Texts and Monographs in Physics

Giovanni Gallavotti

The Elements of Mechanics

With 53 Illustrations

Springer-Science+Business Media, LLC

Giovanni Gallavotti
Istituto di Matematica
Universita degli Studi di Roma
00185, Rome
Italy

Editors

Wolf Beiglböck
Institut für Angewandte Mathematik
Universität Heidelberg
Im Neuenheimer Feld 5
D-6900 Heidelberg 1
Federal Republic of Germany

Elliott H. Lieb
Department of Physics
Joseph Henry Laboratories
Princeton University
P.O. Box 708
Princeton, NJ 08540
U.S.A.

Tullio Regge
Istituto de Fisica Teorica
Universita di Torino
C. so M. d'Azeglio, 46
10125 Torino
Italy

Walter Thirring
Institut für Theoretische Physik
der Universität Wien
Boltzmanngasse 5
A-1090 Wien
Austria

Library of Congress Cataloging in Publication Data
Gallavotti, Giovanni.
The elements of mechanics.
(Texts and monographs in physics)
Includes bibliographical references and indexes.
1. Mechanics. I. Title. II. Series.
QA805.G24 1983 531 83-360

This book was first printed in Italian: *Meccanica Elementare*. ©1980 Editore Boringhieri, Torino.

Originally published by Springer-Verlag New York, Inc. in 1983
Softcover reprint of the hardcover 1st edition 1983

Typeset by Computype, St. Paul, Minnesota.
Printed and bound by R.R. Donnelley & Sons, Harrisonburg, Virginia.

9 8 7 6 5 4 3 2 1

ISBN 978-3-662-00733-4 ISBN 978-3-662-00731-0 (eBook)
DOI 10.1007/978-3-662-00731-0

A Daniela per amore infinito

Preface

The word "elements" in the title of this book does not convey the implication that its contents are "elementary" in the sense of "easy": it mainly means that no prerequisites are required, with the exception of some basic background in classical physics and calculus.

It also signifies "devoted to the foundations". In fact, the arguments chosen are all very classical, and the formal or technical developments of this century are absent, as well as a detailed treatment of such problems as the theory of the planetary motions and other very concrete mechanical problems. This second meaning, however, is the result of the necessity of finishing this work in a reasonable amount of time rather than an a priori choice.

Therefore a detailed review of the "few" results of ergodic theory, of the "many" results of statistical mechanics, of the classical theory of fields (elasticity and waves), and of quantum mechanics are also totally absent; they could constitute the subject of two additional volumes on mechanics.

This book grew out of several courses on *meccanica razionale*, i.e., essentially, theoretical mechanics, which I gave at the University of Rome during the years 1975–1978.

The subjects cover a wide range. Chapter 2, for example, could be used in an undergraduate course by students who have had basic training in classical physics; Chapters 3 and 4 could be used in an advanced course; while Chapter 5 might interest students who wish to delve more deeply into the subject, and it could be used in a graduate course.

My desire to write a self-contained book that gradually proceeds from the very simple problems on the qualitative theory of ordinary differential equations to the more modern theory of stability led me to include arguments of mathematical analysis, in order to avoid having to refer too much to existing textbooks (e.g., see the basic theory of the ordinary differential equations in §2.2–§2.6 or the Fourier analysis in §2.13, etc.).

I have inserted many exercises, problems, and complements which are meant to illustrate and expand the theory proposed in the text, both to avoid excessive size of the book and to help the student to learn how to solve theoretical problems by himself. In Chapters 2–4, I have marked with an asterisk the problems which should be developed with the help of a teacher; the difficulty of the exercises and problems grows steadily throughout the book, together with the conciseness of the discussion.

The exercises include some very concrete ones which sometimes require the help of a programmable computer and the knowledge of some physical data. An algorithm for the solution of differential equations and some data tables are in Appendix O and Appendix P, respectively.

The exercises, problems, and complements must be considered as an important part of the book, necessary to a complete understanding of the theory. In some sense they are even more important than the propositions selected for the proofs, since they illustrate several aspects and several examples and counterexamples that emerge from the proofs or that are naturally associated with them.

I have separated the proofs from the text: this has been done to facilitate reading comprehension by those who wish to skip all the proofs without losing continuity. This is particularly true for the more mathematically oriented sections. Too often students tend to confuse the understanding of a mathematical proposition with the logical contortions needed to put it into an objective, written form. So, before studying the proof of a statement, the student should meditate on its meaning with the help (if necessary) of the observations that follow it, possibly trying to read also the text of the exercises and problems at the end of each section (particularly in studying Chapters 3–5).

The student should bear in mind that he will have understood a theorem only when it appears to be self-evident and as needing no proof at all (which means that its proof should be present in its entirety in his mind, obvious and natural in all its aspects and, if necessary, describable in all details). This level of understanding can be reached only slowly through an analysis of several exercises, problem, examples, and careful thought.

I have illustrated various problems of classical mechanics, guided by the desire to propose always the analysis of simple rather than general cases. I have carefully avoided formulating "optimal" results and, in particular, have always stressed (by using them almost exclusively) my sympathy for the only "functions" that bear this name with dignity, i.e., the C^∞-functions and the elementary theory of integration ("Riemann integration").

I have tried to deal only with concrete problems which could be "constructively" solved (i.e., involving estimates of quantities which could actually be computed, at least in principle, with the described methods), and I hope to have avoided indulging in purely speculative or mathematical considerations. I realize that I have not been entirely successful and I apologize to those readers who agree with this point of view without, at the same time, accepting mathematically nonrigorous treatments.

Finally, let me comment on the conspicuous absence of the basic elements of the classical theory of fluids. The only excuse that I can offer, other than that of nonpertinence (which might seem a pretext to many), is that, perhaps, the contents of this book (and of Chapter 5 in particular) may serve as an introduction to this fascinating topic of mathematical physics.

The final sections, §5.9–§5.12, may be of some interest also to nonstudents since they provide a self-contained exposition of Arnold's version of the Kolmogorov–Arnold–Moser theorem.

This book is an almost faithful translation of the Italian edition, with the addition of many problems and §3.12 and with §5.5, §5.7, and §5.12 rewritten.

I wish to thank my colleagues who helped me in the revision of the manuscript and I am indebted to Professor V. Franceshini for providing (from his files) the very nice graphs of §5.8.

I am grateful to Professor Luigi Radicati for the interest he showed in inviting me to write this book and providing the financial help from the Italian printer P. Boringhieri.

The English translation of this work was partially supported by the "Stiftung Volkswagenwerk" through the IHES.

GIOVANNI GALLAVOTTI

Roma 27 December 1981

Contents

Appendices

CHAPTER 1

Phenomenic Reality and Models

§1.1. Statements

The results of physical experiments are determined from observations which are based on the measurement of various physical entities, i.e., the association of well-defined sequences of numbers with well-defined sequences of events.

The physical entities are "operationally defined". By this we mean that they are defined in terms of the operations used to construct the numbers that provide their "measure".

For instance, we know the sequence of operations necessary to measure the "distance" between two given points P and Q in space. One chooses a particular ruler and places it on the straight line joining points P and Q, starting from P. Taking the endpoint of the ruler as the new starting point, one repeats this procedure n times until the endpoint of the ruler is superimposed on Q. If the distance PQ is not an exact multiple of the length of the ruler, one may, after n such operations, reach a point $Q_n \neq Q$ preceding Q on the line PQ; and after $n+1$ operations one may reach point Q_{n+1} following Q on the line PQ. Then one takes a new ruler "ten times shorter" and puts it on Q_nQ trying to match, as before, the second endpoint with Q. When this turns out to be impossible, one can, in analogy with the first case, define a new point Q_{n_1} on Q_nQ and, then, take a third ruler ten times shorter than the second and repeat the operation.

We, thus, inductively build a number $n + 0.n_1n_2 \ldots$ (in decimal representation) which, by definition, is the measure of the distance between P and Q.

The above sequence of operations appears well defined but, in fact, a careful analysis shows that it does not have the necessary prerequisites to be considered a mathematically precise definition. What, for instance, is "space", what is a "point", what is a "ruler"? Is it possible to "divide" a ruler into parts, and infinitely often?

The physicist is not too concerned (or, rather, he is not at all concerned) with such aspects of the question: he considers a physical entity well defined whenever the empirical procedure necessary for its measurement is clear.

An operational procedure is considered to be clear when every observer is led to the same result when measuring the same physical entity. It should be stressed, however, that this is an empirical criterion perpetually subject to critique; thus, physical entities which today are considered to be well defined may no longer be so in the future.

Hence, the physicist, from his observations of nature, obtains a set of numbers corresponding to the performance of some operations which are considered to be "objectively defined".

Trying to organize such numbers coherently, the physicist often formulates "models": i.e., he associates well-defined mathematical structures with his measures, and he tries to establish a (small) number of mathematical relationships among them. From such relationships new ones logically follow which, reinterpreted through the model used inversely, may serve to predict relations between various empirical measurements.

The belief in the existence of good models motivated Galileo to write: "Philosophy is written in this great book which is always open before our eyes (I mean the universe), but it cannot be understood unless one first learns the language and distinguishes the characters in which it is written. It is a mathematical language and the characters are triangles, circles, and other geometrical figures, without which it cannot be understood by the human mind; without them one would vainly wander through a dark labyrinth".[1]

A mathematical model is considered satisfactory whenever it does not lead to contradictions with the experiments. If a contradiction occurs, the physicist dismisses the model as "wrong"; nevertheless, the mathematical construction built with it remains valid and is witness to an imperfect representation of nature.

Strictly speaking, there is no model which is not wrong: only models which have not yet shown to be wrong exist. However, all "serious" models (such as the dynamics of point masses, the theory of relativity, quantum mechanics, electromagnetism, thermodynamics, statistical mechanics, etc.) have led, and still continue to lead, to the formulation of extremely interesting mathematical problems. Furthermore, it often happens that the analysis of the mathematical properties of a "wrong" model helps in the

[1] G. Galilei, *Il Saggiatore*, p. 232, see References.

formulation of the new "more elaborate" model that the physicist tries to set up as a substitute.

We see, then, how the link between phenomenic reality and mathematics is established through what has been called "a model". However, it would be impossible to give a precise mathematical definition of the notion of a model since it is a rather empirical notion which can only be well understood through the analysis of several concrete cases.

§1.2. An Example of a Model

We will consider the historically particularly important and significant case of the "mechanics of a point mass". Let us see concretely how the mathematical model of one or several point masses subject to forces can be built starting from empirical observations.

The first statement (or "axiom", to use a mathematical term) says that the point masses are in a three-dimensional Euclidean space E^3 in which any point can in fact be represented by its three coordinates with respect to an orthogonal reference system $(0; \mathbf{i}, \mathbf{j}, \mathbf{k})$.

Such an idealization has a clear mathematical meaning, but it appears to be unprovable in mathematical terms: it just renders the following empirical observation.

In practice, a point in space is determined by measuring (often only in principle and with the ruler method described in §1) its distance from three mutually orthogonal walls. It is to be remarked that all such operations are ordinarily considered well defined.

A second statement (or "axiom") concerns "time" which, for the physicist, is the physical entity measured by a "clock" (classically described as a pendulum, although any more modern device will do as well). One assumes that time is an absolute "entity": in other words, one states that, at least in principle, it is possible to associate with every point in space a clock mechanically identical at every point, and, furthermore, to coordinate ("synchronize") the clocks.

This means that if P, P' are two points and t, t' are two chosen time instants with $t < t'$, it is then possible to send a signal from P towards P' leaving P at time t and reaching P' at time t' (as indicated by the local clocks in P and in P', respectively); while, vice versa, if $t > t'$, the above operation should be impossible.

A little thought makes it clear that the operational definition of a "system of synchronized clocks" is based on the empirical fact that it is possible to send signals with arbitrary speed. It is also clear that the notion of time is a purely phenomenological notion, far from being mathematically well posed.

Accepting the point of view so far discussed, we are led to say that the mathematical scheme, or model, representing the space-time continuum, where our observations take place, consists of a four-dimensional space:

each of its points (x, y, z, t) represents a point seen in a Cartesian coordinate frame $(0; \mathbf{i}, \mathbf{j}, \mathbf{k})$ ("laboratory") and observed at the instant t (as measured by the formerly introduced universal clocks).

Empirically, a point mass is any object which, at least as far as our observations are concerned, can be assimilated with a point in space (for instance, a planet or a star in the universe, a stone falling in a ravine, a ship sailing in the ocean, etc.). Such a point preserves its identity over the course of time; hence, we can define its trajectory through a function of time $t \rightarrow \mathbf{x}(t)$, where $\mathbf{x}(t) = (x(t), y(t), z(t))$ is the vector whose components are the point's coordinates at time t, in the chosen reference frame $(0; \mathbf{i}, \mathbf{j}, \mathbf{k})$.

Mathematically, a point mass moving in the reference frame $(0; \mathbf{i}, \mathbf{j}, \mathbf{k})$ observed as t varies over an interval I is represented as a curve C in E^3 by the vector equations $P(t) - 0 = \mathbf{x}(t)$, $t \in I$; and the parameter t has the interpretation of time (i.e., it is called "time").

Given a point mass moving as t varies in I, one can associate with it its velocity at time $t \in I$. Operationally, velocity is defined by fixing $t_0 \in I$, finding the positions $P(t_0)$ and $P(t_0 + \epsilon)$, and setting

$$\mathbf{v}(t_0) = \frac{P(t_0 + \epsilon) - P(t_0)}{\epsilon},$$

where the parameter $\epsilon > 0$ is to be chosen "suitably small" (according to well-defined criteria which, however, depend on the concrete cases).

The mathematical model defines the point mass velocity at time $t_0 \in I$ as the derivative of the function $t \rightarrow \mathbf{x}(t)$ at $t = t_0$.

To complete the mathematical model of a point mass, we still have to define the "force" acting on it.

Operationally, the force acting at a given instant on the point mass consists of three scalar quantities which together define a vector $\mathbf{f}(t)$. The force acting on the point mass moving in E^3 and observed in the frame $(0; \mathbf{i}, \mathbf{j}, \mathbf{k})$ is measured through a "dynamometer" which is an instrument whose use is convenient to describe in a strongly idealized form. It is, basically, a suitably built spring which we shall imagine as a very thin, light segment with a hook.

Consider a point mass moving in E^3 at time t_0 with a velocity (v_x, v_y, v_z) relative to the reference frame $(0; \mathbf{i}, \mathbf{j}, \mathbf{k})$. If we wish to measure the force acting upon it, we hook it to the dynamometer to which we have imparted the same velocity (v_x, v_y, v_z), which will be kept fixed during the measurement. We shall then try to adjust the spring's length and direction so that the acceleration at time $t_0 + \epsilon$ is 0, where $\epsilon > 0$ is chosen "suitably small". (The empirical notion of acceleration and the corresponding mathematical model of it, as the second derivative with respect to t of the point's position, is discussed along the same lines as the notion of velocity.)

The force is then the vector $\mathbf{f}$ whose direction is that of the dynamometer at time $t_0 + \epsilon$, whose orientation is that pointing towards the hook and whose modulus is the measure of the spring's elongation.

What has been said so far can be summarized by saying that a point mass subject to forces and observed in a frame $(0; \mathbf{i}, \mathbf{j}, \mathbf{k})$ in E^3 as time varies within an interval I is, in its mathematical model, described by a curve in seven-dimensional space: one of its points $(t, x, y, z, f_x, f_y, f_z)$ represents a point mass which at time t has coordinates (x, y, z) in $(0; \mathbf{i}, \mathbf{j}, \mathbf{k})$ and, in the same frame, is subject to a force (f_x, f_y, f_z). The curve representing this situation can be parameterized by the parameter t itself, as t varies in some time interval I; it shall also be assumed that in this parametric representation the functions $t \to (x(t), y(t), z(t))$ are twice continuously differentiable so that a mathematical definition of velocity and acceleration is meaningful.

§1.3. The Laws of Mechanics

Once it is established what is meant by a point mass subject to forces and studied in a given frame of reference in E^3 as the time varies in an interval I (briefly, "a point mass subject to forces"), it is possible to complete the mathematical model of the point's mechanics. For this purpose, we have to discuss the laws of dynamics and their mathematical interpretation.

Experimentally, given a point mass, we can observe the existence of a simple relation between its acceleration $\mathbf{a}$ at time t (observed in a given frame of reference) and the force $\mathbf{f}$ acting on it at that time (observed in the same frame). Such a relation is called the Second Law of Mechanics and establishes the existence of a constant $m > 0$, characteristic of the point mass and independent of the frame of reference used for the observations, such that:

$$m\mathbf{a} = \mathbf{f}.$$

This law introduces, via the properties of the differential equations, many relations among the quantities $\mathbf{x}, \mathbf{v}, t$, and such relations can sometimes be experimentally checked.

For instance, if it is known a priori which force will act on the point mass whenever it is at the point (x, y, z) at time t with velocity (v_x, v_y, v_z), then, denoting such force as $\mathbf{f}(v_x, v_y, v_z, x, y, z, t) = \mathbf{f}(\mathbf{v}, \mathbf{x}, t)$, the differential equation

$$m\ddot{\mathbf{x}} = \mathbf{f}(\dot{\mathbf{x}}, \mathbf{x}, t)$$

allows the determination of the motion following an initial state where the velocity $\mathbf{v}_0$ and the position $\mathbf{x}_0$ are given at time t_0, at least for a small time interval around t_0 if $\mathbf{f}$ is a smooth function.

The First Principle of Mechanics postulates the existence of at least one reference frame $(0; \mathbf{i}, \mathbf{j}, \mathbf{k})$, called "inertial frame", in E^3 where a point mass "far" from the other objects in the universe appears to be subjected to a null force in $(0; \mathbf{i}, \mathbf{j}, \mathbf{k})$. Such a frame is experimentally identified with a frame with origin in a fixed star and with axes oriented towards three more fixed stars. It is within such a frame that one usually considers the motion.

Of course the notions of "far" and of "fixed star" are empirical notions rather than mathematical ones.

Mathematically, the first principle is used to grant to a particular frame of reference in the space-time continuum a privileged role and to define the "absolute force" or the "true force" as that acting on the point mass in this frame. This frame has to be chosen once and for all and it is called the "fixed reference frame" (as opposed to "moving reference frame").

It is possible and sometimes convenient to introduce frames whose origin and axes vary with time with respect to the "fixed" frame $(0; \mathbf{i}, \mathbf{j}, \mathbf{k})$: $(0(t); \mathbf{i}(t), \mathbf{j}(t), \mathbf{k}(t))$.

Since $\mathbf{f} = m\mathbf{a}$, it follows that if the moving frame is in uniform rectilinear translatory motion with respect to the fixed frame, then the force acting upon the point is the same whether observed in the fixed frame or in the moving frame: hence, in this moving frame, the "inertia principle", i.e., the first principle, is valid: a point mass which is "very far" from the other objects in the universe is subject to a null force, since the acceleration is the same in the two frames. All frames in rectilinear uniform motion with respect to a fixed frame are called "inertial frames".

The mathematical model of a point mass with mass m subject to forces and obeying the laws of dynamics is then, simply, a point mass subject to forces, in the sense of the preceding section, and such that the relation

$$m\mathbf{a} = \mathbf{f}$$

holds and, furthermore, $\mathbf{f}$ is a function of the point's velocity, position, and time; i.e., the following relation holds:

$$\mathbf{f} = \mathbf{f}(\mathbf{v}, \mathbf{x}, t).$$

Clearly, from such a mathematical viewpoint (where $\mathbf{f}$ is imagined as given a priori), the first principle is deprived of its deep physical meaning.

We conclude the analysis of a mathematical formulation of a model for the mechanics of point masses by a short description of the mathematical model for a "system of N point masses".

Mathematically, such a system consists of N point masses with mass $m_1, \ldots, m_N$ in the above sense, and verifying the Third Principle of Mechanics.

By this we mean that it should be possible to represent the force $\mathbf{f}_i$ acting on the ith point as

$$\mathbf{f}_i = \sum_{j \neq i} \mathbf{f}_{j \to i},$$

where $\mathbf{f}_{j \to i}$ are such that

(a) $\mathbf{f}_{i \to j} = -\mathbf{f}_{j \to i}, j, i = 1, 2, \ldots, N, i \neq j$;
(b) $\mathbf{f}_{j \to i}$ is parallel to $P_i - P_j$, i.e., to the line joining the positions P_i and P_j of the ith and jth points;

(c) $\mathbf{f}_{j\to i}$ depends solely upon the positions and velocities of the ith and jth points and on time:

$$\mathbf{f}_{j\to i} = \mathbf{f}_{j\to i}(\mathbf{v}_j, \mathbf{v}_i, P_j, P_i, t).$$

This assumption corresponds to a precise empirical fact: it is possible to define operationally what should be understood by $\mathbf{f}_{j\to i}$, "the force exerted by i on j".

For instance, the force $\mathbf{f}_{j\to i}$ could be measured as follows: one measures, in the given inertial frame of reference, the force $\mathbf{f}_i$ acting on i and then one measures, after removing the point j from the system, the new force acting on the ith point, obtaining the result $\mathbf{f}_i^{(j)}$; then one sets

$$\mathbf{f}_{j\to i} = \mathbf{f}_i - \mathbf{f}_i^{(j)}.$$

The Third Principle of Mechanics arises from the experimental observation that $\mathbf{f}_{j\to i} = -\mathbf{f}_{i\to j}$, that $\mathbf{f}_{i\to j}$ is parallel to $P_i - P_j$, that the total force acting on a single point mass is the sum of the forces exerted on it by the other system's points (in the sense of vectors addition) if observed in an inertial frame of reference, and, finally, that $\mathbf{f}_{i\to j}$ depends only upon the positions and velocities of the points involved and, possibly, on time.

Physics often places still more requirements and restrictions upon the laws of force which can be used to give a more detailed specification of a mechanical system's model. However, they do not have a general character comparable to the three principles but, rather, are statements explaining which laws of force are to be considered a good model under given circumstances. For instance, two point masses "without structure" (this is, again, an empirical notion which we refrain from elucidating) attract each other with a force of intensity mm'/kr^2, where r is the distance between the points, m and m' are their masses, and k is a universal constant. If the structure of the two points can be summarized by saying that they have an "electric charge e" (a new empirical notion), the mutual force will be the vector sum of the above-described gravitational force and of a repulsive force with intensity $e^2/k'r^2$, where k' is another universal constant.

The principles of mechanics already place enough restrictions upon the nature of the forces admissible in mechanical problems that it is convenient and interesting to examine their implications before passing to the analysis of special models obtained by concretely specifying the "force laws", i.e., the functions giving the forces in terms of the points' positions and velocities and of time.

It should be stressed, and this is a general comment on the mathematical models for physical phenomena, that the mathematical model is always "poorer" than the physical reality that it tries to imitate. We already saw, for instance, that in the above mathematical model for mechanics, the first principle loses its meaning. Another example, implicit in the above discussion, is the following.

Clearly, to give an operational meaning to the notions of position, speed, force, etc., we must be able to repeat "identical" experiments several times (e.g., see the position measurement in §1.1) by repeating the measurement operations. However, time inexorably flows away, and this is impossible.

Physically, this difficulty is avoided by the "principle of homogeneity of space-time" which says that experiments starting at any time in any space location will yield the same results if the points involved are in the same relative positions and situations.

In the mathematical model for mechanics just described, the necessity of understanding the above problems does not arise, nor do many other similar problems which the reader will easily think of.

Usually it is possible to complicate the models in order to imbue them with any given number of physical facts: but an analysis of this type of questions would lead us beyond the scope of this book.

In any case, one always needs to decide where to put a stop to the process of model improvement, which would otherwise hopelessly continue ad infinitum. We must recall that we have the more down-to-earth, and more interesting, problem of obtaining some concrete prediction algorithms for our observations of nature.

§1.4. General Thoughts on Models

In this book we shall meet more abstract schematization processes concerning empirically observed phenomena (e.g., when we discuss the notion of an "observable" or of a "vibrating string"). In such cases, however, we shall not repeat the details of the construction of the mathematical model: this is a very common practice based on the idea that the very words used to designate well-defined mathematical objects will *implicitly* define the model.

It is such a practice, or better, its imperfect understanding, which sometimes causes misunderstandings between physicists and mathematicians and provokes allegations of nonrigorous use of mathematics.

It is important to realize that when the physicist speaks in mathematical terms he is by no means attributing to them the same rigid meaning that a mathematician would assume for them. Rather he is using this language to help himself in the formulation of a model which, once well defined, he shall rigorously treat (since he believes, or at least hopes, that the book of nature is written in mathematical characters).

Possible logically nonrigorous steps or apparently wild mathematical approximations in a physicists' argument should always be interpreted as further complications or, better, refinements of the model that the physicist is trying to build.

In the hectic development of research, a physicist often modifies a model while using it, or he modifies the mathematical meanings of the objects and entities which belong to the model without changing their names (otherwise, a dictionary would not suffice). He does this because his main interest is in the model's construction and only secondarily in its mathematical theory, which he often considers trivial.

To avoid excessively pedantic discussions, we shall adhere, in the following, to the well-established practice of avoiding the physical analysis necessary to the construction of a model and shall leave it to the reader to imagine such an analysis via the suggestive names used for the various mathematical entities (with the exception of a few important cases). In any case, this book is devoted to the mathematical, rather than physical aspects, of mechanical problems.

Bibliographical Comment. It is very useful to study at least the definition and the laws of motion in the *Philosophiae Naturalis Principia Mathematica* by I. Newton (see References) to exactly understand the Newtonian formulation of mechanics and its modernity. To avoid "reading too much", i.e., to avoid interpreting these immortal pages in too modern a way, it is a good idea to read the paper *Essays on the history of mechanics* by C. Truesdell, pp. 85–137 (see References).

The reading of the first two chapters of the work by E. Mach (see References) will be a very useful and stimulating complement to the first three chapters of this book.

CHAPTER 2

Qualitative Aspects of One-Dimensional Motion

§2.1. Energy Conservation

Consider a point mass, with mass m, on the line $\mathbf{R}$ and subject to a force law depending uniquely on its position. Therefore, a force law $\xi \to f(\xi)$ is given, $\xi \in \mathbf{R}$, which we shall suppose to be of class C^∞, associating with every point ξ on the line $\mathbf{R}$ the component $f(\xi)$ of the force acting on the point when it happens to occupy ξ.

A "motion" of the point mass, observed as t varies in an interval I, is a function $t \to x(t)$, $t \in I$, of class $C^\infty(I)$ such that

$$m\ddot{x}(t) = f(x(t)), \qquad \forall t \in I. \tag{2.1.1}$$

The "energy conservation theorem" follows by multiplying Eq. (2.1.1), side by side, by $\dot{x}(t)$:

$$m\dot{x}\ddot{x} = \dot{x}f(x), \tag{2.1.2}$$

omitting, as will often be done, the explicit mention of t-dependence. One then notices that, defining the functions

$$\eta \to T(\eta) = \tfrac{1}{2}m\eta^2, \qquad \xi \to V(\xi) = -\int^{\xi} f(\xi')\,d\xi', \tag{2.1.3}$$

it is

$$\frac{d}{dt}T(\dot{x}) = m\dot{x}\ddot{x}\,\frac{d}{dt}V(x) = -f(x)\dot{x}, \tag{2.1.4}$$

so that Eq. (2.1.2) becomes

$$\frac{d}{dt}(T(\dot{x}) + V(x)) = 0. \tag{2.1.5}$$

This implies that to every motion $t \to x(t)$, $t \in I$, we can associate a constant E, depending on the motion under consideration, such that

$$T(\dot{x}(t)) + V(x(t)) = E, \qquad \forall t \in I. \tag{2.1.6}$$

The expressions $T(\dot{x})$ and $V(x)$ are respectively called the "kinetic energy" and the "potential energy" and Eq. (2.1.6) has to be read as follows: "in every motion developing under the action of a force with potential energy V, the sum of the kinetic energy and potential energy is a constant". This constant is given the name "total energy" of the considered motion.

The "qualitative theory" of Eq. (2.2.1) is concerned with the analysis of the properties of the motion verifying Eq. (2.2.1), which are valid independently of the choice of f, at least for vast classes of functions f.

The energy conservation is a first example of a qualitative property.

Observations. The energy conservation goes back at least to Huygens; afterwards, it was used by J. and D. Bernoulli together with the law of conservation of linear momentum (Descartes) (see C. Truesdell, p. 105 and following).

Equation (2.1.6) implies an expression for the velocity:

$$\dot{x}(t) = \pm\left(\frac{2}{m}(E - V(x(t)))\right)^{1/2}, \qquad t \in I. \tag{2.1.7}$$

This relation, which will be used and discussed in §2.6, allows the reduction of the determination of the evolution law $t \to x(t)$, "time law", to an area-computation problem for a planar figure, "quadature". In fact, supposing $\dot{x} > 0$, it yields:

$$t = \int_{x(0)}^{x(t)} \left(\frac{2}{m}(E - V(\xi))\right)^{-1/2} d\xi \tag{2.1.8}$$

when $I \supset [0, t]$.

Hence, the area under the graph of the curve with equation $\xi \to y(\xi) = ((2/m)(E - V(\xi)))^{-1/2}$ above the interval $[x(0), x(t)]$ is the time that the point needs to reach $x(t)$ starting from $x(0)$ at time 0 with positive speed and energy E, at least for small t (i.e., as long as $\dot{x} > 0$).

Newton "reduced to quadratures" the simplest problems of motion without explicitly using energy conservation (see, for instance, I. Newton, *Principia*, Book I, Propositions XXXIX, XLI, LIII, LVI, etc.).

§2.2. General Existence, Uniqueness, and Regularity Properties of Motion. Uniqueness

In the preceding section, we supposed "having" a motion developing, under the action of a force f, in a time interval I.

We can ask which further properties of a particular motion allow us to select it from among all motions which, in the same time interval I, take place under the action of the same force.

One can even preliminarily ask whether, given an interval I, there exist any motions, i.e., C^∞ solutions of Eq. (2.1.1) thought of as an equation for $t \to x(t)$, $t \in I$.

In view of the importance of such questions, before proceeding in the analysis of Eq. (2.1.1), we shall devote some attention to the general problem of the existence, uniqueness, and regularity of the solutions of differential equations in $\mathbf{R}^d$.

Observe first that Eq. (2.1.1), thought of as a "second-order" differential equation in $\mathbf{R}^1$, is equivalent to a "first-order" differential equation in $\mathbf{R}^2$: it suffices to write it as

$$\begin{cases} \dot{x}(t) = y(t), \\ \dot{y}(t) = f(x(t)), \qquad t \in I, \end{cases} \tag{2.2.1}$$

where Eq. (2.2.1) is an equation for the unknown C^∞ function $t \to (x(t), y(t))$ defined on I and with values in $\mathbf{R}^2$.

More generally, consider an arbitrary "sth-order" differential equation in $\mathbf{R}^d$, $s = 0, 1, \ldots,$ like

$$\frac{d^s\mathbf{x}}{dt^s}(t) = \mathbf{f}\left(\frac{d^{s-1}\mathbf{x}}{dt^{s-1}}(t), \ldots, \frac{d\mathbf{x}}{dt}(t), \mathbf{x}(t), t\right) \tag{2.2.2}$$

with $t \in I$, where $\mathbf{f}$ is an $\mathbf{R}^d$-valued C^∞ function defined on $\mathbf{R}^d \times \mathbf{R}^d \times \cdots \times \mathbf{R}^d \times \mathbf{R}$ and $t \to \mathbf{x}(t)$ is an unknown $\mathbf{R}^d$-valued C^∞ function on I.

This equation may be thought as a first-order equation in $\mathbf{R}^{ds}$ by setting

$$\begin{aligned} \frac{d\mathbf{x}}{dt}(t) &= \mathbf{y}_1(t), \\ \frac{d\mathbf{y}_1}{dt}(t) &= \mathbf{y}_2(t), \\ \frac{d\mathbf{y}_{s-2}}{dt}(t) &= \mathbf{y}_{s-1}(t), \\ \frac{d\mathbf{y}_{s-1}}{dt}(t) &= \mathbf{f}(\mathbf{y}_{s-1}(t), \ldots, \mathbf{y}_1(t), \mathbf{x}(t), t), \end{aligned} \tag{2.2.3}$$

and then considering Eq. (2.2.3) as an equation for the C^∞ function $t \to (\mathbf{x}(t), \mathbf{y}_1(t), \ldots, \mathbf{y}_{s-1}(t))$ defined on the interval I and with values in $\mathbf{R}^d \times \cdots \times \mathbf{R}^d = \mathbf{R}^{ds}$.

Equation (2.2.2) is the most general differential equation that we shall meet in this book. By virtue of the preceding remark, it will then suffice, for our purposes, to study first-order differential equations in $\mathbf{R}^d$ having the form

$$\dot{\mathbf{x}}(t) = \mathbf{F}(\mathbf{x}(t), t). \tag{2.2.4}$$

It will turn out to be convenient to introduce a precise convention about what a differential equation is or about what one of its solutions is.

1 Definition. Given an $\mathbf{R}^d$-valued function $\mathbf{F} \in C^\infty(\mathbf{R}^d \times \mathbf{R})$, the expression (2.2.4), denoted, for short, $\dot{\mathbf{x}} = \mathbf{F}(\mathbf{x}, t)$, will be called a "differential equation on $\mathbf{R}^d$ associated with $\mathbf{F}$".

A "$C^{(k)}$ solution", $k \geqslant 1$, of Eq. (2.2.4) on the interval I, closed or open or semiopen, will be a $C^{(k)}$ function which turns Eq. (2.2.4) into an identity when substituted into it.[1] A "solution" of Eq. (2.2.4) for $t \in I$ is a C^∞ solution. The solutions of Eq. (2.2.4) will often be called "motions".

Let us first examine the uniqueness problem for the solutions of Eq. (2.2.4).

1 Proposition. *Let $(\xi, t) \to \mathbf{F}(\xi, t)$ be an $\mathbf{R}^d$-valued C^∞ function on $\mathbf{R}^d \times \mathbf{R}$. Given $a > 0$, $b > 0$, $t_0 \in \mathbf{R}$, let $t \to \mathbf{x}(t)$ be a $C^{(1)}$ solution of Eq. (2.2.4) on $J = [t_0 - a, t_0 + b]$:*

(i) *the function $t \to \mathbf{x}(t)$ is in $C^\infty(J)$;*
(ii) *if $t \to \mathbf{y}(t)$ is another solution of eq. (2.2.4) on J and if $\mathbf{y}(t_0) = \mathbf{x}(t_0)$, then $\mathbf{x}(t) = \mathbf{y}(t)$, $\forall t \in J$.*

Observations.

(1) This proposition applied to Eq. (2.2.2) via Eq. (2.2.3) tells us that two $C^{(s)}$ solutions of an sth-order differential equation in $\mathbf{R}^d$ for $t \in J$ coincide if and only if at time $t_0 \in J$ ("initial time") they have the same first $(s-1)$ derivatives ("equal initial data").

When Eq. (2.2.2) is the equation governing a motion in $\mathbf{R}^d$, it is $s = 2$; this means that the motion is uniquely determined, if existing at all, by its initial position $\mathbf{x}(t_0)$ and by its initial velocity $\dot{\mathbf{x}}(t_0)$, i.e., as one says, by its initial "act of motion" $(\dot{\mathbf{x}}(t_0), \mathbf{x}(t_0))$.

(2) It would appear that it might be interesting or important to know if, by specifying properties of the solutions of Eq. (2.2.2) other than the just-mentioned initial data at some initial time, one uniquely determines the solution verifying such properties,[2] if existing at all.

The uniqueness criterion that we chose above for illustration purposes, Proposition 1, has been selected only because it quickly leads to a simple answer and because it is one of the uniqueness criteria which are most useful in many applications.

(3) From the proof it will appear that if $\mathbf{F}$ had been only supposed to be of class $C^{(k)}$, $k \geqslant 1$, then uniqueness would have followed in an equal way. The regularity of $t \to \mathbf{x}(t)$, $t \in J$, could also be deduced in this case, but one would only obtain that $t \to \mathbf{x}(t)$ is a $C^{(k+1)}$ function.

[1] We shall see that every $C^{(k)}$ solution, $k \geqslant 1$, is automatically a C^∞ solution.

[2] For instance, we can ask the following question. Consider Eq. (2.2.2) with $s = 2$ and let t_1, t_2 be two times and $\mathbf{x}_1, \mathbf{x}_2 \in \mathbf{R}^d$ be two positions. Is the motion [solution of Eq. (2.2.2)] leading from $\mathbf{x}_1$ to $\mathbf{x}_2$ as time elapses from t_1 to t_2 (assuming that one such motion, at least, exists) unique? We shall see that the answer to this question will, in general, be no.

PROOF. By integrating both sides of Eq. (2.2.4) and by setting $\mathbf{x}_0 = \mathbf{x}(t_0) = \mathbf{y}(t_0)$, we get:

$$\mathbf{x}(t) = \mathbf{x}_0 + \int_{t_0}^{t} \mathbf{F}(\mathbf{x}(\tau), \tau)\, d\tau, \qquad t \in J, \tag{2.2.5}$$

and, similarly, since also $t \to \mathbf{y}(t)$ is a solution of Eq. (2.2.4):

$$\mathbf{y}(t) = \mathbf{x}_0 + \int_{t_0}^{t} \mathbf{F}(\mathbf{y}(\tau), \tau)\, d\tau, \qquad t \in J. \tag{2.2.6}$$

Hence,

$$\mathbf{x}(t) - \mathbf{y}(t) = \int_{t_0}^{t} (\mathbf{F}(\mathbf{x}(\tau), \tau) - \mathbf{F}(\mathbf{y}(\tau), \tau))\, d\tau. \tag{2.2.7}$$

To prove (ii), we must find an estimate of the right-hand side allowing us to infer that the left-hand side is zero.

The procedure that will be followed is very interesting since it obviously goes beyond the particular result that we wish to obtain.

Informally, the argument is the following: the difference $|\mathbf{x}(t) - \mathbf{y}(t)|$ is, by Eq. (2.2.7), about $|t - t_0|\,|\mathbf{F}(\mathbf{x}(t), t) - \mathbf{F}(\mathbf{y}(t), t)|$, if $t \sim t_0$; however, the increment $|\mathbf{F}(\mathbf{x}(t), t) - \mathbf{F}(\mathbf{y}(t), t)|$ is proportional, by the Lagrange theorem, to the increment of the argument of $\mathbf{F}$, i.e., to $C|\mathbf{x}(t) - \mathbf{y}(t)|$, where C is an estimate of the first derivatives of $\mathbf{F}$. Hence, Eq. (2.2.7) implies that $|\mathbf{x}(t) - \mathbf{y}(t)|$ and $C|t - t_0|\,|\mathbf{x}(t) - \mathbf{y}(t)|$ are about equal if $t \sim t_0$, and this, in turn, implies that $|\mathbf{x}(t) - \mathbf{y}(t)| = 0$ for t close to t_0 because for $t \sim t_0$ one has $C|t - t_0| < 1$.

To estimate the integrand of Eq. (2.2.7), we can use Taylor's formula: let $S \subset \mathbf{R}^d$ be a sphere with so large a radius that it contains all the values $\mathbf{x}(\tau), \mathbf{y}(\tau)$, $\forall \tau \in J$, and let

$$M_S = \max_{\xi \in S, \tau \in J} \sum_{i,j=1}^{d} \left| \frac{\partial F^{(i)}}{\partial \xi_j}(\xi, \tau) \right|, \tag{2.2.8}$$

where $F^{(i)}(\xi, t)$ is the ith component of the vector

$$\mathbf{F}(\xi, t) = \left(F^{(1)}(\xi, t), \ldots, F^{(d)}(\xi, t)\right) \in \mathbf{R}^d.$$

Then, from the Taylor formula:

$$|\mathbf{F}(\mathbf{x}(\tau), \tau) - \mathbf{F}(\mathbf{y}(\tau), \tau)| \leqslant M_S |\mathbf{x}(\tau) - \mathbf{y}(\tau)|. \tag{2.2.9}$$

By inserting this inequality into Eq. (2.2.7), we find:

$$|\mathbf{x}(t) - \mathbf{y}(t)| \leqslant M_S \int_{t_0}^{t} |\mathbf{x}(\tau) - \mathbf{y}(\tau)|\, d\tau. \tag{2.2.10}$$

Let $M(t) = \max_{t_0 \leqslant \tau \leqslant t} |\mathbf{x}(\tau) - \mathbf{y}(\tau)|$, $t \in [t_0, t_0 + b]$; then Eq. (2.2.10) implies $|\mathbf{x}(t) - \mathbf{y}(t)| \leqslant M_S M(t) |t - t_0|$, $\forall t \in [t_0, t_0 + b]$.

Since $M(t)$ is monotonic nondecreasing and since this inequality holds for all $t \in [t_0, t_0 + b]$, one easily finds that

$$M(t) \leqslant M_S |t - t_0| M(t), \qquad \forall t \in [t_0, t_0 + b], \tag{2.2.11}$$

which implies $M(t) = 0$ for $|t - t_0| < M_S^{-1}$, $t \in [t_0, t_0 + b]$.

Hence, $\mathbf{x}(t_0 + M_S^{-1}) = \mathbf{y}(t_0 + M_S^{-1})$, if $t_0 + M_S^{-1} < t_0 + b$, and the argument can be repeated, replacing t_0 by $t_0 + M_S^{-1}$, to show that $M(t) = 0$ for $t \in [t_0, t_0 + 2M_S^{-1}]$ if $t_0 + 2M_S^{-1} < t_0 + b$, etc., so that $M(t) = 0$ for $t \in [t_0, t_0 + b]$. For $t \in [t_0 - a, t_0]$, one proceeds likewise.[3]

To check (i), i.e., that $t \to \mathbf{x}(t)$ is a C^∞ function on J, one can remark that if $t \to \mathbf{x}(t)$ is a $C^{(1)}(J)$ function, then Eq. (2.2.4) implies that $t \to \dot{\mathbf{x}}(t)$ is in $C^{(1)}(J)$, being a composition of a C^∞ function with a $C^{(1)}$ function; furthermore, by differentiating Eq. (2.2.4):

$$\ddot{\mathbf{x}}(t) = \sum_{i=1}^{d} \frac{\partial \mathbf{F}}{\partial \xi_i}(\mathbf{x}(t), t) \cdot \dot{\mathbf{x}}^{(i)}(t) + \frac{\partial \mathbf{F}}{\partial t}(\mathbf{x}(t), t), \tag{2.2.12}$$

which, in turn, implies that $t \to \ddot{\mathbf{x}}(t)$ is a $C^{(1)}$ function by the same argument as above. Then, by differentiating Eq. (2.2.12), one finds that $t \to \dddot{\mathbf{x}}(t)$ is a $C^{(1)}$ function on J, etc. mbe

Problems for §2.2

1. If $t \to \mathbf{x}(t)$, $t \geqslant 0$, solves $\dot{\mathbf{x}} = \mathbf{f}(\mathbf{x})$ and $\mathbf{x}(0) = \mathbf{x}(T)$ for some $T \geqslant 0$, then $\mathbf{x}(t) = \mathbf{x}(t+T)$, $\forall t \geqslant 0$; assume $\mathbf{f} \in C^\infty(\mathbf{R}^d)$. Would this also be true if $\mathbf{f} \in C^{(1)}(\mathbf{R}^d)$? (Hint: Use uniqueness.)

2. The property of the preceding problem is not valid when the differential equation's right-hand side is explicitly time dependent (i.e., $\dot{\mathbf{x}} = \mathbf{f}(\mathbf{x}, t)$, and $\partial \mathbf{f}/\partial t \not\equiv 0$, the "nonautonomous case"). Find an example.

3. Let $\mathbf{f}(\mathbf{x}, t)$ be such that $\mathbf{f}(\boldsymbol{\xi}, t) = \mathbf{f}(\boldsymbol{\xi}, t+T)$ for some $T > 0$ and for all $\boldsymbol{\xi} \in \mathbf{R}^d$, $t \in \mathbf{R}$. Suppose that $t \to \mathbf{x}(t)$ is a solution of $\dot{\mathbf{x}} = \mathbf{f}(\mathbf{x}, t)$ such that for some integer $m > 0$, one has $\mathbf{x}(0) = \mathbf{x}(mT)$, then $\mathbf{x}(t) \equiv (t + mT)$, $\forall t \geqslant 0$. (Hint: Use uniqueness.)

4. Consider the equation $\dot{x}(t) = l(t)x(t)$ with $l \in C^\infty(\mathbf{R})$. Show that if $t \to x(t)$ and $t \to y(t)$ are two solutions for $t \in J$ and if $x(t) \not\equiv 0$, there exists a constant A such that $y(t) \equiv Ax(t)$, $\forall t \in J$.

5. If the function l of the Problem 4 is periodic with period $T > 0$ and $t \to x(t) \not\equiv 0$, $t \geqslant 0$, is one of its solutions then also $t \to x(t+T)$ is a solution. Hence, $\exists \lambda \neq 0$ such that $x(t, T) = \lambda x(t)$. Show that $\lambda > 0$. (Hint: Otherwise either $\lambda = 0$ and $x(T) = 0$, hence $x(t) \equiv 0$ (by uniqueness on $[0, +\infty)$), or $\lambda < 0$ and there would be $\bar{t} \in [0, T]$ where $x(\bar{t}) = 0$; hence, again, $x(t) = 0$ by uniqueness.)

6. The most general solution $t \to y(t)$, $t \in \mathbf{R}_+$, of the equation in Problem 4, with l periodic with period T has the form $y(t) = A\lambda^{t/T} z(t)$, where z is a C^∞ function on $\mathbf{R}_+$ with period T.

[3] Alternatively, Eq. (2.2.10) could be iterated n times to yield, if $\mu = \max|\mathbf{x}(\tau) - \mathbf{y}(\tau)|$, $\tau \in [t_0 - a, t_0 + b]$:

$$|\mathbf{x}(t) - \mathbf{y}(t)| \leqslant M_S^n \int_{[t_0, t]} d\tau_1 \int_{[t_0, \tau_1]} d\tau_2 \cdots \int_{[t_0, \tau_{n-1}]} d\tau_n |\mathbf{x}(\tau_n) - \mathbf{y}(\tau_n)|$$

$$\leqslant M_S^n \mu \int_{[t_0, t]} d\tau_1 \int_{[t_0, \tau_1]} d\tau_2 \cdots \int_{[t_0, \tau_{n-1}]} d\tau_n = M_S^n \mu \frac{|t - t_0|^n}{n!} \leqslant M_S^n \mu \frac{(a+b)^n}{n!}$$

so that $\mathbf{x}(t) - \mathbf{y}(t) \equiv 0$ since n is arbitrary and it can be let to $+\infty$.

7.* Consider the equation $\dot{\mathbf{x}} = \mathbf{L}(t)\mathbf{x}$, in $\mathbf{R}^d$, where $t \to \mathbf{L}(t)$, $t \in \mathbf{R}$, is a $d \times d$-matrix-valued C^∞ function. Consider d solutions $\mathbf{x}^{(1)}, \dots, \mathbf{x}^{(d)}$ for $t \in I = [a, b]$ and call them "independent" if $\exists t_0 \in I$ such that the d vectors $\mathbf{x}^{(1)}(t_0), \dots, \mathbf{x}^{(d)}(t_0)$ are linearly independent. Show that if $t \in I$, then also $\mathbf{x}^{(1)}(t), \dots, \mathbf{x}^{(d)}(t)$ are linearly independent whenever they are such for $t = t_0$ and, furthermore, any solution $t \to \mathbf{y}(t)$, $t \in I$, can be represented as $\mathbf{y}(t) = \sum_{j=1}^d A_j \mathbf{x}^{(j)}(t)$, $\forall t \in I$. (Hint: If for $t = \bar{t}$, the d vectors were not independent, we could find constants $\bar{A}_1, \dots, \bar{A}_d$, not all equal to zero, such that $\sum_{j=1}^d \bar{A}_j \mathbf{x}^{(j)}(\bar{t}) = \mathbf{0}$; hence, by linearity and uniqueness, $\sum_{j=1}^d \bar{A}_j \mathbf{x}^{(j)}(t) = \mathbf{0}$, $\forall t \in I$, which contradicts the independence for $t = t_0$.)

8. Show that Problem 7 implies that, given d solutions $t \to \mathbf{x}^{(1)}(t), \dots, \mathbf{x}^{(d)}(t)$, $t \in I$, to the equation $\dot{\mathbf{x}} = \mathbf{L}(t)\mathbf{x}$, then the matrix $W(t)$ ("Wronskian matrix of $\mathbf{x}^{(1)}, \dots, \mathbf{x}^{(d)}$") defined by

$$W_{ij}(t) = x_j^{(i)}(t), \qquad i, j = 1, 2, \dots, d, \quad t \in I$$

has a determinant $w(t)$ nonvanishing for $t \in I$ if and only if $\exists t_0 \in I$ such that $w(t_0) \neq 0$. (Hint: By linear algebra, this is just another way of phrasing Problem 7. Recall that d vectors are linearly independent if and only if the "determinant of their components" is not zero.)

9. Using the determinant differentiation rule, by rows, show that

$$\frac{d}{dt} w(t) \equiv \frac{d}{dt} \det W(t) = \left(\sum_{i=1}^{d} l_{ii}(t) \right) w(t);$$

hence, if $\sum_{i=1}^d l_{ii}(t) = l(t)$, one has $w(t) = w(t_0) \exp \int_{t_0}^t l(\tau)\, d\tau$.

10. In the context of Problem 8, suppose that the matrix function $t \to \mathbf{L}(t)$, $t \in \mathbf{R}$, is periodic with period $T > 0$, i.e., $t \to l_{ij}(t)$, $i, j = 1, \dots, d$, are T-periodic functions. Let $\mathbf{x}^{(1)}, \dots, \mathbf{x}^{(d)}$ be d linearly independent solutions for $t \geqslant 0$. Then there exist d^2 constants $A_j^{(i)}$, $i, j = 1, \dots, d$, such that

$$\mathbf{x}^{(i)}(t + T) = \sum_{j=1}^{d} A_j^{(i)} \mathbf{x}^{(j)}(t), \qquad t \geqslant 0.$$

Show that $\det W(T)/\det W(0) = w(T)/w(0) = \det A \neq 0$.

11. Suppose that the matrix A is similar, via a real nonsingular matrix S, to a real diagonal matrix Λ, $\Lambda_{ij} = \lambda_i \delta_{ij}$, $i, j = 1, \dots, d$: $SAS^{-1} = \Lambda$. Then, in the context of Problem 10, define

$$\mathbf{y}^{(i)}(t) = \sum_{j=1}^{d} S_{ij} \mathbf{x}^{(j)}(t).$$

Show that $\mathbf{y}^{(1)}, \dots, \mathbf{y}^{(d)}$ are linearly independent solutions and

$$\mathbf{y}^{(i)}(t + T) = \lambda_i \mathbf{y}^{(i)}(t), \qquad t \geqslant 0$$

with $\lambda_1, \dots, \lambda_d \neq 0$.

12. Suppose that A is a matrix similar to a diagonal matrix Λ via a complex nonsingular matrix S. Show that $\mathbf{y}^{(1)}, \dots, \mathbf{y}^{(d)}$, defined as in the preceding problem, are complex solutions of $\dot{\mathbf{x}} = \mathbf{L}(t)\mathbf{x}$ and that $\mathbf{y}^{(i)}(t + T) = \lambda_i \mathbf{y}^{(i)}(t)$, $\forall t \geqslant 0$, $i = 1, 2, \dots, d$. (For applications, recall that from linear algebra (see Appendix E), a sufficient condition for the similarity between A and a diagonal matrix Λ, $\Lambda_{ij} = \lambda_i \delta_{ij}$ is that the roots $\lambda_1, \dots, \lambda_d$ of the secular equation $\det(A - \lambda) = 0$ are pairwise different.)

13. Given the assumptions of Problem 11 and supposing $\lambda_1, \ldots, \lambda_d > 0$, show that the most general solution to $\dot{\mathbf{x}} = \mathbf{L}(t)\mathbf{x}$ has the form

$$\mathbf{x}(t) = \sum_{j=1}^{d} \alpha_j \lambda_j^{t/T} \mathbf{z}^{(j)}(t),$$

where the functions $\mathbf{z}^{(1)}, \ldots, \mathbf{z}^{(d)}$ are d C^∞ functions periodic with period T, and $\alpha_1, \ldots, \alpha_d$ are arbitrary constants. (Hint: Let $\mathbf{z}^{(i)}(t) = \lambda_i^{-t/T}\mathbf{y}^{(i)}(t)$.)

14. Suppose that for every nonzero complex number, there exists a C^∞ function $t \to \gamma(t)$, $t \in \mathbf{R}$, such that $\gamma(t + t') = \gamma(t)\gamma(t')$, $\gamma(0) = 1$, $\gamma(T) = \lambda^{-1}$, $\gamma(t) \neq 0$, $\forall t \in \mathbf{R}$; then the conclusions of Problem 13 would hold, setting $\lambda^{-t/T} \equiv \gamma(t)$, without the assumption $\lambda_j > 0$, $j = 1, \ldots, d$, but only with the assumption $\det A \neq 0$. See also the following problem.

15. Let $\lambda \in \mathbf{C}$, $\lambda^{-1} = \rho(\cos\theta + i\sin\theta)$, $\rho > 0$, $\theta \in [0, 2\pi]$. Define $\gamma(t) = \rho^{t/T}(\cos(\theta t/T) + i\sin(\theta t/T))$. Show that $\gamma(0) = 1$, $\gamma(t)\gamma(t') = \gamma(t + t')$, $\gamma(T) = \lambda^{-1}$, $\gamma(t) \neq 0$, $\forall t \in \mathbf{R}$ (e.g., $(-1)^{t/T} = \cos(\pi t/T) + i\sin(\pi t/T)$).

Observations to Problems 8–15.

We shall see that there always exist d linearly independent solutions to $\dot{\mathbf{x}} = \mathbf{L}(t)\mathbf{x}$. However, the existence of S is a restrictive condition. When such an S does not exist, it is possible to show that the most general solution to $\dot{\mathbf{x}} = \mathbf{L}(t)\mathbf{x}$, with $\mathbf{L}$ periodic with period $T > 0$ and C^∞ can be written in the form

$$\mathbf{x}(t) = \sum_{j=1}^{p} \sum_{k=0}^{\delta(j)-1} \alpha_{jk} \lambda_j^{t/T} t^k z^{(j)}(t),$$

where $\sum_{j=1}^{p}\delta(j) = d$, and $\delta(j), \lambda_j$, are suitably chosen, and $t \to \mathbf{z}^{(j)}(t)$, $t \geqslant 0$, are C^∞ functions periodic with period T and possibly complex valued (when λ_j are not positive and $\lambda_j^{t/T}$ is interpreted as explained in Problem 15), and α_{jk} are arbitrary constants (see, for instance, H. Poincaré, Les Methodes Nouvelles de la Mécanique Céléste, Vol. I, pp. 63–68, see References).

16. Consider a differential equation $\ddot{x} + a(t)\dot{x} + b(t)x = 0$, $t \in \mathbf{R}$, $a, b \in C^\infty(\mathbf{R})$. After reducing it to a first-order system of two differential equations in $\mathbf{R}^2$, interpret the results of Problems 7–15 in terms of its solutions. Show first that the matrix $W(t)$ associated to this system is expressed in terms of two of its solutions $t \to x^{(1)}(t)$ and $t \to x^{(2)}(t)$ as

$$W(t) = \begin{pmatrix} x^{(1)}(t) & \dot{x}^{(1)}(t) \\ x^{(2)}(t) & \dot{x}^{(2)}(t) \end{pmatrix}$$

and

$$\dot{w}(t) = a(t)w(t).$$

17.* Like Problem 16 for the sth-order differential equation in $\mathbf{R}$:

$$\frac{d^s x}{dt^s} + \sum_{j=0}^{s-1} a_j(t)\frac{d^j x}{dt^j} = 0, \qquad t \in \mathbf{R}.$$

§2.3. General Existence, Uniqueness, and Regularity Properties of the Motion. Existence

An existence problem for the solutions of Eq. (2.2.4), hence of Eq. (2.2.2), naturally associated with the uniqueness property given in Proposition 1, §2.2, is solved by the following proposition:

2 Proposition. *Let* $\mathbf{F}$ *be an* $\mathbf{R}^d$*-valued function in* $C^\infty(\mathbf{R}^d \times \mathbf{R})$.

Let $\xi_0 \in \mathbf{R}^d$ *and* $t_0 \in \mathbf{R}$. *Let* $S(\xi_0, \rho)$ *be the closed sphere in* $\mathbf{R}^d$ *with center* ξ_0 *and radius* ρ. *Let* $\theta > 0$.

There exists $T_{\rho,\theta} > 0$ *and a solution of Eq.* (2.2.4), *i.e.,* $\dot{\mathbf{x}} = \mathbf{F}(\mathbf{x}, t)$, *defined for* $t \in [t_0 - T_{\rho,\theta}, t_0 + T_{\rho,\theta}]$ *and of class* C^∞ *such that*:

$$\mathbf{x}(t_0) = \xi_0, \qquad \mathbf{x}(t) \in S(\xi_0, \rho), \qquad t \in \left[t_0 - T_{\rho,\theta}, t_0 + T_{\rho,\theta}\right]. \tag{2.3.1}$$

Furthermore, if one defines:

$$M_{\rho,\xi_0,t_0,\theta} = \max_{\substack{\xi \in S(\xi_0,\rho) \\ t \in [t_0 - \theta, t_0 + \theta]}} \mathbf{F}(\xi, t) \equiv M, \tag{2.3.2}$$

one can choose:

$$T_{\rho,\theta} = \rho\theta(\rho + \theta M)^{-1}. \tag{2.3.3}$$

Observations.

(1) By Proposition 1, §2.2, it is enough to show the existence of a $C^{(1)}$ solution verifying Eq. (2.3.1).

(2) The proof that follows is a "constructive one" in the sense that it provides a sequence $t \to \mathbf{x}^{(n)}(t)$, $t \in [T_0 - T_{\rho.\theta}, T_0 + T_{\rho,\theta}]$, of functions approximating (as $n \to +\infty$) the solution and, at the same time, it provides an estimate of the approximation error defined as $\max|x(t) - x^{(n)}(t)|$, where the maximum is taken on the interval $[t_0 - T_{\rho,\theta}, t_0 + T_{\rho,\theta}]$.

(3) It is often useful, in applications, not to follow the solution scheme proposed by the following proof of Proposition 2. It might, in fact, be more convenient to use *ad hoc* procedures based on the particular features of the $\mathbf{F}$ under analysis in a concrete case. Usually, with such procedures one finds much better error estimates than the ones following from general methods where one cannot take into account some special properties of the equations (e.g., symmetry properties, Hamiltonian form, etc.).

(4) It is important to understand informally the bound on the magnitude of the interval of existence.

In fact, during the proof, it appears necessary to have an a priori control of how far $\mathbf{x}(t)$ can travel away from the initial position ξ_0.

The continuity of $\mathbf{F}$ guarantees the boundedness of the maximum of $|\mathbf{F}(\xi, t)|$ for, say, $\xi \in S(\xi_0, \rho)$, $t \in [t_0 - \theta, t_0 + \theta]$. It follows that during the whole time interval $[t_0 - T_{\rho,\theta}, t_0 + T_{\rho,\theta}]$, the point $\mathbf{x}(t)$ stays inside $S(\xi_0, \rho)$ because $\dot{\mathbf{x}}(t) = \mathbf{F}(\mathbf{x}(t), t)$ and the right-hand side of this relation does not exceed M, Eq. (2.3.2): notice, in fact, that $T_{\rho,\theta}$ has been chosen—just to achieve this effect—smaller than both θ and ρM^{-1} (i.e., $T_{\rho,\theta} = (\theta^{-1} + \rho^{-1}M)^{-1}$ so that $MT_{\rho,\theta} < \rho$).

The interval $[t_0 - T_{\rho,\theta}, t_0 + T_{\rho,\theta}]$ is certainly not optimal, at least because the choice of the set $S(\xi_0, \rho) \times [t_0 - \theta, t_0 + \theta]$, where the maximum of $|F|$ is considered, was arbitrary. A better existence interval could be obtained using this arbitrariness and optimizing the result over the possible sets on which one takes the maximum.

Also, once the existence of a solution verifying Proposition 2 has been established, one could apply Proposition 2 and Proposition 1 to the equation with initial datum $\mathbf{x}(t_0 + T_{\rho,\theta})$ at the initial time $t_0 + T_{\rho,\theta}$, thus continuing it beyond $T_{\rho,\theta}$.

(5) Finally one cannot hope, in general, for an infinite existence interval containing $\mathbf{R}_+$: this can be seen through counterexamples. The simplest among them is provided by the equation $\dot{x} = x^2$, $x(0) = 1$, in $\mathbf{R}$.

Proof. Rather than studying $C^{(1)}$ solutions of $\dot{\mathbf{x}} = \mathbf{F}(\mathbf{x}, t)$ verifying the initial conditions (2.3.1), we shall study $\mathbf{R}^d$-valued $C^{(0)}([t_0 - T_{\rho,\theta}, t_0 + T_{\rho,\theta}])$ solutions of the equation:

$$\mathbf{x}(t) = \xi_0 + \int_{t_0}^{t} \mathbf{F}(\mathbf{x}(\tau), \tau)\, d\tau. \tag{2.3.4}$$

It clearly appears that every $C^{(0)}([t_0 - T_{\rho,\theta}, t_0 + T_{\rho,\theta}])$ function verifying Eq. (2.3.4) is a $C^{(1)}$ solution to the original equation also verifying Eq. (2.3.1), and vice versa.

For $t \in [t_0 - T_{\rho,\theta}, t_0 + T_{\rho,\theta}]$ define the sequence of $\mathbf{R}^d$-valued functions $t \to \mathbf{x}^{(n)}(t)$, $n = 0, , \ldots,$ through the following recursive scheme:

$$\begin{aligned}
&\mathbf{x}^{(0)}(t) = \xi_0,\\
&\mathbf{x}^{(1)}(t) = \xi_0 + \int_{t_0}^{t} \mathbf{F}\big(\mathbf{x}^{(0)}(\tau), \tau\big)\, d\tau, \qquad t \in \mathbf{R}\\
&\vdots\\
&\mathbf{x}^{(n)}(t) = \xi_0 + \int_{t_0}^{t} \mathbf{F}\big(\mathbf{x}^{(n-1)}(\tau), \tau\big)\, d\tau,
\end{aligned} \tag{2.3.5}$$

and notice that each such function is in $C^{\infty}(\mathbf{R})$.

Let us show the existence, uniformly in $t \in [t_0 - T_{\rho,\theta}, t_0 + T_{\rho,\theta}]$, of the limit function

$$\lim_{n \to \infty} \mathbf{x}^{(n)}(t) = \mathbf{x}(t). \tag{2.3.6}$$

By the uniformity claimed for the limit of Eq. (2.3.6), the limit function will also be continuous, since $\mathbf{x}^{(n)}$ is such, and we can hope with some assurance that it will be a solution of Eq. (2.3.5).

The existence and uniformity of the limit of Eq. (2.3.6) will be proved by rewriting it as

$$\mathbf{x}^{(0)}(t) + \sum_{k=1}^{\infty} \big(\mathbf{x}^{(k)}(t) - \mathbf{x}^{(k-1)}(t)\big) \tag{2.3.7}$$

and deducing that if

$$\mu_k = \max_{t\in[t_0-T_{\rho,\theta},t_0+T_{\rho,\theta}]} |\mathbf{x}^{(k)}(t) - \mathbf{x}^{(k+1)}(t)|, \tag{2.3.8}$$

the series

$$\sum_{k=0}^{+\infty} \mu_k < +\infty \tag{2.3.9}$$

is convergent. This will mean that the series of Eq. (2.3.7) is uniformly convergent for $t \in [t_0 - T_{\rho,\theta}, t_0 + T_{\rho,\theta}]$ and, therefore, the same can be said of the limit of Eq. (2.3.6).

To estimate μ_k we can refer to Eq. (2.3.5) and notice that for $k = 2, 3, \ldots,$

$$\mathbf{x}^{(k)}(t) - \mathbf{x}^{(k-1)}(t) = \int_{t_0}^{t}\Big(\mathbf{F}\big(\mathbf{x}^{(k-1)}(\tau),\tau\big) - \mathbf{F}\big(\mathbf{x}^{(k-2)}(\tau),\tau\big)\Big)\,d\tau. \tag{2.3.10}$$

Then using the Lagrange theorem in the form

$$\begin{gathered} |\mathbf{F}(\xi,\tau) - \mathbf{F}(\eta,\tau)| \leqslant L|\xi - \eta|, \\ \forall \xi,\eta \in S(\xi_0,\rho), \qquad \forall \tau \in \big[t_0 - T_{\rho,\theta}, t_0 + T_{\rho,\theta}\big] \end{gathered} \tag{2.3.11}$$

where

$$L = \max_{\substack{\xi\in S(\xi_0,\rho)\\ t\in[t_0-T_{\rho,\theta},t_0+T_{\rho,\theta}]}} \sum_{i,j=1}^{d} \left|\frac{\partial F^{(i)}}{\partial \xi_j}(\xi,t)\right|, \tag{2.3.12}$$

we can deduce from Eqs. (2.3.10) and (2.3.11):

$$|\mathbf{x}^{(k)}(t) - \mathbf{x}^{(k-1)}(t)| \leqslant L\int_{[t_0,t]} |\mathbf{x}^{(k-1)}(\tau) - \mathbf{x}^{(k-2)}(\tau)|\,d\tau, \tag{2.3.13}$$

$\forall k = 2, 3, \ldots,$ provided we preliminarily check that for all $k = 0, 1, \ldots,$ the functions $t \to \mathbf{x}^{(k)}(t)$, $t \in [t_0 - T_{\rho,\theta}, t_0 + T_{\rho,\theta}]$, take their values in $S(\xi_0, \rho)$.

This last property can be easily proved inductively starting from Eq. (2.3.5) by virtue of the choice of $T_{\rho,\theta}$ (chosen, as essentially stated in Observation (4), just in such a way to make this property true): by induction suppose that $|\mathbf{x}^{(h)}(t) - \xi_0| \leqslant \rho$, $\forall h = 0, \ldots, k-1$; it is a property which holds for $k = 1$. Let us show that $|\mathbf{x}^{(k)}(t) - \xi_0| \leqslant \rho$. From Eqs. (2.3.5) and (2.3.3) we have:

$$\begin{aligned} |\mathbf{x}^{(k)}(t) - \xi_0| &\leqslant \int_{[t_0,t]} d\tau \big|\mathbf{F}\big(\mathbf{x}^{(k-1)}(\tau),\tau\big)\big| \leqslant M_{\rho,\xi_0,t_0,\theta}|t - t_0| \\ &\leqslant M_{\rho,\xi_0,t_0,\theta} T_{\rho,\theta} < \rho. \end{aligned} \tag{2.3.14}$$

Returning to Eq. (2.3.13), we see that it is implied recursively, and then by Eq. (2.3.14) with $k = 1$: $t \in [t_0 - T_{\rho,\theta}, t_0 + T_{\rho,\theta}]$,

$$|\mathbf{x}^{(k)}(t) - \mathbf{x}^{(k-1)}(t)| \leqslant L^{k-1} \int_{[t_0,t]} d\tau_1 \int_{[t_0,\tau_1]} d\tau_2 \ldots \times \int_{[t_0,\tau_{k-2}]} d\tau_{k-1} |\mathbf{x}^{(1)}(\tau_{k-2}) - \boldsymbol{\xi}_0| \leqslant \frac{L^{k-1} T_{\rho,\theta}^{k-1}}{(k-1)!} \rho \tag{2.3.15}$$

because $T_{\rho,\theta} \geqslant |t - t_0|$.

Obviously Eq. (2.3.15) shows the convergence of the series of Eq. (2.3.9) and, therefore, the limit of Eq. (2.3.6) exists uniformly for $t \in [t_0 - T_{\rho,\theta}, t_0 + T_{\rho,\theta}]$ and defines a function $t \to \mathbf{x}(t)$ on this interval with values in $S(\boldsymbol{\xi}_0, \rho)$. This function verifies Eq. (2.3.4) as it can be immediately seen by taking the $n \to \infty$ limit in Eq. (2.3.5) and by using the uniformity of the limit of Eq. (2.3.6) to exchange the integration with the limit. mbe

Exercises and Problems for §2.3

1. Give a lower estimate for the magnitude of $T_{\rho,\theta}$, the existence interval's amplitude as in Proposition 2, for the following second-order equations, assuming $x(0) = 0$, $\dot{x}(0) = 1$ or $x(0) = 1$, $\dot{x}(0) = 0$ as initial data at $t_0 = 0$:

$$\ddot{x} = x, \qquad \ddot{x} = x + x^3, \qquad \ddot{x} = x - \dot{x} + x^3, \qquad \ddot{x} = -\dot{x}^2, \qquad \ddot{x} = -\sin x.$$

Also estimate $\sup_{\rho,\theta} T_{\rho,\theta}$ from below (Hint: Reduce the equation to first order and then apply Proposition 1.)

2. Solve the equations $\dot{x} = -x^2$, $\dot{x} = \cos x$, $\dot{x} = (\cos x)^2$ with initial datum $x(0) = 1$.

3. Solve the equation $\ddot{x} = x$ with initial datum $x(0) = 1$, $\dot{x}(0) = 0$.

4. Solve the equation $\dot{x} = x + y$, $\dot{y} = -x + 2y$ with initial datum $x(0) = 0$, $y(0) = 1$.

5. Using the "quadrature method", solve the equation $\ddot{x} = 4(x^3 - x)$, $x(0) = 0$, $\dot{x}(0) = \sqrt{2}$ (see §2.1, final comment).

6. As in Problem 5 for $\ddot{x} = -(4x^3 + 6x^2 - 2)$, $x(0) = 0$, $\dot{x}(0) = \sqrt{2}$.

7. Find two linearly independent solutions for the equation in Problem 4.

8.* Compute $w(t)$ for the equation in Problem 4 (see Problem 8, §2.2).

9.* Let $t \to \mathbf{L}(t)$ be a $d \times d$-matrix-valued C^∞ function on $\mathbf{R}$. Show that the equation $\dot{\mathbf{x}}(t) = \mathbf{L}(t)\mathbf{x}(t)$ admits d linearly independent solutions defined for $|t| \leqslant T$ with T small enough. (Hint: Let $\mathbf{x}^{(i)}$ be the solution with initial data $x_j^{(i)}(0) = \delta_{ij}$, $i, j = 1, \ldots d$. Then evaluate an existence interval for such initial data.)

10.* Compute $T_{1,1}$ for the equation in Problem 9 when $|t_0| \leqslant \sigma$ and $\boldsymbol{\xi}_0$ is arbitrary, $\boldsymbol{\xi}_0 = \mathbf{x}(t_0)$; for the symbols, see Proposition 1. Show that $|\boldsymbol{\xi}_0| T_{1,1}$ can be taken to be t_0 and $\boldsymbol{\xi}_0$ independent at a given $\sigma > 0$. Deduce from this that every solution to $\dot{\mathbf{x}} = \mathbf{L}(t)\mathbf{x}$ can be extended to a solution defined for $t \in \mathbf{R}$.

11. Let $\mathbf{L}$ be a $d \times d$ matrix and consider the equation $\dot{\mathbf{x}} = L\mathbf{x}$ in $\mathbf{R}^d$. Suppose that $\mathbf{L}$ has d *pairwise distinct real eigenvalues* (see Appendix E for the eigenvalue notion) $\lambda_1, \ldots, \lambda_d$. Let $\mathbf{v}^{(1)}, \ldots, \mathbf{v}^{(d)}$ be the respective real linearly independent eigenvectors (see Appendix E). Show that the functions $t \to e^{\lambda_i t}\mathbf{v}^{(i)}$ are d linearly independent solutions. Show that any solution $t \to \mathbf{x}(t)$ has the form

$$\mathbf{x}(t) = \sum_{j=1}^{d} \alpha_j e^{\lambda_j t}\mathbf{v}^{(j)}$$

with $(\alpha_1, \ldots, \alpha_d) \in \mathbf{R}^d$.

§2.4. General Existence, Uniqueness, and Regularity Properties of Motion. Regularity

We saw that $C^{(1)}$ solutions of Eq. (2.2.4), $\dot{\mathbf{x}} = \mathbf{F}(\mathbf{x}, t)$, are necessarily C^∞ solutions.

This is the simplest regularity property shown by the solutions of such differential equations with $\mathbf{F}$ in C^∞.

We shall now analyze other regularity properties of the solutions of Eq. (2.2.4).

In applications it often happens that the right-hand side of Eq. (2.2.4) depends on parameters $\boldsymbol{\alpha} \in \mathbf{R}^m$ and that, furthermore, it is important to know how the solutions change as the initial data $\boldsymbol{\xi}_0$ and the parameters $\boldsymbol{\alpha}$ vary in $\mathbf{R}^d$ and $\mathbf{R}^m$, respectively.

A first answer to this question is provided by the following proposition.

3 Proposition. *Let $\boldsymbol{\xi}, t, \boldsymbol{\alpha} \to \mathbf{F}(\boldsymbol{\xi}, t, \boldsymbol{\alpha})$ be a $C^\infty(\mathbf{R}^d \times \mathbf{R} \times \mathbf{R}^m)$ function taking its values in $\mathbf{R}^d$, and consider the equation*

$$\mathbf{x}(t) = \boldsymbol{\xi}_0 + \int_{t_0}^{t} \mathbf{F}(\mathbf{x}(\tau), \tau, \boldsymbol{\alpha}_0)\, d\tau \tag{2.4.1}$$

as an equation for the continuous function $t \to \mathbf{x}(t)$ parametrized by $\boldsymbol{\xi}_0, t_0, \boldsymbol{\alpha}_0 \in \mathbf{R}^d \times \mathbf{R} \times \mathbf{R}^m$.

Given $\rho, \theta, a > 0$ and $(\bar{\boldsymbol{\xi}}, \bar{t}, \bar{\boldsymbol{\alpha}}) \in \mathbf{R}^d \times \mathbf{R} \times \mathbf{R}^m$, there exists $T > 0$ such that:

(i) *Equation (2.4.1) admits a solution for every $(\boldsymbol{\xi}_0, t_0, \boldsymbol{\alpha}_0)$ close enough to $(\bar{\boldsymbol{\xi}}, \bar{t}, \bar{\boldsymbol{\alpha}})$, i.e., such that $|\bar{\boldsymbol{\xi}} - \boldsymbol{\xi}_0| < \rho/2$, $|\bar{t} - t_0| < \theta/2$, $|\bar{\boldsymbol{\alpha}} - \boldsymbol{\alpha}_0| < a$. Such solution will be denoted $t \to S_t(\boldsymbol{\xi}_0; t_0, \boldsymbol{\alpha}_0)$ and it is defined for $t \in [t_0 - T, t_0 + T]$.*

(ii) *The function $S_t(\boldsymbol{\xi}_0; t_0, \boldsymbol{\alpha}_0)$, defined for*

$$|\boldsymbol{\xi}_0 - \bar{\boldsymbol{\xi}}| \leqslant \frac{\rho}{2}, \quad |t_0 - \bar{t}| \leqslant \frac{\theta}{2}, \quad |\boldsymbol{\alpha}_0 - \boldsymbol{\alpha}| \leqslant a, \quad |t - \bar{t}| \leqslant T \tag{2.4.2}$$

takes its values inside the sphere $S(\bar{\boldsymbol{\xi}}; \rho)$ with center $\bar{\boldsymbol{\xi}}$ and radius ρ and it is a C^∞ function of its arguments.

(iii) *The value T can be taken as:*

$$T = \frac{\rho\theta}{2(\rho + \theta \max|\mathbf{F}(\boldsymbol{\xi}, \boldsymbol{\alpha}, t)|)}, \tag{2.4.3}$$

where the maximum is considered on the set $|\boldsymbol{\xi}-\bar{\boldsymbol{\xi}}| \leqslant \rho$, $|\boldsymbol{\alpha}-\bar{\boldsymbol{\alpha}}| \leqslant a$, $|t-\bar{t}| \leqslant \theta$.

Observations.

(1) Equation (2.4.1) is equivalent to

$$\dot{\mathbf{x}}(t)=\mathbf{F}(\mathbf{x}(t),t,\boldsymbol{\alpha}_0), \qquad \mathbf{x}(t_0)=\boldsymbol{\xi}_0, \tag{2.4.4}$$

and, therefore, the above proposition provides a regularity theorem for the solutions of Eq. (2.4.4) as functions of the initial data, of the initial time, of time itself, and of the parameters $\boldsymbol{\alpha}$ on which $\mathbf{F}$ may possibly depend.

The set (2.4.2) and the key estimate (2.4.3) should not be taken too seriously as they are not optimal: they merely show an example of the type of concreteness that can be attained in the formulation of a regularity criterion (see, also, Observation 4, p. 18).

(2) Let $\boldsymbol{\beta}=(\beta_1,\ldots,\beta_{d+m+2})\equiv((\boldsymbol{\xi}_0)_1,\ldots,(\boldsymbol{\xi}_0)_d,(\boldsymbol{\alpha}_0)_1,\ldots,(\boldsymbol{\alpha}_0)_m,t,t_0)$ and let

$$\begin{aligned}\mathbf{x}(t)&=(x_1(t),\ldots,x_d(t))\\&=S_t(\boldsymbol{\xi}_0;t_0,\boldsymbol{\alpha}_0)\equiv((S_t(\boldsymbol{\xi}_0;t_0,\boldsymbol{\alpha}_0))_1,\ldots,(S_t(\boldsymbol{\xi}_0;t_0,\boldsymbol{\alpha}_0))_d).\end{aligned} \tag{2.4.5}$$

A formal differentiation of Eq. (2.4.4) with respect to β_i, $i=1,2,\ldots,m+d$, gives

$$\frac{d}{dt}\frac{\partial\mathbf{x}(t)}{\partial\beta_i}=\sum_{h=1}^{d}\frac{\partial\mathbf{F}}{\partial\xi_h}(\mathbf{x}(t),t,\boldsymbol{\alpha}_0)\frac{\partial x_h(t)}{\partial\beta_i}+\sum_{k=1}^{m}\frac{\partial\mathbf{F}}{\partial\alpha_k}(\mathbf{x}(t),t,\boldsymbol{\alpha}_0)\frac{\partial\alpha_k}{\partial\beta_i} \tag{2.4.6}$$

and

$$\left(\frac{\partial\mathbf{x}(t)}{\partial\beta_i}\right)_{t=t_0}=\frac{\partial\boldsymbol{\xi}_0}{\partial\beta_i}. \tag{2.4.7}$$

One could also obtain analogous equations for the higher-order derivatives.

(3) We shall see from the proof that the above $d(m+d)$ derivatives $\partial x_j(t)/\partial\beta_i$ do actually verify these equations.

(4) The $d(m+d)$ equations (2.4.6) and (2.4.7) can be considered by imagining that $\mathbf{x}(t)$ is a known function [obtained by first solving Eq. (2.4.4)]. Then for each $i=1,\ldots,m+d$, Eq. (2.4.6) can be thought of as a system of d differential equations for the functions of t, $t\to\partial\mathbf{x}(t)/\partial\beta_i$, with initial data at t_0 given by Eq. (2.4.7).

Each such system can be solved by regarding it as an ordinary *linear* system of differential equations of the type (2.2.4) with suitable initial data.

Actually this is a method to compute the derivatives $\partial x_i/\partial\beta_i$ which, as we shall see in several instances, turns out to be quite useful. It is also very useful in numerical analysis.

(5) Similarly, the following equations for the t or t_0 derivatives follow

from Eq. (2.4.4):

$$\frac{\partial \mathbf{x}(t)}{\partial t} = \mathbf{F}(\mathbf{x}(t), t, \boldsymbol{\xi}_0), \qquad \mathbf{x}(t_0) = \boldsymbol{\xi}_0 \tag{2.4.8}$$

and

$$\begin{aligned} \frac{d}{dt}\frac{\partial \mathbf{x}(t)}{\partial t_0} &= \sum_{h=1}^{d} \frac{\partial \mathbf{F}}{\partial \xi_h}(\mathbf{x}(t), t, \boldsymbol{\alpha}_0)\frac{\partial x_h(t)}{\partial t_0}, \\ \left(\frac{\partial \mathbf{x}(t)}{\partial t_0}\right)_{t=t_0} &= -\mathbf{F}(\boldsymbol{\xi}_0, t_0, \boldsymbol{\alpha}_0) \end{aligned} \tag{2.4.9}$$

to which remarks (3) and (4) apply.

(6) Had $\mathbf{F}$ been assumed to be a $C^{(k)}$ function, $k \geqslant 1$, on $\mathbf{R}^d \times \mathbf{R} \times \mathbf{R}^m$ one could still have obtained a regularity result: however, one could only show, with the same proof that follows, that the function $(t, \boldsymbol{\xi}_0, \boldsymbol{\alpha}_0, t_0) \to S_t(\boldsymbol{\xi}_0; t_0, \boldsymbol{\alpha}_0)$ is a $C^{(k)}$ function in the region of Eq. (2.4.2).

(7) Proposition 3 also yields a regularity theorem for the solutions of higher-order differential equations, of the type considered in Eq. (2.2.2), via the reduction to first order described in Eq. (2.2.3). The explicit statement of the corresponding results is left as a problem for the reader.

Proof. This proof is essentially a repetition of the proof of Proposition 2, on the existence property. Here we provide only a sketch of the proof, leaving to the reader the elaboration of the details, if he deems it necessary.

The statement about the existence (and uniqueness) of the solutions $t \to S_t(\boldsymbol{\xi}_0; t_0, \boldsymbol{\alpha}_0)$ follows easily from Proposition 2: Proposition 2 also implies the estimate (2.4.3) for T which follows from Eq. (2.3.3), identifying the parameters ρ, θ of Proposition 2 with $\rho/2, \theta/2$.

Let us show that $(\boldsymbol{\xi}_0, t, t_0, \boldsymbol{\alpha}_0) \to S_t(\boldsymbol{\xi}_0; t_0, \boldsymbol{\alpha}_0)$ is a $C^{(1)}$ function on the set (2.4.2).

Let $\boldsymbol{\beta} = (\beta_1, \ldots, \beta_{m+d+2})$ be defined as in Observation 2. Recall that, as seen in §2.3, we can think of $t \to S_t(\boldsymbol{\xi}_0; t_0, \boldsymbol{\alpha}_0)$ as being obtained via a limit of the functions $t \to \mathbf{x}^{(n)}(t, \boldsymbol{\xi}_0, t_0, \boldsymbol{\alpha}_0)$ recursively defined for $t \in [t_0 - T, t_0 + T]$ by

$$\begin{aligned} &\mathbf{x}^{(0)}(t, \boldsymbol{\xi}_0, t_0, \boldsymbol{\alpha}_0) = \boldsymbol{\xi}_0 \\ &\quad \vdots \\ &\mathbf{x}^{(n)}(t, \boldsymbol{\xi}_0, t_0, \boldsymbol{\alpha}_0) = \boldsymbol{\xi}_0 + \int_{t_0}^{t} \mathbf{F}\big(\mathbf{x}^{(n-1)}(\tau, \boldsymbol{\xi}_0, t_0, \boldsymbol{\alpha}_0), \tau, \boldsymbol{\alpha}_0\big)\, d\tau \end{aligned} \tag{2.4.10}$$

for $n = 1, 2, 3, \ldots$.

The functions $(t, \boldsymbol{\xi}_0, \boldsymbol{\alpha}_0, t_0) \to (\mathbf{x}^{(n)}(t, \boldsymbol{\xi}_0, t_0, \boldsymbol{\alpha}_0))$ are obviously [see Eq. (2.4.10)] C^∞ functions of their arguments, $\forall n$.

Furthermore, by differentiating Eq. (2.4.10) with respect to β_i, i

$= 1, \ldots, m + d + 2$, one finds:

$$\begin{aligned}\frac{\partial \mathbf{x}^{(n)}(t, \boldsymbol{\xi}_0, t_0, \boldsymbol{\alpha}_0)}{\partial \beta_i} &= \frac{\partial \boldsymbol{\xi}_0}{\partial \beta_i} + \int_{t_0}^{t} \left\{ \sum_{j=1}^{d} \frac{\partial \mathbf{F}}{\partial \xi_j} \left(\mathbf{x}^{(n-1)}(\tau, \boldsymbol{\xi}_0, t_0, \boldsymbol{\alpha}_0)\right) \frac{\partial x_j^{(n-1)}(\tau, \boldsymbol{\xi}_0, t_0, \boldsymbol{\alpha}_0)}{\partial \beta_i} \right. \\ &\quad \left. + \sum_{l=1}^{m} \frac{\partial \mathbf{F}}{\partial \alpha_l} \left(\mathbf{x}^{(n-1)}(\tau, \boldsymbol{\xi}_0, t, \boldsymbol{\alpha}_0), \tau, \boldsymbol{\alpha}_0\right) \frac{\partial \alpha_l}{\partial \beta_i} \right\} d\tau \\ &\quad + \mathbf{F}\left(\mathbf{x}^{(n-1)}(t, \boldsymbol{\xi}_0, t_0, \boldsymbol{\alpha}_0), t, \boldsymbol{\alpha}_0\right) \frac{\partial t}{\partial \beta_i} - \mathbf{F}(\boldsymbol{\xi}_0, t_0, \boldsymbol{\alpha}_0) \frac{\partial t_0}{\partial \beta_i},\end{aligned} \tag{2.4.11}$$

where the last two terms arise from the contributions to the derivatives by the integration extremes and, furthermore, the identity $\mathbf{x}^{(n-1)}(t_0, \boldsymbol{\xi}_0, t_0, \boldsymbol{\alpha}_0) \equiv \boldsymbol{\xi}_0$ has been used. This relation between $\partial \mathbf{x}^{(n)}/\partial \beta_i$ and $\partial \mathbf{x}^{(n-1)}/\partial \beta_i$ can be used to estimate the differences $\partial \mathbf{x}^{(n)}/\partial \beta_i - \partial \mathbf{x}^{(n-1)}/\partial \beta_i$ along the lines of the proof of Proposition 2.

By proceeding in the same way and remarking that Eq. (2.4.10), by the choice of T, implies $\forall t \in [t_0 - T, t_0 + T]$ and $\forall n = 0, 1, \ldots,$

$$|\mathbf{x}^{(n)}(t, \boldsymbol{\xi}_0, t_0, \boldsymbol{\alpha}_0) - \boldsymbol{\xi}_0| \leq \frac{\rho}{2}; \tag{2.4.12}$$

and one can deduce from Eqs. (2.4.11) and (2.4.12) that there are two constants $\overline{M}, \overline{L}$ [see Eqs. (2.3.12) and (2.3.15)] such that:

$$\left| \frac{\partial \mathbf{x}^{(n)}}{\partial \beta_i} (t, \boldsymbol{\xi}_0, t_0, \boldsymbol{\alpha}_0) \right| \leq \overline{M}, \tag{2.4.13}$$

$$\left| \frac{\partial \mathbf{x}^{(n)}}{\partial \beta_i} (t, \boldsymbol{\xi}_0, t_0, \boldsymbol{\alpha}_0) - \frac{\partial \mathbf{x}^{(n-1)}}{\partial \beta_i} (t, \boldsymbol{\xi}_0, t_0, \boldsymbol{\alpha}_0) \right| \leq \frac{\overline{L}^{n-1}}{(n-1)!} \overline{M} \tag{2.4.14}$$

hold for all $t, \boldsymbol{\xi}_0, t_0, \boldsymbol{\alpha}_0$ in the region of Eq. (2.4.2) and for all $n = 1, 2, \ldots$.

Then Eqs. (2.4.13) and (2.4.14) imply the existence and the uniformity, in the region of Eq. (2.4.2), of the limit:

$$\begin{aligned}\boldsymbol{\varphi}_i(t, \boldsymbol{\xi}_0, t_0, \boldsymbol{\alpha}_0) &= \lim_{n \to +\infty} \frac{\partial \mathbf{x}^{(n)}}{\partial \beta_i} (t, \boldsymbol{\xi}_0, t_0, \boldsymbol{\alpha}_0) \\ &= \frac{\partial \mathbf{x}^{(0)}}{\partial \beta_i} (t, \boldsymbol{\xi}_0, t_0, \boldsymbol{\alpha}_0) \\ &\quad + \sum_{n=1}^{\infty} \left(\frac{\partial \mathbf{x}^{(n)}}{\partial \beta_i} (t, \boldsymbol{\xi}_0, t_0, \boldsymbol{\alpha}_0) - \frac{\partial \mathbf{x}^{(n-1)}}{\partial \beta_i} (t, \boldsymbol{\xi}_0, t_0, \boldsymbol{\alpha}_0) \right),\end{aligned} \tag{2.4.15}$$

$\forall i = 1, 2, \ldots, m + d + 2$. The above limit is, therefore, a continuous function in the region of Eq. (2.4.2).

Since the limit $\lim_{n \to \infty} \mathbf{x}^{(n)}(t, \boldsymbol{\xi}_0, t_0, \boldsymbol{\alpha}_0)$ exists and equals $S_t(\boldsymbol{\xi}_0; t_0, \boldsymbol{\alpha}_0)$, the uniformity of the limit of Eq. (2.4.15) guarantees the permutability of the limit and of the $\partial/\partial\beta_i$ operations, thereby showing the differentiability of $S_t(\xi_0; t_0, \alpha_0)$ in the region of Eq. (2.4.2). It also shows, *en passant*, via the consideration of the limit as $n\to\infty$ of Eq. (2.4.11), the validity of the statements in Observation 3.

An essentially identical argument can be developed to show that $(t, \xi_0, t_0, \alpha_0) \to S_t(\xi_0; t_0, \alpha_0)$ is in class $C^{(p)}$, $\forall p \geqslant 1$, in the region of Eq. (2.4.2). It will suffice to differentiate Eq. (2.4.11) suitably many times to obtain relations analogous to it for the higher derivatives; such relations will then be used to obtain estimates analogous to Eqs. (2.4.13) and (2.4.14). mbe

Exercises and Problems for §2.4

1. Solve the equation $\ddot{x} - 2\dot{x} + \alpha x = 0$, $\alpha > 1$, with initial data $x(0) = x_0$, $\dot{x}(0) = v_0$, by finding two solutions of the form $t \to Ae^{\lambda t}$. By taking the limit $\alpha \to 1$, find the solution, with the same initial data, to $\ddot{x} - 2\dot{x} + x = 0$ (using Proposition 3).

2. Show that the equations $\ddot{x} = -\epsilon x$, $\epsilon > 0$, and $\ddot{x} = 0$ have, for the same initial conditions, solutions $x_\epsilon(t)$ and $x_0(t)$ such that $\lim_{\epsilon\to 0} x_\epsilon(t) = x_0(t)$, $\forall t \in \mathbf{R}$. However, show that this limit relation is not uniform in t, except for special initial data.

3. Consider the equation $\dot{\mathbf{x}} = \mathbf{F}(\mathbf{x}, t, \alpha)$ and suppose that $\mathbf{F}(\mathbf{0}, t, \alpha) \equiv \mathbf{0}$. Then, given $R > 0$ and fixed $t_0 = 0$, show the existence of $\epsilon > 0$, $\sigma > 0$, $\rho > 0$, such that:

$$(1-\sigma)|\mathbf{w}| \leqslant |S_t\mathbf{w}| \leqslant (1+\sigma)|\mathbf{w}|$$

for all $|t| \leqslant \epsilon$, $|\mathbf{w}| \leqslant \rho$, $|\alpha| \leqslant R$, having denoted $S_t\mathbf{w}$ the solution to the equation with initial datum $\mathbf{w}$ at $t_0 = 0$. (Hint: Apply Lagrange's theorem to estimate $|S_t\mathbf{w} - \mathbf{w}|$ in terms of the maximum of $|\mathbf{F}|$ in a suitable set, and then, likewise, $|S_\tau\mathbf{w} - S_\tau\mathbf{0}| \equiv |S_\tau\mathbf{w}|$, (as $S_\tau\mathbf{0} \equiv \mathbf{0}$), for $|t| \leqslant \epsilon$, $|\mathbf{w}| \leqslant \rho$: use the regularity theorem to bound the derivatives of $t, \mathbf{w}, \alpha \to S_t\mathbf{w}$; see Observations 2–5 to Proposition 3.)

§2.5. Local and Global Solutions of Differential Equations

The theory developed so far for the equation:

$$\dot{\mathbf{x}}(t) = \mathbf{F}(\mathbf{x}(t), t), \qquad \mathbf{x}(t_0) = \xi_0, \tag{2.5.1}$$

where $\mathbf{F}$ is an $\mathbf{R}^d$-valued function in $C^\infty(\mathbf{R}^d \times \mathbf{R})$, is a "local theory"; the existence theorem given in Proposition 2, §2.3, gives, in fact, a solution to Eq. (2.5.1) defined in a finite neighborhood of t_0.

For clear reasons, it is often necessary in applications to have "global solutions", i.e., solutions to Eq. (2.5.1) defined in time intervals containing a neighborhood of $\mathbf{R}_+ = [0, +\infty)$.

To analyze this problem, let us introduce the following definition.

2 Definition. A solution $t \to S_t(\xi_0; t_0)$ of Eq. (2.5.1) defined for $t \in (a, b)$ is called "maximal" if there are no other solutions defined in open intervals properly containing (a, b).

Given two solutions of Eq. (2.5.1) defined in two open intervals I_1 and I_2, it is clear that such solutions coincide in $I_1 \cap I_2$ (see Proposition 1, p. 13); if $I_2 \supset I_1$, one says that the second solution is a "continuation" of the first.

Observations.

(1) A solution to Eq. (2.5.1) is, therefore, maximal if and only if it "cannot be continued".

(2) Clearly, for every initial datum ξ_0 and every initial time t_0, there is a solution which is maximal: the interval of definition of such a solution is the union of all open intervals on which it is possible to define a solution to Eq. (2.5.1).

(3) This maximality definition only involves open intervals; however, this notion would be the same even if we allowed other types of intervals in the definition of maximality. To understand this, just use the existence theorem of §2.3, p. 18, to continue solutions of Eq. (2.5.1) out of closed or half-closed intervals.

The following proposition clarifies the above notion by showing that a solution of a differential equation can be nonglobal in the future (or in the past) if and only if it "diverges in a finite time".

4 Proposition. *Let $t \to S_t(\xi_0; t_0)$ be a maximal solution for Eq.* (2.5.1) *and let (a,b) be the interval on which this solution is defined. If $b < +\infty$, it must be*

$$\limsup_{t \to b^-} |S_t(\xi_0, t_0)| = +\infty; \tag{2.5.2}$$

if $a > -\infty$, it must be

$$\limsup_{t \to a^+} |S_t(\xi_0, t_0)| = +\infty. \tag{2.5.3}$$

PROOF. Assume $b < +\infty$ and that Eq. (2.5.2) does not hold. Then there exists $K < +\infty$ such that

$$|S_t(\xi_0, t_0)| \leqslant K, \qquad \forall t \in [t_0, b). \tag{2.5.4}$$

Using Proposition 2, we can find for every $\tau \in [t_0, b)$ a solution to the equation:

$$\dot{\mathbf{x}}(t) = \mathbf{F}(\mathbf{x}(t), t), \qquad \mathbf{x}(\tau) = S_\tau(\xi_0, t_0) \tag{2.5.5}$$

defined for $t \in [\tau - T_1, \tau + T_1]$, where T_1, by Eq. (2.3.3) with $\rho = \theta = 1$, can be chosen as

$$T_1 = \Bigg(1 + \max_{\substack{|t-\tau| \leqslant 1 \\ |\xi - \mathbf{x}(\tau)| \leqslant 1}} |\mathbf{F}(\xi, t)|\Bigg)^{-1} \geqslant \Bigg(1 + \max_{\substack{|t| \leqslant 1+|a|+|b| \\ |\xi| \leqslant 1+K}} |\mathbf{F}(\xi, t)|\Bigg)^{-1} \equiv \overline{T}_1 \tag{2.5.6}$$

and, obviously, for $t \in (\tau - \overline{T}_1, \tau + \overline{T}_1) \cap (a,b)$, one must have $\mathbf{x}(t) = S_t(\xi_0; t_0)$ by the uniqueness theorem.

The solution under investigation can therefore be extended to a solution defined for

$$t \in \bigcup_{\tau \in (t_0, b)} (\tau - \overline{T}_1, \tau + \overline{T}_1) = (t_0 - \overline{T}_1, b + \overline{T}_1), \tag{2.5.7}$$

manifestly contradicting the supposed maximality of (a, b).

A similar argument holds if $a > -\infty$. mbe

Considering Proposition 4, it is convenient to introduce the following definition.

3 Definition. Consider the differential equation $\dot{\mathbf{x}} = \mathbf{F}(\mathbf{x}, t)$ with $\mathbf{F}$ being an $\mathbf{R}^d$-valued $C^\infty(\mathbf{R}^d \times \mathbf{R})$ function.

Suppose that there is an $\mathbf{R}_+$-valued continuous function defined on $\mathbf{R}^3$: $(r, s, t) \rightarrow \mu(r, s, t)$, such that if $t \rightarrow S_t(\xi_0; t_0)$ is a solution to Eq. (2.5.1) defined for $t \in (a, b)$ then:

$$|S_t(\xi_0, t_0)| \leqslant \mu(|\xi_0|, t_0, t), \qquad \forall \xi_0 \in \mathbf{R}^d, \quad t_0 \in R, \quad t \in \mathbf{R}, \quad t, \geqslant t_0. \tag{2.5.8}$$

We shall say that our differential equation is "normal" (in the future).

If μ can be chosen to be (s, t) independent, we shall say that our equation has "bounded trajectories" (in the future).

Observations.

(1) Equation (2.5.8) is a strong condition on the motions generated by $\dot{\mathbf{x}} = \mathbf{F}(\mathbf{x}, t)$; because of its independence on the existence interval (a, b), it is often called an "a priori estimate" on the motions governed by $\dot{\mathbf{x}} = \mathbf{F}(\mathbf{x}, t)$.

(2) An equation of higher order, like Eq. (2.2.2), will be called normal, or with bounded trajectories, if once reduced to a first-order equation it becomes a normal equation, or an equation with bounded trajectories, in the sense just introduced.

More concretely, this means that it is possible to give an a priori estimate (i.e., independent of the interval of definition) of the moduli of $\mathbf{x}(t)$, $\dot{\mathbf{x}}(t), \ldots, (d^{s-1}\mathbf{x}/dt^{s-1})(t)$ in terms of the observation time t, of the initial time t_0, and of the initial data $\mathbf{x}(t_0), \dot{\mathbf{x}}(t_0), \ldots, (d^{s-1}\mathbf{x}/dt^{s-1})(t_0)$; furthermore, this estimate depends continuously on those parameters.

The importance of the above definition is manifest in the following proposition.

5 Proposition. *If the differential equation* (2.5.1), $\dot{\mathbf{x}} = \mathbf{F}(\mathbf{x}, t)$, *is normal, then it admits a "global solution", i.e., a solution defined in a neighborhood of* $[t_0, +\infty)$, *for any given initial datum* ξ_0 *and initial time* t_0.

Proof. Let (a, b) be a maximal existence interval for a solution to Eq. (2.5.1), and suppose that $b < +\infty$. Then by Definition 3:

$$\limsup_{t \rightarrow b^-} |S_t(\xi_0, t_0)| \leqslant \mu(|\xi_0|, t_0, b) \tag{2.5.9}$$

would hold, contradicting Proposition 4, Eq. (2.5.2). mbe

An example of a normal equation (which also has bounded trajectories) is provided by the following proposition.

6 Proposition. *Consider the differential equation $m\ddot{x} = f(x)$ in* **R**. [*See Eq.* (2.1.1) *describing the motions of a point mass, with mass $m > 0$, on a line and subject to a force depending only on the position, $f \in C^\infty(\mathbf{R})$.*] *Suppose that the potential energy V, see Eq.* (2.1.3), *is bounded below. Then the differential equation is normal. If $\lim_{x\to\pm\infty} V(x) = +\infty$, the differential equation also has bounded trajectories.*

PROOF. If $t \to x(t)$ is a solution to Eq. (2.1.1), defined for $t \in (a,b)$ and verifying $x(t_0) = \xi_0, \dot{x}(t_0) = \eta_0$ for some $t_0 \in (a,b)$, one has the energy conservation (see §2.1):

$$\tfrac{1}{2}m\dot{x}(t)^2 + V(x(t)) = E = \tfrac{1}{2}m\eta_0^2 + V(\xi_0), \qquad \forall t \in (a,b), \quad (2.5.10)$$

and, therefore, if $M = \inf_{\xi\in\mathbf{R}} V(\xi)$:

$$|\dot{x}(t)| = \left(\frac{2}{m}(E - V(x(t)))\right)^{1/2} \leqslant \left(\frac{2}{m}(E - M)\right)^{1/2} \quad (2.5.11)$$

and $M > -\infty$, by assumption. Furthermore,

$$|x(t)| = \left|\xi_0 + \int_{t_0}^{t} \dot{x}(\tau)\,d\tau\right| \leqslant |\xi_0| + \left(\frac{2}{m}(E - M)\right)^{1/2}|t - t_0| \quad (2.5.12)$$

which calling $\mu(|\xi_0|, t, t_0)$ the right-hand side of Eq. (2.5.12), yields an a priori estimate, showing normality.

If $\lim_{\xi\to\pm\infty} V(\xi) = +\infty$, let $\xi \to W(\xi)$ be a symmetric (i.e., $W(\xi) \equiv W(-\xi)$) continuous function which is strictly increasing for $\xi \geqslant 0$ and which is a lower bound to $V(\xi)$: $V(\xi) > W(\xi)$, $\forall \xi \in \mathbf{R}$, and such that $\lim_{\xi\to\infty} W(\xi) = +\infty$. Since V is also bounded below, it is clear that such a function does exist.

Let $\mu(E)$ be the positive solution to $W(\xi) = E$, existing for all $E \geqslant M$, i.e., for all E's of the form (2.5.10). Then the motion with energy E given by the right-hand side of Eq. (2.5.10) must verify $|x(t)| \leqslant \mu(E)$, as $|x(t)| > \mu(E)$ would imply, by the left-hand side of Eq. (2.5.10) and by the choice of W, that $m\dot{x}(t)^2/2 < 0$.

By the assumed continuity and strict monotonicity of W, the function $\mu(E)$ is continuous in E for $E \geqslant M$; hence, we have found a t_0, t-independent a priori bound $|x(t)| \leqslant \mu(E)$. mbe

We conclude this section with an important observation on the "a priori estimates".

In applications one often meets functions $(\mathbf{x}, t) \to \mathbf{F}(\mathbf{x}, t)$ which are C^∞ functions for $(\mathbf{x}, t) \in (\mathbf{R}^d/A) \times \mathbf{R}$, where A is a "singularity set" usually consisting of points, lines, surfaces, or even in a set with interior points; inside $A \times \mathbf{R}$ the **F** might be undefined. In such cases the singularity of **F** means that the model originating the differential equations (2.5.1) is not a good model of the physical phenomenon that it hopes to describe, at least if the initial data or the motion generated by them enter the region A.

For instance, the attractive force exerted by the Sun on the Earth is well described by the formula $k|\mathbf{x}|^{-2}$ only if the distance between the Earth and the Sun is large compared to the Sun's diameter; it is clear that the singularity in $\mathbf{x} = \mathbf{0}$ is purely fictitious and due to an excessive idealization!

It follows that in such cases one is free to modify $\mathbf{F}$ by changing it into a function $\mathbf{F}^{(A)} \in C^{\infty}(\mathbf{R}^d \times \mathbf{R})$ which, outside a small neighborhood of $A \times \mathbf{R}$, coincides with $\mathbf{F}$.

The equation

$$\dot{\mathbf{x}} = \mathbf{F}^{(A)}(\mathbf{x}, t) \tag{2.5.13}$$

will then be an equally good model of the same physical phenomenon.

However, it is obvious that the only interesting motions, among those described by Eq. (2.5.13), will be those evolving outside a neighborhood of $A \times \mathbf{R}$, where, in fact, $\mathbf{F}$ and $\mathbf{F}^{(A)}$ are indistinguishable.

In this book we shall sometimes find equations of the form (2.5.1) with $\mathbf{F}$ singular in some region. However, in all those cases will appear the possibility of an "a priori estimate" guaranteeing the existence of a continuous positive function μ' on $\mathbf{R}^d \times \mathbf{R} \times \mathbf{R}$ such that if $S_t(\xi_0; t_0)$ is a solution to Eq. (2.5.13), defined for $t \in (a, b)$ with initial datum at t_0 given by $\xi_0 \in \{$set of initial data "thought of as interesting"$\} = \tilde{A}$, then

$$d(S_t(\xi_0; t_0), A) \geqslant \mu'(\xi_0, t, t_0), \tag{2.5.14}$$

where $d(\xi, A) =$ (distance of ξ from A) and μ' is positive for $\xi_0 \in \tilde{A}$. Usually one shall fix $\tilde{A} = A^c =$ (complement of A) by possibly enlarging the set A.

By what has been said so far, it appears that if we are interested only in motions starting outside $\tilde{A}$ and $\tilde{A} = A^c$, we shall imagine that such motions verify Eq. (2.5.13) and, therefore, we shall be able to apply to them the various results concerning the differential equations with right-hand side in C^{∞}.

The above elucubrations motivate the following definition:

4 Definition. Let $(\xi, t) \to \mathbf{F}(\xi, t)$ be a C^{∞} function defined on $(\mathbf{R}^d/A) \times \mathbf{R}$ with values in $\mathbf{R}^d$, where $A \subset R^d$ is a closed set. Suppose that:

(i) there exists an $\mathbf{R}^d$-valued function $\mathbf{F}^{(A)} \in C^{\infty}(\mathbf{R}^r \times \mathbf{R})$ coinciding with $\mathbf{F}$ on $(\mathbf{R}^d/A) \times \mathbf{R}$;

(ii) there exists a real valued function μ' on $\mathbf{R}^d \times \mathbf{R} \times \mathbf{R}$, continuous and positive valued on $(\mathbf{R}^d/A) \times \mathbf{R} \times \mathbf{R}$, verifying $\mathbf{x}(t_0) = \xi_0, \dot{\mathbf{x}}(t) = \mathbf{F}(\mathbf{x}(t), t), \forall t \in (a, b)$, then $\forall t \geqslant t_0$ and $t \in (a, b)$:

$$d(S_t(\xi_0, t_0), A) \geqslant \mu'(\xi_0, t, t_0) > 0; \tag{2.5.15}$$

(iii) the differential equation $\dot{\mathbf{x}} = \mathbf{F}^{(A)}(\mathbf{x}, t)$ is normal.

In such a situation we shall say that the "singular differential equation" $\dot{\mathbf{x}} = \mathbf{F}(\mathbf{x}, t)$ is "normal outside A".

It is an exercise to prove the following proposition.

7 Proposition. *Let* $(\xi, t) \to \mathbf{F}(\xi, t)$ *be an* $\mathbf{R}^d$*-valued* C^∞ *function on* $(\mathbf{R}^d/A) \times \mathbf{R}$ *and consider the singular differential equation* $\dot{\mathbf{x}} = \mathbf{F}(\mathbf{x}, t)$: *if this equation is normal outside* A, *every initial datum* $\xi_0 \notin A$ *originates a* C^∞ *solution of*

$$\dot{\mathbf{x}}(t) = \mathbf{F}(\mathbf{x}(t), t); \qquad \mathbf{x}(t_0) = \xi_0; \qquad \mathbf{x}(t) \notin \mathbf{R}^d/A \tag{2.5.16}$$

defined in a neighborhood of $[t_0, +\infty)$, i.e., *a global solution.*

Observation. As the reader will verify when looking at Chapter 4, §4.8 and §4.9, an interesting example of the situation contemplated in Definition 4 and Proposition 7 can be found in the two-body problem: The set A will be, in this case, the closure of a neighborhood of the set of the initial data with vanishing areal velocity.

Such data are those in which the two bodies are heading into or out of a collision and which are, therefore, to be considered singular.

Exercises and Problems for §2.5

1. Formulate the notions of normal differential equation "in the past" or of differential equation with bounded trajectories "in the past", and reformulate all the propositions of §2.5 to deal with the problem of the existence of solutions in intervals like $(-\infty, t_0]$ or $(-\infty, +\infty)$.

2. Consider the equation in R, $\ddot{x} + (d/dx)\log(1 + x^2) = 0$. Determine whether it is normal and with bounded trajectories. Compute $x(1)$ with an approximation of 60% if $x(0) = 0$, $\dot{x}(0) = \frac{1}{2}$.

3. Same as Problem 2 for $\ddot{x} + \sin x = 0$, $x(0) = 0$, $\dot{x}(0) = \frac{1}{4}$.

4. Same as Problems 2 and 3 for the differential equations in Problems 1 and 2 of §2.3.

5.* Same as Problem 2 but with a 1% approximation and using a desk computer together with the error estimate implicit in the existence theorem of §2.3. Alternatively, use the algorithm of Appendix O, together with a desk computer.

6.* Same as Problem 5 but using energy conservation and the relative quadrature formula, together with a desk computer.

7.* Same as Problem 6 but for the equation in Problem 3.

8. Let $t \to \mathbf{x}(t)$ be an $\mathbf{R}^d$-valued $C^\infty(\mathbf{R})$ function such that $\exists M > 0$ for which

$$|\mathbf{x}(t)| \leqslant |\mathbf{x}(0)| + M\int_0^t |\mathbf{x}(\tau)|\, d\tau, \qquad t \geqslant 0.$$

Show that $|\mathbf{x}(t)| \leqslant y(t)$, $t \geqslant 0$, where y is defined as the solution of $y(t) = |x(0)| + M\int_0^t y(\tau)\, d\tau$, $t \geqslant 0$, i.e., $y(t) = |\mathbf{x}(0)| \exp Mt$.

9.* If $\xi \to \varphi(\xi)$ is a continuous positive monotonically increasing function of $\xi \in \mathbf{R}_+$ and if $t \to \mathbf{x}(t)$ is in $C^\infty(\mathbf{R})$ and

$$|\mathbf{x}(t)| \leqslant |\mathbf{x}(0)| + \int_0^t \varphi(|\mathbf{x}(\tau)|)\, d\tau, \qquad t \geqslant 0,$$

show that $|\mathbf{x}(t)| \leqslant y(t)$, $t \geqslant 0$, where y is defined as the solution of $y(t) = |\mathbf{x}(0)| +$

$\int_0^t \varphi(y(\tau))\,d\tau$, i.e., setting $\Phi(y) = \int_{|\mathbf{x}(0)|}^{y}(d\eta/\varphi(\eta))$, as the function verifying $\Phi(y(t)) \equiv t$ (or $y(t) \equiv \Phi^{-1}(t)$).

10.* Given the differential equation $\dot{\mathbf{x}} = \mathbf{f}(\mathbf{x}, t)$ in $\mathbf{R}^d$, define for $T > 0$:

$$\varphi_T(s) = \max_{\substack{t \in [0, T] \\ |\xi| < s}} |\mathbf{f}(\xi, t)|.$$

Show that a sufficient condition for the normality of the equation is that the equation in $\mathbf{R}$:

$$\dot{y} = \varphi_T(y), \qquad y(0) = |\mathbf{x}(0)|$$

admits a global solution (i.e., a solution on $[0, +\infty)$) for all $T > 0$. (Hint: $\mathbf{x}(t) = \mathbf{x}(0) + \int_0^t \mathbf{f}(\mathbf{x}(\tau), \tau)\,d\tau \Rightarrow |\mathbf{x}(t)| \leqslant |\mathbf{x}(0)| + \int_0^t \varphi_T(|\mathbf{x}(\tau)|)\,d\tau$; then apply Problem 9.)

11.* If $t \to \mathbf{L}(t)$ is a matrix-valued $C^\infty(\mathbf{R})$ function with values in the $d \times d$ matrices, the equation $\dot{\mathbf{x}} = \mathbf{L}(t)\mathbf{x}$ is normal in the future (as well as in the past); hence, it has global solutions (Hint: Apply Problem 10.)

12. In the context of Problem 11, show that the differential equation admits d linearly independent global solutions (defined on $(-\infty, +\infty)$). (Hint: Use Problem 11 and Problem 9 of §2.3.)

13. In the context of Problem 11, suppose that $\mathbf{L}(t)$ is a time-independent matrix $\mathbf{L}$. Using the results of Problem 11 of §2.3, p. 21, and supposing that all the eigenvalues of $\mathbf{L}$ are real and pairwise distinct, show that the equation $\dot{\mathbf{x}} = \mathbf{L}\mathbf{x}$ has bounded trajectories if and only if $\lambda_j \leqslant 0$, $j = 1, \ldots, d$.

14.* In the context of Problem 11, let $\mathbf{g} \in C^\infty(\mathbf{R})$ be an $\mathbf{R}^d$-valued function. Show the normality of the equation $\dot{\mathbf{x}} = \mathbf{L}(t)\mathbf{x} + \mathbf{g}(t)$.

15.* Consider a differential equation in $\mathbf{R}^d$, $\dot{\mathbf{x}} = \mathbf{F}(\mathbf{x}, t)$, with $\mathbf{F} \in C^\infty(\mathbf{R}^d \times \mathbf{R})$. Suppose that $|\mathbf{F}(\mathbf{x}, t)| \leqslant \gamma(t)|\mathbf{x}| + \beta(t)$, where $\beta, \gamma \in C^\infty(\mathbf{R})$, $\beta, \gamma \geqslant 0$. Show that the equation is normal by finding an a priori estimate. (Hint: Combine Problems 9 and 4.)

16.* Same as Problem 5 with $|\mathbf{F}(\mathbf{x}, t)| \leqslant \beta(t) + \gamma(t)|\mathbf{x}|\log(e + |\mathbf{x}|)$.

17. Consider a differential equation $\dot{\mathbf{x}} = \mathbf{f}(\mathbf{x}, t, \boldsymbol{\alpha})$ of the type considered in §2.4, $\mathbf{f} \in C^\infty(\mathbf{R}^d \times \mathbf{R} \times \mathbf{R}^m)$. Suppose that this equation admits an a priori estimate like Eq. (2.5.8), for $\forall \boldsymbol{\alpha} \in \mathbf{R}^m$, with an $\boldsymbol{\alpha}$-independent function.

Show that, in this case, the "local regularity theorem", Proposition 3, p. 22, becomes "global", i.e., the function $(t, \xi_0, t_0, \alpha_0) \to S_t(\xi_0; t_0, \alpha_0)$ is a C^∞-function of $\xi_0 \in \mathbf{R}^d$, $\alpha_0 \in \mathbf{R}^m$, $t_0 \in \mathbf{R}$, $t \in \mathbf{R}$, $t \geqslant t_0$.

§2.6. More Generalities about Differential Equations. Autonomous Equations

Before proceeding in the analysis of some applications, it is convenient to set up a few more definitions, mainly as an excuse to illustrate some simple but interesting general remarks about differential equations.

5 Definition. Let $(\xi, \xi^{(1)}, \ldots, \xi^{(s-1)}) \to \mathbf{f}(\xi, \xi^{(1)}, \ldots, \xi^{(s-1)})$ be an $\mathbf{R}^d$-valued C^∞ function on $\mathbf{R}^{sd}$. Consider the equation for the $\mathbf{R}^d$-valued function $t \to \mathbf{x}(t)$ defined for t in an interval I [see Eq. (2.2.2)]:

$$\frac{d^s\mathbf{x}}{dt^s}(t) = \mathbf{f}\left(\mathbf{x}(t), \frac{d\mathbf{x}}{dt}(t), \ldots, \frac{d^{s-1}\mathbf{x}}{dt^{s-1}}(t)\right), \tag{2.6.1}$$

$$\mathbf{x}(t_0) = \xi_0, \quad \frac{d\mathbf{x}}{dt}(t_0) = \xi^{(1)}, \quad \ldots, \quad \frac{d^{s-1}\mathbf{x}}{dt}(t_0) = \xi^{(s-1)}. \tag{2.6.2}$$

We shall say that Eq. (2.6.1) is an "autonomous" differential equation of class C^∞. In other words, Eq. (2.2.2) is said to be autonomous when the right-hand side "does not explicitly depend upon time".

The space $\mathbf{R}^d \times \cdots \times \mathbf{R}^d = \mathbf{R}^{sd}$, thought of as the space of the possible initial data $(\xi, \xi^{(1)}, \ldots, \xi^{(s-1)})$ for Eq. (2.6.1), will be called the "space of the initial data" or the "data space".

Also we set the following.

6 Definition. We shall say that a C^∞ autonomous differential equation like Eq. (2.6.1) is "reversible" if any solution, $t \to \mathbf{x}(t)$, to Eq. (2.6.1) defined for $t \in (-\epsilon, \epsilon)$, $\epsilon > 0$, is such that the function $t \to \mathbf{x}(-t)$, $t \in (-\epsilon, \epsilon)$, is also a solution to Eq. (2.6.1).

Observation. We shall see that many differential equations describing nondissipative dynamical systems are reversible. Basically, f originates a reversible system when s is even and f depends evenly on the odd derivatives.

The interest in autonomous equations lies, from a mathematical point of view, in the validity of the following easy propositions.

8 Proposition. *Consider a normal autonomous first-order*[4] *differential equation in* $\mathbf{R}^d$.

It is possible to define on $\mathbf{R}^d$ *a family* $(S_t)_{t \geqslant 0}$ *of maps, mapping* $\mathbf{R}^d$ *into itself, such that the functions*

$$t \to S_{t-t_0}(\xi_0), \qquad \xi_0 \in \mathbf{R}^d; \quad t, t_0 \in \mathbf{R}; \quad t \geqslant t_0 \tag{2.6.3}$$

are solutions to Eq. (2.6.1) *with initial datum at* $t = t_0$ *given by* ξ_0.

For every $t \geqslant 0$, *the map* S_t *is a* C^∞ *map and*

$$S_t(S_{t'}(\xi)) = S_{t+t'}(\xi), \qquad \forall t, t' \geqslant 0; \quad \forall \xi \in \mathbf{R}^d. \tag{2.6.4}$$

Furthermore, the maps S_t *are* C^∞ *regular jointly in* ξ_0 *and* t: i.e., *the function* $(t, \xi_0) \to S_t(\xi_0), (t, \xi_0) \in \mathbf{R}_+ \times \mathbf{R}^d$ *are in* $C^\infty(\mathbf{R}_+ \times \mathbf{R}^d)$.

Proof. Let $t \to S_t(\xi_0; t_0)$ be the solution to Eqs. (2.6.1) and (2.6.2) with $s = 1$, defined for $t \geqslant t_0$. It does exist since Eq. (2.6.1) is now supposed to be a normal equation. Let:

$$S_t(\xi_0) = S_t(\xi_0, 0) \qquad \text{for} \quad t \geqslant 0. \tag{2.6.5}$$

[4] I.e., $s = 1$ in Eq. (2.6.1).

In §2.5 we saw that S_t is a C^∞ map of $\mathbf{R}^d$ into itself for each $t \in \mathbf{R}_+$ and, also, that $(t, \xi_0) \to S_t(\xi_0)$, $(t, \xi_0) \in \mathbf{R}_+ \times \mathbf{R}^d$ is in $C^\infty(\mathbf{R}_+ \times \mathbf{R}^d)$.

For $t \geqslant t_0$, let $\mathbf{x}(t) = S_{t-t_0}(\xi_0)$. Since "$\mathbf{f}$ does not explicitly depend on time", one finds:

$$\begin{aligned} \frac{d\mathbf{x}}{dt}(t) &= \frac{d}{dt} S_{t-t_0}(\xi_0) = \frac{d}{dt} S_{t-t_0}(\xi_0, 0) \\ &= f(S_{t-t_0}(\xi_0, 0)) = f(S_{t-t_0}(\xi_0)) = f(\mathbf{x}(t)). \end{aligned} \tag{2.6.6}$$

Hence $t \to S_{t-t_0}(\xi_0)$ is a solution to Eq. (2.6.1) for $t \geqslant t_0$. Furthermore,

$$S_{t_0-t_0}(\xi_0) = S_0(\xi_0) = S_0(\xi_0, 0) \equiv \xi_0 \tag{2.6.7}$$

which, by the uniqueness theorem, Proposition 1, p. 13 gives

$$S_{t-t_0}(\xi_0) \equiv S_t(\xi_0, t_0), \qquad t \geqslant t_0.$$

Similarly, one checks that $t \to S_t(S_{t'}(\xi_0))$ is a solution to Eq. (2.6.1) with initial datum at $t = 0$ equal to $S_{t'}(\xi_0)$; such is also $t \to S_{t+t'}(\xi_0)$; hence, Eq. (2.6.4) is also proved. mbe

9 Corollary. *Consider an autonomous equation of order s, as in Eq.* (2.6.1), *and suppose that it is normal.*

It is possible to define, on the data space $\mathbf{R}^{ds}$, *a family* $(S_t)_{t \geqslant 0}$ *of* C^∞ *maps of* $\mathbf{R}^{ds}$ *into itself such that the function*

$$t \to S_{t-t_0}(\xi_0, \xi^{(1)}, \ldots, \xi^{(s-1)}) = \left(\mathbf{x}(t), \mathbf{x}^{(1)}(t), \ldots, \mathbf{x}^{(s-1)}(t)\right) \tag{2.6.8}$$

is a solution to the equations

$$\begin{aligned} &\dot{\mathbf{x}}(t) = \mathbf{x}^{(1)}(t), \quad \dot{\mathbf{x}}^{(1)}(t) = \mathbf{x}^{(2)}(t), \ldots, \dot{\mathbf{x}}^{(s-2)}(t) = \mathbf{x}^{(s-1)}(t), \\ &\dot{\mathbf{x}}^{(s-1)}(t) = f\left(\mathbf{x}(t), \mathbf{x}^{(1)}(t), \ldots, \mathbf{x}^{(s-1)}(t)\right) \end{aligned} \tag{2.6.9}$$

[*equivalent to Eq.* (2.6.1)] *and verifies the initial data*

$$\mathbf{x}(t_0) = \xi_0, \quad \mathbf{x}^{(1)}(t_0) = \xi^{(1)}, \ldots, \mathbf{x}^{(s-1)}(t_0) = \xi^{(s-1)}. \tag{2.6.10}$$

Furthermore,

$$S_t S_{t'} = S_{t+t'}, \qquad \forall t, t' \geqslant 0 \tag{2.6.11}$$

and the maps S_t *are* C^∞ *regular also, jointly in t and* $\xi_0, \ldots, \xi_{s-1}$: i.e., *the map* $(t, \xi_0, \ldots, \xi_{s-1}) \to S_t(\xi_0, \ldots, \xi_{s-1})$, $(t, \xi_0, \ldots, \xi_{s-1}) \in \mathbf{R}_+ \times \mathbf{R}^d \times \cdots \times \mathbf{R}^d$ *is in* $C^\infty(\mathbf{R}_+ \times \mathbf{R}^d \times \cdots \times \mathbf{R}^d)$.

Proof. It is an immediate consequence of the equivalence between Eqs. (2.6.1), (2.6.2) and Eqs. (2.6.9), (2.6.10) and of Proposition 8. mbe

7 Definition. Given a normal sth-order autonomous differential equation on $\mathbf{R}^d$, the family $(S_t)_{t \geqslant 0}$ of maps of the data space into itself, defined in Proposition 8, will be called the "flow" on $\mathbf{R}^{ds}$ which "solves Eqs. (2.6.1) and (2.6.2)".

Observations.

(1) Because of Eq. (2.6.11), the flow $(S_t)_{t\geq 0}$ is, in mathematical language, a "semigroup".

When Eq. (2.6.1) is also normal in the past, it becomes possible to define S_t for $t \leq 0$, and the family $(S_t)_{t\in\mathbf{R}}$ forms a group, i.e., it verifies Eq. (2.6.11) for all $t, t' \in \mathbf{R}$ (exercise).

(2) All the normal reversible equations are also normal in the past (exercise); hence, such a class of equations provides an important instance when the solution flow is a group.

An interesting remark about autonomous equations, already met in Problem 3, §2.2, is the following proposition.

10 Proposition. *Consider a normal* sth-*order autonomous differential equation on* $\mathbf{R}^d$, *like Eq.* (2.6.1).

Suppose that the initial datum $(\xi_0^{(0)}, \xi_0^{(1)}, \ldots, \xi_0^{(s-1)}) \in \mathbf{R}^{sd}$ *is such that there is some* $T > 0$ *for which*

$$S_T(\xi_0^{(0)}, \ldots, \xi_0^{(s-1)}) = (\xi_0^{(0)}, \ldots, \xi_0^{(s-1)}); \qquad (2.6.12)$$

then the motion generated by $(\xi_0^{(0)}, \ldots, \xi_0^{(s-1)})$ *is a "periodic motion" with period* T, *i.e., it is a periodic solution of Eq.* (2.6.1) *with period* T.

PROOF. The function $t \to S_{t+T}(\xi_0^{(0)}, \ldots, \xi_0^{(s-1)})$, $t \geq 0$, where $(S_t)_{t\geq 0}$ is the solution flow to Eq. (2.6.1), is again a solution to Eq. (2.6.1) and, for $t = 0$, verifies the initial condition $(\xi_0^{(0)}, \ldots, \xi_0^{(s-1)})$ by our assumption (2.6.2). Hence, by uniqueness, it coincides with $t \to S_t(\xi_0^{(0)}, \ldots, \xi_0^{(s-1)})$. This means that $t \to S_t(\xi_0^{(0)}, \ldots, \xi_0^{(s-1)})$ is periodic with period T. mbe

Observation. More generally, it is clear that $t \to S_{t+T}(\xi^{(0)}, \ldots, \xi^{(s-1)})$ is a solution to Eq. (2.6.9) for $t \geq 0$: i.e., if $t \to \mathbf{x}(t)$ is a solution to an autonomous equation, $t \to \mathbf{x}(t + T)$ is also a solution for $T \in \mathbf{R}$.

Exercises and Problems for §2.6

Show that the following differential equations are normal both in the past and in the future and:

1. Draw the trajectories of the flow $(S_t)_{t\in\mathbf{R}}$ in the data space $\mathbf{R}^2$ for the equation $\ddot{x} = -g$, $g \in \mathbf{R}$.

2. Same as Problem 1 for $\ddot{x} = -\omega^2 x$, $\omega^2 > 0$.

3. Same as Problem 1 for $\ddot{x} = -g - \lambda\dot{x}$, $g, \lambda \in \mathbf{R}$.

4. Same as Problem 1 for $\ddot{x} = \omega^2 x$, $\omega^2 > 0$.

5. Describe the trajectories of the flow $(S_t)_{t\in\mathbf{R}}$ in the data space $\mathbf{R}^6$ for the equations in $\mathbf{R}^3$: $\ddot{\mathbf{x}} = -\mathbf{g}$ or $\ddot{\mathbf{x}} = -\omega^2\mathbf{x}$.

6. Same as Problem 5 for the equation, in $\mathbf{R}^2$, $\dot{x} = ax + y, \dot{y} = -x + ay$, discussing the result in terms of $a \in \mathbf{R}$.

7. If Eq. (2.6.1) is normal and reversible, prove that it is also normal in the past. Show that the flow $(S_t)_{t \geqslant 0}$, solving it for $t \geqslant 0$, can be extended to a flow $(S_t)_{t \in \mathbf{R}}$, solving Eq. (2.6.1) for all $t \in \mathbf{R}$: one can define $S_{-t} = (S_t)^{-1}$, $\forall t \geqslant 0$. In this case, the family $(S_t)_{t \in \mathbf{R}}$ forms a group of maps of $\mathbf{R}^{ds}$ onto itself.

§2.7. One-Dimensional Conservative Periodic and Aperiodic Motions

Having completed a general analysis of the existence, uniqueness, and regularity properties of the ordinary differential equations, let us go back to the qualitative theory of the motions developing under the action of a purely positional force f considered in §2.1 [see Eq. (2.1.1)].

For such motions, in §2.1, we found the energy conservation theorem (so that they are called "conservative motions" generated by "conservative forces").

We shall now analyze under which circumstances the motions, solutions of a differential equation like Eq. (2.1.1), are periodic or aperiodic.

Let V be the potential energy generating the force f (i.e., $f(\xi) = -(dV/d\xi)(\xi)$) and, in order to have motions defined for all times ("globally defined motions"), suppose that V is bounded below (see Proposition 6).

Let $(\eta_0, \xi_0) \in \mathbf{R}^2$, $t_0 \in \mathbf{R}$, and let $t \to x(t)$, $t \geqslant t_0$, be the solution to Eq. (2.1.1) with data:

$$\dot{x}(t_0) = \eta_0, \qquad x(t_0) = \xi_0. \tag{2.7.1}$$

If $E = (m/2)\eta_0^2 + V(\xi_0)$, we can represent graphically the initial datum and the potential as in Fig. 2.1.

Because of energy conservation, it is clear that, if ξ_0, as in the picture, is between two distinct solutions $x_-(E) < x_+(E)$ of $V(\xi) = E$ and if $(dV/d\xi)(x_-(E)) < 0$, $(dV/d\xi)(x_+(E)) > 0$, then the motion $t \to x(t)$ will never leave the interval $[x_-(E), x_+(E)]$. In fact V would be strictly larger than E to the left of $x_-(E)$ and to the right of $x_+(E)$ and, therefore, the motion with energy E would have to have negative kinetic energy when occupying such a position.

Such a trapped motion will be periodic if and only if it takes a finite time for it to run from $x_-(E)$ to $x_+(E)$. This amount of time can be estimated easily by the quadrature formula (2.1.8), p. 11.

If $x_-(E)$ or $x_+(E)$ or both do not exist, the above argument says that the motion may be unbounded. The above argument also does not give any precise predictions when the derivative of V vanishes in at least one of the two points $x_-(E), x_+(E)$.

The following proposition shows a general result and in its proof all the above problems are implicitly or explicitly solved.

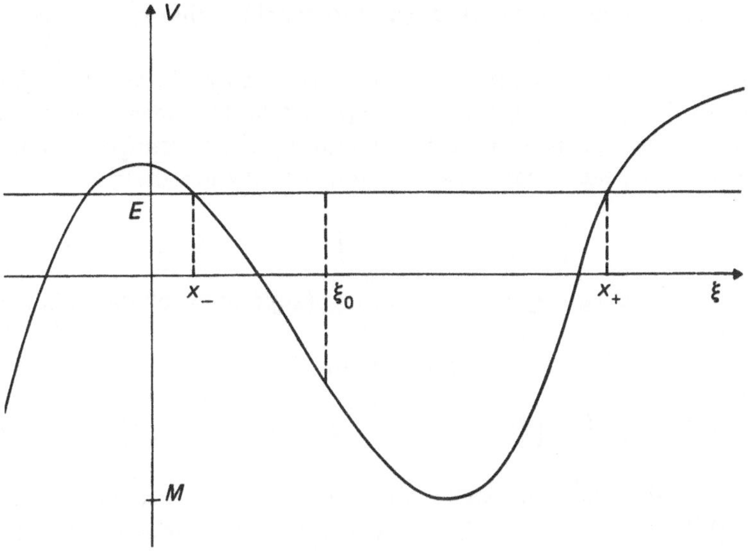

Figure 2.1.

11 Proposition. *The motion*

$$t \to x(t), \qquad t \in [t_0, +\infty), \qquad \text{of} \quad m\ddot{x} = -(dV/dx)(x)$$

with initial datum (2.7.1) *is periodic with a positive minimal period if and only if* ξ_0 *is between two adjacent roots* $x_- < x_+$ *of* $V(\xi) = E$, *where the derivative of* V *is, respectively, negative or positive.*

Proof. Suppose that $V(x_\pm) = E$, $-V'(x_-)$ and $V'(x_+) > 0$, and $V(\xi) < E$ for $\xi \in (x_-, x_+)$ and let $\xi_0 \in [x_-, x_+]$. As already noticed it must be that $\forall t \geqslant t_0$, $x(t) \in [x_-, x_+]$.

Suppose, first, that $\eta_0 > 0$ and define $t_+ = \{$supremum of the values $t > t_0$ such that $\dot{x}(\tau) > 0$ for all $\tau \in [t_0, t)\}$.

From energy conservation, one deduces:

$$\dot{x}(t) = \left(\frac{2}{m}(E - V(x(t)))\right)^{1/2}, \qquad t_0 \leqslant t < t_+, \tag{2.7.2}$$

where the choice of the sign in front of the square root comes from the continuity of $\dot{x}(t)$ and from $\eta_0 > 0$. To estimate t_+, observe that Eq. (2.7.2) implies:

$$t - t_0 = \int_{\xi_0}^{x(t)} \left(\frac{2}{m}(E - V(\xi))\right)^{-1/2} d\xi, \qquad t_0 \leqslant t < t_+, \tag{2.7.3}$$

If we show that $\lim_{t \to t_+} x(t) = x_+$, it will follow from Eq. (2.7.3) that

$$t_+ - t_0 = \int_{\xi_0}^{x_+} \left(\frac{2}{m}(E - V(\xi))\right)^{-1/2} d\xi \tag{2.7.4}$$

which can be used to estimate t_+ and to conclude that $t_+ < +\infty$: i.e., the point reaches x_+ in a finite time.

Once the point reaches x_+, it cannot stay there since $f(x_+) = -(dV/dx)(x_+) < 0$ and, therefore, $\ddot{x}(t_+) < 0$. This means that $\dot{x}(t) < 0$, $x(t) < x_+$ in a right-hand neighborhood of t_+, by Lagrange's theorem. We can then repeat the already used argument to deduce that:

$$\dot{x}(t) = -\left(\frac{2}{m}(E - V(x(t)))\right)^{-1/2}, \qquad \forall t \in [t_+, t_-), \tag{2.7.5}$$

where t_- is analogously defined as $t_- = \{$supremum of the values $t > t_+$ such that $\dot{x}(\tau) < 0$ for all $\forall \tau \in [t_+, t)\}$.

Proceeding as before, we shall show that

$$t_- - t_+ = \int_{x_+}^{x_-} \left(\frac{2}{m}(E - V(\xi))\right)^{-1/2} d\xi \quad \text{and} \quad t_- - t_+ < +\infty. \tag{2.7.6}$$

It is clear that the same arguments can be again repeated and, therefore, after a suitable time $t' - t_-$, the point will again go through ξ_0 with positive velocity η_0 and

$$t' - t_- = \int_{x_-}^{\xi_0} \left(\frac{2}{m}(E - V(\xi))\right)^{-1/2} d\xi. \tag{2.7.7}$$

By Proposition 10, p. 35, it is clear that from now on the motion will identically repeat itself: i.e., $x(t+T) \equiv x(t)$, $\forall t \geqslant t_0$, if T is the sum of the time intervals of Eqs. (2.7.4), (2.7.6), and (2.7.7):

$$T = 2\int_{x_-}^{x_+} \left(\frac{2}{m}(E - V(\xi))\right)^{-1/2} d\xi, \tag{2.7.8}$$

hence, the motion will be periodic and T will be, by construction, its minimal period.

It remains to show that $\lim_{t\to t_+} x(t) = x_+$ and that $t_+ < +\infty$.

In fact, since $\dot{x}(t) \geqslant 0$, $\forall t \in [t_0, t_+)$, it is clear that the limit $\lim_{t\to t_+} x(t) = \bar{x}$ exists and that it is approached monotonically. Then, if $\bar{x} < x_+$, it would follow that

$$\bar{v} = \lim_{t\to t_+} \dot{x}(t) = \left(\frac{2}{m}(E - V(\bar{x}))\right)^{-1/2} > 0;$$

hence, $\dot{x}(t)$ would be >0 in the right-hand neighborhood of t_+, if $t_+ < +\infty$, against the very definition of t_+ or, if $t_+ = +\infty$, this would mean that $\bar{x} = +\infty$ against $\bar{x} \leqslant x_+$. Hence, $\bar{x} = x_+$ and Eq. (2.7.4) holds.

To show that Eq. (2.7.4) also implies $t_+ < +\infty$, apply Lagrange's theorem to infer that there is a point $\tilde{x} \in (\xi_0, x_+)$ such that for all $\xi \in (\tilde{x}, x_+)$:

$$E - V(\xi) \geqslant E - V(x_+) - \frac{1}{2}\frac{dV}{d\xi}(x_+)(\xi - x_+) \equiv \frac{f(x_+)(\xi - x_+)}{2}, \tag{2.7.9}$$

since $E = V(x_+)$ and $f(x_+) = -(dV/d\xi)(x_+) < 0$ and $(E - V(\xi)) - (E - V(x_+)) - f(x_+)(\xi - x_+)$ is infinitesimal of higher order in $(\xi - x_+)$ as $\xi \to x_+$. Therefore,

$$t_+ - t_0 \leqq \int_{\xi_0}^{\tilde{x}} \left(\frac{2}{m}(E - V(\xi)) \right)^{-1/2} d\xi$$
$$+ \int_{\tilde{x}}^{x_+} \left(\frac{2}{m} \frac{f(x_+)}{2} (\xi - x_+) \right)^{-1/2} d\xi < +\infty \tag{2.7.10}$$

since the first integral is finite because $\max(2(E - V(\xi))/m)^{-1} < +\infty$ in $[\xi_0, \tilde{x}]$, while the second integral is also finite since this is seen by explicit computation.

To conclude the proof, one has to consider the alternatives intially set aside, $\eta_0 < 0$ or $\eta_0 = 0$ (i.e., $\xi_0 = x_\pm$). Such cases are easily reduced to the one just treated. Finally, we must also show that if $f(x_+) = 0$, or $f(x_-) = 0$, or $f(x_+) = f(x_-) = 0$, or x_+, or x_- do not exist, then the motion is not periodic or it has period 0. This last case is realized if (and only if) $\eta_0 = 0$ and $f(\xi_0) = 0$: one says that ξ_0 is an "equilibrium point". Among the remaining cases, we shall treat, as an example, the case $\eta_0 > 0$, $f(x_+) = -(dV/d\xi)(x_+) = 0$.

Proceeding as before, we show that t_+ is given by Eq. (2.7.4). This time, however, to estimate t_+ we must improve Eq. (2.7.9) using Taylor's formula to second order since $f(x_+) = 0$. If $f'(x_+) = (df/d\xi)(x_+)$, we find:

$$E - V(\xi) = \tfrac{1}{2} f'(x_+)(\xi - x_+)^2 + o\big((\xi - x_+)^2\big) \tag{2.7.11}$$

because the left-hand side vanishes together with its first derivative in x_+. Hence, there is $\tilde{x}' \in [\xi_0, x_+)$ such that, if $f'(x_+) > 0$, as we suppose for simplicity:

$$E - V(\xi) \leqslant f'(x_+)(\xi - x_+)^2, \qquad \xi \in (\tilde{x}', x_+). \tag{2.7.12}$$

Then, if $f'(x_+) > 0$, we deduce from Eqs. (2.7.4) and (2.7.2):

$$t_+ - t_0 \geqslant \int_{\tilde{x}'}^{x_+} \left(\frac{2}{m} f'(x_+)(\xi - x_+)^2 \right)^{-1/2} d\xi = +\infty. \tag{2.7.13}$$

The case $f'(x_+) = 0$ is treated likewise, as, in this case, $E - V(\xi)$ is infinitesimal of higher than second order in $\xi - x_+$ and an inequality like Eq. (2.7.12) holds, therefore, with $f'(x_+)$ replaced, say, by 1.

The case $f'(x_+) < 0$ is impossible if $\eta_0 > 0$ (since this would mean that x_+ is a minimum for V) (exercise). mbe

For future reference let us state the following obvious proposition.

12 Proposition. *If $\xi_0 \in \mathbf{R}$, the constant function $t \to x(t) \equiv \xi_0$ is a solution to Eq.* (2.1.1) *if and only if ξ_0 is a stationary point for the potential energy V.*

Exercises and Problems for §2.7

1. Estimate the period of the motions indicated below with an error rigorously bounded by 60%:

$$\ddot{x} = x(x-1), \qquad x(0) = 0, \qquad \dot{x}(0) = \tfrac{1}{2},$$
$$\ddot{x} = -2x - 4x^3, \qquad x(0) = 1/\sqrt{2}, \qquad \dot{x}(0) = 0,$$
$$\ddot{x} = -x^3, \qquad x(0) = 0, \qquad \dot{x}(0) = 1,$$
$$\ddot{x} = -x/(1+x), \qquad x(0) = 0, \qquad \dot{x}(0) = \tfrac{1}{2},$$
$$\ddot{x} = \log(1+x), \qquad x(0) = \tfrac{1}{2}, \qquad \dot{x}(0) = 0.$$

2. Find, if they exist, values of E to which correspond aperiodic motions for the equations in Problem 1, and for:

$$\ddot{x} = -xe^{-x^2}; \qquad \ddot{x} = -\sin x.$$

3. Same as Problem 1 with an error rigorously bounded by 10% or 1%, using a desk computer.

4. Find whether the motions associated with the second equation in Problem 1 admit a motion with period $T = 10$ and, if it exists, estimate within 20% the amplitude of such a motion.

5. Show that the period of the motion of total energy E verifying $\ddot{x} = -x^3$ has a period $T(E)$ proportional to $E^{-1/4}$ if the potential energy is defined as $V(\xi) = \xi^4/4$. Show that the proportionality constant is $2\int_{-1}^{1}(1-\xi^4)^{-1/2}\,d\xi$. (Hint: Write the formula of quadrature for T, Eq. (2.7.8), and change variable as $\xi \to \xi E^{-1/4}$.)

6.* Show that the period of the motion with energy E verifying $m\ddot{x} = -(dV/dx)(x)$, with V such that $V(0) = 0$, $V'(0) = 0$, $V''(0) > 0$, is such that $\lim_{E\to 0^+} T(E) = 2\pi(m/V''(0))^{1/2}$.

7. Let $\xi \to V(\xi)$ be a C^∞ convex even function vanishing at the origin. Let

$$\bar{V}(\xi) = \tfrac{1}{2}\sigma\xi^2 \equiv \tfrac{1}{2}(\sup_{\xi'} V''(\xi'))\xi^2.$$

Consider a motion, associated to the potential energy V, having total energy E. Show that its period is larger than the period of the motions with potential energy $\bar{V}$.

8. Suppose that $V(\xi) = |\xi|^\alpha/\alpha$, $\alpha > 1$, and show that the period of the motion with energy E is proportional to $E^{1/\alpha - 1/2}$ (see Problem 5).

9. Find the limit as $E \to +\infty$ of the period of the motion with energy E developing with potential energy $V(\xi) = \xi^2/2 + \xi^4/4$.

10. Same as Problem 9 with V such that $V(\xi) = V(-\xi)$, $\lim_{\xi\to\infty} V(\xi)/\xi^2 = +\infty$.

11. Same as Problem 9 with V such that $V(\xi) = V(-\xi)$, $\lim_{\xi\to\infty} V(\xi)/\xi^2 = 0$, $\lim_{\xi\to\infty} V(\xi) = +\infty$.

§2.8. Equilibrium: Stability in the Absence of Friction

We already noted, Proposition 12, p. 39, that the stationary solutions of $m\ddot{x} = f(x)$, i.e., the solutions like $t \to \xi_0 = \text{constant}$, correspond to the stationary points of the potential energy function V. In such positions, "equilibrium positions", the exerted force vanishes.

We now wish to distinguish the equilibrium points on the basis of a qualitative property: the stability of their equilibria. Obviously it is first necessary to establish what should be understood by the term "stable equilibrium". This is an empirical notion susceptible to assuming different precise meanings, depending on the particular problem where it appears necessary to study the stability of an equilibrium point.

It is perhaps useful to provide several different definitions of stability for an equilibrium point, leaving to the imagination of the reader the identification of different types of problems for which such types of notions might be relevant.

A deeper analysis of the stability notion will be found in Chapter V which is entirely devoted to stability theory.

In the following, x_0 shall denote an equilibrium point for $m\ddot{x} = f(x)$ under the assumption that f is generated by a C^∞ potential V bounded from below so that the equation of motion is normal (see Proposition 6).

8 Definition. x_0 is a stable equilibrium position if there is a function $\epsilon \to a(\epsilon) < +\infty$ defined for $\epsilon \geqslant 0$ and infinitesimal as $\epsilon \to 0$, such that every motion following an initial condition $x(0) = x_0$, $|\dot{x}(0)| \leqslant \epsilon$ has the property:

$$|x(t) - x_0| < a(\epsilon), \qquad \forall t \geqslant 0. \tag{2.8.1}$$

Observations.

(1) In other words, x_0 is a stable equilibrium position if a point mass placed in x_0 with small velocity stays indefinitely close to x_0; and the smaller $\dot{x}(0)$, the closer it will stay.

(2) The fact that $a(\epsilon)$ might be $+\infty$ means that we admit the possibility that initial data whose velocity $\dot{x}(0)$ is too large may originate motions which travel indefinitely far from x_0. Equation (2.8.1) is really a condition which is relevant only for ϵ small.

(3) The choice of $t_0 = 0$ as initial time is irrelevant since the motion's equation is autonomous.

In most applications it is by no means sufficient to know that x_0 is a stable equilibrium position in the sense of Definition 8. For instance, it is sometimes necessary that the function $a(\epsilon)$, which for obvious reasons could be called the "tolerance" function, has a preassigned structure. This leads to the following definition:

9 Definition. Given a function of the variable $\epsilon \geqslant 0$, $\epsilon \to b(\epsilon) \leqslant +\infty$ (not necessarily infinitesimal as $\epsilon \to 0$), one says that x_0 is a stable equilibrium position "with tolerance b" if the motion $t \to x(t)$, $t \geqslant 0$, following an initial condition $x(0) = x_0$, $|\dot{x}(0)| \leqslant \epsilon$ is such that

$$|x(t) - x_0| \leqslant b(\epsilon), \qquad \forall t \geqslant 0. \tag{2.8.2}$$

Observations.

(1) Definition 9 differs from Definition 8 because $\epsilon \to b(\epsilon)$ is a priori given and also because $b(\epsilon)$ is not necessarily infinitesimal as $\epsilon \to 0$.

(2) Obviously one can also give other analogous definitions where the "perturbed" initial data look like $x(0) = x_0 + \epsilon$, $\dot{x}(0) = 0$, or some other.

Without formalizing the possibilities hidden in Observation 2, we continue by stating a well-known simple stability criterion for stability in the sense of Definition 8. We shall afterwards discuss a third stability definition involving the introduction of novel interesting ideas.

The following proposition holds.

13 Proposition. *If x_0 is a strict minimum for the potential energy function V, then x_0 is a stable equilibrium point in the sense of the Definition 8.*

PROOF. Let $E_\epsilon = m\epsilon^2/2 + V(x_0)$ be the total energy of the initial datum $x(0) = x_0$, $\dot{x}(0) = \epsilon$. Since, by assumption, x_0 is a point of strict minimum for V, see Fig. 2.2, i.e., $V(\xi) > V(x_0)$ if $\xi \neq x_0$ and $|\xi - x_0|$ is small enough, we can define the positions $x_{-,\epsilon}$ and $x_{+,\epsilon}$ which are the first root of $E - V(\xi) = 0$ to the left or to the right of x_0, respectively. It is also easy to check that the strict minimum assumption also implies that

$$\lim_{\epsilon \to 0} x_{\pm,\epsilon} = x_0 \tag{2.8.3}$$

and that, furthermore, $x_{+,\epsilon}$ and $x_{-,\epsilon}$ are, respectively, monotonically increasing and decreasing. For large ϵ, it might happen that $E_\epsilon - V(\xi)$ does not have one of the two roots $x_{-,\epsilon}$ or $x_{+,\epsilon}$ or both. We interpret this by setting $x_{-,\epsilon} = -\infty$ or $x_{+,\epsilon} = +\infty$.

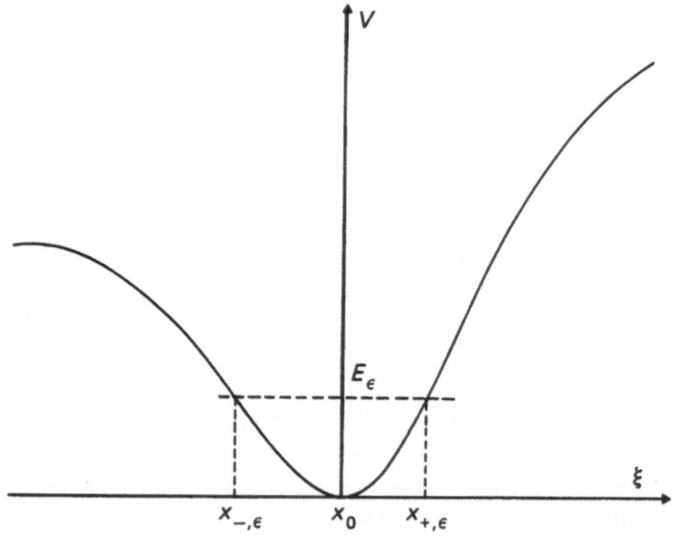

Figure 2.2.

Then if one sets

$$a(\epsilon) = \max_{\sigma = \pm} |x_{\sigma,\epsilon} - x_0|, \tag{2.8.4}$$

one realizes that Eq. (2.8.1) is verified for the motions $t \to x(t)$ such that $x(0) = 0$, $|\dot{x}(0)| < \epsilon$, using the arguments of Proposition 11, p. 37. mbe

Observations.

(1) The proof method of Proposition 13 allowed us to define, in fact, the minimal tolerance function; i.e., Eq. (2.8.4).

It is therefore easy to provide stability criteria in the sense of Definition 9 by using the preceding proof, under the assumptions of Proposition 13.

(2) Note that if $(d^2V/d\xi^2)(x_0) > 0$, the function $a(\epsilon)$ in Eq. (2.8.4) is of $O(\epsilon)$.

Exercises and Problems for §2.8

1. Determine the stable equilibrium positions in the sense of Definition 9 with tolerance functions:

$$b(\epsilon) = \tfrac{1}{2} + \epsilon \quad \text{or} \quad \begin{cases} b(\epsilon) = 3\epsilon & \text{for } \epsilon < \tfrac{1}{5}, \\ b(\epsilon) = +\infty & \text{for } \epsilon > \tfrac{1}{5}, \end{cases}$$

for a unit mass point acted upon by a force with potential energy

$$V(\xi) = \xi(\xi - 1), \text{ or } \log(1 + \xi^2), \text{ or } -\sin\xi, \text{ or } \tfrac{1}{2}\xi^2 e^{-\xi^2}.$$

2. Show that not all the stable equilibrium positions for $V(\xi) = (\sin\xi^2)e^{-\xi^2}$ have tolerance $b(\epsilon) = \frac{1}{2}$ if $\epsilon < 1$ and $b(\epsilon) = +\infty$ if $\epsilon > 1$.

3.* Show that the potential energy V defined by

$$V(\xi) = e^{-1/|\xi|}\left(\xi^2 + \left(\sin\frac{1}{\xi}\right)^2\right), \qquad \xi \neq 0$$

$$= 0, \qquad \xi = 0$$

has infinitely many stable equilibrium positions in the sense of Definition 8. (Hint: Show that $V'(\xi)$ is infinitely many times positive and negative near zero.)

§2.9. Stability and Friction

A further alternative definition of an equilibrium point x_0 for a force law $f \in C^\infty(\mathbf{R})$, with potential energy V bounded from below, comes from the remark that, in practice, when x_0 is a stable equilibrium position, then, under a small perturbation of the equilibrium state, the point mass moves away from x_0 to return eventually to x_0 with essentially zero velocity. Think of what really happens when a pendulum is slightly deflected from its equilibrium position.

To give a mathematically precise meaning to the stability criterion that seems to emerge from these considerations, we have to formulate a precise definition of the term "friction".

An accurate analysis of the friction phenomenon could be found in physics and engineering textbooks: it will be enough for us to observe that, empirically, a friction force acts "against the motion"; then one understands why a mathematical model for a friction force is that of a force law depending on the position x and, mainly, on the velocity $\dot{x}$ of the point mass in such a way to systematically have a sign opposite to that of $\dot{x}$.

The simplest model describes the friction force A in terms of a non-negative C^∞ function $(\eta, \xi) \to \alpha(\eta, \xi)$ defined on $\mathbf{R}^2$ as:

$$A(\dot{x}, x) = -\dot{x}\alpha(\dot{x}, x) \tag{2.9.1}$$

with α verifying the further property that $\alpha(\eta, \xi) \neq 0$ for $\eta \neq 0$; i.e., friction is absent only if the point is standing still. There are, however, phenomena for which this is not a good model, like the so-called "static friction" cases (which are modeled by discontinuous friction forces).

Remarkable examples are: "linear friction",

$$A(\dot{x}, x) = -\lambda\dot{x}, \qquad \lambda > 0; \tag{2.9.2}$$

"cubic" friction,

$$A(\dot{x}, x) = -\lambda\dot{x}(1 + \lambda'\dot{x}^2), \qquad \lambda, \lambda' > 0; \tag{2.9.3}$$

and "quadratic friction",

$$A(\dot{x}, x) = -\lambda\dot{x}(1 + \lambda'\dot{x}^2)^{1/2}, \qquad \lambda, \lambda' > 0. \tag{2.9.4}$$

We can then formulate the following stability notion.

10 Definition. If x_0 is an equilibrium point for $m\ddot{x} = f(x)$, we shall say that it is "strongly stable" if for small enough ϵ the motions $t \to x(t)$, $t \geqslant 0$, with initial data $x(0) = x_0$, $\dot{x}(0) = \epsilon$ and described by the (normal) equation

$$m\ddot{x} = -\lambda\dot{x} + f(x) \tag{2.9.5}$$

are such that

$$\lim_{t \to +\infty} x(t) = x_0, \qquad \forall \lambda > 0. \tag{2.9.6}$$

Observation. In other words, this means that x_0 is strongly stable if, in the presence of an arbitrarily small friction, an initial datum $x(0) = x_0$, $\dot{x}(0) = \epsilon$ produces a motion returning asymptotically to x_0 at least if ϵ is not too large.

The following is a stability criterion in the new sense.

14 Proposition. *Let x_0 be an equilibrium point for $m\ddot{x} = -(dV/dx)(x)$, with $V \in C^\infty(\mathbf{R})$ bounded from below. Suppose that for $\xi - x_0 \neq 0$ and* *small enough, the derivative* $-f'(\xi)=(d^2V/d\xi^2)(\xi)$ *is positive ("strict convexity of* V *at* x_0*"); then* x_0 *is a strongly stable equilibrium point.*

Observations

(1) The condition on V is verified if, for instance, V has a strict minimum in x_0 and not all its derivatives vanish in x_0 (exercise).

(2) The function V defined to be 0 for $\xi=0$ and, for $\xi\neq 0$:

$$V(\xi)=e^{-1/|\xi|}\left(\xi^2+\left(\sin\frac{1}{\xi}\right)^2\right) \tag{2.9.7}$$

is a potential energy function to which the criterion of Proposition 14 cannot be applied. One can see that, actually, Eq. (2.9.7) provides a counterexample to the thought that might flash that the above strong stability notion is equivalent to the one of Definition 8. The origin is, in fact, a stable equilibrium position because of Proposition 13, p. 42, but it is not a strongly stable equilibrium point.

(3) The proof of Proposition 14 is a particular case of quite general technique adaptable to the analysis of various stability problems as we shall see again in Chapter 5.

PROOF. Intuitively we can probably expect that in the presence of friction, energy is no longer conserved: we shall, indeed, show that the energy of the motion $t\to x(t)$, solution to Eq. (2.9.5), defined as $E(t)=\frac{1}{2}m\dot{x}(t)^2+V(x(t))$, $t\geqslant 0$, is a nonconstant function of t such that $\lim_{t\to+\infty}E(t)=E_0=V(x_0)$. Since $V(\xi)\geqslant V(x_0)=E_0$ and in the vicinity of x_0 there is just one point, namely x_0, where $V(\xi)=E_0$, it must necessarily follow that $\lim_{t\to+\infty}x(t)=x_0$, if ϵ is small.

To study the energy variation, as time elapses, of a motion verifying Eq. (2.9.5), we compute its derivative:

$$\frac{d}{dt}E(t)=\frac{d}{dt}\left(\frac{m\dot{x}(t)^2}{2}+V(x(t))\right)=\dot{x}(m\ddot{x}-f(x))=-\lambda\dot{x}^2\leqslant 0 \tag{2.9.8}$$

which shows that, in presence of linear friction, the energy is monotonically nonincreasing (and strictly decreasing when the velocity does not vanish).

Therefore, the limit

$$E_\infty=\lim_{t\to+\infty}E(t)\geqslant\inf_{\xi\in\mathbf{R}}V(\xi)>-\infty \tag{2.9.9}$$

exists.

Since x_0 is, by the assumption on f', a strict minimum point, there are (if ϵ is small enough) two points $x_{+,\epsilon}$ and $x_{-,\epsilon}$ to the right of x_0 and to the left of x_0, respectively, that cannot be bypassed by the motion with Eq. (2.9.5) and initial datum $x(0)=x_0$, $\dot{x}(0)=\epsilon$, because $E(t)\leqslant E(0)$, $\forall t\geqslant 0$.

Figure 2.3 eloquently illustrates this, making it unnecessary to expound further details.

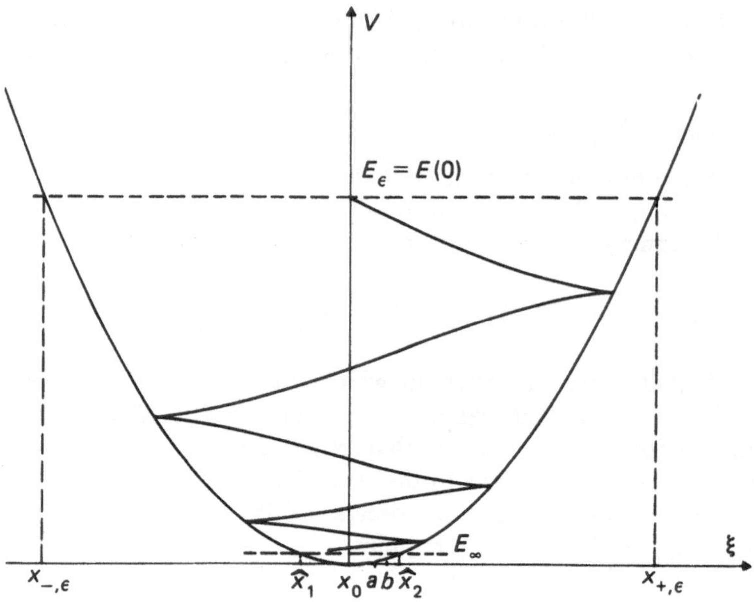

Figure 2.3.

Suppose that ϵ has been chosen so small that, as in Fig. 2.3, $f'(\xi) \neq 0$ if $\xi \neq x_0$, $x_{-,\epsilon} \leqslant \xi \leqslant x_{+,\epsilon}$: this is possible by the supposed structure of the minimum of V in x_0.

Suppose now, per absurdum, $E_\infty > E_0$, as in Fig. 2.3. Observe first that as $t \to +\infty$, $\lim_{t\to+\infty} x(t) = \hat{x}$ must exist.

Otherwise if $\bar{x} = \liminf_{t\to+\infty} x(t) < \bar{\bar{x}} = \limsup_{t\to+\infty} x(t)$, there would be an interval $[a,b] \subset (\bar{x}, \bar{\bar{x}})$, where $\min_{\xi\in[a,b]}(E_\infty - V(\xi)) > 0$. Such an interval would have to be run infinitely many times as $t \to +\infty$, since $\bar{x}$ and $\bar{\bar{x}}$ are limit points for $x(t)$; furthermore, when the point mass is in $[a,b]$, its velocity is neither too small nor too large:

$$|\dot{x}(t)| = \left(\frac{2}{m}(E(t) - V(x(t))\right)^{1/2} \geqslant \left(\frac{2\gamma}{m}\right)^{1/2} \tag{2.9.10}$$

for some $\gamma > 0$, and

$$|\dot{x}(t)| \leqslant \left(\frac{2}{m}(E(0) - E_0)\right)^{1/2}. \tag{2.9.11}$$

Therefore, every time the point mass enters $[a,b]$, it spends therein at least a time T:

$$T = (b-a)\left(\frac{2}{m}(E(0) - E_0)\right)^{-1/2} \tag{2.9.12}$$

by Eq. (2.9.11) and, therefore [see Eq. (2.9.8)], it loses an amount of energy given, at least, by

$$-\lambda \frac{2}{m}\gamma T. \tag{2.9.13}$$

Hence, after infinitely many passages through $[a,b]$, the energy should become $E_\infty = -\infty$, against $E_\infty \geqslant E_0$. We conclude that, indeed, the limit $\hat{x} = \lim_{t\to+\infty} x(t)$ exists. Clearly $\hat{x}$ must be one of the two abscissae of the intersections of E_∞ with the graph of V, i.e., in Fig. 2.3, one of the two points $\hat{x}_1$ or $\hat{x}_2$; otherwise,

$$\lim_{t\to+\infty} \dot{x}(t) = \pm\left(\frac{2}{m}(E_\infty - V(\hat{x}))\right)^{1/2} \neq 0$$

and $x(t)$ could not have a finite limit.[5] This, in turn, implies that $\lim_{t\to+\infty} \dot{x}(t) = 0$.

The last property is, however, in contradiction with the equations of motion (2.9.5) which would imply that

$$\lim_{t\to+\infty} \ddot{x}(t) = \frac{f(\hat{x})}{m} \neq 0, \tag{2.9.14}$$

i.e., that the limit as $t \to +\infty$ of $\dot{x}(t)$ could not be finite while we proved it to be zero.

Hence, E_∞ cannot be larger than E_0, and, then, as already remarked at the beginning of this proof, $\lim_{t\to+\infty} x(t) = x_0$. mbe

Exercises and Problems for §2.9

1. Show that the equation for the energy variation versus the position is, for the motions verifying $m\ddot{x} + \lambda\dot{x} + V'(x) = 0$, given by:

$$\frac{dE}{dx}(x) = \pm\lambda\sqrt{\frac{2}{m}(E(x) - V(x))}\ .$$

2.* Consider the motions described by the equations:

$$m\ddot{x} + \lambda\dot{x} + V'(x) = 0,$$
$$m\ddot{x} + \lambda x + W'(x) = 0$$

with $\dot{x}(0) = \dot{y}(0) = 0$, $x(0) = y(0) = x_0$ and suppose that for $x_0 \leqslant \xi \leqslant x_1$, one has $0 < -W'(\xi) < -V'(\xi)$.

Denote $v_x(\xi)$ and $v_y(\xi)$ the velocity of the motions x and y, respectively, at their passage through $\xi \in [x_0, x_1]$ and suppose also that it is known that $\dot{x}(t)$, $\dot{y}(t)$ are non-negative for all the times preceding the (respective) time of first passage through x_1.

Show that $v_x(\xi) \geqslant v_y(\xi)$, $\forall \xi \in [x_0, x_1]$. (Hint: Use the result of Problem 1 to deduce from $v_x(\xi) = \sqrt{2(E(\xi) - V(\xi))/m}$:

$$\frac{d}{d\xi}\left(v_x(\xi)^2 - v_y(\xi)^2\right) = \frac{2}{m}\left(-\lambda(v_x(\xi) - v_y(\xi)) - V'(\xi) + W'(\xi).\right)$$

This proves that $(d/d\xi)(v_x(\xi)^2 - v_y(\xi)^2) > 0$ for $\xi > x_0$ and close enough to x_0; hence, for such ξ's, $v_x(\xi) > v_y(\xi)$. If there existed $\bar{\xi} \in (x_0, x_1]$ where $v_x(\xi) = v_y(\xi)$ we

[5] Exercise: If $\lim_{t\to+\infty} f(t)$ and $\lim_{t\to+\infty} f'(t)$ exist, then $\lim_{t\to+\infty} f'(t) = 0$ (denoting f' the derivative of f).

could consider the smallest among them: still call it $\bar{\xi}$. Then $(d/d\xi)(v_x(\bar{\xi})^2 - v_y(\bar{\xi})^2) \leqslant 0$, since $\bar{\xi}$ is the first point where $v_x(\xi) = v_y(\xi)$; but this contradicts the above equation for $v_x(\xi)^2 - v_y(\xi)^2$ since $v_x(\xi) = v_y(\bar{\xi})$ while $-V'(\bar{\xi}) + W'(\bar{\xi}) > 0$.)

3.* Consider the case analogous to the one in Problem 2 with initial datum $\dot{x}(0) = \dot{y}(0) = v_0 > 0$.

4.* Formulate and prove results analogous to Problems 2 and 3, when $v_0 < 0$, $0 < W'(\xi) < V'(\xi)$.

5. Consider the equation $\ddot{x} + \lambda\dot{x} - f(x) = 0$, $x(0) = 0$, $\dot{x}(0) = 1$ or $\dot{x}(0) = -\sqrt{2/15}$. Determine the limit, as $t \to +\infty$, of $x(t)$ for $\lambda = 50$ and for f with potential energy $V(\xi) = \xi^2(1+\xi)^2$.

6. Same as Problem 5 for $\lambda = 10$ and $x(0) = 0$, $\dot{x}(0) = 10$.

7. Same as Problem 5 for $V(\xi) = (\xi^2 - 1)(\xi + 2)$, $x(0) = 3/2$ and $\dot{x}(0) = 0$, $\lambda = 4$; or $V(\xi) = \xi^2(\xi + 1)(\xi + 2)$, $\lambda = 1$, $x(0) = 0$, $\dot{x}(0) = -\sqrt{2}$.

8. How large should λ be so that the motion verifying $\ddot{x} = -\dot{x} + V'(x)$, with $V(\xi) = \frac{1}{2}\xi^2 e^{-\xi^2}$, $x(0) = 0$, $\dot{x}(0) = 10$, is attracted by the origin? (Find a lower bound only.)

9. Show that for λ small enough, the motion in Problem 8 "runs away", i.e., $\lim_{t\to+\infty} x(t) = +\infty$. For such a motion, after an arbitrary choice of λ, estimate the time necessary to reach the point with abscissa $\xi = 10$. (Find an upper and a lower bound.)

§2.10. Period and Amplitude: Harmonic Oscillators

In this section we shall consider the motions of a point mass with mass m subject to a force law f generated by a C^∞ potential energy V such that

$$\begin{aligned} &\text{(i)} \quad V(\xi) = V(-\epsilon), \\ &\text{(ii)} \quad \frac{dV}{d\xi}(\xi) \neq 0, \qquad \xi \neq 0, \\ &\text{(iii)} \quad \lim_{\xi\to+\infty} V(\xi) = +\infty. \end{aligned} \tag{2.10.1}$$

In §2.7 it was proved that all such motions are periodic with period $T < +\infty$.

We now ask whether there exist potential energy functions V verifying Eq. (2.10.1) and generating motions with energy-independent (or amplitude-independent) period.

It is well known that the "elastic energy" $V(\xi) = V(0) + \frac{1}{2}k\xi^2$ generates "isochronous" motions of period $T = 2\pi(m/k)^{1/2}$, which remain constant as the total energy varies ("harmonic oscillations").

We now show that this isochrony is a characteristic property of the harmonic oscillators in the class (2.10.1), (Landau–Lifshitz, see References).

15 Proposition. *If all the motions developing under the action of a force with potential energy V verifying Eq.* (2.10.1) *have the same period, there is*

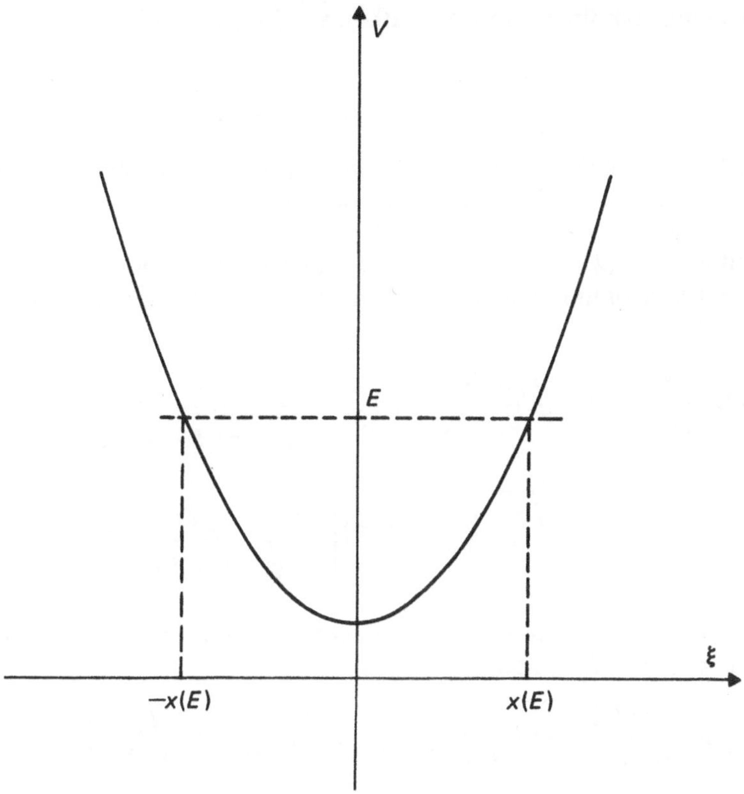

Figure 2.4.

$k > 0$ *such that*

$$V(\xi) = \tfrac{1}{2}k\xi^2 + V(0). \tag{2.10.2}$$

Observation. Using the idea involved in the following proof, it is also possible to treat the case when V does not verify (i). See the observations following Corollary 16 below.

Proof. Let E be the energy of the motion associated with the potential energy of Eq. (2.10.1) and let $x(E)$ be the corresponding amplitude ($x(E) = x_+$ with the notations of §2.7, see Fig. 2.4).

This motion's period is [see Eq. (2.7.8)]

$$T(E) = 4\int_0^{x(E)}\left(\frac{2}{m}(E - V(\xi))\right)^{-1/2} d\xi. \tag{2.10.3}$$

Since V is monotonically increasing in ξ for $\xi \geqslant 0$, we can construct the inverse function to the function V. We denote it by $v \to \xi(v)$, defined for $v \in [V(0), +\infty)$ and such that $V(\xi(v)) \equiv v$, $\forall v \in [V(0), +\infty)$. The second relation in Eq. (2.10.1) implies that $v \to \xi(v)$ is in C^∞ $(V(0), +\infty)$, say, by the implicit function theorem (see Appendix G).

Changing coordinates in Eq. (2.10.4), $\xi \to \xi(v)$, we find:

$$T(E) = 4\int_{V(0)}^{E} \frac{\xi'(v)\,dv}{\left[2(E-v)/m\right]^{1/2}}, \tag{2.10.4}$$

where $\xi'(v)$ is the derivative of $\xi(v)$ with respect to v. Supposing $E \to T(E)$ known for $E \in (V(0), +\infty)$; Eq. (2.10.4) becomes an equation for $\xi(v)$ which can be solved through the following artifice.

Multiply Eq. (2.10.4) by $(b-E)^{-1/2}$ and integrate both sides between $V(0)$ and b (assuming that the arbitrary parameter b is chosen larger than $V(0)$:

$$\begin{aligned}\int_{V(0)}^{b} dE\,\frac{T(E)}{\sqrt{b-E}} &= 4\sqrt{\frac{m}{2}}\int_{V(0)}^{b} dE\int_{V(0)}^{E}\frac{\xi'(v)\,dv}{\sqrt{(E-v)(b-E)}}\\ &= 4\sqrt{\frac{m}{2}}\int_{V(0)}^{b} dv\,\xi'(v)\left[\int_{v}^{b}\frac{dE}{\sqrt{(E-v)(b-E)}}\right].\end{aligned} \tag{2.10.5}$$

The integral in the last parenthesis can be explicitly computed and its value is π, $(\forall v, b!)$. Hence,

$$\int_{V(0)}^{b} dE\,\frac{T(E)}{\sqrt{b-E}} = 4\pi\sqrt{\frac{m}{2}}\,(\xi(b)-\xi(V(0))) = 4\pi\sqrt{\frac{m}{2}}\,\xi(b). \tag{2.10.6}$$

This formula is interesting in itself since it provides the expression of the potential energy "as a function of the period" for all V's verifying Eq. (2.10.1).

When $T(E) = T =$ constant, $\forall E \in (V(0), +\infty)$, one deduces, from Eq. (2.10.6),

$$\xi(b) = \frac{T}{4\pi}\sqrt{\frac{2}{m}}\,2\sqrt{b-V(0)} \tag{2.10.7}$$

which, remembering the definition of $\xi(b)$, means that

$$V(\xi) = \frac{1}{2}m\left(\frac{T}{2\pi}\right)^{-2}\xi^2 + V(0). \tag{2.10.8}$$

mbe

The remark after Eq. (2.10.6) provides the following corollary.

16 Corollary. *Let $E \to T(E)$, $E \in (V(0), +\infty)$, be the period of the motions with energy E developing under the action of a potential verifying Eq.* (2.10.1) *and let $V(0) = 0$. Then V is given by*

$$\int_{0}^{V(\xi)} \frac{T(E)\,dE}{\sqrt{V(\xi)-E}} = 4\pi\sqrt{\frac{m}{2}}\,\xi. \tag{2.10.9}$$

Observations.

(1) In the above proof, it is necessary that $V(\xi) = V(-\xi)$: if V verifies only (ii) and (iii) of Eq. (2.10.1), then Eq. (2.10.3) is no longer correct and should be replaced by

$$T(E) = 2\int_{x_-(E)}^{x_+(E)} \left(\frac{2}{m}(E - V(\xi))\right)^{-1/2} d\xi, \qquad (2.10.10)$$

where $x_+(E), x_-(E)$ are the roots of $E - V(\xi) = 0$ [uniquely defined by Eq. (2.10.1), (ii) and (iii)]. Proceeding as in the proof of Propositions 15 and 16, after splitting Eq. (2.10.10) into two integrals like Eq. (2.10.3) between $x_-(E)$ and 0 and between 0 and $x_+(E)$, one finds:

$$2\pi\sqrt{\frac{m}{2}}\,(x_+(b) - x_-(b)) = \int_0^b dE\,\frac{T(E)}{\sqrt{b-E}}\,, \qquad (2.10.11)$$

determining $x_+(b) - x_-(b)$ in terms of the period function.

(2) We see, therefore, that because of Observation (1), there are infinitely many C^∞ functions $\xi \to V(\xi)$ verifying (ii) and (iii) and originating motions with energy-independent period. They can be visualized by saying that their graphs are obtained by horizontally deforming the parabolae of Eq. (2.10.8), keeping fixed the distances between the values $x_-(E)$ and $x_+(E)$ such that $V(x_\pm(E)) = E$.

A necessary and sufficient condition that V, verifying (ii) and (iii) of Eq. (2.10.1), generates isochronous periodic motions is that for all $E > V(0)$,

$$x_+(E) - x_-(E) = k'\sqrt{E - V(0)} \qquad (2.10.12)$$

for some $k' > 0$.

(3) Note that Eq. (2.10.9) does not, in general, imply that there is a $V \in C^\infty(\mathbf{R})$ verifying it for arbitrarily given $E \to T(E)$ (see the problems below).

Exercises and Problems for §2.10

1. Determine the potential V verifying Eq. (2.10.1) and $V(0) = 0$ such that $T(E)$ is $(1 + E)$ or $(1 + E^2)$ or $\log(1 + E)$; check whether $V \in C^\infty(\mathbf{R})$ or $V \in C^\infty(\mathbf{R}/\{0\})$.

2. Let $E \to T(E) > 0$ be a C^∞ function defined for $E \geqslant 0$. Suppose that $T(E) = T_0(1 + \Sigma_{k=1}^{\infty}\tau_k E^k)$ for E small enough and suppose $|\tau_k| \leqslant \rho^k$ for some $\rho > 0$, $\forall k > 0$. Show that $\xi(b)$ in Eq. (2.10.6) is given by

$$\xi(b) = T_0\sqrt{b}\left(1 + \sum_{k=1}^{\infty} b^k \tau_k \int_0^1 \frac{x^k\,dx}{(1-x)^{1/2}}\right)$$

for b small enough.

3. In the context of Problem 2, using the implicit functions theorem (see Appendix G) to invert, the function

$$\xi^2 = T_0^2 V\left(1 + \sum_{k=1}^{\infty} V^k \tau_k \int_0^1 \frac{x^k\,dx}{(1-x)^{1/2}}\right)^2$$

to obtain V as a function of ξ^2 for ξ^2 small, show that there is a $V \in C^\infty(\mathbf{R})$ verifying Eq. (2.10.1) and producing motions with energy E whose period is $T(E)$ for all E small enough.

4.* Let $E \to T(E) > 0$ be a C^∞ function defined for $E \geqslant 0$. Show that given $N > 0$, there is a C^∞ function A_N such that the function $\xi(b)$ in Eq. (2.10.6) can be expressed as

$$\xi(b) = T_0\sqrt{b}\left(1 + \sum_{k=1}^{N} \tilde{\tau}_k b^k + b^{N+1}A_{N+1}(b)\right),$$

where $\tilde{\tau}_1, \ldots, \tilde{\tau}_N$ are suitably chosen constraints and $T = T(0)$. (Hint: Use the Lagrange–Taylor expansion to order N on the left-hand side of Eq. (2.10.6) to express $T(E)$ (see Appendix B).)

5.* Using the result of Problem 4, indicate which, among the following functions $E \to T(E)$, cannot be the period function describing the periods of the motions with energy E of some even C^∞ potential: $E \to (1 + E)$, $E \to 1 + (\cos E)^2$, $E \to 1 + (\sin\sqrt{E}\,/\sqrt{E}\,)^2$, $E \to 1 + \frac{1}{2}(\sin\sqrt{E}\,)^2$, $E \to 1 + \operatorname{sh}\sqrt{E}$, $E \to 1 + \log(1 + E)$, $E \to 1 + \log(1 + \sqrt{E}\,)$. (Hint: The problem is essentially whether the function (2.10.6) really can be used to obtain b (i.e., V) as a function of ξ which, also, is C^∞.)

6.* Let $V \to \xi(V)$ be defined by

$$\xi(V) = \frac{1}{4\pi}\sqrt{\frac{2}{m}}\,\sqrt{V}\int_0^1 \frac{T(xV)\,dx}{\sqrt{1-x}}, \qquad V \geqslant 0$$

obtained from Eq. (2.10.6) by setting $b = V$, $V(0) = 0$, and changing the integration variable. Assume that $E \to T(E)$ is a C^∞ function of $E \in [0, +\infty)$. Show that a necessary and sufficient condition for the existence of potential V verifying Eq. (2.10.1) and producing motions with energy $E \geqslant 0$ with period $T(E)$ is that $\xi(V) \to_{V \to +\infty} +\infty$, $\xi'(V) > 0$, $\forall V > 0$. Show also that this happens if $T'(E) > 0$, $\forall E \geqslant 0$, and $T(E)$ is bounded for $E \geqslant 0$; show, however, that such conditions are only sufficient conditions.

7. Let $V \in C^{(0)}(\mathbf{R})$ verify (i), (ii), and (iii) of Eq. (2.10.1) and suppose that $V \in C^\infty((-\infty, 0) \cup (0, +\infty))$ and $V(0) = 0$. Define $t \to x(t)$, $t \geqslant 0$, to be a motion generated by V if x is a $C^{(1)}$ function verifying $\dot{x}(t)^2/2 + V(x(t)) = E > 0$ and $\dot{x}(t)$ changes sign to the right and to the left of any time t when $\dot{x}(t) = 0$.

Show that any initial datum $x(0), \dot{x}(0)$ gives rise to a unique motion generated by V and respecting the datum, if $E > 0$.

8.* Show that if in Problem 6 one only drops the condition $T(E) > 0$ replacing it by $T(E) \geqslant 0$, one has the same results, provided V is allowed to vary in the class of potentials considered in Problem 7.

9. Find a calculation algorithm for the tabulation of a function $\xi \to V(\xi)$ which generates motions with period $\log(1 + \sqrt{E}\,)$ with 30% accuracy as E varies in the interval $4 \leqslant E \leqslant 10$. (Hint: Define $T(E)$ "arbitrarily for $E \notin [4, 10]$ and use Eq. (2.10.6).)

10. Using a desk computer connected with a plotter, actually perform the calculations in Problem 9, drawing the graph of V (without tabulating it) and the graph of the amplitude $x(E)$, $E \in [4, 10]$.

§2.11. The Damped Oscillator: Euler's Formulae

In §2.10 we saw that the harmonic oscillator is a system with the absolutely remarkable property of exhibiting only periodic motions with the same period.

In this section, and in the following, we shall examine other important properties of harmonic oscillators before dedicating some attention to the study of the stability of such properties with respect to "small" modifications of the force law.

Consider a point mass with mass $m > 0$ whose motions are described by the equation

$$m\ddot{x}(t) = -kx(t) - \lambda\dot{x}(t) + \varphi(t), \tag{2.11.1}$$

where $k > 0$, $\lambda > 0$, and $\varphi \in C^\infty(\mathbf{R})$ is a preassigned function. Equation (2.11.1) is a normal differential equation (see §2.5), as it can be readily verified by multiplying it by $\dot{x}(t)$ and obtaining

$$\frac{d}{dt}E(t) = -\lambda\dot{x}(t)^2 + \dot{x}(t)\varphi(t) \leqslant \max_{\eta\in\mathbf{R}}\big(-\lambda\eta^2 + \eta\varphi(t)\big) = \frac{\varphi(t)^2}{4\lambda} \tag{2.11.2}$$

if $E(t) = m\dot{x}(t)^2/2 + kx(t)^2/2$. Hence, for all $t \geqslant 0$, we find the a priori estimate

$$E(t) = \frac{1}{2}m\dot{x}(t)^2 + \frac{1}{2}kx(t)^2 \leqslant E(0) + \int_0^t \frac{\varphi(\tau)^2}{4\lambda}\,d\tau \tag{2.11.3}$$

which implies normality by Proposition 5, p. 28.

The motions described by Eq. (2.11.1) are called the "forced oscillations" of a linearly damped harmonic oscillator. In this section we shall study the case $\varphi \equiv 0$, i.e., the equation

$$m\ddot{x} = -\lambda\dot{x} - kx \tag{2.11.4}$$

describing the linearly damped oscillators.

The arguments used to prove the strong stability criterion, Proposition 14, p. 45, can be easily adapted to the particular case of Eq. (2.11.4) and allow us to conclude that its motions have a trivial asymptotic behavior as $t \to +\infty$: $\lim_{t\to+\infty} x(t) = 0$.

Actually Eq. (2.11.4) can be "explicitly" solved and from the solution's formulae one gets a very detailed description of the motions.

We shall prove the following proposition.

17 Proposition. *Given* $(\eta_0, \xi_0, t_0) \in \mathbf{R}^3$, *there exist* A_0, A_0' *in* $\mathbf{R}$ *such that the solution of Eq.* (2.11.4) *with initial datum*

$$\dot{x}(t_0) = \eta_0, \qquad x(t_0) = \xi_0 \tag{2.11.5}$$

can be written as

$$x(t) = e^{-(\lambda/2m)(t-t_0)}\Big(A_o e^{(\lambda/2m)\sqrt{1-(4mk/\lambda^2)}\,(t-t_0)} + A_0' e^{-(\lambda/2m)\sqrt{1-(4mk/\lambda^2)}\,(t-t_0)}\Big) \tag{2.11.6}$$

if $\lambda^2 > 4mk$; *or as*

$$x(t) = e^{-(\lambda/2m)(t-t_0)}\Big(A_0 \cos\sqrt{(k/m)(1-\lambda^2/4mk)}\,(t-t_0) + A_0' \sin\sqrt{(k/m)(1-\lambda^2/4mk)}\,(t-t_0)\Big) \tag{2.11.7}$$

if $\lambda^2 < 4mk$; *or as*

$$x(t) = e^{-(\lambda/2m)(t-t_0)}(A_0 + A_0'(t-t_0)) \tag{2.11.8}$$

if $\lambda^2 = 4mk$.

Observations.

(1) Notice that $\lim_{t\to+\infty} x(t) = 0$ exponentially fast for all solutions.

(2) Clearly there are two time scales in the motions described above (they coincide if $\lambda^2 = 4mk$). For small λ (compared with $\sqrt{4mk}$), one time scale is $2m/\lambda$ and the other is $\sim 2\pi\sqrt{m/k}$ and $2m/\lambda \gg 2\pi\sqrt{m/k}$. The first time scale controls the damping ("friction time scale") and the other controls the oscillatory motion ("proper time scale") [see Eq. (2.11.7)].

(3) The above solutions can be continued to solutions of Eq. (2.11.4) on the entire time range. However, $\limsup_{t\to-\infty} |x(t)| = +\infty$ unless $x(t) \equiv 0$.

PROOF. A possible proof is by direct verification, i.e., by inserting Eqs. (2.11.6)–(2.11.8) into Eq. (2.11.4) and by checking that in each case the initial data can be satisfied by suitably choosing A_0, A_0'. We present a more instructive proof which illustrates a general method and allows to introduce some new mathematical notions.

Look for solutions of Eq. (2.11.4) having the form

$$x(t) = Ae^{\alpha t}, \qquad A \neq 0. \tag{2.11.9}$$

By inserting Eq. (2.11.9) into Eq. (2.11.4), we see that in order that Eq. (2.11.9) be a solution it must be

$$m\alpha^2 + \lambda\alpha + k = 0; \tag{2.11.10}$$

hence, $\alpha = \alpha_+$ or $\alpha = \alpha_-$ with

$$\alpha_\pm = -\frac{\lambda}{2m}\left[1 \pm \sqrt{-1\,\frac{4mk}{\lambda^2}}\right]. \tag{2.11.11}$$

If $\lambda^2 > 4mk$, there are no problems. For $t \in \mathbf{R}$, setting

$$x(t) = A_0 e^{\alpha_+(t-t_0)} + A_0' e^{\alpha_-(t-t_0)}, \tag{2.11.12}$$

one obtains a solution of Eq. (2.11.4) for all $A_0, A'_0 \in \mathbf{R}$, since Eq. (2.11.4) is a linear homogeneous equation. Imposing the initial conditions yields the system

$$\xi_0 = A_0 + A'_0, \qquad \eta_0 = \alpha_+ A_0 + \alpha_- A'_0 \tag{2.11.13}$$

whose determinant is

$$\alpha_+ - \alpha_- = \frac{\lambda}{m}\left(1 - \frac{4mk}{\lambda^2}\right)^{1/2} \neq 0.$$

This proves the proposition if $\lambda^2 > 4mk$.

The case $\lambda^2 = 4mk$ can be easily obtained by first letting $\lambda^2 > 4mk$, solving Eqs. (2.11.4) and (2.11.5), and taking the limit $\lambda^2 \to 4mk$ and using the regularity theorem, Proposition 3, for differential equations.

The determination of A_0 and A'_0 from Eq. (2.11.13) gives, for $\lambda^2 > 4mk$,

$$x(t) = \eta_0 \frac{e^{\alpha_+(t-t_0)} - e^{\alpha_-(t-t_0)}}{\alpha_+ - \alpha_-} - \xi_0 \frac{\alpha_- e^{\alpha_+(t-t_0)} - \alpha_+ e^{\alpha_-(t-t_0)}}{\alpha_+ - \alpha_-} \tag{2.11.14}$$

which, as $\lambda^2 \to 4mk$, gives

$$\left(\xi_0 + \left(\eta_0 + \frac{\lambda}{2m}\xi_0\right)(t - t_0)\right)e^{-(\lambda/2m)(t-t_0)}. \tag{2.11.15}$$

For $\lambda^2 < 4mk$, the roots $\alpha_\pm$ are complex and Eq. (2.11.9) does not directly make sense.

However, if we could give a meaning to the exponential of a complex number in such a way that the function $t \to e^{zt}$ has the properties

$$\frac{d}{dt}e^{zt} = ze^{zt}, \qquad \forall z \in \mathbf{C} \tag{2.11.16}$$

and, of course, $e^z = \sum_{k=0}^{\infty} z^k/k!$ for z real, we could still take Eq. (2.11.4) as the solution to Eqs. (2.11.4) and (2.11.5).

It is natural to define $\forall z \in \mathbf{C}$,

$$e^z = \sum_{k=0}^{\infty} \frac{z^k}{k!} \tag{2.11.17}$$

since the series is absolutely convergent even if z is complex.

It is then possible to check Eq. (2.11.16) by series differentiation of Eq. (2.11.17) with z replaced by zt: in fact, such a series can be differentiated term by term, as is easy to see.

Some remarkable properties of e^z are

(i)

$$e^z e^{z'} = e^{z+z'}, \qquad e^{\bar{z}} = \overline{e^z}, \tag{2.11.18}$$

where the bar denotes complex conjugation. This property can be checked by *series multiplication*, as for z real, and by conjugation of the series.

(ii)

$$e^{x+iy} = e^x(\cos y + i \sin y), \qquad \forall x, y \in \mathbf{R} \tag{2.11.19}$$

which is checked by recalling the Taylor series for the sine and cosine:

$$e^{x+iy} = e^x e^{iy} = e^x \sum_{k=0}^{\infty} \frac{(iy)^k}{k!} = e^x \left(\sum_{k=0}^{\infty} \frac{(-1)^k y^{2k}}{(2k)!} + i \sum_{k=0}^{\infty} \frac{(-1)^k y^{2k+1}}{(2k+1)!} \right)$$
$$= e^x(\cos y + i \sin y) \tag{2.11.20}$$

(iii) By Eq. (2.11.19), one has

$$\cos y = \frac{e^{iy} + e^{-iy}}{2}, \qquad \sin y = \frac{e^{iy} - e^{-iy}}{2i}. \tag{2.11.21}$$

Hence, we see that Eq. (2.11.14) gives a solution to Eqs. (2.11.4) and (2.11.5), even if $\lambda^2 < 4mk$, by interpreting the complex exponentials as given by Eqs. (2.11.17) and (2.11.19). Note that Eq. (2.11.14) defines a real function of t, as $\alpha_+ = \bar{\alpha}_-$ and the coefficients of η_0, ξ_0 in Eq. (2.11.14) are therefore real because of the second relation in Eq. (2.11.18).

Since, by (2.11.18):

$$\begin{aligned} \operatorname{Re} e^{\alpha_+ t} &= e^{-(\lambda/2m)t} \cos\sqrt{(k/m)(1-\lambda^2/4mk)}\; t, \\ \operatorname{Im} e^{\alpha_+ t} &= e^{-(\lambda/2m)t} \sin\sqrt{(k/m)(1-\lambda^2/4mk)}\; t, \end{aligned} \tag{2.11.22}$$

it is clear that Eq. (2.11.6) follows from Eqs. (2.11.14) and (2.11.22). mbe

Observations.

(1) Using the representation (2.11.14) and the complex exponentials, the two cases $\lambda^2 > 4mk$ and $\lambda^2 < 4mk$ are formally unified. This is the first instance, among several that we shall meet, where the use of complex-valued functions appears useful and simplifies formulae and calculations even in problems in which one is eventually only interested in "real-valued results".

(2) The formula:

$$e^{x+iy} = e^x(\cos y + i \sin y), \qquad x, y \in \mathbf{R} \tag{2.11.23}$$

is called "Euler's formula" and it will be widely used in the following.

Notice the remarkable facts that the polar representation of a complex number $z = \rho(\cos\theta + i\sin\theta)$ becomes, because of Eq. (2.11.23):

$$z = \rho e^{i\theta}, \tag{2.11.24}$$

and also $|e^{iy}| \equiv 1$, $\forall y \in \mathbf{R}$, is true and more generally:

$$|e^{x+iy}| = e^x, \qquad \forall x, y \in \mathbf{R} \tag{2.11.25}$$

so that $e^z \neq 0$, $\forall z \in \mathbf{C}$.

Exercises and Problems for §2.11

1. Through Euler's formulae prove the "De Moivre formula": i.e., show that for $\forall n \geqslant 0$ and n an integer, $(\cos\theta + i\sin\theta)^n = (\cos n\theta + i \sin n\theta)$.

2. Through Euler's formulae and the Newton binomial, show that for $n \geqslant 0$ and integer:

$$(\cos\theta)^n = \left(\frac{e^{i\theta}+e^{-i\theta}}{2}\right)^n = \frac{1}{2^n}\sum_{k=0}^{n}\binom{n}{k}\cos(n-2k)\theta.$$

3. Study the analogue to Problem 2 for $(\sin\theta)^n$.

4. Via Euler's formulae, compute $\Sigma_{j=0}^{n}\cos j\theta$ using the addition formula for geometric series.

5. Compute

$$\int_0^{2\pi} e^{in\theta}\frac{d\theta}{2\pi}, \qquad \int_0^{2\pi}(\cos\theta)^n\frac{d\theta}{2\pi}, \qquad n \in \mathbf{Z}_+,$$

using Euler's formulae and Problem 2.

6. Compute

$$\int_0^{2\pi}(\sin\theta)^n\frac{d\theta}{2\pi}, \qquad \int_0^{2\pi}(\sin\theta)^n(\cos\theta)^m\frac{d\theta}{2\pi}, \qquad n \in \mathbf{Z}_+.$$

7. Find two linearly independent solutions of $\ddot{x}+\dot{x}+x=0$ and compute their determinant $w(t)$, $t \geqslant 0$ (see Problem 16 in §2.2).

8. Consider the system of equations in $\mathbf{R}^d$: $\dot{\mathbf{x}} = L\mathbf{x}$, where L is a $d\times d$ matrix $L=(l_{ij})_{i,j=1,\ldots,d}$, with constant coefficients. Determime whether there are solutions having the form $\mathbf{x}(t)=e^{\alpha t}\mathbf{x}(0)$. Which algebraic equation does α satisfy? Which equation does $\mathbf{x}(0)$ have to verify? (See also Appendices E and F.)

9. Apply the method suggested in Problem 8 to find two linearly independent solutions of $\dot{x} = ax+y$, $\dot{y} = -x+ay$ and describe the flow $(S_t)_{t\geqslant 0}$ in the data space as a varies.

10. Compute the time interval between the nth and the $(n+1)$th passage through the origin of the solutions of $\ddot{x}+\dot{x}+x=0$ and $\ddot{x}+\frac{3}{2}\dot{x}+x=0$.

§2.12. Forced Harmonic Oscillations in the Presence of Friction

We now consider Eq. (2.11.1) with $\varphi \neq 0$. Its motions are the "linearly damped harmonic oscillations with forcing term φ".

An obvious but important remark about Eq. (2.11.1) is that its most general solution can be written as the sum of a particular solution $t \to x_{\text{part}}(t)$, $t \geqslant 0$, of Eq. (2.11.1) and of a solution of Eq. (2.11.4), i.e., of the homogeneous equation associated with Eq. (2.11.1). In fact, the linearity of this equation provides that the difference between two of its solutions is a solution of Eq. (2.11.4).

Hence, in formulae, a solution $t \to x(t)$, $t \geqslant 0$, of Eq. (2.11.1) can be written:

$$x(t) = x_{\text{part}}(t) + x_0(t), \tag{2.12.1}$$

where $t \to x_0(t)$ is a solution of Eq. (2.11.4).

In §2.11 we saw that $\lim_{t\to+\infty} x_0(t) = 0$ and, furthermore, we found explicit expressions for the most general solution $t \to x_0(t)$. Hence, the discussion of the properties of the motions described by Eq. (2.11.1) is reduced to that of a particular solution of the same equation which we can choose as convenience suggests. This remark is particularly relevant whenever one is interested in the "asymptotic behavior" as $t \to +\infty$, where $t \to x_0(t)$ is infinitesimal.

Let us now describe a method for the construction of a particular solution to Eq. (2.11.1) valid in the interesting though special case when φ is periodic with period $T > 0$.

18 Proposition. *Let $\varphi \in C^\infty(\mathbf{R})$ be a real-valued periodic function with period $T > 0$. Then Eq.* (2.11.1) *admits a solution with the same period.*

Observation.

Consequently, we can say that all the motions described by Eq. (2.11.1) with a periodic forcing term are "asymptotically periodic": this means that there is a periodic solution $t \to x_{\text{per}}(t)$, $t \in \mathbf{R}_+$ of Eq. (2.11.1) such that any other solution $t \to x(t)$ has the property $|x(t) - x_{\text{per}}(t)| \to_{t\to+\infty} 0$.

PROOF. First consider the apparently special cases

$$\varphi(t) = \hat{\varphi} \cos \frac{2\pi}{T} t \quad \text{or} \quad \varphi(t) = \hat{\varphi} \sin \frac{2\pi}{T} t, \qquad \hat{\varphi} \in \mathbf{R} \tag{2.12.2}$$

and notice that they can be treated simultaneously by solving the equation

$$m\ddot{x} + \lambda\dot{x} + kx = \hat{\varphi} e^{(2\pi i/T)t}. \tag{2.12.3}$$

In fact, the real and imaginary parts of a solution to Eq. (2.12.3) are solutions to Eq. (2.11.1) with φ given, respectively, by the first or the second solution of Eq. (2.11.2) as a result of Euler's formulae $\operatorname{Re} e^{(2\pi i/T)t} = \cos(2\pi/T)t$, $\operatorname{Im} e^{(2\pi i/T)t} = \sin(2\pi/T)t$.

On the other hand, remembering the properties of the complex exponentials (i.e., $(d/dt)e^{zt} = ze^{zt}$), it is clear that Eq. (2.12.3) admits a particular periodic solution

$$x_{\text{per}}(t) = \frac{\hat{\varphi} e^{(2\pi i/T)t}}{-m(2\pi/T)^2 + i\lambda(2\pi/T) + k}. \tag{2.12.4}$$

Hence, the particular cases (2.12.2) are solved by the real and imaginary parts of Eq. (2.12.4), respectively.

To analyze more general cases, we shall again use the linearity of Eq. (2.11.1). If this equation is considered with right-hand side $\varphi \in C^\infty(\mathbf{R})$ or $\psi \in C^\infty(\mathbf{R})$ and if $t \to x_\varphi(t)$ and $t \to x_\psi(t)$, $t \in \mathbf{R}_+$, are particular solutions of it, then $t \to x_\varphi(t) + x_\psi(t)$, $t \in \mathbf{R}_+$, is a particular solution of Eq. (2.11.1) with right-hand side $\varphi + \psi$.

Consider, then, the case:

$$\varphi(t)=\sum_{n=0}^{N}\hat{\varphi}_n^{(1)}\cos\frac{2\pi}{T}nt+\sum_{n=1}^{N}\hat{\varphi}_n^{(2)}\sin\frac{2\pi}{T}nt, \tag{2.12.5}$$

where $\hat{\varphi}_n^{(1)},\hat{\varphi}_n^{(2)}$, $n=0,1,2,\ldots,N$, are real constants.

By using Euler's formulae, we can rewrite Eq. (2.11.5) as:

$$\varphi(t)=\sum_{n=-N}^{N}\hat{\varphi}_n e^{(2\pi i/T)nt}, \tag{2.12.6}$$

where $\hat{\varphi}_n$ is defined by

$$\hat{\varphi}_n=\overline{\hat{\varphi}}_{-n}=\frac{\hat{\varphi}_n^{(1)}-i\hat{\varphi}_n^{(2)}}{2},\qquad n>0;\qquad \hat{\varphi}_0=\hat{\varphi}_0^{(1)}. \tag{2.12.7}$$

Hence, a particular solution of Eq. (2.11.1) with φ given by Eq. (2.12.5) [or Eq. (2.12.6)] is

$$x_{\text{per}}(t)=\sum_{n=-N}^{N}\frac{\hat{\varphi}_n e^{(2\pi i/T)nt}}{-m((2\pi/T)n)^2+i(2\pi/T)n\lambda+k} \tag{2.12.8}$$

which is real since the addends in Eq. (2.12.8) with index n and $-n$ are complex conjugates because of Eq. (2.12.7).

So the proposition is proved when φ is given by Eq. (2.12.5) or Eqs. (2.12.6) and (2.12.7).

Obviously the same methods permit us to solve the case when φ is given by:

$$\varphi(t)=\sum_{n=-\infty}^{+\infty}\hat{\varphi}_n e^{(2\pi i/T)nt},\qquad t\in\mathbf{R} \tag{2.12.9}$$

with

$$\hat{\varphi}_n=\overline{\hat{\varphi}}_{-n},\qquad n=0,1,\ldots, \tag{2.12.10}$$

provided the series (2.12.9) converges well enough so that the function $t\to x_{\text{per}}(t)$, $t\in\mathbf{R}$, defined by

$$x_{\text{per}}(t)=\sum_{n=-\infty}^{+\infty}\hat{\varphi}_n\frac{e^{(2\pi i/T)nt}}{-m((2\pi/T)n)^2+i\lambda(2\pi/T)n+k} \tag{2.12.11}$$

is of class C^∞ and its first and second derivatives (at least) can be computed by summing the corresponding derivatives of the functions in Eq. (2.12.11).

A simple sufficient condition for these properties is that there is a constant c_p such that

$$|\hat{\varphi}_n|\leqslant c_p(1+|n|^p)^{-1},\qquad n=0,\pm1,\pm2,\ldots \tag{2.12.12}$$

for all $p\geqslant 0$ or, equivalently:

$$\lim_{n\to\infty}|\hat{\varphi}_n|(1+|n|^p)=0,\qquad \forall p\geqslant 0. \tag{2.12.13}$$

If Eq. (2.12.12) holds, the series (2.12.9) is uniformly convergent together with the derivative series obtained by differentiating Eq. (2.12.9) term by term an arbitrary number of times. For instance, the series of the kth derivatives of Eq. (2.12.9) is

$$\sum_{n=-\infty}^{+\infty} \hat{\varphi}_n \left(\frac{2\pi i}{T} n \right)^k e^{(2\pi i/T)t} \tag{2.12.14}$$

and its nth term has a modulus bounded by

$$|\hat{\varphi}_n| \left(\frac{2\pi}{T} |n| \right)^k \leqslant \left(\frac{2\pi}{T} \right)^k c_p \frac{|n|^k}{(1+|n|)^p}, \tag{2.12.15}$$

by Eq. (2.12.12) and by $|\exp(2\pi i/T)nt| \equiv 1$. The right-hand side of Eq. (2.12.15) is t independent and can also be summed over n if one chooses the (arbitrary) parameter $p > k+1$.

Then the series differentiation theorems guarantee that Eq. (2.12.9) is a C^∞ function whose derivatives can be computed by "series differentiation".

Hence, the proposition is proved also when φ is given by Eqs. (2.12.9) and (2.12.10) with $\hat{\varphi}_n$ verifying Eq. (2.12.12), $\forall p \geqslant 0$, i.e., with $\hat{\varphi}_n$ decreasing faster than any power as $n \to \infty$.

At this point we use the following very important proposition which tells us that the last case considered is, actually, the most general and therefore completes our proof.

19 Proposition. *Let $T > 0$ and $\varphi \in C^\infty(\mathbf{R})$ be a periodic function with period T. There exists a unique sequence $(\hat{\varphi}_n)_{n\in\mathbf{Z}}$ of complex numbers such that*

$$\text{(i)} \quad \hat{\varphi}_n = \bar{\hat{\varphi}}_{-n}, \qquad n = 0,1,2,\dots; \tag{2.12.16}$$

$$\text{(ii)} \quad \lim_{n\to\infty} (1+|n|)^p |\hat{\varphi}_n| = 0, \qquad \forall p \in \mathbf{Z}_+; \tag{2.12.17}$$

$$\text{(iii)} \quad \varphi(t) = \sum_{n=-\infty}^{+\infty} \hat{\varphi}_n e^{(2\pi i/T)nt}, \qquad \forall t \in \mathbf{R}. \tag{2.12.18}$$

The numbers $\hat{\varphi}_n$ are called the "harmonics" of φ with respect to the period T and

$$\text{(iv)} \quad \hat{\varphi}_n = \frac{1}{T} \int_0^T \varphi(t) e^{-(2\pi i/T)nt}\, dt, \qquad \forall n \in \mathbf{Z}, \tag{2.12.19}$$

and, finally, $\forall s = 0,1,\dots$:

$$\text{(v)} \quad \frac{d^s\varphi}{dt^s}(t) = \sum_{n=-\infty}^{+\infty} \left(\frac{2\pi i}{T} n \right)^s \hat{\varphi}_n e^{(2\pi i/T)nt}, \qquad \forall t \in \mathbf{R}. \tag{2.12.20}$$

Observations.

(1) Equation (2.12.17) can also be read as: the sequence $(\hat{\varphi}_n)_{n\in\mathbf{Z}}$ approaches zero, as $n \to \infty$, "faster than any power". It is equivalent to Eq. (2.12.12).

(2) Proposition 19 implies, via Eqs. (2.12.18) and (2.12.19), that two C^∞ functions periodic with the same period $T > 0$ coincide if and only if all their harmonics relative to the period T coincide.

(3) Proposition 19 is a beautiful "structure theorem" on the C^∞-periodic functions on $\mathbf{R}$: it is the "Fourier series theorem".

The proof of this proposition will be given in the next section and it will also conclude the proof of Proposition 18.

§2.13. Fourier's Series for C^∞-Periodic Functions

Preliminary to the proof of Proposition 19, p. 60, let us remark that if a function $t \to \varphi(t)$, $t \in \mathbf{R}$, is defined by Eq. (2.12.18) with $(\hat{\varphi}_n)_{n\in Z}$ verifying Eqs. (2.12.16) and (2.12.17), then φ is necessarily a C^∞ function, by the series differentiation theorem [see, also, the considerations concerning Eqs. (2.12.14) and (2.12.15)]. Furthermore, since Eq. (2.12.18) is, in this case, uniformly convergent:

$$\int_0^T e^{-(2\pi i/T)nt}\varphi(t)\,\frac{dt}{T} = \sum_{k=-\infty}^{+\infty} \hat{\varphi}_k \int_0^T e^{-(2\pi i/T)(n-k)t}\,\frac{dt}{T} \tag{2.13.1}$$

by the interchangeability of the integration and the summation operations in unformly convergent series.

However:

$$\int_0^T e^{-(2\pi i/T)(n-k)t}\,\frac{dt}{T} = \begin{cases} 1 & \text{if } n = k, \\ 0 & \text{if } n \neq k, \end{cases} \tag{2.13.2}$$

as can be seen by explicit calculation of the integral. Relation (2.13.2) is often written

$$\int_0^T e^{-(2\pi i/T)(n-k)t}\,\frac{dt}{T} = \delta_{nk}, \tag{2.13.3}$$

$n, k = 0, \pm 1, \pm 2, \ldots$ with $\delta_{nn} \equiv 1$, $\delta_{nk} \equiv 0$ if $k \neq n$.

By substituting Eq. (2.13.2) into Eq. (2.13.1), one deduces

$$\frac{1}{T}\int_0^T e^{-(2\pi i/T)nt}\varphi(t)\,dt = \hat{\varphi}_n, \qquad n \in \mathbf{Z} \tag{2.13.4}$$

which shows that if φ has the form of Eq. (2.12.18) with $(\hat{\varphi}_n)_{n\in\mathbf{Z}}$ verifying Eq. (2.12.17), then the numbers $\hat{\varphi}_n$ are uniquely determined by Eq. (2.12.19). If φ is real, then Eq. (2.12.19) [or Eq. (2.13.4)] implies Eq. (2.12.16).

The above considerations show the validity of an "inverse" proposition to Proposition 19 and motivate the validity of Eq. (2.12.19). They are also useful since they allow the introduction of the fundamental relation (2.13.3).

We now give the proof of Proposition 19 §2.12, ("Fourier's theorem").

PROOF. Let $\varphi \in C^\infty(\mathbf{R})$ be a real periodic function with period $T > 0$. Define

$$\hat{\varphi}_n = \frac{1}{T}\int_0^T e^{-(2\pi i/T)nt}\varphi(t)\,dt, \qquad n \in \mathbf{Z}. \tag{2.13.5}$$

It is evident that $\hat{\varphi}_n = \bar{\hat{\varphi}}_{-n}$, $\forall n \in \mathbf{Z}$, because φ is real. Hence, Eq. (2.12.16) holds.

To study the asymptotic behavior of $\hat{\varphi}_n$ as $n \to \infty$, integrate Eq. (2.13.5) by parts.

$$\begin{aligned}\hat{\varphi}_n &= \frac{1}{T}\left[\frac{e^{(2\pi i/T)nt}}{-(2\pi i/T)n}\varphi(t)\right]_0^T - \frac{1}{T}\int_0^T \frac{e^{-(2\pi i/T)nt}}{-(2\pi i/T)n}\varphi'(t)\,dt \\ &= \frac{1}{T}\frac{1}{((2\pi i/T)n)}\int_0^T e^{-(2\pi i/T)nt}\varphi'(t)\,dt,\end{aligned} \tag{2.13.6}$$

where φ' denotes the first derivative of φ, and the periodicity of φ has been used to eliminate the first term in the intermediate relation.

Since φ' is obviously a T-periodic C^∞ function and so are the higher derivatives $\varphi'', \varphi''', \ldots, (d^p\varphi/dt) \equiv \varphi^{(p)}$, we can iterate Eq. (2.13.6) by again integrating by parts.

After p steps, $p = 0, 1, 2, \ldots,$ one finds:

$$\hat{\varphi}_n = \frac{1}{((2\pi i/T)n)^p}\frac{1}{T}\int_0^T e^{-(2\pi i/T)tn}\varphi^{(p)}(t)\,dt. \tag{2.13.7}$$

Hence, if

$$\tilde{c}_p = \max_{0<t<T}|\varphi^{(p)}(t)|, \tag{2.13.8}$$

one has, $\forall p = 0, 1, \ldots$:

$$|\hat{\varphi}_n| < \left(\frac{T}{2\pi|n|}\right)^p \tilde{c}_p, \qquad \forall n \in \mathbf{Z} \tag{2.13.9}$$

which is obviously equivalent to Eq. (2.12.17).

It remains to prove Eq. (2.12.18) with $\hat{\varphi}_n$, $n \in \mathbf{Z}$, given by Eq. (2.13.5). In fact, the relation (2.12.20) is, as already remarked, a consequence of Eqs. (2.12.18) and (2.12.17).

Consider the order N approximant to the series (2.12.18); we elaborate it by using Eq. (2.13.5):

$$\begin{aligned}\sum_{n=-N}^{N} e^{(2\pi i/T)nt}\hat{\varphi}_n &= \sum_{n=-N}^{N} e^{(2\pi i/T)nt}\int_0^T e^{-(2\pi i/T)n\tau}\varphi(\tau)\frac{d\tau}{T} \\ &= \int_0^T\left(\sum_{n=-N}^{N} e^{(2\pi i/T)(t-\tau)}\right)\varphi(\tau)\frac{d\tau}{T}\end{aligned} \tag{2.13.10}$$

which is an identity, $\forall \varphi \in C^\infty(\mathbf{R})$.

The summation in the parenthesis in Eq. (2.13.10) is obviously a C^∞ function in t and τ, periodic in both variables with period T, and it has the value $2N+1$ if $\tau = t + mT$, with m an integer. It also has the property

$$\frac{1}{T}\int_0^T\left(\sum_{n=-N}^{N} e^{(2\pi i/T)n(t-\tau)}\right)d\tau \equiv 1, \qquad \forall t \in \mathbf{R} \tag{2.13.11}$$

which immediately follows from Eq. (2.13.3) by changing $t-\tau$ into t' and by using the mentioned periodicity. Furthermore, the function in parenthesis in Eqs. (2.13.11) and (2.13.10) can be written as

$$1+\sum_{n=1}^{N} e^{(2\pi i/T)n(t-\tau)}+\sum_{n=1}^{N} e^{-(2\pi i/T)n(t-\tau)} \tag{2.13.12}$$

and the two sums can be "explicitly" summed as geometric sums with ratios $\exp \pm 2\pi i(t-\tau)$.

After a few steps, the result is

$$\sum_{n=-N}^{N} e^{(2\pi i/T)n(t-\tau)}=\begin{cases} \dfrac{\sin(N+\frac{1}{2})(2\pi/T)(t-\tau)}{\sin\frac{1}{2}(2\pi/T)(t-\tau)}, & \tau\neq t+mT,\ m \text{ an integer}, \\ 1+2N & \tau=t+mT,\ m \text{ an integer}. \end{cases} \tag{2.13.13}$$

Coming back to Eq. (2.13.10) and using Eqs. (2.13.13) and (2.13.11):

$$\begin{aligned} \sum_{n=-N}^{N} \hat{\varphi}_n e^{(2\pi i/T)nt} &= \frac{1}{T}\int_0^T \frac{\sin(N+\frac{1}{2})(2\pi/T)(t-\tau)}{\sin\frac{1}{2}(2\pi/T)(t-\tau)}\varphi(\tau)\,d\tau \\ &= \frac{1}{T}\int_0^T \frac{\sin(N+\frac{1}{2})(2\pi/T)(t-\tau)}{\sin\frac{1}{2}(2\pi/T)(t-\tau)} \\ &\quad \times(\varphi(t)+\varphi(\tau)-\varphi(t))\,d\tau \\ &= \varphi(t)+\frac{1}{T}\int_0^T \frac{\sin(N+\frac{1}{2})(2\pi/T)(t-\tau)}{\sin\frac{1}{2}(2\pi/T)(t-\tau)} \\ &\quad \times(\varphi(\tau)-\varphi(t))\,d\tau. \end{aligned} \tag{2.13.14}$$

Hence, to show Eq. (2.12.8), we have to show that

$$\lim_{N\to\infty}\frac{1}{T}\int_0^T \frac{\sin(N+\frac{1}{2})(2\pi/T)(t-\tau)}{\sin\frac{1}{2}(2\pi/T)(t-\tau)}(\varphi(\tau)-\varphi(t))\,d\tau=0. \tag{2.13.15}$$

The reason why this is true is related to the remark that the function

$$\begin{aligned} \tau\to\psi_t(\tau) &= \frac{\varphi(\tau)-\varphi(t)}{\sin\frac{1}{2}(2\pi/T)(t-\tau)}\cos\frac{1}{2}\frac{2\pi}{T}(t-\tau) && \text{if } \tau\neq t+mT m \text{ an integer} \\ &= \frac{T}{\pi}\varphi'(t) && \text{if } \tau=t+mT m \text{ an integer} \end{aligned} \tag{2.13.16}$$

is a C^∞ function with period T. This is a remark whose proof is left as an exercise (see Problems 1–4 of this section).

By the trigonometric addition formulae, the integral in Eq. (2.13.15) then becomes

$$\frac{1}{T}\int_0^T\left\{\psi_t(\tau)\sin\frac{2\pi}{T}N(t-\tau)+(\varphi(\tau)-\varphi(t))\cos\frac{2\pi}{T}N(t-\tau)\right\}d\tau. \tag{2.13.17}$$

It appears that this expression, via Euler's formulae, is a linear combination of four harmonics of order $\pm N$ of the function of the τ variable $\tau\to\psi_t(\tau)$ and $\tau\to\varphi(t)-\varphi(\tau)$ which, as discussed above, are C^∞ functions, periodic with period T.

Hence, the integral in Eq. (2.13.17) must tend toward zero faster than any power of N as $N-\infty$: in fact, formula (2.13.9) is valid for an arbitrary T-periodic C^∞ function. The same, then, occurs for Eq. (2.13.15), and (2.12.18) is proved. mbe

Exercises and Problems for §2.13

1. Let $f\in C^\infty(\mathbf{R})$, $f(0)=0$. Define $\psi(t)=f(t)/t$, $t\neq 0$, and $\psi(0)=f'(0)$. By applying the Taylor–Lagrange theorem (see Appendix B), show that $\psi\in C^\infty(\mathbf{R})$.

2. In the context of problem 1, show that for $k=0,1,\ldots,$

$$\psi^{(k)}(t)=\sum_{h=0}^{k}\frac{(-t)^{k-h}}{(k-h)!}f^{(k-h)}(t)\frac{(-1)^k k!}{t^{k+1}},\qquad \forall t\neq 0,$$

$$\psi^{(k)}(0)=\frac{f^{(k+1)}(0)}{(k+1)},$$

where the superscript k denotes the kth derivative.

3. Show that if f, $g\in C^\infty(\mathbf{R})$ and $g(t_0)=0$, $g'(t_0)\neq 0$, the function

$$\psi(t)=\frac{f(t)-f(t_0)}{g(t)},\qquad t\neq t_0,$$

$$\psi(t_0)=\frac{f'(t_0)}{g'(t_0)}$$

is a C^∞ function in the vicinity of t_0.

4. If $f\in C^\infty(\mathbf{R})$ and is periodic with period T, the function

$$\psi_t(\tau)=\frac{(f(\tau)-f(t))\cos(\pi/T)(t-\tau)}{\sin(\pi/T)(t-\tau)}\qquad \text{if } \tau\neq T+mT,\ m \text{ an integer}$$

$$=\frac{T}{\pi}f'(t)\qquad \text{if } \tau=t+mT,\ m \text{ an integer}$$

is a C^∞ function of τ and it is periodic with period T. (Hint: Use Problem 3.)

5. Using Eq. (2.12.19), compute the Fourier coefficient of order $0, 1, -1$ for the function $f(t)=(1-\frac{1}{2}\cos t)^{-1}$, thinking of it as a periodic function with period 2π or 4π.

6. Using the Taylor series for the function $(1-\xi)^{-\alpha}$, compute the Fourier series' coefficients of the complex-valued functions with period 2π:

$$f(t) = \left(1 - \tfrac{1}{2}e^{it}\right)^{-1} \quad \text{or} \quad f(t) = \left(1 - \tfrac{1}{2}e^{it}\right)^{-m/n}$$

with $m, n \in \mathbf{Z}$.

7. Let, $\forall z \in C$:

$$\sin z = \frac{e^{iz} - e^{-iz}}{2i}, \qquad \cos z = \frac{e^{iz} + e^{-iz}}{2}.$$

Using the Taylor series for the exponential [see Eq. (2.11.17)], determine the Fourier series' coefficients of $f(t) = \sin e^{it}$ or of $f(t) = \cos e^{it}$, $t \in \mathbf{R}$, as 2π-periodic functions or as 4π-periodic functions.

8. Let, $\forall z \in \mathbf{C}$, $|z| < 1$:

$$\log(1+z) = \sum_{n=1}^{\infty} \frac{(-1)^{n+1}}{n} z^n.$$

By using the series expansion for the exponential, Eq. (2.11.17), show that $\exp(\log(1+z)) \equiv 1 + z$ and compute the Fourier transform of the 2π-periodic function $f(t) = -\log(1 - \frac{1}{2}e^{it})$.

9. Same as Problem 7 for $f(t) = (1 - \frac{1}{2}\cos t)^{-1}$, $t \in \mathbf{R}$. Estimate $\hat{f}_2$ up to 10%, i.e., find an expression for $\hat{f}_n$, but estimate it only for $n = 2$.

10. Compute the Fourier transform of $f(t) = -\log(1 - \frac{1}{2}\cos t)$ as a 2π-periodic function. Estimate $\hat{f}_3$ up to 30%.

11. Same as Problem 10 for $f(t) = \exp\cos t$. Estimate up to 1% $\hat{f}_0, \hat{f}_{\pm 1}$.

12.* Show that all the functions in Problems 5–11 have an exponentially decaying Fourier transform. In each case give an estimate of the decay constant.

13. Give an example of a C^∞ function, periodic with period 2π, whose Fourier transform does not decay exponentially (Hint: First define the transform and then the function, as its sum.)

14. Show that the function $f(t) = \sum_{n=-\infty}^{+\infty} e^{int}/(1+n^4)$ is continuous, term by term differentiable, periodic together with its derivative and with period 2π, but not C^∞.

15.* Analyze critically the proof of §2.13 to deduce that if f is a T-periodic continuous function which is piecewise differentiable with continuous bounded derivatives in each piece, then

$$f(t) = \lim_{N\to+\infty} \sum_{n=-N}^{N} \hat{f}_n e^{(2\pi i/T)nt}$$

with $\hat{f}_n$ given by $\int_0^T f(t) e^{-(2\pi i/T)nt}(dt/T)$, $n \in \mathbf{Z}$.

If f is discontinuous but piecewise continuous with derivatives bounded and continuous in each piece, the preceding formula holds in every continuity point. In the discontinuity points, if $f(t^\pm) = \lim_{\tau\to t^\pm} f(\tau)$, the series' sum is $(f(t^+) + f(t^-))/2$ (considering 0 and T as the same point from the point of view of the discontinuities). (Hint: To reduce the second part to the first, show the truth of the second part in the case of a function which takes just two values (i.e., which has only two discontinuities being otherwise constant). Then show that any function of the second type is a sum of a function of the first type plus a finite number of piecewise constant functions.) (Recall that a function f defined on the interval a, b is piecewise continuous if a, b can be represented as a union of n closed intervals $[a_1, b_1], [a_2, b_2], \ldots, [a_n, b_n]$ and, for every $i = 1, 2, \ldots, n$, the function f coincides in the interior of $[a_i, b_i]$ with a function f_i continuous on the entire interval $[a_i, b_i]$: f may take arbitrary values at the extremes of each interval $[a_i, b_i]$).

§2.14. Nonlinear Oscillations. The Pendulum and Its Forced Oscillations. Existence of Small Oscillations

In the preceding sections we saw that the period of the forced oscillations of a damped harmonic oscillator is identical to that of the forcing term (§2.12).

However, the notion of "linear" or "harmonic" oscillator is too rigid a notion and, in applications, a linear oscillator can only appear as a simplified model of some more complex entity.

For instance, very often a linear oscillator appears as a model for the "small oscillations" of a system governed by a nonlinear equation. The prototype of these nonlinear systems is the pendulum.

It is natural to ask the question of the stability of the properties of the solutions to certain classes of equations with respect to the variations of the equations themselves: in fact, it is clear that in applications one shall only "trust" those models' predictions which do not change by "slightly" changing the models themselves. This is because, as stressed in Chapter 1, there is no "absolutely valid model".

As an example of a motion-stability problem in the above sense, we shall now treat some questions concerning the pendulum's forced motion; i.e., the motion governed by the (normal) equation:

$$m\ddot{x}(t) + \lambda\dot{x}(t) + k\sin x(t) = f(t), \qquad t \in \mathbf{R}_+ \tag{2.14.1}$$

with λ, m, and $k > 0$ and where $f \in C^\infty(\mathbf{R})$ is a periodic function with period $T > 0$.

In the following, it will be necessary to compare several motions, i.e., functions of t, and to fix the ideas we shall adopt, as a measure of the magnitude on $[a,b] \subset I$ of a function $\varphi \in C^\infty(I)$, the quantity[6]

$$\|\varphi\|_{[a,b]} = \sup_{t \in [a,b]} |\varphi(t)|. \tag{2.14.2}$$

We now ask if the motions of Eq. (2.14.1) have the following properties:

(1) If $t \to x(t)$, $t \in \mathbf{R}_+$, is a motion described by Eq. (2.14.1) and if $x(0)$, $\dot{x}(0)$, $\|f\|_{\mathbf{R}_+}$ are small enough, the motion is also small ("existence of the small oscillations").

[6] Obviously, there are other possible magnitude measures. Usually the "good" one is determined from the needs of the particular applications. Examples of other measures are

$$\int_a^b |\varphi(t)|\,dt, \qquad \sup_{t \in (a,b)} (|\varphi(t)| + |\dot{\varphi}(t)|), \qquad \left(\int_a^b |\varphi(t)|^2\,dt\right)^{1/2}.$$

(2) When $\|f\|_{\mathbf{R}_+}$ is sufficiently small, Eq. (2.14.1) admits a solution with the same period of f.
(3) As $t \to +\infty$, every solution can be asymptotically confused with the periodic solution, in (2) above, provided such a solution exists and the data $x(0)$, $\dot{x}(0)$ are small enough.

In other words, we ask if the above three properties, which have been explicitly or implicitly checked for the forced linear oscillations without restrictions on $x(0)$, $\dot{x}(0)$, $\|f\|_{\mathbf{R}_+}$, are still true in a nonlinear case, at least in the small oscillations' regime.

In this section we analyze problem (1) and introduce the following proposition which "solves" it:

20 Proposition. *There exist constants* $\gamma, \gamma' > 0$ *such that if* $f \in C^\infty(\mathbf{R})$ *(not necessarily periodic) and* $(x_0, v_0) \in \mathbf{R}^2$, *the motion* $t \to x(t)$ *described by Eq.* (2.14.1) *and following the initial data* $x(0) = x_0$, $\dot{x}(0) = v_0$, *verifies*

$$\|x\|_{\mathbf{R}_+} \leqslant \gamma(|x_0| + |v_0| + \|f\|_{\mathbf{R}_+}) \qquad \text{if} \quad |x_0| + |v_0| + \|f\|_{\mathbf{R}_+} < \gamma'. \tag{2.14.3}$$

Observations.

(1) Equation (2.14.1) is just one example of a nonlinear equation, chosen among others for its historical and romantic importance. The results and methods that follow apply to much more general equations. The reader will recognize that in the proof the key point is that $k\sin\xi - k\xi$ is infinitesimal, as $\xi \to 0$, of higher order in ξ. As an exercise, the reader can, with the obvious modifications, repeat the proof that follows to investigate the validity of the statement identical to Proposition 20 for the equation $m\ddot{x} + \lambda\dot{x} + k\psi(x) = f(t)$, $\lambda > 0$, under the sole assumptions that $\psi \in C^\infty(\mathbf{R})$, $\psi(0) = 0$, $\psi'(0) = (d\psi/d\xi)(0) > 0$.

(2) It is relevant to realize the necessity of the restriction on $\|f\|_{\mathbf{R}_+}$: this is made obvious by considering the equation

$$m\ddot{x} + \lambda\dot{x} + \sin x = \lambda\omega + \sin\omega t, \tag{2.14.4}$$

whose solution, among others, $t \to \omega t$ is unbounded. However, restrictions on x_0 and v_0 are not necessary. In other words, in Proposition 20, one could replace $|x_0| + |v_0| + \|f\|_{\mathbf{R}_+} < \gamma'$ with $\|f\|_{\mathbf{R}_+} < \gamma'$. We have imposed them only for the purpose of simplifying the proof.

(3) The proof's idea is to "compare" the solution of Eq. (2.14.1) with the solution of a similar equation where $\sin x$ is replaced by its first-order approximation, namely x. Such comparison will not be "direct", but it will take place by rewriting $k\sin x$ as $kx + k(\sin x - x)$ and considering the function $t \to k(\sin x(t) - x(t))$ as a known function bounded by $k|x(t)|^3/6$, because of the inequality $0 \leqslant \xi - \sin\xi \leqslant \xi^3/6$, $\forall \xi \in \mathbf{R}_+$.

In this way, one gets a linear equation with forcing term $f(t) - k(\sin x(t) - x(t))$. Solving it "explicitly" (see the following proof), one finds a t-independent relation between the amplitude $M(t) = \max_{0 \leqslant \tau \leqslant t} |x(\tau)| \equiv \|x\|_{[0,t]}$ and its cube which, as we shall see, implies that $M(t)$ must stay bounded, $\forall t \in \mathbf{R}_+$.

This method of proof is a particular case of a general method to obtain a priori estimates on the solutions of nonlinear equations close to linear ones and, sometimes, it is called the "self-consistency" method. The reader should meditate on the reason for this name after reading the following proof. The self-consistency method will be again used in this book, for instance in the proof of the Lyapunov stability criterion (see §5.4).

Proof. Assume, for simplicity, $\lambda^2 \neq 4mk$. Before analyzing Eq. (2.14.1), it is useful to remark that the equation

$$m\ddot{x} + \lambda\dot{x} + kx = F(t), \qquad t \in \mathbf{R}_+ \tag{2.14.5}$$

with $F \in C^\infty(\mathbf{R})$, admits, among its solutions defined for $t \geqslant 0$, the solution

$$p_0(t) = \int_0^t \frac{e^{\alpha_+(t-\tau)} - e^{\alpha_-(t-\tau)}}{\alpha_+ - \alpha_-} F(\tau) \frac{d\tau}{m}, \tag{2.14.6}$$

where α_+ and α_- are the two roots of $m\alpha^2 + \lambda\alpha + k = 0$, i.e.,

$$\alpha_\pm = -\frac{\lambda}{2m}\left[1 \pm \sqrt{1 - \frac{4mk}{\lambda^2}}\right]. \tag{2.14.7}$$

This property can be checked directly by inserting Eq. (2.14.6) into Eq. (2.14.5), and it is a special case of a general property of the linear differential equations which will be illustrated further through exercises and problems at the end of this section.[7]

As already remarked in §2.12, the most general solution to Eq. (2.14.5) will have the form

$$x(t) = \bar{x}(t) + p_0(t), \tag{2.14.8}$$

where $t \to \bar{x}(t)$, $t \in \mathbf{R}_+$, solves Eq. (2.14.5) with $F \equiv 0$.

Note also that Eq. (2.14.6) implies that p_0 is real valued, even when $\alpha_\pm$ are complex, provided F is real; also,

$$p_0(0) = 0, \qquad \dot{p}_0(0) = 0. \tag{2.14.9}$$

Coming back to Eq. (2.14.1) with initial conditions $x(0) = x_0$, $\dot{x}(0) = v_0$, we rewrite it as

$$m\ddot{x}(t) + \lambda\dot{x}(t) + kx(t) \equiv f(t) + k(x(t) - \sin x(t)), \tag{2.14.10}$$

and by the preceding remarks, pretending that the right-hand side is a "known function" of t, we can say

$$x(t) = \bar{x}(t) + \int_0^t \frac{e^{\alpha_+(t-\tau)} - e^{\alpha_-(t-\tau)}}{\alpha_+ - \alpha_-} \left[f(\tau) + k(x(\tau) - \sin x(\tau))\right] \frac{d\tau}{m}, \tag{2.14.11}$$

[7] Note also that if F is periodic, Eq. (2.14.6) will not be so, in general. Hence, this method for obtaining particular solutions to Eq. (2.14.5) is different from the one in §2.12, valid for periodic F's and based on the Fourier series.

where $t \to \bar{x}(t)$ is a solution to Eq. (2.14.5) with $F \equiv 0$ and verifying [see Eq. (2.14.9)]

$$\bar{x}(0) = x_0, \qquad \dot{\bar{x}}(0) = v_0. \tag{2.14.12}$$

From §2.12, it follows that

$$\bar{x}(t) = \frac{v_0 - \alpha_- x_0}{\alpha_+ - \alpha_-} e^{\alpha_+ t} - \frac{v_0 - \alpha_+ x_0}{\alpha_+ - \alpha_-} e^{\alpha_- t}, \tag{2.14.13}$$

and since $\operatorname{Re}\alpha_- \leqslant \operatorname{Re}\alpha_+ < 0$, it follows that $|\exp \alpha_\pm t| \leqq \exp \operatorname{Re}\alpha_+ t \leqslant 1$, $\forall t \geqslant 0$: hence,

$$\|\bar{x}\|_{\mathbf{R}_+} \leqslant \left(\frac{|\alpha_+| + |\alpha_-|}{|\alpha_+ - \alpha_-|} |x_0| + \frac{2}{|\alpha_+ - \alpha_-|} |v_0| \right). \tag{2.14.14}$$

Setting $M(t) = \|x\|_{[0,t]}$, we deduce from Eq. (2.14.11), using the inequality

$$0 \leqslant \xi - \sin\xi \leqslant \frac{\xi^3}{6}, \qquad \forall \xi \in \mathbf{R}_+, \tag{2.14.15}$$

that, $\forall t \geqslant 0$:

$$|x(t)| \leqslant \|\bar{x}\|_{\mathbf{R}_+} + \left(\|f\|_{\mathbf{R}_+} + \frac{k}{6} M(t)^3 \right) \int_0^t \frac{d\tau}{m} \cdot 2 \frac{e^{\operatorname{Re}\alpha_+(t-\tau)}}{|\alpha_+ - \alpha_-|}. \tag{2.14.16}$$

Hence, by integration,

$$|x(t)| \leqslant \|\bar{x}\|_{\mathbf{R}_+} + \frac{2m^{-1}}{|\operatorname{Re}\alpha_+||\alpha_+ - \alpha_-|} \left(\|f\|_{\mathbf{R}_+} + \frac{k}{6} M(t)^3 \right) \tag{2.14.17}$$

which implies, by Eq. (2.14.14),

$$|x(t)| \leqslant A + BM(t)^3, \qquad t \geqslant 0 \tag{2.14.18}$$

with

$$A = \frac{|\alpha_+| + |\alpha_-|}{|\alpha_+ - \alpha_-|} |x_0| + \frac{2}{|\alpha_+ - \alpha_-|} |v_0| + \frac{2\|f\|_{\mathbf{R}_+}}{m|\operatorname{Re}\alpha_+||\alpha_+ - \alpha_-|}, \tag{2.14.19}$$

$$B = \frac{2k}{6m|\operatorname{Re}\alpha_+||\alpha_+ - \alpha_-|}. \tag{2.14.20}$$

It is then immediately seen from Eq. (2.14.18) that the continuity and monotonicity of $M(t) = \|x\|_{[0,t]}$ and the arbitrariness of $t \geqslant 0$ imply

$$M(t) \leqslant A + BM(t)^3, \qquad \forall t \in \mathbf{R}_+, \tag{2.14.21}$$

and from Eq. (2.14.19), it also follows that

$$M(0) = |x(0)| = |x_0| < A. \tag{2.14.22}$$

It is now easy to complete the proof. The graph of the function $M \to A + BM^3 - M$ has the form illustrated in Fig. 2.5 if $27BA^2 < 4$.

Hence, if $|x_0|, |v_0|, \|f\|_{\mathbf{R}_+}$ are small enough so that this inequality involving A and B holds [see Eqs. (2.14.19) and (2.14.20)], we see that the equation $A + BM^3 - M = 0$ has three real roots $\mu_1(A)$, $\mu_2(A)$, $\mu_3(A)$, with

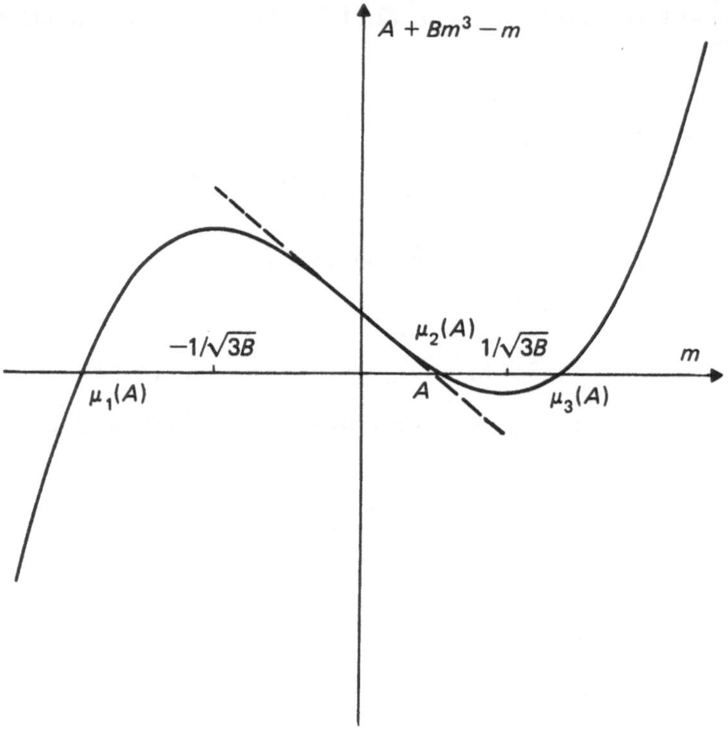

Figure 2.5.

$\mu_1(A) < 0$, $0 < \mu_2(A) < (3B)^{-1/2} < \mu_3(A)$. Furthermore, $A + BM^3 - M \geqslant A - M$ for all $M \geqslant 0$: hence, $\mu_2(A) > A$. Also, if $M \geqslant 0$, $M < (3B)^{-1/2}$, it follows that $0 \leqslant A + BM^3 - M \leqslant A + B(1/3B)M - M = A - \frac{2}{3}M$, i.e., $\mu_2(A) \leqslant \frac{3}{2}A$. So, concluding:

$$A < \mu_2(A) < \frac{3A}{2}. \qquad (2.14.23)$$

Since the function $t \to M(t)$, $t \in \mathbf{R}_+$, is continuous and verifies Eqs. (2.14.21) and (2.14.22) and $M(t) \geqslant 0$, it is clear that

$$M(t) \leqslant \mu_2(A) \leqslant \frac{3A}{2}, \qquad \forall t \in \mathbf{R}_+ \qquad (2.14.24)$$

which concludes the proof. The constant γ' is determined by the condition $27BA^2 < 4$ and γ by Eq. (2.14.24) recalling Eq. (2.14.19). mbe

Exercises and Problems for §2.14

1. Consider the differential equation $\dot{x} = ax + f(t)$ and show that $p(t) = \int_0^t e^{a(t-\tau)} f(\tau)\, d\tau$ is a solution to it with initial datum $p(0) = 0$, $\forall f \in C^\infty(\mathbf{R})$, $a \in \mathbf{R}$.

2. Let L be a $d \times d$ matrix with constant coefficients and consider the differential equation $\dot{\mathbf{x}} = L\mathbf{x} + \mathbf{f}(t)$, where $\mathbf{f} \in C^\infty(\mathbf{R})$ is an $\mathbf{R}^d$-valued function. Assume that L has d distinct eigenvalues $\lambda_1, \ldots, \lambda_d$ with respective eigenvectors $\mathbf{v}^{(1)}, \mathbf{v}^{(2)}$,

$\ldots, \mathbf{v}^{(d)}$. Show that if for $\mathbf{w} \in \mathbf{R}^d$, we denote $\alpha_1(\mathbf{w}), \ldots, \alpha_d(\mathbf{w})$ the components of $\mathbf{w}$ on the basis $\mathbf{v}^{(1)}, \ldots, \mathbf{v}^{(d)}$ (see Appendix E), then

$$\mathbf{p}(t) = \int_0^t \sum_{j=1}^d e^{\lambda_j(t-\tau)} \alpha_j(\mathbf{f}(\tau)) \mathbf{v}^{(j)}\, d\tau$$

is a particular solution to the equation, with $\mathbf{p}(0) = 0$. (Hint: Note that $\Sigma_{j=1}^d \alpha_j(\mathbf{f}(t)) \mathbf{v}^{(j)} \equiv \mathbf{f}(t)$ and check the validity of the equation by substitution.)

3.* In the context of Problem 2, Let $\mathbf{x}^{(1)}, \ldots, \mathbf{x}^{(d)}$ be d linearly independent solutions of the equation $\dot{\mathbf{x}} = L\mathbf{x}$ with initial data $x_j^{(i)}(0) = \delta_{ij}$, $i, j = 1, \ldots, d$. Let $W_{ij}(t) = x_j^{(i)}(t)$, $i, j = 1, \ldots, d$ be the matrix already introduced in Problems 7–9, §2.2 ("wronskian matrix"), verifying that $dW/dt = LW$. (Hint: Use the differential equation verified by each row of W.)

4.* In the context of Problems 2 and 3, show that

$$t \to \mathbf{p}(t) = \int_0^t W(t-\tau)\mathbf{f}(\tau)\, d\tau$$

is a special solution to $\dot{\mathbf{x}} = L\mathbf{x} + f(t)$ with initial datum $\mathbf{x}(0) = 0$; i.e., it coincides with the one in Problem 2.

5. Apply the method of Problem 2 to find particular solutions to the equation $\dot{x} = -x + y + f_1(t)$, $\dot{y} = -x - y + f_2(t)$.

6. Same as Problem 4 for $m\ddot{x} + \lambda\dot{x} + kx = f(t)$, after reducing it to a first-order system of equations. Consider the case $f(t) = t$. Show that the solution thus constructed verifies $x(0) = 0$, $\dot{x}(0) = 0$.

7. Same as Problem 4 for $d^4x/dt^4 - d^2x/dt^2 + x = t$, after reducing it to a first-order system. Show that the solution thus constructed verifies $x(0) = 0$, $\dot{x}(0) = 0$, $\ddot{x}(0) = 0$, $\dddot{x}(0) = 0$.

§2.15. Damped Pendulum: Small Forced Oscillations

We shall now show that the pendulum, as the damped linear oscillator, also admits periodic motions isochronous with the forcing term, at least if the oscillations are small. This solves the problem (2) posed in p. 67, §2.14.

Again, we treat the pendulum for definiteness. The theory developed below is valid for equations obtained from Eq. (2.14.1) by changing $\sin x$ into $\psi(x)$, where ψ is an arbitrary C^∞ function such that $\psi(0) = 0$, $\psi'(0) > 0$.

Consider the normal equation

$$m\ddot{x} + \lambda\dot{x} + k\sin x = \gamma f(t), \qquad t \in \mathbf{R}_+, \tag{2.15.1}$$

$\gamma \in \mathbf{R}$, $\lambda, m, k > 0$, $\lambda^2 \neq 4mk$ (for simplicity's sake), and let us prove the following proposition.

21 Proposition. *Let $t \to f(t)$, $t \in \mathbf{R}$, be a C^∞-periodic function with period $T > 0$. There exists a periodic motion with period T verifying Eq.* (2.15.1), *provided γ is small enough.*

Observation. The proof below is based on a very general method used to treat such questions and relying on the implicit function theorem (see Appendix G).

Together with Eq. (2.15.1), one considers the "linearized equation"

$$m\ddot{x} + \lambda\dot{x} + kx = \gamma f, \tag{2.15.2}$$

which, as already shown in §2.12, admits a periodic solution isochronous with f:

$$t \to \bar{x}_\gamma(t) \equiv \gamma\tilde{x}(t) = \gamma \sum_{n=-\infty}^{+\infty} \frac{\hat{f}_n e^{(2\pi i/T)nt}}{-m(2\pi/T)^2 n^2 + k + (2\pi i n/T)\lambda}, \tag{2.15.3}$$

where $(\hat{f}_n)_{n\in\mathbf{Z}}$ are the harmonics of f.

We then look for a periodic solution to Eq. (2.15.1) having the form

$$t \to x(t) = \gamma\tilde{x}(t) + y(t), \qquad t \in \mathbf{R}_+ \tag{2.15.4}$$

with initial data:

$$y(0) = \epsilon, \qquad \dot{y}(0) = \eta \tag{2.15.5}$$

hoping to be able to show that ϵ, η, and y exist and are infinitesimal as $\gamma \to 0$, of higher order in γ(i.e., hoping that $\gamma\tilde{x}(t)$ is a very good approximation to $x(t)$ for small γ).

The function $t \to x(t)$, solution of Eqs. (2.15.1) and (2.15.5), depends on ϵ, η, γ [in a C^∞ way, by the regularity theorem (see Proposition 3 and Problem 17 of §2.5)] and we shall set

$$\begin{aligned} x(T) &= \gamma\tilde{x}(T) + a(\epsilon, \eta, \gamma), \\ \dot{x}(T) &= \gamma\dot{\tilde{x}}(T) + b(\epsilon, \eta, \gamma); \end{aligned} \tag{2.15.6}$$

i.e., $y(T) = a(\epsilon, \eta, \gamma)$, $\dot{y}(T) = b(\epsilon, \eta, \gamma)$ [see Eq. (2.15.4)].

Therefore, the condition that Eq. (2.15.1) admits a periodic solution with period T can be written (see Proposition 12) as

$$\begin{aligned} a(\epsilon, \eta, \gamma) &= \epsilon, \\ b(\epsilon, \eta, \gamma) &= \eta, \end{aligned} \tag{2.15.7}$$

since $\tilde{x}(0) = \tilde{x}(T)$, $\dot{\tilde{x}}(0) = \dot{\tilde{x}}(T)$ by the periodicity of $\tilde{x}$.

So the problem of proving Proposition 21 is equivalent to proving the solubility of the implicit functions problem of expressing, from Eq. (2.15.7), ϵ and η as functions of γ for γ small.

Proof. Note that the functions $f_1(\epsilon, \eta, \gamma) = a(\epsilon, \eta, \gamma) - \epsilon$ and $f_2(\epsilon, \eta, \gamma) = b(\epsilon, \eta, \gamma) - \eta$ are C^∞ functions.

To study Eq. (2.15.7), we write the equation verified by $t \to y(t)$ defined in Eq. (2.15.4):

$$m\ddot{y}(t) + \lambda\dot{y}(t) + ky(t) = k(\gamma\tilde{x}(t) + y(t) - \sin(\gamma\tilde{x}(t) + y(t)), \tag{2.15.8}$$

$$y(0) = \epsilon, \qquad \dot{y}(0) = \eta.$$

This equation and the uniqueness theorem for differential equations show that if $\epsilon = 0$, $\eta = 0$, $\gamma = 0$, it follows that $y(t) \equiv 0$, $t \in \mathbf{R}_+$, and, therefore,

$$f_1(0,0,0) = 0, \qquad f_2(0,0,0) = 0. \tag{2.15.9}$$

It is then natural to look for solutions of Eq. (2.15.7) near $\gamma = 0$ through the implicit functions theorem (see Appendix G).

The solubility condition of Eq. (2.15.7) for small γ is that the jacobian matrix

$$J = \begin{bmatrix} \dfrac{\partial f_1}{\partial \epsilon}(0,0,0) & \dfrac{\partial f_1}{\partial \eta}(0,0,0) \\ \dfrac{\partial f_2}{\partial \epsilon}(0,0,0) & \dfrac{\partial f_2}{\partial \eta}(0,0,0) \end{bmatrix} \tag{2.15.10}$$

has nonvanishing determinant.

To compute the derivatives in Eq. (2.15.10), we recall that

$$a(\epsilon,\eta,\gamma) = y(T), \qquad b(\epsilon,\eta,\gamma) = \dot{y}(T), \tag{2.15.11}$$

where $t \to y(t)$ solves Eq. (2.15.8).

We can pretend that the right-hand side of Eq. (2.15.8) is a known function of $t \in \mathbf{R}_+$ and write

$$y(t) = \bar{y}(t) + \int_0^t \frac{e^{\alpha_+(t-\tau)} - e^{\alpha_-(t-\tau)}}{\alpha_+ - \alpha_-} \times k(\gamma\tilde{x}(\tau) + y(\tau) - \sin(\gamma\tilde{x}(\tau) + y(\tau)))\frac{d\tau}{m} \tag{2.15.12}$$

along the same lines as the proof in the preceding section where $t \to \bar{y}(t)$ is a solution to

$$m\ddot{\bar{y}} + \lambda\dot{\bar{y}} + k\bar{y} = 0, \qquad \bar{y}(0) = \epsilon, \quad \dot{\bar{y}}(0) = \eta, \tag{2.15.13}$$

[see Eq. (2.14.13)]:

$$\bar{y}(t) = \frac{\eta - \alpha_-\epsilon}{\alpha_+ - \alpha_-}e^{\alpha_+ t} + \frac{\alpha_+\epsilon - \eta}{\alpha_+ - \alpha_-}e^{\alpha_- t}. \tag{2.15.14}$$

Hence,

$$a(\epsilon,\eta,\gamma) = \eta\frac{e^{\alpha_+ T} - e^{\alpha_- T}}{\alpha_+ - \alpha_-} + \epsilon\frac{\alpha_+ e^{\alpha_- T} - \alpha_- e^{\alpha_+ T}}{\alpha_+ - \alpha_-} + \int_0^T \frac{e^{\alpha_+(T-\tau)} - e^{\alpha_-(T-\tau)}}{\alpha_+ - \alpha_-} \times k(\gamma\tilde{x}(\tau) + y(\tau) - \sin(\gamma\tilde{x}(\tau) + y(\tau)))\frac{d\tau}{m}, \tag{2.15.15}$$

and a similar expression can be found for b by differentiating Eq. (2.15.12) with respect to t and setting $t = T$.

From Eq. (2.15.15), we can compute the partial derivatives of a with respect to ϵ, η, γ in $(0,0,0)$, without really knowing $y(t)$ (remarkably enough).

For instance:

$$\frac{\partial a}{\partial \epsilon}(0,0,0) = \frac{\alpha_+ e^{\alpha_- T} - \alpha_- e^{\alpha_+ T}}{\alpha_+ - \alpha_-} + \int_0^T \frac{d\tau}{m} \left\{ \frac{e^{\alpha_+(T-\tau)} - e^{\alpha_-(T-\tau)}}{\alpha_+ - \alpha_-} \cdot k(1 - \cos y(\tau)) \frac{\partial y}{\partial \epsilon}(\tau) \right\}$$
$$\equiv \frac{\alpha_+ e^{\alpha_- T} - \alpha_- e^{\alpha_+ T}}{\alpha_+ - \alpha_-}, \tag{2.15.16}$$

where $\tau \to y(\tau)$, $\tau \geqslant 0$, is the solution to Eq. (2.15.8) with $\epsilon = \eta = \gamma = 0$, i.e., $y(\tau) \equiv 0$. Note that $(\partial y/\partial \epsilon)(\tau)$ is unknown but is multiplied by zero and, therefore, it is not necessary to know it.

Similarly:

$$\frac{\partial a}{\partial \eta}(0,0,0) = \frac{e^{\alpha_+ T} - e^{\alpha_- T}}{\alpha_+ - \alpha_-},$$
$$\frac{\partial b}{\partial \epsilon}(0,0,0) = \alpha_+ \alpha_- \frac{e^{\alpha_- T} - e^{\alpha_+ T}}{\alpha_+ - \alpha_-}, \tag{2.15.17}$$
$$\frac{\partial b}{\partial \eta}(0,0,0) = \frac{\alpha_+ e^{\alpha_+ T} - \alpha_- e^{\alpha_- T}}{\alpha_+ - \alpha_-},$$

and, hence, it is possible to write the matrix J and, with some patience, the algebraic calculations lead to

$$\det J = (e^{\alpha_+ T} - 1)(e^{\alpha_- T} - 1) \neq 0. \tag{2.15.18}$$

This completes the proof since the implicit functions theorem (Appendix G) implies that Eq. (2.15.7) can be uniquely solved for small γ with $\epsilon(\gamma)$, $\eta(\gamma)$ of the order $O(\gamma)$.

Actually, the implicit functions theorem implies that the derivatives of $\epsilon(\gamma)$, $\eta(\gamma)$, with respect to γ at $\gamma = 0$ are proportional to the derivatives of f_1 and f_2 with respect to γ in $\epsilon = \eta = \gamma = 0$. Since such derivatives can be computed in the same way as those in Eqs. (2.15.16) and (2.15.17) and they turn out to be zero, it also follows that $\epsilon(\gamma)$, $\eta(\gamma)$ are of the order $O(\gamma^2)$ as expected.

mbe

Problems for §2.15

1. Show that the oscillator $\ddot{x} + \dot{x} + x + x^3 = f(t)$, $f \in C^\infty(\mathbf{R})$, has small oscillations in the sense of Proposition 20. Show that if f has the form $f(t) = \gamma\varphi(t)$, $\gamma \in \mathbf{R}$, and φ periodic with period $T > 0$, then for γ small enough the equation admits a periodic solution with period T (Hint: Go through the proof of Proposition 21, replacing $\sin x$ by $x + x^3$ everywhere.)

2. Show that the motion $\ddot{x} + \dot{x} + x^3 = 0$, $x(0) = 1$, $\dot{x}(0) = 0$ never goes through the origin as $t \to +\infty$. How does this result depend on the datum?

3. Same as Problem 2 for $\ddot{x} + 3\dot{x} + x + x^3 = 0$, $x(0) = \frac{1}{2}$, $\dot{x}(0) = 0$.

4.* Consider the oscillator $\ddot{x} + \dot{x} + x + x^3 = 0$ and find the limit T_∞, as $t \to +\infty$, of the time $T(t)$ elapsing between the two consecutive passages through the origin with positive speed taking place after t. Show that it does not depend on the initial datum (Answer: $T_\infty = 4\pi/\sqrt{3}$).

5.* Same as Problem 4 for $\ddot{x} + \dot{x} + \tanh x = 0$; discover why T_∞ is the same as that in Problem 4.

6. Examine critically the proof of Proposition 21 to see under which assumptions its conclusions remain valid when $\lambda = 0$ (Answer: If and only if $\det J \neq 0$, i.e., if and only if the forcing period T is not an integer multiple of the "proper period" $T_0 = 2\pi\sqrt{m/k}$.)

§2.16. Small Damping: Resonances

We shall not study the problem (3), p. 67, in detail since, in the next few sections, we shall conclude our analysis of the damped motions (in one dimension) by an application where a similar, but more difficult, problem is analyzed. Let us simply formulate a result about problem (3), without proof:

22 Proposition. *Let $f \in C^\infty(\mathbf{R})$ be a periodic function with period $T > 0$ and consider the forced pendulum of Eq.* (2.15.1). *If γ is small enough, the equation admits one periodic solution $t \to x_p(t)$, $t \in \mathbf{R}_+$, with period T, and every other solution $t \to x(t)$, $t \in \mathbf{R}_+$, with initial datum (x_0, v_0) with $|x_0| + |v_0|$ small enough approaches, exponentially fast, the periodic solution: i.e., there are $C > 0$, $\mu > 0$ such that*

$$|x(t) - x_p(t)| \leqslant Ce^{-\mu t}, \qquad \forall t \in \mathbf{R}_+ . \tag{2.16.1}$$

Observations.

(1) The proof of this proposition is very similar to that of Proposition 25 on the theory of the clock. The reader will reconstruct it from that proof.

(2) Hence, the small oscillations of the nonlinear damped oscillators are qualitatively very similar to those of the damped linear oscillators, at least if one is only concerned with properties (1), (2), and (3) selected for discussion at §2.14.

So far, the presence of friction has revealed itself to be essential to the theory (see, however, Problem 6, §2.15). In fact something "goes wrong" as

$\lambda \to 0$. This can be easily seen for the linear oscillators, as we shall briefly see in the following. However, this time we consider only harmonic oscillators not just "for simplicity", but because only in this case will it be possible to obtain something without excessive conceptual and technical difficulties.

In the nonlinear case, the discussion is, surprisingly at first sight, much more involved (and interesting) and, also, the results are unfortunately less detailed and complete than desirable for applications. Some basic ideas and technical tools will be developed in §5.9–§5.12 of Chapter 5.

Actually, contrary to what is sometimes believed, the motion of mechanical systems is much simpler and stable when friction is present than when it is absent. When friction vanishes, the motion becomes very sensitive to the details of the equations of motion and to the initial data, as far as asymptotic behavior is concerned, in this way introducing new difficulties and peculiarly new phenomena. Also, from the mathematical point of view, the frictionless motion theory appears to be deep and rich with connections to the most diverse fundamental problems[8] in analysis and geometry: from number theory to topology to probability theory.

Our discussion of the small friction case will be based on the following two linear (normal) equations:

$$m\ddot{x} + \lambda\dot{x} + kx = f \tag{2.16.2}$$

with $\lambda > 0$, $k > 0$, $m > 0$ and

$$m\ddot{x} + kx = f, \tag{2.16.3}$$

where f is a C^∞ function periodic with period $T > 0$. The discussion will be restricted to the following simple proposition (for the time being).

23 Proposition. *Given $x_0, v_0 \in \mathbf{R}$, and setting $t \to x_\lambda(t)$, $t \in \mathbf{R}_+$, the solution to Eq. (2.16.2) with initial data $x_\lambda(0) = x_0$, $\dot{x}_\lambda(0) = v_0$ and calling $t \to x_0(t)$, $t \in \mathbf{R}_+$, the solution to Eq. (2.16.3) with data $x(0) = x_0$, $\dot{x}(0) = v_0$, the following results hold:*

(i)
$$\lim_{\lambda \to 0} x_\lambda(t) = x_0(t), \qquad \forall t \in \mathbf{R}_+. \tag{2.16.4}$$

(ii) *The preceding limit is "uniform as $\lambda t \to 0$": i.e., given $\epsilon > 0$, there exist $\delta_\epsilon > 0$, $\lambda_\epsilon > 0$ such that*

$$|x_\lambda(t) - x_0(t)| < \epsilon \qquad \forall \lambda < \lambda_\epsilon, \quad \forall t < \delta_\epsilon \lambda^{-1}. \tag{2.16.5}$$

(iii) *If T is not an integer multiple of the "proper period" $T_0 = 2\pi\sqrt{m/k}$ of the undamped free harmonic oscillator, one has*

$$x_0(t) = A_0 \cos\left(\frac{2\pi t}{T_0} + \varphi_0\right) + \sum_{n=-\infty}^{+\infty} \frac{\hat{f}_n e^{(2\pi i/T)nt}}{-m(2\pi/T)^2 n^2 + k}, \tag{2.16.6}$$

where A_0, φ_0 are suitable constants and $(\hat{f}_n)_{n \in \mathbf{Z}}$ are the harmonics of f on the period T: this is the "nonresonant case".

[8] However, at a deeper level of understanding, similar statements could also be made for dissipative systems: a glimpse of how complex they may become is given in §5.8.

(iv) *If $T = \bar{n}T_0$ for some integer $\bar{n}$:*

$$x_0(t) = A_0 \cos\left(\frac{2\pi t}{T_0} + \varphi_0\right) + \sum_{\substack{n=-\infty \\ n \neq \pm \bar{n}}}^{+\infty} \frac{\hat{f}_n e^{(2\pi i n/T)t}}{-m(2\pi/T)^2 n^2 + k}$$
$$+ 2t \operatorname{Re} \frac{\hat{f}_{\bar{n}} e^{(2\pi i/T_0)t}}{2i(2\pi/T_0)m}. \tag{2.16.7}$$

This is the "resonant case".

Observations.

(1) (ii) is particularly significant and says that the smaller the friction, the longer the time during which the friction-driven motion coincides, within a given approximation ϵ, with the frictionless motion (this time being at least δ_ϵ/λ). Hence, (ii) strengthens and implies (i).

(2) The above proposition also illustrates the "resonance phenomenon". By what has been seen in §2.12, the solution to Eq. (2.16.2) of interest to us is

$$x_\lambda(t) = A_+ e^{\alpha_+ t} + A_- e^{\alpha_- t} + \sum_{n=-\infty}^{+\infty} \frac{\hat{f}_n e^{(2\pi i/T)nt}}{-m(2\pi/T)^2 n^2 + (2\pi i/T)\lambda n + k}, \tag{2.16.8}$$

where $v_0 = \dot{x}_\lambda(0)$, $x_0 = x_\lambda(0)$ have to be used to determine the constants A_+, A_- and

$$\alpha_\pm = -\frac{\lambda}{2m}\left[1 \pm i\sqrt{\frac{4mk}{\lambda^2} - 1}\right], \qquad \operatorname{Re}\alpha_\pm = -\frac{\lambda}{2m}. \tag{2.16.9}$$

From Eq. (2.16.8), it immediately follows that as $t \to +\infty$, the asymptotic motion is T periodic and it is given by

$$\bar{x}_\lambda(t) = \sum_{n=-\infty}^{+\infty} \frac{\hat{f}_n e^{(2\pi i/T)nt}}{-m(2\pi/T)^2 n^2 + k + (2\pi i/T)n\lambda}, \tag{2.16.10}$$

provided the first two "transient terms" in Eq. (2.16.8) are very small: i.e., provided $\lambda t/2m \gg 1$.

If there is $\bar{n}$ such that $T = \bar{n}T_0$, we select the two terms in Eq. (2.16.10) with $n = \pm\bar{n}$ and rewrite them as

$$\bar{x}_\lambda(t) = 2\operatorname{Re}\frac{\hat{f}_{\bar{n}} e^{(2\pi i/T_0)t}}{(2\pi i/T_0)\lambda} + \sum_{|n| \neq \bar{n}} \frac{\hat{f}_n e^{(2\pi i n/T)t}}{-m(2\pi/T)^2 n^2 + k + (2\pi i/T)n\lambda}. \tag{2.16.11}$$

Letting $\hat{f}_{\bar{n}} = \rho_{\bar{n}} e^{i\delta_{\bar{n}}}$, $\rho_{\bar{n}} \geq 0$, $\delta_{\bar{n}} \in \mathbf{R}$, the first term becomes

$$2\rho_n \frac{\sin((2\pi/T_0)t + \delta_n)}{(2\pi/T_0)\lambda}, \tag{2.16.12}$$

while the series in Eq. (2.16.11) can be majorized uniformly in λ by

$$\sum_{|n|\neq \bar{n}} \frac{|\hat{f}_n|}{|-m(2\pi/T)^2 n^2 + k|}. \tag{2.16.13}$$

So we see that if $T = \bar{n}T_0$, for some integer $\bar{n}$, and if the force f is arbitrarily small but such that $\hat{f}_{\bar{n}} \neq 0$, the motion impressed by f to the oscillator may attain an enormous amplitude, as Eqs. (2.16.11)–(2.16.13) show, for small λ.

Obviously if T/T_0 is not integer but almost such, $(T/T_0 \simeq \bar{n} \in \mathbf{Z})$, it will happen that the series of Eq. (2.16.10) will contain terms (those with $n = \pm \bar{n}$) with denominators which, even though not vanishing as $\lambda \to 0$, will become very small producing two contributions to Eq. (2.16.10) that could "dominate" the others.

(3) It should be stressed that resonance manifests itself only when the terms $A_\pm \exp \alpha_\pm t$ in Eq. (2.16.8) are small and, therefore, only if $\lambda t/2m \gg 1$. Hence, although it is true that in the resonating linear oscillator $(T = \bar{n}T_0,\ \bar{n} \in \mathbf{Z}_+)$ a very small force can produce huge oscillations (proportional to λ^{-1}), it is also true that the time it takes for this to happen is very large (proportional to λ^{-1}).

Note also that

$$A_+ = \bar{A}_- = \frac{(v_0 - \dot{\bar{x}}_\lambda(0)) - \alpha_-(x_0 - \bar{x}_\lambda(0))}{\alpha_+ - \alpha_-} \tag{2.16.14}$$

becomes singular as $\lambda \to 0$, in resonance cases, because such are $\bar{x}_\lambda(0)$, $\dot{\bar{x}}_\lambda(0)$ [see Eqs. (2.16.10) and (2.16.12)].

(4) Equations (2.16.6) and (2.16.7) give the most general solutions to Eq. (2.16.3) as A_0, φ_0 vary arbitrarily. They show that when $\lambda = 0$, the linear oscillator's motions are not longer periodic but, rather, are "sums" of two periodic motions with respective periods T_0 and T equal to the "proper" period of the free oscillator and to the period of the forcing force provided that T/T_0 is not an integer. If T/T_0 is integer, and if the harmonic component of order T/T_0 of the force f does not vanish, the asymptotic motion is even unbounded: "undamped resonance".

Furthermore, in every case, the asymptotic motion depends on the initial datum (through A_0, φ_0). It is clear that the initial datum dependence surviving in the asymptotic regime is due to the absence of friction: analytically this appears via the fact that $\exp \alpha_\pm t \not\to 0$, since $\operatorname{Re} \alpha_\pm = 0$ if $\lambda = 0$.

(5) The proof of Proposition 23 is a simple discussion of the limit as $\lambda \to 0$ of the expressions (2.16.8) and (2.16.14). No problem arises in the absence of resonance. In the resonant case, the limit is most conveniently discussed by collecting together the first two terms in Eq. (2.16.8) and the two resonant terms in the series (2.16.8) (i.e., those with $n = \pm T/T_0$). The calculations are straightforward and are left to the reader.[9]

[9]Note that (i) would also directly follow from the regularity theorem (Proposition 3, 22, and problem 17, p. 32).

Exercises for §2.16

1. Determine up to 20%, the asymptotic amplitude of the oscillations of the oscillator: $\ddot{x}+\dot{x}+x=f(t)$, $f(t)=(1-\cos 2\pi t/T)^{-1}$, for $T=1, 4\pi, \sqrt{2}$. Which, in each case, are the resonant harmonics? (Call "resonance" a harmonic of order $n \in \mathbf{Z}$ if the function $\xi \to |(2\pi/T)^2\xi^2-(2\pi/T_0)^2|$ takes its minimum between n and $n+1$.)

2. Determine the asymptotic amplitude of the motion described by $\ddot{x}+x=f(t)$ with f given as in Problem 1.

3. Estimate how small λ has to be taken so that the amplitude of the asymptotic oscillations described by $\ddot{x}+\lambda\dot{x}+x=f(t)$, with $f(t)=10^{-3}(1-10^{-2}\cos t)^{-1}$, is not smaller than $A=1, 10, 10^2, 10^6$.

4. Same as Problem 3 with $f(t)=10^{-3}(1-0.99\cos t)^{-1}$.

5. Write a computer program for the empirical solution (i.e., without error estimates) by a desk computer of the equation in Problem 1 with the purpose of drawing graphs, in the data space, of the trajectories corresponding to the various choices of the initial datum (using a computer with plotter or screen and always avoiding the tabulation of results).

§2.17. An Application: Construction of a Rigorously Periodic Oscillator in the Presence of Friction. The Anchor Escapement Feedback Phenomena

Che l'una parte l'altra tira e urge
Tin tin sonando con sì dolce nota
Che 'l ben disposto spirto d'amor turge.[10]

In §2.12 we saw that a damped harmonic oscillator can move exactly periodically with a period equal to that of the forcing term. Furthermore, any of its motions differs from this periodic one by an amount which becomes exponentially small as $t \to +\infty$ It is natural to try to use this property to build a clock, i.e., a mechanism moving in a rigorously periodic fashion despite friction. However, it is clear that the difficulty of producing a rigorously periodic force seems to be, at least, of the same order of magnitude as that of producing a periodic motion.

The anchor escapement is a contrivance in a timepiece which controls the motion of the train of wheelwork and through which the energy of the weight is delivered to the pendulum by means of impulses which keep the latter in vibration (Webster).

This mechanism simultaneously solves the two problems of building a rigorously periodic force and of inducing a rigorously periodic motion. It takes advantage of the presence of friction to cause the oscillator to move

[10] In basic English:
That every part pulls another
tin tin singing, so sweetly:
that the well inclined spirit is filled with love.
(Dante, *Paradiso*, Canto X).

asymptotically in a periodic way in the sense that the difference between the actual oscillator's position $x(t)$, at time t, and that of a certain ideal periodic motion's position $x_{per}(t)$ tends exponentially to zero as $t \to +\infty$.

A very schematic empirical description of the anchor escapement is the following.

The "anchor" is a device set in motion by the oscillator as it passes through the point $x_0 = 0$, for instance, with positive velocity. At this instant, a notched wheel connected to a weight is liberated from a brake and starts moving. A little later, the notch of the wheel reaches the oscillator and accompanies it for a short while, exerting a push on it. Then the notched wheel loses contact with the oscillator, which remains free, allowing the wheel to return to its original position by continuing its rotation. In this simplified scheme, the wheel has just one notch instead of the usual few dozens.

In the meantime, the oscillator, now free, continues its (damped) oscillation, and the entire process starts afresh at the new passage through x_0 with positive speed.

We can try to schematize the just-described mechanical system through a motion governed by the equation:

$$m\ddot{x} + \lambda\dot{x} + kx = 0, \qquad x \leqslant 0, \tag{2.17.1}$$

$$m\ddot{x} + \lambda\dot{x} + kx = f(\dot{x}_0, \tau), \qquad x \geqslant 0, \tag{2.17.2}$$

where the action of the notched wheel is schematized by a force $f(\dot{x}_0, \tau)$ depending upon the velocity $\dot{x}_0$ of the last passage through $x_0 = 0$ with positive speed and upon the time τ elapsed since then.

Note that Eqs. (2.17.1) and (2.17.2) are equations of motion quite different from the ones considered in the preceding sections. The force appearing in Eq. (2.17.2) not only depends on time but also upon the past history of the motion itself. Therefore it is not a differential equation in the sense of §2.2, Definition 1.

Consequently, we do not even know yet whether Eqs. (2.17.1) and (2.17.2) have a solution, i.e., a C^∞ function $t \to x(t)$, $t \in \mathbf{R}_+$, which turns Eqs. (2.17.1) and (2.17.2) into an identity (not even if f is a C^∞ function of its arguments).

To study Eqs. (2.17.1) and (2.17.2), it is useful to place some restrictions on the form of f which we intend to consider: i.e., it is useful to further specialize the model. This is done to avoid problems too complex from a technical point of view, as well as to avoid developing a theory for too general an f, which may not correspond to a force law that is reasonable for our problem.

For the sake of example, let us assume that $f(\dot{x}_0, \tau)$ vanishes whenever $\dot{x}_0$ does not belong to an interval $[v_-, v_+]$, $0 < v_- < v_+$:

$$\dot{x}_0 \notin [v_-, v_+] \to f(\dot{x}_0, \tau) = 0. \tag{2.17.3}$$

This assumption corresponds to the fact that when the oscillator sweeps through $x_0 = 0$ too fast, it is never reached by the wheel's notch; while if it sweeps too slowly, the amplitude of oscillation is too small to allow the oscillator to touch the notch.

Assume, also, that once $\dot{x}_0 \in [v_-, v_+]$, the force on the oscillator only depends on the time τ elapsed since the last passage through $x_0 = 0$ with positive speed; i.e.,

$$f(\dot{x}_0, \tau) = P\chi(\dot{x}_0)\, g(\tau) \tag{2.17.4}$$

where $\chi(\dot{x}_0) = 1$ if $\dot{x}_0 \in [v_-, v_+]$, $\chi(\dot{x}_0) = 0$ otherwise, and $t \to g(t) \geqslant 0$ is a $C^\infty(\mathbf{R})$ function vanishing outside an interval $[\alpha, T_g]$, $\alpha > 0$, $T_g > 0$ with a maximum equal to 1. The constant P, which we shall take as a positive adjustable parameter, models the "intensity" of the force. Physically, one can imagine that P depends on the weight moving the notched wheel, while g is a detailed description of the wheel's action mechanism.

Therefore, Eq. (2.17.4) will be considered a mathematical model of the force generated by the anchor escapement. Such a model is only a schematization, where some of the properties of any real mechanism are certainly oversimplified. Nevertheless, as we shall see, it is a model presenting some interesting characteristics such as, primarily, the "self-control" or "feedback" mechanism providing that the system (2.17.1), (2.17.2), and (2.17.4) "searches automatically", in certain circumstances, for a situation of motion that allows it to move periodically.

The function g has a graph like that in Fig. 2.6.

Some further properties which we must impose on g should be that g vanishes for $\tau < \alpha$, for some $\alpha > 0$, or for $\tau > T_g > \alpha > 0$, and T_g should be small compared to the time necessary for an elongation of the oscillator from the position $x_0 = 0$ to the position of maximum distance from $x_0 = 0$.

The time $\alpha > 0$ is a mechanical constant representing the delay between the beginning of the wheel's motion and the actual oscillator-notch contact. Clearly the $\dot{x}_0$ independence of α is a strong idealization.

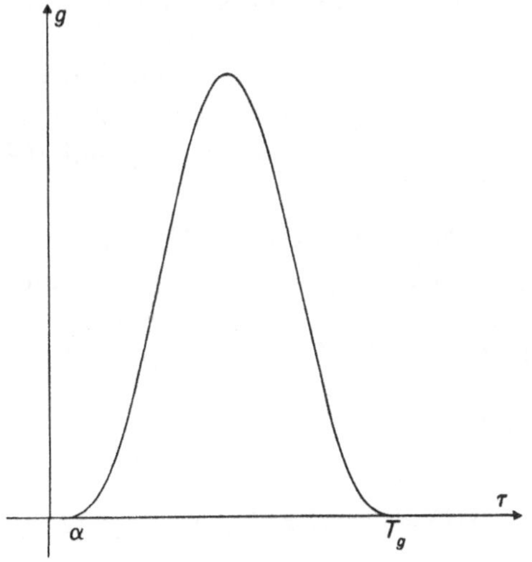

Figure 2.6.

The physically obvious requirement on T_g can be translated into mathematical terms by requiring $T_g \ll T_0/4$, where $T_0 = 2\pi\sqrt{m/k}$ is the ideal oscillator period. This attempts to translate the fact that the notch has to detach itself from the oscillator before it starts swinging back toward the origin. Empirically, the condition $T_g \ll T_0/4$ should guarantee this fact, at least if the friction is small so that it produces negligible effects for times of the order of T_0; i.e., as we saw in §2.12, if $\lambda^2 < 4mk$.

As a conclusion to the above considerations, let us assume as a model for the anchor escapement Eqs. (2.17.1), (2.17.2), and (2.17.4) with g as in Fig. 2.6 with $0 < \alpha < T_g < T_0/4$ and with $\lambda^2 < 4mk$, and let us prove the following proposition, which begins the theory of the model.

24 Proposition. *Under suitable compatibility conditions between the parameters P, v_-, v_+, Eqs.* (2.17.1), (2.17.2), *and* (2.17.4) *admit a periodic C^∞ solution defined for $t \in \mathbf{R}_+$.*

Observations. In the upcoming section we shall discuss the compatibility conditions by showing that they can be easily satisfied when λ is small enough. Later, in §2.19, we shall also show that when λ is small enough and the compatibility conditions are fulfilled, the motions with initial data close enough to those of the periodic motion become close to such motion exponentially fast (see Figs. 2.7 and 2.8).

PROOF. Let $t \to x(t)$ be a given periodic motion with period $T > T_g$ and such that $x(0) = 0$.

It is clear that the function defined, for $t \geqslant 0$:

$$\varphi(t) = f(v_0, \tau) = P\chi(v_0)\, g(\tau), \tag{2.17.5}$$

where v_0 is the velocity of the given motion at its last passage through the origin, with positive speed and before time t, and τ is the time elapsed since such time, is a C^∞-periodic function of t, provided the time necessary to return to 0 with positive speed is equal to T itself.[11]

Assuming, then, that $t \to x(t)$, $t \in \mathbf{R}_+$, is a C^∞-periodic motion verifying Eqs. (2.17.1), (2.17.2), and (2.17.4) and period T equal to its first return time to the origin with positive speed and assuming that $v_0 = \dot{x}(0) \in [v_-, v_+]$ we shall have, $\forall t \geqslant 0$,

$$m\ddot{x}(t) + \lambda\dot{x}(t) + kx(t) = \varphi(t). \tag{2.17.6}$$

If we note that $\varphi(t) \equiv Pg(t)$, $\forall t \in [0, T_g]$, and if we set

$$\hat{g}_n = \int_0^T g(\tau)e^{-(2\pi i/T)n\tau}\frac{d\tau}{T} \equiv \int_0^{T_g} g(\tau)e^{-(2\pi/T)in\tau}\frac{d\tau}{T}, \tag{2.17.7}$$

recalling that $T_g < T$, we shall have (see (§2.12))

$$x(t) = P\sum_{n=-\infty}^{+\infty}\frac{\hat{g}_n e^{(2\pi i/T)nt}}{-m(2\pi/T)^2n^2 + k + (2\pi/T)in\lambda}, \tag{2.17.8}$$

[11] It could a priori happen that the motion sweeps through the origin more than twice (even infinitely many times) in an interval of time equal to the period T.

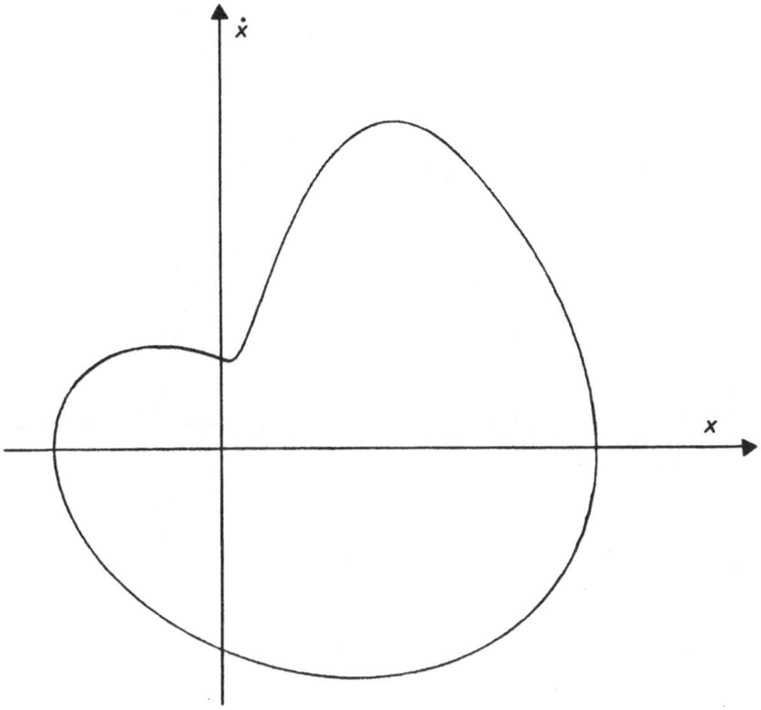

Figure 2.7. Graph of a periodic solution $t \to (x(t), \dot{x}(t))$ of Eqs. (2.14.1), (2.14.2), and (2.14.4) with convenient choices of the arbitrary parameters and of the function g.

and the series is uniformly convergent because $\hat{g}_n$ approaches zero as $n \to \infty$ faster than any power (being the Fourier transform of the C^∞-periodic function φ).

Of course, we still have to determine T and to check that for such T, Eq. (2.17.8) is really a solution to Eqs. (2.17.1), (2.17.2), and (2.17.4). In other words, we must impose the condition that Eq.(2.17.8) is such that

(i) $$x(0) = 0; \tag{2.17.9}$$

(ii) $$\dot{x}(0) \in [v_-, v_+]; \tag{2.17.10}$$

(iii) $$T > T_g; \tag{2.17.11}$$

(iv) T is the first return time in 0 with positive velocity. (2.17.12)

Relation (2.17.9) is an equation for the period T:

$$0 = \sum_{n=-\infty}^{+\infty} \hat{g}_n \left(-m \left(\frac{2\pi}{T} \right)^2 n^2 + k + \frac{2\pi i}{T} \lambda n \right)^{-1}, \tag{2.17.13}$$

and it should be noted that in this equation, T also appears in the coefficients $\hat{g}_n$ [see Eq. (2.17.7)].

Then if the parameters v_-, v_+, P, T_g are such that Eq. (2.17.13) admits at least one solution T and if with this choice of T Eq. (2.14.8) verifies the

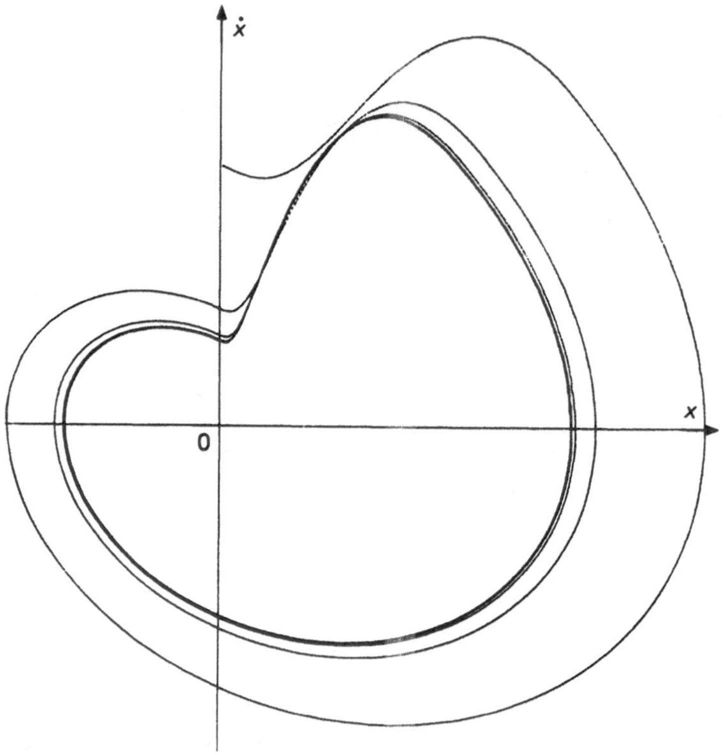

Figure 2.8. Graph of a solution $t \to (x(t), \dot{x}(t))$ with initial datum chosen arbitrarily: it becomes indistinguishable from the periodic solution of Fig. 2.7 within a few oscillations (three in the precision of the drawing).

compatibility conditions (2.17.10)–(2.17.12), it follows that Eq. (2.17.8) is a T-periodic solution to Eqs. (2.17.1), (2.17.2), and (2.17.4). mbe

Exercises for §2.17

1. Choose arbitrarily a function g and $m, k, \lambda, v_-, v_+ > 0$ and write a computer program providing a heuristic (i.e., without error estimate) solution to Eqs. (2.17.1), (2.17.2), and (2.17.4) in which P and the datum $\dot{x}(0)$ are left as free parameters. The output of the program should be a graph like those in Figs. 2.7 and 2.8.

2. Run the above program on a desk computer with plotter or screen finding, by trial and error, a value of P yielding a nontrival periodic motion.

§2.18. Compatibility Conditions in the Theory of the Anchor Escapement

This section, as well as the next, will suppose some maturity on the reader's part and, therefore, on first reading it would be appropriate to skip the proof of this section and to read the next section only up to the beginning of the proof of Proposition 26.

As promised in the previous section, we shall now show that if λ is small enough, given v_-, v_+, g with $T_g < T_0/4$, it is possible to fix P so that Eq. (2.17.8) actually verifies the four compatibility conditions (2.17.9)–(2.17.12) and, therefore, is a periodic solution to Eqs. (2.17.1), (2.17.2), and (2.17.4), i.e., to the equation for the anchor-escapement model.

Consider Eq. (2.17.13) as an equation for T^{-1} parametrized by $\lambda > 0$, and let us find some of its solutions having the form

$$T^{-1} = T_0^{-1}(1 + \lambda\beta) \tag{2.18.1}$$

suggested by the idea that for small λ the oscillator may oscillate with a periodic motion with period close to the period $T_0 = 2\pi\sqrt{m/k}$ of the frictionless oscillator.

The equation for T, Eq. (2.17.13), then becomes, after explicitly separating out of the series' sum the two complex conjugate terms with $n = \pm 1$:

$$\begin{aligned} 0 = 2\,\mathrm{Re}\left\{ \hat{g}_1\left(-m\left(\frac{2\pi}{T_0}\right)^2(1+\lambda\beta)^2 + k + \frac{2\pi i}{T_0}\lambda(1+\lambda\beta)\right)^{-1}\right\} \\ + \sum_{\substack{n=-\infty \\ n\neq\pm 1}}^{+\infty} \hat{g}_n\left(-m\left(\frac{2\pi}{T}\right)^2 n^2 + k + \frac{2\pi i}{T}\lambda n\right)^{-1} \end{aligned} \tag{2.18.2}$$

which, using $T_0 = 2\pi\sqrt{m/k}$, becomes

$$\begin{aligned} 0 = 2\,\mathrm{Re}\left\{ \hat{g}_1\left(-k(2\beta + \beta^2\lambda)\lambda + \frac{2\pi i}{T_0}\lambda(1+\lambda\beta)\right)^{-1}\right\} \\ + \sum_{\substack{n=-\infty \\ n\neq\pm 1}}^{+\infty} \hat{g}_n\left(-m\left(\frac{2\pi}{T}\right)^2 n^2 + k + \frac{2\pi i}{T}\lambda n\right)^{-1}. \end{aligned} \tag{2.18.3}$$

We see that for small λ the first of the above two addends shows a small denominator. To avoid having to study an equation with small denominators, let us multiply Eq. (2.18.3) by λ. Then, $\forall(\lambda, \beta) \in \mathbf{R}^2$, $\lambda\beta \geqslant -\frac{1}{2}$ (so that $T_0^{-1}(1+\lambda\beta) > 0$), and we can define

$$\begin{aligned} \Phi(\lambda, \beta) = 2\,\mathrm{Re}\left\{ \hat{g}_1\left(-k(2\beta + \beta^2\lambda) + \frac{2\pi i}{T_0}(1+\lambda\beta)\right)^{-1}\right\} \\ + \lambda \sum_{n=-\infty}^{+\infty} \hat{g}_n\left(-m\left(\frac{2\pi}{T}\right)^2 n^2 + k + \frac{2\pi i}{T}\lambda n\right)^{-1}, \end{aligned} \tag{2.18.4}$$

where $T^{-1} = T_0^{-1}(1+\lambda\beta)$ and it is perhaps worth recalling that $\hat{g}_n$ are also T dependent.

We now rewrite Eq. (2.18.3) as

$$\Phi(\lambda, \beta) = 0 \tag{2.18.5}$$

with the additional restriction $\lambda\beta > -\frac{1}{2}$ [to be amply sure that the denominators in Eq. (2.18.4) do not vanish].

To study Eq. (2.18.5), note that the equation $\Phi(0, \beta) = 0$ leads to

$$2\,\mathrm{Re}\,\frac{\hat{g}_1}{-2k\beta + 2\pi i/T_0} = 0 \tag{2.18.6}$$

which, if we set

$$b(T^{-1}) = \mathrm{Re}\,\hat{g}_1, \qquad c(T^{-1}) = \mathrm{Im}\,\hat{g}_1, \tag{2.18.7}$$

has as a solution the quantity β_0:

$$\beta_0 = -\frac{1}{2k}\frac{2\pi}{T_0}\frac{c(T_0^{-1})}{b(T_0^{-1})}, \tag{2.18.8}$$

provided that $b(T_0^{-1}) \neq 0$.

Note also that $b(T_0^{-1}) \neq 0$ as it follows from [see Eq. (2.17.7)]

$$\begin{aligned} b(T_0^{-1}) &= \int_0^{T_g} \frac{d\tau}{T_0}\, g(\tau)\cos\frac{2\pi}{T_0}\tau, \\ c(T_0^{-1}) &= \int_0^{T_g} \frac{d\tau}{T_0}\, g(\tau)\sin\frac{2\pi}{T_0}\tau \end{aligned} \tag{2.18.9}$$

thanks to the assumption $T_g < T_0/4$ which implies that for $\tau \in (0, T_g)$, the sine and cosine in Eq. (2.18.9) are positive.

The above remarks give us hope for the existence of a solution to Eq. (2.18.5) having the form

$$\beta(\lambda) = \beta_0 + O(\lambda), \tag{2.18.10}$$

at least if λ is small.

If this were true, we could compute the velocity $\dot{x}(0)$ from Eq. (2.17.8):

$$\begin{aligned} \dot{x}(0) &= P\sum_{n=-\infty}^{+\infty} \frac{2\pi i n}{T}\,\hat{g}_n\left(-\left(\frac{2\pi}{T}\right)^2 n^2 m + k + \frac{2\pi}{T} i\lambda n\right)^{-1} \\ &= \frac{P}{\lambda}\left[2\,\mathrm{Re}\,\frac{2\pi i}{T}\,\hat{g}_1\left(-k(2\beta + \lambda\beta^2) + \frac{2\pi i}{T}\right)^{-1}\right. \\ &\qquad \left. + \lambda\sum_{\substack{n=-\infty \\ n\neq\pm 1}}^{+\infty} \frac{2\pi i}{T} n\hat{g}_n\left(-\left(\frac{2\pi}{T}\right)^2 n^2 m + k + \frac{2\pi i}{T}\lambda n\right)^{-1}\right] \end{aligned} \tag{2.18.11}$$

Hence, Eqs. (2.18.11) and (2.18.10) would imply, with some algebra (and patience),

$$\begin{aligned} \dot{x}(0) &= \frac{P}{\lambda}\left[2\,\mathrm{Re}\left\{\frac{2\pi}{T_0}\frac{ib(T_0^{-1}) - c(T_0^{-1})}{-2k\beta_0 + 2\pi i/T_0}\right\} + O(\lambda)\right] \\ &= \frac{2P}{\lambda}\left(b(T_0^{-1}) + O(\lambda)\right), \end{aligned} \tag{2.18.12}$$

having used Eq. (2.18.8).

Therefore, if λ is so small that $|O(\lambda)| < b(T_0^{-1})$, we have $\dot{x}(0) \neq 0$, and we can choose P so that $\dot{x}(0) \in [v_-, v_+]$. Note that $P \to 0$ proportionally to λ, as $\lambda \to 0$, if one imposes $\dot{x}(0) \in [v_-, v_+]$: this agrees with the obvious empirical observation that the "weight" necessary to move the oscillator must be small in proportion to friction.

Similarly, starting from Eqs. (2.18.7) and (2.18.10), we could check Eq. (2.17.12) for small λ. It can in fact be seen that it is enough to verify Eq. (2.17.12) by replacing $x(t)$ in Eq. (2.17.8) with the only contributions to the series (2.17.8) coming from the $n = \pm 1$ terms (which in the preceding discussion seem to be the only important ones for small λ as far as the computation of T and of $\dot{x}(0)$ are concerned). For such an approximation to $x(t)$, the statement of Eq. (2.17.12) is, however, obvious since such an approximate motion is a harmonic motion with period T. We leave to the reader the elaboration of the details.

Finally, Eq. (2.17.11) would also immediately follow from Eqs. (2.18.1) and (2.18.10) for small λ.

The above analysis can be summarized in the following proposition.

25 Proposition. *If Eq.* (2.18.5), *as an equation for β parametrized by λ, admits a solution having the form of Eq.* (2.18.10) *for λ small enough, then the equation for the anchor-escapement model* [*Eqs.* (2.17.1), (2.17.2), *and* (2.17.4)] *admits a periodic solution with period T such that*:

$$T^{-1} = T_0^{-1}(1 + \beta_0\lambda + O(\lambda)) \tag{2.18.13}$$

if λ is sufficiently small and if P is suitably chosen.

Therefore, to complete the solution of our question, it only remains to verify that Eq. (2.18.5) does indeed admit a solution β like Eq. (2.18.10) for small λ.

Since we already know a pair (λ, β) verifying Eq. (2.18.5), namely the pair $(0, \beta_0)$, it is natural to try to treat Eq. (2.18.5) through the implicit function theorem (see Appendix G). By this theorem, it will be enough to check that the function Φ, defined in the open set of $\mathbf{R}^2$ containing the points (λ, β) such that $\lambda\beta > -\frac{1}{2}$, is of class C^∞ in its domain of definition and has a first-order derivative with respect to β such that

$$\frac{\partial \Phi}{\partial \beta}(0, \beta_0) \neq 0. \tag{2.18.14}$$

In this case, Eq. (2.18.5) will admit a solution β for λ small enough, like

$$\beta = \beta_0 - \frac{(\partial\Phi/\partial\lambda)(0, \beta_0)}{(\partial\Phi/\partial\epsilon)(0, \beta_0)}\lambda + o(\lambda). \tag{2.18.15}$$

To see that Φ is a C^∞ function near $(0, \beta_0)$, one shows that from expression (2.17.7) and from estimates (2.13.7) it follows that

$$\hat{g}_n = \frac{1}{((2\pi i/T)n)^k} \int_0^{T_g} \frac{d^k g}{d\tau^k}(\tau) e^{-(2\pi i/T)n\tau} \frac{d\tau}{T} \tag{2.18.16}$$

and, by Newton's formula for the pth derivative of a product:

$$\frac{\partial^p \hat{g}_n}{\partial (T^{-1})^p} = \frac{1}{(2\pi i n)^k} \int_0^{T_g} d\tau \frac{d^k g}{d\tau^k}(\tau) e^{-(2\pi i/T)n\tau} \times \sum_{p=0}^{p} \binom{p}{j} (-2\pi i n\tau)^{p-j} (-k+1)(-k) \cdots \times (-k-j+2)(T^{-1})^{-k+1-j}; \tag{2.18.17}$$

hence,

$$\left| \frac{\partial^p \hat{g}_n}{\partial (T^{-1})^p} \right| \leqslant \max_{\substack{0<\tau<T_g \\ 0<j<p}} \left[(k+p)^j (2\pi |n| T_g)^{p-j} (T^{-1})^{-k+1-j} \left| \frac{d^k g}{d\tau^k}(\tau) \right| \right] \tag{2.18.18}$$

which clearly implies that as long as $T < +\infty$ (i.e., $1 + \beta\lambda > 0$), the function in Eq. (2.18.4) is a C^∞ function of β and λ.

The last three relations also imply that the derivatives of Φ with respect to λ and β can be computed by term-by-term differentiation of the series defining Φ, in the region $\lambda\beta > -\frac{1}{2}$. After a brief computation, such a term-by-term differentiation evaluated at $(0, \beta_0)$ yields:

$$\frac{\partial \Phi}{\partial \beta}(0, \beta_0) = \left(\frac{T_0}{\pi} \right)^2 \frac{kb(T_0^{-1})^3}{b(T_0^{-1})^2 + c(T_0^{-1})^2} > 0 \tag{2.18.19}$$

and this check of Eq. (2.18.14) concludes the discussion of the compatibility conditions showing that they can indeed be satisfied for small enough λ.

§2.19. Encore about the Anchor Escapement: Stability of the Periodic Motion

In the preceding sections we showed that the anchor-escapement model [Eqs. (2.17.1), (2.17.2), and (2.17.4)] admits a periodic solution for small enough friction if the intensity of weight P is suitably chosen.

Imagining to have fixed P conveniently in terms of λ, we shall denote such a periodic motion $t \to \bar{x}(t)$, $t \in \mathbf{R}_+$. Clearly, the existence of the motion $\bar{x}$ is not interesting in itself, for applications. In fact, to put the system in this state of motion, one would have to impress exactly the velocity v_0 at $t = 0$, with v_0 defined by Eq. (2.18.11) after putting the oscillator in $x_0 = 0$. In fact, these are the initial data of the periodic motion corresponding to the a priori given λ and P.[12]

[12]One should also show that to such an initial datum an actually periodic motion does follow: i.e., one should prove a uniqueness theorem, at least for the initial data under examination. This is possible, as well as it is also possible to show a uniqueness property on the perturbed motions that we shall meet in this section. However, we shall not enter into the proof of the validity of the uniqueness properties that interest us: the reader should do this as a problem. Note that Proposition 1, p. 13, does not directly apply here, since the equations do not have the form contemplated in §2.2.

The periodic motion studied in the preceding sections is interesting for applications only if it is "stable", i.e., only if starting the system in an initial state $x(0) = 0$, $\dot{x}(0) = v_0 + \eta$, perturbed with respect to that which would generate a periodic motion, would produce a motion $t \to x_\eta(t)$, $t \in \mathbf{R}_+$, according to Eqs. (2.17.1), (2.17.2), and (2.17.4), which exists and is unique, at least for small η, and, furthermore,

$$|x_\eta(t) - \bar{x}(t - \tau_\eta)| \underset{t \to +\infty}{\to} 0 \tag{2.19.1}$$

if τ_η is suitably chosen.

In applications, one would like to require more: for instance, one would wish that the limit (2.19.1) is attained with an exponential speed with a halving time of the order of the period T of the periodic motion. In such a case, after a "few" oscillations, the motion would be identical to the rigorously periodic one, for all practical purposes. This is what actually occurs in the pendulum clock.

To examine the stability problem, we prove the following proposition.

26 Proposition. *The periodic motion of the anchor-escapement model, $t \to \bar{x}(t)$, $t \in \mathbf{R}_+$, built in* §2.17 *and* §2.18, *is stable in the sense expressed in Eq.* (2.19.1) *if λ is small enough. The limit* (2.19.2) *is reached exponentially with a halving time $T_{1/2}$ of the order of magnitude*

$$T_{1/2} \cong \max(T, 2m\lambda^{-1}). \tag{2.19.2}$$

Observations.

(1) During the proof, the role of friction and its importance will clearly appear. It is a rather general rule that the dissipative motions are more stable than the corresponding frictionless motions, as long as the friction is not too strong. The price paid for this stability, of obvious and essential importance in applications, is naturally the necessity of a moving force to maintain the motion itself.

(2) One could require, and prove, stability with respect to initial data that are more general and realistic than those considered in Proposition 26. For instance, with respect to initial data like $x(0) = \epsilon$, $\dot{x}(0) = v_0 + \eta$, the theory and results would be essentially the same.

(3) Proposition 26 concludes our theory of the anchor escapement. One should clearly bear in mind that the mathematical equations (2.17.1), (2.17.2), and (2.17.4) are just a model, in some respects not very satisfactory. For instance, $\dot{x}_0$ independence of the force $f(\dot{x}_0, \tau)$, once $\dot{x}_0 \in [v_-, v_+]$, is clearly unrealistic.

However, the model considered performs perfectly one of the typical functions of models and clarifies the possibility of the existence of an important mechanism which would also have to be present in more refined models: the possibility of a motion controlling itself via a feedback reaction inducing it to move periodically after a short while. This self-control, understood and practically realized at a time when the field of mechanics was new, is a phenomenon which appears in many models concerning the most diverse physical systems. The design and construction of the most precise machines are based on it, as well as the very possibility of their existence.

PROOF. Define [see Eq. (2.19.1) and the preceding lines for notation]:

$$x_\eta(t) = \bar{x}(t) + \xi(t) \tag{2.19.3}$$

and let us show the existence of a C^∞ solution of Eqs. (2.17.1), (2.17.2), and (2.17.4) verifying the initial conditions $x_\eta(0) = 0$, $\dot{x}_\eta(0) = \dot{\bar{x}}(0) + \eta$, provided η is small enough and the values of P, λ are such that the periodic motion $t \to \bar{x}(t)$, $t \geqslant 0$, exists. Call T the period of x. First, note that if $t \to \xi_1(t)$ is the solution of the equation

$$\begin{aligned} &m\ddot{\xi}_1 + \lambda\dot{\xi}_1 + k\xi_1 = 0, \qquad t \in \mathbf{R}_+ ,\\ &\xi_1(0) = 0, \qquad \dot{\xi}_1(0) = \eta \end{aligned} \tag{2.19.4}$$

and if $\overline{T}_1$ is the first positive time when the motion $t \to \bar{x}(t) + \xi_1(t)$ passes through the origin with positive speed, then the motion solves Eqs. (2.17.1), (2.17.2), and (2.17.4) for $t \in [0, \overline{T}_1]$ if η is small.

To understand this property, note that the solution of Eq. (2.19.4) is

$$\xi_1(t) = \eta e^{-(\lambda/2m)t} \frac{\sin\sqrt{(k/m - \lambda^2/4m^2)}\, t}{\sqrt{(k/m - \lambda^2/4m^2)}} ; \tag{2.19.5}$$

hence, $\forall t \geqslant 0$:

$$\begin{aligned} |\xi_1(t)| &\leqslant |\eta| \left(\frac{k}{m} - \frac{\lambda^2}{4m^2} \right)^{-1/2},\\ |\dot{\xi}_1(t)| &\leqslant |\eta| \left(1 + \left(\frac{k}{m} - \frac{\lambda^2}{4m^2} \right)^{-1/2} \right) \left(\frac{k}{m} - \frac{\lambda^2}{4m^2} \right)^{-1/2}. \end{aligned} \tag{2.19.6}$$

Then, if $|\eta|$ is small enough, to fix the ideas $|\eta| < \delta_\lambda$ with δ_λ suitably chosen, it is clear that Eq. (2.19.3) with $\xi_1(t)$ replacing $\xi(t)$ verifies Eqs. (2.17.1), (2.17.2), and (2.17.4).

To estimate a choice for δ_λ, the following conditions must be imposed: (1) $T < \overline{T}_1 < T + \alpha, \forall |\eta| < \delta_\lambda$; (2) the velocity at the first passage through the origin is negative and at the second passage is positive.

Such conditions are true for the reference motions $\bar{x}$ if λ is small enough since, in such a case, as already mentioned and used in §2.18, the reference motion is almost a harmonic motion of period $\sim T_0$ for which the conditions under analysis manifestly hold. Therefore, by continuity, they must remain true for the motion $t \to \bar{x}(t) + \xi_1(t)$ if η is small. We leave the elaboration of the details to the reader.

The fact that $t \to \bar{x}(t) + \xi_1(t)$ is a solution for $t \in [0, \overline{T}_1]$ will not, in general, remain true for $t > \overline{T}_1$ because in Eq. (2.17.4) the time τ is now counted beginning at $\overline{T}_1$, and $T \neq \overline{T}_1$ in general.

To study the motions for times following $\overline{T}_1$, we can proceed as follows. Set

$$\eta_1 = \dot{\bar{x}}(\overline{T}_1) + \dot{\xi}_1(\overline{T}_1) - \dot{\bar{x}}(0); \tag{2.19.7}$$

then if $|\eta_1| < \delta_\lambda$, we can define, as we already saw, the function $t \to (x(t) - \bar{x}(t - \overline{T}_1)) = \xi_2(t)$ where $\xi_2(t)$ is defined for t between $\overline{T}_1$ and the first instant $\overline{T}_2$ successive to $\overline{T}_1$, when the motion sweeps through 0 with positive speed for the first time, as the solution of Eq. (2.19.4) with initial datum:

$$\xi_2(\overline{T}_1) = 0, \qquad \dot{\xi}_2(\overline{T}_1) = \eta_1 . \tag{2.19.8}$$

Repeating the argument, we can indefinitely continue to define $\eta_2, \eta_3, \ldots$ *provided* $|\eta_i| < \delta_\lambda$, $i = 1, 2, \ldots,$ thus obtaining the definition of the times $\overline{T}_1, \overline{T}_2, \overline{T}_3, \ldots$ corresponding to the successive passages through 0 with positive speed.

It is clear that the stability property asserted in the proposition will have been proven once we shall have shown the existence of two constants $c_\lambda > 0$, $0 < \theta_\lambda < 1$ such that for all η, $|\eta| < \delta_\lambda$:

$$|\overline{T}_{p+1} - (\overline{T}_p + T)| \leqslant c_\lambda \theta_\lambda^p \qquad p = 1, 2, \ldots \tag{2.19.9}$$

and

$$|\eta_p| \leqslant \theta_\lambda^p |\eta|, \qquad p = 1, 2, \ldots, \tag{2.19.10}$$

at least for small λ. Setting $\overline{T}_0 = 0$, the constant τ_η [see Eq. (2.19.1)] will then be

$$\tau_\eta = \sum_{i=1}^{\infty} \left(\overline{T}_i - (T + \overline{T}_{i-1})\right). \tag{2.19.11}$$

Let us then show the validity of Eqs. (2.19.9) and (2.19.10).

If $T_0 = 2\pi\sqrt{m/k}$, the value for θ_λ that we shall find will have the form

$$\theta_\lambda = e^{-(\lambda T_0/2m)}(1 + o(\lambda)), \qquad T_0 = 2\pi\sqrt{\frac{m}{k}} \tag{2.19.12}$$

for λ small enough (note that $\theta_\lambda < 1$ as soon as λ is so small that $(-T_0/2m + \frac{1}{2}\lambda T_0^2/4m^2 + e^{-\lambda T_0/2m} o(\lambda)/\lambda) < 0$). This will also clearly prove Eq. (2.19.2) and, neglecting the infinitesimal $o(\lambda)$ in Eq. (2.19.12), we see that the larger the friction (compatibly with the supposed $\lambda^2 < 4mk$ and with the existence of $\bar{x}$), the faster the motion tends to become periodic.

To discuss Eqs. (2.19.9) and (2.19.10), one has to find a more concrete expression for $\overline{T}_1$ and, in general, for $\overline{T}_i$, $i \geqq 1$.

Let $\overline{T}_1 = T + \kappa_1$: the equation for κ_1 is

$$x_\eta(T + \kappa_1) = 0 \tag{2.19.13}$$

with the added condition that $T + \kappa_1$ should be the first positive time when the oscillator passes again through the origin with positive speed.

For $\kappa, \eta \in \mathbf{R}^2$, $\kappa > -T$, define

$$\psi(\kappa, \eta) = \bar{x}(T + \kappa) + \xi_1(T + \kappa) \tag{2.19.14}$$

and Eq. (2.19.13) becomes

$$\psi(\kappa,\eta)=0. \tag{2.19.15}$$

Since, as we saw in the preceding sections, $t\to\bar{x}(t)$, $t\in\mathbf{R}_+$, is a C^∞ function and, obviously, so is $(\eta,t)\to\xi_1(t)$, we can say that ψ is a C^∞ function on its domain of definition, $\kappa>-T$.

Furthermore, it is easy to see, by Eqs. (2.19.14) and (2.19.5), that

$$\psi(0,0)=0,\quad\text{and}\quad\frac{\partial\psi}{\partial\kappa}(0,0)=v_0,$$
$$\frac{\partial\psi}{\partial\eta}(0,0)=e^{-(\lambda/2m)T}\frac{\sin(k/m-\lambda^2 4m^2)^{1/2}T}{(k/m-\lambda^2/4m^2)^{1/2}}. \tag{2.19.16}$$

Then, by the implicit function theorem (see Appendix G), there is, for small η, a unique small solution of Eq. (2.19.15) which we denote $\kappa_1(\eta)$ and

$$\kappa_1(\eta)=-\frac{e^{-(\lambda T/2m)}}{v_0}\frac{\sin(k/m-\lambda^2/4m^2)^{1/2}T}{(k/m-\lambda^2/4m^2)^{1/2}}\eta+o_\lambda(\eta), \tag{2.19.17}$$

where the index λ in $o_\lambda(\eta)$ recalls that the infinitesimal depends also on λ.

By taking Eq. (2.17.14) into account:

$$T=T_0(1-\beta_0\lambda+o(\lambda)), \tag{2.19.18}$$

and using $\sin\sqrt{(k/m)}\,T_0=0$, one finds, with simple steps, from Eqs. (2.19.17) and (2.19.18), that

$$\kappa_1(\eta)=\frac{\beta_0 T_0}{v_0}(\lambda+o'(\lambda))\eta+o_\lambda(\eta). \tag{2.19.19}$$

where $o'(\lambda)$ is a suitable infinitesimal of order λ^2.

It then becomes possible to compute η_1:

$$\eta_1=\dot{\bar{x}}(T+\kappa_1)+\dot{\xi}_1(T+\kappa_1)-v_0=\ddot{\bar{x}}(T)\kappa_1+\dot{\xi}_1(T+\kappa_1)+\tilde{o}_\lambda(\kappa_1), \tag{2.19.20}$$

where $\tilde{o}_\lambda$ is a λ-dependent second-order infinitesimal: this expression arises just by expanding $\dot{\bar{x}}$ in Taylor series near T and using $\dot{\bar{x}}(T)=v_0$.

The equations of motion (2.17.1) imply that $\ddot{\bar{x}}(T)=\ddot{\bar{x}}(0)=-(\lambda/m)v_0$ and Eq. (2.19.20) implies [via Eqs. (2.19.18) and (2.19.19) and some patience]:

$$\eta_1=\eta e^{-\lambda T_0/2m}(1+\tilde{o}(\lambda)+\tilde{O}_\lambda(\eta)), \tag{2.19.21}$$

where $\tilde{o}$ is an infinitesimal of higher order in λ while $\tilde{O}_\lambda(\eta)$ is a λ-dependent infinitesimal of the same order as η. It is then clear that there is a $\delta'_\lambda<\delta_\lambda$ sufficiently small so that $|\tilde{O}_\lambda(\eta)|\leqslant|\tilde{o}(\lambda)|$, $\forall|\eta|<\delta'_\lambda$; then Eq. (2.19.21) implies that

$$|\eta_1|\leqslant\theta_\lambda|\eta|,\qquad\forall|\eta|<\delta'_\lambda \tag{2.19.22}$$

with θ_λ given by Eq. (2.19.12).

Hence, if λ is small enough one finds that $|\eta| < \delta'_\lambda$ implies $|\eta_1| < \delta'_\lambda$ and the argument can be indefinitely repeated to successively estimate $|\eta_1|$, $|\eta_2|, \ldots$. It is clear that from Eqs. (2.19.22) and (2.19.19), one deduces Eqs. (2.19.9) and (2.19.10), and Proposition 26 is thus proven. mbe

Problems for §2.19

1. Investigate heuristically the stability of the solutions of Eqs. (2.17.1), (2.17.2), and (2.17.4), using the computer program of problem 1, §2.17. For each value of λ let v_0, x_0 be the data, at time zero, of a periodic motion verifying Eqs. (2.17.1), (2.17.2) and (2.17.4); let the computer draw, with its plotter, the graph of the periodic motion superimposed with the graph of the motion of a harmonic oscillator with the same mass and elastic constant (but no friction or forcing term). Repeat this operation as λ varies using it to compare visually the two motions.

2. Same as Problem 1, replacing kx by $k \sin x$ in Eqs. (2.17.1) and (2.17.2) (i.e., replacing the basic oscillator with a pendulum).

§2.20. Frictionless Forced Oscillations: Quasi-Periodic Motions

In §2.16 we saw that under the action of a periodic force, a frictionless harmonic oscillator moves with a motion "sum" (or "superposition") of two periodic motions with respective periods equal to the proper oscillator's period T_0 and to the forcing term's period T, provided T/T_0 is not an integer.

In this section we discuss a proposition that allows us to visualize some remarkable properties of such motions. One of them appears by representing them as motions on the data space (see §2.6), i.e., on the plane $\mathbf{R}^2$ thought of as the space of the initial velocities and positions. This means that the motion $t \to x(t)$, $t \in \mathbf{R}_+$, the solution of Eq. (2.16.3), is represented by a curve $t \to (\dot{x}(t), x(t))$, $t \in \mathbf{R}_+$. This is a representation of the motion which we have not yet used: it is somewhat redundant because once $t \to x(t)$ is given, its t derivative is automatically given. On the other hand, every point of the curve $t \to (\dot{x}(t), x(t))$, $t \in \mathbf{R}_+$, completely determines the motion. Also it may sometimes be useful to know which are the pairs $(\dot{x}, x)$ which can appear during the evolution of a given motion. In such a case, this information can be directly extracted from the geometric locus described in $\mathbf{R}^2$ by $t \to (\dot{x}(t), x(t))$, $t \in \mathbf{R}_+$, without having to know explicitly which values of t correspond to the various points of the locus.

Therefore, in the data space, a periodic motions appears as a closed curve. A motion like those met in §2.12, asymptotically periodic, appears as a curve spiraling around the closed curve representing the periodic motion and becoming indefinitely closer to it.

The structure of a superposition of two periodic motions in the data-space representation is of particular interest: it is elucidated by the following well-known proposition (Euler).

27 Proposition. *Let $f, g \in C^{\infty}(\mathbf{R})$ be two periodic functions with minimal period 2π and let f', g' be their first derivatives. Given $\omega, \omega_0 > 0$, consider the motion in $\mathbf{R}^2$ described by $t \to (\eta(t), \xi(t))$:*

$$\begin{aligned} \eta(t) &= \omega f'(\omega t) + \omega_0 g'(\omega_0 t), \\ \xi(t) &= f(\omega t) + g(\omega_0 t). \end{aligned} \tag{2.20.1}$$

Such a motion[13] *is periodic if and only if ω/ω_0 is a rational number. If ω/ω_0 is irrational, the curve $t \to (\eta(t), \xi(t))$, $t \geqslant t_0$, $\forall t_0 \in \mathbf{R}_+$ densely fills the region $\Omega_{f,g}$:*

$$\begin{aligned} \Omega_{f,g} = \{(\eta, \xi) \mid (\eta, \xi) \in \mathbf{R}^2 : \eta &= \omega f'(\alpha) + \omega_0 g'(\beta), \\ \xi &= f(\alpha) + g(\beta); \alpha, \beta \in [0, 2\pi]\}. \end{aligned} \tag{2.20.2}$$

Observations.

(1) The region $\Omega_{f,g}$ can be easily visualized. Consider the curve Γ_f in the (η, ξ)-plane, having equations

$$\eta = \omega f'(\alpha), \qquad \xi = f(\alpha), \qquad \alpha \in [0, 2\pi]. \tag{2.20.3}$$

By the periodicity of f, this is a closed curve [see Fig. 2.9]. Given $\alpha \in [0, 2\pi]$, consider the curve $\Gamma_g(\alpha)$ with equations

$$\eta = \omega f'(\alpha) + \omega_0 g'(\beta), \qquad \xi = f(\alpha) + g(\beta) \qquad \beta \in [0, 2\pi] \tag{2.20.4}$$

which, since g, too, is periodic, is a closed curve "around" $(\omega f'(\alpha), f(\alpha))$.

As α varies in $[0, 2\pi]$, the curve $\Gamma_g(\alpha)$ "glides along Γ_f" and "sweeps" the region $\Omega_{f,g}$. A simple case is illustrated in Fig. 2.9.

(2) The relevance of this proposition for the harmonic nonresonant forced oscillations is obvious after the discussion of §2.16 (see (iii) in

[13] Notice that $\eta(t) = \dot{\xi}(t)$.

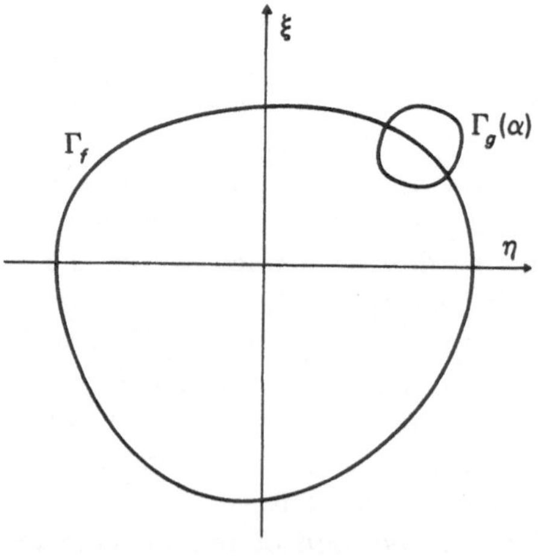

Figure 2.9.

Proposition 23). It shows that such oscillations, when T and T_0 have an irrational ratio, are not periodic although they come back as close as desired to the initial datum, provided one waits long enough.

(3) Also, for the purpose of future applications, it is interesting to give a geometric interpretation to Proposition 27 when ω/ω_0 is irrational. In this case, the analytic expression of the trajectory density in $\Omega_{f,g}$ is: given $\sigma > 0$ and $t_0 \in \mathbf{R}_+$, for all $(\alpha, \beta) \in [0, 2\pi]^2$, there is $t_\sigma(\alpha, \beta) > t_0$ such that

$$\begin{aligned} &|(\alpha - \omega t_\sigma(\alpha, \beta)) \bmod 2\pi| < \sigma, \\ &|(\beta - \omega_0 t_\sigma(\alpha, \beta)) \bmod 2\pi| < \sigma, \end{aligned} \tag{2.20.5}$$

i.e., there are two integers $m_\sigma(\alpha, \beta)$ and $n_\sigma(\alpha, \beta)$ such that

$$\begin{aligned} &|\alpha - \omega t_\sigma(\alpha, \beta) - 2\pi m_\sigma(\alpha, \beta)| < \sigma, \\ &|\beta - \omega_0 t_\sigma(\alpha, \beta) - 2\pi n_\sigma(\alpha, \beta)| < \sigma. \end{aligned} \tag{2.20.6}$$

Now think of the plane $\mathbf{R}^2$ as being paved with squares with side size 2π and with corners at $(2\pi r, 2\pi s)$, r and s being integers. In this plane, consider the straight line through the origin with slope ω_0/ω:

$$y = \omega_0 t, \qquad x = \omega t, \qquad t \in \mathbf{R} \tag{2.20.7}$$

and the half-line corresponding to $t \geqslant t_0$.

Next, identify the plane's points whose coordinates differ by integer multiples of 2π (see Fig. 2.10). The just-described line can now be thought of as a set of segments in the square $[0, 2\pi]^2$, where corresponding points on opposite sides are identified (topologically, we can say that we regard the square $[0, 2\pi]^2$ as a two dimensional torus).

Equation (2.20.6) says that at least one of the segments associated with the line of Eq. (2.20.7) with $t \geqslant t_0$ cuts the square neighborhood of side 2σ around (α, β) (see Fig. 2.11).

In other words, the half-line of Eq. (2.20.7) with $t \geqslant t_0$, brought back inside $[0, 2\pi]^2$ through the identification of the plane's points mod 2π (i.e., thought of as a coil around the torus) densely fills $[0, 2\pi]^2$.

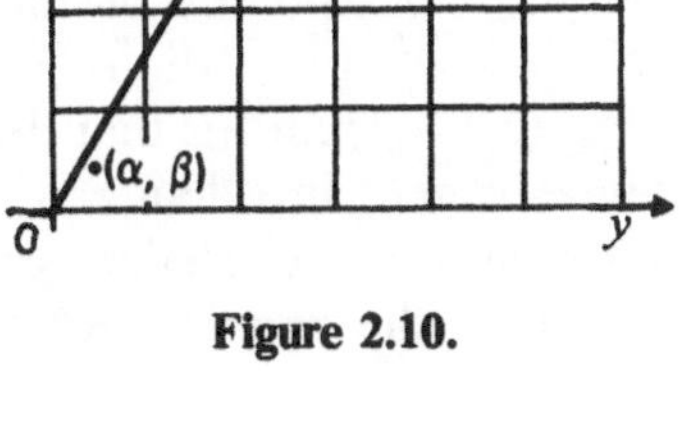

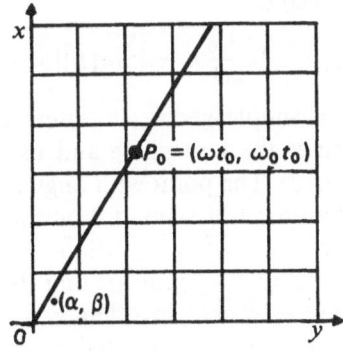

Figure 2.10.

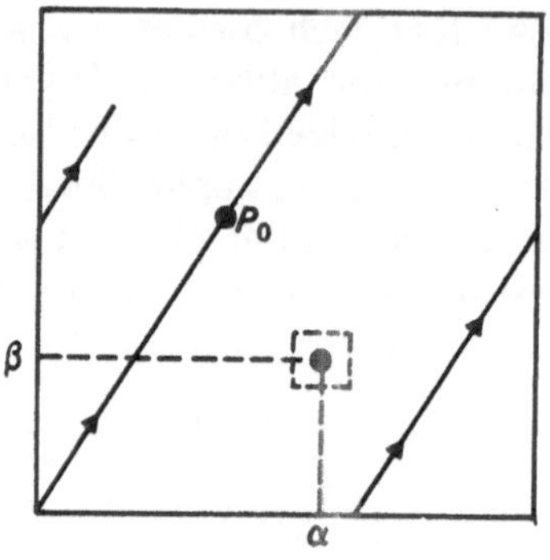

Figure 2.11.

PROOF. If $\omega/\omega_0 \equiv T_0/T = p/q =$ (ratio of two relatively prime integers), where we define $T = 2\pi/\omega$, $T_0 = 2\pi/\omega_0$, then the motion of Eq. (2.20.1) is obviously periodic with period $T' = pT = qT_0$. As an exercise, the reader can show that in the geometric interpretation of Fig. 2.11, this means that the line becomes a finite set of segments (forming a closed curve if $[0, 2\pi]^2$ is thought of as a torus).

Suppose now that ω/ω_0 is irrational. Define for every integer n the number τ_n as

$$\alpha - \omega\tau_n + 2\pi n = 0 \Rightarrow \tau_n = \frac{\alpha + 2\pi n}{\omega}. \tag{2.20.8}$$

To check Eq. (2.20.5) and, therefore, the validity of the Proposition, it will suffice to show that given $n_0 \in \mathbf{Z}$ and $\sigma > 0$ arbitrarily, there exists $n \in \mathbf{Z}$, $n \geqslant n_0$, and $m(n) \in \mathbf{Z}$ such that

$$|\beta - \omega_0\tau_n - 2\pi m(n)| < \sigma. \tag{2.20.9}$$

It is useful for the reader to understand (along the lines of observation 3) the geometrical meaning of Eqs. (2.20.8) and (2.20.9) (exercise).

By substituting τ_n, given by Eq. (2.20.8), in Eq. (2.20.9) one finds

$$|\beta - \frac{\omega_0}{\omega}\alpha - 2\pi\frac{\omega_0}{\omega}n - 2\pi m(n)| < \sigma, \tag{2.20.10}$$

i.e., setting $\varphi_0 = \beta - (\omega_0/\omega)\alpha$:

$$|\varphi_0 - 2\pi\frac{\omega_0}{\omega}n - 2\pi m(n)| < \sigma. \tag{2.20.11}$$

Equation (2.20.11) has a simple geometric interpretation which is convenient to illustrate: consider the unit circle and its rotation R by an angle $\theta = 2\pi(\omega_0/\omega)$ (see Fig. 2.12). The point with angular coordinate $2\pi(\omega_0/\omega)n$ can be interpreted as the image of a point 0 on the circle under the action of the rotation R^n, i.e., of n successive rotations R. If φ_0 is also interpreted as a point on the circle, Eq. (2.20.11) means that the rotation R^n brings the origin to an angular distance from φ_0 less than σ.

Then our problem is to show the existence, given $\sigma > 0$, of infinitely many integers $n > 0$ such that the rotation R^n brings 0 to an angular

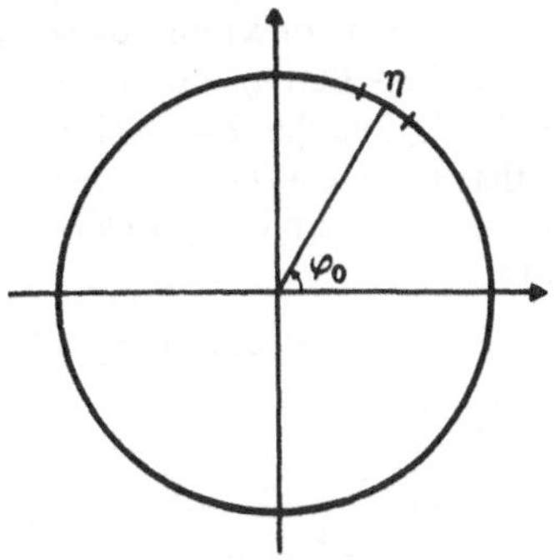

Figure 2.12

distance of less than σ from φ_0. It is clear that in order to show this, it will be enough to show that there is $\tilde{n} > 0$ such that $R^{\tilde{n}}$ displaces the point 0 by a nonvanishing quantity ϵ with modulus less than σ. In fact, when this happens, it is manifest that with a rotation R^n, $n = k\tilde{n}$, $k = 0, 1, 2, \ldots,$ one successively displaces 0 by $\epsilon, 2\epsilon, 3\epsilon, \ldots, k\epsilon, \ldots$ and therefore, sooner or later (and infinitely often), one arrives at the situation that the origin falls inside a neighborhood of φ_0 with angular amplitude σ.

To show the existence of $\tilde{n}$, note that the sequence $(2\pi(\omega_0/\omega)k)_{k\in\mathbf{Z}_+}$, thought of as a sequence of angular coordinates on the circle, corresponds to a family of points which are pairwise distinct since

$$2\pi\frac{\omega_0}{\omega}k_1 = 2\pi\frac{\omega_0}{\omega}k_2 + 2\pi\mu \tag{2.20.12}$$

with k_1, k_2, μ integers would imply, if $k_1 \neq k_2$, that $\omega_0/\omega = \mu/(k_1 - k_2)$ = (rational number). Then, if $\bar{\varphi}$ is an accumulation point of the above family of points on the circle, there must exist two distinct points in such a sequence closer to $\bar{\varphi}$ than $\sigma/2$; i.e., there exist $k_1, k_2 > 0$ such that

$$\left|\left(2\pi\frac{\omega_0}{\omega}k_1 - 2\pi\frac{\omega_0}{\omega}k_2\right)\bmod 2\pi\right| < \sigma, \tag{2.20.13}$$

and this means that the rotation $R^{k_1-k_2}$ displaces the point 0 on the circle at a point whose angular distance from 0 is ϵ and $0 < |\epsilon| < \sigma$ [note that $\epsilon \neq 0$, by the remark related to Eq. (2.20.12)]. Hence, one can take $\tilde{n} = k_1 - k_2$ if $k_1 > k_2$ or $\tilde{n} = k_2 - k_1$ if $k_2 > k_1$. mbe

Exercises and Problems for §2.20

Problems (1)–(9) are inspired by Kintchin, *Continued Fractions*, see references, and they aim at providing tools for studying the remaining problems.

1. Let $r > 0$ be an irrational number represented by its continued fraction

$$r = a_0 + \cfrac{1}{a_1 + \cfrac{1}{a_2 + \cfrac{1}{a_3 + \cdots}}} \equiv \{a_0, a_1, a_2, \ldots\}$$

defined by setting $[x] =$ (integral part of x) and $a_0 = [r]$, $r_1 = (r - a_0)^{-1}$, $a_1 = [r_1]$, $r_2 = (r_1 - a_1)^{-1}$, $a_2 = [r_2]$, etc. Show that $a_j > 0$, $\forall j > 0$. Compute $a_0, a_1, a_2 \ldots$ for $r =$ golden section $= (\sqrt{5} - 1)/2$ (note that $r = 1/(1 + r)$), $r = (1 + \sqrt{5})/2$ (note that $r = 1 + 1/r$), $r = \sqrt{2}$ (note that $r = 1 + 1/(1 + r)$), $r = \pi$ (recall $\pi = 3.141592653589 \ldots$ and using a pocket computer to find empirically (i.e., without error estimates) $a_0, a_1, a_2, \ldots, a_8$, one finds

$$a_0 = 3,\ a_1 = 7,\ a_2 = 15,\ a_3 = 1,\ a_4 = 291, \ldots).$$

2. In the context of Problem 1, let

$$R_k = a_0 + \cfrac{1}{a_1 + \cfrac{1}{a_2 + \cdots \cfrac{}{\cdots + \cfrac{1}{a_k}}}} \equiv \{a_0, a_1, \ldots, a_k\}.$$

Show that $R_{2k} < r < R_{2k+1}$, $\forall k \geqslant 0$.

3. In the context of Problems 1 and 2, note that if $\{a_1, a_2, \ldots, a_k\} = p'/q'$ with $p', q' \in \mathbf{Z}_+$, then $\{a_0, a_1, \ldots, a_k\} = (a_0 p' + q')/p'$. Deduce from this that a vector $\mathbf{v}_k = (p_k, q_k) \in \mathbf{Z}_+^2$ such that $R_k = p_k/q_k$ is the one given by

$$\mathbf{v}_k = \begin{pmatrix} a_0 & 1 \\ 1 & 0 \end{pmatrix}\begin{pmatrix} a_1 & 1 \\ 1 & 0 \end{pmatrix} \cdots \begin{pmatrix} a_k & 1 \\ 1 & 0 \end{pmatrix}\begin{pmatrix} 1 \\ 0 \end{pmatrix}.$$

4. Deduce from Problem 3 that $\mathbf{v}_k = a_k \mathbf{v}_{k-1} + \mathbf{v}_{k-2}$, i.e.,

$$p_k = a_k p_{k-1} + p_{k-2} \Rightarrow p_k \geqslant p_{k-1}, \qquad \forall k > 1,$$
$$q_k = a_k q_{k-1} + q_{k-2} \Rightarrow q_k \geqslant q_{k-1}, \qquad \forall k > 1.$$

(Hint: $\left(\begin{smallmatrix} a_k & 1 \\ 1 & 0 \end{smallmatrix}\right)\left(\begin{smallmatrix} 1 \\ 0 \end{smallmatrix}\right) = \left(\begin{smallmatrix} a_k \\ 1 \end{smallmatrix}\right) = a_k\left(\begin{smallmatrix} 1 \\ 0 \end{smallmatrix}\right) + \left(\begin{smallmatrix} 0 \\ 1 \end{smallmatrix}\right)$ *and* $\left(\begin{smallmatrix} a_{k-1} & 1 \\ 1 & 0 \end{smallmatrix}\right)\left(\begin{smallmatrix} 0 \\ 1 \end{smallmatrix}\right) = \left(\begin{smallmatrix} 1 \\ 0 \end{smallmatrix}\right)$; use this to eliminate the last matrix in the product of matrices appearing in Problem 3.)

5. From the recursion relation in Problem 4, deduce that

$$q_k p_{k-1} - p_k q_{k-1} = -(q_{k-1} p_{k-2} - p_{k-1} q_{k-2}) = (-1)^k, \qquad \forall k \geqslant 2,$$
$$q_k p_{k-2} - p_k q_{k-2} = a_k (q_{k-1} p_{k-2} - p_{k-1} q_{k-2}) = (-1)^{k-1} a_k, \qquad \forall k \geqslant 2.$$

Hence:

$$\frac{p_{k-1}}{q_{k-1}} - \frac{p_k}{q_k} = \frac{(-1)^k}{q_k q_{k-1}},$$

$$\frac{p_{k-2}}{q_{k-2}} - \frac{p_k}{q_k} = \frac{(-1)^{k-1}}{q_k q_{k-2}} a_k.$$

(Hint: Multiply the first equation of the recursive formula in Problem 4 by q_{k-1} and the second by p_{k-1} and subtract, etc.)

6. From Problem 5 deduce that

$$\frac{p_0}{q_0} < \frac{p_2}{q_2} \cdots < r < \cdots < \frac{p_3}{q_3} < \frac{p_1}{q_1}$$

and

$$\left| r - \frac{p_k}{q_k} \right| < \frac{1}{q_k q_{k+1}}.$$

7. Show that $q_k \geqslant 2^{(k-1)/2}$, $k \geqslant 0$ and $p_k \geqslant 2^{(k-2)/2}$, $k \geqslant 1$. (Hint: Note that $a_k \geqslant 1$, $\forall k \geqslant 1$, and use the recursive relation in Problem 4 and $p_j, q_j \geqslant 1$.)

8. Show that

$$\frac{1}{q_k(q_k+q_{k+1})} < \left| r - \frac{p_k}{q_k} \right| < \frac{1}{q_k q_{k+1}} .$$

(Hint: If $a/b < c/d$, then $(a+sc)/(b+sd)$ increases with s for $s \geqslant 0$, while if $a/b > c/d$ it decreases. Hence, if k is even,

$$\frac{p_{k-2}+sp_{k-1}}{q_{k-2}+sq_{k-1}}$$

increases with s and for $s = a_k$ it becomes p_k/q_k which is such that $p_k/q_k < r < p_{k-1}/q_{k-1}$. Therefore,

$$\frac{p_{k-2}}{q_{k-2}} \leqslant \frac{p_{k-2}+p_{k-1}}{q_{k-2}+q_{k-1}} < r$$

and so

$$\left| r - \frac{p_{k-2}}{q_{k-2}} \right| > \left| \frac{p_{k-2}+p_{k-1}}{q_{k-2}+q_{k-1}} - \frac{p_{k-2}}{q_{k-2}} \right| \equiv \frac{1}{q_{k-2}(q_{k-2}+q_{k-1})}$$

by Problem 5. For k odd, one argues similarly.

9. If $r = \{a_0, a_1, \ldots\}$ and if $c = \min_{i \geqslant 1} a_i \geqslant 1$, $c_k = \max_{1 \leqslant i \leqslant k} a_i \geqslant 1$, it follows from Problem 4 that

$$(1+c)^{(k-1)/2} \leqslant p_k \leqslant (1+c_k)^{(k-1)/2},$$

$$(1+c)^{(k-1)/2} \leqslant q_k \leqslant (1+c_k)^{(k-1)/2}.$$

10. Given $\sigma > 0$, estimate the first time τ when $|(\omega\tau - \alpha) \bmod 2\pi| < \sigma$, $|(\omega_0\tau - \beta) \bmod 2\pi| < \sigma$ in terms of the continued fraction's coefficients of ω/ω_0, assuming it irrational. (Hint: Use Problems 1–9; let $\sigma < 2\pi$ and define k so that $q_{k+1}^{-1} < \sigma/2\pi < q_k^{-1}$. Note that, if $\tilde{\tau} = 2\pi q_k/\omega_0$, the point $x = (\omega\tilde{\tau}, \omega_0\tilde{\tau})$ will come close to the origin by $\omega\tilde{\tau}$ (mod 2π). Since $|2\pi\omega q_k/\omega_0 - 2\pi p_k| \equiv 2\pi q_k|(\omega/\omega_0) - (p_k/q_k)| = \epsilon \leqslant 2\pi q_k/q_k q_{k+1} < \sigma$, we see that x is closer than σ but not closer than $\epsilon \geqslant 2\pi/2q_{k+1}$ (see Problem 7). This implies an estimate for τ as $\tau < \tilde{\tau}2\pi/\epsilon \leqslant 4\pi q_k q_{k+1}/\omega_0$. It is also easy to see, if one uses the fact that p_k/q_k is the "best rational approximation" to ω/ω_0, among those of the form p/q with $q \leqslant q_k$, that there is a point (α, β) for which the time τ can be estimated to be $\tau \geqslant (2\pi q_k/\omega_0) - (\sigma/\omega)$. The bounds in Problems 7 and 9 can be used to obtain concrete estimates for τ by replacing q_k by $2^{(k-1)/2}$.)

11. Determine the region Ω densely covered by the data-space trajectory of the motion

$$\ddot{x} + x = 3\cos\omega t, \qquad x(0) = \dot{x}(0) = 0$$

when ω is irrational.

12. For ω = golden section (see Problem 1), estimate the minimum time τ necessary for the trajectory of the oscillator in Problem 11 to cover Ω so that any point in Ω has a distance from the trajectory $t \to (\dot{x}(t), x(t))$, $t \in [0, \tau]$, not exceeding $\sigma = 2\pi/2^4$.

13. Same as Problem 12 for $\omega = \sqrt{2}$ and for $\omega = \pi$; (use for π an empirically computed continued fraction; see Problem 1).

14. Let $\tilde{\omega} = \{a_0, a_1, \ldots, a_k\}$, $a_i \geqslant 1$, $i > 0$, be a rational number. In terms of k, estimate the maximum distance of a point in Ω from the trajectory of the oscillator in Problem 11 with $\tilde{\omega}$ replacing ω.

§2.21. Quasi-Periodic Functions. Multiperiodic Functions. Tori and the Multidimensional Fourier Theorem

The considerations of §2.20 suggest the following definition.

11 Definition. A function $f \in C^\infty(\mathbf{R})$ is "quasi-periodic with pulsations $\omega_1, \dots, \omega_d$" if there exists a $\varphi \in C^\infty(\mathbf{R}^d)$ such that

$$\varphi(\xi_1, \dots, \xi_i, \dots, \xi_d) = \varphi(\xi_1, \dots, \xi_i + 2\pi, \dots, \xi_d), \tag{2.21.1}$$
$$\forall \xi \in \mathbf{R}^d, \quad \forall i = 1, 2, \dots, d$$

and

$$f(t) = \varphi(\omega_1 t, \dots, \omega_d t), \qquad t \in \mathbf{R}. \tag{2.21.2}$$

The numbers $T_1 = 2\pi/\omega_1, \dots, T_d = 2\pi/\omega_d$ are called the "periods" of f, while $\gamma_1 = T_1^{-1}, \dots, \gamma_d = T_d^{-1}$ are the "frequencies" of f.

Observations.

(1) It is therefore possible to say that the motion of a harmonic oscillator with pulsation ω_0 forced by a periodic force with pulsation ω is, in the absence of resonances, a quasi-periodic function with pulsations ω_0 and ω [see Eq. (2.16.6) and §2.20].

(2) The above definition of a quasi-periodic function is more restrictive than the one usually found in mathematical literature: it is, however, sufficiently general for our purposes.

(3) It is useful to note that given f, there exist several choices of d, $\omega_1, \dots, \omega_d$ and φ allowing us to represent f as in Eq. (2.21.2). A trivial example is provided by the consideration of a function $\varphi \in C^\infty(\mathbf{R})$, periodic with period 2π, and of the functions of $\xi \in \mathbf{R}$ or of $(\xi_1, \xi_2) \in \mathbf{R}^2$ defined as $\psi(\xi) = \varphi(2\xi)$ or $\bar{\psi}(\xi_1, \xi_2) = \varphi(2\xi_1 + 3\xi_2)$ which, via the formulae

$$f(t) = \psi\left(\frac{\omega}{2} t\right) \equiv \varphi(\omega t), \tag{2.21.3}$$

$$f(t) = \bar{\psi}\left(\frac{\omega}{4} t, \frac{\omega}{6} t\right) \equiv \varphi(\omega t), \tag{2.21.4}$$

allow a representation of f as a quasi-periodic function with angular velocities $\omega/2$ or ω or $(\omega/4, \omega/6)$.

(4) The pulsations (or "angular velocities") in Definition 11 need not necessarily all be positive: some may be zero or negative.

The functions φ used to introduce the notion of quasi-periodic function are remarkable in themselves, and it is convenient to set up the following definition.

12 Definition. Given $L_1, \dots, L_d > 0$, consider the pavement of $\mathbf{R}^d$ whose tesserae are the parallelepiped $[0, L_1] \times \cdots \times [0, L_d]$ and the parallelepipeds obtained by translating it by $(n_1 L_1, \dots, n_d L_d)$, $n_1, \dots, n_d$ integers.

Two points $\xi, \eta \in \mathbf{R}^d$ will be declared equivalent if they are "equally located" in the pavement's tesserae, i.e., if there are d integers $n_1, \dots, n_d$

such that $\xi_i - \eta_i = n_i L_i$, $i = 1, \ldots, d$. We shall denote $\mathbf{T}^d(L_1, \ldots, L_d)$ the set of the equivalence classes thus obtained and we shall set

$$d(\{\boldsymbol{\xi}\}, \{\boldsymbol{\eta}\}) = \min_{\substack{\boldsymbol{\xi}' \in \{\boldsymbol{\xi}\} \\ \boldsymbol{\eta}' \in \{\boldsymbol{\eta}\}}} |\boldsymbol{\xi}' - \boldsymbol{\eta}'| \tag{2.21.5}$$

if $\{\boldsymbol{\xi}\}, \{\boldsymbol{\eta}\} \in \mathbf{T}^d(L_1, \ldots, L_d)$ and $\{\boldsymbol{\xi}\}$ denotes the equivalence class containing $\boldsymbol{\xi}$.

The set $\mathbf{T}^d(L_1, \ldots, L_d)$, regarded as a metric space with the distance function defined by Eq. (2.21.5) ("distance on the torus"), will be called a "d-dimensional torus" with sides $L_1, \ldots, L_d$. If $L_1 = L_2 = \ldots L_d = 2\pi$, this torus will be denoted $\mathbf{T}^d$, simply.

Observations.

(1) The above definition, in spite of its apparent complexity, is simple and can be informally summarized by saying that the torus $\mathbf{T}^d(L_1, \ldots, L_d)$ is obtained by "identifying the opposite sides" of the parallelepiped $[0, L_1] \times \cdots \times [0, L_d]$ of $\mathbf{R}^d$. For this reason it is customary to describe the points of $\mathbf{T}^d(L_1, \ldots, L_d)$ through the Cartesian coordinates in $\mathbf{R}^d$ of one of the corresponding representatives without explicit mention of the equivalence relation: when $L_1 = \cdots = L_d = 2\pi$, such coordinates are called the "natural angular coordinates" or "flat coordinates" on $\mathbf{T}^d$. In general, the distance [Eq. (2.21.5)] is called the distance between $\boldsymbol{\xi}$ and $\boldsymbol{\eta}$ "on the torus $\mathbf{T}^d(L_1, \ldots, L_d)$".

(2) Clearly $\mathbf{T}^d$ can be regarded as the product of d unit circles. If $(\varphi_1, \ldots, \varphi_d)$ are the natural angular coordinates of $\boldsymbol{\varphi} \in \mathbf{T}^d$, a natural one-to-one continous mapping of $\mathbf{T}^d$ into $S \times \cdots \times S$, where $S =$ (unit circle in the complex plane), is

$$\boldsymbol{\varphi} = (\varphi_1, \ldots, \varphi_d) \leftrightarrow \mathbf{z} = (z_1, \ldots, z_d) = (e^{i\varphi_1}, \ldots, e^{i\varphi_d}) \tag{2.21.6}$$

and the distance (2.21.5) turns out to be equivalent to the distance on $S \times \cdots \times S$ as a subset of $\mathbf{C}^d$.

Therefore, the d-dimensional torus $\mathbf{T}^d$ can be regarded as a subset of the d-dimensional complex space $\mathbf{C}^d$. This representation is more intrinsic since it does not involve coordinates defined mod 2π. It will turn out to be a deep and very useful representation (see Chapter V, §5.10, 5.12).

13 Definition. The set $C^\infty(\mathbf{T}^d(L_1, \ldots, L_d))$ is, by definition, the set of the functions f defined on $\mathbf{T}^d(L_1, \ldots, L_d)$ such that setting

$$\tilde{f}(\boldsymbol{\xi}) = f(\{\boldsymbol{\xi}\}), \qquad \forall \boldsymbol{\xi} \in \mathbf{R}^d, \tag{2.21.7}$$

the function $\tilde{f}$ is in $C^\infty(\mathbf{R}^d)$.

The set of functions on $\mathbf{R}^d$ having the form of Eq. (2.21.7) is the set of the "multiperiodic functions on $\mathbf{R}^d$" with periods $L_1, \ldots, L_d$ or with pulsations $2\pi/L_1, \ldots, 2\pi/L_d$.

When $\tilde{f}$ has the form of Eq. (2.21.7) with $f \in C^\infty(\mathbf{T}^d(L_1, \ldots, L_d))$, the same happens for the partial derivatives of $\tilde{f}$ since the derivatives of a

C^∞-multiperiodic function are still multiperiodic; i.e., given d nonnegative integers $n_1, \ldots, n_d$, there is $\varphi_{n_1,\ldots,n_d} \in C^\infty(\mathbf{T}^d(L_1, \ldots, L_d))$ such that

$$\frac{\partial^{n_1+\cdots+n_d}\tilde{f}}{\partial\xi_1^{n_1}\ldots\partial\xi_d^{nd}}(\boldsymbol{\xi}) = \varphi_{n_1,\ldots,n_d}(\{\boldsymbol{\xi}\}). \tag{2.21.8}$$

and, obviously, we shall set

$$\frac{\partial^{n_1+\cdots+n_d}f}{\partial\xi_1^{n_1}\ldots\partial\xi_d^{nd}}(\{\boldsymbol{\xi}\}) = \varphi_{n_1,\ldots,n_d}(\{\boldsymbol{\xi}\}). \tag{2.21.9}$$

Depending on the circumstances, we shall choose to think or not to think of a C^∞-multiperiodic function with periods $L_1, \ldots, L_d$ and its partial derivatives as an element of $C^\infty(\mathbf{T}^d(L_1, \ldots, L_d))$.

Observations.

(1) Another natural definition of $C^\infty(\mathbf{T}^d)$, for $L_1 = \cdots = L_d = 2\pi$, could be related to observation (2) to Definition 12: one could say that $f \in C^\infty(\mathbf{T}^d)$ if $f(\boldsymbol{\varphi}) = F(\mathbf{z})$, where F is a C^∞ function on C^d[14] and $\mathbf{z}$ is given by Eq. (2.21.6). This would in fact be an equivalent definition, as could be shown; see Problems (6)–(10) at the end of this section.

(2) In an obvious way, after Definition 13, one can define the classes $C^\infty(V \times \mathbf{T}^d)$, where V is an open set in $\mathbf{R}^q$, and the derivatives of their elements. One can also define $W \times T^l$-valued functions in $C^\infty(V \times \mathbf{T}^d)$, and their derivatives, as the $C^\infty(V \times \mathbf{T}^d)$ functions with values in $\mathbf{R}^s \times \mathbf{R}^l$ whose last l components are thought of as angular coordinates on $\mathbf{T}^l$ (for $W \subset \mathbf{R}^s$ and $V \subset \mathbf{R}^q$ open sets).

(3) An obvious example of a multiperiodic function on $\mathbf{R}^d$ with pulsations $\omega_1, \ldots, \omega_d$ is given by the sum of the series

$$f(\xi_1, \ldots, \xi_d) = \sum_{n_1,\ldots,n_d}^{n_j \in \mathbf{Z}} c_{n_1,\ldots,n_d} e^{i(n_1\omega_1\xi_1 + \cdots + n_d\omega_d\xi_d)}; \tag{2.21.10}$$

when the coefficients $c_{n_1,\ldots,n_d} \in \mathbf{C}$ verify the relation

$$c_{n_1,\ldots,n_d} = \bar{c}_{-n_1,\ldots,-n_d} \tag{2.21.11}$$

and $\forall s = 0, 1, \ldots,$ there exists $\gamma_s > 0$ such that

$$(1 + |n_1|)^s \ldots (1 + |n_d|)^s |c_{n_1,\ldots,n_d}| \leqslant \gamma_s. \tag{2.21.12}$$

The partial derivatives of such an f, in the sense of Definition 12, can be computed by series differentiation, as a result of Eq. (2.21.12).

It is important to realize that, vice versa, Eqs. (2.21.10), (2.21.11), and (2.21.12) provide the most general example. This is essentially the content of the following proposition ("multidimensional Fourier theorem").

[14] A C^∞ function on $\mathbf{C}^d$ is a function on $\mathbf{C}^d$ which is C^∞ in the real and imaginary parts of the coordinates.

28 Proposition. *If f is a C^∞-multiperiodic function on $\mathbf{R}^d$ with periods $L_1, \ldots, L_d > 0$, then it is possible to represent f by formula (2.21.10) with coefficients $c_{n_1,\ldots,n_d}$ verifying Eqs. (2.21.11) and (2.21.12) and given by*

$$c_{n_1,\ldots,n_d} = \int_0^{L_1} \frac{d\xi_1}{L_1} \int_0^{L_2} \frac{d\xi_2}{L_2} \cdots \times \int_0^{L_d} \frac{d\xi_d}{L_d} e^{-i(\omega_1 n_1 \xi_1 + \cdots + \omega_d n_d \xi_d)} f(\xi_1, \ldots, \xi_d), \tag{2.21.13}$$

where $\omega_i = 2\pi / L_i$, $i = 1, \ldots, d$.

Observations.

(1) If $d = 1$, this proposition coincides with the Fourier development theorem for periodic functions (see Proposition 19).

(2) Since $\tilde{f}(\xi_1, \ldots, \xi_d) = f(\xi_1/\omega_1, \ldots, \xi_d/\omega_d)$ is multiperiodic with periods 2π, it will suffice to prove the above proposition when $\omega_1 = \cdots = \omega_d = 1$.

PROOF (Case $\omega_1 = \cdots = \omega_d = 1$). The proof of this proposition can be developed by induction. Since we are already aware of its truth for $d = 1$ (see Proposition 19, §2.12), let us assume its validity for $d = 1, 2, \ldots, k$ and then consider the case $d = k + 1$.

Let $f \in C^\infty(\mathbf{T}^{k+1})$ and contemplate the function $\psi_{\xi_{k+1}}$ parametrized by $\xi_{k+1} \in \mathbf{R}$ and defined on $\mathbf{R}^k$:

$$\psi_{\xi_{k+1}}(\xi_1, \ldots, \xi_d) = f(\xi_1, \ldots, \xi_d, \xi_{k+1}). \tag{2.21.14}$$

Such a function, $\forall \xi_{k+1} \in \mathbf{R}$, is a C^∞-2π-multiperiodic function on $\mathbf{R}^k$.

The inductive hypothesis implies

$$f(\xi_1, \ldots, \xi_k, \xi_{k+1}) = \sum_{(n_1, \ldots, n_k) \in \mathbf{Z}^k} \hat{\psi}_{n_1,\ldots,n_k}(\xi_{k+1}) e^{i(n_1\xi_1 + \cdots + n_k\xi_k)} \tag{2.21.15}$$

with

$$\hat{\psi}_{n_1,\ldots,n_k}(\xi_{k+1}) = \int_0^{2\pi} \cdots \int_0^{2\pi} \frac{d\xi_1 \ldots d\xi_k}{(2\pi)^k} f(\xi_1, \ldots, \xi_k, \xi_{k+1}) e^{-i(n_1\xi_1 + \cdots + n_k\xi_k)}. \tag{2.21.16}$$

On the other hand, Eq. (2.21.16) immediately implies that $\hat{\psi}_{n_1,\ldots,n_k}(\xi_{k+1})$ is a C^∞ function, periodic with period 2π, of ξ_{k+1} for all choices of $(n_1, \ldots, n_k) \in \mathbf{Z}^k$. We can then apply the Fourier theorem for $d = 1$ to infer

$$\hat{\psi}_{n_1,\ldots,n_d}(\xi_{k+1}) = \sum_{n_{k+1} \in \mathbf{Z}} e^{in_{k+1}\xi_{k+1}} \int_0^{2\pi} \hat{\psi}_{n_1,\ldots,n_d}(\xi') e^{-in_{k+1}\xi'} \frac{d\xi'}{2\pi}, \tag{2.21.17}$$

i.e., using Eq. (2.21.13) as defined of $c_{n_1,\ldots,n_{k+1}}$ and inserting Eq. (2.21.16) in the right-hand side of Eq. (2.21.17), we find

$$\hat{\psi}_{n_1,\ldots,n_k}(\xi_{k+1}) = \sum_{n_{k+1}\in\mathbf{Z}} c_{n_1,\ldots,n_k,n_{k+1}} e^{in_{k+1}\xi_{k+1}}. \tag{2.21.18}$$

Substituting Eq. (2.21.18) into Eq. (2.21.15), one obtains Eq. (2.21.10), provided Eq. (2.21.12) holds (which implies that the series on n_{k+1} and on $n_1, \ldots, n_k$ can be unconditionally summed and interchanged since they are absolutely convergent).

It is therefore necessary to check Eq. (2.21.12) in order to complete the proof. In fact, Eq. (2.21.11) is evident from Eq. (2.21.13) which has now become, temporarily, a definition of $c_{n_1,\ldots,n_{k+1}}$. One can proceed as in the analogous situation met in the one-dimensional case: one integrates Eq. (2.21.13) by parts. By integrating σ times with respect to ξ_j by parts, we find, if $n_j \neq 0$:

$$c_{n_1,\ldots,n_{k+1}} = \frac{1}{(in_j)^\sigma} \int_0^{2\pi} \cdots \int_0^{2\pi} \frac{d\xi_1 \ldots d\xi_{k+1}}{(2\pi)^{k+1}} \frac{\partial^\sigma f}{\partial \xi_j^\sigma}(\xi_1, \ldots, \xi_{k+1}) e^{-\sum_{r=1}^{k+1} in_r\xi_r} \tag{2.21.19}$$

and, if $F'_\sigma = \max_{\xi, j} |(\partial^\sigma f/\partial\xi_j^\sigma)(\boldsymbol{\xi})|$, this yields

$$|c_{n_1,\ldots,n_{k+1}}| \leqslant \frac{F'_\sigma}{|n_j|^\sigma}. \tag{2.21.20}$$

Since it is obvious from Eq. (2.21.13) that, $\forall n_1, \ldots, n_{k+1}$ (zero or not), $c_{n_1,\ldots,n_{k+1}}$ is bounded by the maximum F'_0 of $|f|$, Eq. (2.21.20) implies the existence of some $F_\sigma > 0$ such that for $\forall j = 1, \ldots, k+1$, $\forall \sigma \in \mathbf{Z}_+$, $\forall (n_1, \ldots, n_{k+1}) \in \mathbf{Z}^{k+1}$:

$$|c_{n_1,\ldots,n_{k+1}}| \leqslant \frac{F_\sigma}{(1+|n_j|)^\sigma} \tag{2.21.21}$$

(take, for instance, $F_\sigma = F'_0 + F'_\sigma$). Hence, multiplying Eq. (2.21.21) on j as j varies between 1 and $k+1$ and then taking the $(k+1)$th root, side by side, of the result, we find

$$|c_{n_1,\ldots,n_{k+1}}| \leqslant F_\sigma\big[(1+|n_1|) \ldots (1+|n_{k+1}|)\big]^{-(\sigma/k+1)}, \tag{2.21.22}$$

implying Eq. (2.21.12) by the arbitrariness of $\sigma \geqslant 0$. mbe

It is useful to explicitly state the following obvious corollary of Proposition 28 and Definition 11.

29 Corollary. *If f is a C^∞-quasi-periodic function with pulsations $\omega_1, \ldots, \omega_d > 0$, then it admits a representation of the type*

$$f(t) = \sum_{\mathbf{n}\in\mathbf{Z}^d} c_{\mathbf{n}} e^{i\mathbf{n}\cdot\omega t}, \tag{2.21.23}$$

where $\omega = (\omega_1, \ldots, \omega_d)$, $\mathbf{n} = (n_1, \ldots, n_d)$, *and the constants* $c_{\mathbf{n}} = c_{n_1, \ldots, n_d}$ *verify Eqs.* (2.21.11) *and* (2.21.12).

It is remarkable that in some cases, *given* $\omega_1, \ldots, \omega_d$, the representation (2.21.23) is unique.

30 Proposition. *Let* $f \in C^\infty(\mathbf{R})$ *be quasi-periodic with pulsations* $\omega_1, \ldots, \omega_d > 0$. *If the pulsations are rationally independent,*[15] *the coefficients of the representation* (2.21.23) *are given by*

$$c_{\mathbf{n}} = \lim_{t \to +\infty} t^{-1} \int_0^t e^{-i\mathbf{n}\cdot\omega\tau} f(\tau)\, d\tau \tag{2.21.24}$$

and, therefore, the representation (2.21.23) *is unique, given* $\omega = (\omega_1, \ldots, \omega_d)$.

Proof. Taking into account the decay properties of $c_{\mathbf{n}}$ as $\mathbf{n} \to \infty$ expressed by Eq. (2.21.12), the integral in Eq. (2.21.24) can be computed by the series from Eq. (2.21.23):

$$t^{-1} \int_0^t e^{-i\mathbf{n}\cdot\omega\tau} f(\tau)\, d\tau = \sum_{\mathbf{m} \in \mathbf{Z}^d} c_{\mathbf{m}} t^{-1} \int_0^t e^{-i(\mathbf{n}-\mathbf{m})\cdot\omega\tau}\, d\tau. \tag{2.21.25}$$

The right-hand integral divided by t has a modulus $\leqslant 1$ (as an average of a function with modulus 1). Therefore, Eq. (2.21.12) shows that the series in Eq. (2.21.25) is a series uniformly convergent with respect to t and that the limit of Eq. (2.21.24) can be computed in Eq. (2.21.25) by interchanging it with the series. If $\mathbf{n} \neq \mathbf{m}$, the integral in Eq. (2.21.25) is

$$\frac{1}{t} \frac{\exp i(\mathbf{n}-\mathbf{m})\cdot\omega t - 1}{i(\mathbf{n}-\mathbf{m})\cdot\omega} \underset{t\to\infty}{\to} 0 \tag{2.21.26}$$

because $(\mathbf{n}-\mathbf{m})\cdot\omega \neq 0$ by the rational independence assumption on ω.

Hence all the terms in Eq. (2.21.25) with $\mathbf{n} \neq \mathbf{m}$ do not contribute to the limit, as $t \to +\infty$, of Eq. (2.21.15) itself. The term with $\mathbf{n} = \mathbf{m}$, on the other hand, only contributes $c_{\mathbf{n}}$; hence, the proposition is proved. mbe

For the sake of completeness, we also wonder about what can be said in the other cases when $\omega_1, \ldots, \omega_d$ are rationally dependent. As an example, we discuss the following proposition.

31 Proposition. *Let* $f \in C^\infty(\mathbf{R})$ *be quasi-periodic with rationally dependent pulsations* $\omega_1, \ldots, \omega_d > 0$. *There exist* $p < d$ *and* p *rationally independent numbers* $\tilde{\omega}_1, \ldots, \tilde{\omega}_p$ *and a multiperiodic function* $\tilde{\varphi} \in C^\infty(\mathbf{T}^p)$ *such that*

$$f(t) = \tilde{\varphi}(\tilde{\omega}_1 t, \ldots, \tilde{\omega}_p t), \qquad \forall t \in \mathbf{R}. \tag{2.21.27}$$

[15] A family $\Omega = (\omega_1, \omega_2, \ldots)$ of real numbers is said to consist of rationally independent numbers when every finite subset $(\omega_{j_1}, \ldots, \omega_{j_p})$ has the property that the relation $\sum_{k=1}^{p} n_k \omega_{j_k} = 0$, with $n_1, \ldots, n_p$ integers, implies $n_1 = \cdots = n_p = 0$.

Observation. Therefore, if $\omega_1, \ldots, \omega_d$ are rationally dependent, it is possible to reduce the complexity of the representation (2.21.23) by reducing the dimension of the multiple series appearing in it.

PROOF. Consider all the subsets of $\omega_1, \ldots, \omega_d$ built with rationally independent numbers and let $\{\bar\omega_1, \ldots, \bar\omega_p\}$ be a maximal one among them (i.e., such that $\{\bar\omega_1, \ldots, \bar\omega_p, \omega'\}$ is not built with rationally independent numbers no matter which $\omega' \in \{\omega_1, \ldots, \omega_d\}$ is chosen).

Without loss of generality, we suppose $\omega_1 = \bar\omega_1, \ldots, \omega_p = \bar\omega_p$. It is then clear that for every $j = p+1, \ldots, d$, there are p rational numbers $\Gamma_1^{(j)}, \ldots, \Gamma_p^{(j)}$, all with the same denominator N, as we may and shall assume, such that

$$\omega_j = \sum_{k=1}^{p} \Gamma_k^{(j)} \omega_k \qquad j = p+1, \ldots, d. \tag{2.21.28}$$

Hence, setting $\Gamma_k^{(j)} = m_k^{(j)}/N$, $m_k^{(j)}$ integer, $j = p+1, \ldots, d$, $k = 1, \ldots, p$, and $\tilde\omega_j = \omega_j/N$, we see that

$$\omega_j = \sum_{k=1}^{p} m_k^{(j)} \tilde\omega_k \qquad j = 1, \ldots, d, \tag{2.21.29}$$

defining $m_k^{(j)} = N\delta_{jk}$ for $j \leqslant p$.

We can now use Proposition 29:

$$\begin{aligned} f(t) &= \sum_{n_1, \ldots, n_d} c_{n_1, \ldots, n_d} e^{i\sum_{h=1}^{d} n_h \omega_h t} \\ &= \sum_{n_1, \ldots, n_d} c_{n_1, \ldots, n_d} e^{i\sum_{h=1}^{d} n_h (\sum_{k=1}^{p} m_k^{(h)} \tilde\omega_k) t} \\ &= \sum_{n_1, \ldots, n_d} c_{n_1, \ldots, n_d} e^{i\sum_{k=1}^{p} \tilde\omega_k (\sum_{h=1}^{d} m_k^{(h)} n_h) t} \\ &= \sum_{q_1, \ldots, q_p} \Bigg(\sum_{\substack{n_1, \ldots, n_d \\ \sum_{h=1}^{d} m_k^{(h)} n_h = q_k}} c_{n_1, \ldots, n_d} \Bigg) e^{i\sum_{k=1}^{p} \tilde\omega_k q_k t}, \end{aligned} \tag{2.21.30}$$

Therefore, we set

$$\tilde c_{q_1, \ldots, q_p} = \sum_{\substack{n_1, \ldots, n_d \\ \sum_{h=1}^{d} m_i^{(h)} n_h = q_k}} c_{n_1, \ldots, n_d}, \tag{2.21.31}$$

and we easily see, from Eq. (2.21.11), that $\tilde c_{q_1, \ldots, q_p} = \bar{\tilde c}_{-q_1, \ldots, -q_p}$. Furthermore, since $|q_j| \leqslant M(\sum_{k=1}^{d} |n_k|)$ with $M = \max_{k,j} |m_j^{(k)}| \geqslant 1$, we see easily that

$$\begin{aligned} &(1 + |q_1|)^s \ldots (1 + |q_p|)^s |\tilde c_{q_1, \ldots, q_p}| \\ &\leqslant M^{ps} \sum_{\substack{n_1, \ldots, n_d \\ \sum_{h=1}^{d} m_k^{(h)} n_h = q_k}} (1 + |n_1|)^{sp} \ldots (1 + |n_d|)^{sp} |c_{n_1, \ldots, n_d}|. \end{aligned} \tag{2.21.32}$$

The series on the right-hand side of Eq. (2.21.32) can be bounded with the help of Eq. (2.21.12) as

$$\sum_{n_1,\ldots,n_d\in\mathbf{Z}^d}(1+|n_1|)^{sp}\ldots(1+|n_p|)^{sp}\frac{\gamma_{sp+2}}{(1+|n_1|)^{sp+2}\ldots(1+|n_d|)^{sp+2}} \tag{2.21.33}$$

$$\leqslant \gamma_{sp+2}\left(\sum_{n=-\infty}^{+\infty}(1+|n|)^{-2}\right)^d$$

Hence, Eqs. (2.21.32) and (2.21.33) mean that the constants $\tilde{c}_{q_1,\ldots,q_p}$ verify an inequality like Eq. (2.21.12) (with p instead of d) and the proposition is now proved since, by Eq. (2.21.30), we can define

$$\tilde{\varphi}(\xi_1,\ldots,\xi_p)=\sum_{q_1,\ldots,q_p}\tilde{c}_{q_1,\ldots,q_p}e^{i\Sigma_{j=1}^{p}\tilde{\omega}_j q_j \xi_j t}. \tag{2.21.34}$$

mbe

Exercise and Problems for §2.21

1. Compute the Fourier coefficients $\hat{f}_{0,0}$, $\hat{f}_{0,1}$, $\hat{f}_{1,0}$ of the function $f(\xi_1,\xi_2) = (1-\frac{1}{4}(\cos\xi_1+\cos\xi_2))$ with an approximation of 50%.

2. Same as Problem 1 for $f(\xi_1,\xi_2)=-\log(1-\frac{1}{4}(\cos\xi_1+\cos\xi_2))$.

3. Show that if $f(\xi_1,\xi_2)=\Sigma_{k=0}^{\infty}4^{-k}C_k(\cos\xi_1+\cos\xi_2)^k$ with $|C_k|\leqslant D$, there exist $C>0$, $\epsilon>0$ such that $|\hat{f}_{n_1,n_2}|\leqslant C\exp-\epsilon(|n_1|+|n_2|)$. Estimate C and ϵ in terms of D.

4. If ω_1/ω_2 is irrational, show that, for $\forall\varphi\in C^\infty(\mathbf{T}^2)$, the closure of the set of the values taken as $t\in\mathbf{R}_+$, by $f(t)=\varphi(\omega_1 t,\omega_2 t)$ coincides with $\varphi(\mathbf{T}^2)=\varphi$-image of $\mathbf{T}^2$ (Hint: See Proposition 27.)

5.* Same as Problem 4 when $\varphi\in C^\infty(\mathbf{T}^d)$, $f(t)=\varphi(\omega_1 t,\ldots,\omega_d t)$ and $\omega_1,\ldots,\omega_d$ are rationally independent.

6. On the complex plane $\mathbf{C}/\{0\}$, define the function $I(z)=e^{i\varphi}$ if $z=\rho e^{i\varphi}\neq 0$. Show that I is a C^∞ function of $\operatorname{Re}z=x$ and $\operatorname{Im}z=y$.

7. In the context of Problem 6, show that

$$\left|\frac{\partial^k I(z^n)}{\partial x^h\partial y^{k-h}}\right|\leqslant|n|^k C_k,\qquad \forall n\in\mathbf{Z},\quad \forall k\in\mathbf{Z}_+$$

for a suitably chosen C_k, for all z such that $\frac{1}{2}<|z|<2$.

8. If $f\in C^\infty(\mathbf{R})$ and f is 2π periodic and if $\hat{f}_n$ are the Fourier coefficients of f, consider the function of $z=x+iy$, $x,y\in\mathbf{R}$, defined for $\frac{1}{2}<|z|<2$ by

$$F(z)=\sum_{n=-\infty}^{+\infty}\hat{f}_n I(z^n).$$

Using Problem 7, show that this function F, as a function of x, y, is C^∞ in the region $\frac{1}{2}<z<2$ and on the unit circle coincides with f:

$$f(\varphi)=F(e^{i\varphi}).$$

9. Using Problem 8, show the validity of the equivalence claimed in observation (1) to Definition 13, p. 102, in the case $d = 1$.

10. Same as Problem 9 in the case $d > 1$. (Hint: If $f \in C^\infty(\mathbf{T}^d)$ and if $\hat{f}_{n_1, \ldots, n_d}$ are its Fourier coefficients, let $\mathbf{z} = (z_1, \ldots, z_d) \in \mathbf{C}^d$ and

$$F(\mathbf{z}) = \sum_{n_1, \ldots, n_d \in Z^d} \hat{f}_{n_1, \ldots, n_d} I(z_1^{n_1}) \ldots I(z_d^{n_d});$$

then show that F is a C^∞ function of $x_i = \operatorname{Re} z_i$, $y_i = \operatorname{Im} z_i$, $i = 1, \ldots, d$, in a neighborhood of the torus $S \times \cdots \times S$, where $S = \{\text{unit circle in } \mathbf{C}\}$ identified with $\mathbf{T}^d$ via Eq. (2.21.6).)

§2.22. Observables and Their Time Averages

To understand the ideas of observables and their time averages, consider an autonomous differential equation

$$m\ddot{x} = f(\dot{x}, x), \tag{2.22.1}$$

where $(\eta, \xi) \to f(\eta, \xi)$ is in $C^\infty(\mathbf{R}^2)$ and $m > 0$.

Suppose, also, that Eq. (2.22.1) is normal. According to Definition 7, we shall denote by $(S_t)_{t \in \mathbf{R}_+}$ the flow which solves Eq. (2.22.1); i.e., if $(\eta, \xi) \in \mathbf{R}^2$, the function

$$t \to (\dot{x}(t), x(t)) = S_t(\eta, \xi), \qquad t \in \mathbf{R}_+ \tag{2.22.2}$$

will be such that $t \to x(t)$, $t \in \mathbf{R}_+$, is a solution of Eq. (2.22.1) with initial datum (η, ξ). Recall, also, that the map $(t, \eta, \xi) \to S_t(\eta, \xi)$, defined on $\mathbf{R}_+ \times \mathbf{R} \times \mathbf{R}$ and with values in $\mathbf{R} \times \mathbf{R}$, is a C^∞ map and

$$S_{t+t'} = S_t S_{t'}, \qquad \forall t, t' \in \mathbf{R}_+ ; \tag{2.22.3}$$

see Corollary 9.

In the above context, we introduce the following concepts.

14 Definition. The set of C^∞ functions on $\mathbf{R}^2$, thought of as the space of the initial data of Eq. (2.22.1), will be called the set of instantaneous "observables" for the point mass described by Eq. (2.22.1).

If $t \to S_t(\eta, \xi)$, $t \geqslant 0$, is a motion of Eq. (2.22.1) and if F is an observable, we shall define the observable's "value" at time $t \in \mathbf{R}_+$ on the motion with initial datum (η, ξ) as

$$F(\dot{x}(t), x(t)) = F(S_t(\eta, \xi)), \tag{2.22.4}$$

and the function $t \to F(S_t(\eta, \xi))$, $t \in \mathbf{R}_+$, is the "history" of the observable F on the motion departing from (η, ξ).

Observations.

(1) The reason for the above terminology is clear. What perhaps needs a few words of comment is why one defines an observable as a function of

velocity and position only, see Eq. (2.22.4), rather than, more generally, as a function of acceleration and higher derivatives as well.

Actually, such a definition would not be more general since, via Eq. (2.22.1) and by what it has been observed in § 2.4, it is possible to compute all the derivatives of $t \to x(t)$ successive to the first by repeatedly differentiating both sides of Eq. (2.22.1), once $x(t)$ and $\dot{x}(t)$ are known.

(2) Therefore, the observables correspond to the physical entities which can be measured by observing the velocity and the position of the point mass at a given instant: they are a mathematical model of such entities.

Given an observable F and a motion $t \to S_t(\eta, \xi)$, $t \in \mathbf{R}_+$, one can raise several questions about the results of the observations of F at various times. As an example, we shall examine the notion of the average value of an observable on a given motion.

It is important to remember and to stress that, concerning the notion of the average value of an observable, it is possible to repeat what has already been said about the notion of the stability of equilibrium. It makes no sense to provide an absolute definition of average value of an observable as time elapses. In fact, it is possible to give several meanings to this concept, each corresponding to different needs that may naturally emerge in applications.

In this and in the following sections, we shall discuss only a few interesting examples of the definition of the time averages, leaving it to the reader to imagine in which applications a definition may appear as a relevant one. The reader should also try to imagine other definitions and the corresponding situations to which they naturally apply. The methods explained below could then be used to elucidate their properties.

15 Definition. Let $F \in C^{\infty}(\mathbf{R}^2)$ be an observable for the motions described by (2.22.1) and let $T > 0$. We define the continuous average value of F on the motion with initial datum $(\eta, \xi) \in \mathbf{R}^2$ and on the time interval $[0, T]$ as the quantity

$$M_T(F; \eta, \xi) = \frac{1}{T} \int_0^T F(S_t(\eta, \xi))\, dt. \tag{2.22.5}$$

The "continuous average value" of F on the motion with initial datum (η, ξ) will be, whenever it exists, the limit

$$\bar{F}(\eta, \xi) = \lim_{T \to +\infty} M_T(F; \eta, \xi). \tag{2.22.6}$$

Similarly, one could define the average value with observation step $a > 0$:

16 Definition. If $F \in C^{\infty}(\mathbf{R}^2)$ is an observable for the motions described by Eq. (2.22.1) and if N is a positive integer, we define the discrete average value with observation step a of F on the motion, leaving the initial datum

(η, ξ) and relative to N observations, as the quantity

$$M_N^{(a)}(F;\eta,\xi) = \frac{1}{N}\sum_{j=0}^{N-1} F(S_{ja}(\eta,\xi)). \tag{2.22.7}$$

The "discrete average value" with step a of F on the motion $t \to S_t(\eta, \xi)$, $t \in \mathbf{R}_+$, is defined by the limit

$$\bar{F}^{(a)}(\eta,\xi) = \lim_{N\to\infty} M_N^{(a)}(F;\eta,\xi) \tag{2.22.8}$$

whenever it exists.

Why should one refrain from considering the following more general notion?

17 Definition. If $\varphi \in C^\infty(\mathbf{R}_+)$ and if $T > 0$, $N \in \mathbf{Z}_+$, $a > 0$ one sets

$$\begin{aligned} \mathfrak{M}_T(\varphi) &= \frac{1}{T}\int_0^T \varphi(t)\,dt, \\ \mathfrak{M}_N^{(a)}(\varphi) &= \frac{1}{N}\sum_{j=0}^{N-1}\varphi(ja), \\ \bar{\varphi} &= \lim_{T\to\infty}\mathfrak{M}_T(\varphi), \\ \bar{\varphi}^{(a)} &= \lim_{N\to\infty}\mathfrak{M}_N^{(a)}(\varphi) \end{aligned} \tag{2.22.9}$$

whenever the limits exist.

From the top down, the quantities defined in Eq. (2.22.9) will be called the "continous average of φ on $[0, T]$", the "discrete average of φ on N observations with step a", the "continous average of φ", and the "discrete average of φ with step a".

Observations.

(1) If φ is constant, $\bar{\varphi} \equiv \bar{\varphi}^{(a)} \equiv \varphi$.

(2) If $\lambda = \lim_{t\to+\infty}\varphi(t)$ exists, then $\bar{\varphi} = \bar{\varphi}^{(a)} = \lambda$: in fact, note that $\mathfrak{M}_T(\varphi) - \lambda = \mathfrak{M}_T(\varphi - \lambda)$ and if T_ϵ is such that, $\forall t \geqslant T_\epsilon$, $|\varphi(t) - \lambda| < \epsilon$, one has

$$\mathfrak{M}_T(\varphi - \lambda) = \frac{1}{T}\int_0^{T_\epsilon}(\varphi(\tau) - \lambda)\,d\tau + \frac{1}{T}\int_{T_\epsilon}^{T}(\varphi(\tau) - \lambda)\,d\tau \tag{2.22.10}$$

and the first term in the right-hand side of Eq. (2.22.10) goes to zero as $T \to \infty$, while the second is bounded by $T^{-1}(T - T_\epsilon)\epsilon < \epsilon$. Hence, $\lim_{T\to\infty}\mathfrak{M}_T(\varphi - \lambda) = 0$ by the arbitrariness of ϵ, and $\bar{\varphi} = \lambda$. Similarly, one checks that $\bar{\varphi}^{(a)} = \lambda$.

(3) If $\varphi \in C^\infty(\mathbf{R})$ is periodic with period $T_\varphi > 0$,

$$\lim_{T\to\infty}\mathfrak{M}_T(\varphi) = \bar{\varphi} = \frac{1}{T_\varphi}\int_0^{T_\varphi}\varphi(\tau)\,d\tau. \tag{2.22.11}$$

In fact, if we write $T = nT_\varphi + \theta$ with n integer and $\theta \in [0, T_\varphi]$, it follows that $T\to\infty \Leftrightarrow n\to\infty$ and

$$\mathfrak{M}_T(\varphi)=\frac{1}{nT_\varphi+\theta}\left(n\int_0^{T_\varphi}\varphi(\tau)\,d\tau+\int_0^{\theta}\varphi(\tau)\,d\tau\right), \tag{2.22.12}$$

evidently implying Eq. (2.22.11).

(4) If $\varphi\in C^\infty(\mathbf{R})$ is a periodic function with period $T_\varphi>0$ and if $a>0$ is such that $T_\varphi/a=p/q$ with p and q relatively prime integers (i.e., if T_φ/a is rational), it follows that

$$\begin{aligned}\bar{\varphi}^{(a)}&=\sum_{m=-\infty}^{+\infty}\hat{\varphi}_{mp},\\ \bar{\varphi}^{(a)}&=\frac{1}{p}\sum_{j=0}^{p-1}\varphi(ja)\end{aligned} \tag{2.22.13}$$

where $\hat{\varphi}_n$ are the harmonics of φ relative to the period T_φ. The first relation in Eq. (2.22.13) can be proved as in (3) above. To prove the second, note that

$$\begin{aligned}\mathfrak{M}_N^{(a)}(\varphi)&=\frac{1}{N}\sum_{j=0}^{N-1}\varphi(ja)=\frac{1}{N}\sum_{j=0}^{N-1}\sum_{n\in\mathbf{Z}}\hat{\varphi}_n e^{(2\pi i n/T_\varphi)ja}\\ &=\sum_{n\in\mathbf{Z}}\hat{\varphi}_n\left(\frac{1}{N}\sum_{j=0}^{N-1}\exp\frac{2\pi i n}{T_\varphi}ja\right)\end{aligned} \tag{2.22.14}$$

and the term in brackets has modulus $\leqslant 1$ (as an average of numbers with modulus not exceeding 1). Hence, the series in Eq. (2.22.14) is uniformly convergent in N and the limit as $N\to\infty$ can be taken term by term. As already remarked elsewhere, if $\exp(2\pi i n a/T_\varphi)\neq 1$, one finds

$$\frac{1}{N}\sum_{j=0}^{N-1}\exp\frac{2\pi i n a}{T_\varphi}j=\frac{1}{N}\,\frac{e^{(2\pi i n a/T_\varphi)N}-1}{e^{2\pi i n a/T_\varphi}-1}\xrightarrow[N\to\infty]{}0, \tag{2.22.15}$$

while if $\exp(2\pi i n a/T_\varphi)\equiv 1$, i.e., if na/T_φ is an integer (i.e., $n=mp$ for some $m\in\mathbf{Z}$), the sum (2.22.15) is clearly 1, identical to $\forall N$. Hence, by taking the limit as $N\to\infty$ in Eq. (2.22.14), Eq. (2.22.13) follows.

(5) If φ is periodic with period $T_\varphi>0$, $\varphi\in C^\infty(\mathbf{R})$, and if T_φ/a is irrational, then

$$\bar{\varphi}^{(a)}=\bar{\varphi}=\hat{\varphi}_0=\frac{1}{T_\varphi}\int_0^{T_\varphi}\varphi(\tau)\,d\tau. \tag{2.22.16}$$

This is true because, in the present case, in the series (2.22.14), all the terms tend to zero except the one with $n=0$ (as $\exp(2\pi i n a/T_\varphi)\neq 1$, $\forall n\neq 0$ [see, also, Eq. (2.22.15)]).

(6) If $\varphi\in C^\infty(\mathbf{R})$ is periodic with period $T_\varphi>0$, let $a>0$ vary so that T_φ/a is rational, but if $T_\varphi/a=p/q$, with p and q relatively prime integers, then $p\to\infty$.[16] Then it follows from Eq. (2.22.13) and from the decay

[16]The number p measures the number of times it is necessary to repeat a to reach a multiple of T_φ, i.e., it measures the "commensurability" of T_φ with respect to a.

properties of the Fourier coefficients of φ that $\bar{\varphi}^{(a)} \to \bar{\varphi}$. Hence, the "less T_φ is commensurable with a", the closer the discrete average $\bar{\varphi}^{(a)}$ is to the continous average $\bar{\varphi}$.

As a consequence of the above remarks and as an example of questions related to the above definitions, we formulate the following proposition.

32 Proposition. *Let $V \in C^\infty(\mathbf{R})$ be bounded below and consider the motions associated with Eq. (2.22.17):*

$$m\ddot{x}(t) = -\frac{\partial V}{\partial \xi}(x(t)), \qquad t \in \mathbf{R}_+ \tag{2.22.17}$$

If F is an observable "with bounded support" (i.e., if $F(\eta, \xi) \equiv 0$ when $|\eta| + |\xi|$ is large enough), every initial datum $(\eta, \xi) \in \mathbf{R}^2$ gives rise to a motion on which both the continuous and the discrete averages with step $a > 0$ are defined.

If $\lim_{\xi \to \pm\infty} V(\xi) = +\infty$, every observable (whether with bounded support or not) has well-defined average values, continuous and discrete. In the preceding cases, the continuous and discrete averages with step $a > 0$ coincide an all motions, with the possible exception of the periodic motions with period commensurable with a.

PROOF. From Proposition 11, p. 37, it follows that the motions described by Eq. (2.22.17) either approach infinity or tend toward a well-defined limit (i.e., $\lim_{t \to +\infty} S_t(\eta, \xi) = (0, x_0)$) or are periodic.

In the first two cases, the above proposition follows from Observation 2 to Definition 17, while in the third case, it follows from Observations 3 and 4. The assumption on the support of F is needed to deal with the case when $S_t(\eta, \xi) \to \infty$: this case cannot occur, according to the law of conservation of energy, when V diverges at infinity; hence, in this case, no restriction on F is necessary. mbe

Exercises and Problems for §2.22

1. Compute the continuous average along the motions $\ddot{x} + x = 0$, $x(0) = 0$, and $\dot{x}(0) = 1$ of the kinetic energy and of the squared elongation (i.e., of the observables $f(\eta, \xi) = \frac{1}{2}\eta^2$ or $g(\eta, \xi) = \xi^2$).

2. Compute the difference between the continuous average of the kinetic energy and that of the potential energy in the oscillations of $m\ddot{x} = -kx$ with energy E. Compute their values as functions of E.

3. Compute the discrete average of the kinetic energy for the motion $\ddot{x} + x = 0$, $x(0) = 0$, $\dot{x}(0) = 1$ for $a = 2\pi$, 4π, $\pi/2$, 1, 2, $\frac{17}{13}$.

4. Same as Problem 1 for the motion $\ddot{x} + \sin x = 0$, $x(0) = 0$, $\dot{x}(0) = \frac{1}{2}$ with 60% accuracy.

5. Same as Problem 3 for the motion in Problem 4 with 60% accuracy.

6. Same as Problems 4 and 5 with 1% accuracy (using a computer).

7. Same as Problem 1 for the motion in Problem 4. Estimate the accuracy needed in the computations to see a difference between the linear-oscillator and pendulum results.

8. Compute the average value of the elongation and of the square of the elongation in the motion $\ddot{x}+x=\cos \pi t$, $x(0)=0$, $\dot{x}(0)=0$ in the continuous case and in the discrete case with step $a=\pi, \frac{17}{13}, \sqrt{2}$.

9. Show that if $\varphi, \psi \in C^{\infty}(\mathbf{R})$ and $\lim_{t\to+\infty}|\varphi(t)-\psi(t)|=0$ and if φ has an average value of any type, then ψ has the same average value.

10. Apply Problem 9 to calculate the continuous average of the squared elongation in the motion of the oscillator $\ddot{x}+\dot{x}+x=\cos t$, $x(0)=0$, $\dot{x}(0)=0$. How does this average change by changing the initial datum? (Answer: It does not change.)

11. Arbitrarily choose a definition of average and estimate the average work done by the friction force ("dissipation per unit time", i.e., average of the observable $w(\eta,\xi)=-\eta^2$) in the motions of the oscillator in Problem 10.

12. In the context of Problem 11, compare the average work per unit time done by the friction force and that done by the forcing force. Interpret the value of their difference.

13. Compute, in general, the continuous average value of the work done by the forcing force and by the friction force in the motions of the oscillators $m\ddot{x}+\lambda\dot{x}+kx=f(t)$ with $m,\lambda,k>0$ and $f(t)=F\cos\omega t$, $F,\omega\in\mathbf{R}$. Also compute the continuous average value of the potential or kinetic energy.

14.* Same as Problem 13 but with a generic $2\pi/\omega$-periodic C^{∞} forcing force f. Express the results by means of the harmonics of f and of the parameters m,λ,k.

15. In the context of Problem 13, find the value of ω to which corresponds maximum average work done by the forcing term ("resonant pulsation").

16.* If $f\in C^{\infty}(\mathbf{R})$ is a quasi-periodic function in the sense of Definition 11, then the average values of f exist both in the continuous and the discrete sense. Find expressions for such quantities and show that if the pulsations of f are $\omega_1,\ldots,\omega_d$ and if $\{\omega_1,\ldots,\omega_d, 2\pi/a\}$ are $(d+1)$ rationally independent numbers, then the discrete average of f with step $a>0$ and the continuous average of f coincide. (Hint: Use the representation of Eq. (2.21.23) and proceed as in Observation 4, Eq. (2.22.14).)

17. Find an example of a function in $C^{\infty}(\mathbf{R})$ which does not have a continuous average.

18. Up to 60%, estimate the average kinetic energy in the motion with energy $E=10$ of the oscillator $\ddot{x}+x^3=0$.

19.* Same as Problem 18 with 1% accuracy (using a computer).

20.* Show that if a potential energy produces periodic motions with period $T(E)$ which, as E varies in $[E_0,E_1]$, is such that $T'(E)>0$, then the discrete average with step $a=1$ and the continuous average of an arbitrary observable coincide for a dense set of values of E in $[E_0,E_1]$, while they do not coincide, in general, on another dense set. The second set is, however, denumerable. (Hint: By the implicit functions theorem, deduce that $T(E)/a=T(E)$ is irrational for all but countably many values of $E\in[E_0,E_1]$).

21. Show that the same results of Problem 20 hold if $T(E)$ is strictly monotonically increasing in $[E_0,E_1]$. They also hold if $T'(E)=0$ only finitely many times in $[E_0,E_1]$.

§2.23. Time Averages on Sequences of Times Known up to Errors. Probability and Stochastic Phenomena

... or mi di' anche:
Questa Fortuna di che tu mi tocche,
Che è, che i ben del mondo ha sì tra branche?[17]

The continuous averages as well as the step-a discrete averages are, as is easily understood, very idealized mathematical notions, even when T or N are $< +\infty$. To be really measured, the continuous averages would demand an infinity of measurements of f, one per each time, and we do not need to underline the degree of abstraction that we must assume to imagine such a sequence of measurements.

Only at first sight are the discrete averages "more concrete" notions. It is in fact unthinkable to be able to perform measurements at time intervals exactly equal to a, because, of course, of the unavoidable errors of time measurement.

Obviously, the considerations of the measurement errors could have been brought up in correspondence with almost every question studied so far or it could be brought up in correspondence with any future questions. Arbitrarily, we decide to discuss it now in connection with the analysis of the averages of functions or observables.

The methods and ideas involved in the effort of making precise the notion of error in the time average computations present the greatest interest and are very general: they could be applied to the consideration of errors in the context of other problems, and the reader could try some of these applications by himself.

A very naive schematization of the process of data accumulation for the calculation of an average is the following: one measures[18] f, the function that we want to average, at the initial time $\tau_0 \simeq 0$; then we wait a time interval $\tau_1 \simeq a$ and repeat, again, the measurement of f, and subsequently the operation is repeated after waiting times $\tau_2, \tau_3, \ldots,$ etc: every τ_i, $i = 1, 2, \ldots,$ is approximately equal to a, though not exactly because of the errors made in the measurement of the time intervals. Afterwards, the average of f will be defined as the "average of the results thus obtained". Such an average, instead of being

$$\lim_{N \to \infty} \mathfrak{M}_N^{(a)}(f), \tag{2.23.1}$$

[17] In basic English:

... now tell me also:
This Fortune of whom you speak
What is she, that the world's goods keep so firmly in her hands?
(Dante, *Inferno*, Canto VII).

[18] For the sake of simplicity, we shall suppose that we are able to perform exact measurements so that the only source of error is that coming from the measurement of the time intervals.

will be

$$\lim_{N\to\infty} \frac{1}{N} \sum_{j=0}^{N-1} f(\tau_0 + \tau_1 + \cdots + \tau_j). \tag{2.23.2}$$

We can give a further idealization of the time measurement errors by imagining that

$$\tau_0 = \epsilon_0, \qquad \tau_i = a + \epsilon_i, \qquad i = 1, 2, \ldots, \tag{2.23.3}$$

and $\epsilon_j = \pm\epsilon$ with $\epsilon > 0$ fixed, $\epsilon \ll a$, and the sign of ϵ_j is "randomly chosen".

One can think of a simple mechanism producing a sequence of errors like those in Eq. (2.23.3). Assume that we are also able to perform perfect time measurements, but that we deliberately proceed as follows: at the initial time we toss a coin and perform a measurement of f at time $\tau_0 = \epsilon_0$, where $\epsilon_0 = \epsilon$ if the result is "heads", while $\epsilon_0 = -\epsilon$ if the result is "tails".

At time τ_0, we again toss the coin and perform the measurement of f at time $\tau_0 + \tau_1$, where $\tau_1 = a + \epsilon_1$ and $\epsilon_1 = \pm\epsilon$ according to the result[19], etc.

One can debate at length on which would be the best mathematical model allowing a satisfactory translation into mathematically clear terms of the just-described sequence of operations.

The most interesting mathematical scheme is based on the notion of probability.

18 Definition. Let $\mathcal{E}$ be a finite set of elements which we call "possible events". On $\mathcal{E}$, let it be defined as a function p which associates $p(e) \geqslant 0$ to $e \in \mathcal{E}$ such that

$$\sum_{e\in\mathcal{E}} p(e) = 1. \tag{2.23.4}$$

The pair $(\mathcal{E}, p)$ will be called a "probability distribution" on $\mathcal{E}$.

If $A \subset \mathcal{E}$ is a subset of $\mathcal{E}$, we set

$$p(A) = \sum_{e\in A} p(e) \tag{2.23.5}$$

and we say that $p(A)$ is the probability of A with respect to the distribution $(\mathcal{E}, p)$.

The above notion of probability is precise from a mathematical point of view, but its connection with reality is far less evident.

It is clear that the relation between this definition and the empirical world cannot be established on a deductive basis in the same way as it is not possible to establish deductively the relation between solutions to differential equations and motions of point masses.

The theory of the motion of a point mass, if identified with the theory of a class of differential equations, appears to us as natural only after long

[19]In other words, instead of leaving the "coin tossing" to the measurement instruments, we "do it ourselves".

practice and experience in comparing the relations between the mathematical model of a point mass and the corresponding empirical, i.e., experimental, properties of the "real" point masses. In this comparison, one refines both the mathematical intuition on the structure of the solutions of some differential equations and the physical intuition about the nature of motion.

Even a superficial knowledge of the theory of differential equations has the consequence that one cannot avoid observing, perhaps unconsciously, the motions more and more closely to unveil in them those properties which are suggested by their analytical interpretation through the model of the point mass as a differential equation.

Similarly, the notion of probability allows the formulation of mathematical models of stochastic phenomena and the quantitative evaluation of the probability of classes of events, reaching results such as "that class of events has large probability" or "probability $\frac{1}{7}$", etc. In terms of empirical interpretation, the meaning to attribute to such results becomes clearer and more refined while one proceeds in the applications, and this allows us to think of them again in more intuitive terms, more immediately expressible in an empirical language and in empirical prescriptions.

The key to the empirical interpretation of the notion of probability is the following: consider a "stochastic phenomenon" developing "following the aim of Her, which is as hidden as a snake in the grass",[20] which we imagine "reproducible" and whose possible events form a certain set $\mathcal{E}$. To say that a mathematical model for such a phenomenon is given by the probability distribution $(\mathcal{E}, p)$ means to formulate a law (on an empirical basis) stating that the number of times that in "n trials", or "repetitions of the event's production", the event $e \in \mathcal{E}$[21] will happen *about* $p(e)n$ times, if n is large, and the deviations from this value are very small, $\ll p(e)n$, except in particularly unlucky situations which can be disregarded "for all practical purposes".

One can wonder about what could be the predictive power of such a law. This power, in fact, is enormous when it is formulated a priori, i.e., without having first measured the occurrence frequencies of every event of $\mathcal{E}$ over a large number of "trials". The laws of dynamics have the same extent of power when they are applied to cases to which they are believed to be applicable, but for which the actual applicability has not been checked a priori and will be checked only a posteriori (think of the microscopic gas theory or of the planetary system theory).

Obviously sometimes a formulated law may be wrong, i.e., the distribution $(\mathcal{E}, p)$ may not be a good model of the stochastic phenomenon in the preceding sense. This may happen for two reasons.

[20] "Seguendo lo giudicio di costei/ che è occulto come in erba l'angue" (Dante, *Inferno*, Canto VII).

[21] $\mathcal{E}$ could be the six faces of a dice and a "try" could be one tossing of the dice (after suitable "shaking"); and the produced event would be the upper face of the dice after tossing, if the dice is "fair", $p(e) = \frac{1}{6}$.

First, the phenomenon may be stochastic but the empirical law on the existence of a well-defined frequency of realization of every possible event may not hold, in the limit of a large number of trials. In mechanics an analogous situation would occur if we discovered a point mass for which one could find, after a few direct measurements of force and corresponding acceleration, that these two physical entities are not proportional.

Alternatively, it might happen that the probability law $(\mathscr{E}, p)$, assumed as modelling the phenomenon under analysis, foresees occurrency frequencies different from the observed ones: this circumstance would have the analogue, in the mechanics of a point mass, of a case where we had "forgotten" to list some force f among the forces acting on a point mass.

Here let us end the discussion on the notion of probability and on its empirical interpretation. One could continue it for much longer at the risk of making the issue and the content of the analysis increasingly nebulous. In fact it is more useful and constructive to illustrate the content of Definition 18 via a few applications to the problems which interest us.

To have at hand a more flexible language, it is convenient to agree on a few more "simple" definitions. First comes the notion of "random variable".

19 Definition. Let $(\mathscr{E}, p)$ be a probability distribution and let f be a real function on $\mathscr{E}$.

(i) f will be called a "random variable".

(ii) If $a_1, a_2, \ldots, a_{n(f)}$ are pairwise distinct values taken by $f(e)$ as e varies in $\mathscr{E}$, we shall call $E_1, E_2, \ldots, E_{n(f)}$ the corresponding sets of events of $\mathscr{E}$; i.e., E_i consists of those elements $e \in \mathscr{E}$ such that $f(e) = a_i$, $i = 1, 2, \ldots, n(f)$. The sets $(E_1, \ldots, E_{n(f)})$ are pairwise disjoint and their union is $\mathscr{E}$. Therefore, they form a "partition" P_f of $\mathscr{E}$, which we call the "partition of $\mathscr{E}$ associated with f".

(iii) The "probability distribution" of the random variable f is the probability distribution (I_f, P_f), where I_f has as elements the $n(f)$ sets $E_1, \ldots, E_{n(f)}$ and

$$P_f(E_i) = p(E_i) = \sum_{e \in E_i} p(e). \tag{2.23.6}$$

(iv) More generally, if $\mathscr{P}$ is a partition of $\mathscr{E}$ into n sets $(E_1, \ldots, E_n)$, we shall define $(\mathscr{E}_{\mathscr{P}}, p_{\mathscr{P}})$ the "probability distribution associated with $\mathscr{P}$" as being the probability distribution in which the elements of $\mathscr{E}_{\mathscr{P}}$ are the sets constituting the partition $\mathscr{P}$ and, if $E \in \mathscr{P}$,

$$p_{\mathscr{P}}(E) = p(E) = \sum_{e \in E} p(e). \tag{2.23.7}$$

Observation. The notion of the probability distribution of a random variable is a relevant one when we are only interested in the random event $e \in \mathscr{E}$ via the value $f(e)$ that it implies for a given random variable f. It is in fact clear that we can identify all the events $e \in \mathcal{E}$ giving rise to the same value of $f(e)$ and call "event" such a collection of events.

Suppose, for instance, that we have to perform a measurement of a quantity g and that such a measurement is affected by an error which can be thought of as due to N "causes", all independent from each other and each producing an additive error on the value of g which is $\pm\epsilon$ with equal probability. A complete description of the error is therefore a N-tuple $\epsilon = (\epsilon_1, \ldots, \epsilon_N)$ of numbers which take the values $\epsilon_i = \pm 1$; the hypothesis of independence and equal probability of the various errors will be translated into a model by saying that all the N-tuples ϵ are equally probable; i.e., on the space $\mathcal{E}$ of the 2^N sequences $\epsilon = (\epsilon_1, \ldots, \epsilon_N)$, with $\epsilon_i = \pm\epsilon$, the probability distribution[22] $p(\epsilon) = 2^{-N}$ is defined.

Suppose, however, that we are not interested in knowing the details of the error's structure but just the total error:

$$f(\epsilon) = \sum_{i=1}^{N} \epsilon_i . \tag{2.23.8}$$

This is a random variable on $\mathcal{E}$. It can take the values $N\epsilon, (N-2)\epsilon, \ldots, (-N+2)\epsilon, -N\epsilon$, and the value $(N-2k)\epsilon$ is taken on all the sequences ϵ containing exactly k minus signs: call E_k the set of all such sequences.

Then the set I_f, in this example, consists of $N+1$ elements E_0, $E_1, \ldots, E_N$ and

$$p_f(E_i) = p(E_i) = \sum_{e \in E_i} \frac{1}{2^N} = \frac{1}{2^N}\binom{N}{i}. \tag{2.23.9}$$

We can obviously regard (I_f, P_f) as a model for the total error without explicit reference to the elementary errors.

The preceding definition gives us a method for building new probability distributions, starting from a given probability distribution. It is useful, in this respect, also to give the following definition providing another way of constructing new probability distributions starting from a given one $(\mathcal{E}, p)$, as suggested by the above observation.

20 Definition. Let $(\mathcal{E}, p)$ be a probability distribution. Let N be a positive integer. We shall denote $(\mathcal{E}, p)^N$ as the probability distribution on $\mathcal{E}^N$ associating with the event $\mathbf{e} = (e_1, \ldots, e_N) \in \mathcal{E}^N$ the probability $p^{(N)}(\mathbf{e})$:

$$p^{(N)}(\mathbf{e}) = p(e_1)p(e_2)\ldots p(e_N). \tag{2.23.10}$$

The distribution $(\mathcal{E}, p)^N$ will be called the "distribution of N events independently extracted with distribution $(\mathcal{E}, p)$".

Let us conclude this series of definitions, necessary to establish a concise and suggestive language for the formulation of some interesting proposi-

[22] This is a celebrated error model. It was used by Gauss for his mathematical theory of errors, one of the first grandiose applications of probability theory.

tions, by describing the important notion of a sequence of random variables converging in probability to a constant limit.

21 Definition. Let $(\mathcal{E}_N, p_N)$, $N = 1,2,\ldots,$ be a sequence of probability distributions and let f_N be a random variable defined on $\mathcal{E}_N$, $N = 1,2,\ldots$. We shall say that the sequence $(f_N)_{N=1}^{\infty}$ of random variables "converges in probability" to a limit $l \in \mathbf{R}$ as $N \to \infty$, if[23]

$$\lim_{N\to\infty} p_N(\{e \mid e \in \mathcal{E}_N, |f_N(e) - l| > \epsilon\}) = 0. \tag{2.23.11}$$

for all $\epsilon > 0$.

Let us provide some examples.

33 Proposition. *Let $(\mathcal{E}, p)$ be a probability distribution and let f be a random variable on $(\mathcal{E}, p)$. Define the random variable f_N on $(\mathcal{E}, p)^N$ as*

$$f_N(\mathbf{e}) = \sum_{i=1}^{N} \frac{f(e_i)}{N} \tag{2.23.12}$$

if $\mathbf{e} = (e_1, \ldots, e_N) \in \mathcal{E}^N$. Then the sequence f_N converges in probability to $\bar{f} = \Sigma_{e\in\mathcal{E}}\, p(e)f(e)$ as $N \to \infty$.

Observations.

(1) This proposition ("law of the large numbers") tells us that the average value of a sum of N independent random variables is "almost constant" if N is large or, better, that the probability that such an average value differs from a certain constant $\bar{f}$ by more than a given quantity ϵ approaches 0 as $N \to \infty$ [see Eq. (2.23.12)].

(2) This proposition clarifies why the quantity $\Sigma_{e\in\mathcal{E}}\, p(e)f(e)$ is called the "average value" of the random variable f with respect to the probability distribution $(\mathcal{E}, p)$.

The proof of Proposition 33 relies on a very elementary but very important inequality ("the Chebysčev inequality") which underlies many probabilistic estimates.

34 Proposition. *Let f be a random variable with respect to the probability distribution $(\mathcal{E}, p)$. Define the "kth moment" of f as*

$$\mu_k(f) = \sum_{e\in\mathcal{E}} |f(e)|^k p(e), \qquad k \in \mathbf{Z}_+ . \tag{2.23.13}$$

Then for $k \in \mathbf{Z}_+$ and $\delta > 0$,

$$p(\{\mathbf{e} \mid \mathbf{e} \in \mathcal{E}, |f(\mathbf{e})| > \delta\}) \leqslant \frac{\mu_k(f)}{\delta^k} . \tag{2.23.14}$$

[23] We use the convention that $\{\mathbf{e} \mid \mathbf{e} \in A, f(\mathbf{e}) \in B\}$ means "subset of A consisting in those e's such that $f(\mathbf{e}) \in B$".

PROOF. By Eq. (2.22.13),

$$\mu_k(f) \geq \sum_{\substack{e\in\mathscr{E}\\ |f(e)|>\delta}} |f(e)|^k p(e) \geq \delta^k \sum_{\substack{e\in\mathscr{E}\\ |f(e)|>\delta}} p(e)$$
$$= \delta^k p(\{e\,|\,e\in\mathscr{E}, |f(e)|>\delta\}). \tag{2.23.15}$$

mbe

PROOF OF PROPOSITION 33. By applying the Chebysčev inequality to the random variable $f_N - f$, one finds

$$p_N\Big(\{\mathbf{e}\,|\,\mathbf{e}\in\mathscr{E}^N, |f_N(\mathbf{e})-\bar f|>\delta\Big) \leq \frac{\mu_2}{\delta^2}, \tag{2.23.16}$$

where

$$\mu_2 = \sum_{\mathbf{e}\in\mathscr{E}^N} p_N(\mathbf{e})\big(f_N(\mathbf{e})-\bar f\big)^2$$
$$= \sum_{e_1,\ldots,e_N} p(e_1)p(e_2)\ldots p(e_N)\Big(\frac{1}{N}\sum_{i=1}^N\big(f(e_i)-\bar f\big)\Big)^2$$
$$= \frac{1}{N^2}\sum_{i,j=1}^N\sum_{e_1,\ldots,e_N} p(e_1)\ldots p(e_N)\times\big(f(e_i)-\bar f\big)\big(f(e_j)-\bar f\big) \tag{2.23.17}$$
$$= \frac{1}{N^2}\sum_{i=1}^N\sum_{e_1,\ldots,e_N} p(e_1)\ldots p(e_N)\big(f(e_i)-\bar f\big)^2,$$

since all the terms with $i\neq j$ vanish because, since $\Sigma_e p(e)=1$, one has

$$\sum_{e_1,\ldots,e_N} p(e_1)\ldots p(e_N)\big(f(e_i)-\bar f\big)\big(f(e_j)-\bar f\big)$$
$$= \sum_{e_i,e_j} p(e_i)p(e_j)\big(f(e_i)-\bar f\big)\big(f(e_j)-\bar f\big) = \Big(\sum_e p(e)\big(f(e)-\bar f\big)\Big)^2 = 0 \tag{2.23.18}$$

by the definition of $\bar f = \Sigma_{e\in\mathscr{E}} p(e)f(e)$, if $i\neq j$.

The last member of Eq. (2.23.17) can be similarly computed yielding

$$\mu_2 = \frac{1}{N^2}N\Big(\sum_e p(e)\big(f(e)-\bar f\big)^2\Big) = \frac{\sigma^2}{N}, \tag{2.23.19}$$

where $\sigma^2 = \Sigma_e p(e)(f(e)-\bar f)^2$ and the proposition is proved as $\mu_2 \to_{N\to\infty} 0$ [see Eq. (2.23.16)].

mbe

Observation. Note that Eqs. (2.23.16) and (2.23.19) show more: they imply that the probability of the event $|f_N(\mathbf{e})-\bar f|>\delta_N$ tends to zero as $N\to\infty$, provided the sequence δ_N is such that $N\delta_N^2\to_{N\to\infty}\infty$, i.e., provided δ_N does not go to zero faster or as $N^{-1/2}$.

Also the problem of the determination of the average value of an observable over a sequence of times succeeding each other at time intervals $a \pm \epsilon$, where the choice of the sign $\pm$ is a random choice in the sense informally discussed at the beginning of this section, can be easily dealt with by the above techniques.

35 Proposition. *Let $f \in C^\infty(\mathbf{R})$ be a periodic function with period $T > 0$. Consider the probability distribution $(\mathcal{E}^N, p_N)$ on the space $\mathcal{E}^N$ of the N-tuples $\epsilon = (\epsilon_0, \ldots, \epsilon_{N-1})$, $\epsilon_i = \pm\epsilon$, $i = 0, 1, \ldots, N-1$, where*

$$p_N(\epsilon) = 2^{-N}, \qquad \forall \epsilon \in \mathcal{E}_N . \tag{2.23.20}$$

Given $a > \epsilon$ such that a/ϵ is irrational, consider the random variable on $(\mathcal{E}^N, p_N)$,

$$\tilde{\mathfrak{M}}_N(\epsilon) = \frac{1}{N} \sum_{j=0}^{N-1} f(ja + \epsilon_1 + \cdots + \epsilon_N). \tag{2.23.21}$$

Then

$$\lim_{N\to\infty} \tilde{\mathfrak{M}}_N(\epsilon) = \frac{1}{T} \int_0^T f(\tau)\, d\tau = \bar{f} \tag{2.23.22}$$

in probability.

Observations.

(1) This proposition is interesting because it shows that even if some measurement errors involving the successive timing of the observations are present, the average value of f, computed using the data successively obtained, has a large probability of being close to the "ideal" average value, i.e., to the continuous average, independent of a and ϵ, provided a/ϵ is irrational.

(2) The coincidence of the stochastic average with the continuous average depends upon the irrationability of a/ϵ, but not on the value of T: it is therefore a property of the structure of the measurement (through the parameters a and ϵ) and does not depend on the characteristic properties of the observable f, as was the case in the comparison between the two ideal notions of the average (continuous and discrete with step a) where the rationality of T/a was relevant (see Proposition 32).

(3) With the same methods of proof, the above proposition could be extended to the case when ϵ_i takes more than two values: $\epsilon_i = \pm\alpha_1, \pm\alpha_2, \ldots, \pm\alpha_k$ and $p(\alpha_j) \equiv p(-\alpha_j)$. In this case, the condition "ϵ/a irrational" will be replaced by the condition "there is at least one value $\bar{\alpha}$ among the values of α_j such that $\bar{\alpha}/a$ is irrational".

(4) Finally, always with the same technique of proof, one could treat the case ϵ/a rational, and this would lead one to conclude that $\mathfrak{M}_N(\epsilon)$ still converges in probability to a well-defined limit easily expressible in terms of the Fourier transform of f [with a result analogous to Eq. (2.22.13)

generally involving T as well; see Problem 17 at the end of this section]. The difference between this new limit and the continuous average could be measured by "the commensurability of a with respect to ϵ" [see Observation 6 to Definition 17, p. 111 (for an analogous comment) and Problem 18 at the end of this section].

When the error takes more than one value, as in Observation 3 above, this difference depends on the maximum degree of commensurability between a and the values of the various errors. It is sufficient that among the various errors there is one with respect to which a is "little" commensurable to imply that $\tilde{\mathfrak{M}}_N(\epsilon)$ converges in probability to a value very close to the continuous average of f.

For this reason, it is rare that the stochastic average sensibly deviates from the continuous average: in the concrete situations, there are always several causes of errors and, correspondingly, ϵ_j can take very many different values. Necessarily, a will not be too commensurable with respect to many of them.

PROOF. The proof is basically a simple check founded on the Chebysčev inequality.

Consider the Fourier representation of f:

$$f(t) = \sum_{n \in \mathbf{Z}} \hat{f}_n e^{(2\pi i/T)nt}, \tag{2.23.23}$$

and take into account that $\hat{f}_0 = \bar{f} = T^{-1}\int_0^T f(\tau)\,d\tau$:

$$\begin{aligned}
\tilde{\mathfrak{M}}_N(\epsilon) - \bar{f} &= \frac{1}{N}\left(\sum_{j=0}^{N-1} f(ja + \epsilon_0 + \epsilon_1 + \cdots + \epsilon_j)\right) - \bar{f} \\
&\equiv \frac{1}{N}\sum_{j=0}^{N-1}\left(f(ja + \epsilon_0 + \cdots + \epsilon_j) - \bar{f}\right) \\
&= \sum_{\substack{n \in \mathbf{Z} \\ n \neq 0}} \hat{f}_n\left(\frac{1}{N}\sum_{j=0}^{N-1} \exp\frac{2\pi i}{T} n(ja + \epsilon_0 + \cdots + \epsilon_j)\right).
\end{aligned} \tag{2.23.24}$$

Hence, to apply the Chebysčev inequality, we compute the second moment of $\tilde{\mathfrak{M}}_N(\epsilon) - \bar{f}$, using Eq. (2.23.24):

$$\begin{aligned}
\mu_2(N) &= \sum_{\epsilon}\left(\tilde{\mathfrak{M}}_N(\epsilon) - \bar{f}\right)^2 p_N(\epsilon) = \frac{1}{2^N}\sum_{\epsilon}\left(\mathfrak{M}_N(f) - \bar{f}\right)^2 \\
&= \sum_{\substack{n_1, n_2 \in \mathbf{Z} \\ n_1 \neq 0, n_2 \neq 0}} \hat{f}_{n_1}\hat{f}_{n_2}\left\{\frac{1}{N^2}\sum_{j_1, j_2}^{0, N-1}\frac{1}{2^N}\sum_{\epsilon}\exp\frac{2\pi i}{T}(j_1 n_1 a + j_2 n_2 a)\right. \\
&\qquad \times \exp\frac{2\pi i}{T}\left[(\epsilon_0 + \cdots + \epsilon_{j_1})n_1\right. \\
&\qquad\qquad \left.\left. + (\epsilon_0 + \cdots + \epsilon_{j_2})n_2\right]\right\}.
\end{aligned} \tag{2.23.25}$$

The series over n_1, n_2 is term-by-term bounded by the convergent series $\Sigma_{n_1,n_2}|\hat{f}_{n_1}||\hat{f}_{n_2}|$: in fact, the factor within curly brackets is a sum of $N^2 2^N$ addends each with modulus $1/N^2 2^N$ and, therefore, its modulus does not exceed 1. Hence, the series in Eq. (2.23.25) is uniformly convergent in N and we can compute its limit as $N \to \infty$ under the summation sign (i.e., term by term). We shall in fact show that all the terms in curly brackets in Eq. (2.23.25) tend to zero as $N \to \infty$; hence, $\mu_2(N) \to_{N\to\infty} 0$ which, by the Chebysčev inequality, will imply Proposition 35.

The contribution to the sum inside the curly brackets in Eq. (2.23.25) coming from the terms with $j_1 = j_2$ involves $N 2^N$ terms with modulus $1/N^2 2^N$. Hence, it tends to zero as $N \to \infty$. Therefore, it will be enough to consider the terms with $j_1 < j_2$ and to show that their contribution to the sum is also infinitesimal as $N \to \infty$. The terms with $j_1 > j_2$ can be similarly treated.

Suppose $j_2 > j_1$: the contribution to the curly bracket term from such addends is

$$\frac{1}{N^2} \sum_{j_1=0}^{N-2} \sum_{j_2=j_1+1}^{N-1} \left(\exp \frac{2\pi i a}{T}(n_1 j_1 + n_2 j_2)\right)$$
$$\times \frac{1}{2^N} \sum_{\epsilon} \exp \frac{2\pi i}{T}(n_1+n_2)(\epsilon_0 + \cdots + \epsilon_{j_1}) \exp \frac{2\pi i}{T} n_2(\epsilon_{j_1} + \cdots + \epsilon_{j_2}) \tag{2.23.26}$$

which, by successively performing the summations over $\epsilon_{N-1}, \ldots, \epsilon_0$, becomes

$$\frac{1}{N^2} \sum_{j_1=0}^{N-2} \sum_{j_1=j_1+1}^{N-1} \left(\exp \frac{2\pi i a}{T}(n_1 j_1 + n_2 j_2)\right)$$
$$\times \left(\cos \frac{2\pi}{T} \epsilon n_2\right)^{j_2 - j_1} \left(\cos \frac{2\pi}{T} \epsilon (n_1 + n_2)\right)^{j_1} \tag{2.23.27}$$
$$= \frac{1}{N^2} \sum_{j_1=0}^{N-1} \sum_{j_2=j_1+1}^{N-1} \left(\left(\exp \frac{2\pi i a}{T}(n_1+n_2)\right) \cos \frac{2\pi}{T} \epsilon (n_1+n_2)\right)^{j_1}$$
$$\times \left(\left(\exp \frac{2\pi i a}{T} n_2\right) \cos \frac{2\pi}{T} \epsilon n_2\right)^{j_2 - j_1}.$$

We can now perform the summation over j_2, noting that if $n_2 \neq 0$,

$$\lambda = e^{(2\pi i/T) n_2 a} \cos \frac{2\pi}{T} \epsilon n_2 \neq 1 \tag{2.23.28}$$

because, as is easily seen, $|\lambda| \leqslant 1$; and if $|\lambda| = 1$, the number ϵ/a would have to be rational, regardless of T.

The result of the sum in Eq. (2.23.27) over j_2 is then

$$\frac{1}{N^2} \sum_{j_1=0}^{N-2} \left(e^{(2\pi i/T) a (n_1+n_2)} \cos \frac{2\pi}{T} \epsilon (n_1+n_2)\right)^{j_1} \lambda \frac{\lambda^{N-j_1-1} - 1}{\lambda - 1}, \tag{2.23.29}$$

and this sum involves $(N-1)$ addends each with modulus bounded by $N^{-2}2/|\lambda-1|$. Hence, it tends to zero as $N\to\infty$. mbe

Exercises and Problems for §2.23

1. Consider the equitable probability distribution $(\mathcal{E}, p)$ on a set of six events $\mathcal{E}=\{1,2,\ldots,6\}$, $p(j)=\frac{1}{6}$ ("perfect dice"). Compute the probability distribution for the following random variables (see Definition 20):

$$f_1(i)=\begin{cases} 1 & \text{if } i \text{ is even,}\\ -1 & \text{if } i \text{ is odd,}\end{cases}$$

$$f_1(i)=\begin{cases} 1 & \text{if } i=1,2,3,\\ -1 & \text{if } i=4,5,6.\end{cases}$$

2. Let $\mathcal{E}$ consist of two elements $+1$ and -1 and let $p(\pm 1)=\frac{1}{2}$. Compute, in $(\mathcal{E},p)^N$, the moment μ_2 of the random variables $f(\epsilon)=(\epsilon_1+\cdots+\epsilon_N)$, $\epsilon\in\mathcal{E}^N$, $\epsilon=(\epsilon_1,\ldots,\epsilon_N)$, and $f'(\epsilon)=\sin(\epsilon_1+\cdots+\epsilon_N)$.

3. Same as Problem 2 with $p(+1)=\frac{2}{3}$, $p(-1)=\frac{1}{3}$.

4. Compute the limit in probability of the random variables $(\epsilon_1+\cdots+\epsilon_N)/N$, as $N\to+\infty$, in $(\mathcal{E},p)^N$, where $\mathcal{E}=\{-1,+1\}$, $p(-1)=\frac{1}{3}$, $p(+1)=\frac{2}{3}$.

5. Consider the stochastic average with step $a=\sqrt{2}$ with respect to an error distribution with the scheme $\mathcal{E}=\{-\epsilon,\epsilon\}$, $p(\pm\epsilon)=\frac{1}{2}$, $\epsilon=1/10$ for the observable "kinetic energy" on the energy 1 motion of the oscillator $\ddot{x}+x=0$. Estimate the number of measures N needed for finding that the average over N observations deviates from the stochastic average (i.e., from the case $N=+\infty$) by 10% at most with a probability of 99%.

6. Same as Problem 5 with error scheme $\mathcal{E}=\{-\epsilon,0,\epsilon\}$, $p(\pm\epsilon)=\frac{1}{3}$, $p(0)=\frac{1}{3}$, using the observable "potential energy."

7. Same as Problem 5 for the motions of the oscillators $\ddot{x}+\dot{x}+x=(1-\cos t)^2$ for the observable "work done per unit time by the forcing force."

8. Interpret $\Sigma_{i=1}^N \log i \equiv \log n!$ as an approximation for the integral between 1 and $n+1$ of the function $\xi\to\log\xi$ and, using this interpretation, show that

$$0<\log n! - n(\log n - 1) < 1+\frac{1}{n}+\log n,$$

i.e.,

$$1<\frac{n!}{n^n e^{-n}}<ne^{1+1/n}.$$

9.* Using the "Stirling formula" (see Problem 14):

$$n! = n^n e^{-n}\sqrt{2\pi n}\left(1+O\left(\frac{1}{n}\right)\right),$$

show that the probability that

$$f_N(\epsilon)=\frac{\epsilon_1+\cdots+\epsilon_N}{\sqrt{N}}\in[a,b]$$

with respect to the probability distribution $(\mathscr{E}, p)^N$, where $\mathscr{E} = \{-1, +1\}$, $p(\pm 1) = \frac{1}{2}$, converges to

$$\int_a^b e^{-(x^2/2)} \frac{dx}{\sqrt{2\pi}}$$

as $N \to +\infty$ ("Gauss' theorem"). (Hint: Recall Eq. (2.23.9) to see that the probability that $(\epsilon_1 + \cdots + \epsilon_N)/\sqrt{N}$ takes the value $(N-2k)/\sqrt{N}$, $k = 0, 1, \ldots, N$, is given by $2^{-N}\binom{N}{k}$; then express the factorials in $\binom{N}{k}$ via the Stirling formula, recalling that k must be such that $a \leqslant (N-2k)/\sqrt{N} \leqslant b$, etc.)

10.* Show that the statement in Problem 9 implies that the sequence of random variables f_N considered there does not converge in probability as $N \to \infty$.

11. Assuming the result in Problem 9, show that the sequence

$$f_N^{(\alpha)}(\epsilon) = \frac{\epsilon_1 + \cdots + \epsilon_N}{N^{\alpha/2}}$$

of random variables with respect to the probability distribution considered in Problem 9 converges to zero in probability if $\alpha > 1$, does not converge if $\alpha = 1$ (see Problem 10), and diverges if $\alpha < 1$ (in the sense that the probability that $|f_N^{(\alpha)}(\epsilon)| < a$ approaches zero, as $N \to \infty$, $\forall a \in [0, +\infty)$).

12. Show that the probability p_N that the random variable $f_N(\epsilon)$ introduced in Problem 9 is positive approaches $\frac{1}{2}$ as $N \to +\infty$ (Hint: Distinguish N even and N odd.)

13. In the context of Problems 9 and 12, estimate how fast $|p_N - \frac{1}{2}| \to 0$ as $N \to \infty$.

14.* Prove Stirling's formula with a constant Γ instead of $\sqrt{2\pi}$, refining the argument in Problem 8. (Hint:

$$\log n! = \sum_{i=2}^{n} \log i = \sum_{i=2}^{n} \int_{i-1}^{i} \log i \, dx = \sum_{i=2}^{n} \int_{i-1}^{i} \log(x - (x-i)\, dx$$

$$= \sum_{i=2}^{n} \int_{i-1}^{i} \left[\log x + \log\left(1 - \frac{x-i}{x}\right)\right] dx = \int_1^n \log x \, dx$$

$$+ \sum_{i=2}^{n} \int_{i-1}^{i} \left[\log\left(1 - \frac{x-i}{x}\right) + \frac{x-i}{x}\right] dx + \sum_{i=2}^{n} \int_{i-1}^{i} - \frac{x-i}{x} \, dx$$

$$= n(\log n - 1) + \sum_{i=2}^{n} \gamma_i + \sum_{i=2}^{n} - \left(1 - i \log \frac{i}{i-1}\right),$$

where γ_i denotes the second integral in the intermediate step. Then $|\gamma_i| \leqslant \text{const}\, i^{-2}$ and $-1 + i \log[i/(i-1) = 1/2i + 1/3i^2 + \cdots$ so that

$$\log n! = n(\log n - 1) + \frac{1}{2} \sum_{i=2}^{n} \frac{1}{i} + \sum_{i=2}^{n} \tilde{\gamma}_i$$

with $|\tilde{\gamma}_i| \leqslant \text{const}\, i^{-2}$. Since $\Sigma_{i=2}^{n} 1/i = \log n - \tilde{C} + O(1/n)$ with $\tilde{C}$ suitably chosen (see next exercise), it follows that

$$\log n! = n(\log n - 1) + \log\sqrt{n} - \tilde{C} - \sum_{i=2}^{n} \tilde{\gamma}_i + O\left(\frac{1}{n}\right);$$

so if $\Gamma = \exp(\tilde{C} + \Sigma_{i=2}^{\infty}\tilde{\gamma}_i)$, it follows that

$$n! = n^n e^{-n}\sqrt{n}\,\Gamma\left(1 + O\left(\frac{1}{n}\right)\right).$$

15. Show that $\Sigma_{i=1}^{n} 1/i = \log n - C + O(1/n)$, where C is a suitable constant ("Euler–Mascheroni constant") (Hint:

$$\begin{aligned}
\sum_{i=1}^{n} \frac{1}{i} &= \sum_{i=1}^{n} \int_i^{i+1} \frac{1}{i}\,dx = \sum_{i=1}^{n} \int_i^{i+1} \frac{1}{x-(i-x)}\,dx \\
&= \sum_{i=1}^{n} \int_i^{i+1} \frac{dx}{x}\,\frac{1}{1-(i-x)/x} \\
&= \sum_{i=1}^{n} \int_i^{i+1} \frac{dx}{x}\left[\frac{1}{1-(i-x)/x} - 1 - \frac{i-x}{x}\right] \\
&\quad + \sum_{i=1}^{n} \int_i^{i+1} \frac{dx}{x}\left(1 + \frac{i-x}{x}\right) \\
&= \sum_{i=1}^{n} \bar{\gamma}_i + \log(n+1) + \sum_{i=1}^{n} i\left(\frac{1}{i^2} - \frac{1}{(i+1)^2}\right) \\
&= \sum_{i=1}^{n} \bar{\gamma}_i + \log(n+1) + \sum_{i=1}^{n} \frac{2i+1}{i(i+1)^2} \equiv \sum_{i=1}^{n} \tilde{\gamma}_i + \log(n+1)
\end{aligned}$$

and show that $|\bar{\gamma}_i|, |\tilde{\gamma}_i| < \text{const}\, i^{-2}$).

16.* Complete the derivation of the Stirling formula begun in Problem 14 by showing that $\Gamma = \sqrt{2\pi}$. (Hint: Use Problem 9, with Γ instead of $\sqrt{2\pi}$, which says that the random variables $f_N(\epsilon)$ lie in $[-A, A]$ with a probability converging to $\int_{-A}^{A} e^{-x^2/2}dx/\Gamma$ (if one does not suppose $\Gamma = \sqrt{2\pi}$ yet). Then, by estimating the factorials in $\binom{N}{k}2^{-N}$ using the Stirling formula with Γ instead of $\sqrt{2\pi}$, see Problem 14, show that

$$\sum_{|N-2k/\sqrt{N}|>A} 2^{-N}\binom{N}{k} \xrightarrow[n\to\infty]{} 0$$

uniformly in N: this implies that $\int_{-\infty}^{+\infty} e^{-x^2/2}dx/\Gamma = 1$; hence, $\Gamma = \sqrt{2\pi}$. The estimate on the $\Sigma 2^{-N}\binom{N}{k}$ is quite delicate and should be decomposed into two estimates: the first for $|k/N - \frac{1}{2}| \in [A/\sqrt{N}, \frac{1}{10}]$, and the second for $|k/N - \frac{1}{2}| \in [\frac{1}{10}, \frac{1}{2}]$.)

17. Show that if a/ϵ is not assumed to be irrational, Eq. (2.23.22) becomes

$$\lim_{N\to\infty} \mathfrak{M}_N(\epsilon) = \sum_n{}^{*} f_n \equiv \tilde{f},$$

where f_n are the harmonics of f and Σ_n^* is a sum running over the n's such that $e^{(2\pi i/T)na}\cos(2\pi/T)n\epsilon = 1$.

18. Deduce from Problem 17 that the limit in Eq. (2.23.22) coincides in probability with the continuous average not only if a/ϵ is irrational, but also if T is irrational with respect to either ϵ or a. Also if $\epsilon/a = p/q$, with p and q relatively prime integers, and if a is varied so that $p\to\infty$, $a\to\bar{a} > 0$, then $\tilde{f}\to\bar{f}$.

§2.24. Extremal Properties of Conservative Motion: Action and Variational Principle

Since the construction of the entire universe is absolutely perfect and is due to a Creator with infinite knowledge, nothing exists in the world which does not exhibit some property of maximum or minimum. Therefore, there cannot be any doubt whatsoever about the possibility that all the effects are determined by their final aims with the help of the maxima method, in the same way in which they are also determined by the initial causes.

The equilibrium positions of a point mass on a line are identified with the points where the potential energy is stationary.

Thinking of equilibrium as a particular form of motion, we can ask whether the other possible motions of the point mass, developing under the action of a conservative force with potential energy V, can be characterized by similar stationarity properties. This analysis will also be useful as a first illustration of the content of the above quoted proposition of Euler. A deeper analysis will be the object of Chapter 3.

Consider a point mass with mass $m > 0$ moving in the time interval $[t_1, t_2]$ from the position ξ_1 to the position ξ_2: such a motion is a C^∞ function $t \to x(t)$, $t \in [t_1, t_2]$, such that $x(t_1) = \xi_1$, $x(t_2) = \xi_2$.

Let $\mathfrak{M}_{t_1,t_2}(\xi_1, \xi_2)$ be the set of all C^∞ motions $t \to x(t)$, $t \in [t_1, t_2]$, such that $x(t_1) = \xi_1$, $x(t_2) = \xi_2$. If $V \in C^\infty(\mathbf{R})$ is a given function bounded from below, it makes sense to consider the motions of the point mass which take place under the influence of the force generated by the potential energy V. It is clear that such motions are a very restricted class in $\mathfrak{M}_{t_1,t_2}(\xi_1, \xi_2)$, possibly empty.

We now inquire whether there is a real-valued function A defined on $\mathfrak{M}_{t_1,t_2}(\xi_1, \xi_2)$ which takes a minimum value or, at least, is stationary on the motions going from ξ_1 to ξ_2 as t goes from t_1 to t_2 under the influence of the force with potential energy V.

The meaning of this question has to be clarified by a preliminary discussion on the meaning of "extremality" of a function defined on a set of motions, i.e., on a set of other functions.

We shall restrict our attention to special functions defined on $\mathfrak{M}_{t_1,t_2}(\xi_1, \xi_2)$: those having the form

$$A(\mathbf{x}) = \int_{t_1}^{t_2} \mathfrak{L}(\dot{x}(t), x(t), t)\, dt, \tag{2.24.1}$$

where $\mathfrak{L} \in C^\infty(\mathbf{R}^3)$ associates (η, ξ, t) with $\mathfrak{L}(\eta, \xi, t)$.

It is clear that Eq. (2.24.1) associates a well defined real number with every $\mathbf{x} \in \mathfrak{M}_{t_1,t_2}(\xi_1, \xi_2)$. This number is called the "action of the motion $\mathbf{x}$ with respect to the Lagrangian function $\mathfrak{L}$".[24]

The notion of "stationarity" or "extremality" of A is very natural in terms of the related notion of "varied motions".

[24] For the origin of this name, see the remarks on pp. 164 and 242.

22 Definition. Given $\mathbf{x} \in \mathfrak{M}_{t_1,t_2}(\xi_1,\xi_2)$ and a real function $(t,\epsilon) \to y(t,\epsilon)$ in $C^\infty([t_1,t_2] \times (-1,1))$ such that

(i) $$y(t,0) = x(t), \qquad \forall t \in [t_1,t_2], \tag{2.24.2}$$

(ii) $$y(t_1,\epsilon) = \xi_1, \qquad y(t_2,\epsilon) = \xi_2, \qquad \forall \epsilon \in (-1,1), \tag{2.24.3}$$

We shall say that the function y is a "variation of $\mathbf{x}$" inside $\mathfrak{M}_{t_1,t_2}(\xi_1,\xi_2)$ parametrized by $\epsilon \in (-1,1)$. The set of all the variations will be denoted by $\mathcal{V}_{\mathbf{x}}$.

More generally, if $\mathfrak{M}$ is a subset of $\mathfrak{M}_{t_1,t_2}(\xi_1,\xi_2)$, we shall denote $\mathcal{V}_{\mathbf{x}}(\mathfrak{M})$ the set of the variations of $\mathbf{x}$ such that, $\forall \epsilon \in (-1,1)$, the function

$$t \to y_\epsilon(t) = y(t,\epsilon), \qquad t \in [t_1,t_2] \tag{2.24.4}$$

is in $\mathfrak{M}$.

Observations.

(1) We can imagine that a varied motion y is a bundle of motions with equal initial and final data (see Fig. 2.13).

(2) Occasionally it will be useful to think of a variation of $\mathbf{x}$ in $\mathfrak{M} \subset \mathfrak{M}_{t_1,t_2}(\xi_1,\xi_2)$ as a "regular curve" in the space $\mathfrak{M}$: for every $\epsilon \in (-1,1)$, one has a point $\mathbf{y}_\epsilon \in \mathfrak{M}$ and $\mathbf{y}_0 = \mathbf{x}$ [see Eq. (2.24.4)].

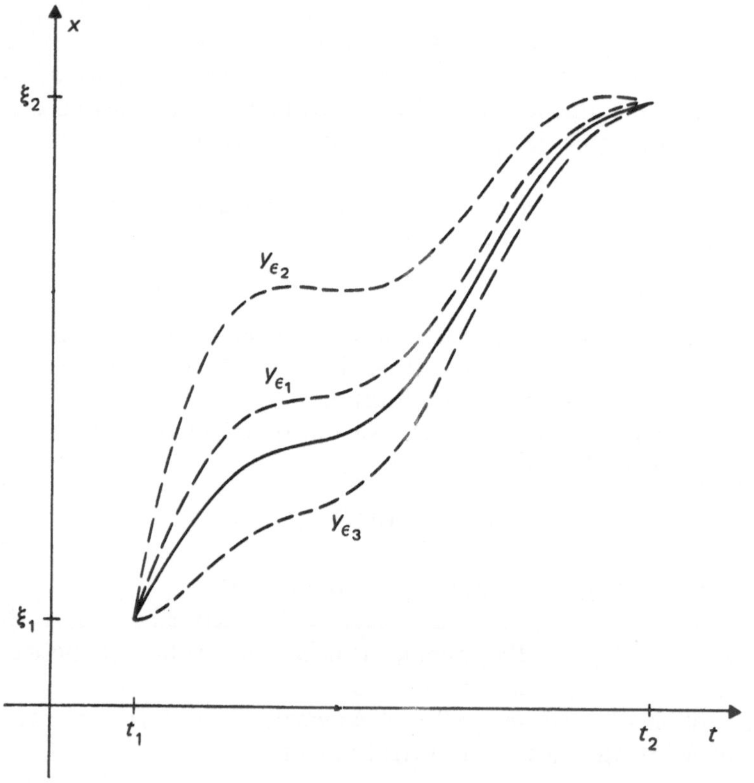

Figure 2.13.

(3) If F is a function on $\mathfrak{M} \subset \mathfrak{M}_{t_1,t_2}(\xi_1, \xi_2)$ and $y \in \mathcal{V}_{\mathbf{x}}(\mathfrak{M})$, it will make sense to consider the function of $\epsilon \in (-1, 1)$: $\epsilon \to F(\mathbf{y}_\epsilon)$, "value of F along the curve y through $\mathbf{x}$ in the point parametrized by ϵ".

It is now possible to give a precise definition of stationarity.

23 Definition. Let $\mathfrak{M} \subset \mathfrak{M}_{t_1,t_2}(\xi_1, \xi_2)$ and let A be a function on $\mathfrak{M}$ having the form of Eq. (2.24.1). We shall say that $\mathbf{x} \in \mathfrak{M}$ is a "stationarity point" for A in $\mathfrak{M}$ if for every $y \in \mathcal{V}_{\mathbf{x}}(\mathfrak{M})$ the function [see Eq. (2.24.4)]

$$\epsilon \to A(\mathbf{y}_\epsilon), \qquad \epsilon \in (-1, 1) \tag{2.24.5}$$

has a stationarity point in $\epsilon = 0$, i.e.,

$$\frac{d}{d\epsilon} A(\mathbf{y}_\epsilon)\bigg|_{\epsilon=0} = 0, \qquad \forall y \in \mathcal{V}_{\mathbf{x}}(\mathfrak{M}). \tag{2.24.6}$$

Observations.

(1) In other words, $\mathbf{x}$ is a stationarity point for A in $\mathfrak{M}$ if on every regular curve $y \in \mathcal{V}_{\mathbf{x}}(\mathfrak{M})$ through $\mathbf{x}$, the function A, thought of as a function of the parameter ϵ parametrizing the curve, has a stationarity point in $\epsilon = 0$, i.e., "in $\mathbf{x}$".

(2) Note that in the theory of the maxima and minima of functions $F \in C^\infty(\mathbf{R}^d)$, there are various equivalent definitions of the stationarity points; for instance,

(a) $(\partial F/\partial x_i)(\mathbf{x}) = 0$, $i = 1, 2, \ldots, d$.
(b) On every C^∞ curve $\epsilon \to \mathbf{y}_\epsilon$, $\epsilon \in (-1, 1)$, through $\mathbf{x} \equiv \mathbf{y}_0$, the function $\epsilon \to F(\mathbf{y}_\epsilon)$ has zero derivative with respect to ϵ in $\epsilon = 0$.

Definition (b) is clearly the "finite-dimensional" analogue by which Definition 23 is inspired: intuitively, one can in fact think of $\mathbf{x} \in \mathfrak{M}_{t_1,t_2}(\xi_1, \xi_2)$ as a vector with infinitely many components $x_t \equiv x(t)$, $t \in [t_1, t_2]$, not independent, however, since they are constrained by the condition that $t \to x_t$ is in $C^\infty([t_1, t_2])$ and $x_{t_1} = \xi_1$, $x_{t_2} = \xi_2$.

(3) Strictly speaking, one should prove that Eq. (2.24.6) makes sense, i.e., that $\epsilon \to A(\mathbf{y}_\epsilon)$ is differentiable in ϵ. However, this is an immediate consequence of the differentiation rules for integrals. Actually, it is easy to find explicit expressions for the derivatives of A. For instance, from Eqs. (2.24.1) and (2.24.5), it follows that

$$\begin{aligned} \frac{d}{d\epsilon} A(\mathbf{y}_\epsilon) &= \frac{d}{d\epsilon} \int_{t_1}^{t_2} \mathcal{L}\left(\frac{\partial y}{\partial t}(t,\epsilon), y(t,\epsilon), t \right) dt \\ &= \int_{t_1}^{t_2} dt \left\{ \frac{\partial \mathcal{L}}{\partial \eta} \left(\frac{\partial y}{\partial t}(t,\epsilon), y(t,\epsilon), t \right) \cdot \frac{\partial^2 y}{\partial \epsilon \partial t}(t,\epsilon) \right. \\ &\qquad \left. + \frac{\partial \mathcal{L}}{\partial \xi} \left(\frac{\partial y}{\partial t}(t,\epsilon), y(t,\epsilon), t \right) \cdot \frac{\partial y}{\partial \epsilon}(t,\epsilon) \right\}, \end{aligned} \tag{2.24.7}$$

and shortening the notations for $(\partial y/\partial t)(t,\epsilon)$ in $\partial y/\partial t$ and for $(\partial^2 y/\partial\epsilon\partial t)(t,\epsilon)$ in $\partial^2 y/\partial\epsilon\partial t$, for $y(t,\epsilon)$ in y, etc. Eq. (2.24.7) can be rewritten:

$$\frac{d}{d\epsilon}A(\mathbf{y}_\epsilon)=\int_{t_1}^{t_2}dt\left\{\frac{\partial \mathcal{L}}{\partial\eta}\left(\frac{\partial y}{\partial t},y,t\right)\frac{\partial^2 y}{\partial\epsilon\partial t}+\frac{\partial \mathcal{L}}{\partial\xi}\left(\frac{\partial y}{\partial t},y,t\right)\frac{\partial y}{\partial\epsilon}\right\}. \quad (2.24.8)$$

Then, avoiding the explicit indication of the arguments $\partial y/\partial t, y, t$ in $\mathcal{L}$ and in its derivatives, a straightforward computation yields

$$\begin{aligned}\frac{d^2}{d\epsilon^2}A(\mathbf{y}_\epsilon)=\int_{t_1}^{t_2}dt\bigg\{&\frac{\partial^2\mathcal{L}}{\partial\eta^2}\left(\frac{\partial^2 y}{\partial\epsilon\partial t}\right)^2+\frac{\partial^2\mathcal{L}}{\partial\xi\partial\eta}\frac{\partial^2 y}{\partial\epsilon\partial t}\frac{\partial y}{\partial\epsilon}\\&+\frac{\partial\mathcal{L}}{\partial\eta}\frac{\partial^3 y}{\partial\epsilon^2\partial t}+\frac{\partial^2\mathcal{L}}{\partial\eta\partial\epsilon}\frac{\partial^2 y}{\partial\epsilon\partial t}\frac{\partial y}{\partial\epsilon}\qquad (2.24.9)\\&+\frac{\partial^2\mathcal{L}}{\partial\xi^2}\left(\frac{\partial y}{\partial\epsilon}\right)^2+\frac{\partial\mathcal{L}}{\partial\xi}\frac{\partial^2 y}{\partial\epsilon^2}\bigg\}\end{aligned}$$

The higher derivatives could be evaluated with similar procedures; i.e., $\epsilon\to A(\mathbf{y}_\epsilon)$ is a C^∞ function.

As in the case of the functions on $\mathbf{R}^d$, it is convenient to distinguish between stationary points and points of "local" or "relative" minimum.

24 Definition. If $\mathbf{x}\in\mathfrak{M}_{t_1,t_2}(\xi_1,\xi_2)$, we say that $\mathbf{x}$ is a "local" minimum for A defined by Eq. (2.24.1) on $\mathfrak{M}$ if for all varied motions $y\in\mathcal{V}_{\mathbf{x}}(\mathfrak{M})$, the function $\epsilon\to A(\mathbf{y}_\epsilon)$ [see Eq. (2.24.5)] has a relative minimum in $\epsilon=0$.

Observations.

(1) A has a local minimum in $\mathbf{x}$ on $\mathfrak{M}$ if on every regular curve y through $\mathbf{x}$ lying on $\mathfrak{M}$, it has a local minimum in $\mathbf{x}$.

(2) Obviously, a necessary condition for A to have a local minimum relative to $\mathfrak{M}$ in $\mathbf{x}\in\mathfrak{M}$ that is $\mathbf{x}$ is a stationarity point for A on $\mathfrak{M}$.

(3) A necessary condition in order that a stationarity point for A on $\mathfrak{M}$ is a local minimum on $\mathfrak{M}$ is that

$$\left.\frac{d^2}{d\epsilon^2}A(\mathbf{y}_\epsilon)\right|_{\epsilon=0}\geqslant 0,\qquad \forall y\in\mathcal{V}_{\mathbf{x}}(\mathfrak{M}) \qquad (2.24.10)$$

if $\mathbf{x}$ is the point of stationarity.

(4) If $\mathbf{x}\in\mathfrak{M}$ is an absolute minimum point for A on $\mathfrak{M}$, i.e., if $A(\mathbf{x}')\geqslant A(\mathbf{x})$, $\forall\mathbf{x}'\in\mathfrak{M}$, it is clear that $\mathbf{x}$ is also a local minimum point for A on $\mathfrak{M}$.

(5) If A has a local minimum in $\mathbf{x}$ relative to $\mathfrak{M}$, it must be that given $y\in\mathcal{V}_{\mathbf{x}}(\mathfrak{M})$, there is $\eta>0$ such that if $\epsilon\in[-\eta,\eta]$, then $A(\mathbf{y}_\epsilon)\geqslant A(\mathbf{x})$: this value of η may, however, depend on the choice of y.

(6) One could be tempted to define a local minimum by requiring that $A(\mathbf{x}) \leqslant A(\mathbf{x}')$, $\forall \mathbf{x}' \in \mathfrak{M}$ and "close enough" to $\mathbf{x}$. But the meaning of "close enough" would be unclear.

Let us now look for a necessary and sufficient stationarity criterion which is as "simple" as the one usually considered in the case of the stationarity of functions on $\mathbf{R}^d$ and concerning the vanishing of the gradient (see Observation 2 (a), to Definition 23).

36 Proposition. *The motion* $\mathbf{x} \in \mathfrak{M}_{t_1,t_2}(\xi_1, \xi_2)$ *is a stationary point for Eq.* (2.24.1) *on all of* $\mathfrak{M}_{t_1,t_2}(\xi_1, \xi_2)$ *if and only if*

$$\frac{d}{dt}\frac{\partial \mathcal{L}}{\partial \eta}(\dot{x}(t), x(t), t) = \frac{\partial \mathcal{L}}{\partial \xi}(\dot{x}(t), x(t), t), \qquad \forall t \in [t_1, t_2]. \quad (2.24.11)$$

Observations.

(1) In this proposition, it is essential that the set $\mathfrak{M} \subset \mathfrak{M}_{t_1,t_2}(\xi_1, \xi_2)$ on which stationarity is considered coincides with $\mathfrak{M}_{t_1,t_2}(\xi_1, \xi_2)$ itself.

(2) Equation (2.24.11) can be thought of as a differential equation for the function $t \to x(t)$, $t \in [t_1, t_2]$, i.e., as an equation for the determination of the stationarity points of A on the entire set $\mathfrak{M}_{t_1,t_2}(\xi_1, \xi_2)$. When Eq. (2.24.11) is viewed in this way, it is called the "Euler–Lagrange" equation for A or $\mathcal{L}$. As it clearly emerges from the proof, it is analogous to the condition of vanishing of the gradient in the stationarity problem for functions on $\mathbf{R}^d$ (see Observation 2 (a) to Definition 23, p. 129).

(3) It is important to stress that, in general, Eq. (2.24.11) is not a differential equation in the sense of Definition 1, p. 13: In important cases, however, Eq. (2.24.11) is equivalent to a differential equation in that sense (see Problems 4–6 at the end of this section).

PROOF. It reduces to a simple check. Let $y \in \mathcal{V}_{\mathbf{x}}$ and set

$$z(t) = \frac{\partial y}{\partial \epsilon}(t, 0), \qquad t \in [t_1, t_2], \quad (2.24.12)$$

$$\dot{z}(t) = \frac{\partial^2 y}{\partial t \partial \epsilon}(t, 0), \qquad t \in [t_1, t_2] \quad (2.24.13)$$

and note that Eq. (2.24.2) implies, $\forall t \in [t_1, t_2]$:

$$y(t, 0) = x(t), \qquad \frac{\partial y}{\partial t}(t, 0) = \dot{x}(t) \quad (2.24.14)$$

and Eq. (2.24.3) implies

$$z(t_1) = z(t_2) = 0. \quad (2.24.15)$$

Then, with the above notations, Eq. (2.24.8) becomes

$$\frac{d}{d\epsilon}A(\mathbf{y}_\epsilon)\bigg|_{\epsilon=0} = \int_{t_1}^{t_2} dt \left\{ \frac{\partial \mathcal{L}}{\partial \eta}(\dot{x}(t), x(t), t)\dot{z}(t) + \frac{\partial \mathcal{L}}{\partial \xi}(\dot{x}(t), x(t), t) z(t) \right\}. \quad (2.24.16)$$

As $\mathbf{y}_\epsilon$ varies in $\mathcal{V}_{\mathbf{x}}$, the function $\mathbf{z}$ defined by Eq. (2.24.12) spans the entire set $\mathfrak{M}_{t_1,t_2}(0,0)$. In fact, Eq. (2.24.15) shows that $\mathbf{z} \in \mathfrak{M}_{t_1,t_2}(0,0)$; furthermore, given arbitrary $\bar{\mathbf{z}} \in \mathfrak{M}_{t_1,t_2}(0,0)$ and setting

$$\bar{y}(t,\epsilon) = x(t) + \epsilon \bar{z}(t) \tag{2.24.17}$$

for $\epsilon \in (-1,1)$, $t \in [t_1,t_2]$, one constructs a varied motion $y \in \mathcal{V}_{\mathbf{x}}$ which, via Eq. (2.24.12), exactly generates $\bar{\mathbf{z}}$.

We can then use the wide arbitrariness of $\mathbf{z}$ in Eq. (2.24.16) to deduce conditions on $\mathbf{x}$. For this purpose it is convenient to eliminate $\dot{z}(t)$ from Eq. (2.24.16) by integrating the first term in the integral by parts; one finds:

$$\begin{aligned} \frac{d}{d\epsilon} A(\mathbf{y}_\epsilon)\Big|_{\epsilon=0} &= \left[\frac{\partial}{\partial \eta}(\dot{x}(t), x(t)) z(t) \right]_{t_1}^{t_2} \\ &- \int_{t_1}^{t_2} \left\{ \frac{d}{dt}\left(\frac{\partial \mathcal{L}}{\partial \eta}(\dot{x}(t), x(t), t) \right) - \frac{\partial \mathcal{L}}{\partial \xi}(\dot{x}(t), x(t)) \right\} z(t)\, dt \end{aligned} \tag{2.24.18}$$

which, by Eq. (2.24.15) and by the preceding remark on the arbitrariness of z, shows that $(dA/d\epsilon)(\mathbf{y}_\epsilon)|_{\epsilon=0} = 0$, $\forall y \in \mathcal{V}_{\mathbf{x}}$, becomes:

$$\begin{aligned} 0 = \int_{t_1}^{t_2} dt\, z(t) \left\{ \frac{d}{dt}\left(\frac{\partial \mathcal{L}}{\partial \eta}(\dot{x}(t), x(t), t) \right) - \frac{\partial \mathcal{L}}{\partial \xi}(\dot{x}(t), x(t), t) \right\}, \\ \forall \mathbf{z} \in \mathfrak{M}_{t_1,t_2}(0,0). \end{aligned} \tag{2.24.19}$$

The equivalence between Eqs. (2.24.19) and (2.24.11) is implied by the principle of vanishing integrals (see Appendix D). mbe

As a consequence of Proposition 36, it is possible to answer the question raised at the beginning of this section. In fact, if one defines for $\mathbf{x} \in \mathfrak{M}_{t_1,t_2}(\xi_1, \xi_2)$:

$$A(\mathbf{x}) = \int_{t_1}^{t_2} \left(\frac{1}{2} m \dot{x}(t)^2 - V(x(t)) \right) dt, \tag{2.24.20}$$

one realizes that the following proposition holds.

37 Proposition. *The motion $\mathbf{x}$ of a point mass with mass $m > 0$ developing from ξ_1 to ξ_2 in the time interval $[t_1, t_2]$ under the influence of a force with potential energy $V \in C^\infty(\mathbf{R})$ makes the action of Eq. (2.24.20) on $\mathfrak{M}_{t_1,t_2}(\xi_1, \xi_2)$ stationary, i.e., it makes the action with Lagrangian density*

$$\mathcal{L}(\eta, \xi, \epsilon) = \tfrac{1}{2} m \eta^2 - V(\xi). \tag{2.24.21}$$

stationary.

Proof. In fact, Eq. (2.24.11) becomes

$$\frac{d}{dt} m \dot{x}(t) = -\frac{\partial V}{\partial \xi}(x(t)), \qquad t \in [t_1, t_2] \tag{2.24.22}$$

which is the equation of motion. mbe

Furthermore, the following interesting proposition holds.

38 Proposition. *Let $t \to \bar{x}(t)$, $t \in \mathbf{R}$, be a motion of a point mass with $m > 0$ developing under the action of a force with potential energy $V \in C^\infty(\mathbf{R})$, bounded from below. Given $t_1 \in \mathbf{R}$, there exists $\bar{t} > t_1$ such that if $t_2 \in [t_1, \bar{t}]$, the motion $t \to \bar{x}(t)$ observed for $t \in [t_1, t_2]$,* i.e., *as an element of $\mathfrak{M}_{t_1,t_2}(\bar{x}(t_1), \bar{x}(t_2))$, not only is a stationarity point for the action with Lagrangian (2.24.21), but is also a local minimum for it in $\mathfrak{M}_{t_1,t_2}(\bar{x}(t_1), \bar{x}(t_2))$.*

Observation. This proposition motivates the name "principle of the least action" occasionally given to the Propositions 37 and 38.

PROOF. By the Observation 5 to Definition 24, p. 130, given $y \in \mathcal{V}_{\bar{\mathbf{x}}}$, we must find a η_y such that $A(\mathbf{y}_\epsilon) \geqslant A(\bar{\mathbf{x}})$, $\forall \epsilon \in [-\eta_y, \eta_y]$.

Given $t_2 > t_1$ and $y \in \mathcal{V}_{\bar{\mathbf{x}}}(\mathfrak{M}_{t_1,t_2}(\bar{x}(t_1), \bar{x}(t_2)))$, define η_y so that $|y_\epsilon(t) - \bar{x}(t)| \leqslant 1$, $\forall t \in [t_1, t_2]$, $\forall \epsilon \in [-\eta_y, \eta_y]$. The comparison of $A(\mathbf{y}_\epsilon)$ with $A(\bar{\mathbf{x}})$ yields

$$A(\mathbf{y}_\epsilon) - A(\bar{\mathbf{x}}) = \int_{t_1}^{t_2} \Big\{ \frac{m}{2}\Big(\big(\dot{\bar{x}}(t) + \dot{z}(t)\big)^2 - \dot{\bar{x}}(t)^2 \Big) - \Big(V\big(\bar{x}(t) + z(t)\big) - V\big(\bar{x}(t)\big)\Big) \Big\}\, dt, \tag{2.24.23}$$

where we set $z(t) = y_\epsilon(t) - \bar{x}(t)$, $t \in [t_1, t_2]$.

Obviously, the function z has the property

$$z(t_1) = z(t_2) = 0 \tag{2.24.24}$$

and it is ϵ dependent. To show that Eq. (2.24.23) is $\geqslant 0$, we apply the Taylor–Lagrange formula (see Appendix B):

$$V(\xi') - V(\xi) = \frac{\partial V}{\partial \xi}(\xi)(\xi' - \xi) + \varphi(\xi', \xi)\frac{(\xi' - \xi)^2}{2}, \tag{2.24.25}$$

where $\varphi \in C^\infty(\mathbf{R}^2)$ is a suitable function.

Then Eq. (2.24.23) becomes

$$A(\mathbf{y}_\epsilon) - A(\bar{\mathbf{x}}) = \int_{t_1}^{t_2} \Bigg\{ \bigg[m\frac{\dot{z}(t)^2}{2} - \varphi\big(\bar{x}(t) + z(t), \bar{x}(t)\big)\frac{z(t)^2}{2} \bigg] + \bigg[m\dot{\bar{x}}(t)\dot{z}(t) - \frac{\partial V}{\partial \xi}\big(\bar{x}(t)\big)z(t) \bigg] \Bigg\}\, dt. \tag{2.24.26}$$

Integrating the first term in the second set of square brackets by parts and using the equation of motion for $\bar{\mathbf{x}}$, Eqs. (2.24.22) and (2.24.24), one realizes that the integral of the term within the second set of square brackets in Eq. (2.24.26) vanishes. Therefore, if

$$M = \max_{\substack{t \in [t_1, t_1 + 1] \\ |\zeta| < 1}} |\varphi(\bar{x}(t) + \zeta, \bar{x}(t))|, \tag{2.24.27}$$

one sees that, if $|\epsilon| < \eta_y$, Eq. (2.24.26) implies

$$A(\mathbf{y}_\epsilon) - A(\bar{\mathbf{x}}) \geq \frac{m}{2}\int_{t_1}^{t_2} \dot{z}(t)^2\,dt - \frac{M}{2}\int_{t_1}^{t_2} z(t)^2\,dt; \tag{2.24.28}$$

if $t_2 - t_1 < 1$, which is a condition that can be implemented by supposing $t_2 < \bar{t}$ and, without loss of generality,

$$\bar{t} < t_1 + 1. \tag{2.24.29}$$

On the other hand, since $z(t_1) = 0$,

$$z(t) = \int_{t_1}^{t} \dot{z}(\tau)\,d\tau, \tag{2.24.30}$$

and applying the Cauchy–Schwartz inequality (see Appendix A) which generally looks like $\forall f, g \in C^\infty([t_1, t_2])$:

$$\left|\int_{t_1}^{t_2} f(\tau)g(\tau)\,d\tau\right| \leq \left(\int_{t_1}^{t_2} f(\tau)^2\,d\tau\right)^{1/2}\left(\int_{t_1}^{t_2} g(\tau)^2\,d\tau\right)^{1/2} \tag{2.24.31}$$

one finds,

$$\begin{aligned}\int_{t_1}^{t_2} z(t)^2\,dt &= \int_{t_1}^{t_2} dt \left|\int_{t_1}^{t} \dot{z}(\tau)\cdot 1\,d\tau\right|^2 \leq \int_{t_1}^{t_2} dt \left(\int_{t_1}^{t} \dot{z}(\tau)^2\,d\tau\right)\left(\int_{t_1}^{t} d\tau\right)\\ &\leq \int_{t_1}^{t_2} dt\,(t - t_1)\left(\int_{t_1}^{t_2} \dot{z}(\tau)^2\,d\tau\right) = \frac{(t_2 - t_1)^2}{2}\int_{t_1}^{t_2} \dot{z}(\tau)^2\,d\tau\end{aligned} \tag{2.24.32}$$

from Eq. (2.24.30). Hence Eqs. (2.24.28) and (2.24.32) mean

$$A(\mathbf{y}_\epsilon) - A(\bar{\mathbf{x}}) \geq \frac{1}{2}\left(m - \frac{M}{2}(t_2 - t_1)^2\right)\int_{t_1}^{t_2} \dot{z}(\tau)^2\,d\tau \tag{2.24.33}$$

which implies $A(\mathbf{y}_\epsilon) - A(\mathbf{x}) \geq 0$ if $t_2 \in [t_1, \bar{t}]$ and if $\bar{t}$ is close enough to t_1 (precisely so that $\bar{t} - t_1 < 1$ and $2m - M(\bar{t} - t_1)^2 > 0$), $\forall y \in \mathcal{V}_{\mathbf{x}}$, $\forall \epsilon \in [-\eta_y, \eta_y]$.
mbe

In the context of Proposition 38, one can wonder about what happens when the interval $[t_1, t_2]$ is not small. In this case it is easy to realize that it is always possible to cut the interval $[t_1, t_2]$ into finitely many small intervals such that the action is locally minimal on the variations of the restrictions of **x** to such intervals.

This situation is strongly reminiscent of the properties of the geodesics on curved surfaces. For instance, on the surface of a sphere, a line joining two points along a great circle ("geodesic of the sphere") has the property of being the line shortest among all those joining the two points and lying on the sphere's surface provided their distance, measured along the line itself, is small enough. However, if the two points are not close enough, it is generally no longer true that such a line is the shortest ("close enough" here means closer than πR if R is the sphere's radius).

Finally, let us present an important comment: we wish to stress the fact that the stationarity (or minimality) of A is an "intrinsic property", i.e., it is independent of the way the motion is described. To make this precise, let $\xi \to \gamma(\xi)$ be a C^∞ function defining a "nonsingular" change of variables (i.e., such that $\gamma'(\xi) \equiv (d\gamma/d\xi)(\xi) \neq 0$, $\forall \xi \in \mathbf{R}$). We can then use as a coordinate for the point ξ the quantity $\gamma(\xi)$.

Let Γ be the inverse function to γ defined on the open interval $I = \gamma(R) = \{\gamma\text{-image of } \mathbf{R}\}$. Suppose, for simplicity, $\gamma(\mathbf{R}) = I = \mathbf{R}$.

If $t \to x(t)$, $t \in [t_1, t_2]$, is a motion in $\mathbf{R}$, we can describe it through the function $t \to s(t) = \gamma(x(t))$, $t \in [t_1, t_2]$. We shall say that such a function describes the motion $\mathbf{x}$ in the system of coordinates on $\mathbf{R}$ associated with the function γ.

Clearly, there is a one-to-one correspondence B between the motions $\mathbf{x} \in \mathfrak{M}_{t_1,t_2}(\xi_1, \xi_2)$ and the motions $\mathbf{x} \in \mathfrak{M}_{t_1,t_2}(\gamma(\xi_1), \gamma(\xi_2))$ established by the relations

$$s(t) = \gamma(x(t)), \qquad x(t) = \Gamma(s(t)), \qquad t \in [t_1, t_2]. \tag{2.24.34}$$

We shall describe the correspondence of Eq. (2.24.34), denoting it by $\mathbf{s} = B\mathbf{x}$.

If we denote Γ' the derivative of Γ, we see that if $\mathbf{s} = B\mathbf{x}$,

$$\dot{x}(t) = \Gamma'(s(t))\dot{s}(t), \tag{2.24.35}$$

and we can note that the Lagrangians

$$\mathfrak{L}(\eta, \xi, t) = \frac{m\eta^2}{2} - V(\xi), \tag{2.24.36}$$

$$\tilde{\mathfrak{L}}(\eta, \xi, t) = m\frac{\Gamma'(\xi)^2}{2}\eta^2 - V(\Gamma(\xi)) \tag{2.24.37}$$

attribute the same action to the motions $\mathbf{x} \in \mathfrak{M}_{t_1,t_2}(\xi_1, \xi_2)$ and $\mathbf{s} = B\mathbf{x} \in \mathfrak{M}_{t_1,t_2}(\gamma(\xi_1), \gamma(\xi_2))$, respectively; i.e., if $\mathbf{s} = B\mathbf{x}$,

$$\begin{aligned} A(\mathbf{x}) &= \int_{t_1}^{t_2} dt \left(\frac{m\dot{x}(t)^2}{2} - V(x(t)) \right) \\ &\equiv \tilde{A}(\mathbf{s}) = \int_{t_1}^{t_2} \left(\frac{m\Gamma'(s(t))^2 \dot{s}(t)^2}{2} - V(\Gamma(s(t))) \right) dt, \end{aligned} \tag{2.24.38}$$

since this immediately follows from Eqs. (2.24.34) and (2.24.35).

If $y \in \mathfrak{V}_{\mathbf{x}}(\mathfrak{M})$, it is natural to associate with y the element By of $\mathfrak{V}_{\mathbf{s}}(B\mathfrak{M})$ defined by

$$(By)(t, \epsilon) = \gamma(y(t, \epsilon)), \qquad (t, \epsilon) \in [t_1, t_2] \times (-1, 1) \tag{2.24.39}$$

and $B\mathfrak{M} \subset \mathfrak{M}_{t_1,t_2}(\gamma(\xi_1), \gamma(\xi_2))$ is the image of $\mathfrak{M}$ via the map of Eq. (2.24.39).

It is then an immediate consequence of Definitions 23 and 24 that if A is stationary or locally minimal on $\mathbf{x} \in \mathfrak{M}_{t_1,t_2}(\xi_1, \xi_2)$ in $\mathfrak{M}$, then $\tilde{A}$ also is stationary or locally minimal on $\mathbf{s} = B\mathbf{x} \in \mathfrak{M}_{t_1,t_2}(\gamma(\xi_1), \gamma(\xi_2))$ in $B\mathfrak{M}$ and vice versa.

In particular, this means that the equations

$$\frac{d}{dt}\left(\frac{\partial \mathfrak{L}}{\partial \eta}(\dot{x}(t), x(t), t)\right) = \frac{\partial \mathfrak{L}}{\partial \xi}(\dot{x}(t), x(t), t),$$
$$\frac{d}{dt}\left(\frac{\partial \tilde{\mathfrak{L}}}{\partial \eta}(\dot{s}(t), s(t), t)\right) = \frac{\partial \tilde{\mathfrak{L}}}{\partial \xi}(\dot{s}(t), s(t), t) \tag{2.24.40}$$

are "equivalent" if $\mathfrak{L}$ and $\tilde{\mathfrak{L}}$ are given by Eqs. (2.24.36) and (2.24.37).

As we shall see, this invariance property of the stationarity (or of the local minimality) with respect to changes of coordinates is perhaps the most interesting aspect of all the considerations of this section. We shall meet some of its very remarkable applications in the theory of systems with many degrees of freedom.

Concluding Remarks

(1) In the analysis of this section we always dealt with conservative systems. In fact, it is not possible to give a simple formulation of the stationary action principle for dissipative motions without introducing singular Lagrangians (see Problems 12–15 at the end of this section).

(2) The action of a motion x with Lagrangian (2.24.36) can be thought of as the product of $(t_2 - t_1)$ times the difference between the average value, in $[t_1, t_2]$, of the kinetic energy and the average value of the potential energy:

$$\frac{A(\mathbf{x})}{t_2 - t_1} = \frac{1}{t_2 - t_1}\int_{t_1}^{t_2} \frac{m\dot{x}(t)^2}{2}\, dt - \frac{1}{t_2 - t_1}\int_{t_1}^{t_2} V(x(t))\, dt. \tag{2.24.41}$$

It is for this reason that one can say that the motion developing for $t \in [t_1, t_2]$ between t_1 and t_2 under the influence of a force of given potential energy V is the one that minimizes the difference between the average kinetic energy and the average potential energy in every short enough time interval in $[t_1, t_2]$.

We leave it to the reader to elaborate his own philosophical considerations on this beautiful mathematical property. The interested reader could go through the history of the variational principles in mechanics and, more generally, in physics, to understand how subjective considerations (as we would call them today) have influenced the formulation of the variational principles themselves and the recognition of their equivalence to the Newtonian equations of motion; see also the comments on p. 164 and p. 242 and the Euler's quotation at the beginning of this section.

Exercises and Problems for §2.24

1. Compute the action between $t_1 = 0$ and $t_2 = 2\pi/\omega$ of the motions of a harmonic oscillator with mass $m > 0$ and pulsation ω.

2. Same as Problem 1 with t_2 arbitrary $(t_2 \neq 2\pi/\omega)$.

3. Compute the action between $t_1 = 0$ and arbitrary t_2 of the motions of a point mass with mass $m > 0$ subject to the force $f = -mg$, $g > 0$.

4. Let $\mathcal{L} \in C^\infty(\mathbf{R}^2)$ be such that the correspondence $(\eta, \xi) \to (p, \xi) = ((\partial\mathcal{L}/\partial\eta)(\eta, \xi), \xi)$ can be inverted in class C^∞ as a mapping of $\mathbf{R}^2$ onto $\mathbf{R}^2$ and let $(p, \xi) \to (f(p, \xi), \xi)$ be the inverse map. Set

$$H(p, \xi) = pf(p, \xi) - \mathcal{L}(f(p, \xi), \xi) \equiv [p\eta - \mathcal{L}(\eta, \xi)]_{\eta = f(p,\xi)}$$

and check that the Lagrange equations

$$\dot{\xi} = \eta,$$

$$\frac{d}{dt}\frac{\partial\mathcal{L}}{\partial\eta}(\eta, \xi) = \frac{\partial\mathcal{L}}{\partial\xi}(\eta, \xi)$$

are equivalent to the "Hamilton equations"

$$\dot{p} = -\frac{\partial H}{\partial\xi}(p, \xi),$$

$$\dot{\xi} = \frac{\partial H}{\partial p}(p, \xi).$$

The motion described in terms of p and ξ, $t \to (p(t), \xi(t))$, is a solution of this differential equation and any of its solutions is called a "Hamiltonian motion" and the space $\mathbf{R}^2$, thought of as the space of the initial data for the above equations, is called a "phase space". (Hint: Note that since, by definition of p and f, one has $p \equiv (\partial/\partial\eta)(f(p, \xi), \xi)$, it follows that

$$\begin{aligned}\frac{\partial H}{\partial p}(p, \xi) &= f(p, \xi) + p\frac{\partial f}{\partial p}(p, \xi) - \frac{\partial\mathcal{L}}{\partial\eta}(f(p, \xi), \xi)\frac{\partial f}{\partial p}(p, \xi)\\ &\equiv f(p, \xi) = \eta\end{aligned}$$

and

$$\begin{aligned}\frac{\partial H}{\partial\xi}(p, \xi) &= p\frac{\partial f}{\partial\xi}(p, \xi) - \frac{\partial\mathcal{L}}{\partial\eta}(f(p, \xi), \xi)\frac{\partial f}{\partial\xi}(p, \xi) - \frac{\partial\mathcal{L}}{\partial\xi}(f(p, \xi), \xi)\\ &\equiv -\frac{\partial\mathcal{L}}{\partial\xi}(f(p, \xi), \xi) = -\frac{\partial\mathcal{L}}{\partial\xi}(\eta, \xi),\end{aligned}$$

having used the definition of H.)

5. The function H in Problem 4 can be expressed in terms of $\mathcal{L}$ and vice versa as

$$H(p, \xi) = \max_{\eta\in\mathbf{R}}(p\eta - \mathcal{L}(\eta, \xi),$$

$$\mathcal{L}(\eta, \xi) = \max_{p\in\mathbf{R}}(p\eta - H(p, \xi))$$

("Legendre duality"), if the maximum is attained at a unique point $\bar{\eta}$ or $\bar{p}$, respectively, and, furthermore, if these are the only stationarity points of the functions in brackets as functions of η or p, respectively. (Hint: Write the stationarity conditions for $p\eta - \mathcal{L}(\eta, \xi)$ and those for $p\eta - H(p, \xi)$ with respect to η or, respectively, to p. Then use the definition of H in Problem 4.)

6. The "Hamilton equations"

$$\dot{p} = -\frac{\partial H}{\partial\xi}(p, \xi), \qquad \dot{\xi} = \frac{\partial H}{\partial p}(p, \xi)$$

with Hamiltonian $H \in C^\infty(\mathbf{R}^2)$ can be obtained by imposing the stationarity of the quantity

$$S=\int_{t_1}^{t_2}(p(t)\dot{x}(t)-H(p(t),x(t)))dt$$

in the space $\mathfrak{M}_{t_1,t_2}((\pi_1,\xi_1),(\pi_2,\xi_2))$ of the $C^\infty([t_1,t_2])$ functions $t\to(p(t),q(t))\in\mathbf{R}^2$ such that $p(t_1)=\pi_1$, $p(t_2)=\pi_2$, $x(t_1)=\xi_1$, $x(t_2)=\xi_2$, ("Hamilton's principle"). (Hint: Apply Proposition 36, Eq. (2.24.1), with $\mathcal{L}(\dot{p},\dot{x},p,x,t)=p\dot{x}-H(p,x)$.)

7. In the context of Problem 6, show that the same Hamilton equations can be obtained by imposing the stationarity of S on the larger space $\mathfrak{M}_{t_1,t_2}(\xi_1,\xi_2)$ of the $C^\infty([t_1,t_2])$ function $t\to(p(t),q(t))\in\mathbf{R}^2$ such that $x(t_1)=\xi_1$, $x(t_2)=\xi_2$. (Hint: Go through the proof of Proposition 36 using the special form of the Lagrangian $\mathcal{L}(\dot{p},\dot{x},p,x)=p\dot{x}-H(p,x)$.)

8. Let $t\to(p(t),x(t))\in\mathbf{R}^2$ be a motion *verifying* the Hamilton equations of Problems 4 and 6. Show that the quantity S defined in Problem 6 coincides with $\int_{t_1}^{t_2}\mathcal{L}(\dot{x}(t),x(t))\,dt$, i.e., with the action of the same motion (of course, if $\mathcal{L}$ and H are related as in Problem 4).

9. Extend Problems 4 and 7 to the case when H and $\mathcal{L}$ depend explicitly on time.

10.* Let H be as in Problem 4 and let $S_t(p,x)=(p(t),x(t))$, $t\geqslant 0$, be the solution of the Hamilton equations (as in Problem 14), supposed normal, with (p,x) as initial datum at $t=0$.

Let $A\subset\mathbf{R}^2$ be a (Riemann) measurable region. Show that area $(S_tA)=$ area A, $\forall t\geqslant 0$, if $S_tA=\{$set of points of the form $S_t(p,x)$, with $(p,x)\in A\}$ ("Liouville's theorem"). (Hint: In general, let $\dot{\mathbf{x}}=\mathbf{f}(\mathbf{x})$ be an autonomous normal differential equation in $\mathbf{R}^d$. Set $\mathbf{y}=S_t\mathbf{x}$, for $t\geqslant 0$. Then

$$\text{volume } S_tA=\int_{S_tA}d\mathbf{x}=\int_A\left|\det\left(\frac{\partial S_{-t}}{\partial\mathbf{y}}(\mathbf{y})\right)\right|d\mathbf{y},$$

where $\partial S_{-t}(\mathbf{y})/\partial\mathbf{y}$ denotes the Jacobian matrix of the coordinate transformation $\mathbf{x}=S_{-t}(\mathbf{y})$. This formula shows that if $|\det(\partial S_{-t}(\mathbf{y})/\partial\mathbf{y})|>0$, the modulus symbol is irrelevant and $t\to$ volume (S_tA) is a C^∞ function, and

$$\frac{d}{dt}\text{volume } S_tA\Big|_{t=\tau}=\int_A\left[\frac{d}{dt}\left(\det\frac{\partial S_{-t}(\mathbf{y})}{\partial\mathbf{y}}\right)\right]_{t=\tau}d\mathbf{y},$$

but (see §2.6) $S_{t+\tau}=S_tS_\tau$; hence, the last expression is equal to:

$$\begin{aligned}&\int_A\left[\frac{d}{dt}\det\frac{\partial S_{-t-\tau}}{\partial\mathbf{y}}(\mathbf{y})\right]_{t=0}dy\\&=\int_A\left[\frac{d}{dt}\det\frac{\partial S_{-t}(S_{-\tau}y)}{\partial\mathbf{y}}\right]_{t=0}dy\\&=\int_A\left[\frac{d}{dt}\det\frac{\partial S_{-t}(S_{-\tau}\mathbf{y})}{\partial(S_{-\tau}\mathbf{y})}\right]_{t=0}\det\left(\frac{\partial S_{-\tau}(\mathbf{y})}{\partial y}\right)dy\end{aligned}$$

by the composite function differentiation rule and by the determinant rules.

It is then sufficient to check that, under suitable circumstances, the derivative

$$\left[\frac{d}{dt}\det\frac{\partial S_{-t}(\mathbf{x})}{\partial\mathbf{x}}\right]_{t=0}\equiv 0,\qquad\forall\mathbf{x}\in\mathbf{R}^d$$

to infer the volume conservation under the same circumstances. If $\dot{\mathbf{x}}=\mathbf{f}(\mathbf{x})$, it follows that

$$S_t\mathbf{x} = \mathbf{x} + t\mathbf{f}(\mathbf{x}) + t^2\boldsymbol{\varphi}(\mathbf{x}, t)$$

by the Taylor–Lagrange formula (see Appendix B), where φ is a C^∞ function of $\mathbf{x}$ and t. Hence,

$$\det \frac{\partial S_t(\mathbf{x})}{\partial \mathbf{x}} = \det\left(1 + t\left(\frac{\partial f^{(i)}}{\partial x_j}(\mathbf{x})\right)\right) + t^2\left(\frac{\partial \varphi^{(i)}}{\partial x_j}(\mathbf{x}, t)\right);$$

hence, by developing the determinant

$$\det \frac{\partial S_t(\mathbf{x})}{\partial \mathbf{x}} = 1 + t\sum_{j=1}^{d} \frac{\partial f^{(j)}}{\partial x_j}(\mathbf{x}) + t^2\psi(\mathbf{x}, t),$$

where ψ is a suitable C^∞ function of $\mathbf{x}$, t. Hence, the derivative of $\det(\partial S_t(\mathbf{x})/\partial \mathbf{x})$ for $t = 0$ is

$$\sum_{j=1}^{d} \frac{\partial f^{(j)}}{\partial x_j}(\mathbf{x}) \equiv \operatorname{div} \mathbf{f}(\mathbf{x}),$$

wherein the right-hand side is the notation used in physics for the left-hand side ("divergence of f").

Therefore, if $\operatorname{div}\mathbf{f} = 0$, the flow S_t generated by $\dot{\mathbf{x}} = \mathbf{f}(\mathbf{x})$ preserves the volume (this also motivates the name "divergence" given to $\operatorname{div}\mathbf{f}(\mathbf{x})$ since it measures the rate of increase of volume under the transformation S_t). In fact, it easily follows from the above considerations that $\det(\partial S_{-t}(\mathbf{x})/\partial \mathbf{x}) \equiv 1$, being constant and equal to 1 for $t = 0$. Then note that the Hamilton equations are divergenceless.)

11.* Let $\dot{\mathbf{x}} = \mathbf{f}(\mathbf{x})$ be an autonomous normal differential equation in $\mathbf{R}^d$, $f \in C^\infty(\mathbf{R}^d)$, and suppose $\operatorname{div}\mathbf{f}(\mathbf{x}) \equiv \sum_{j=1}^{d}(\partial f^{(j)}/\partial x_j)(\mathbf{x}) = 0$, $\forall \mathbf{x} \in \mathbf{R}^d$. So, by the hint to Problem 10, it follows that the solution flow $(S_t)_{t\in R_+}$ preserves the volume: volume $S_tA \equiv$ volume A.

Suppose that the solution flow maps a bounded open set $\Omega \subset \mathbf{R}^d$ into itself: $S_t\Omega \subset \Omega$, $\forall t \in \mathbf{R}_+$. Show that given $\mathbf{x}_0 \in \Omega$, $t_0 > 0$, and a neighborhood $U \subset \Omega$ of $\mathbf{x}_0$, there exists $t \geqslant t_0$ such that $S_tU \cap U \neq \emptyset$; i.e., close to any point $\mathbf{x}_0 \in \Omega$, there is another point which comes as close to $\mathbf{x}_0$ after a given arbitrarily large time ("Poincaré's recurrence theorem"). (Hint: Suppose $S_{t_0}U \cap U = \emptyset$, otherwise $t = t_0$; then consider $S_{2t_0}U$. If $S_{2t_0}U \cap U = \emptyset$, show that the three sets U, $S_{t_0}U$, $S_{2t_0}U$ must be pairwise disjoint; if $S_{2t_0}U \cap U \neq \emptyset$, then $t = 2t_0$. In the first case, consider $S_{3t_0}U$: if $S_{3t_0}U \cap U \neq \emptyset$, the four sets U, $S_{t_0}U$, $S_{2t_0}U$, and $S_{3t_0}U$ must be pairwise disjoint; if not, take $t = 3t_0$, etc. The result could fail only if the sequence U, $S_{t_0}U$, $S_{2t_0}U, \ldots, S_{kt_0}U, \ldots$ is an infinite sequence of pairwise disjoint sets. However, in such a case, volume $\Omega \geqslant \sum_{k=0}^{\infty}$ volume $S_{kt_0}U = \sum_{k=0}^{\infty}$ volume $U = +\infty$, since U is open, which is absurd since U is a bounded set.)

12. Show that the equation $\ddot{x} + \gamma\dot{x} = 0$, $\gamma > 0$ describing a free particle moving under the action of linear friction is the Euler–Lagrange equation associated with the Lagrangian

$$\mathcal{L}(\dot{x}, x) = \dot{x}\log\dot{x} + \gamma x$$

in the region $\dot{x} > 0$ (Kobussen). (Define the Euler–Lagrange equations as $(d/dt)\,\partial\mathcal{L}/\partial\dot{x} = \partial\mathcal{L}/\partial x$.)

13. Let $V \in C^\infty(\mathbf{R})$ be bounded below. Show that if $F \in C^\infty(\mathbf{R})$ has a nonvanishing derivative, the equations $\ddot{x} = -(dV/dx)(x)$ can be described by the Lagrangian function

$$\mathfrak{L}(\eta, \xi) = \eta \int_1^{\eta} \frac{F(\frac{1}{2}y^2 + V(\xi))dy}{y^2}$$

in the region $\dot{x} > 0$, i.e., $\eta > 0$.

What does $\mathfrak{L}$ become if $F(e) \equiv e$, $\forall e \in \mathbf{R}$? (Kobussen)

14. Consider the damped oscillator $\ddot{x} + \dot{x} + \omega^2 x = 0$ and let $\alpha = (4\omega^2 - \gamma)^{-1/2}$, $\gamma > 0$. Show that in the region $\eta > 0$, $\xi > 0$, the Lagrangian

$$\mathfrak{L}(\eta, \xi) = -\frac{1}{2}\log(\eta^2 + \gamma\eta\xi + \omega^2\xi) + \alpha\left(2\frac{\eta}{\xi} + \gamma\right)\operatorname{arctg}\alpha\left(2\frac{\eta}{\xi} + \gamma\right)$$

has, as Euler–Lagrange equations, the damped oscillator equations (Kobussen).

15. Let $\ddot{x} = g(\dot{x}, x)$ be a differential equation. Show that in order that a function $\mathfrak{L}$ on a subset A of $\mathbf{R}^2$ generates, (via the Euler–Lagrange equations) the equation $\ddot{x} = g(\dot{x}, x)$ for the motions developing in A, it must be

$$\frac{\partial \mathfrak{L}}{\partial \xi} - \eta \frac{\partial^2 \mathfrak{L}}{\partial \xi \partial \eta} - g(\eta, \xi) \frac{\partial^2 \mathfrak{L}}{\partial \eta^2} = 0, \qquad \forall (\eta, \xi) \in A.$$

(Hint: Write the Euler–Lagrange equations substituting $\ddot{x}$ with $g(\dot{x}, x)$ (Kobussen).)[25]

[25]The last four problems are taken from a paper by J. Kobussen (see references). The equation for $\mathfrak{L}$ in Problem 15 allows one to find many Lagrangians for the same equation. Note, however, that such Lagrangians will generally be singular somewhere in $\mathbf{R}^2$, always so, probably, if the equation $\ddot{x} = g(\dot{x}, x)$ is nonconservative. So, strictly speaking, this confirms the fact that a Lagrangian description, in the sense of §2.24, with $\mathfrak{L} \in C^{\infty}(\mathbf{R}^2)$ can only be found for conservative systems.

CHAPTER 3

Systems with Many Degrees of Freedom. Theory of the Constraints. Analytical Mechanics

§3.1. Systems of Point Masses

We begin with some definitions which are perhaps obvious from the considerations of Chapters 1 and 2, but are nevertheless necessary.

We begin with a notational convention that shall allow important formal simplifications. If $M = m_1 + m_2 + \cdots + m_p$ is a sum of p positive integers, we shall be able to think of the space $\mathbf{R}^M$ as identical with the space $\mathbf{R}^{m_1} \times \mathbf{R}^{m_2} \times \cdots \times \mathbf{R}^{m_p}$. A point $\boldsymbol{\xi} = (\xi_1, \ldots, \xi_M) \in \mathbf{R}^M$ will be identified with the p-tuple of vectors $(\boldsymbol{\xi}^{(1)}, \ldots, \boldsymbol{\xi}^{(p)}) \in \mathbf{R}^{m_1} \times \cdots \times \mathbf{R}^{m_p}$, where $\boldsymbol{\xi}^{(i)} = (\xi_{m_1 + \cdots + m_{i-1} + 1}, \ldots, \xi_{m_1 + \cdots + m_i})$ for $i = 1, 2, \ldots, p$.

Very often such a decomposition of $\boldsymbol{\xi}$ into $(\boldsymbol{\xi}^{(1)}, \ldots, \boldsymbol{\xi}^{(p)})$ will be "natural" in the context of the discussion. For instance, if a point in $\mathbf{R}^{3N}$ represents a configuration of a system of N point masses, it will be "natural" to think of $\boldsymbol{\xi}$ as $(\boldsymbol{\xi}^{(1)}, \boldsymbol{\xi}^{(2)}, \ldots, \boldsymbol{\xi}^{(N)})$, where $\boldsymbol{\xi}^{(i)} \in \mathbf{R}^3$, $i = 1, \ldots, N$, represents the position in $\mathbf{R}^3$ of the ith point mass. Every time that it will appear useful, when a natural decomposition of $\boldsymbol{\xi} \in \mathbf{R}^M$ into $(\boldsymbol{\xi}^{(1)}, \ldots, \boldsymbol{\xi}^{(p)})$, $\boldsymbol{\xi}^{(i)} \in \mathbf{R}^{m_i}$, $i = 1, \ldots, p$, emerges from the context, we shall regard $\boldsymbol{\xi}$ as a p-tuple of vectors in $\mathbf{R}^{m_1} \times \cdots \times \mathbf{R}^{m_p}$.

Such an identification will be made without explicit mention, provided no real ambiguities arise. Thus, a $\mathbf{R}^{3N}$-valued function $t \to \boldsymbol{\varphi}(t)$ defined on $\mathbf{R}$ will be written, if this is natural within the context, as $t \to (\boldsymbol{\varphi}^{(1)}(t), \ldots, \boldsymbol{\varphi}^{(N)}(t))$ with $t \to \boldsymbol{\varphi}^{(i)}(t)$, $i = 1, \ldots, N$, an $\mathbf{R}^3$-valued function, etc.

If $\mathbf{F}$ is an $\mathbf{R}^d$-valued C^∞ function on $\mathbf{R}^M = \mathbf{R}^{m_1} \times \cdots \times \mathbf{R}^{m_p}$, we shall denote the Jacobian matrix

$$\left(\frac{\partial F^{(i)}}{\partial \xi_j} (\boldsymbol{\xi}) \right)_{\substack{i=1,\ldots,d, \\ j=1,\ldots,M}}$$

with the symbol $(\partial \mathbf{F}/\partial \boldsymbol{\xi})(\boldsymbol{\xi})$. If $\boldsymbol{\xi} = (\boldsymbol{\xi}^{(1)}, \ldots, \boldsymbol{\xi}^{(p)}) \in \mathbf{R}^{m_1} \times \cdots \times \mathbf{R}^{m_p}$, the symbol $(\partial \mathbf{F}/\partial \boldsymbol{\xi}^{(s)})(\boldsymbol{\xi})$ will denote the Jacobian matrix $(\partial F^{(i)}/\partial \xi_l)(\boldsymbol{\xi})$, where $i = 1, \ldots, d$ and l varies in the set of indices corresponding to the coordinates of $\boldsymbol{\xi}^{(s)}$ (i.e., $l = m_1 + \cdots + m_{s-1} + 1, \ldots, m_1 + \cdots + m_s$).

We can now set up the following definition.

1 Definition. A "motion" of a system of N point masses in $\mathbf{R}^d$, observed as the time varies in the interval I, is a C^∞ function $t \to \mathbf{x}(t) = (\mathbf{x}^{(1)}(t), \ldots, \mathbf{x}^{(N)}(t))$ defined for $t \in I$ and taking values in $\mathbf{R}^{Nd} = \mathbf{R}^d \times \cdots \times \mathbf{R}^d$.

We shall say that a motion $\mathbf{x}$ of a system of N point masses with respective masses $m_1, \ldots, m_N > 0$ is governed by a "force law $\mathbf{F}$" or "develops under the influence" of a force law $\mathbf{F}$ if:

(i) $\mathbf{F}$ is an $\mathbf{R}^{dN}$-valued C^∞ function defined on $\mathbf{R}^{2dN+1}$ which we shall write $\mathbf{F} = (\mathbf{f}^{(1)}, \ldots, \mathbf{f}^{(N)})$ with $\mathbf{f}^{(i)}$ being an $\mathbf{R}^d$-valued C^∞ function on $\mathbf{R}^{2dN+1}$, $i = 1, 2, \ldots, N$.

(ii) For $i = 1, \ldots, N$, $\forall t \in I$:

$$m_i \ddot{\mathbf{x}}^{(i)}(t) = f^{(i)}(\dot{\mathbf{x}}^{(1)}(t), \ldots, \dot{\mathbf{x}}^{(N)}(t), \mathbf{x}^{(1)}(t), \ldots, \mathbf{x}^{(N)}(t), t). \quad (3.1.1)$$

(iii) Equation (3.1.1), thought of as a differential equation, is normal for all values of $m_1, \ldots, m_N > 0$ (see Definition 3, §2.5).

Observations. Requirement (iii) is a restriction of "physical nature" on the force laws $\mathbf{F}$ that we shall consider. Such laws will often be subject to other restrictions and, always (beginning with the next section), to the condition of verifying the third principle of dynamics (see Chapter 1, §1.3).

A particularly important role will be played by the "conservative force laws" which deserve a formal definition and to which we dedicate the rest of the section.

2 Definition. A force law for a system of N point masses in $\mathbf{R}^d$, i.e., a function $\mathbf{F} \in C^\infty(\mathbf{R}^{2dN+1})$ with values in $\mathbf{R}^{dN}$, verifying (i) and (iii) of Definition 1, is called "conservative" if:

(i) it depends solely on the system's configuration, i.e., there exist N $\mathbf{R}^d$-valued C^∞ functions defined on $\mathbf{R}^{dN}$, $\tilde{\mathbf{f}}^{(1)}, \ldots, \tilde{\mathbf{f}}^{(N)}$, such that

$$f^{(i)}(\boldsymbol{\eta}^{(1)}, \ldots, \boldsymbol{\eta}^{(N)}, \boldsymbol{\xi}^{(1)}, \ldots, \boldsymbol{\xi}^{(N)}, t) \equiv \tilde{\mathbf{f}}^{(i)}(\boldsymbol{\xi}^{(1)}, \ldots, \boldsymbol{\xi}^{(N)}); \quad (3.1.2)$$

(ii) there is a real-valued function $V \in C^\infty(\mathbf{R}^{dN})$ such that for $i = 1, \ldots, N$:

$$\tilde{\mathbf{f}}^{(i)}(\boldsymbol{\xi}^{(1)}, \ldots, \boldsymbol{\xi}^{(N)}) = -\frac{\partial V}{\partial \boldsymbol{\xi}^{(i)}}(\boldsymbol{\xi}^{(1)}, \ldots, \boldsymbol{\xi}^{(N)}) \tag{3.1.3}$$

which will be called the "potential energy" of the force law $\mathbf{F}$.

The interest of this definition lies in the fact that the majority of force models are described by conservative force laws, i.e., by force laws that can be expressed as in Eq. (3.1.3) which, according to the conventions set up at the beginning of this section, means:

$$\left(\tilde{\mathbf{f}}^{(i)}(\boldsymbol{\xi}^{(1)}, \ldots, \boldsymbol{\xi}^{(N)})\right)_j = -\frac{\partial V}{\partial (\boldsymbol{\xi}^{(i)})_j}(\boldsymbol{\xi}^{(1)}, \ldots, \boldsymbol{\xi}^{(N)}), \tag{3.1.4}$$

where $i = 1, \ldots, N$ and $j = 1, \ldots, d$.

Furthermore, the energy conservation theorem can be easily extended to systems of N point masses subject to conservative forces.

Given a motion $\mathbf{x}$ of a system of N point masses with respective mass $m_1, \ldots, m_N > 0$, we define the "kinetic energy" at time t as the quantity

$$T(t) = \frac{1}{2} \sum_{i=1}^{N} m_i \left(\dot{\mathbf{x}}^{(i)}(t)\right)^2, \tag{3.1.5}$$

and the "potential energy" at time t of the force $\mathbf{F}$ governing the motion, and supposed conservative with potential energy function $V \in C^\infty(\mathbf{R}^{dN})$, will be defined as

$$V(t) = V\left(\mathbf{x}^{(1)}(t), \ldots, \mathbf{x}^{(N)}(t)\right). \tag{3.1.6}$$

One then notes that

$$\frac{d}{dt} T(t) = \sum_{i=1}^{N} m_i \dot{\mathbf{x}}^{(i)}(t) \cdot \ddot{\mathbf{x}}^{(i)}(t), \tag{3.1.7}$$

$$\frac{d}{dt} V(t) = \sum_{i=1}^{N} \frac{\partial V}{\partial \boldsymbol{\xi}^{(i)}}(\mathbf{x}(t)) \cdot \dot{\mathbf{x}}^{(i)}(t); \tag{3.1.8}$$

hence, by Eqs. (3.1.1), (3.1.2), and (3.1.3):

$$\frac{d}{dt}(T(t) + V(t)) = \sum_{i=1}^{N} \dot{\mathbf{x}}^{(i)}(t) \cdot \left(m_i \ddot{\mathbf{x}}^{(i)}(t) + \frac{\partial V}{\partial \boldsymbol{\xi}^{(i)}}(\mathbf{x}(t)) \right) = 0. \tag{3.1.9}$$

Therefore, the following proposition holds.

1 Proposition. *If $t \to \mathbf{x}(t) = (\mathbf{x}^{(1)}(t), \ldots, \mathbf{x}^{(N)}(t))$, $t \in I$, is the motion of a system of N point masses, governed by a conservative force law with potential energy V, there is a constant E, "total energy" of the motion, equal at all times to the sum of the kinetic energy and the potential energy:*

$$T(t) + V(t) = E, \qquad \forall t \in I \tag{3.1.10}$$

with $T(t)$ and $V(t)$ defined in Eqs. (3.1.5) *and* (3.1.6).

Observation. It is worth stressing that here we are meeting a first but very important difference between one-dimensional and multi-dimensional motions: in the case of the motion of a single point in one dimension, every purely positional force law is conservative. If $d > 1$ or $N > 1$, there are purely positional force laws which are not conservative in the above sense. For instance, if $N = 1$, $d = 2$, the force law $f_1(\xi_1, \xi_2) = 0$, $f_2(\xi_1, \xi_2) = \xi_1$ is not conservative since $\partial f_1/\partial \xi_2 \neq \partial f_2/\partial \xi_1$, while it is clear that these two derivatives should coincide if $\mathbf{f}$ were conservative (since they would be the mixed second-order derivatives of the same function V).

Exercises for §3.1

1. Let $\mathbf{f}$ be a $C^\infty(R^3/\{\mathbf{0}\})$ function with values in $\mathbf{R}^3$ having the form $\mathbf{f}(\mathbf{x}) = \varphi(|\mathbf{x}|) \cdot \mathbf{x}/|\mathbf{x}|$ with $\varphi \in C^\infty(R_+/\{0\})$. Consider the force law for a system of N point masses given by

$$\begin{aligned}\mathbf{f}^{(i)}(\boldsymbol{\xi}^{(1)}, \ldots, \boldsymbol{\xi}^{(N)}) &= \sum_{j \neq i} \varphi(|\boldsymbol{\xi}^{(i)} - \boldsymbol{\xi}^{(j)}|) \frac{\boldsymbol{\xi}^{(i)} - \boldsymbol{\xi}^{(j)}}{|\boldsymbol{\xi}^{(i)} - \boldsymbol{\xi}^{(j)}|} \\ &\equiv \sum_{j \neq i} \mathbf{f}(\boldsymbol{\xi}^{(i)} - \boldsymbol{\xi}^{(j)}).\end{aligned}$$

This force law is defined for configurations such that $\boldsymbol{\xi}^{(i)} \neq \boldsymbol{\xi}^{(j)}$, $\forall i \neq j$, and strictly speaking is, therefore, a generalization of the force law notion of Definitions 1 and 2 (requiring the force to be defined for every configuration $(\boldsymbol{\xi}^{(1)}, \ldots, \boldsymbol{\xi}^{(N)})$). It will be called conservative if there is a function V of class C^∞ on the configurations such that $\boldsymbol{\xi}^{(i)} \neq \boldsymbol{\xi}^{(j)}$, $\forall i \neq j$, and such that Eq. (3.1.4) holds.

In this extended sense, show that the above force law is conservative and

$$V(\boldsymbol{\xi}^{(1)}, \ldots, \boldsymbol{\xi}^{(N)}) = \sum_{j < j'} \Phi(|\boldsymbol{\xi}^{(j)} - \boldsymbol{\xi}^{(j')}|),$$

where $r \to \Phi(r)$, $r \in \mathbf{R}_+$, is a primitive function to φ: $\Phi(r) = \int^r \varphi(r')\, dr'$.

Find sufficient conditions on φ so that the above force law can be extended by continuity to all configurations becoming a conservative force law in the sense of Definition 2.

2. Let $\Phi_{j,j'}(r)$, $j, j' = 1, \ldots, N$, $j < j'$, be $N(N-1)$ $C^\infty(\mathbf{R})$ functions. Write the force law with potential energy function

$$V(\boldsymbol{\xi}^{(1)}, \ldots, \boldsymbol{\xi}^{(N)}) = \sum_{j < j'} \Phi_{j,j'}(|\boldsymbol{\xi}^{(j)} - \boldsymbol{\xi}^{(j')}|)$$

Find sufficient conditions on Φ so that the force law is of class C^∞.

§3.2. Work. Linear and Angular Momentum

One can wonder whether it is possible to extend the energy conservation theorem so that it could be applied to systems subject to nonconservative force laws.

The answer is, in some sense, affirmative and it is known as the "alive forces theorem". To formulate this simple theorem, one needs the notion of "work of a force" on a given motion.

3 Definition. (i) A $\mathbf{R}^{dN}$-valued $C^\infty(\mathbf{R}^{2dN+1})$ function $\mathbf{F}$ verifying properties (i) and (iii) of Definition 1 will be called a "force law" for a system of N point masses.

(ii) If $\mathbf{x}$ is a motion, defined for $t \in I$, of a system of N point masses and if $\mathbf{\Phi}$ is a force law for it, not necessarily coinciding with the force law generating the motion $\mathbf{x}$ [i.e. not necessarily verifying Eq. (3.1.1)], one defines the "work" of the force $\mathbf{\Phi}$ in the time interval $[t_1, t_2] \subset I$ as the quantity

$$L_{t_1,t_2}(\mathbf{\Phi}, \mathbf{x}) = \sum_{i=1}^{N} \int_{t_1}^{t_2} \boldsymbol{\varphi}^{(i)}(\dot{\mathbf{x}}(t), \mathbf{x}(t), t) \cdot \dot{\mathbf{x}}^{(i)}(t)\, dt, \tag{3.2.1}$$

where, following the conventions of §3.1, we set $\mathbf{\Phi} = (\boldsymbol{\varphi}^{(1)}, \ldots, \boldsymbol{\varphi}^{(N)})$, $\mathbf{x} = (\mathbf{x}^{(1)}, \ldots, \mathbf{x}^{(N)})$.

Observations.

(1) Let $\mathbf{\Phi}$ be a purely positional force law, i.e., $\forall (\boldsymbol{\eta}, \boldsymbol{\xi}, t) \in \mathbf{R}^{2dN+1}$, $i = 1, \ldots, N$:

$$\boldsymbol{\varphi}^{(i)}(\boldsymbol{\eta}^{(1)}, \ldots, \boldsymbol{\eta}^{(N)}, \boldsymbol{\xi}^{(1)}, \ldots, \boldsymbol{\xi}^{(N)}, t) = \tilde{\boldsymbol{\varphi}}^{(i)}(\boldsymbol{\xi}^{(1)}, \ldots, \boldsymbol{\xi}^{(N)}). \tag{3.2.2}$$

Then

$$L_{t_1,t_2}(\mathbf{\Phi}, \mathbf{x}) = \int_{t_1}^{t_2} \sum_{i=1}^{N} \tilde{\boldsymbol{\varphi}}^{(i)}(\mathbf{x}(t)) \cdot \dot{\mathbf{x}}^{(i)}(t)\, dt \tag{3.2.3}$$

and one recognizes in the above integral a line integral of the differential form

$$\sum_{i=1}^{N} \tilde{\boldsymbol{\varphi}}^{(i)}(\boldsymbol{\xi}) \cdot d\boldsymbol{\xi}^{(i)} \tag{3.2.4}$$

on the curve $\mathfrak{T}(\mathbf{x})$ described in $\mathbf{R}^{Nd}$ by the point $\mathbf{x}(t)$ as t varies in $[t_1, t_2]$ ("trajectory of $\mathbf{x}$"). Formula (3.2.4) is usually read by saying that the work done by a force on a point which undergoes a displacement is the "scalar product of the force times the displacement".

(2) From Observation 1, it follows that the work done by a purely positional force law $\mathbf{\Phi}$ in a given time interval during which the system is displaced from the configuration $\mathbf{x}(t_1)$ to $\mathbf{x}(t_2)$ along a certain trajectory $\mathfrak{T}$ solely depends upon the trajectory $\mathfrak{T}$ and does not depend on the time law governing the motion along $\mathfrak{T}$.

(3) If $\mathbf{\Phi}$ is a conservative force with potential energy V [see Eq. (3.1.3)], one sees that the differential form of Eq. (3.2.4) coincides with the differen-

tial of $-V$:

$$\sum_{i=1}^{N} \tilde{\varphi}^{(i)}(\xi) \cdot d\xi^{(i)} = -\sum_{i=1}^{N} \frac{\partial V(\xi)}{\partial \xi^{(i)}} \cdot d\xi = -dV; \tag{3.2.5}$$

hence, from Eq. (3.2.3), it follows that

$$L_{t_1,t_2}(\Phi, \mathbf{x}) = -V(\mathbf{x}(t_2)) + V(\mathbf{x}(t_1)), \tag{3.2.6}$$

showing that the work performed in a given time interval by a conservative force on a motion $\mathbf{x}$ depends solely on the initial and final configurations of the motion, i.e., it is also independent of the trajectory followed by the motion.

The "theorem of the alive forces" can now be formulated.

2 Proposition. *Let $t \to \mathbf{x}(t)$, $t \in I$, be a motion of a system of N point masses with masses $m_1, \ldots, m_N > 0$ developing in $\mathbf{R}^d$ under the action of a force law $\mathbf{F}$.*

Then the variation of the kinetic energy, or "alive force",[1] *between the times $t_1, t_2 \in I$, is equal to the work performed in $[t_1, t_2]$ by $\mathbf{F}$ on the motion $\mathbf{x}$:*

$$T(t_2) - T(t_1) = L_{t_1,t_2}(\mathbf{F}, \mathbf{x}). \tag{3.2.7}$$

Observations. By Eq. (3.2.6), Eq. (3.2.7) becomes the already discussed energy conservation theorem, Proposition 1, whenever $\mathbf{F}$ is conservative.

PROOF. By Definition 1, p. 142, of motion developing under the action of a force $\mathbf{F} = (\mathbf{f}^{(1)}, \ldots, \mathbf{f}^{(N)})$, we have

$$m_j \ddot{\mathbf{x}}^{(j)} = \mathbf{f}^{(j)}(\dot{\mathbf{x}}^{(1)}, \ldots, \dot{\mathbf{x}}^{(N)}, \mathbf{x}^{(1)}, \ldots, \mathbf{x}^{(N)}, t), \tag{3.2.8}$$

$\forall j = 1, 2, \ldots, N$. Multiplying both sides scalarly by $\dot{\mathbf{x}}^{(j)}$ and summing over j:

$$\sum_{j=1}^{N} m_j \ddot{\mathbf{x}}^{(j)} \cdot \dot{\mathbf{x}}^{(j)} = \sum_{j=1}^{N} \mathbf{f}^{(j)} \cdot \dot{\mathbf{x}}^{(j)}, \tag{3.2.9}$$

and integrating both sides with respect to t between t_1 and t_2, one finds Eq. (3.2.7).

mbe

The interest of Proposition 2 lies in its generality as a consequence of Eq. (3.1.1).

There are other immediate consequences of Eq. (3.1.1) valid under the additional assumption that the force law $\mathbf{F}$ governing the motion verifies the third principle of dynamics: they are the so called "cardinal equations" of dynamics, whose interest is also due to their great generality.

As discussed in Chapter 1, the hypothesis that a force law $\mathbf{F}$ for a system of N point masses verifies the third law of dynamics means several things mathematically. First, if $\mathbf{F} = (\mathbf{f}^{(1)}, \ldots, \mathbf{f}^{(N)})$, the function $\mathbf{f}^{(j)}$, $j = 1, \ldots,$

[1] In the ancient times the alive force was actually defined to be twice the kinetic energy.

N, can be represented as

$$\begin{aligned}&\mathbf{f}^{(j)}(\boldsymbol{\eta}^{(1)},\ldots,\boldsymbol{\eta}^{(N)},\boldsymbol{\xi}^{(1)},\ldots,\boldsymbol{\xi}^{(N)},t)\\&\quad=\mathbf{f}^{(j)e}(\boldsymbol{\eta}^{(j)},\boldsymbol{\xi}^{(j)},t)+\sum_{\substack{i=1\\i\neq j}}^{N}\mathbf{f}^{(i\to j)}(\boldsymbol{\eta}^{(i)},\boldsymbol{\eta}^{(j)},\boldsymbol{\xi}^{(i)},\boldsymbol{\xi}^{(j)},t),\end{aligned}\tag{3.2.10}$$

where $\mathbf{f}^{(j)}\in C^{\infty}(\mathbf{R}^{2d+1})$, $\mathbf{f}^{(i\to j)}\in C^{\infty}(\mathbf{R}^{4d+1})$ are suitable $\mathbf{R}^d$-valued functions, $\forall i,j=1,\ldots,N$.

For reasons discussed in Chapter 1, the function $\mathbf{f}^{(j)e}$ is called the "external force" acting upon the jth point mass and $\mathbf{f}^{(i\to j)}$ is called the "force exerted by the ith point on the jth one".

Second, one assumes that

$$\mathbf{f}^{(i\to j)}(\boldsymbol{\eta},\boldsymbol{\eta}',\boldsymbol{\xi},\boldsymbol{\xi}',t)=-\mathbf{f}^{(j\to i)}(\boldsymbol{\eta}',\boldsymbol{\eta},\boldsymbol{\xi}',\boldsymbol{\xi},t)\tag{3.2.11}$$

and, finally,

$$\mathbf{f}^{(i\to j)}(\boldsymbol{\eta},\boldsymbol{\eta}',\boldsymbol{\xi},\boldsymbol{\xi}',t)\text{ is parallel to }(\boldsymbol{\xi}'-\boldsymbol{\xi}).\tag{3.2.12}$$

Equations (3.2.10)–(3.2.12) are the analytic form taken, in our notations, by the third principle of dynamics for the force law $\mathbf{F}$ acting on the system of point masses under consideration (see, also, Chapter 1).

4 Definition. A force law for a system of N point masses in $\mathbf{R}^d$ verifies the third principle of dynamics if it admits a representation like Eq. (3.2.10) verifying Eqs. (3.2.11) and (3.2.12). In this case, the quantity

$$\mathbf{R}^{(e)}(\boldsymbol{\eta}^{(1)},\ldots,\boldsymbol{\eta}^{(N)},\boldsymbol{\xi}^{(1)},\ldots,\boldsymbol{\xi}^{(N)},t)=\sum_{i=1}^{N}\mathbf{f}^{(j)e}(\boldsymbol{\eta}^{(j)},\boldsymbol{\xi}^{(j)},t)\tag{3.2.13}$$

thought of as an $\mathbf{R}^d$-valued $C^{\infty}(\mathbf{R}^{2dN+1})$ function takes the name of "total external force" of the force law $\mathbf{F}$.

If $d=3$, the quantity

$$\begin{aligned}&\mathbf{M}^{(e)}_{\boldsymbol{\alpha}}(\boldsymbol{\eta}^{(1)},\ldots,\boldsymbol{\eta}^{(N)},\boldsymbol{\xi}^{(1)},\ldots,\boldsymbol{\xi}^{(N)},t)\\&\quad=\sum_{j=1}^{N}(\boldsymbol{\xi}^{(j)}-\boldsymbol{\alpha})\wedge\mathbf{f}^{(j)e}(\boldsymbol{\eta}^{(j)},\boldsymbol{\xi}^{(j)},t)\end{aligned}\tag{3.2.14}$$

is called the "total momentum of the external forces" of $\mathbf{F}$ with respect to the point $\boldsymbol{\alpha}\in\mathbf{R}^d$.

Observation. If $d\neq 3$, it is still possible to define the momentum of the forces with respect to a point: however, it cannot be naturally thought of as a vector in $\mathbf{R}^d$. To avoid complications, rather than on the shaky grounds that the "physical case" is $d=3$, we do not deal with this question.

The following proposition holds ("cardinal equations of dynamics").

3 Proposition. *Given a motion $t\to\mathbf{x}(t)$, $t\in\mathbf{R}_+$, of N point masses in $\mathbf{R}^3$ with masses $m_1,\ldots,m_N>0$, define the "linear momentum" at time t and* *the "angular momentum", with respect to* $\boldsymbol{\alpha} \in \mathbf{R}^3$, *at time* t *as the quantities*

$$\mathbf{Q}(t) = \sum_{j=1}^{N} m_j \dot{\mathbf{x}}^{(j)}(t),$$
$$\mathbf{K}_\alpha(t) = \sum_{j=1}^{N} m_j(\boldsymbol{\xi}^{(j)} - \boldsymbol{\alpha}) \wedge \dot{\mathbf{x}}^{(j)}(t). \tag{3.2.15}$$

If the motion develops under the action of a force law $\mathbf{F}$ *verifying the third principle of dynamics and if one shortens* $\mathbf{R}^{(e)}(\dot{\mathbf{x}}^{(1)}(t), \ldots, \dot{\mathbf{x}}^{(N)}(t), \mathbf{x}^{(1)}(t), \ldots, \mathbf{x}^{(N)}(t), t)$ *as* $\mathbf{R}^{(e)}(t)$ *and, likewise,* $\mathbf{M}_\alpha^{(e)}(\dot{\mathbf{x}}^{(1)}(t), \ldots, \mathbf{x}^{(N)}(t), t)$ *as* $\mathbf{M}_\alpha^{(e)}(t)$ [*see Eqs.* (3.2.13) *and* (3.2.14)], *then*

$$\frac{d}{dt}\mathbf{Q}(t) = \mathbf{R}^{(e)}(t),$$
$$\frac{d}{dt}\mathbf{K}_\alpha(t) = \mathbf{M}_\alpha^{(e)}(t). \tag{3.2.16}$$

Observations.

(1) Sometimes the linear momentum is called the "quantity of motion", while the angular momentum is called "momentum of the quantity of motion". The cardinal equations (3.2.16) show that their time variation depends only upon the external forces acting on the system.

(2) The first cardinal equation in (3.2.16) is often called the "baricenter theorem" or the "center of mass theorem". To understand this name's origin, associate with the motion in $\mathbf{R}^{3N}$, $t \to \mathbf{x}(t) = (\mathbf{x}^{(1)}(t), \ldots, \mathbf{x}^{(N)}(t))$, $t \in I$, the motion in $\mathbf{R}^3$, $t \to \mathbf{x}_G(t)$, where

$$\mathbf{x}_G(t) = \frac{\sum_{i=1}^{N} m_i \mathbf{x}^{(i)}(t)}{\sum_{i=1}^{N} m_i}. \tag{3.2.17}$$

If we call the motion $t \to \mathbf{x}_G(t)$, $t \in I$, the "baricenter motion" and if we set $M = \sum_{i=1}^{N} m_i$ ("total mass of the system"), the first relations in Eqs. (3.2.15) and (3.2.16) become, respectively,

$$\mathbf{Q}(t) = M\dot{\mathbf{x}}_G(t), \tag{3.2.18}$$

$$M\ddot{\mathbf{x}}_G(t) = \mathbf{R}^{(e)}(t) \tag{3.2.19}$$

and we can read Eq. (3.2.19) as "the baricenter of a system of N point masses moves as if it were a single point mass subject to the action of a force equal to the total external force acting on the system".

If the external force has the form $\mathbf{f}^{(i)e} = m_i\mathbf{g}$, $\mathbf{g} \in \mathbf{R}^3$, "gravity force," the point G has many other nice properties which motivate its name: they are discussed below.

(3) Note that, in general, Eq. (3.2.19) is not a "closed equation": the right-hand side cannot, in fact, be computed without already knowing the locations and the speeds of all the system's particles. Nevertheless, there are

some exceptional particular cases of special importance. For instance, if the external force acting on the jth point is independent of its position and velocity: this is the case of the gravity force.

(4) It is worth stressing that, in general, it is not true that the momentum of the external forces can be computed by imagining the total force as applied to the baricenter; i.e., as $(\mathbf{x}_G - \boldsymbol{\alpha}) \wedge \mathbf{R}^{(e)}$. Neither is it generally true that the derivative of the baricenter's angular momentum, i.e., of $M(\mathbf{x}_G - \boldsymbol{\alpha}) \wedge \dot{\mathbf{x}}_G$, is the momentum of the total external forces.

(5) However, in the special case

$$\mathbf{f}^{(j)e} = m_j \mathbf{g}, \tag{3.2.20}$$

where $\mathbf{g}$ is a fixed vector ("gravity force"), one finds

$$\mathbf{R}^{(e)} = \left(\sum_{j=1}^{N} m_j \right) \mathbf{g} = M\mathbf{g} \tag{3.2.21}$$

and by Eq. (3.2.17),

$$\begin{aligned} \mathbf{M}_{\boldsymbol{\alpha}}^{(e)} &= \sum_{j=1}^{N} (\mathbf{x}^{(j)} - \boldsymbol{\alpha}) \wedge m_j \mathbf{g} = \left(\sum_{j=1}^{N} m_j (\mathbf{x}^{(j)} - \boldsymbol{\alpha}) \right) \wedge \mathbf{g} \\ &= M(\mathbf{x}_G - \boldsymbol{\alpha}) \wedge \mathbf{g} = (\mathbf{x}_G - \boldsymbol{\alpha}) \wedge \mathbf{R}^{(e)}. \end{aligned} \tag{3.2.22}$$

Furthermore,

$$\begin{aligned} \frac{d}{dt}(\mathbf{x}_G - \boldsymbol{\alpha}) \wedge M\dot{\mathbf{x}}_G &= (\mathbf{x}_G - \boldsymbol{\alpha}) \wedge M\ddot{\mathbf{x}}_G + \dot{\mathbf{x}}_G \wedge M\dot{\mathbf{x}}_G \\ &\equiv (\mathbf{x}_G - \boldsymbol{\alpha}) \wedge M\ddot{\mathbf{x}}_G = (\mathbf{x}_G - \boldsymbol{\alpha}) \wedge \mathbf{R}^{(e)} = \mathbf{M}_{\boldsymbol{\alpha}}^{(e)}; \end{aligned} \tag{3.2.23}$$

i.e., in the case of the gravity forces, the most daring thoughts are allowed: Eqs. (3.2.22) and (3.2.23) show the uniqueness of the gravity force case with respect to the cardinal equations and they explain why the point defined in Eq. (3.2.17) is given the name of "center of gravity", or "center of mass" or "baricenter".

PROOF. From Eq. (3.2.8), by summing both sides over j, it follows that

$$\sum_{j=1}^{N} m_j \ddot{\mathbf{x}}^{(j)} = \sum_{j=1}^{N} \mathbf{f}^{(j)}, \tag{3.2.24}$$

but Eqs. (3.2.10) and (3.2.11) and the first of Eqs. (3.2.15) imply $\sum_{j=1}^{N} m_j \ddot{\mathbf{x}}^{(j)} = \mathbf{R}^{(e)}$, i.e., the first of Eqs. (3.2.16).

Similarly, by externally multiplying both sides of Eq. (3.2.8) by $(\mathbf{x}^{(j)}(t) - \tau)$, $\boldsymbol{\alpha} \in \mathbf{R}^3$, and summing:

$$\begin{aligned} \sum_{j=1}^{N} m_j (\mathbf{x}^{(j)}(t) - \boldsymbol{\alpha}) \wedge \ddot{\mathbf{x}}^{(j)}(t) &= \sum_{j=1}^{N} (\mathbf{x}^{(j)}(t) - \boldsymbol{\alpha}) \wedge \mathbf{f}^{(j)} \\ &= \sum_{j=1}^{N} (\mathbf{x}^{(j)}(t) - \boldsymbol{\alpha}) \wedge \mathbf{f}^{(j)e} = \mathbf{M}_{\boldsymbol{\alpha}}^{(e)}, \end{aligned} \tag{3.2.25}$$

having used Eqs. (3.2.10) and (3.2.11) and, particularly, Eq. (3.2.12) in the third step to eliminate the contribution of the internal forces:

$$\begin{aligned}\sum_{j=1}^{N}(\mathbf{x}^{(j)}(t)-\boldsymbol{\alpha})\wedge\sum_{\substack{i=1\\ i\neq j}}^{N}\mathbf{f}^{(i\to j)} &= \sum_{i\neq j}(\mathbf{x}^{(j)}(t)-\boldsymbol{\alpha})\wedge\mathbf{f}^{(i\to j)}\\ &= \frac{1}{2}\sum_{i\neq j}\left\{(\mathbf{x}^{(j)}(t)-\boldsymbol{\alpha})\wedge\mathbf{f}^{(i\to j)}+(\mathbf{x}^{(i)}(t)-\boldsymbol{\alpha})\wedge\mathbf{f}^{(j\to i)}\right\}\\ &= \frac{1}{2}\sum_{i\neq j}(\mathbf{x}^{(j)}(t)-\mathbf{x}^{(i)}(t))\wedge\mathbf{f}^{(i\to j)}=\mathbf{0}.\end{aligned} \tag{3.2.26}$$

Furthermore,

$$\begin{aligned}(\mathbf{x}^{(j)}(t)-\boldsymbol{\alpha})\wedge\ddot{\mathbf{x}}^{(j)}(t) &\equiv (\mathbf{x}^{(j)}(t)-\boldsymbol{\alpha})\wedge\frac{d}{dt}\dot{\mathbf{x}}^{(j)}(t)\\ &= \frac{d}{dt}\left\{(\mathbf{x}^{(j)}(t)-\boldsymbol{\alpha})\wedge\dot{\mathbf{x}}^{(j)}(t)\right\}\\ &\quad -\left\{\frac{d}{dt}(\mathbf{x}^{(j)}(t)-\boldsymbol{\alpha})\right\}\wedge\dot{\mathbf{x}}^{(j)}(t)\\ &= \frac{d}{dt}\left\{(\mathbf{x}^{(j)}(t)-\boldsymbol{\alpha})\wedge\dot{\mathbf{x}}^{(j)}(t)\right\}-\dot{\mathbf{x}}^{(j)}(t)\wedge\dot{\mathbf{x}}^{(j)}(t)\\ &\equiv \frac{d}{dt}\left\{(\mathbf{x}^{(j)}(t)-\boldsymbol{\alpha})\wedge\dot{\mathbf{x}}^{(j)}(t)\right\}.\end{aligned} \tag{3.2.27}$$

Hence,

$$\sum_{j=1}^{N}m_j(\mathbf{x}^{(j)}(t)-\boldsymbol{\alpha})\wedge\ddot{\mathbf{x}}^{(j)}(t)=\frac{d}{dt}\sum_{j=1}^{N}m_j(\mathbf{x}^{(j)}(t)-\boldsymbol{\alpha})\wedge\dot{\mathbf{x}}^{(j)}(t)=\frac{d}{dt}\mathbf{K}_{\boldsymbol{\alpha}} \tag{3.2.28}$$

which, together with Eq. (3.2.25), proves the second equation in (3.2.16). mbe

Exercises and Problems for §3.2

1. In Appendix P, there is a table of the masses of the nine main planets and of their distance from the Sun. The Sun's mass and radius can also be found there. Find the configuration of the planets in which the center of mass of the above ten heavenly bodies is farthest from the Sun's center and compute the ratio of this distance to the Sun's radius. (Assume that the planets move in circular orbits around the Sun.)

2. Same as Problem 1, not counting the Sun.

3. From the data in Appendix P, find the position of the Earth-Moon center of mass relative to the Earth and compare its distance from the center of the Earth with the Earth's radius. (Assume the distance between the Earth and Moon to be equal to the maximal or to the minimal distance.)

4. Find the value of the Earth-Moon system's angular momentum with respect to the Sun's center, assuming that the latter is fixed in a reference frame with axes fixed with the fixed stars. Assume also that the configuration Moon-Earth-Sun is that of a lunar eclipse and neglect the Moon's orbital inclination. Should the angular momentum be time independent? If not, indicate what should be neglected to make it time independent. (Hint: The attraction of the Sun on the Earth and on the Moon has vanishing momentum with respect to the Sun's center, while the Earth-Moon forces are internal forces to the Earth-Moon system.)

5. If $V \in C^\infty(\mathbf{R}^{Nd})$ is bounded below the force law $\mathbf{F} = (\mathbf{f}^{(1)}, \ldots, \mathbf{f}^{(N)})$:

$$\mathbf{f}^{(i)} = -\frac{\partial V}{\partial \boldsymbol{\xi}^{(i)}}(\boldsymbol{\xi}), \qquad i = 1, \ldots, N$$

is actually a force law in the sense of Definition 3 (i). Show the validity of this statement. (Hint: One need only check condition (iii) of Definition 1. This is obtained by finding an a priori estimate in the sense of §2.5, using energy conservation. Proceed along the lines of the analogous one-dimensional case, §2.5, Proposition 6, p. 29).

§3.3. The Least Action Principle

The least action principle seen in §2.24 has an obvious extension to systems of N point masses in $\mathbf{R}^d$ subject to conservative forces.

Let us give the following definition.

5 Definition. (i) Let $\mathfrak{M}_{t_1,t_2}(\boldsymbol{\xi}_1, \boldsymbol{\xi}_2)$ be the set of the motions $t \to \mathbf{x}(t) = (\mathbf{x}^{(1)}(t), \ldots, \mathbf{x}^{(N)}(t))$ in $C^\infty([t_1, t_2])$ such that $\mathbf{x}(t_1) = \boldsymbol{\xi}_1, \mathbf{x}(t_2) = \boldsymbol{\xi}_2$, with $\boldsymbol{\xi}_1, \boldsymbol{\xi}_2, \mathbf{x}(t) \in \mathbf{R}^{Nd}$.

(ii) If $\mathbf{x} \in \mathfrak{M} \subset \mathfrak{M}_{t_1,t_2}(\boldsymbol{\xi}_1, \boldsymbol{\xi}_2)$, we shall denote $\mathcal{V}_{\mathbf{x}}(\mathfrak{M})$ the space of the "variations" of the motion $\mathbf{x}$ in $\mathfrak{M}$: it is the set of the $\mathbf{R}^{Nd}$-valued functions $\mathbf{y} \in C^\infty([t_1, t_2] \times (-1, 1))$, $(t, \epsilon) \to \mathbf{y}(t, \epsilon)$ such that:

(a) $\mathbf{y}(t, 0) \equiv \mathbf{x}(t)$, $\forall t \in [t_1, t_2]$; (3.3.1)

(b) $\mathbf{y}(t_1, \epsilon) \equiv \boldsymbol{\xi}_1$, $\mathbf{y}(t_2, \epsilon) \equiv \boldsymbol{\xi}_2$, $\forall \epsilon \in (-1, 1)$; (3.3.2)

(c) for all $\epsilon \in (-1, 1)$, the function $t \to \mathbf{y}_\epsilon(t)$,

$$\mathbf{y}_\epsilon(t) = \mathbf{y}(t, \epsilon), \qquad t \in [t_1, t_2] \tag{3.3.3}$$

is a motion $\mathbf{y}_\epsilon \in \mathfrak{M}$. We shall set $\mathcal{V}_{\mathbf{x}}(\mathfrak{M}_{t_1,t_2}(\boldsymbol{\xi}_1, \boldsymbol{\xi}_2)) \equiv \mathcal{V}_{\mathbf{x}}$.

(iii) If $\mathfrak{L} \in C^\infty(\mathbf{R}^{2Nd+1})$ is a real-valued function, we shall define the action with Lagrangian density $\mathfrak{L}$ as the real-valued function A on $\mathfrak{M}_{t_1,t_2}(\boldsymbol{\xi}_1, \boldsymbol{\xi}_2)$ defined as

$$A(\mathbf{x}) = \int_{t_1}^{t_2} \mathfrak{L}(\dot{\mathbf{x}}(t), \mathbf{x}(t), t)\, dt. \tag{3.3.4}$$

(iv) The action A in Eq. (3.3.4) is said to be stationary or locally minimal on the motion $\mathbf{x} \in \mathfrak{M} \subset \mathfrak{M}_{t_1,t_2}(\boldsymbol{\xi}_1, \boldsymbol{\xi}_2)$ on $\mathfrak{M}$ if the function $\epsilon \to A(\mathbf{y}_\epsilon)$, $\epsilon \in (-1, 1)$, is stationary or locally minimal for $\epsilon = 0$ and for all $\mathbf{y} \in \mathcal{V}_{\mathbf{x}}(\mathfrak{M})$.

The stationarity condition of Eq. (3.3.4) on all of $\mathfrak{M}_{t_1,t_2}(\boldsymbol{\xi}_1,\boldsymbol{\xi}_2)$ in $\mathbf{x}$ is deduced exactly along the same lines and patterns followed to prove the analogous condition seen in Proposition 36, §2.24, p. 131, through the principle of the vanishing integrals (Appendix D). Therefore, we shall leave as an exercise for the reader the detailed proof of the following proposition.

4 Proposition. *A motion* $\mathbf{x} \in \mathfrak{M}_{t_1,t_2}(\boldsymbol{\xi}_1,\boldsymbol{\xi}_2)$ *is a stationary point in* $\mathfrak{M}_{t_1,t_2}(\boldsymbol{\xi}_1,\boldsymbol{\xi}_2)$ *for the action of Eq.* (3.3.4) *if and only if*

$$\frac{d}{dt}\left(\frac{\partial \mathcal{L}}{\partial \boldsymbol{\eta}^{(i)}}(\dot{\mathbf{x}}(t),\mathbf{x}(t),t)\right) = \frac{\partial \mathcal{L}}{\partial \boldsymbol{\xi}^{(i)}}(\dot{\mathbf{x}}(t),\mathbf{x}(t),t) \tag{3.3.5}$$

for all $\forall t \in [t_1,t_2]$, $\forall i = 1,2,\ldots,N$.

Observations.

(1) In Eq. (3.3.5) we use the notation on the derivatives introduced in §3.1.

(2) Often, with an obvious abuse of notation, Eq. (3.3.5) is compactly written as

$$\frac{d}{dt}\frac{\partial \mathcal{L}}{\partial \dot{\mathbf{x}}} = \frac{\partial \mathcal{L}}{\partial \mathbf{x}}\,. \tag{3.3.6}$$

An immediate corollary to Proposition 4 is the following.

5 Proposition. *Given a real-valued* $C^\infty(\mathbf{R}^{Nd})$ *function* V, *bounded from below,*[2] *the motion* $\mathbf{x} \in \mathfrak{M}_{t_1,t_2}(\boldsymbol{\xi}_1,\boldsymbol{\xi}_2)$ *makes the action with Lagrangian density*

$$\mathcal{L}(\boldsymbol{\eta},\boldsymbol{\xi},t) = \sum_{i=1}^{N} \frac{m_i}{2}(\boldsymbol{\eta}^{(i)})^2 - V(\boldsymbol{\xi}), \tag{3.3.7}$$

$m_j > 0$, $j = 1,\ldots,N$, *stationary on* $\mathfrak{M}_{t_1,t_2}(\boldsymbol{\xi}_1,\boldsymbol{\xi}_2)$ *if and only if* $\mathbf{x}$ *is a motion of a system of* N *point masses in* $\mathbf{R}^d$ *with masses* $m_1,\ldots,m_N > 0$ *which for* $t \in [t_1,t_2]$ *develops subject to influence of the force law* $\mathbf{F}$ *with potential energy* V.

Proof. It is enough to substitute Eq. (3.3.7) into Eq. (3.3.5) to see that, in this case, Eq. (3.3.5) becomes Eq. (3.1.1) with $\mathbf{F}$ given by Eq. (3.1.3), i.e.,

$$m_j\ddot{\mathbf{x}}^{(j)}(t) = -\frac{\partial V}{\partial \boldsymbol{\xi}^{(j)}}(\mathbf{x}(t)), \qquad j = 1,\ldots,N \tag{3.3.8}$$

if $\mathbf{x}(t) = (\mathbf{x}^{(1)}(t),\ldots,\mathbf{x}^{(N)}(t))$, $t \in [t_1,t_2]$. mbe

The following generalization of Proposition 38, §2.24, is also valid, but we leave the proof as a problem for the reader (since it is an essentially word-by-word repetition of that of Proposition 38, p. 133).

[2] See Problem 5, §3.2.

6 Proposition. *Let $t \to \mathbf{x}(t)$, $t \in \mathbf{R}_+$, be a motion of a system of N point masses in $\mathbf{R}^d$ with masses $m_1, \ldots, m_N > 0$ developing under the action of a conservative force with potential energy $V \in C^\infty(\mathbf{R}^{Nd})$. Given $t_1 \in \mathbf{R}_+$ and $t_2 > t_1$, if $t_2 - t_1$ is small enough, the motion $t \to \mathbf{x}(t)$ considered in the time interval $[t_1, t_2]$ is a point of local minimum in $\mathfrak{M}_{t_1,t_2}(\mathbf{x}(t_1), \mathbf{x}(t_2))$ for the action with Lagrangian* (3.3.7).

The comments seen at the end of Chapter 2, pp. 134–136, extend, in an obvious way, to the contents of this section. It is, in particular, quite important that the reader extends to the case of a system of point masses the observations made in §2.24, concerning the representations of motions in coordinates other than Cartesian coordinates and concerning the invariance of the Lagrange equations (3.3.6) with respect to changes in coordinates (see §2.24, p. 134 and following).

In the following sections and in their exercises, we shall see some interesting applications of the "Lagrangian formulation" (3.3.6) of the equations of motion as a "change of coordinates invariant" formulation of such equations. Among these will be the theory of perfect constraints.

§3.4. Introduction to the Constrained Motion Theory

Elli avien cappe con cappucci bassi
Dinanzi a li occhi, fatte della taglia
Che in Clugnì per li monaci fassi.
Di fuor dorate son, sì ch'elli abbaglia;
Ma dentro tutte piombo, e gravi tanto . . . [3]

The principle of least action inspires the following somewhat trivial considerations. Let $\mathbf{x} \to A(\mathbf{x})$ be the action of Eqs. (3.3.4) and (3.3.7) defined on the motions in $\mathfrak{M}_{t_1,t_2}(\boldsymbol{\xi}_1, \boldsymbol{\xi}_2)$ of a system of N point masses system subject to a conservative force $\mathbf{F}$.

Suppose that we know, a priori, that the force law is such that the motion $\mathbf{x}$ that develops under its influence from $\boldsymbol{\xi}_1$ to $\boldsymbol{\xi}_2$, within times t_1 and t_2, verifies some properties like $|\mathbf{x}(t)| \leqslant S$ or $|\dot{\mathbf{x}}(t)| \leqslant P$ or $\mathbf{x}^{(1)}(t) \equiv \mathbf{0}$, etc. Then it is clear that the research of $\mathbf{x}$ in $\mathfrak{M}_{t_1,t_2}(\boldsymbol{\xi}_1, \boldsymbol{\xi}_2)$ can be restricted to the subset $\mathfrak{M}$ of the motions in $\mathfrak{M}_{t_1,t_2}(\boldsymbol{\xi}_1, \boldsymbol{\xi}_2)$ verifying the properties under consideration.

Very often it happens that a system of point masses is subject to "constraints", i.e., to force laws that allow only a "few" motions among

[3] In basic English:
They had capes with low hoods
in front of the eyes, made in the fashion
that in Cluny is used for the monks.
Golden they are outside, so that they dazzle
but inside they are all leaden and heavy a lot . . .
(Dante, *Inferno*, Canto XXIII).

those a priori possible, at least for vast classes of initial data. Think of a point mass constrained to remain on a surface: in this case, the surface acts on the point with a force systematically such as to forbid the abandonment of the surface itself by the point mass, whenever the initial data $(\boldsymbol{\eta}, \boldsymbol{\xi})$ have $\boldsymbol{\xi}$ on the surface and $\boldsymbol{\eta}$ tangent to it.

Think, also, of a rigid system of N point masses. Now the ith point will exert on the jth point a force $\mathbf{f}^{(i \to j)}$ systematically such that the two points keep each other at a fixed distance.

By taking into account the constraints, the allowable motions in $\mathfrak{M}_{t_1,t_2}(\boldsymbol{\xi}_1, \boldsymbol{\xi}_2)$ will generally be parameterizable with l coordinates, often $l \ll Nd$, and consequently, we shall be able to imagine a description of the system's motions in terms of l functions of time. Therefore, the Lagrangian and the action will also be expressible in terms of the same l functions.

Hence, the action of a motion $\mathbf{x}$ allowed by the constraints will take the form

$$A(\mathbf{x}) = \int_{t_1}^{t_2} dt\, \tilde{\mathcal{L}}(\dot{a}_1(t), \ldots, \dot{a}_l(t), a_1(t), \ldots, a_l(t), t) \tag{3.4.1}$$

if $t \to (a_1(t), \ldots, a_l(t))$, $t \in [t_1, t_2]$, is the description of the motion $\mathbf{x}$ in the l "essential coordinates".

To be less vague, assume that there are N $\mathbf{R}^d$-valued functions in $C^\infty(\mathbf{R}^l)$:

$$\boldsymbol{\alpha} = (\alpha_1, \ldots, \alpha_l) \to \mathbf{X}^{(i)}(\boldsymbol{\alpha}) = \mathbf{X}^{(i)}(\alpha_1, \ldots, \alpha_l), \tag{3.4.2}$$

$i = 1, \ldots, N$, such that the set of the motions $t \to \mathbf{x}(t) = (\mathbf{x}^{(1)}(t), \ldots, \mathbf{x}^{(N)}(t))$, $t \in [t_1, t_2]$, which are "constrained" or "allowed" by the constraints is simply the set of the motions which is the image of the motions in $\mathbf{R}^l$ via the transformation (3.4.2). Thus, given a motion $t \to \mathbf{a}(t)$, $t \in [t_1, t_2]$, in $\mathbf{R}^l$ one describes, via Eq. (3.4.2), the constrained motion $t \to \mathbf{x}(t)$, $t \in [t_1, t_2]$, where

$$\mathbf{x}^{(i)}(t) = \mathbf{X}^{(i)}(\mathbf{a}(t)), \qquad i = 1, 2, \ldots, N \tag{3.4.3}$$

which we shorten as $\mathbf{x}(t) = \mathbf{X}(\mathbf{a}(t))$.

In other words, let us admit that the conservative force law $\mathbf{F}$ for the system of N point masses under consideration is such that the motions in $\mathfrak{M}_{t_1,t_2}(\boldsymbol{\xi}_1, \boldsymbol{\xi}_2)$ that can actually develop under its influence starting from a given class of initial data are necessarily contained in the class of the motions having the form of Eq. (3.4.3) with $\mathbf{a} \in \mathfrak{M}_{t_1,t_2}(\boldsymbol{\xi}_1, \boldsymbol{\xi}_2)$, where $\boldsymbol{\alpha}_1, \boldsymbol{\alpha}_2 \in \mathbf{R}^l$ and $\mathbf{X}(\boldsymbol{\alpha}_1) = \boldsymbol{\xi}_1$, $\mathbf{X}(\boldsymbol{\alpha}_2) = \boldsymbol{\xi}_2$.

If $\mathbf{x}$ is a constrained motion in the sense just discussed, its action, Eq. (3.3.4), with respect to the Lagrangian (3.3.7), where V is the potential energy of $\mathbf{F}$, can be written as in Eq. (3.4.1) if $\tilde{\mathcal{L}} \in C^\infty(\mathbf{R}^{2l+1})$ is the function

$$\tilde{\mathcal{L}}(\beta_1, \ldots, \beta_l, \alpha_1, \ldots, \alpha_l, t) = \sum_{i=1}^{N} \frac{m_i}{2} \left(\sum_{j=1}^{N} \frac{\partial \mathbf{X}^{(i)}}{\partial \alpha_j}(\boldsymbol{\alpha}) \beta_j \right)^2 - V(\mathbf{X}(\boldsymbol{\alpha})). \tag{3.4.4}$$

In fact, this is true because $\dot{\mathbf{x}}^{(i)}(t)$ can be computed, by differentiating Eq. (3.4.3), as

$$\dot{\mathbf{x}}^{(i)}(t) = \sum_{j=1}^{l} \frac{\partial \mathbf{X}^{(i)}}{\partial \alpha_j}(\mathbf{a}(t))\dot{a}_j(t), \qquad j = 1, 2, \ldots, N, \tag{3.4.5}$$

whenever $\mathbf{x}$ is the constrained motion image of $\mathbf{a}$: $\mathbf{x} = \mathbf{X}(\mathbf{a})$.

Hence, if $\mathbf{x} \in \mathfrak{M}_{t_1,t_2}(\boldsymbol{\xi}_1, \boldsymbol{\xi}_2)$ is the motion that actually develops under the influence of the force $\mathbf{F}$ and if $\mathbf{x}$ is the image via Eq. (3.4.3) of $\mathbf{a}$, it is clear that the action A with Lagrangian (3.3.7) is stationary in $\mathfrak{M}_{t_1,t_2}(\boldsymbol{\xi}_1, \boldsymbol{\xi}_2)$ on $\mathbf{x}$, while the action $\tilde{A}$ with Lagrangian given by Eq. (3.4.4) is stationary on $\mathbf{a}$ in $\mathfrak{M}_{t_1,t_2}(\boldsymbol{\alpha}_1, \boldsymbol{\alpha}_2)$. This property is an immediate consequence of the fact that if A is stationary on a motion $\mathbf{x}$ in $\mathfrak{M}_{t_1,t_2}(\boldsymbol{\xi}_1, \boldsymbol{\xi}_2)$, it is also stationary on $\mathbf{x}$ in any smaller set $\mathfrak{M}' \subset \mathfrak{M}_{t_1,t_2}(\boldsymbol{\xi}_1, \boldsymbol{\xi}_2)$ (provided $\mathbf{x} \in \mathfrak{M}'$). In our case, through Eq. (3.4.3), $\mathfrak{M}'$ would be the set of the motions which is the image of the motions in $\mathfrak{M}_{t_1,t_2}(\boldsymbol{\alpha}_1, \boldsymbol{\alpha}_2)$.

By the Proposition 4, §3.3, the stationarity condition for $\tilde{A}$, i.e., for the action on $\mathfrak{M}_{t_1,t_2}(\boldsymbol{\alpha}_1, \boldsymbol{\alpha}_2)$ with Lagrangian density (3.4.4), is

$$\frac{d}{dt}\frac{\partial \tilde{\mathfrak{L}}}{\partial \beta_i}(\dot{\mathbf{a}}(t), \mathbf{a}(t), t) = \frac{\partial \tilde{\mathfrak{L}}}{\partial \alpha_i}(\dot{\mathbf{a}}(t), \mathbf{a}(t), t), \tag{3.4.6}$$

$i = 1, 2, \ldots, 1$, $\forall t \in [t_1, t_2]$.

The importance of the above considerations is easily realized: Eq. (3.4.6) is already the equation of motion after the elimination of the parameters describing the system, necessary a priori but made "useless" or "redundant" by the presence of the constraints which allow one to reduce the number of the coordinates necessary to describe the actually "possible" configurations, from Nd down to l via (3.4.2) and (3.4.3).

Therefore, the idea occurs that the mechanism for the elimination of the redundant coordinates in conservative systems subject to simple constraints, like Eqs. (3.4.2) and (3.4.3), might be particularly simple: it will be enough to rewrite the Lagrangian density of the action only in terms of the essential coordinates through Eq. (3.4.2) and, then, deduce Eq. (3.4.6).

However, the principle of the conservation of the difficulties makes it clear that there must be some serious obstacle to the actual applications of such a shining but simplistic vision.

The true constraints are, in fact, generated by forces that, as we shall see shortly, generally are neither simple nor conservative (in the sense of Definition 2, p. 142, §3.1) but depend on the velocities of the points as well as on their positions.

In such situations, the above considerations become essentially useless since they are not applicable to the simplest and most interesting motions constrained in the sense that they are parametrizable as in Eqs. (3.4.2) and (3.4.3), by l coordinates.

To understand better what has just been said, let us consider the case of a point mass constrained to stay on a curve $\Gamma \subset \mathbf{R}^3$ with intrinsic parametric equations given by

$$s \to \boldsymbol{\xi}(s), \qquad s \in \mathbf{R}, \tag{3.4.7}$$

where s is the curvilinear abscissa on Γ (which we suppose to be a simple curve, i.e., without double points and open). We suppose that the curve Γ exerts a force on the point mass which keeps it on Γ for all motions starting from initial data $(\boldsymbol{\eta}, \boldsymbol{\xi})$ with $\boldsymbol{\xi} = \boldsymbol{\xi}(s_0)$, $\boldsymbol{\eta} = (d\boldsymbol{\xi}/ds)(s_0)\dot{s}_0$ (i.e., with $\boldsymbol{\xi} \in \Gamma$ and $\boldsymbol{\eta}$ tangent to it), with $(\dot{s}_0, s_0) \in \mathbf{R}^2$.

If $\boldsymbol{\tau}(s)$, $\mathbf{n}(s)$ denote, respectively, the tangent and the principal normal versors to Γ at the point with curvilinear abscissa s and if $r(s)$ denotes the curvature radius at the same point, it is well known that

$$\boldsymbol{\tau}(s) = \frac{d\boldsymbol{\xi}(s)}{ds}, \qquad \frac{\mathbf{n}(s)}{r(s)} = \frac{d\boldsymbol{\tau}(s)}{ds}. \tag{3.4.8}$$

Then if $t \to s(t)$, $t \in \mathbf{R}$, is a motion on Γ described by the time variation of the curvilinear abscissa, we find

$$\frac{d}{dt}\boldsymbol{\xi}(s(t)) = \dot{s}(t)\boldsymbol{\tau}(s(t)) \tag{3.4.9}$$

and

$$\frac{d^2}{dt^2}\boldsymbol{\xi}(s(t)) = \ddot{s}(t)\boldsymbol{\tau}(s(t)) + \frac{\dot{s}(t)^2}{r(s(t))}\mathbf{n}(s(t)). \tag{3.4.10}$$

If the point is subject to a force which is the sum of the constraint's reaction $\mathbf{R}(\dot{s}, s)$ and of an external force $\mathbf{f}(s)$, then

$$m\ddot{\mathbf{x}} = \mathbf{f} + \mathbf{R} \tag{3.4.11}$$

if $m > 0$ is the point's mass and $\mathbf{x}(t) = \boldsymbol{\xi}(s(t))$ denotes the point's motion in $\mathbf{R}^3$.

By Eq. (3.4.10), Eq. (3.4.11) becomes

$$\begin{aligned} m\ddot{s} &= \mathbf{f} \cdot \boldsymbol{\tau} + \mathbf{R} \cdot \boldsymbol{\tau}, \\ m\frac{\dot{s}^2}{r} &= \mathbf{f} \cdot \mathbf{n} + \mathbf{R} \cdot \mathbf{n} \end{aligned} \tag{3.4.12}$$

and from the second equation, we deduce that the normal component of the constraint's reaction is

$$\mathbf{R} \cdot \mathbf{n} = m\frac{\dot{s}^2}{r(s)} - \mathbf{f}(s) \cdot \mathbf{n}(s) \tag{3.4.13}$$

at the point of Γ with coordinate s when it is occupied by a point mass with speed along Γ given by $\dot{s}$.

From Eq. (3.4.13), we see that $\mathbf{R}(\dot{s}, s)$ is necessarily $\dot{s}$ dependent if $0 < r(s) < +\infty$, as we shall suppose, and therefore the constraint's reaction cannot be conservative in the very restrictive sense of §3.1.

Nevertheless, we shall see that the essence of the idea which arose in connection with Eq. (3.4.6) will be saved: it will, however, be necessary to go through a long analysis which, as is to be expected, involves a deeper physicomathematical discussion of the notion of constraint. Such a discussion will be aimed at clarifying the definition of constraint, i.e., the physical phenomenon mathematically modeled as a "constraint".

In the next section we shall first give a general definition of constraint, stressing its main mathematical properties and delaying until the following sections a deeper discussion showing how the empirical notion of a frictionless constraint is naturally schematized by the mathematical structures introduced in the next section.

Exercises and Problems for §3.4

1. Let Γ be a circle in $\mathbf{R}^3$ with radius r. Find $\boldsymbol{\tau}(s)$, $\mathbf{n}(s)$, $r(s)$ [see Eq. (3.4.8)].

2. Let Γ be an ellipse with equations $z=0$, $x^2/a^2+y^2/b^2=1$, $a,b>0$. Find $\boldsymbol{\tau}$, $\mathbf{n}$, r at the point $(x,y,0)$.

3. Show that the force law

$$\mathbf{R}(\dot{\mathbf{x}},\mathbf{x}) = -\frac{m\dot{\mathbf{x}}^2}{r^2}\mathbf{x}, \qquad (\dot{\mathbf{x}},\mathbf{x}) \in \mathbf{R}^2 \times \mathbf{R}^2$$

produces a constraint for the motions of a point mass, with mass $m>0$, with initial data $(\boldsymbol{\eta},\boldsymbol{\xi})$ with $\boldsymbol{\eta}\cdot\boldsymbol{\xi}=0$, $|\boldsymbol{\xi}|=r$. The constraint is to the circle $\Gamma=\{\boldsymbol{\xi}\,|\,\boldsymbol{\xi}\in\mathbf{R}^2, |\boldsymbol{\xi}|=r\}$. (Hint: Show that the circular uniform motion verifies the equations of motions and use the uniqueness theorem.)

4. Same as Problem 3 in $\mathbf{R}^3$, replacing the circle Γ with the surface of a sphere.

5. Same as Problem 3, using Archimedes' spiral (with equations $\rho=a\theta$, $a>0$, in polar coordinates), finding an appropriate force $\mathbf{R}$ producing a constraint to the spiral.

6. Find an appropriate force $\mathbf{R}$ producing a constraint to Γ, as defined in Problems 3–5, if the point mass is also subject to a conservative force with potential energy $V=|\mathbf{x}|^2/2$.

7. Show that no purely positional force law $\mathbf{R}$ can force every motion with initial data $(\boldsymbol{\eta},\boldsymbol{\xi})$ with $\boldsymbol{\eta}\cdot\boldsymbol{\xi}=0$, $|\boldsymbol{\xi}|=1$ to move on the unit circle in $\mathbf{R}^2$, regardless of the mass m of the point. (Hint: Let $(\boldsymbol{\eta}_0,\boldsymbol{\xi}_0)$ be an initial datum at $t=0$ producing a motion which stays on the circle. Consider the motion with initial datum $(2\boldsymbol{\eta}_0,\boldsymbol{\xi}_0)$ and show that it must abandon the unit circle, for $t>0$ and small, by using the Lagrange–Taylor theorem or, alternatively, by using Eq. (3.4.13).)

§3.5 Ideal Constraints as Mathematical Entities

The following is a rather general mathematical definition of a constrained motion for a system of N point masses.

6 Definition. Given s real-valued C^∞ ($\mathbf{R}^{2Nd+1}$)-functions $\psi^{(1)},\ldots,\psi^{(s)}$, we shall say that a system of N point masses with masses $m_1,\ldots,m_N>0$ subject to a force law $\mathbf{F}$ is constrained by the constraints $\psi^{(1)},\ldots,\psi^{(s)}$ if $\mathbf{F}$

is such that the motions $t \to \mathbf{x}(t) = (\mathbf{x}^{(1)}(t), \ldots, \mathbf{x}^{(N)}(t))$, $t \in \mathbf{R}_+$, developing under its influence identically verify the s relations, $i = 1, \ldots, s$:

$$\psi^{(i)}\big(\dot{\mathbf{x}}^{(1)}(t), \ldots, \dot{\mathbf{x}}^{(N)}(t), \mathbf{x}^{(1)}(t), \ldots, \mathbf{x}^{(N)}(t), t\big) = 0, \qquad (3.5.1)$$

$\forall t \in \mathbf{R}_+$, provided there is a time t (e.g., $t = 0$) when Eq. (3.5.1) holds.

Examples

(1) If $V \in C^{\infty}(\mathbf{R}^{Nd})$ and $E \in \mathbf{R}$, the function

$$\begin{aligned}\psi(\boldsymbol{\eta}^{(1)}, \ldots, \boldsymbol{\eta}^{(N)}, \boldsymbol{\xi}^{(1)}, \ldots, \boldsymbol{\xi}^{(N)}, t)\\ = \frac{1}{2}\sum_{i=1}^{N} m_i (\boldsymbol{\eta}^{(i)})^2 - V(\boldsymbol{\xi}^{(1)}, \ldots, \boldsymbol{\xi}^{(N)}) - E\end{aligned} \qquad (3.5.2)$$

is a constraint for the motions of a system of N point masses with masses $m_1, \ldots, m_N > 0$ subject to a force law with V as a potential energy.

(2) Given a system of N point masses, with masses $m_1, \ldots, m_N > 0$ subject to a force law $\mathbf{F}$ verifying the third principle of dynamics and with zero external forces, let $\mathbf{q}, \mathbf{m} \in \mathbf{R}^3$. Define the six functions on $\mathbf{R}^{2Nd+1}$ (actually independent on the last coordinate):

$$\begin{aligned}\psi(\boldsymbol{\eta}^{(1)}, \ldots, \boldsymbol{\eta}^{(N)}, \boldsymbol{\xi}^{(1)}, \ldots, \boldsymbol{\xi}^{(N)}, t) &= \sum_{i=1}^{N} m_i \boldsymbol{\eta}^{(i)} - \mathbf{q},\\ \psi'(\boldsymbol{\eta}^{(1)}, \ldots, \boldsymbol{\eta}^{(N)}, \boldsymbol{\xi}^{(1)}, \ldots, \boldsymbol{\xi}^{(N)}, t) &= \sum_{i=1}^{N} m_i \boldsymbol{\xi}^{(i)} \wedge \boldsymbol{\eta}^{(i)} - \mathbf{m};\end{aligned} \qquad (3.5.3)$$

then the above six functions provide six constraints for the system.

(3) More generally, every conservation law may be interpreted as a constraint.

(4) The above examples may be pushed to some extreme analogues: given $(\boldsymbol{\eta}_0, \boldsymbol{\xi}_0) \in \mathbf{R}^{2Nd}$ and calling S_t the evolution flow associated with a time independent force law $\mathbf{F}$ acting on a system of N point masses, the $2Nd$ functions:

$$\psi(\boldsymbol{\eta}, \boldsymbol{\xi}, t) = S_{-t}(\boldsymbol{\eta}, \boldsymbol{\xi}) - (\boldsymbol{\eta}_0, \boldsymbol{\xi}_0) \qquad (3.5.4)$$

are constraints for the system.

(5) Consider a point mass with mass $m > 0$ in $\mathbf{R}^3$ subject to a force law given by

$$\mathbf{F}(\boldsymbol{\eta}, \boldsymbol{\xi}) = -m\frac{(\boldsymbol{\eta})^2 \boldsymbol{\xi}}{r^2}, \qquad (3.5.5)$$

where $r > 0$ is constant.

Then the following two functions:

$$\begin{aligned}\psi_1(\boldsymbol{\eta}, \boldsymbol{\xi}) &= (\boldsymbol{\xi}^2 - r^2)^2 + (\boldsymbol{\eta} \cdot \boldsymbol{\xi})^2\\ \psi_2(\boldsymbol{\eta}, \boldsymbol{\xi}) &= \xi_3^2 + \eta_3^2\end{aligned} \qquad (3.5.6)$$

are constraints for the system (see Problem 3, §3.4).

Observation. The above examples of constraints may leave the reader a bit perplexed, particularly Example 4. In some sense it shows that all the motions can be considered as constrained motions.

As we shall see, the constraints become interesting only when they can actually be "constructed", so that they can be used to reduce the number of degrees of freedom or of parameters necessary to describe the motions. A constraint of the type in the Example 4 is of little use in practice since it can be constructed only when all the motions of the system are perfectly understood (i.e., when S_t is a "known transformation"). However, this is usually the aim of the theory and it cannot be considered as a starting point.

Particularly interesting are the velocity-independent and time-independent constraints.

7 Definition. In the context of Definition 6, assume that there exist s real-valued functions in $C^\infty(\mathbf{R}^{Nd})$, $\varphi^{(1)}, \ldots, \varphi^{(s)}$ such that, $\forall i = 1, \ldots, s$,

$$\psi^{(i)}(\boldsymbol{\eta}^{(1)}, \ldots, \boldsymbol{\eta}^{(N)}, \boldsymbol{\xi}^{(1)}, \ldots, \boldsymbol{\xi}^{(N)}, t) \equiv \varphi^{(i)}(\boldsymbol{\xi}_1, \ldots, \boldsymbol{\xi}^{(N)}) \tag{3.5.7}$$

for all $(\boldsymbol{\eta}, \boldsymbol{\xi}, t) \in \mathbf{R}^{2Nd+1}$.

We shall say that the system is "subject to s holonomous constraints $\varphi^{(1)}, \ldots, \varphi^{(s)}$".[4] We shall denote $\mathfrak{M}_{t_1,t_2}(\boldsymbol{\xi}_1, \boldsymbol{\xi}_2 \,|\, \varphi^{(1)}, \ldots, \varphi^{(s)})$ the subset of the motions in $\mathfrak{M}_{t_1,t_2}(\boldsymbol{\xi}_1, \boldsymbol{\xi}_2)$ consisting of the motions $\mathbf{x} \in \mathfrak{M}_{t_1,t_2}(\boldsymbol{\xi}_1, \boldsymbol{\xi}_2)$ of the system such that

$$\varphi^{(i)}(\mathbf{x}(t)) \equiv 0, \qquad \forall i = 1, \ldots, s. \tag{3.5.8}$$

This set will be called the set of the motions "subject to" or "compatible with" the constraints $\varphi^{(1)}, \ldots, \varphi^{(s)}$.

Finally, if $\boldsymbol{\xi}, \boldsymbol{\eta} \in \mathbf{R}^{Nd}$, we shall say that $\boldsymbol{\xi}$ is a configuration "compatible with the constraints" if $\varphi^{(j)}(\boldsymbol{\xi}) = 0$, $j = 1, \ldots, s$, and we shall say that $\boldsymbol{\eta}$ is a velocity "compatible with the constraints in $\boldsymbol{\xi}$" if there is a motion $t \to \mathbf{x}(t)$, defined for t near zero, such that $\mathbf{x}(0) = \boldsymbol{\xi}$, $\dot{\mathbf{x}}(0) = \boldsymbol{\eta}$ and $\mathbf{x}(t)$ is compatible with the constraints for all t.

Observations.

(1) By the assumed time invariance of the constraints [see Eq. (3.5.7)], it is clear that the choice of time $t = 0$ in the last part of Definition 7 has no special meaning.

(2) In Problem 2 at the end of this section, we mention that when the vectors $(\partial\varphi^{(j)}/\partial\boldsymbol{\xi})(\boldsymbol{\xi})$, $j = 1, \ldots, s$, are s linearly independent vectors in $\mathbf{R}^{Nd}$, the constraint compatibility condition for a velocity $\boldsymbol{\eta}$ can simply be analytically expressed as

$$\frac{\partial\varphi^{(j)}}{\partial\boldsymbol{\xi}}(\boldsymbol{\xi}) \cdot \boldsymbol{\eta} = 0, \qquad j = 1, \ldots, s \tag{3.5.9}$$

which has a clear geometrical meaning.

[4] Holonomous simply means "depending on the site".

(3) Given $\boldsymbol{\xi} \in \mathbf{R}^{Nd}$ compatible with the constraints, the set of the velocity vectors $\boldsymbol{\eta}$ compatible with the constraints in $\boldsymbol{\xi}$ is always nonempty since it contains $\boldsymbol{\eta} = \mathbf{0}$.

Our first task will now be that of setting up a precise definition of a "perfect holonomous constraint".

A possible definition is inspired by Eq. (3.4.12): in that case, we shall naturally call the constraint to the line Γ "ideal" if $\mathbf{R} \cdot \boldsymbol{\tau} = 0$, i.e., if the "only effect" of the constraint to the line Γ is that of keeping the motion on Γ; in fact, the equation of motion simply becomes

$$m\ddot{s} = \mathbf{f}(s) \cdot \boldsymbol{\tau}(s) \tag{3.5.10}$$

which can be read "the acceleration along Γ is proportional to the projection on Γ of the active force", i.e., of the part of the force distinct from the constraint's reaction.

The relation $\mathbf{R} \cdot \boldsymbol{\tau} = 0$ can be interpreted as meaning that the reaction acts orthogonally to Γ. However, it is not immediately clear what should be meant by the reaction being orthogonal to the constraint in the case of the general constraints considered in Definition 7.

After some thought, the following notion appears natural: the constraint's reaction or, more generally, a force law $\mathbf{R}(\boldsymbol{\eta}, \boldsymbol{\xi})$ acting on a system of N points masses in $\mathbf{R}^d$, when it occupies the configuration $\boldsymbol{\xi}$ with velocity $\boldsymbol{\eta}$ (both constraint compatible), is "orthogonal to the constraint" if, calling $\boldsymbol{\eta}'$ the velocity of any other constraint compatible motion at the time when it occupies the configuration $\boldsymbol{\xi}$, it is

$$\mathbf{R}(\boldsymbol{\eta}, \boldsymbol{\xi}) \cdot \boldsymbol{\eta}' = 0 \tag{3.5.11}$$

which, more explicitly, is

$$\sum_{j=1}^{N} \boldsymbol{\eta}'^{(j)} \cdot \mathbf{R}^{(j)}(\boldsymbol{\eta}, \boldsymbol{\xi}) = 0. \tag{3.5.12}$$

One could argue and debate about this extension. However, in this section we shall first investigate its mathematical meaning, delaying the discussion of its deep and interesting physical interpretation until later on. Let us therefore establish the following definition.

8 Definition. Let $\mathbf{F}$ be a time-independent force law for a system of N point masses in $\mathbf{R}^d$. Assume that $\mathbf{F}$ produces s holonomous constraints $\varphi^{(1)}, \ldots, \varphi^{(s)}$.

Given a positional force law $\mathbf{F}^{(a)} \in C^\infty(\mathbf{R}^{Nd})$ for the system, we define the "constraints' reaction" with respect to the "active force" $\mathbf{F}^{(a)}$ as the quantity $\mathbf{R} = \mathbf{F} - \mathbf{F}^{(a)}$. Furthermore, we shall say that the constraints' system is "ideal" with respect to the pair $(\mathbf{R}, \mathbf{F}^{(a)})$ if for all $\boldsymbol{\xi} \in \mathbf{R}^{Nd}$ compatible with the constraints, i.e., such that $\varphi^{(j)}(\boldsymbol{\xi}) \equiv 0, j = 1, \ldots, s$ (see Definition 7), one has

$$\mathbf{R}(\boldsymbol{\eta}_1, \boldsymbol{\xi}) \cdot \boldsymbol{\eta}_2 = 0 \tag{3.5.13}$$

for all choices of constraint compatible velocity vectors $\boldsymbol{\eta}_1, \boldsymbol{\eta}_2$ (in the sense of Definition 7).

We shall refer to this situation by using the shortened locution "the system of point masses is subject to the active force $\mathbf{F}^{(a)}$ and to s holonomous ideal constraints $\varphi^{(1)}, \ldots, \varphi^{(s)}$".

Observations.

(1) Therefore, the last sentence means that the system is subject to a time-independent force law $\mathbf{F}$ producing the constraints $\varphi^{(1)}, \ldots, \varphi^{(s)}$ which are ideal with respect to the active force $\mathbf{F}^{(a)}$ and to the "reaction" $\mathbf{R} = \mathbf{F} - \mathbf{F}^{(a)}$.

Strictly speaking, the last sentence of Definition 8 should be subject to a consistency check: in terms of the information contained in it, it should be possible to reconstruct the equations of motion at least as far as the constrained motions are concerned; i.e., given $\mathbf{F}^{(a)}$ and the constraints it should be possible to reconstruct $\mathbf{F}(\boldsymbol{\eta}, \boldsymbol{\xi})$ for all constraint compatible $(\boldsymbol{\eta}, \boldsymbol{\xi})$. This is actually possible and basically, it is the content of Proposition 8 (to follow) and of the first observation to it (see, also, Problem 2 at the end of this section).

(2) It is important to stress that the decomposition $\mathbf{F} = \mathbf{F}^{(a)} + \mathbf{R}$ of the force as a sum of an "active force" and of an "ideal constraint's reaction" is certainly not unique, if it exists at all. For instance, if $\tilde{\mathbf{F}}$ is a conservative force field for our system whose potential energy $\tilde{V} \in C^\infty(\mathbf{R}^{Nd})$ is constant on the region of $\mathbf{R}^{Nd}$, where

$$\varphi^{(1)}(\boldsymbol{\xi}) = \cdots = \varphi^{(s)}(\boldsymbol{\xi}) = 0, \tag{3.5.14}$$

then the decomposition

$$\mathbf{F} = (\mathbf{F}^{(a)} + \tilde{\mathbf{F}}) + (\mathbf{R} - \tilde{\mathbf{F}}) \tag{3.5.15}$$

can be shown to be another decomposition of $\mathbf{F}$ into an "active" part $\mathbf{F}^{(a)} + \tilde{\mathbf{F}}$ and into a "reaction" $\mathbf{R}' = \mathbf{R} - \tilde{\mathbf{F}}$ verifying Eq. (3.5.13).

In fact, if $t \to \mathbf{x}(t)$, $t \in \mathbf{R}$, is a motion compatible with the constraints and passing through $\boldsymbol{\xi}$ at $t = 0$ with speed $\boldsymbol{\eta}_2$, we shall have $\tilde{V}(\mathbf{x}(t)) = \text{constant}$; hence,

$$\begin{aligned} 0 &= -\left.\frac{d\tilde{V}(\mathbf{x}(t))}{dt}\right|_{t=0} = -\frac{\partial \tilde{V}}{\partial \boldsymbol{\xi}}(\boldsymbol{\xi}) \cdot \dot{\mathbf{x}}(0) = -\frac{\partial \tilde{V}}{\partial \boldsymbol{\xi}}(\boldsymbol{\xi}) \cdot \boldsymbol{\eta}_2 \\ &= \tilde{\mathbf{F}}(\boldsymbol{\xi}) \cdot \boldsymbol{\eta}_2 . \end{aligned} \tag{3.5.16}$$

(3) The ambiguity seen in Observation 2 has a physical interpretation: it is generally ambiguous to talk about the constraint's reactions before having specified which are the other forces "not due to the constraints". Think of a point constrained to glide on a horizontal plane: we can always look at it as if it were subject to a force orthogonal to the plane and of arbitrary intensity G, besides the vertical downward gravity force mg. The point will not change its motion, at least in absence of friction, but the

table's reaction will be mg upwards in the first case and $mg + G$ upwards in the second.

(4) Hence, on the basis of the above definition of ideality, the ideality of a constraints' system depends on the choice of $\mathbf{F}^{(a)}$: only once both $\mathbf{F}$ and $\mathbf{F}^{(a)}$ are given it is possible to define $\mathbf{R}$ and check Eq. (3.5.13). Therefore, the ideality of a constraint is not a property that can be described only in terms of the total force $\mathbf{F}$ producing it.

Translating into a mathematical model concrete problems, it often happens that one is given the constraints' equations $\varphi^{(1)}, \ldots, \varphi^{(s)}$ and, separately, the active forces $\mathbf{F}^{(a)}$ and the "reaction of the constraints" $\mathbf{R}$. In fact, in applications it is often possible to distinguish operationally between the forces due to the constraints ("constraint reactions") and those due to other causes ("active forces"). In such cases, $\mathbf{R}$ is a priori given, or at least some of its properties are a priori given.

(5) Equation (3.5.13) is often called the "symbolic equation of dynamics" or "D'Alembert's principle" or "virtual works principle". The last name is usually given to Eq. (3.5.13) in its applications to statics where it is considered with $\boldsymbol{\eta}_1 = 0$ (see, also, the next comment and Observation 2, p. 164, and the concluding remarks, p. 242.

(6) Equation (3.5.13) is also read "the virtual work of an ideal constraint's reaction always vanishes". This is perhaps the most suggestive way of reading this equation since it stresses the fact that the velocity vector $\boldsymbol{\eta}_2$ is not the same as that, $\boldsymbol{\eta}_1$, provoking the reaction in $\boldsymbol{\xi}$. It is, in fact, the velocity of another possible motion through $\boldsymbol{\xi}$ ("virtual motion"). The word "work" is naturally a reference to the fact that $\mathbf{R}(\boldsymbol{\eta}_1, \boldsymbol{\xi}) \cdot \boldsymbol{\eta}_2$ is the work per unit time that the constraint's reaction to the motion $\mathbf{x}$ (passing at a given time through $\boldsymbol{\xi}$ with speed $\boldsymbol{\eta}_1$) performs on another motion passing, at the same time, through $\boldsymbol{\xi}$ with speed $\boldsymbol{\eta}_2$.

In the upcoming sections, we will analyze the physical meaning of Definition 8, i.e., we shall discuss the physical circumstances in which it becomes a relevant definition. Before that analysis, let us examine some remarkable consequences of this definition.

The first consequence is the following proposition: the "theorem of energy conservation for ideally constrained systems".

7 Proposition. *Let $t \to \mathbf{x}(t)$, $t \in I$, be a motion of a system of N point masses in $\mathbf{R}^d$ with masses $m_1, \ldots, m_N > 0$ subject to a system of s ideal holonomous constraints and to a conservative active force $\mathbf{F}^{(a)}$ with potential energy $V^{(a)} \in C^\infty(\mathbf{R}^{Nd})$. Assume that $\mathbf{x}$ respects the constraints. Then there is a constant E such that*

$$T(t) + V^{(a)}(t) = E, \qquad t \in I, \tag{3.5.17}$$

where $V^{(a)}(t) = V^{(a)}(\mathbf{x}(t))$ and $T(t)$ is the kinetic energy of the motion $\mathbf{x}$ at time t.

Observation. The main point is that the above proposition does not assume that the constraint's reaction is conservative in the sense of §3.1, but "only" that it is ideal, i.e., that it verifies Eq. (3.5.13).

PROOF. It is an immediate consequence of the theorem of alive forces, Proposition 2, §3.2, p. 146, that the variation of kinetic energy between two times t_1 and t_2 is equal to the sum of the work performed on the motion $\mathbf{x}$ by the force $\mathbf{F}^{(a)}$, i.e., $V^{(a)}(t_1) - V^{(a)}(t_2)$, and by the reaction $\mathbf{R}$, i.e.,

$$\sum_{j=1}^{N} \int_{t_1}^{t_2} \mathbf{R}^{(j)}(\dot{\mathbf{x}}(t), \mathbf{x}(t)) \cdot \dot{\mathbf{x}}^{(j)}(t)\, dt. \tag{3.5.18}$$

However, by assumption, the motion $\mathbf{x}$ respects the constraints and, also, Eq. (3.5.13) holds. Using Eq. (3.5.13) with $\boldsymbol{\xi} = \mathbf{x}(t)$, $\boldsymbol{\eta}_1 = \boldsymbol{\eta}_2 = \dot{\mathbf{x}}(t)$, we see that the work in Eq. (3.5.18) vanishes; hence, $T(t_2) - T(t_1) = V^{(a)}(t_1) - V^{(a)}(t_2)$, implying Eq. (3.5.17). mbe

Far more interesting is the following proposition: the "least-action principle for ideally constrained systems".

8 Proposition. *Consider a system of N point masses in* $\mathbf{R}^d$, *with masses* $m_1, \ldots, m_N > 0$, *subject to s holonomous ideal constraints* $\varphi^{(1)}, \ldots, \varphi^{(s)}$ *and to the active force* $\mathbf{F}^{(a)}$, *conservative and with potential energy* $V^{(a)} \in C^\infty(\mathbf{R}^{Nd})$. *Denote by* $\mathbf{R}$ *the constraint's reaction.*

The action with Lagrangian

$$\mathcal{L}(\boldsymbol{\eta}, \boldsymbol{\xi}, t) = \frac{1}{2} \sum_{j=1}^{N} m_j (\boldsymbol{\eta}^{(j)})^2 - V^{(a)}(\boldsymbol{\xi}) \tag{3.5.19}$$

is stationary in $\mathfrak{M}_{t_1,t_2}(\boldsymbol{\xi}_1, \boldsymbol{\xi}_2 \,|\, \varphi^{(1)}, \ldots, \varphi^{(s)})$ *on the motions which are generated by the force* $\mathbf{F}^{(a)} + \mathbf{R} = \mathbf{F}$.

Furthermore, let $t \to \mathbf{x}(t)$, $t \in \mathbf{R}_+$, *be a motion of the system developing under the action of the force* $\mathbf{F}$ *and respecting the constraints. Given* $t_1 \in \mathbf{R}_+$, *there exists* $\bar{t} > t_1$ *such that if* $t_2 \in [t_1, \bar{t}]$, *the action with Lagrangian* (3.5.19) *is locally minimal in* $\mathfrak{M}_{t_1,t_2}(\mathbf{x}(t_1), \mathbf{x}(t_2) \,|\, \varphi^{(1)}, \ldots, \varphi^{(s)})$ *on the motion* $\mathbf{x}$ *observed for* $t \in [t_1, t_2]$ *and thought of as an element of* $\mathfrak{M}_{t_1,t_2}(\mathbf{x}(t_1), \mathbf{x}(t_2) \,|\, \varphi^{(1)}, \ldots, \varphi^{(s)})$.

Observations.

(1) The importance of the above proposition lies in the fact that, if wisely used, it allows one to "eliminate" the degrees of freedom redundant because of the constraints.

Suppose one is able to find N C^∞ functions on $\mathbf{R}^l$ taking values in $\mathbf{R}^d$:

$$\boldsymbol{\alpha} = (\alpha_1, \ldots, \alpha_l) \to \mathbf{X}^{(i)}(\boldsymbol{\alpha}) = \mathbf{X}^{(i)}(\alpha_1, \ldots, \alpha_l), \tag{3.5.20}$$

$i = 1, \ldots, N$, such that $\forall \boldsymbol{\alpha} \in \mathbf{R}^l$:

$$\varphi^{(j)}(\mathbf{X}(\boldsymbol{\alpha})) = 0, \qquad j = 1, \ldots, s, \tag{3.5.21}$$

i.e., such that the image of $\mathbf{R}^l$ via the transformation (3.5.20) is a subset of $\mathbf{R}^{Nd}$ which automatically "verifies the constraints".[5]

Also, suppose one knows that the motion $\hat{\mathbf{x}} \in \mathfrak{M}_{t_1,t_2}(\xi_1, \xi_2 | \varphi^{(1)}, \ldots, \varphi^{(s)})$ that we are studying and which develops under the action of the force $\mathbf{F}$, is the image in $\mathbf{R}^{Nd}$ of a motion $\hat{\mathbf{a}} \in \mathfrak{M}_{t_1,t_2}(\hat{\mathbf{a}}(t_1), \hat{\mathbf{a}}(t_2))$ in $\mathbf{R}^l$ via the transformation (3.5.20).

The above assumptions mean that we have a good understanding of the constraint's structure so that we can find an explicit parametric representation of a class of configurations satisfying it.

It is clear that the action $A(\mathbf{x})$ with Lagrangian (3.5.19) can be computed on the motions $\mathbf{x} \in \mathfrak{M}_{t_1,t_2}(\xi_1, \xi_2 | \varphi^{(1)}, \ldots, \varphi^{(s)})$ that are images via Eq. (3.5.21) of motions $\mathbf{a} \in \mathfrak{M}_{t_1,t_2}(\hat{\mathbf{a}}(t_1), \hat{\mathbf{a}}(t_2))$ in $\mathbf{R}^l$ as $\tilde{A}(\mathbf{a})$, where $\tilde{A}$ is the action on $\mathfrak{M}_{t_1,t_2}(\mathbf{a}(t_1), \mathbf{a}(t_2))$ with (t-independent) Lagrangian

$$\tilde{\mathfrak{L}}(\boldsymbol{\beta}, \boldsymbol{\alpha}) = \frac{1}{2} \sum_{j=1}^{N} m_j \left(\sum_{k=1}^{l} \frac{\partial \mathbf{X}^{(j)}}{\partial \alpha_k}(\boldsymbol{\alpha}) \beta_k \right)^2 - V^{(a)}(\mathbf{X}(\boldsymbol{\alpha})) \tag{3.5.22}$$

[see, also, §3.4, Eq. (3.4.5), where this is derived].

Since by Proposition 8 A is stationary in $\hat{\mathbf{x}}$ in $\mathfrak{M}_{t_1,t_2}(\xi_1, \xi_2 | \varphi^{(1)}, \ldots, \varphi^{(s)})$, we have that $\tilde{A}$ is stationary in $\hat{\mathbf{a}}$ on the entire set $\mathfrak{M}_{t_1,t_2}(\hat{\mathbf{a}}(t_1), \hat{\mathbf{a}}(t_2))$ and, therefore, by Proposition 4, §3.3, this means that

$$\frac{d}{dt}\left(\frac{\partial \tilde{\mathfrak{L}}}{\partial \beta_i}(\dot{\hat{\mathbf{a}}}(t), \hat{\mathbf{a}}(t)) \right) = \frac{\partial \tilde{\mathfrak{L}}}{\partial \alpha_i}(\dot{\hat{\mathbf{a}}}(t), \hat{\mathbf{a}}(t)) \tag{3.5.23}$$

for $i = 1, 2, \ldots, l$ and $t \in [t_1, t_2]$.

Equation (3.5.23) provide l equations for the l unknown functions $t \to a_i(t)$, $i = 1, 2, \ldots, l$, $t \in [t_1, t_2]$. These are the equations of motion for the essential coordinates once the degrees of freedom which have become inessential because of the constraints have been eliminated.

It is of fundamental importance to realize the difference between this section's considerations and those, apparently alike, of §3.4, pp. 154–155. Those, in fact, had been developed assuming that the force $\mathbf{F}$ was conservative in the sense of §3.1. In the present case, as the example of §3.4 p. 156 shows, the force will generally be velocity dependent.

After a few exercises the reader will understand how great a simplification Eq. (3.5.23) implies in the deduction of the equations of motion, with respect to the alternative procedure of writing the equations of motions in the ordinary Cartesian coordinates followed by the elimination of the constraint reactions [remarkably absent in Eq. (3.5.23)] and of the redundant coordinates. In many instances, for example think of a rigid body, N can be large but l very small.

(2) It is convenient to say a few words to explain why the name "principle" is granted to the Proposition 8 as well as to several other propositions or definitions already met (D'Alembert's principle, virtual

[5]For instance, in the case of the point constrained on a line (§3.4), one can take $l = 1$ and $\alpha \to \mathbf{x}(\alpha) \equiv \xi(\alpha)$, $\alpha \in \mathbf{R}$, and the parameter α has the meaning of a curvilinear abscissa on the curve.

work principle, etc.). Such names have interesting historical origins: the reader should not believe that the discussion of the laws of mechanics and the treatment of all the mechanical problems by the application of the equation $\mathbf{f} = m\mathbf{a}$, together with the two other laws of dynamics, to the point masses into which a system can be ideally decomposed has always been obvious and natural since the work of Newton.

As already remarked, Newton himself did not arrive in a very clear way at such a conclusion. For instance, in his study of rigid motions he had recourse to arguments quite different from modern methods based on the cardinal equations (i.e., on Newton's laws).

Both before and after Newton, philosophers were accustomed to studying mechanical problems on the basis of special assumptions, "principles", which were deduced by them through more or less general considerations often a bit obscure.

Newton's principles can be thought of as belonging to the above class of principles, and, initially, they were used particularly in the theory of the motions of heavenly bodies. Together with the three principles first formulated by Newton, there already existed, more or less clearly formulated at least in particular cases, the energy conservation principle for simple systems (Huygens), the principle of the linear momentum conservation (going back at least to Descartes), the virtual works principle (which was used in the solution of problems in statics by Del Monte, Galilei, Stevin, etc.), the inertia principle (Galilei), and to these principles many others can be added: they were invented, even in years following Newton's time, to treat complex mechanical problems.

With Euler's work, the synthesis of all the different principles began through the realization that they could all be unified and deduced from Newton's and, what is often not sufficiently clearly stated, equivalent to Newton's if suitably interpreted (probably beyond the intentions and meanings the inventors attributed to them) (see, also, the concluding remarks to Chapter 3, p. 242.

Let us go back to the simple proof of Proposition 8.

PROOF. Let $\mathbf{x} \in \mathfrak{M}_{t_1,t_2}(\xi_1, \xi_2 | \varphi^{(1)}, \ldots, \varphi^{(s)}) \equiv \mathfrak{M}$ be a motion developing under the influence of $\mathbf{F} = \mathbf{F}^{(a)} + \mathbf{R}$. The action of $\mathbf{x}$ with respect to the Lagrangian (3.5.19) is

$$A(\mathbf{x}) = \int_{t_1}^{t_2} \left\{ \sum_{j=1}^{N} \frac{m_j}{2} \left(\dot{\mathbf{x}}^{(j)}(t)\right)^2 - V^{(a)}(\mathbf{x}(t)) \right\} dt. \tag{3.5.24}$$

Let $\mathbf{y} \in \mathcal{V}_{\mathbf{x}}(\mathfrak{M})$; let us compute the derivative with respect to ϵ of the function $A(\mathbf{y}_\epsilon)$ in $\epsilon = 0$. If we set (see Definition 5, §3.3) $\mathbf{z}(t) = (\partial \mathbf{y}/\partial \epsilon)(t, 0) = (\mathbf{z}^{(1)}(t), \ldots, \mathbf{z}^{(N)}(t))$, $t \in [t_1, t_2]$, we have $\mathbf{z}(t_1) = \mathbf{z}(t_2) = 0$ and

$$\frac{d}{d\epsilon} A(\mathbf{y}_\epsilon)\Big|_{\epsilon=0} = \int_{t_1}^{t_2} \sum_{j=1}^{N} \left\{ m_j \dot{\mathbf{x}}^{(j)}(t) \cdot \dot{\mathbf{z}}^{(j)}(t) - \frac{\partial V^{(a)}}{\partial \xi^{(j)}}(\mathbf{x}(t)) \cdot \mathbf{z}^{(j)}(t) \right\} dt. \tag{3.5.25}$$

By integrating the terms containing $\dot{\mathbf{z}}^{(j)}(t)$ by parts and using $\mathbf{z}(t_1) = \mathbf{z}(t_2)$

$= \mathbf{0}$, one deduces, as usual,

$$\frac{d}{d\epsilon}A(\mathbf{y}_\epsilon)\bigg|_{\epsilon=0} = -\int_{t_1}^{t_2}\sum_{j=1}^{N}\left\{m_j\ddot{\mathbf{x}}^{(j)}(t) + \frac{\partial V^{(a)}}{\partial \boldsymbol{\xi}^{(j)}}(\mathbf{x}(t))\right\}\cdot \mathbf{z}^{(j)}(t)\,dt. \qquad (3.5.26)$$

The equations of motion for $\mathbf{x}$ are, by assumption,

$$m_j\ddot{\mathbf{x}}^{(j)}(t) = -\frac{\partial V^{(a)}}{\partial \boldsymbol{\xi}^{(j)}}(\mathbf{x}(t)) + \mathbf{R}^{(j)}(\dot{\mathbf{x}}(t),\mathbf{x}(t)); \qquad (3.5.27)$$

hence, we cannot conclude that the right-hand side of Eq. (3.5.26) vanishes, but only that it is equal to

$$\frac{d}{d\epsilon}A(\mathbf{y}_\epsilon)\bigg|_{\epsilon=0} = -\int_{t_1}^{t_2}\sum_{j=1}^{N}\mathbf{R}^{(j)}(\dot{\mathbf{x}}(t),\mathbf{x}(t))\cdot \mathbf{z}^{(j)}(t)\,dt. \qquad (3.5.28)$$

However, Eq. (3.5.13) will allow us to infer that, $\forall \tau \in (t_1, t_2)$:

$$\sum_{j=1}^{N}\mathbf{R}^{(j)}(\dot{\mathbf{x}}(\tau),\mathbf{x}(\tau))\cdot \mathbf{z}^{(j)}(\tau) \equiv 0 \qquad (3.5.29)$$

if we show that $(\mathbf{x}(\tau), \mathbf{z}(\tau))$ are a position-velocity pair compatible with the constraints; i.e., if we show the existence of a motion defined for t near τ and constraint compatible, which at $t = \tau$ is in $\mathbf{x}(\tau)$ with velocity $\mathbf{z}(\tau)$.

Recalling the definition of $\mathbf{z}$, one sees that such a motion indeed exists. To build it, one simply defines $t \to \mathbf{y}(\tau, t-\tau)$ for $t - \tau \in (-1, 1)$, i.e., for t close to τ. This function of t has, for $t = \tau$, velocity $\mathbf{z}(\tau)$ and, furthermore, verifies the constraints and has a value $\mathbf{x}(\tau)$, for $t = \tau$, since $\mathbf{y} \in \mathcal{V}_{\mathbf{x}}(\mathfrak{M})$.

We shall not explicitly prove the local minimum property: its (long) proof is entirely analogous to the proof of Proposition 37, §2.24, and should not present particular difficulties to the reader. mbe

Problems for §3.5

1. Give an example of a holonomous constraint for a system of N point masses in $\mathbf{R}^3$ for which the only constraint compatible velocity $\boldsymbol{\eta}$ is $\boldsymbol{\eta} = 0$. (Hint: Find a constraint φ such that $\varphi(\boldsymbol{\xi}) = 0$ determines an isolated point.)

2.* Let $\varphi^{(1)}, \dots, \varphi^{(s)}$ be s holonomous constraints for a system of N point masses in $\mathbf{R}^3$. Given $\boldsymbol{\xi} \in \mathbf{R}^{3N}$ constraint compatible, assume that the s vectors $(\partial\varphi^{(j)}/\partial\boldsymbol{\xi})(\boldsymbol{\xi}) \in \mathbf{R}^{3N}$, $j = 1, \dots, s$, are linearly independent. Show that $\boldsymbol{\eta}$ is a constraint compatible velocity if and only if Eq. (3.5.9) holds. (Hint: The necessity is obvious. Conversely, consider the conditions on a constraint compatible motion of the form

$$t \to \boldsymbol{\xi} + t\boldsymbol{\eta} + \sum_{j=1}^{s}\delta_j(t)\frac{\partial\varphi^{(j)}}{\partial\boldsymbol{\xi}}$$

given by

$$\varphi^{(k)}\left(\boldsymbol{\xi} + t\boldsymbol{\eta} + \sum_{j=1}^{s}\delta_j\frac{\partial\varphi^{(j)}}{\partial\boldsymbol{\xi}}(\boldsymbol{\xi})\right) = 0, \qquad k = 1, \dots, s$$

which are regarded as equations for δ_j parametrized by t and solved, for $t=0$, by $\delta_j=0$. We now regard the left-hand side as a function of $t,\delta_1,\ldots,\delta_s$, and call it $\Phi^{(k)}(t,\delta_1,\ldots,\delta_s)$ and we try to define $\delta_j(t)$, $j=1,\ldots,s$, for t near zero, applying the implicit functions theorem (Appendix G). The Jacobian matrix for $t=0$, $\delta_1=\cdots=\delta_s=0$, is

$$M_{kh}=\frac{\partial\Phi^{(k)}}{\partial\delta_h}(\mathbf{0})=\sum_{p=1}^{3N}\frac{\partial\varphi^{(k)}}{\partial\xi_p}(\boldsymbol{\xi})\frac{\partial\varphi^{(h)}}{\partial\xi_p}(\boldsymbol{\xi}),\qquad h,k=1,\ldots,s$$

which has rank s, by the supposed linear independence of the vectors $(\partial\varphi^{(k)}/\partial\boldsymbol{\xi})(\boldsymbol{\xi})$. In fact, the linear independence means that, $\forall\mathbf{c}\in\mathbf{R}^s$

$$\sum_{k=1}^{s}c_k\frac{\partial\varphi^{(k)}}{\partial\boldsymbol{\xi}}(\boldsymbol{\xi})\neq\mathbf{0}$$

unless $\mathbf{c}=\mathbf{0}$; therefore, if $\mathbf{c}\neq\mathbf{0}$:

$$\begin{aligned}\sum_{h,k=1}^{s}c_hc_kM_{hk}&=\sum_{p=1}^{3N}\sum_{h=1}^{s}\sum_{k=1}^{s}c_hc_k\frac{\partial\varphi^{(k)}}{\partial\xi_p}(\boldsymbol{\xi})\frac{\partial\varphi^{(h)}}{\partial\xi_p}(\boldsymbol{\xi})\\&=\sum_{p=1}^{3N}\left(\sum_{h=1}^{s}c_h\frac{\partial\varphi^{(h)}}{\partial\xi_p}(\boldsymbol{\xi})\right)^2>0\end{aligned}$$

and, since $M_{hk}=M_{kh}$, this means that the matrix M is positive definite; hence, $\det M>0$ (see Appendix F).

Hence, by the implicit functions theorem, there exist s C^∞ functions $t\to\delta_j(t)$, $j=1,\ldots,s$, defined near $t=0$ such that the motion

$$t\to\mathbf{x}(t)=\boldsymbol{\xi}+\boldsymbol{\eta}t+\sum_{j=1}^{s}\delta_j(t)\frac{\partial\varphi^{(j)}}{\partial\boldsymbol{\xi}}(\boldsymbol{\xi})$$

verifies the constraints and, furthermore, $\delta_j(0)=0$ and, by the implicit functions theorem:

$$\delta_k(t)=-t\sum_{h=1}^{s}(M^{-1})_{kh}\frac{\partial\varphi^{(h)}}{\partial\boldsymbol{\xi}}(\boldsymbol{\xi})\cdot\boldsymbol{\eta}+t^2\tilde{\delta}_k(t)$$

for some C^∞ functions $\tilde{\delta}_k$ defined near $t=0$. By the assumption on $\boldsymbol{\eta}$, the t-linear term vanishes: hence, $\dot{\delta}_k(0)=0$, i.e., $\dot{\mathbf{x}}(0)=\boldsymbol{\eta}\ldots$.)

3.* Given a system of N point masses in $\mathbf{R}^3$, with masses $m_1,\ldots,m_N>0$, subject to ideal holonomous constraints $\varphi^{(1)},\ldots,\varphi^{(s)}$ and to active force $\mathbf{F}^{(a)}$, show the possibility of an explicit expression for the reaction $\mathbf{R}$ acting on a constraint compatible motion, at a time t when $\mathbf{x}(t)=\boldsymbol{\xi}$, $\dot{\mathbf{x}}(t)=\boldsymbol{\eta}$. (Hint: Let $m_1=\cdots=m_N=1$ for simplicity and suppose also that the s vectors in $\mathbf{R}^{3N}$, $(\partial\varphi^{(j)}/\partial\boldsymbol{\xi})(\boldsymbol{\xi})$, $j=1,\ldots,s$, are independent, again for simplicity. From $\varphi^{(j)}(\mathbf{x}(t))\equiv0$, $j=1,\ldots,s$, deduce by two-fold differentiation

$$\frac{\partial\varphi^{(j)}}{\partial\boldsymbol{\xi}}(\mathbf{x}(t))\cdot\ddot{\mathbf{x}}(t)+\sum_{p,q=1}^{N}\frac{\partial^2\varphi^{(j)}}{\partial\xi_p\partial\xi_q}(\mathbf{x}(t))\dot{x}_p(t)\dot{x}_q(t)\equiv0$$

and then combine this equation with the equation of motion $\ddot{\mathbf{x}}=\mathbf{F}^{(a)}+\mathbf{R}$ to obtain

$$\frac{\partial\varphi^{(j)}}{\partial\boldsymbol{\xi}}(\boldsymbol{\xi})\cdot\mathbf{R}(\boldsymbol{\eta},\boldsymbol{\xi})=-\left\{\frac{\partial\varphi^{(j)}}{\partial\boldsymbol{\xi}}(\boldsymbol{\xi})\cdot\mathbf{F}^{(a)}(\boldsymbol{\xi})+\sum_{p,q=1}^{3N}\frac{\partial^2\varphi^{(j)}}{\partial\xi_p\partial\xi_q}(\boldsymbol{\xi})\eta_p\eta_q\right\},$$

$j=1,\ldots,s$. But, by the ideality assumption, $\mathbf{R}(\boldsymbol{\eta},\boldsymbol{\xi})$ has to be orthogonal to every

η' such that $(\partial\varphi^{(j)}/\partial\xi)(\xi)\cdot\eta'=0$ [see Problem 2 and Eq. (3.5.13)]. Hence, $\mathbf{R}$ has to be a linear combination of the s vectors $(\partial\varphi^{(j)}/\partial\xi)(\xi)$, $j=1,\ldots,s$, and the coefficients can be determined by the scalar products $(\partial\varphi^{(j)}/\partial\xi)(\xi)\cdot\mathbf{R}$, $j=1,\ldots,s$, since the s vectors $(\partial\varphi^{(j)}/\partial\xi)(\xi)$ are linearly independent. Deal also with the general case: different masses and linearly dependent vectors $\partial\varphi^{(j)}/\partial\xi$.)

4. A "constraint" of the form $\varphi(\xi)\geqslant 0$ for a system of N point masses in $\mathbf{R}^d$ is called "unilateral". Show that such constraints are not more general than those considered in Definition 7. (Hint: Let $\alpha\to\chi(\alpha)$ be a C^∞ function, strictly positive if $\alpha<0$ and zero if $\alpha\geqslant 0$; then consider the constraint $\psi(\xi)=\chi(\varphi(\xi))=0$, etc.)

5. Show that any velocity is compatible with a unilateral constraint $\varphi\geqslant 0$ in the positions ξ, where $\varphi(\xi)>0$.

6. Which are the velocities η compatible with a unilateral constraint $\varphi(\xi)\geqslant 0$ in a position ξ, where $\varphi(\xi)=0$? Suppose $(\partial\varphi/\partial\xi)(\xi)\neq\mathbf{0}$ (Answer: Those such that $\eta\cdot(\partial\varphi/\partial\xi)(\xi)=0$, i.e., the same as those for the constraint $\varphi(\xi)=0$!)

7. Extend the notion of velocity η compatible, in a configuration ξ, with some holonomous (unilateral or not) constraints $\varphi^{(1)},\ldots,\varphi^{(s)}$ by saying that η is constraint compatible at ξ if there is a constraint compatible motion $t\to\mathbf{x}(t)$, defined for $t\geqslant 0$ small enough (rather than for $t\gtreqless 0$ and small enough) such that $\dot{\mathbf{x}}(0)=\eta$, $\mathbf{x}(0)=\xi$. Show that there are cases where η can be constraint compatible in this new sense without being so in the old one of Definition 7. We call the velocities which are constraint compatible in the new sense "(+)-compatible velocities". (Hint: The two notions will differ when ξ is such that $\varphi(\xi)=0$ in the case of a unilateral constraint $\varphi\geqslant 0$. Give a physical interpretation of such extra velocities in terms of "collision velocities" with the constraint.)

8.* Show that the smoothness requirements on $\mathbf{F}$ and $\mathbf{F}^{(a)}$ used for the Definition 8 of an ideal constraint cannot generally hold if the system is subject to a unilateral constraint $\varphi\geqslant 0$ (thought of as a constraint via the construction of Problem 4); i.e., a unilateral constraint cannot, in general, be ideal in the sense of Definition 8. (Hint: There are, in general, motions starting in the region $\varphi>0$ which in a finite time reach a point ξ_0, where $\varphi(\xi_0)=0$, "collision with the constraint", with a speed which is not (+)-constraint compatible in the sense of Problem 7. At this point the speed must have a discontinuity against the assumption that $\mathbf{F}\in C^\infty(\mathbf{R}^{3N})$ and the regularity theorem for the differential equations.)

§3.6. Real and Ideal Constraints

The discussion of §3.5 is largely unsatisfactory.

The notion of constraints used there has been given on a purely mathematical basis and it is quite unclear which is the physical phenomenon mathematically modeled by the constraints, ideal or not, of the preceding sections.

In this and in the following sections, we will radically modify the point of view and we will show that an ideal constraint for a system of N point masses can also be thought of as a limiting case of suitable very strong conservative force fields which oblige the systems' trajectories to lie on certain surfaces in $\mathbf{R}^{Nd}$ or in their vicinity.

From a physical viewpoint, one always imagines a constraint as a complex of forces acting on a system of point masses and due to their tendency to deform some obstacles. Such a tendency provokes imperceptibly small (at least as far as our observations are concerned[6]) deformations of the obstacles. Think of a point constrained on a rail or on a surface or think of a rigid system.

Note, also, that in the above concrete cases, the elegant theory of §3.5 is totally useless: the constraints now constrain in an approximate sense only and, therefore, they are not of the type considered there.

The question which is more interesting for us in this context is whether or not the solutions of the equations obtained by minimizing the Lagrangian (3.5.19) on the motions constrained by s holonomous constraints $\varphi^{(1)}, \ldots, \varphi^{(s)}$ (see Definition 8, §3.5) provide good approximations to the real motion under the influence of the real constraints, which necessarily constrain only in an approximate sense.

This is a really interesting problem in physics and applications, in contrast with the question underlying §3.5 which, abstractly, asked for a definition of a perfect constraint that would give rise to a sufficient condition in order that the equations of motion could be deduced from the least-action principle associated with the Lagrangian (3.5.19), for the $(\varphi^{(1)}, \ldots, \varphi^{(s)})$-constrained motions (see §3.5, Proposition 8, p. 163).

To understand better the spirit and the meaning of the various definitions that will follow, it is convenient to analyze a simple but significant example.

Consider a point mass with mass $m > 0$ in $\mathbf{R}^2$ and suppose that it is subject to an elastic force with potential energy $V^{(a)} = (m\omega^2/2)(x^2 + y^2)$ and to a restoring conservative force toward the $y = 0$ axis with potential energy

$$\lambda W(x, y) = \tfrac{1}{2} m\lambda y^2, \qquad \lambda > 0. \tag{3.6.1}$$

Consider the point's motions under the action of the force with potential energy

$$V^{(a)} + \lambda W = \tfrac{1}{2} m\omega^2(x^2 + y^2) + \tfrac{1}{2} m\lambda y^2. \tag{3.6.2}$$

It is clear that if λ is very large, such a force simulates a constraint to the line $y = 0$ in a sense which has still to be precisely understood. To understand this statement, let us study the motions which start on the $y = 0$ axis and develop under the influence of the force with potential energy of Eq. (3.6.2). The equations of motion are

$$\begin{aligned} m\ddot{x} &= -\omega^2 m x, \\ m\ddot{y} &= -\omega^2 m y - m\lambda y, \end{aligned} \tag{3.6.3}$$

$$x(0) = x_0, \qquad \dot{x}(0) = v_0, \qquad y(0) = 0, \qquad \dot{y}_0 = w_0,$$

and, to be definite, suppose $x_0 > 0$.

[6]When they can be appreciated, one no longer speaks of a constraint.

Because of their extreme simplicity, Eqs. (3.6.3) can be elementarily solved: if $t \to (x(t), y(t))$, $t \in \mathbf{R}$, denotes the solution of Eqs. (3.6.3):

$$x_\lambda(t) = \left(x_0^2 + \frac{v_0^2}{\omega^2} \right)^{1/2} \cos(\omega t + \varphi_0),$$

$$y_\lambda(t) = \frac{w_0}{(\lambda + \omega^2)^{1/2}} \sin(\lambda + \omega^2)^{1/2} t, \tag{3.6.4}$$

$$\varphi_0 = \operatorname{arctg} - \frac{v_0}{\omega x_0} .$$

One then sees that the limit as $\lambda \to \infty$ of the motion of Eqs. (3.6.4) is the motion $t \to (x(t), y(t))$ with

$$x(t) = \left(x_0^2 + \frac{v_0^2}{\omega^2} \right)^{1/2} \cos(\omega t + \varphi_0),$$

$$y(t) \equiv 0, \tag{3.6.5}$$

for all w_0. This is, as is easy to check, exactly the solution of the equations obtained by imposing stationarity on the motions constrained by the ideal holonomous constraint $\xi_2 = 0$ of the action with Lagrangian:

$$\tilde{\mathcal{L}}(\eta_1, \eta_2, \xi_1, \xi_2) = \frac{m(\eta_1^2 + \eta_2^2)}{2} - m\omega^2 \frac{(\xi_1^2 + \xi_2^2)}{2} . \tag{3.6.6}$$

On the basis of Observation (1) to Proposition 8 of §3.5, these equations coincide with those for the motions $t \to x(t)$, $t \in \mathbf{R}$, in $\mathbf{R}^1$ which make stationary the "constrained Lagrangian":

$$\frac{1}{2} m\eta_1^2 - m \frac{\omega^2}{2} \xi_1^2 \tag{3.6.7}$$

which, in our case, is what Eq. (3.6.6) becomes by imposing the constraint $\xi_2 = 0$.

It is interesting to note that the more "rigid" the approximate constraint realized by Eq. (3.6.2), the smaller the deviations from a motion respecting the constraints ($|y_\lambda| \lesssim 1/\sqrt{\lambda}$) for λ large [see Eqs. (3.6.4)]. At the same time, however, the coordinate $y_\lambda(t)$, which simply represents the constraint's violation, oscillates more and more rapidly: in fact, its vibrations have a frequency:

$$\nu_\lambda = \frac{(\lambda + \omega^2)^{1/2}}{2\pi} . \tag{3.6.8}$$

These very small but very-high-frequency vibrations ("fatigue vibrations" of the constraint) provide a good intuitive representation of the effect of an approximate ideal constraint on the motion of a system. In general, it is possible to think that a system of N point masses subject to an approximate ideal constraint moves as if it were on the surface $\Sigma \subset \mathbf{R}^{Nd}$ defined by the constraint, with some very small elongations orthogonal to Σ described by oscillatory motions with very small amplitude and very large frequency.

On the basis of the above heuristic discussion, the following definition should appear quite natural.

9 Definition. Given a system of N point masses, with masses $m_1, \ldots, m_N > 0$, in $\mathbf{R}^d$, let $\Sigma \subset \mathbf{R}^{Nd}$ be a closed set and let W be a real $C^\infty(\mathbf{R}^{Nd})$ function vanishing on Σ and having there a strict minimum; i.e., Σ is the set of the points $\boldsymbol{\xi} \in \mathbf{R}^{Nd}$ where $W(\boldsymbol{\xi}) = 0$ and for all $\boldsymbol{\xi} \notin \Sigma$ it is $W(\boldsymbol{\xi}) > 0$. We shall say that the conservative force law with potential energy

$$\boldsymbol{\xi} \rightarrow \lambda W(\boldsymbol{\xi}) \geqslant 0, \qquad \lambda > 0 \tag{3.6.9}$$

is a "model of conservative approximate constraint to the region Σ with structure W and rigidity λ".

We shall denote such a force law by (Σ, W, λ). If Σ is a regular surface with dimension $l < Nd$, we shall say that the constraint model (Σ, W, λ) is a "bilateral approximate conservative constraint with dimension l" or with "codimension $Nd - l$", (see also Definition 10 below).

Observation. In general, Σ may contain interior points: in this case, one says that Eq. (3.6.9) is a model for a "unilateral" approximate constraint.

It is convenient to recall the definition of a regular surface in $\mathbf{R}^d$.

10 Definition. Let $\boldsymbol{\beta} \rightarrow \Xi(\boldsymbol{\beta})$ be a C^∞ function defined on a convex open set $\Omega \subset \mathbf{R}^d$, taking its values in a neighborhood $U \subset \mathbf{R}^d$. Suppose that Ξ is invertible, i.e., one to one, as a map of Ω onto U and, furthermore, assume that Ξ is nonsingular, i.e., that its Jacobian matrix defined by

$$J_{ij}(\boldsymbol{\beta}) = \frac{\partial \Xi_i}{\partial \beta_j}(\boldsymbol{\beta}), \qquad i, j = 1, \ldots, d \tag{3.6.10}$$

has a nonvanishing determinant, $\forall \boldsymbol{\beta} \in \Omega$.

We shall say that Ξ "establishes a regular system of coordinates on U" and, if $\boldsymbol{\xi} = \Xi(\boldsymbol{\beta})$, we shall say that $\boldsymbol{\beta} \in \Omega$ is the coordinate of $\boldsymbol{\xi}$ in the coordinate system on U associated with Ξ.

We shall denote the just-described coordinate system by (U, Ξ) and Ω will be called the "basis" of the coordinate system.

A closed set $\Sigma \subset \mathbf{R}^d$ will be called a "regular l-dimensional surface" if for all $\boldsymbol{\xi}_0 \in \Sigma$ it is possible to find a neighborhood U_0 and a regular system of local coordinates (U_0, Ξ) with basis Ω_0 such that the points of $\Sigma \cap U_0$ are all those with coordinates $\boldsymbol{\beta} = (\beta_1, \ldots, \beta_d)$ such that

$$\beta_1 = \beta_2 = \cdots = \beta_{d-l} = 0. \tag{3.6.11}$$

We say that (U_0, Ξ) is a local system of coordinates "adapted to Σ".

Sometimes a regular s-dimensional surface is called a regular surface with codimension $d - l$.

Let us go back to the definition of the approximate conservative constraint. From now on, we will confine our attention to the approximate bilateral conservative constraints with dimension l, or, as it is customary to say, with "l degrees of freedom", $0 < l < Nd$.

Consider a system of N point masses in $\mathbf{R}^d$, with masses $m_1, \ldots, m_N > 0$, subject to the action of a conservative force with potential energy $V^{(a)} \in C^\infty(\mathbf{R}^{Nd})$, bounded from below, and to a model of approximate conservative constraint (Σ, W, λ), bilateral and l-dimensional.

Suppose that Σ is defined by equations of the type

$$\varphi^{(1)}(\xi) = \cdots = \varphi^{(s')}(\xi) = 0 \tag{3.6.12}$$

with $\varphi^{(i)} \in C^\infty(\mathbf{R}^{Nd})$ (the number s' need not be $(Nd - l)$, although this will often be the case).

It is natural to study the motions $t \to \mathbf{x}(t)$, $t \in \mathbf{R}_+$, developing under the action of the conservative force with potential energy

$$\xi \to V^{(a)}(\xi) + \lambda W(\xi), \tag{3.6.13}$$

following an initial datum $\mathbf{x}_\lambda(0) = \xi_0$, $\dot{\mathbf{x}}(0) = \eta_0$ with ξ_0 compatible with the constraints

$$\xi_0 \in \Sigma. \tag{3.6.14}$$

We ask if there exists a limit

$$\mathbf{x}(t) = \lim_{\lambda \to +\infty} \mathbf{x}_\lambda(t), \qquad t \in \mathbf{R}_+ . \tag{3.6.15}$$

Furthermore, we ask if $t \to \mathbf{x}(t)$, $t \in \mathbf{R}_+$, coincides (when existing) with a motion developing under the action of the ideal constraints of Eq. (3.6.12) and of the active force with potential $V^{(a)}$, in conformity with the Definition 8, p. 160, and Proposition 8, p. 163.

It is easy to realize that there cannot be a positive answer if the problem is posed in the above generality.

Just reconsider the point mass in $\mathbf{R}^2$, with mass $m > 0$, constrained to the line $\xi_2 = 0$ with the new constraint model (Σ, W, λ) with $\Sigma = \{\xi_1 - \text{axis}\}$ and, if $\xi_1 \equiv x$, $\xi_2 \equiv y$,

$$W(x, y) = \frac{m}{2} y^2(1 + x^2). \tag{3.6.16}$$

and, also, subject to the same active force with potential energy $V^{(a)} = \frac{1}{2} m\omega^2(x^2 + y^2)$.

The equations of motion, similar to Eqs. (3.6.3), now become

$$\begin{gathered} m\ddot{x} = -m\omega^2 x - \lambda m y^2 x, \\ m\ddot{y} = -\lambda m y(1 + x^2) - m\omega^2 y, \\ x(0) = x_0, \qquad \dot{x}(0) = v_0, \qquad y(0) = 0, \qquad \dot{y}(0) = w_0 . \end{gathered} \tag{3.6.17}$$

These equations are more complex than Eqs. (3.6.3), and we shall discuss them only in a heuristic, nonrigorous way.

Let $t \to (x(t), y(t))$, $t \in \mathbf{R}_+$, be the solution of Eqs. (3.6.17). Energy conservation implies that if

$$E = m\frac{v_0^2 + w_0^2}{2} + \frac{m\omega^2}{2} x_0^2, \tag{3.6.18}$$

one has, $\forall t \geqslant 0$,

$$E = m\frac{\dot{x}_\lambda(t)^2 + \dot{y}_\lambda(t)^2}{2} + m\omega^2\frac{x_\lambda(t)^2 + y_\lambda(t)^2}{2} + \frac{\lambda m}{2}\, y_\lambda(t)^2(1 + x_\lambda(t)^2). \tag{3.6.19}$$

Then Eq. (3.6.19) implies

$$|\dot{x}_\lambda(t)|,\ |\dot{y}_\lambda(t)| \leqslant \left(\frac{2E}{m}\right)^{1/2}, \tag{3.6.20}$$

$$|x_\lambda(t)| \leqslant \left(\frac{2E}{m\omega^2}\right)^{1/2}, \tag{3.6.21}$$

$$|y_\lambda(t)| \leqslant \left(\frac{2E}{m\lambda}\right)^{1/2} \tag{3.6.22}$$

which follow by observing that all the addends in Eq. (3.6.19) are non-negative.

Fix a finite time interval $[0, T]$ and note that the first of Eqs. (3.6.17) together with Eqs. (3.6.20), (3.6.21), and (3.6.22) implies that the function $\ddot{x}_\lambda(t)$ has a uniformly bounded modulus for all $\lambda > 1$ and for $t \in [0, T]$. Hence, if T is "small", one can think that the function $t \to x_\lambda(t)$, $t \in [0, T]$, is practically constant together with its first derivative and, then, we can heuristically set $x_\lambda(t) = x_0$, $t \in [0, T]$, in the second equation of Eqs. (3.6.17).

Within this "approximation", Eqs. (3.6.17) becomes "elementarily soluble":

$$y_\lambda(t) \cong w_0\big(\omega^2 + \lambda(1 + x_0^2)\big)^{-1/2}\sin\big(\omega^2 + \lambda(1 + x_0^2)\big)^{1/2}t \tag{3.6.23}$$

which shows that, at least if $t \in [0, T]$, $y_\lambda(t)$ varies very quickly if λ is large, with a frequency $\nu_\lambda \simeq (\omega^2 + \lambda(1 + x_0^2))^{1/2}/2\pi$ remaining, nevertheless, small by Eq. (3.6.22).

Since, as just seen, $x_\lambda(t)$ varies slowly (essentially λ independently), we can substitute in the first of Eqs. (3.6.17) $\lambda y_\lambda(t)^2$ with its average value $\lambda \overline{y_\lambda(t)^2}$ between 0 and t, if t is large compared to the characteristic time T_λ of y's variation $T_\lambda \simeq 2\pi O(\lambda^{-1/2})$:

$$\begin{aligned}\lambda\,\overline{y_\lambda(t)^2} &\cong \frac{\lambda}{t}\int_0^t \frac{w_0^2}{\big(\omega^2 + \lambda(1 + x_0^2)\big)}\Big(\sin\big(\omega^2 + \lambda(1 + x_0^2)\big)^{1/2}\tau\Big)^2 d\tau \\ &= \frac{\lambda w_0^2}{\omega^2 + \lambda(1 + x_0^2)}\left[\frac{1}{t\big(\omega^2 + \lambda(1 + x_0^2)\big)^{1/2}} \times \int_0^{t(\omega^2 + \lambda(1 + x_0^2))^{1/2}} (\sin^2\theta)\, d\theta\right],\end{aligned} \tag{3.6.24}$$

where Eq. (3.6.23) has been used and θ has been defined as

$$\theta = \tau\left(\omega^2 + \lambda(1 + x_0^2)\right)^{1/2} = \frac{2\pi\tau}{T_\lambda}\,.$$

Since we assume $t \gg T_\lambda = 2\pi(\omega^2 + \lambda(1 + x_0^2))^{-1/2}$, the integral in square brackets can be replaced by

$$\lim_{R\to\infty} \frac{1}{R}\int_0^R (\sin\theta)^2\,d\theta = \frac{1}{2}\,. \tag{3.6.25}$$

Hence,

$$\lambda\,\overline{y_\lambda(t)^2} \simeq \frac{1}{2}\,\frac{\lambda w_0^2}{\omega^2 + \lambda(1 + x_0^2)} \sim \frac{1}{2}\,\frac{w_0^2}{(1 + x_0^2)} \tag{3.6.26}$$

if λ is large enough. Then performing the substitution $\lambda y_\lambda(t)^2 \to \lambda\overline{y_\lambda(t)^2}$ in the first of Eqs. (3.6.17), one finds

$$m\ddot{x}_\lambda = -m\omega^2 x_\lambda - \frac{w_0^2 m}{2(1 + x_0^2)}\,x_\lambda \tag{3.6.27}$$

for t near 0 (but $t \gg T_\lambda$) and λ large (note that $T_\lambda \to 0$ as $\lambda \to +\infty$).

For arbitrary values of t, a similar argument suggests that, in general, the acceleration $\ddot{x}_\lambda$ should verify the equation (when $\lambda = +\infty$):

$$m\ddot{x} = -m\omega^2 x - \frac{m w_0^2}{2}\,\frac{x}{1 + x^2}\,. \tag{3.6.28}$$

Hence, the model of constraint to the line $y = 0$ with the structure of Eq. (3.6.16) does not give rise to the motions that develop under the action of an ideal constraint to the line $y = 0$ and of an active force with potential energy $V^{(a)}(x) = \frac{1}{2} m\omega^2 x^2$, when $\lambda \to +\infty$, as one could have naively expected.

Rather one should think that the limit motion, for $\lambda \to +\infty$, of x_λ is a motion subject to an ideal constraint to $y = 0$ and to the active force whose potential energy is

$$V'^{(a)}(x) = \frac{m\omega^2}{2}\,x^2 + \frac{w_0^2 m}{4}\,\log(1 + x^2), \tag{3.6.29}$$

which, unfortunately, is dependent on the initial velocity w_0 transversal to the constraint (and, of course, on the particular form of the structure function W).

It is then possible to think, in general, that in the limit of infinite rigidity, the model of a conservative bilateral approximate constraint generates motions which respect the constraints and develop as if they were ideal but under the influence of an active force modified with respect to the one with potential energy $V^{(a)}$ which naively could be thought to be the force "not due to the constraints". In general, the structure W of the constraints has some influence, even for λ large, and contributes to the active forces in a way that may also depend on the initial data or, better, on the "initial stresses on the constraints", as in the case of the last example, where the active force depends also on the initial velocity component w_0 orthogonal to the constraint. This conjecture also sheds some light on the slightly formal distinction in §3.5 between the active force and the constraint's reaction.

In the following section we will deal with questions related to the following problems.

(1) Which further condition is it necessary to place on an approximate constraint model (Σ, W, λ) to imply that the motion $t \to x_\lambda(t)$, $t \geqslant 0$, developing under the action of the force with potential energy

$$V^{(a)} + \lambda W, \tag{3.6.30}$$

following the initial datum

$$\mathbf{x}(0) = \boldsymbol{\xi}_0 \in \Sigma, \qquad \dot{\mathbf{x}}(0) = \boldsymbol{\eta}_0 \tag{3.6.31}$$

is well approximated by the motion that takes place under the action of the active force with potential energy $V^{(a)}$ and of the ideal constraints $\varphi^{(1)}, \ldots, \varphi^{(s')}$ and follows the initial datum $\mathbf{x}(0) = \boldsymbol{\xi}_0$, $\dot{\mathbf{x}}(0) = \boldsymbol{\eta}_0^{\Sigma}$, where $\boldsymbol{\eta}_0^{\Sigma}$ is a suitable "projection of $\boldsymbol{\eta}_0$ on Σ", assuming that Σ is determined by the equations $\varphi^{(1)}(\boldsymbol{\xi}) = \cdots = \varphi^{(s')}(\boldsymbol{\xi}) = 0$? (See Definitions 7 and 8, §3.5.)

In other words the question is: when does an approximate conservative constraint appear as well approximated by an ideal constraint model in the sense of Definition 8, §3.5, with the "naive" identification of the active forces?

(2) If (Σ, W, λ) is a model of an approximate conservative constraint, is it true that the motion developing under the action of a force with the potential energy of Eq. (3.6.30) and following the initial datum of Eq. (3.6.31) is well approximated, as $\lambda \to +\infty$, by a motion developing under the influence of the ideal constraints $\varphi^{(1)}, \ldots, \varphi^{(s')}$ (determining Σ) and of a conservative active force with potential energy $V'^{(a)}$, possibly different from $V^{(a)}$ and $\boldsymbol{\eta}_0$-dependent?

(3) In the same situation as that of question (2) and if $\boldsymbol{\eta}_0$ is suitable, i.e., $\boldsymbol{\eta}_0 \equiv \boldsymbol{\eta}_0^{\Sigma}$, where $\boldsymbol{\eta}_0^{\Sigma}$ is as in question (1), is it true that $V'^{(a)} = V^{(a)}$? [This seems to be true in the example, heuristically explicitly studied above, about the constraint to the line $y = 0$ generated by Eq. (3.6.16), when $w_0 = 0$; see Eqs. (3.6.17) and (3.6.29)].

Actually, we shall really study in detail question (1) only, which we shall refer to as the problem of the determination of "sufficient perfection conditions for approximate bilateral conservative constraints".

It will be useful and necessary to analyze in some deeper way the kinematics of the system of point masses subject to constraints: this is a purely geometric analysis, very suggestive for its relationship with differential geometry.

The following section is mainly devoted to this task.

Exercises and Problems for §3.6

1. Show that the polar coordinates are a regular system of coordinates in various regions $U \subset \mathbf{R}^2$ or $U \subset \mathbf{R}^3$.

2. Show that the surface of a sphere is a regular surface in the sense of Definition 10, p. 171.

3. Show that the surface of the paraboloid $z = (x^2 + y^2)/2$ is a regular two-dimensional surface in the sense of Definition 10, p. 171. Treat similarly the hyperboloid and ellipsoid cases.

4. Consider the ellipsoid surface $x^2/a + y^2/b + z^2/c = 1$, $a, b, c > 0$, and show that the system of coordinates

$$x \equiv (1+\beta_1)\sqrt{a\frac{(\beta_2 - a)(\beta_3 - a)}{(b-a)(c-a)}} ,$$

$$y \equiv (1+\beta_1)\sqrt{b\frac{(\beta_2 - b)(\beta_3 - b)}{(c-b)(a-b)}} ,$$

$$z \equiv (1+\beta_1)\sqrt{c\frac{(\beta_2 - c)(\beta_3 - c)}{(a-c)(b-c)}}$$

is a local system of regular coordinates in the vicinity of various points of the ellipsoid's surface if $a < b < c$. This system is adapted to the surface itself, which has equations $\beta_1 = 0$ ("Jacobi's coordinates").

5. Let $V \in C^\infty(\mathbf{R})$, $V'(\xi) \equiv (dV/d\xi)(\xi) \neq 0$, $\forall \xi \neq 0$, $\lim_{\xi \to \pm\infty} V'(\xi) = +\infty$, $V(0) = 0$. Given a point $(\eta, \xi) \in \mathbf{R}^2$, $(\eta, \xi) \neq (0,0)$, define

$$E = \frac{\eta^2}{2} + V(\xi), \qquad T(E) = 2\int_{x_-(E)}^{x_+(E)} \frac{d\xi'}{\sqrt{2(E - V(\xi'))}} ,$$

where $x_-(E) < x_+(E)$ are the two roots of $E - V(\xi) = 0$. Define

$$\varphi(\xi, \eta) = \frac{2\pi}{T(E)}\left\{\begin{array}{l}\text{time necessary to the motion } x \text{ such that} \\ \ddot{x} = -dV/dx \text{ with initial datum } (0, x_-(E)) \text{ to} \\ \text{reach } (\eta, \xi)\end{array}\right\}$$

which defines $\varphi(\text{mod } 2\pi)$ (as every motion with initial velocity η and position ξ is periodic and, sooner or later, visits $x_-(E)$ with velocity 0).

Show that the coordinates (E, φ) are a regular system of coordinates near any point in $\mathbf{R}^2$ except the origin and that they generalize the polar coordinates (ρ, θ) with E being analogous to $\rho^2/2$ and φ to θ. (Hint: First consider the case

$V(\xi) = \xi^2/2$ and explicitly find $\varphi(\eta, \xi)$. Draw the qualitative form of the curves $\eta^2/2 + V(\xi) = E$ as E varies.)

§3.7. Kinematics of the Quasi-constrained Systems. Reformulation of the Perfection Criteria for Approximate Conservative Constraints

In this section we shall regard $\mathbf{R}^{Nd}$ as a vector space in which the scalar product between two vectors $\boldsymbol{\eta} = (\boldsymbol{\eta}^{(1)}, \ldots, \boldsymbol{\eta}^{(N)})$ and $\boldsymbol{\chi} = (\boldsymbol{\chi}^{(1)}, \ldots, \boldsymbol{\chi}^{(N)})$ is defined by

$$\sum_{i=1}^{N} m_i \boldsymbol{\eta}^{(i)} \cdot \boldsymbol{\chi}^{(i)}, \tag{3.7.1}$$

where $m_1, \ldots, m_N$ are given positive numbers.

The length of a vector is then

$$\|\boldsymbol{\eta}\| = \left(\sum_{i=1}^{N} m_i (\boldsymbol{\eta}^{(i)})^2 \right)^{1/2}. \tag{3.7.2}$$

The strange convention above allows one to say that the kinetic energy of a motion $t \to \mathbf{x}(t)$, $t \geqslant t_0$, of a system of N point masses, with masses $m_1, \ldots, m_N$, is

$$T(t) = \tfrac{1}{2} \|\dot{\mathbf{x}}(t)\|^2, \tag{3.7.3}$$

i.e., it is one-half the square of the velocity of the point representing the system's configuration in $\mathbf{R}^{Nd}$ without explicit reference to the masses (which of course are now hidden in the definition of length given by Eq. (3.7.2)).

Let $(U, \boldsymbol{\Xi})$ be a local system of regular coordinates in $\mathbf{R}^{Nd}$ with basis $\Omega \subset \mathbf{R}^{Nd}$ (see Definition 10, p. 171) and let $t \to \mathbf{x}(t)$, $t \geqslant t_0$, be a motion of a system of N point masses, with masses $m_1, \ldots, m_N > 0$, taking place for $t \in [t_1, t_2]$ inside U (i.e., $\mathbf{x}(t) \in U$, $\forall t \in [t_1, t_2]$).

We can then consider the motion $t \to \mathbf{b}(t)$, $t \in [t_1, t_2]$, "image" in the basis Ω of the motion $t \to \mathbf{x}(t)$, $t \in [t_1, t_2]$, via the coordinate system $(U, \boldsymbol{\Xi})$, i.e., the motion such that

$$\mathbf{x}(t) = \boldsymbol{\Xi}(\mathbf{b}(t)), \qquad t \in [t_1, t_2]. \tag{3.7.4}$$

It is obviously possible to express the kinetic energy of the motion $\mathbf{x}$ in terms of the kinematical properties of its image motion $\mathbf{b}$.

In fact, if $\boldsymbol{\Xi} = (\boldsymbol{\Xi}^{(1)}, \ldots, \boldsymbol{\Xi}^{(N)})$, differentiating Eq. (3.7.4) with respect to t:

$$\dot{\mathbf{x}}^{(j)}(t) = \sum_{l=1}^{Nd} \frac{\partial \boldsymbol{\Xi}^{(j)}}{\partial \beta_l} (\mathbf{b}(t)) \dot{b}_l(t), \qquad j = 1, \ldots, N \tag{3.7.5}$$

where $\mathbf{x}(t) = (\mathbf{x}^{(1)}(t), \ldots, \mathbf{x}^{(N)}(t))$, $\mathbf{b}(t) = (b_1(t), \ldots, b_{Nd}(t))$, with the usual notations. Hence,

$$\begin{aligned} T(t) &= \frac{1}{2} \sum_{j=1}^{N} m_j (\dot{\mathbf{x}}^{(j)}(t))^2 \\ &= \sum_{l',l''=1}^{Nd} \sum_{j=1}^{N} \frac{m_j}{2} \left(\frac{\partial \Xi^{(j)}}{\partial \beta_{l'}} (\mathbf{b}(t)) \frac{\partial \Xi^{(j)}}{\partial \beta_{l''}} (\mathbf{b}(t)) \right) \dot{b}_{l'}(t) \dot{b}_{l''}(t) \end{aligned} \tag{3.7.6}$$

which we shall write as

$$T(t) = \frac{1}{2} \sum_{l',l''=1}^{Nd} g_{l'l''}(\mathbf{b}(t)) \dot{b}_{l'}(t) \dot{b}_{l''}(t), \tag{3.7.7}$$

having set, $\forall \boldsymbol{\beta} \in \Omega$, $\forall l', l'' = 1, \ldots, Nd$:

$$g_{l'l''}(\boldsymbol{\beta}) = \sum_{j=1}^{N} m_j \frac{\partial \Xi^{(j)}}{\partial \beta_{l'}} (\boldsymbol{\beta}) \cdot \frac{\partial \Xi^{(j)}}{\partial \beta_{l''}} (\boldsymbol{\beta}) = g_{l''l'}(\boldsymbol{\beta}). \tag{3.7.8}$$

It is convenient to establish a general definition in connection with Eqs. (3.7.7) and (3.7.8) because of the generality of Eq. (3.7.7) itself.

11 Definition. The function $\boldsymbol{\beta} \to g(\boldsymbol{\beta})$ defined by Eq. (3.7.8) on Ω and with values in the $Nd \times Nd$ matrices will be called the "kinetic matrix" for the scalar product of Eq. (3.7.1) or, equivalently, for a system of N point masses in $\mathbf{R}^d$, with masses $m_1, \ldots, m_N > 0$, "relative to the local system of regular coordinates (U, Ξ) in $\mathbf{R}^{Nd}$".

Observations.

(1) Via Eq. (3.7.7), the kinetic matrix allows one to compute the kinetic energy in arbitrary local coordinates; hence, its name.

(2) We shall later list and discuss some of the properties of the kinetic matrix (at the end of the section). For the moment, note that $g(\boldsymbol{\beta})$ is a symmetric matrix whose elements, thought of as functions on Ω, are in $C^\infty(\Omega)$.

For the study of the kinematics of the quasi-constrained systems, the following purely geometrical definition is useful.

12 Definition. Let Σ be a regular surface in $\mathbf{R}^{Nd}$ with codimension s and let U be a neighborhood of a point $\boldsymbol{\xi}_0 \in \Sigma$ on which a system (U, Ξ) of local regular coordinates, with basis Ω, is defined.

Assume that the coordinate system is "adapted" to Σ (see Definition 10, p. 171) i.e., that $\Sigma \cap U$ is described in (U, Ξ) by

$$\beta_1 = \beta_2 = \cdots = \beta_s = 0. \tag{3.7.9}$$

(a) We shall say that (U, Ξ) is "well adapted" to Σ if the kinetic matrix, for the scalar product of Eq. (3.7.1), associated with (U, Ξ) has the first principal submatrix which is constant on the plane of Eq. (3.7.9); i.e., if for

$\boldsymbol{\beta} = (0, \ldots, 0, \beta_{s+1}, \ldots, \beta_{Nd}) \in \Omega$ it is, $\forall l', l'' = 1, \ldots, s$:

$$g_{l'l''}(\boldsymbol{\beta}) = g_{l'l''}(0, \ldots, 0, \beta_{s+1}, \ldots, \beta_{Nd}) = \gamma_{l'l''}, \tag{3.7.10}$$

where γ is an $s \times s$ $\boldsymbol{\beta}$-independent matrix.

(b) We shall say that $(U, \boldsymbol{\Xi})$ is "orthogonal" on Σ with respect to the scalar product of Eq. (3.7.1) if $\forall \boldsymbol{\beta} = (0, \ldots, 0, \beta_{s+1}, \ldots, \beta_{Nd})$:

$$g_{lk}(\boldsymbol{\beta}) = 0, \qquad l = 1, 2, \ldots, s; \quad k = s+1, \ldots, Nd. \tag{3.7.11}$$

Observations.

(1) Let $t \to \mathbf{x}(t)$, $t \in \mathbf{R}_+$, be a motion of N point masses in $\mathbf{R}^d$, with masses $m_1, \ldots, m_N > 0$, which at some time $\bar{t}$ happens to be in $U \cap \Sigma$ with velocity $\dot{\mathbf{x}}(\bar{t})$ "purely transversal" to Σ in the coordinate system $(U, \boldsymbol{\Xi})$, i.e., such that the motion $\mathbf{b}$, image of $\mathbf{x}$ in Ω for t close to $\bar{t}$, has velocity $\dot{\mathbf{b}}(\bar{t})$ with components vanishing "along Σ".

$$\dot{\mathbf{b}}(\bar{t}) = \big(\dot{b}_1(\bar{t}), \ldots, \dot{b}_s(\bar{t}), 0, \ldots, 0\big). \tag{3.7.12}$$

If the coordinates system is well adapted, then the kinetic energy of $\mathbf{x}$ at time t depends only on $\dot{\mathbf{b}}(\bar{t})$ but not on the particular position $\mathbf{b}(\bar{t})$.

(2) If the coordinate system $(U, \boldsymbol{\Xi})$ is orthogonal on Σ, the kinetic energy of a motion $t \to \mathbf{x}(t)$ which for $t = \bar{t}$ crosses Σ is in this instant a sum of two terms: one depending only on $\dot{b}_1(\bar{t}), \ldots, \dot{b}_s(\bar{t})$, besides $\mathbf{b}(\bar{t})$, and the other depending only on $\dot{b}_{s+1}(\bar{t}), \ldots, \dot{b}_{Nd}(\bar{t})$, besides $\mathbf{b}(\bar{t})$:

$$\begin{aligned} T_1(\bar{t}) &= \sum_{l',l''}^{1,s} g_{l'l''}\big(\mathbf{b}(\bar{t})\big)\dot{b}_{l'}(\bar{t})\dot{b}_{l''}(\bar{t}), \\ T_2(\bar{t}) &= \sum_{l',l''}^{s+1,Nd} g_{l'l''}\big(\mathbf{b}(\bar{t})\big)\dot{b}_{l'}(\bar{t})\dot{b}_{l''}(\bar{t}), \end{aligned} \tag{3.7.13}$$

and $T(t) = T_1(t) + T_2(t)$. In other words, one can say that, in such a system of coordinates, for $t = \bar{t}$ the kinetic energy is the sum of the kinetic energies of the component of the motion orthogonal to Σ and of the component of the motion parallel to it.

(3) Thinking of this, it should become geometrically evident that if Σ is a regular surface in $\mathbf{R}^{Nd}$ and $\boldsymbol{\xi}_0 \in \Sigma$, it will always be possible to construct a system of local coordinates in a neighborhood U of $\boldsymbol{\xi}_0$ which is well adapted and orthogonal to Σ (see Proposition 12 to follow).

The equations of motion can immediately be written in an arbitrary coordinate system, either in the absence or in the presence of constraints, using the following propositions.

9 Proposition. *Let $V \in C^\infty(\mathbf{R}^{Nd})$ be a real-valued function bounded from below. Let $t \to \mathbf{x}(t)$, $t \geqslant t_0$ be a motion of N point masses, with masses $m_1, \ldots, m_N > 0$, in $\mathbf{R}^d$ developing under the influence of the force $\mathbf{F}$ with potential energy V. Suppose that for $t \in [t_1, t_2]$, the motion $\mathbf{x}$ takes place in* *a neighborhood $U \subset \mathbf{R}^{Nd}$, where a local system of regular coordinates (U, Ξ) is established with basis $\Omega \subset \mathbf{R}^{Nd}$. Call $\mathbf{b}$ the motion in Ω, image of the considered motion $\mathbf{x}$, for $t \in [t_1, t_2]$, via the coordinate transformation Ξ.*

Then the motion $\mathbf{b}$ verifies the Lagrange equations[7] *associated with the Lagrangian:*

$$(\boldsymbol{\alpha}, \boldsymbol{\beta}) \to \mathcal{L}(\boldsymbol{\alpha}, \boldsymbol{\beta}) = \sum_{l', l''=1}^{Nd} \frac{1}{2} g_{l'l''}(\boldsymbol{\beta}) \alpha_{l'} \alpha_{l''} - V(\Xi(\boldsymbol{\beta})), \tag{3.7.14}$$

where $\boldsymbol{\alpha} \in \mathbf{R}^{Nd}$, $\boldsymbol{\beta} \in \Omega$ and g is the kinetic matrix of Eq. (3.7.1) *relative to (U, Ξ). Explicitly, such equations are, $\forall l = 1, \ldots, Nd$:*

$$\frac{d}{dt}\left(\sum_{l'=1}^{Nd} g_{ll'}(\mathbf{b}(t)) \dot{b}_{l'}(t)\right) = -\left(\frac{\partial V(\Xi(\boldsymbol{\beta}))}{\partial \beta_l}\right)_{\boldsymbol{\beta}=\mathbf{b}(t)} + \frac{1}{2} \sum_{l', l''=1}^{Nd} \frac{\partial g_{l'l''}}{\partial \beta_l}(\mathbf{b}(t)) \dot{b}_{l'}(t) \dot{b}_{l''}(t). \tag{3.7.15}$$

PROOF. This is an easy exercise based on the definition of Lagrangian equations and on the least-action principle as in Proposition 4, §3.3. It will be left to the reader.

10 Proposition. *Consider N point masses in $\mathbf{R}^d$, with masses $m_1, \ldots, m_N > 0$, and let $\mathbf{F}^{(a)}$ be a conservative force for it with potential energy $V^{(a)}$, bounded from below.*

Let $\Sigma \subset \mathbf{R}^{Nd}$ be a codimension-s regular surface and suppose that Σ is the set of points $\boldsymbol{\xi} \in \mathbf{R}^{Nd}$ such that

$$\varphi^{(1)}(\boldsymbol{\xi}) = \cdots = \varphi^{(s')}(\boldsymbol{\xi}) = 0, \tag{3.7.16}$$

where $\varphi^{(1)}, \ldots, \varphi^{(s')} \in C^\infty(\mathbf{R}^{Nd})$.

Let $t \to \mathbf{x}(t) \in \Sigma$, $t \in [t_1, t_2]$, be a motion developing in a neighborhood U of $\boldsymbol{\xi}_0 \in \Sigma$ under the influence of the force $\mathbf{F}^{(a)}$ and of the holonomous ideal constraint $\varphi^{(1)}, \ldots, \varphi^{(s')}$ in the sense of Definition 8, §3.5, *p.* 160.

Suppose that (U, Ξ) is a local system of regular coordinates adapted to Σ and with basis Ω (see Definition 10, §3.6) *and let $\mathbf{b}$ be the image motion of $\mathbf{x}$ on Ω:*

$$\mathbf{b}(t) = \Xi^{-1}(\mathbf{x}(t)), \qquad t \in [t_1, t_2]. \tag{3.7.17}$$

Clearly, $\mathbf{b}$ is such that

$$b_1(t) = \cdots = b_s(t) = 0, \qquad t \in [t_1, t_2]. \tag{3.7.18}$$

[7] If $(\boldsymbol{\alpha}, \boldsymbol{\beta}) \to \mathcal{L}(\boldsymbol{\alpha}, \boldsymbol{\beta})$ is a real C^∞ function defined on an open set $W \subset \mathbf{R}^{2M}$, we shall say that a C^∞ function $t \to \mathbf{b}(t)$, $t \in [t_1, t_2]$, such that $(\dot{\mathbf{b}}(t), \mathbf{b}(t)) \in W$, $\forall t \in [t_1, t_2]$, verifies the Lagrange equations associated with $\mathcal{L}$ if

$$\frac{d}{dt}\left(\frac{\partial \mathcal{L}}{\partial \alpha_i}(\dot{\mathbf{b}}(t), \mathbf{b}(t))\right) = \frac{\partial \mathcal{L}}{\partial \beta_i}(\dot{\mathbf{b}}(t), \mathbf{b}(t)), \qquad i = 1, 2, \ldots, M$$

even when W does not coincide with $\mathbf{R}^{2M}$, as usually supposed so far in connection with the Lagrange equations.

where $\mathbf{0}$ is the origin in $\mathbf{R}^s$ and $\mathbf{b}^{(s)}(t)$ is a suitable $\mathbf{R}^{Nd-s}$-valued function.

(ii$_1$) $t \to \mathbf{b}(t)$ is a $C^\infty([t_1, t_2])$ function verifying Eqs. (3.7.18) and (3.7.20), $\forall t \in [t_1, t_2]$.

Condition (iii) is equivalent to:

(iii$_1$) If $\xi_0 \in U$ and (U, Ξ) is orthogonal on Σ:

$$\Xi(\mathbf{b}(0)) = \xi_0, \qquad \dot{b}_i(0) = 0, \qquad i = 1, 2, \ldots, s. \tag{3.7.28}$$

In the next section we shall discuss an important sufficient perfection criterion for approximate conservative constraints and the perfection will be checked in the form of (i$_1$), (ii$_1$), and (iii$_1$) above.

We conclude this section by stating some simple properties of the kinetic matrices on $\mathbf{R}^{Nd}$ associated with the scalar product of Eq. (3.7.1) in a local system of regular coordinates. We shall also sketch the proof of some geometric properties of the regular surfaces (i.e., the existence of well-adapted and orthogonal coordinates).

11 Proposition.

(i) *The matrix $g(\boldsymbol{\beta})$ defined in Eq.* (3.7.6) *on Ω is, $\forall \boldsymbol{\beta} \in \Omega$, symmetric and positive definite.*

(ii) *The matrix elements of $g(\boldsymbol{\beta})$, as well as the matrix elements of the matrices inverting $g(\boldsymbol{\beta})$ or any of its principal submatrices, are in $C^\infty(\Omega)$ as functions of $\boldsymbol{\beta} \in \Omega$.*

(iii) *There exists a positive continuous function $\boldsymbol{\beta} \to C(\boldsymbol{\beta})$, defined on Ω, such that if $\mu(\boldsymbol{\beta})$ is a $q \times q$ principal submatrix of $g(\boldsymbol{\beta})$ or an inverse to such a matrix, then $\forall \boldsymbol{\sigma} = (\sigma_1, \ldots, \sigma_q) \in \mathbf{R}^q$:*

$$C(\boldsymbol{\beta}) \sum_{l=1}^{q} \sigma_l^2 \leqslant \sum_{l',l''}^{1,q} \mu_{l'l''}(\boldsymbol{\beta}) \sigma_{l'} \sigma_{l''} \leqslant C(\boldsymbol{\beta})^{-1} \sum_{l=1}^{q} \sigma_l^2 . \tag{3.7.29}$$

Observation. We recall that a $q \times q$ matrix μ is called positive definite if it is symmetric and

$$\sum_{l',l''}^{1,q} \mu_{l'l''} \sigma_{l'} \sigma_{l''} > 0, \qquad \forall \boldsymbol{\sigma} \in \mathbf{R}^q, \ \boldsymbol{\sigma} \neq \mathbf{0} \tag{3.7.30}$$

(see Appendix F).

PROOF. The symmetry is obvious and has already been remarked on [see Eq. (3.7.8)]. The positivity of $g(\boldsymbol{\beta})$ follows from its kinematic interpretation: in fact, $\frac{1}{2} \sum_{l',l''=1}^{Nd} g_{l'l''}(\boldsymbol{\beta}) \sigma_{l'} \sigma_{l''}$ is the kinetic energy of a motion which at some time happens to be in $\Xi(\boldsymbol{\beta})$ with velocity

$$\dot{\mathbf{x}}^{(j)} = \sum_{l=1}^{Nd} \frac{\partial \Xi^{(j)}}{\partial \beta_l} (\boldsymbol{\beta}) \sigma_l , \qquad j = 1, \ldots, N$$

[see Eq. (3.7.5)].

If $\boldsymbol{\sigma} \neq \mathbf{0}$, then $\dot{\mathbf{x}} \neq \mathbf{0}$ because the coordinate system is regular [see Eq. (3.6.10)]. Hence,

$$\sum_{l',l''}^{1,Nd} g_{l'l''}(\boldsymbol{\beta})\sigma_{l'}\sigma_{l''} = \sum_{j=1}^{N} m_j(\dot{\mathbf{x}}^{(j)})^2 > 0.$$

From algebra, it is well known that a matrix $g(\boldsymbol{\beta})$ is positive definite if and only if all its principal submatrices are positive definite: in this case also their inverse matrices are all positive definite and all the mentioned matrices have a positive determinant. Furthermore, if μ is a $q \times q$ positive-definite matrix, there is a positive continuous function of its matrix elements such that

$$C_1(\mu)\sum_{l=1}^{q}\sigma_l^2 < \sum_{l',l''}^{1,q}\mu_{l'l''}\sigma_{l'}\sigma_{l''} \leqslant C_1(\mu)^{-1}\sum_{l=1}^{q}\sigma_l^2 \tag{3.7.31}$$

(see, also, Appendix F, Corollary 3 and related exercises).

It is then clear that the above proposition is an immediate consequence of these algebraic properties and of the observation that all the mentioned matrix elements are in $C^{\infty}(\Omega)$: in fact they are obtained by taking products and sums of matrix elements of $g(\boldsymbol{\beta})$ and possibly dividing the results by products of determinants of some principal submatrices of the matrix $g(\boldsymbol{\beta})$, which are in turn positive by what has just been mentioned. mbe

The following proposition concerns the existence of a local system of regular coordinates $(U, \boldsymbol{\Xi})$ well adapted and orthogonal to a regular surface Σ in the vicinity of one of its points $\boldsymbol{\xi}_0$.

12 Proposition. *Given a regular surface $\Sigma \subset \mathbf{R}^{Nd}$ with codimension s and given the scalar product of Eq.* (3.7.1) *and $\boldsymbol{\xi}_0 \in \Sigma$, it is always possible to find a neighborhood U of $\boldsymbol{\xi}_0$ on which it is possible to define a local regular system of coordinates well adapted and orthogonal to Σ. It is even possible to construct it so that the kinetic matrix is given, in the basis points corresponding to $\Sigma \cap U$, by*

$$g_{ll'} = \gamma\delta_{ll'}, \qquad l, l' = 1, 2, \ldots, s, \quad \gamma > 0 \tag{3.7.32}$$

("Fermi coordinates" on Σ).

Proof. We only give a sketch, leaving to the reader experienced in analysis and analytic geometry the task of completing the proof (see, also, exercises and problems at the end of this section).

Upon a first reading, the student should avoid trying to complete the proof, although it is convenient that he gets some ideas about the geometrical meaning of Proposition 12 by reading and trying to understand the following.

Let $\boldsymbol{\xi}_0 \in \Sigma$ and let U be a bounded neighborhood of $\boldsymbol{\xi}_0$ on which a local system of regular coordinates adapted to Σ, $(U, \boldsymbol{\Xi})$, is established. Suppose that $\boldsymbol{\Xi}^{-1}(\boldsymbol{\xi}_0) = \mathbf{0}$, say. We shall build our orthogonal and well-adapted system $(\hat{U}, \hat{\boldsymbol{\Xi}})$ by suitably choosing $\hat{U} \subset U$.

This system is geometrically illustrated in the case $d = 2$, $s = 1$, in Fig. 3.1.

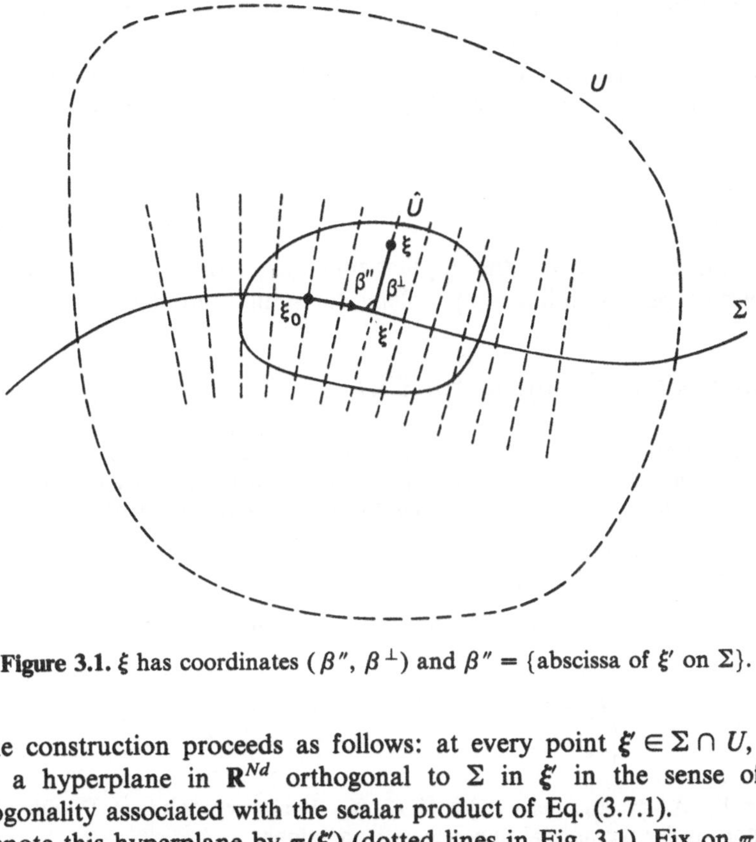

Figure 3.1. ξ has coordinates $(\beta'', \beta^{\perp})$ and $\beta'' = \{$abscissa of ξ' on $\Sigma\}$.

The construction proceeds as follows: at every point $\xi' \in \Sigma \cap U$, consider a hyperplane in $\mathbf{R}^{Nd}$ orthogonal to Σ in ξ' in the sense of the orthogonality associated with the scalar product of Eq. (3.7.1).

Denote this hyperplane by $\pi(\xi')$ (dotted lines in Fig. 3.1). Fix on $\pi(\xi')$ a system of Cartesian mutually orthogonal axes with versors $\mathbf{e}_1(\xi'), \ldots, \mathbf{e}_s(\xi')$, the orthogonality being in the sense of the scalar product of Eq. (3.7.1) and the length of the axes being measured in the same sense.

Choose the above unit vectors so that the points $\xi' + \mathbf{e}_i(\xi')$, $i = 1, \ldots, s$, have coordinates which are C^∞ functions of the $Nd - s$ nontrivial coordinates of ξ' in (U, Ξ); i.e., choose the Cartesian axes "so that they are C^∞ functions of $\xi' \in \Sigma \cap U$".

There is a neighborhood U' of ξ_0, $U' \subset U$ such that every point $\xi \in U'$ is on a unique plane $\pi(\xi')$ with ξ' suitably chosen on $\Sigma \cap U$.

To every $\xi \in U'$, we then associate Nd coordinates $\boldsymbol{\beta}$: the first s of them, denoted $\hat{\boldsymbol{\beta}}^{\perp} = (\hat{\beta}_1, \ldots, \hat{\beta}_s)$, are the coordinates of ξ in the Cartesian frame chosen on the plane $\pi(\xi')$ containing ξ; the remaining $Nd - s$ coordinates $\hat{\boldsymbol{\beta}}'' = (\hat{\beta}_{s+1}, \ldots, \hat{\beta}_{Nd})$ are the coordinates with $\hat{\beta}_i = \beta_i$, $i = s+1, \ldots, Nd$, if ξ' has in (U, Ξ) coordinates $(0, \ldots, 0, \beta_{s+1}, \ldots, \beta_{Nd})$.

Setting $\xi = \hat{\Xi}(\hat{\boldsymbol{\beta}})$, one can check that as ξ varies in U', the point $\hat{\boldsymbol{\beta}}$ varies in some open set $\Omega' \subset \mathbf{R}^{Nd}$. Furthermore, the function $\hat{\Xi}$ is a C^∞-invertible map of Ω' onto U' which is not singular, i.e., its Jacobian matrix [see Eq. (3.5.10)] has a never-vanishing determinant if U' is small enough.

Let $\hat{S}^{Nd}$ be a sphere contained in Ω' centered at $\Xi^{-1}(\xi_0) = \mathbf{0}$ and set $\hat{U} = \hat{\Xi}(\hat{S}^{Nd})$. It is clear from the construction that the pair $(\hat{U}, \hat{\Xi})$ is a coordinate system which is of Fermi type on Σ with basis $\hat{S}^{Nd}$ and $\gamma = 1$.

The difficulty of the rigorous proof consists of justifying the actual possibility of the various "choices" involved in the above descriptive argument and in checking the validity of the statements claimed about the uniqueness of the plane $\pi(\xi')$ through ξ and on the nonsingularity of the Jacobian matrix

$$\hat{J}(\hat{\beta}) = \frac{\partial \hat{\Xi}}{\partial \hat{\beta}}(\hat{\beta}), \qquad \hat{\beta} \in \hat{\mathbf{S}}^{Nd}. \tag{3.7.33}$$

The main idea is to use the implicit function theorem to check the above properties (See the following problems for some more details.)

Exercises and Problems for §3.7

1. Establish an orthogonal regular system of coordinates well adapted to the circle $\Gamma \subset \mathbf{R}^2$, with radius 1 and center at the origin, with respect to the scalar product $\eta \cdot \chi = \eta_1\chi_1 + \eta_2\chi_2$, in the neighborhood of a generic point $\xi_0 \in \Gamma$.

2. Same as Problem 1, replacing Γ with the parabola $y = x^2$, the hyperbola $xy = 1$, or the ellipse $x^2/a^2 + y^2/b^2 = 1$, $a, b > 0$.

3.* Let $\Gamma \subset \mathbf{R}^2$ be a simple C^∞ curve in $\mathbf{R}^2$ parametrized in terms of its curvilinear abscissa $s \in \mathbf{R}$ as:

$$\begin{cases} \xi_1 = X_1(s), & X_1'(s)^2 + X_2'(s)^2 = 1, \\ \xi_2 = X_2(s), & \lim\limits_{s \to \pm\infty} X_1(s)^2 + X_2(s)^2 = +\infty, \end{cases}$$

where X_1', X_2' are the derivatives of X_1, X_2. For every point on Γ with abscissa $s \in R$, consider the normal line $n(s)$ with equations

$$\xi_1 X_1'(s) + \xi_2 X_2'(s) = 0.$$

Show that given $R > 0$, there is $\delta > 0$ such that the segments of length 2δ cut around $(X_1(s), X_2(s))$ on the line $n(s)$ are pairwise disjoint, $\forall s$, $|s| < R$. (Hint: Define for $|s| < R$, $|\sigma| < \delta$:

$$F_1(s,\sigma) = X_1(s) + \sigma X_2'(s), \qquad F_2(s,\sigma) = X_2(s) - \sigma X_1'(s)$$

and note that the equations

$$F_1(s,\sigma) = F_1(\bar{s},\bar{\sigma}), \qquad F_2(s,\sigma) = F_2(\bar{s},\bar{\sigma})$$

thought of as equations for (s,σ) parametrized by $\bar{s},\bar{\sigma}$ have $s = \bar{s}$, $\sigma = \bar{\sigma}$ as a unique solution near $\bar{s}$, $\bar{\sigma}$ if $\bar{\sigma}$ is small (using the implicit function theorem). Then, by using the possibility of choosing δ small, show that they have a unique solution in the entire region $|s| < R$, $|\sigma| < \delta$, etc.)

4.* In the context of Problem 3, show that there is $\delta' \leqslant \delta$ such that the image via (F_1, F_2) of $(-R, R) \times (-\delta', \delta')$ is a neighborhood U of $(F_1(0,0), F_2(0,0)) \in \Gamma$, where the map $(s,\sigma) \leftrightarrow (F_1(s,\sigma), F_2(s,\sigma))$ is invertible, C^∞ and with nonvanishing Jacobian, i.e., the map $(U, \mathbf{F})$ is an adapted system of local regular coordinates.

5.* In the context of Problem 4, compute the kinetic matrix $g(s,\sigma)$ and show that $(U, \mathbf{F})$ is a well-adapted orthogonal coordinate system for Γ, with respect to the scalar product in Problem 1, and g is on Γ a 2×2 diagonal matrix.

6.* Let $z = s + i\sigma$, $(s,\sigma) \in \mathbf{R}^2$, and let f be a function on $\mathbf{C}$ admitting a representation

$$f(z) = \sum_{n=0}^{\infty} c_n z^n$$

with $c_n \to_{n\to\infty} 0$ so fast that the series has an infinite radius of convergence.

Also suppose that $f'(z_0) = \sum_{n=0}^{\infty} n c_n z_0^{n-1} \neq 0$ for some $z_0 \in \mathbf{C}$.

Let γ_1, γ_2 be two segments of regular curves crossing at $z_0 \in \mathbf{C}$ and there forming an angle φ_0. Show that the f images of γ_1 and γ_2, $f(\gamma_1)$ and $f(\gamma_2)$, cross at $f(z_0)$ forming the same angle φ_0. (Hint: Let $\{dz_1\}$ and $\{dz_2\}$ be two infinitesimal segments in z_0 directed along γ_1 and γ_2. Show that $f(\{dz_1\}) = f'(z_0)\{dz_1\}$, $f(dz_2) = f'(z_0)\{dz_2\}$, where $f'(z_0) = \sum_{n=0}^{\infty} n c_n z^{n-1} = \rho_0 e^{i\theta_0}$, and interpret this as saying that dz_1 and dz_2 are transformed into two infinitesimal segments emerging from $f(z_0)$, elongated by a factor ρ_0, and rotated by an angle θ_0 ("conformal mapping property").)

7.* In the context of Problem 6, suppose that $z \to f(z)$ is one to one near z_0 and that $f'(z_0) \neq 0$. Call U a neighborhood of z_0 where this happens. Show that the map $(s,\sigma) \in U \to (s',\sigma') = (\operatorname{Re} f(z), \operatorname{Im} f(z)$ (i.e., $z \to f(z)$) maps U onto a neighborhood V of $(Re\, f(z_0), \operatorname{Im} f(z_0))$ and establishes a local system of regular coordinates on V.

Let U be a disk around z_0 and $z_0 = 0$. Consider the curve in V whose equations are $s' = \operatorname{Re} f(s)$, $\sigma' = \operatorname{Im} f(s)$ for $(s,0) \in U$. Show that the above coordinate system is orthogonal on the curve Γ image of the points in U of the form $(s,0)$.

8.* Without use of complex functions, extend the argument of Problem 3 to a regular surface Σ in $\mathbf{R}^d$ with Σ and d arbitrary, using the ordinary scalar product in $\mathbf{R}^d$. (Hint: Follow the pattern of the sketch of the proof of Proposition 12 and of the Problem 3, using the implicit function theorem.)

§3.8. A Perfection Criterion for Approximate Constraints

This section is devoted to the analysis of the following interesting proposition, "Arnold's theorem", see historical note on p. 211.

13 Proposition. *Consider N point masses, with masses $m_1, \ldots, m_N > 0$, and a model of bilateral conservative approximate constraint* (Σ, W, λ) *with codimension s.*

Suppose that for every $\xi \in \Sigma$, there is a neighborhood U admitting a system of local regular coordinates (U, Ξ) with basis Ω, well adapted and orthogonal on Σ with respect to the scalar product of Eq. (3.7.1) *and such that*

$$W(\Xi(\boldsymbol{\beta})) = \overline{W}(\beta_1, \beta_2, \ldots, \beta_s), \tag{3.8.1}$$

where $\overline{W}$ is a real $C^{\infty}(\mathbf{R}^s)$ function, vanishing at the origin and having a strict minimum there, and $\boldsymbol{\beta} = (\beta_1, \ldots, \beta_s, \beta_{s+1}, \ldots, \beta_{Nd})$.

Then the constraint model (Σ, W, λ) *is an ideal approximate constraint, in the sense of Definition* 13, *p.* 181.

Observations.

(1) We already noted that it is always possible to find a neighborhood U of ξ_0 on which a local system of regular coordinates, well adapted and orthogonal on Σ, can be established.

In general, however, the functions $\boldsymbol{\beta} \to W(\boldsymbol{\Xi}(\boldsymbol{\beta}))$ will depend on *all* the Nd coordinates of $\boldsymbol{\beta} \in \Omega$ and not only on the first s (one can say that W will not, in general, be "purely orthogonal" to the constraint).

(2) Before proceeding to the proof of Proposition 13, let us discuss the following example.

Example.

Consider a two point mass system, with masses $m_1, m_2 > 0$, in $\mathbf{R}^3$ and the constraint model defined by

$$\Sigma = \{\boldsymbol{\xi} \,|\, \boldsymbol{\xi} = (\boldsymbol{\xi}^{(1)}, \boldsymbol{\xi}^{(2)}), |\boldsymbol{\xi}^{(1)} - \boldsymbol{\xi}^{(2)}| = l\}, \tag{3.8.2}$$

$$W(\boldsymbol{\xi}) = \left((\boldsymbol{\xi}^{(1)} - \boldsymbol{\xi}^{(2)})^2 - l^2\right)^2, \tag{3.8.3}$$

where $l > 0$ is given.

Let us check that the approximate conservative constraint model (Σ, W, λ) verifies the assumptions of Proposition 13.

Define the following Baricentral-Polar coordinates:

$$\begin{aligned} &\beta_1 = |\boldsymbol{\xi}^{(1)} - \boldsymbol{\xi}^{(2)}| - l \equiv \rho - l, \qquad \beta_2 = \theta, \qquad \beta_3 = \varphi, \\ &\beta_4 = (\boldsymbol{\xi}_G)_1, \qquad \beta_5 = (\boldsymbol{\xi}_G)_2, \qquad \beta_6 = (\boldsymbol{\xi}_G)_3, \end{aligned} \tag{3.8.4}$$

where (ρ, θ, φ) are the polar coordinates of the vector $\boldsymbol{\xi}^{(2)} - \boldsymbol{\xi}^{(1)}$ and $\boldsymbol{\xi}_G$ is the vector determining the baricenter's position:

$$\boldsymbol{\xi}_G = \frac{m_1 \boldsymbol{\xi}^{(1)} + m_2 \boldsymbol{\xi}^{(2)}}{m_1 + m_2}. \tag{3.8.5}$$

Through Eq. (3.8.4), one can easily establish a regular local system of coordinates $(U, \boldsymbol{\Xi})$ in the vicinity of any point $\boldsymbol{\xi}_0 \in \Sigma$ such that $\theta \in (0, \pi)$, $\varphi \in (0, 2\pi)$ (which, by the arbitrariness of the choice of Cartesian axes, is not a real restriction). This reference system is adapted to Σ, and Σ is given by $\beta_1 = 0$.

Let us show that it is well adapted and orthogonal for the scalar product of Eq. (3.7.1). We need the kinetic matrix: it can be computed by noting that

$$m_1(\dot{\boldsymbol{\xi}}^{(1)})^2 + m_2(\dot{\boldsymbol{\xi}}^{(2)})^2 = (m_1 + m_2)\dot{\boldsymbol{\xi}}_G^2 + \frac{2m_1 m_2}{m_1 + m_2}(\dot{\boldsymbol{\xi}}^{(1)} - \dot{\boldsymbol{\xi}}^{(2)})^2 \tag{3.8.6}$$

which follows immediately from the relations

$$\begin{aligned} \boldsymbol{\xi}^{(1)} &= \boldsymbol{\xi}_G + \frac{m_2}{m_1 + m_2}(\boldsymbol{\xi}^{(1)} - \boldsymbol{\xi}^{(2)}), \\ \boldsymbol{\xi}^{(2)} &= \boldsymbol{\xi}_G - \frac{m_1}{m_1 + m_2}(\boldsymbol{\xi}^{(1)} - \boldsymbol{\xi}^{(2)}) \end{aligned} \tag{3.8.7}$$

by differentiation and some algebra.

Since

$$(\dot{\boldsymbol{\xi}}^{(1)} - \dot{\boldsymbol{\xi}}^{(2)})^2 = \dot{\rho}^2 + \rho^2\dot{\theta}^2 + \rho^2 \sin^2\theta \dot{\varphi}^2, \tag{3.8.8}$$

which is clear if one recalls that the line element in polar coordinates is

$d\rho^2 + \rho^2 d\theta^2 + \rho^2(\sin\theta)^2 d\varphi^2$ and that (ρ, θ, φ) are just the polar coordinates of $(\boldsymbol{\xi}^{(1)} - \boldsymbol{\xi}^{(2)})$, we see that Eqs. (3.8.6) and (3.8.8) give

$$\begin{aligned} m_1(\dot{\boldsymbol{\xi}}^{(1)})^2 + m_2(\dot{\boldsymbol{\xi}}^{(2)})^2 = (m_1 + m_2)(\dot{\beta}_4^2 + \dot{\beta}_5^2 + \dot{\beta}_6^2) \\ + \frac{2m_1 m_2}{m_1 + m_2}(\dot{\beta}_1^2 + (l + \beta_1)^2\dot{\beta}_2^2 + (l + \beta_1)^2(\sin\beta_2)^2\dot{\beta}_3^2), \end{aligned} \tag{3.8.9}$$

showing that the coordinates of Eq. (3.8.4) are well adapted and orthogonal on Σ (in fact, $g_{11}(\boldsymbol{\beta}) \equiv 2m_1 m_2/(m_1 + m_2)$ and the quadratic form of Eq. (3.8.9) does not contain the mixed terms $\dot{\beta}_1\dot{\beta}_l$, $l > 1$).

In this coordinate system, the constraint's structure function W of Eq. (3.8.3) is simply $(\beta_1^2 + 2l\beta_1)^2$, i.e., it depends only on β_1.

A further example is the model (Σ, W, λ) for a single point in $\mathbf{R}^3$ bound to a regular surface $\sigma : \Sigma = \{\boldsymbol{\xi} \,|\, \boldsymbol{\xi} \in \sigma\}$, and W is a C^∞ function of $\boldsymbol{\xi}$, positive outside Σ and having, for $\boldsymbol{\xi}$ close enough to Σ, the form

$$W(\boldsymbol{\xi}) = \left(\mathbf{n}_{\bar{\boldsymbol{\xi}}} \cdot (\boldsymbol{\xi} - \bar{\boldsymbol{\xi}})\right)^2 \tag{3.8.10}$$

where $\bar{\boldsymbol{\xi}}$ is the point on σ closest to $\boldsymbol{\xi}$ and $\mathbf{n}_{\bar{\boldsymbol{\xi}}}$ is a unit vector normal to σ in $\bar{\boldsymbol{\xi}}$. (The proof is left to the reader as a problem.)

PROOF (Proposition 13). Let $(\boldsymbol{\eta}_0, \boldsymbol{\xi}_0)$ be an initial datum for the given system of point masses, $\boldsymbol{\xi}_0 \in \Sigma$.

Fix $\lambda \geqslant 1$ and a function $V^{(a)} \in C^\infty(\mathbf{R}^{Nd})$ bounded from below. Let U be a neighborhood of the point $\boldsymbol{\xi}_0$ where it is possible to define a local system of regular coordinates $(U, \boldsymbol{\Xi})$ with basis Ω, well adapted and orthogonal on Σ and such that Eq. (3.8.1) holds in this system. Suppose that $\boldsymbol{\Xi}^{-1}(\boldsymbol{\xi}_0) = \mathbf{0}$.

We also suppose, for the sake of simplicity, that W has a rather special form:

$$\overline{W}(\beta_1, \ldots, \beta_s) = \frac{1}{2}\sum_{i=1}^{s} \beta_i^2. \tag{3.8.11}$$

In spite of the particularity of Eq. (3.8.11), this is an assumption that can easily be eliminated through some formal complications which would only make the true difficulties of the problem and the solution's method more obscure (see problems at the end of this section).

Denote $t \to \mathbf{x}_\lambda(t)$, $t \in \mathbf{R}_+$, the motion that the N point masses perform under the influence of the force with potential energy $V^{(a)} + \lambda W$ starting from the initial datum $(\boldsymbol{\xi}_0, \boldsymbol{\eta}_0)$.

By energy conservation, this will be

$$\sum_{i=1}^{N} \frac{m_i}{2}\left(\dot{\mathbf{x}}_\lambda^{(i)}(t)\right)^2 + V^{(a)}(\mathbf{x}_\lambda(t)) + \lambda W(\mathbf{x}_\lambda(t)) = E, \tag{3.8.12}$$

where

$$E = \sum_{i=1}^{N} \frac{m_i}{2}\left(\boldsymbol{\eta}_0^{(i)}\right)^2 + V^{(a)}(\boldsymbol{\xi}_0) \tag{3.8.13}$$

is λ independent because $\boldsymbol{\xi}_0 \in \Sigma$ and W vanishes on Σ.

Then Eq. (3.8.12) and the assumed boundedness of $V^{(a)}$ imply that

$$\sum_{i=1}^{N} \frac{m_i}{2}\left(\dot{\mathbf{x}}_\lambda^{(i)}(t)\right)^2 \leqslant E + \sup\left(-V^{(a)}(\boldsymbol{\xi})\right) = C < +\infty. \qquad (3.8.14)$$

If S_ρ denotes a closed sphere with radius ρ and center $\boldsymbol{\xi}_0$, contained in U, Eq. (3.8.14) will imply that the motion $t \to \mathbf{x}_\lambda(t)$ will develop remaining inside S_ρ for $t \in [0, T]$, i.e., $\mathbf{x}_\lambda(t) \in S_\rho$, $t \in [0, T]$, if T is chosen so that

$$T\sqrt{\frac{2C}{\min_j m_j}} \leqslant \rho. \qquad (3.8.15)$$

We then fix T verifying Eq. (3.8.15) and study the motion $t \to \mathbf{x}_\lambda(t)$, $t \in [0, T]$.

We will be concerned with showing the existence of the limit $\lim_{\lambda \to +\infty} \mathbf{x}_\lambda(t) = \mathbf{x}(t)$ and the validity of Eqs. (3.7.18) and (3.7.20) only for $t \in [0, T]$.

The treatment of the general case (t arbitrarily large) contains some additional difficulties and we will not discuss it in detail. Such difficulties have a geometrical character and depend on the fact that $(U, \boldsymbol{\Xi})$ is generally a local coordinate system and not a global one for all of Σ, see Problem 5 at the end of this section.

We first need to show that the motion $t \to \mathbf{x}_\lambda(t)$, $t \in [0, T]$, tends to evolve on Σ as $\lambda \to +\infty$.

This is a simple consequence of energy conservation and of the positivity (only) of W: it does not depend on the special hypothesis on the constraint's nature [Eq. (3.8.11)], but it would be valid even for general approximate conservative constraints.

Let $t \to \mathbf{b}_\lambda(t)$, $t \in [0, T]$, be the image of the motion $\mathbf{x}_\lambda$, observed for $t \in [0, T]$, in the basis Ω of the coordinate system: $\mathbf{b}_\lambda(t) = \boldsymbol{\Xi}^{-1}(\mathbf{x}_\lambda(t))$, $t \in [0, T]$.

Let us rewrite the energy conservation equation in the local coordinates $(U, \boldsymbol{\Xi})$ by using the kinetic matrix $g_{l,l'}$ in this reference system:

$$\frac{1}{2}\sum_{l',l''}^{1,Nd} g_{l'l''}(\mathbf{b}_\lambda(t))\dot{b}_{\lambda l'}(t)\dot{b}_{\lambda l''}(t) + V^{(a)}(\boldsymbol{\Xi}(b_\lambda(t))) + \frac{\lambda}{2}\sum_{i=1}^{s} b_{\lambda i}(t)^2 = E, \qquad (3.8.16)$$

having used Eq. (3.8.11) to express W.

The first of the above three addends is non-negative (being the kinetic energy; see, also, Proposition 11, §3.7, p. 183). Hence, Eq. (3.8.16) implies

$$|b_{\lambda i}(t)| \leqslant \left(\frac{2C}{\lambda}\right)^{1/2} \qquad i = 1, 2, \ldots, s. \qquad (3.8.17)$$

if C is defined by Eq. (3.8.14).

From the examples discussed in §3.6, it is expected that the motion $t \to \mathbf{b}_\lambda(t)$, although squeezed on Σ, will very quickly oscillate transversally to Σ. It will therefore be useful to estimate the velocities $\dot{b}_{\lambda i}(t)$, $i = 1, \ldots, s$, of the vanishing coordinates.

Equation (3.8.16) also provides such estimates: by Proposition 11, p. 183, we can say that there is a constant $g^{-1} = \{$minimum of $C(\boldsymbol{\beta})$ in Eq. (3.7.29) for $\boldsymbol{\Xi}(\boldsymbol{\beta}) \in S_\rho\}$ for which

$$\frac{1}{2} g^{-1} \sum_{l=1}^{Nd} \left(\dot{b}_{\lambda l}(t)\right)^2 \leqslant \frac{1}{2} \sum_{l',l''=1}^{Nd} g_{l'l''}(\mathbf{b}_\lambda(t)) \dot{b}_{\lambda l'}(t) \dot{b}_{\lambda l''}(t), \tag{3.8.18}$$

$\forall t \in [0, T]$, because for such t's and by the choice of T, $\mathbf{x}_\lambda(t) = \boldsymbol{\Xi}(\mathbf{b}_\lambda(t)) \in S_\rho$.

Then Eqs. (3.8.18) and (3.8.16) imply, $\forall \lambda \geqslant 0$,

$$|\dot{b}_{\lambda l}(t)| \leqslant \sqrt{2Cg}\,, \qquad l = 1, \ldots, Nd; \quad t \in [0, T]. \tag{3.8.19}$$

We now use the orthogonality and adaptation properties of the coordinate system $(U, \boldsymbol{\Xi})$; if one sets $\boldsymbol{\beta} = (\boldsymbol{\beta}_v, \boldsymbol{\beta}_n)$ with $\boldsymbol{\beta}_v = (\beta_1, \ldots, \beta_s) \in \mathbf{R}^s$, $\boldsymbol{\beta}_n = (\beta_{s+1}, \ldots, \beta_{Nd}) \in \mathbf{R}^{Nd-s}$, one has

$$g_{l'l''}(\mathbf{0}, \boldsymbol{\beta}_n) \equiv 0, \qquad l' = 1, \ldots, s; \quad l'' = s+1, \ldots, Nd \tag{3.8.20}$$

(orthogonality) and

$$g_{l'l''}(\mathbf{0}, \boldsymbol{\beta}_n) = \gamma_{ll'}, \qquad l, l' = 1, \ldots, s, \tag{3.8.21}$$

(good adaptation), where γ is a constant $s \times s$ matrix.

Since the functions $g_{l'l''}(\boldsymbol{\beta})$, $\boldsymbol{\beta} \in \Omega$, are C^∞ functions, one deduces from the Taylor–Lagrange theorem (see Appendix B), $\forall (\boldsymbol{\beta}_v, \boldsymbol{\beta}_n) \in \Omega$, $\forall l = 1, \ldots, s$; $l' = s+1, \ldots, Nd$:

$$g_{ll'}(\boldsymbol{\beta}_v, \boldsymbol{\beta}_n) = \sum_{j=1}^{s} g_{l,l',j}(\boldsymbol{\beta}_v, \boldsymbol{\beta}_n) \beta_j, \tag{3.8.22}$$

and $\forall l, l' = 1, \ldots, s$:

$$g_{ll'}(\boldsymbol{\beta}_v, \boldsymbol{\beta}_n) = \gamma_{ll'} + \sum_{j=1}^{s} g_{l,l',j}(\boldsymbol{\beta}_v, \boldsymbol{\beta}_n) \beta_j, \tag{3.8.23}$$

where $g_{l,l',j}(\boldsymbol{\beta})$, $\boldsymbol{\beta} \in \Omega$, are suitable $C^\infty(\Omega)$ functions.

Equations (3.8.22) and (3.8.23) can be used to "more explicitly" write the equations of motion (3.7.15) for the "nonconstrained coordinates", i.e., for the β_j's with $j = s+1, \ldots, Nd$. Using Eq. (3.8.1), one finds, for $l = s+1, \ldots, Nd$:

$$\begin{aligned}
&\frac{d}{dt}\left\{\left[\sum_{l'=1}^{s} \sum_{j=1}^{s} g_{l',l,j}(\mathbf{b}_\lambda(t)) b_{\lambda j}(t) \dot{b}_{\lambda l'}(t)\right] + \sum_{l'=s+1}^{Nd} g_{ll'}(\mathbf{b}_\lambda(t)) \dot{b}_{\lambda l'}(t)\right\} \\
&\quad = -\sum_{k=1}^{N} \frac{\partial V^{(a)}}{\partial \boldsymbol{\xi}^{(k)}}(\boldsymbol{\Xi}(\mathbf{b}_\lambda(t))) \cdot \frac{\partial \boldsymbol{\Xi}^{(k)}}{\partial \beta_l}(\mathbf{b}_\lambda(t)) \\
&\qquad + \frac{1}{2} \sum_{l',l''}^{s+1,Nd} \frac{\partial g_{l'l''}}{\partial \beta_l}(\mathbf{b}_\lambda(t)) \dot{b}_{\lambda l'}(t) \dot{b}_{\lambda l''}(t) \qquad\qquad (3.8.24) \\
&\qquad + \left[\sum_{l'=1}^{s} \sum_{l''=s+1}^{Nd} \sum_{j=1}^{s} \frac{\partial g_{l',l'',j}}{\partial \beta_l}(\mathbf{b}_\lambda(t)) b_{\lambda j}(t) \dot{b}_{\lambda l'}(t) \dot{b}_{\lambda l''}(t)\right. \\
&\qquad \left. + \frac{1}{2} \sum_{l',l''=1}^{s} \sum_{j=1}^{s} \frac{\partial g_{l',l'',j}}{\partial \beta_l}(\mathbf{b}_\lambda(t)) b_{\lambda j}(t) \dot{b}_{\lambda l'}(t) \dot{b}_{\lambda l''}(t)\right],
\end{aligned}$$

where we place in square brackets the terms which should vanish, as $\lambda \to +\infty$, in order that Eq. (3.8.24) could reduce at least formally to Eq. (3.7.20) as wished on the basis of Definition 13, §3.7, p. 181 and Observation (4) to Definition 13 p. 182.

Note that in Eq. (3.8.24) every term in square brackets contains factors proportional to one of the first s coordinates which, by Eq. (3.8.17), vanish as $\lambda \to +\infty$ uniformly in $t \in [0, T]$. The coefficients in Eq. (3.8.24) of such coordinates are uniformly bounded in λ, as the motion takes place in S_ρ, for $t \in [0, T]$, and there the $g \ldots$ are C^∞ functions and, therefore, bounded together with their derivatives; furthermore, Eq. (3.8.19) provides λ-independent bounds for $\dot{b}_{\lambda l}(t)$.

To understand in a rigorous way that the above formal convergence of Eqs. (3.8.17) and (3.8.24) to Eqs. (3.7.8) and (3.7.20) implies that, uniformly in $t \in [0, T]$, the functions $t \to b_{\lambda l}(t)$, $l = s+1, \ldots, Nd$, converge to limits $b_l(t)$ verifying Eq. (3.7.20) with the desired initial conditions (3.7.24), some more work is still necessary.

Integrate both sides of Eq. (3.8.20) with respect to t, $\forall l = s+1, \ldots, Nd$:

$$\begin{aligned}
&\sum_{l'=s+1}^{Nd} g_{ll'}(\mathbf{b}_\lambda(t))\dot{b}_{\lambda l'}(t) + \left[\sum_{l'=1}^{s}\sum_{j=1}^{s} g_{l',l,j}(\mathbf{b}_\lambda(t))b_{\lambda j}(t)\dot{b}_{\lambda l'}(t)\right] \\
&\quad - \sum_{l'=s+1}^{Nd} g_{ll'}(\mathbf{b}(0))\dot{b}_{l'}(0) \\
&= \int_0^t \left\{ -\sum_{k=1}^{N} \frac{\partial V^{(a)}}{\partial \boldsymbol{\xi}^{(k)}}(\Xi(\mathbf{b}_\lambda(t'))) \cdot \frac{\partial \Xi^{(k)}}{\partial \beta_l}(\mathbf{b}_\lambda(t') \right. \\
&\qquad \left. + \frac{1}{2}\sum_{l',l''}^{s+1,Nd} \frac{\partial g_{l'l''}}{\partial \beta_l}(\mathbf{b}_\lambda(t'))\dot{b}_{\lambda l'}(t'))\dot{b}_{\lambda l''}(t')\right\} dt' \qquad (3.8.25) \\
&\quad + \left[\int_0^t dt' \left(\sum_{l'=1}^{s}\sum_{l''=s+1}^{Nd}\sum_{j=1}^{Nd} \frac{\partial g_{l',l'',j}}{\partial \beta_l}(\mathbf{b}_\lambda(t'))b_{\lambda j}(t')\dot{b}_{\lambda l'}(t')\dot{b}_{\lambda l''}(t') \right.\right. \\
&\qquad \left.\left. + \frac{1}{2}\sum_{l',l''=1}^{s}\sum_{j=1}^{s} \frac{\partial g_{l',l'',j}}{\partial \beta_l}(\mathbf{b}_\lambda(t'))b_{\lambda j}(t')\dot{b}_{\lambda l'}(t')b_{\lambda l''}(t')\right)\right],
\end{aligned}$$

where in the second line we used the hypothesis that $b_{\lambda l}(0) = 0$, if $l = 1, \ldots, s$, and that the initial data $b_{\lambda l}(0), \dot{b}_{\lambda l}(0)$, $l = 1, \ldots, Nd$, are λ independent.

Now bring the second and third addends to the right-hand side and consider the resulting equations as $(Nd - s)$ linear equations in the $(Nd - s)$ unknowns $\dot{b}_{\lambda l}(t)$, $l = s+1, \ldots, Nd$, pretending that the right-hand side is known. The matrix of the coefficients is the last $(Nd - s) \times (Nd - s)$ principal submatrix g_s of the kinetic matrix g: $(g_s)_{ij} = g_{ij}(\mathbf{b}_\lambda(t))$, $i, j = s+1, \ldots, Nd$. By Proposition 11, §3.7, p. 183, g_s admits an inverse matrix

$g_s^{-1}(\mathbf{b}_\lambda(t))$ (making explicit its $\mathbf{b}_\lambda(t)$ dependence). Therefore, we can express $\dot{b}_{\lambda l}(t)$, $l = s+1, \ldots, Nd$, in terms of the right-hand side. We get

$$\begin{aligned}
\dot{b}_{\lambda \bar{l}}(t) = &\left[- \sum_{l=s+1}^{Nd} (g_s^{-1}(\mathbf{b}_\lambda(t)))_{\bar{l}l} \sum_{l'=1}^{s} \sum_{j=1}^{s} g_{l',l,j}(\mathbf{b}_\lambda(t)) b_{\lambda j}(t) \dot{b}_{\lambda l}(t) \right] \\
&+ \sum_{l=s+1}^{Nd} (g_s^{-1}(\mathbf{b}_\lambda(t)))_{\bar{l}l} \sum_{l'=s+1}^{Nd} g_{l,l'}(\mathbf{b}(0)) \dot{\mathbf{b}}_{l'}(0) \\
&+ \sum_{l=s+1}^{Nd} (g_s^{-1}(\mathbf{b}_\lambda(t)))_{\bar{l}l} \\
&\times \int_0^t \Bigg\{ \Bigg(- \sum_{k=1}^{N} \frac{\partial V^{(a)}}{\partial \xi^{(k)}} (\Xi(\mathbf{b}_\lambda(t'))) \cdot \frac{\partial \Xi^{(k)}}{\partial \beta_l} (\mathbf{b}_\lambda(t')) \\
&\qquad + \frac{1}{2} \sum_{l',l''}^{s+1,Nd} \frac{\partial g_{l'l''}}{\partial \beta_l} (\mathbf{b}_\lambda(t')) \dot{b}_{\lambda l'}(t') \dot{b}_{\lambda l''}(t') \Bigg) \Bigg\} dt' \qquad (3.8.26) \\
&+ \Bigg[\sum_{l=s+1}^{Nd} (g_s^{-1}(\mathbf{b}_\lambda(t)))_{\bar{l}l} \\
&\quad \times \int_0^t \Bigg\{ \sum_{l'=1}^{s} \sum_{l''=s+1}^{Nd} \sum_{j=1}^{s} \frac{\partial g_{l',l'',j}}{\partial \beta_l} (\mathbf{b}_\lambda(t')) b_{\lambda j}(t') \dot{b}_{\lambda l'}(t') \dot{b}_{\lambda l''}(t') \\
&\qquad + \frac{1}{2} \sum_{l',l''=1}^{s} \sum_{j=1}^{s} \frac{\partial g_{l',l'',j}}{\partial \beta_l} (\mathbf{b}_\lambda(t')) b_{\lambda j}(t') \dot{b}_{\lambda l'}(t') \dot{b}_{\lambda l''}(t') \Bigg\} dt' \Bigg].
\end{aligned}$$

It is now obvious to remark that the terms in square brackets vanish uniformly in $t \in [0, T]$ as $\lambda \to +\infty$ because of Eqs. (3.8.17) and (3.8.19) and because of the uniform boundedness in $\Xi^{-1}(S_\rho)$ of the $g\ldots$ functions and of their derivatives.

Furthermore, we shall try to prove the convergence to a limit, as $\lambda \to +\infty$, of the terms which are not in square brackets in Eq. (3.8.26): call $\delta_{\lambda \bar{l}}(t)$, $t \in [0, T]$, their sum and let us first show that a subsequence $\lambda_n \to +\infty$, extracted from an arbitrary diverging sequence, exists such that $\delta_{\lambda_n \bar{l}}(t)$ converges to a limit $\delta_{\bar{l}}(t)$ uniformly in $t \in [0, T]$, $\bar{l} = s+1, \ldots, Nd$.

This will be shown by proving that the family of functions on $[0, T]$ parameterized by λ and $\bar{l}$: $(\delta_{\lambda l})_{\lambda \geq 1,\, l=s+1,\ldots,Nd}$ is an equicontinuous and equibounded family of functions on $[0, T]$, and then applying the Ascoli–Arzelà theorem (see Appendix H).

Finally, we will prove the actual existence of the limit as $\lambda \to +\infty$ of $\delta_{\lambda l}(t)$ by showing that every limit of the converging subsequences verifies a certain differential equation with given initial conditions, whatever the subsequence is, and applying the uniqueness theorem for the differential equations: this differential equation will essentially turn out to coincide with Eq. (3.7.20) and the proof will then be easily completed.

The equiboundedness (see Appendix H) of the functions is clear from Eqs. (3.8.17) and (3.8.19).

The equicontinuity of the contribution to $\delta_{\lambda\bar{l}}$ coming from the integral of Eq. (3.8.26) and that coming from the part outside the integral can be separately shown. They follow from the remarks:

(i) Consider a family $(\mu_\alpha)_{\alpha\in A}$ of functions on $[0, T]$ given by

$$\mu_\alpha(t) = \int_0^t \nu_\alpha(\tau)\, d\tau, \tag{3.8.27}$$

where $(\nu_\alpha)_{\alpha\in a}$ is a family of equibounded continuous functions on $[0, T]$, i.e., a family of functions bounded as $|\nu_\alpha(t)| \leqslant B$, $\forall t \in [0, T]$, with a suitable B, $\forall \alpha \in A$. Then the family $(\mu_\alpha)_{\alpha\in A}$ is equicontinuous; in fact,

$$|\mu_\alpha(t) - \mu_\alpha(t')| = \left| \int_{t'}^t \nu_\alpha(\tau)\, d\tau \right| \leqslant B|t - t'|, \qquad \forall \alpha \in A. \tag{3.8.28}$$

(ii) Families of functions obtained by composing a given $C^\infty(\mathbf{R}^h)$ function and a family of equicontinuous equibounded $\mathbf{R}^h$-valued functions on $[0, T]$ form equicontinuous equibounded families of functions (exercise) $(\forall h > 0)$.

It is then clear by suitably combining the criteria (i) and (ii) and Eqs. (3.8.17), and (3.8.19) and the fact that $(g_s^{-1})_{l'l''}$ are C^∞ functions on $\Xi^{-1}(S_\rho)$ and $t \to \mathbf{b}_\lambda(t)$ is an equicontinuous family [by (i) and by Eq. (3.8.19)], that $\delta_{\lambda\bar{l}}$ form an equicontinuous equibounded family of functions on $[0, T]$ parameterized by $\lambda \geqslant 1$, $\bar{l} = s + 1, \ldots, Nd$.

Then we can apply the Ascoli–Arzelà criterion (see Appendix H). From every diverging sequence of positive numbers, it is possible to extract a subsequence $(\lambda_n)_{n\in\mathbf{Z}_+}$ diverging and such that the limit

$$\lim_{n\to\infty} \delta_{\lambda_n\bar{l}}(t) = \delta_{\bar{l}}(t) \tag{3.8.29}$$

exists uniformly for $t \in [0, T]$, $\bar{l} = s + 1, \ldots, Nd$.

Equation (3.8.26) then implies (since it has already been observed that the terms in square brackets in the right-hand side vanish uniformly as $\lambda \to +\infty$, $\forall t \in [0, T]$) that

$$\lim_{n\to\infty} \dot{b}_{\lambda_n\bar{l}}(t) = \delta_{\bar{l}}(t). \tag{3.8.30}$$

So

$$\lim_{n\to\infty} b_{\lambda_n\bar{l}}(t) \equiv \lim_{n\to\infty} \left(b_{\lambda_n\bar{l}}(0) + \int_0^t \dot{b}_{\lambda_n\bar{l}}(\tau)\, d\tau \right) = \lim_{n\to\infty} \left(b_{\bar{l}}(0) + \int_0^t \dot{b}_{\lambda_n\bar{l}}(\tau)\, d\tau \right)$$

$$= b_{\bar{l}}(0) + \int_0^t \delta_{\bar{l}}(\tau)\, d\tau \equiv b_{\bar{l}}(t), \quad \bar{l} = s + 1, \ldots, Nd, \tag{3.8.31}$$

uniformly in $t \in [0, T]$, since the initial datum is λ independent, and $b_{\bar{l}}(t)$ is defined by the last identity. Of course, by changing the subsequence λ_n, we cannot yet be sure that $\delta_{\bar{l}}$ and $b_{\bar{l}}$, thus defined, do not change.

The functions $t \to b_{\bar{l}}(t)$, $t \in [0, T]$, defined in Eq. (3.8.31) are, by Eq. (3.8.31) itself, once differentiable and

$$\dot{b}_{\bar{l}}(t) = \delta_{\bar{l}}(t), \qquad \forall t \in [0, T], \quad \forall \bar{l} = s+1, \ldots, Nd. \tag{3.8.32}$$

Coming back to Eq. (3.8.25) with $\lambda = \lambda_n$ and using Eqs. (3.8.17), (3.8.31), (3.8.30), and (3.8.32), we find that as $n \to \infty$, $\forall l = s+1, \ldots, Nd$:

$$\begin{aligned} \sum_{l'=s+1}^{Nd} g_{ll'}(\mathbf{b}(t))\dot{b}_{l'}(t) = {} & \sum_{l'=s+1}^{Nd} g_{ll'}(\mathbf{b}(0))\dot{b}_{l'}(0) \\ & + \int_0^t \left\{ -\sum_{k=1}^{N} \frac{\partial V^{(a)}}{\partial \xi^{(k)}}(\Xi(\mathbf{b}(t'))) \cdot \frac{\partial \Xi^{(k)}}{\partial \beta_l}(\mathbf{b}(t')) \right. \\ & \left. + \frac{1}{2} \sum_{l',l''=s+1}^{Nd} \frac{\partial g_{l'l''}}{\partial \beta_l}(\mathbf{b}(t'))\dot{b}_{l'}(t')\dot{b}_{l''}(t') \right\} dt', \end{aligned} \tag{3.8.33}$$

having set $\mathbf{b}(t) = (0, \ldots, 0, b_{s+1}(t), \ldots, b_{Nd}(t))$ [recall that $b_l(t)$ is defined by Eq. (3.8.31) only for $l = s+1, \ldots, Nd$].

We see that $t \to \mathbf{b}(t)$ verifies the wanted initial conditions at $t = 0$, Eq. (3.7.24), as well as Eq. (3.7.18) and, by differentiating Eq. (3.8.33), also Eq. (3.7.20). It is also true that $\mathbf{b} \in C^\infty([0, T])$. In fact, pretending that the right-hand side of Eq. (3.8.33) is known, we interpret Eq. (3.8.33) as a linear system in the unknowns $\dot{b}_l(t)$: its coefficients form the already-met nonsingular matrix $g_s(\mathbf{b}(t))$. Proceeding as in Eq. (3.8.26), we find

$$\begin{aligned} \dot{b}_{\bar{l}}(t) = {} & \sum_{l=s+1}^{Nd} \left(g_s^{-1}(\mathbf{b}(t))\right)_{\bar{l}l} \sum_{l'=s+1}^{Nd} g_{ll'}(\mathbf{b}(0))\dot{b}_{l'}(0) + \sum_{l=s+1}^{Nd} \left(g_s^{-1}(\mathbf{b}(t))\right)_{\bar{l}l} \\ & \times \int_0^t \left\{ -\sum_{k=1}^{N} \frac{\partial V^{(a)}}{\partial \xi^{(k)}}(\Xi(\mathbf{b}(t'))) \cdot \frac{\partial \Xi^{(k)}}{\partial \beta_l}(\mathbf{b}(t')) \right. \\ & \left. + \frac{1}{2} \sum_{l',l''=s+1}^{Nd} \frac{\partial g_{l'l''}}{\partial \beta_l}(\mathbf{b}(t'))\dot{b}_{l'}(t')\dot{b}_{l''}(t') \right\} dt', \end{aligned} \tag{3.8.34}$$

and since the right-hand side is obviously once differentiable, it follows that $b_{\bar{l}}$ is twice differentiable. Differentiating both sides, we find an expression for $\ddot{b}_{\bar{l}}$ in terms of $\mathbf{b}$, $\dot{\mathbf{b}}$, and some integrals: hence, $\ddot{\mathbf{b}}$ is differentiable, etc. So $\mathbf{b}$ is in $C^\infty([0, T])$.

It remains to show that the limit as $\lambda \to +\infty$ of $b_{\lambda\bar{l}}(t)$ exists, $\forall \bar{l} = s+1, \ldots, Nd$, without "passing to subsequences". It suffices to show that every divergent subsequence $\lambda_n \to +\infty$ for which the limit $\lim_{n\to\infty} b_{\lambda_n\bar{l}}(t)$, $\bar{l} = s+1, \ldots, Nd$, exists uniformly has to converge to the same limit.

It is clearly enough to show that there is only one function $t \to \mathbf{b}(t)$ verifying Eq. (3.8.33) and in $C^\infty([0, T])$ and such that $\mathbf{b}(t) \in \Xi^{-1}(S_\rho)$, $\forall t \in [0, T]$, because every limit of a uniformly convergent subsequence has to verify Eq. (3.8.33).

The following trick, which will be sublimated in §3.11 and §3.12, can be used.

Set, $\forall l = s+1, \ldots, Nd$:

$$p_l(t) = \sum_{l'=s+1}^{Nd} g_{ll'}(\mathbf{b}(t))\dot{b}_{l'}(t) \tag{3.8.35}$$

or, for short,

$$\mathbf{p} = g_s(\mathbf{b})\dot{\mathbf{b}}, \tag{3.8.36}$$

where $g_s(\boldsymbol{\beta})$ is the last principal matrix of $g(\boldsymbol{\beta})$ of order $Nd - s$. Then by differentiation with respect to t, Eq. (3.8.33) yields the following equations, $\forall l = s+1, \ldots, Nd$:

$$\begin{aligned}\dot{p}_l = &- \sum_{k=1}^{N} \frac{\partial V^{(a)}}{\partial \boldsymbol{\xi}^{(k)}}(\boldsymbol{\Xi}(\mathbf{b}))\frac{\partial \boldsymbol{\Xi}^{(k)}}{\partial \beta_l}(\mathbf{b}) \\ &+ \frac{1}{2}\sum_{l',l''=s+1}^{Nd} \frac{\partial g_{l'l''}}{\partial \beta_l}(\mathbf{b})\big(g_s(\mathbf{b})^{-1}\mathbf{p}\big)_{l'}\big(g_s(\mathbf{b})^{-1}\mathbf{p}\big)_{l''}, \\ \dot{b}_l = &\big(g_s(\mathbf{b})^{-1}\mathbf{p}\big)_l\end{aligned} \tag{3.8.37}$$

with the notations of Eq. (3.8.33), having dropped the t dependence from $p(t), \mathbf{b}(t)$ and having deduced the second equation from Eq. (3.8.36). In Eq. (3.8.37), $\mathbf{b}$ means $(0, \ldots, 0, b_{s+1}, \ldots, b_{Nd})$.

Note that $p_l(0), b_l(0)$, $l = s+1, \ldots, Nd$, are, by Eq. (3.8.36) or by assumption, independent of the sequence used to construct $\mathbf{b}$.

So every sequence $\lambda_n \to +\infty$ for which $b_{\lambda_n l}(t)$ is uniformly convergent in $t \in [0, T]$ to a limit, $\forall l = s+1, \ldots, Nd$, can be used to construct a solution of the differential equation (3.8.37) for $t \to (p_l(t), b_l(t))_{l=s+1,\ldots,Nd}$ verifying the initial condition $(p_l(0), b_l(0))_{l=s+1,\ldots,Nd}$. Unfortunately, Eq. (3.8.37) is not quite a differential equation of the type considered in the uniqueness theorem, Proposition 1, §2.2, p. 13, since the right-hand side of Eq. (3.8.37) is defined only for $\mathbf{b} \in \boldsymbol{\Xi}^{-1}(S_\rho)$ as a function of the p_l's, b_l's.

However, all the functions $\mathbf{b}(t)$, $t \in [0, T]$, which can be built via the above construction are such that $\mathbf{b}(t) \in \boldsymbol{\Xi}^{-1}(S_\rho)$, $t \in [0, T]$. It is then easy to deduce from Proposition 1, p. 13, as a corollary, that every solution to Eq. (3.8.37) $t \to (\mathbf{p}(t), \mathbf{b}(t))$, $t \in [0, T]$, verifying $\mathbf{b}(t) \in \boldsymbol{\Xi}^{-1}(S_\rho)$, $\forall t \in [0, T]$, must be identical to every other with this property. mbe

Problems for §3.8

1. Let $\Sigma \subset \mathbf{R}^{Nd}$ be a regular surface with codimension s. Let $(U, \boldsymbol{\Xi})$ be a regular system of local coordinates well adapted and orthogonal on Σ with respect to the scalar product of Eq. (3.7.1). Denote $\boldsymbol{\beta} \in \Omega$ the coordinates of $\boldsymbol{\xi} = \boldsymbol{\Xi}(\boldsymbol{\beta}) \in U$. Set $\boldsymbol{\beta} = (\boldsymbol{\beta}_v, \boldsymbol{\beta}_n) \in \mathbf{R}^s \times \mathbf{R}^{Nd-s}$. Show that the change of coordinates $(\boldsymbol{\beta}_v, \boldsymbol{\beta}_n) \to (\Lambda\boldsymbol{\beta}_v, \tilde{\Lambda}\boldsymbol{\beta}_n)$, with Λ and $\tilde{\Lambda}$ being two $s \times s$ and $(Nd - s) \times (Nd - s)$ constant matrices, allows us to define a new system of local coordinates which is still well adapted and orthogonal on Σ.

2. In the context of Problem 1, let $\overline{W}(\beta_1, \ldots, \beta_s)$ be a C^∞ function of $(\beta_1, \ldots, \beta_s, \beta_{s+1}, \ldots, \beta_{Nd})$ independent of the last $(Nd - s)$ coordinates. Suppose that $M_{ij} = (\partial \overline{W}/\partial\beta_i\partial\beta_j)(\mathbf{0})$, $i, j = 1, \ldots, s$, is a $s \times s$ matrix which is positive definite. Show that there is a change of coordinates $\boldsymbol{\beta} \to \boldsymbol{\beta}'$ of the linear type considered in Problem 1 changing $\overline{W}$ into a function such that $M_{ij} = \delta_{ij}$, $i, j = 1, \ldots, s$. (Hint: Use $\tilde{\Lambda} = 1$, $\Lambda = J = \{$orthogonal matrix diagonalizing $M\}$ (see Appendix F, §F.4). Then make a further change of coordinates of the same type with $\tilde{\Lambda} = 1$ and $\Lambda = \{$diagonal matrix with diagonal elements $(w_1^{-1/2}, \ldots, w_s^{-1/2})\}$, where $(w_1, \ldots, w_s)$ are the s eigenvalues of M.)

3. Show that there is essentially no change in the proof of Proposition 13 if Eq. (3.8.11) is changed into

$$\overline{W}(\beta_1, \ldots, \beta_s) = \frac{1}{2}\sum_{i=1}^{s} \beta_i^2 + o(\beta_1^2 + \cdots + \beta_s^2).$$

4.* Alternatively to Problem 3, but with the same assumptions, show that there is a change of coordinates which, possibly restricting the size of U to $U' \subset U$, changes $\boldsymbol{\beta}$ into $\boldsymbol{\beta}'$, retaining the orthogonality and good adaptation of the β' coordinates, changing $\tilde{W}$ into $\frac{1}{2}(\sum_{j=1}^{s}(\beta_j')^2)$. (Hint: For $\boldsymbol{\beta} \in \Omega$, $\Omega \equiv \{$basis of $(U, \Xi)\}$, let $W(\boldsymbol{\beta}) = \boldsymbol{\beta}^2/2 + \sum_{i,j,l}^{1,s}\gamma_{ijl}(\boldsymbol{\beta})\beta_i\beta_j\beta_l$, where γ_{ijl} are suitable $C^\infty(\Omega)$ functions symmetric in the indices ijl (this assumption is not restrictive because of Problem 2 and of the Taylor–Lagrange theorem, Appendix B). Then define

$$\beta_l' = \beta_l + \sum_{j,k=1}^{s} f_{ljk}(\boldsymbol{\beta})\beta_j\beta_k$$

with f symmetric in l, j, k and of class C^∞ in $\boldsymbol{\beta}$ and impose

$$\begin{aligned}\frac{1}{2}(\boldsymbol{\beta})^2 + \sum_{jkl}\gamma_{jkl}(\boldsymbol{\beta})\beta_j\beta_k\beta_l &\equiv \frac{1}{2}(\boldsymbol{\beta}')^2\\ &= \frac{1}{2}(\boldsymbol{\beta})^2 + \sum_{ljk}\beta_l f_{ljk}(\boldsymbol{\beta})\beta_j\beta_k\\ &\quad + \frac{1}{2}\sum_{l}\sum_{\substack{jk\\ j'k'}} f_{ljk}(\boldsymbol{\beta}) f_{lj'k'}(\boldsymbol{\beta})\beta_l^2\beta_j\beta_k\beta_{j'}\beta_{k'}\end{aligned}$$

Therefore, γ_{jkl} has to be equal to

$$f_{jkl}(\boldsymbol{\beta}) + \frac{1}{2}\sum_{\{l_1,j_1,k_1,j_1',k_1'\}\supset^*\{j,k,l\}} f_{l_1j_1k_1}(\boldsymbol{\beta}) f_{l_1j_1'k_1'}(\boldsymbol{\beta})\left[\frac{\beta_{l_1}^2\beta_{j_1}\beta_{k_1}\beta_{j_1'}\beta_{k_1'}}{\beta_j\beta_k\beta_l}\right],$$

where $\supset^*$ means that the monomial in square brackets has to "simplify" so that all the terms in the denominator cancel with some in the numerator. Show that for small $\boldsymbol{\beta}$, by the implicit functions theorem, the above relation allows us to determine f in terms of γ and $\boldsymbol{\beta}$. Then, again by the implicit functions theorem, invert the relation between $\boldsymbol{\beta}$ and $\boldsymbol{\beta}'$ to complete the change of coordinates).

5.* Let $t \to \mathbf{x}(t)$, $t \in \mathbf{R}_+$, be a motion of N point masses, with masses $m_1, \ldots, m_N > 0$, which develops under the influence of an active force $\mathbf{F}^{(a)}$, conservative with potential energy $V^{(a)} \in C^\infty(\mathbf{R}^{Nd})$ bounded from below, and of an ideal constraint to a regular surface $\Sigma \subset \mathbf{R}^{Nd}$ with a codimension s.

Let $\boldsymbol{\xi}_0 = \mathbf{x}(0)$, $\boldsymbol{\eta}_0 = \dot{\mathbf{x}}(0)$, and let $\tilde{\boldsymbol{\eta}}_0$ be any velocity vector such that $\tilde{\boldsymbol{\eta}}_0^{\Sigma} = \boldsymbol{\eta}_0$ and call $t \to \mathbf{x}_\lambda(t)$ the motion with initial datum $(\tilde{\boldsymbol{\eta}}_0, \boldsymbol{\xi}_0)$ of the same system moving under the influence of the same active force and of an approximate constraint model (Σ, W, λ) verifying the assumptions of Proposition 13. Call $T_0 = \{$supremum of the T's such that $\lim_{\lambda \to +\infty} \mathbf{x}_\lambda(t) = \mathbf{x}(t)$ uniformly for $t \in [0, T]\}$.

Show that $T_0 = +\infty$. (Hint: The part of Proposition 13 proved in this section says that $T_0 > 0$. Suppose $T_0 < +\infty$. Let $\hat{\xi}_0 = \mathbf{x}(T_0)$, $\hat{\eta}_0 = \dot{\mathbf{x}}(T_0)$ and note that the energies of the motions $\mathbf{x}_\lambda$ and $\mathbf{x}$ are λ independent and coincide. Discuss the system's motion in the coordinate system (U, Ξ) around $\hat{\xi}_0$ which is well adapted and orthogonal to Σ and in which W admits a representation like Eq. (3.8.11). Show that in (U, Ξ), $\hat{\mathbf{x}}_\lambda$ verifies equations like Eqs. (3.8.24), (3.8.25), (3.8.26), and (3.8.31) with some slight but obvious changes which do not affect the conclusion that $\lim_{\lambda \to +\infty} \mathbf{x}_\lambda(t) = \mathbf{x}(t)$ as long as $\mathbf{x}_\lambda(t)$ stays inside U at a positive distance from ∂U. Since the conservation of energy implies the bound of Eq. (3.8.14) on speed, it is clear that for large λ, $\mathbf{x}_\lambda(t)$ and $\mathbf{x}(t)$ will stay at a positive distance from ∂U for all times in a neighborhood of T_0. Hence, $T_0 = +\infty$.)

6.* Show that a system of N point masses, with masses $m_1, \ldots, m_N > 0$, bound by an ideal bilateral constraint to a regular surface $\Sigma \subset \mathbf{R}^{Nd}$ and also subject to a conservative active force with inferiorly bounded potential energy $V^{(a)}$, has (for all $\xi_0 \in \Sigma$, η_0 tangent to Σ) a unique global motion $t \to \mathbf{x}(t)$, $t \in \mathbf{R}_+$, such that $\dot{\mathbf{x}}(0) = \eta_0$, $\mathbf{x}(0) = \xi_0$; i.e., a motion verifying Eq. (3.7.20) in every local system of regular coordinates (Hint: Use the energy conservation and the existence and uniqueness theorems for Eq. (3.7.20) following from its transformation into Eq. (3.8.37): energy conservation together with the semiboundedness of $V^{(a)}$ gives an a priori estimate.)

§3.9. Application to the Rigid Motion. The König's Theorem

The general perfection criterion for approximate constraints discussed in §3.8 is interesting because it permits us to show the perfection of some classes of constraint models.

In this section we wish to show how, on the basis of the results of §3.8, Proposition 13, it becomes possible to show that a natural model of the rigidity constraint is approximately ideal.

Consider the following model (Σ, W, λ), which is one of the most important constraint models for N point masses. Let $l_{ij} > 0$ be given numbers defined for $(i, j) \in S = \{$subset of the set of pairs of different points in $\{1, \ldots, N\}\}$; let σ_i, $i \in T \subset \{1, \ldots, N\}$, be a family of regular surfaces in $\mathbf{R}^3$; then define

$$\Sigma = \Big\{\xi \,|\, \xi = (\xi^{(1)}, \ldots, \xi^{(N)}) \in \mathbf{R}^{3N}, |\xi^{(i)} - \xi^{(j)}| = l_{ij} \text{ for } i, j \in S; \quad \xi^{(i)} \in \sigma_i \text{ for } i \in T\Big\}, \tag{3.9.1}$$

$$W(\xi) = \sum_{i,j \in S} \psi_{ij}\big(|\xi^{(i)} - \xi^{(j)}|^2 - l_{ij}^2\big) + \sum_{i \in T} \psi_i\big(|\xi^{(i)} - \sigma_i|^2\big), \tag{3.9.2}$$

where $\psi_{ij}, \psi_i \in C^\infty(\mathbf{R})$, $\psi_{ij}(0) = \psi_i(0) = 0$, and have a strict minimum at zero; the notation $|\xi^{(i)} - \sigma_i|^2$ denotes a C^∞ function on $\mathbf{R}^3$ positive outside

σ_i and, near σ_i, equal to the square of the distance between ξ and σ_i. Here σ_i may also be a single point.

Clearly, (Σ, W, λ) is a natural model of rigidity for some system's pairs (those in S) and for permanence on a surface or on a point (if σ_i is zero dimensional) for some system's points (those in T).

In applications, it is quite common to meet only constraints for which the above is a good model, when friction is neglected.

It is not completely trivial to show that Eqs. (3.9.1) and (3.9.2) are an approximate ideal model in the sense of Definition 13, §3.7.

In this case, we shall examine, for simplicity, only the case in which Eq. (3.9.1) is a "total rigidity" constraint, i.e., the case when S contains so many pairs (e.g., all) to allow only configurations which can be obtained by rigid motions of a single one or, at most, finitely many. Nevertheless, we formulate the general result.

14 Proposition. *The model* (Σ, W, λ) *defined by Eqs.* (3.9.1) *and* (3.9.2) *is a model of an approximate ideal constraint for a system of N point masses with (arbitrary) masses* $m_1, \ldots, m_N > 0$.

PROOF. (Case $T = \emptyset$, S such that Σ is a total rigidity constraint). The surface Σ, in the case under examination, decomposes into a finite number of connected parts, each representing a rigid system in the usual sense of the word.[9]

We shall suppose $N \geq 3$, the $N = 2$ case having been already discussed in the Example in §3.8, p. 188. We shall also suppose that the points 1, 2, and 3 are not aligned in the configurations of Σ: the degenerate case of N aligned points could be treated likewise.

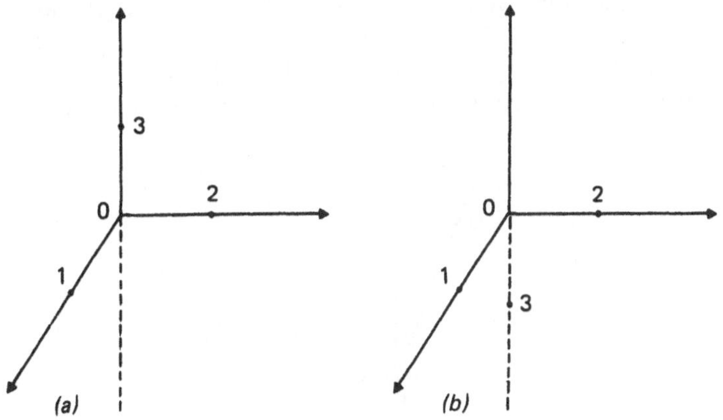

Figure 3.2.

[9] Σ may consist of several connected parts: for instance, if $N = 4$ and the distances of the points 0, 1, 2, and 3 are $d(0,1) = 1$, $d(0,2) = 1$, $d(1,2) = \sqrt{2}$, $d(1,3) = \sqrt{2}$, $d(2,3) = \sqrt{2}$, respectively, then Σ contains two connected parts. The first consists of the configurations obtained by rotations and translations of the configuration in Fig 3.2(a) and the other of those in Fig. 3.2(b)).

The configurations $\xi' \in \Sigma$ located on the same connected component of Σ shall be uniquely determined by the position G of the system's baricenter in the "fixed" Cartesian reference frame $(0; \mathbf{i}, \mathbf{j}, \mathbf{k})$ and by three orthogonal unit vectors $\mathbf{i}_1, \mathbf{i}_2, \mathbf{i}_3$ fixed with the system ("co-moving") and finally by the positions $P_1, \ldots, P_N$ of N point masses in the reference frame $(G; \mathbf{i}_1, \mathbf{i}_2, \mathbf{i}_3)$. By the rigidity constraint, the points $P_1, \ldots, P_N$ will have coordinates $(P_i - G)_l$, $l = 1,2,3$, $i = 1,2, \ldots, N$, which are given constants, $\forall \xi'$ in the same connected component of Σ, in the frame $(G; \mathbf{i}_1, \mathbf{i}_2, \mathbf{i}_3)$.

We suppose to have fixed $\mathbf{i}_3$ parallel to $(P_2 - P_1)$ and $\mathbf{i}_2$ parallel to the plane (P_1, P_2, P_3).

To prove Proposition 14, it will be sufficient to build a system of coordinates, local near $\xi_0 \in \Sigma$, regular, well adapted, and orthogonal to Σ with respect to the scalar product of Eq. (3.7.1) and with the extra property that W, the constraint's structure function, has the property of Eq. (3.8.1).

Without loss of generality, we shall suppose that the plane of the first two axes in the co-moving frame $(G_0; \mathbf{i}_1^0, \mathbf{i}_2^0, \mathbf{i}_3^0)$ associated with $\xi_0 \in \Sigma$, i.e., the plane $(\mathbf{i}_1^0, \mathbf{i}_2^0)$, is not parallel to the plane $(\mathbf{i}, \mathbf{j})$.

Let ξ be a configuration close to Σ: in general, $\xi \notin \Sigma$.

We associate with ξ, $3N$ coordinates which we obtain through the following construction.

It is clear that ξ can be determined by assigning a configuration $\xi' \in \Sigma$ and the vectors $\boldsymbol{\kappa}^{(1)}, \ldots, \boldsymbol{\kappa}^{(N)}$ providing the deviations of the points in ξ with respect to the corresponding points $P_1, \ldots, P_N$ in ξ'. The $(3N + 6)$ coordinates necessary to determine the $3N$ components of $\boldsymbol{\kappa}^{(1)}, \ldots, \boldsymbol{\kappa}^{(N)}$ in the frame $(G; \mathbf{i}_1, \mathbf{i}_2, \mathbf{i}_3)$ fixed with ξ' and the six coordinates giving the position and orientation in space of $(G; \mathbf{i}_1, \mathbf{i}_2, \mathbf{i}_3)$, i.e., of ξ', are clearly too many.

The coordinates that we shall use to determine $(G; \mathbf{i}_1, \mathbf{i}_2, \mathbf{i}_3)$ are the three Cartesian coordinates of G in the fixed frame $(0; \mathbf{i}, \mathbf{j}, \mathbf{k})$ and the three "Euler angles" (θ, φ, ψ). The angles $\theta = \mathbf{i}_3 \hat{} \mathbf{k}$, $\varphi = \mathbf{i} \hat{} \mathbf{n}$, $\psi = \mathbf{n} \hat{} \mathbf{i}_1$ are illustrated in Fig. 3.3. $\mathbf{n}$ is the unit vector along the intersection between the plane $(\mathbf{i}, \mathbf{j})$

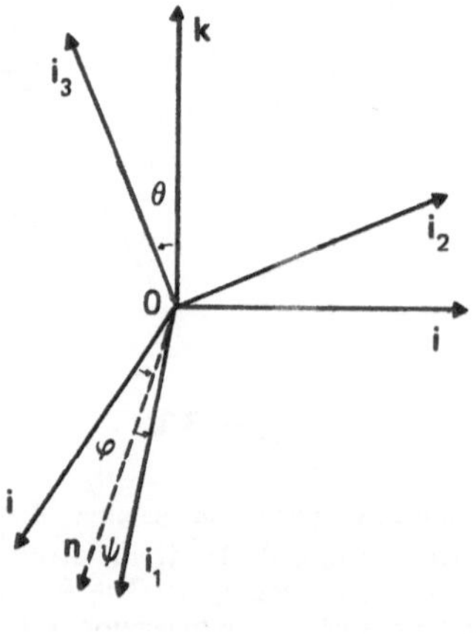

Figure 3.3. The Euler angles.

and the plane $(\mathbf{i}_1, \mathbf{i}_2)$, arbitrarily oriented (the "nodes line"). Note that the components in $(0; \mathbf{i}_1, \mathbf{i}_2, \mathbf{i}_3)$ of $\mathbf{k}$, $\mathbf{i}_3$, $\mathbf{n}$ are, respectively:

$$\begin{aligned} \mathbf{k} &= (\sin\theta \sin\psi, \sin\theta\cos\psi, \cos\theta), \\ \mathbf{i}_3 &= (0, 0, 1), \\ \mathbf{n} &= (\cos\psi, -\sin\psi, 0) \end{aligned} \tag{3.9.3}$$

which will be useful in the following.

To obtain a local system of regular coordinates near ξ_0, we shall remove from the $3N+6$ abundant coordinates just introduced six among them by imposing the following six restrictions:

$$\sum_{i=1}^{N} m_i \boldsymbol{\kappa}^{(i)} = \mathbf{0}, \tag{3.9.4}$$

$$\sum_{i=1}^{N} (P_i - G) \wedge m_i \boldsymbol{\kappa}^{(i)} = \mathbf{0} \tag{3.9.5}$$

which signify that G is actually the baricenter of ξ as well as that of ξ' and that the configuration $\xi' \in \Sigma$ is so chosen that the system $(m_i \boldsymbol{\kappa}^{(i)})_{i=1}^{N}$ of vectors ("quantities of deviation") have a vanishing "angular momentum". The above restrictions should be thought of as restrictions on the choice of the reference configuration $\xi' \in \Sigma$, a priori arbitrary.

The six coordinates that we eliminate via Eqs. (3.9.4) and (3.9.5) are, for instance, the first two components of $\boldsymbol{\kappa}^{(1)}$ in $(G; \mathbf{i}_1, \mathbf{i}_2, \mathbf{i}_3)$, the three components of $\boldsymbol{\kappa}^{(2)}$, and the first component of $\boldsymbol{\kappa}^{(3)}$ still in $(G; \mathbf{i}_1, \mathbf{i}_2, \mathbf{i}_3)$. The "free" coordinates $\boldsymbol{\beta} = (\beta_1, \ldots, \beta_{3N})$ will then be (orderly enumerated):

$$\left(\kappa_3^{(1)}, \kappa_2^{(3)}, \kappa_3^{(3)}, \kappa_1^{(4)}, \kappa_2^{(4)}, \kappa_3^{(4)}, \ldots, \kappa_3^{(N)}, \theta, \varphi, \psi, (\xi_G)_1, (\xi_G)_2, (\xi_G)_3\right),$$

where (θ, φ, ψ) are the Euler angles of $(\mathbf{i}_1, \mathbf{i}_2, \mathbf{i}_3)$ with respect to $(\mathbf{i}, \mathbf{j}, \mathbf{k})$ and $(\xi_G)_1, (\xi_G)_2, (\xi_G)_3$ are the coordinates of G in $(0; \mathbf{i}, \mathbf{j}, \mathbf{k})$, while $\kappa_j^{(i)}$ are the components in $(G; \mathbf{i}_1, \mathbf{i}_2, \mathbf{i}_3)$ of the deviations $\boldsymbol{\kappa}^{(i)}$.

Given the $3N$ coordinates $\boldsymbol{\beta}$, the configuration $\xi = \Xi(\boldsymbol{\beta})$ is built as follows:

(i) $\xi_G = (\beta_{3N-2}, \beta_{3N-1}, \beta_{3N}) \equiv \boldsymbol{\beta}_G$ determines the baricenter G.

(ii) $\boldsymbol{\beta}_{\text{rot}} = (\beta_{3N-5}, \beta_{3N-4}, \beta_{3N-3}) = (\theta, \varphi, \psi)$ determine the orientation of the axes $\mathbf{i}_1, \mathbf{i}_2, \mathbf{i}_3$. Therefore, the positions $P_1, \ldots, P_N$ of the N points of the auxiliary configuration, called ξ' above, are determined.

(iii) The coordinates $\boldsymbol{\beta}_V = (\beta_1, \ldots, \beta_{3N-6})$ determine $\boldsymbol{\kappa}^{(4)}, \ldots, \boldsymbol{\kappa}^{(N)}$ and, hence, the positions in $(G; \mathbf{i}_1, \mathbf{i}_2, \mathbf{i}_3)$ of the points labeled $4, 5, \ldots, N$ and, furthermore, the coordinates $\kappa_3^{(1)}$ and $\kappa_2^{(3)}, \kappa_3^{(3)}$ of $\boldsymbol{\kappa}^{(1)}, \boldsymbol{\kappa}^{(3)}$.

(iv) The coordinates of $\boldsymbol{\kappa}^{(2)}$, as well as the remaining coordinates of $\boldsymbol{\kappa}^{(1)}, \boldsymbol{\kappa}^{(3)}$, are determined from Eqs. (3.9.4) and (3.9.5). From Eq. (3.9.4), one finds

$$\boldsymbol{\kappa}^{(2)} = -\frac{m_1}{m_2}\boldsymbol{\kappa}^{(1)} - \sum_{i=3}^{N} \frac{m_i}{m_2}\boldsymbol{\kappa}^{(i)} \tag{3.9.6}$$

which, inserted into Eq. (3.9.5), yields

$$m_1(P_1 - P_2) \wedge \boldsymbol{\kappa}^{(1)} + \sum_{i=3}^{N} m_i(P_i - P_2) \wedge \boldsymbol{\kappa}^{(i)} = 0. \tag{3.9.7}$$

By scalar multiplication of Eq. (3.9.7) by $(P_1 - P_2)$, we find

$$\sum_{i=3}^{N} m_i(P_i - P_2) \wedge \boldsymbol{\kappa}^{(i)} \cdot (P_1 - P_2) = 0 \tag{3.9.8}$$

which determines the value of $\kappa_1^{(3)}$. In fact, recalling that $\mathbf{i}_1$ is orthogonal to the plane P_1, P_2, P_3 and that these three points are not aligned, it is clear that $(P_3 - P_2) \wedge \mathbf{i}_1 \cdot (P_1 - P_2) \neq 0$, so that Eq. (3.9.8) is a linear equation for $\kappa_1^{(3)}$ (with non-zero coefficient in front of $\kappa_1^{(3)}$).

Once $\boldsymbol{\kappa}^{(3)}$ is completely determined, Eq. (3.9.7) unambiguously provides the first two components of $\boldsymbol{\kappa}^{(1)}$, since Eq. (3.9.7) only leaves the component of $\boldsymbol{\kappa}^{(1)}$ along $(P_1 - P_2)$ undetermined, which, however, is just $\kappa_3^{(1)}$ (recall that $\mathbf{i}_3$ is parallel, by construction, to $(P_1 - P_2)$, i.e., it is already known to be β_1).

Finally, once $\boldsymbol{\kappa}^{(1)}$ and $\boldsymbol{\kappa}^{(3)}$ are completely known, $\boldsymbol{\kappa}^{(2)}$ emerges from Eq. (3.9.6).

It is now easy to check the invertibility, near $\boldsymbol{\xi}_0$, of the transformation associating with $\boldsymbol{\beta} = (\boldsymbol{\beta}_V, \boldsymbol{\beta}_{\text{rot}}, \boldsymbol{\beta}_G)$ the configuration $\boldsymbol{\Xi}(\boldsymbol{\beta}) = \boldsymbol{\xi}$ built following rules (i)–(iv). Such a transformation is also of class C^∞ with nonvanishing Jacobian matrix near $\boldsymbol{\xi}_0$. However, we do not enter into the laborious analysis of the check of the regularity, invertibility, and nonsingularity of $\boldsymbol{\Xi}$: it does not present any conceptual problem.

Hence, the transformation $\boldsymbol{\Xi}$ establishes a regular system of local coordinates in some small enough neighborhood U of $\boldsymbol{\xi}_0 \in \Sigma$.

Clearly, the points in $\Sigma \cap U$ are those described by $\beta_1 = \cdots = \beta_{3N-6} = 0$; i.e., $(U, \boldsymbol{\Xi})$ is adapted to Σ. Actually, $(U, \boldsymbol{\Xi})$ is well adapted and orthogonal on Σ, with respect to the scalar product of Eq. (3.7.1). To show this, we have to find the kinetic matrix associated with $(U, \boldsymbol{\Xi})$ and Eq. (3.7.1). Therefore, we must express the kinetic energy of a motion $t \to \mathbf{x}(t)$ of N point masses through its image motion $t \to \mathbf{b}(t) = (\mathbf{b}_V(t), \mathbf{b}_{\text{rot}}(t), \mathbf{b}_G(t)) = \boldsymbol{\Xi}^{-1}(\mathbf{x}(t))$, assuming that the motion $\mathbf{x}$ takes place inside U for $t \in [t_1, t_2]$.

By the definition of the coordinates $\boldsymbol{\beta}$, one has, for $t \in [t_1, t_2]$:

$$\mathbf{x}^{(i)}(t) = \mathbf{b}_G(t) + \sum_{l=1}^{3} \left(\kappa_l^{(i)}(t) + (P_i - G)_l\right)\mathbf{i}_l(t), \tag{3.9.9}$$

and by differentiation one finds

$$\dot{\mathbf{x}}^{(i)}(t) = \dot{\mathbf{b}}_G(t) + \sum_{l=1}^{3} \dot{\kappa}_l^{(i)}(t)\mathbf{i}_l(t) + \sum_{l=1}^{3} \left(\kappa_l^{(i)}(t) + (P_i - G)_l\right)\frac{d\mathbf{i}_l(t)}{dt}. \tag{3.9.10}$$

We will now use a well-known kinematic formula giving a simple expression to the time derivative of three mutually orthogonal unit vectors which are time dependent:

$$\frac{d\mathbf{i}_l(t)}{dt} = \boldsymbol{\omega}(t) \wedge \mathbf{i}_l(t), \qquad l = 1,2,3, \tag{3.9.11}$$

where $\boldsymbol{\omega}(t)$ is a suitable vector called "angular velocity" of the triple $(\mathbf{i}_1, \mathbf{i}_2, \mathbf{i}_3)$.[10]

[10]To understand Eq. (3.9.11), note that, in general, the space orientation of three mutually orthogonal axes $\mathbf{i}_1, \mathbf{i}_2, \mathbf{i}_3$ imagined as emerging from a fixed point Ω can only vary if its three Euler angles (θ, φ, ψ) with respect to a fixed triple $(\mathbf{i}, \mathbf{j}, \mathbf{k})$ change. If only θ varies, it means that the reference frame $(\Omega; \mathbf{i}_1, \mathbf{i}_2, \mathbf{i}_3)$ rotates around the node line (see Fig. 3.3), and it is then clear that every point P co-moving with $(\Omega; \mathbf{i}_1, \mathbf{i}_2, \mathbf{i}_3)$ has velocity $\mathbf{V}_P = \dot{\theta}\mathbf{n} \wedge (P - \Omega)$. This holds, in particular, for the extremities of $\mathbf{i}_1, \mathbf{i}_2, \mathbf{i}_3$; hence:

$$\frac{d\mathbf{i}_l}{dt} = \dot{\theta}\mathbf{n} \wedge \mathbf{i}_l, \qquad l = 1, 2, 3.$$

A similar argument shows that if $\mathbf{i}_1, \mathbf{i}_2, \mathbf{i}_3$ move because only φ or ψ vary, then

$$\frac{d\mathbf{i}_l}{dt} = \dot{\varphi}\mathbf{k} \wedge \mathbf{i}_l \quad \text{or} \quad \frac{d\mathbf{i}_l}{dt} = \dot{\psi}\mathbf{i}_3 \wedge \mathbf{i}_l, \qquad l = 1, 2, 3.$$

More generally, if $\mathbf{i}_1, \mathbf{i}_2, \mathbf{i}_3$ vary because θ, φ, ψ simultaneously vary, it will be (by the differentiation rule of composed functions):

$$\frac{d\mathbf{i}_l}{dt} = \boldsymbol{\omega} \wedge \mathbf{i}_l, \qquad l = 1, 2, 3$$

with $\boldsymbol{\omega}$ given by $\dot{\theta}\mathbf{n} + \dot{\varphi}\mathbf{k} + \dot{\psi}\mathbf{i}_3$, i.e., Eq. (3.9.11).

In connection with Eq. (3.9.11), it is natural to note one of its famous consequences: the relation between a motion $t \to P(t)$ in a frame $(0; \mathbf{i}, \mathbf{j}, \mathbf{k})$ and the same motion in a frame $(\Omega(t); \mathbf{i}_1(t)\mathbf{i}_2(t), \mathbf{i}_3(t))$, time dependent. From the vector relation

$$P(t) - 0 = (P(t) - \Omega(t)) + (\Omega(t) - 0)$$

written componentwise as

$$x(t)\mathbf{i} + y(t)\mathbf{j} + z(t)\mathbf{k} = x_1(t)\mathbf{i}_1(t) + x_2(t)\mathbf{i}_2(t) + x_3(t)\mathbf{i}_3(t) + x^{\Omega}(t)\mathbf{i} + y^{\Omega}(t)\mathbf{j} + z^{\Omega}(t)\mathbf{k}$$

with obvious notations, it follows, by differentiation, that

$$\mathbf{V}^{(a)} = \mathbf{V}^{(r)} + x_1(t)\frac{d\mathbf{i}_1}{dt}(t) + x_2(t)\frac{d\mathbf{i}_2}{dt}(t) + x_3(t)\frac{d\mathbf{i}_3}{dt}(t) + \mathbf{V}_{\Omega},$$

where $\mathbf{V}^{(a)} = \dot{x}(t)\mathbf{i} + \dot{y}(t)\mathbf{j} + \dot{z}(t)\mathbf{k}$ is the velocity of $t \to P(t)$ in $(0; \mathbf{i}, \mathbf{j}, \mathbf{k})$, $\mathbf{V}^{(r)}$ is the velocity of the same motion "relative" to $(\Omega(t); \mathbf{i}_1(t), \mathbf{i}_2(t), \mathbf{i}_3(t))$, i.e., $\mathbf{V}^{(r)} = \dot{x}_1(t)\mathbf{i}_1(t) + \dot{x}_2(t)\mathbf{i}_2(t) + \dot{x}_3(t)\mathbf{i}_3(t)$ and $\mathbf{V}_{\Omega}$ is the velocity of the motion $t \to \Omega(t)$ in $(0; \mathbf{i}, \mathbf{j}, \mathbf{k})$, i.e., $\mathbf{V}_{\Omega} = \dot{x}^{\Omega}(t)\mathbf{i} + \dot{y}^{\Omega}(t)\mathbf{j} + \dot{z}^{\Omega}(t)\mathbf{k}$.

Then, by using Eq. (3.9.11):

$$\mathbf{V}^{(a)} = \mathbf{V}^{(r)} + \boldsymbol{\omega} \wedge (x_1(t)\mathbf{i}_1(t) + x_2(t)\mathbf{i}_2(t) + x_3(t)\mathbf{i}_3(t)) + \mathbf{V}_{\Omega} = \mathbf{V}^{(r)} + (\boldsymbol{\omega} \wedge (P - \Omega) + \mathbf{V}_{\Omega}).$$

The term in parentheses has the obvious interpretation of the "drag velocity" that the point P would have if it were fixed in $(\Omega; \mathbf{i}_1, \mathbf{i}_2, \mathbf{i}_3)$; hence, the above formula reads "the absolute speed equals the sum of the relative speed plus the drag speed". Furthermore, the velocity of a point P fixed in a moving frame $(\Omega; \mathbf{i}_1, \mathbf{i}_2, \mathbf{i}_3)$ is given by

$$\mathbf{V}_P = \mathbf{V}_{\Omega} + \boldsymbol{\omega} \wedge (P - \Omega),$$

where $\mathbf{V}_P$ is the velocity in $(0; \mathbf{i}, \mathbf{j}, \mathbf{k})$ of P, $\boldsymbol{\omega}$ is the angular velocity of the triplet $(\mathbf{i}_1, \mathbf{i}_2, \mathbf{i}_3)$ in $(0; \mathbf{i}, \mathbf{j}, \mathbf{k})$, and $\mathbf{V}_{\Omega}$ is the speed of Ω in $(0; \mathbf{i}, \mathbf{j}, \mathbf{k})$. The last relation is obviously of great interest in the theory of rigid motion.

A useful expression for $\boldsymbol{\omega}$ is, in terms of the Euler angles (see p. 203 and footnote 10):

$$\boldsymbol{\omega} = \dot{\theta}\mathbf{n} + \dot{\varphi}\mathbf{k} + \dot{\psi}\mathbf{i}_3 . \tag{3.9.12}$$

Coming back to Eq. (3.9.10), we shall rewrite it by using Eq. (3.9.11):

$$\dot{\mathbf{x}}^{(i)} = \dot{\mathbf{b}}_G + \overset{*}{\boldsymbol{\kappa}}^{(i)} + \boldsymbol{\omega} \wedge \left(\boldsymbol{\kappa}^{(i)} + (P_i - G)\right), \tag{3.9.13}$$

where we set $\overset{*}{\boldsymbol{\kappa}}^{(i)}(t) = \sum_{l=1}^{3} \overset{*}{\kappa}_l^{(i)}(t)\mathbf{i}_l(t)$ (which differs from $\dot{\boldsymbol{\kappa}}^{(i)}(t)$; it is, in fact, the velocity of the ith point relative to the moving frame, while $\dot{\boldsymbol{\kappa}}^{(i)}(t)$ is its velocity relative to the fixed frame), and $(P_i - G) = \sum_{l=1}^{3}(P_i - G)_l\mathbf{i}_l(t)$.

It is now possible to compute the kinetic energy, using Eq. (3.9.13):

$$\begin{aligned}
T &= \frac{1}{2}\sum_{i=1}^{N} m_i(\dot{\mathbf{x}}^{(i)})^2 = \frac{1}{2}\sum_{i=1}^{N} m_i\left(\dot{\mathbf{b}}_G + \boldsymbol{\omega} \wedge \left(\boldsymbol{\kappa}^{(i)} + (P_i - G)\right) + \overset{*}{\boldsymbol{\kappa}}^{(i)}\right)^2 \\
&= \frac{1}{2}\left(\sum_{i=1}^{N} m_i\right)(\dot{\mathbf{b}}_G)^2 + \frac{1}{2}\sum_{i=1}^{N} m_i\left(\boldsymbol{\omega} \wedge \left(\boldsymbol{\kappa}^{(i)} + (P_i - G)\right)\right)^2 \\
&\quad + \frac{1}{2}\sum_{i=1}^{N} m_i(\overset{*}{\boldsymbol{\kappa}}^{(i)})^2 + \sum_{i=1}^{N} m_i\dot{\mathbf{b}}_G \cdot \boldsymbol{\omega} \wedge \left(\boldsymbol{\kappa}^{(i)} + (P_i - G)\right) \\
&\quad + \sum_{i=1}^{N} m_i\dot{\mathbf{b}}_G \cdot \overset{*}{\boldsymbol{\kappa}}^{(i)} + \sum_{i=1}^{N} m_i\boldsymbol{\omega} \wedge \left(\boldsymbol{\kappa}^{(i)} \wedge (P_i - G)\right) \cdot \overset{*}{\boldsymbol{\kappa}}^{(i)} .
\end{aligned} \tag{3.9.14}$$

The fourth and fifth terms in the right-hand side vanish identically, which follows by taking the constant vectors out of the summations and by recalling the definition of the baricenter (by which $\sum_{i=1}^{N} m_i(P_i - G) = \mathbf{0}$) as well as Eq. (3.9.4).

To study the second and the sixth terms of the right-hand side Eq. (3.9.14), we will use the formula

$$(\mathbf{a} \wedge \mathbf{b}) \cdot \mathbf{c} = (\mathbf{b} \wedge \mathbf{c}) \cdot \mathbf{a} = (\mathbf{c} \wedge \mathbf{a}) \cdot \mathbf{b}, \qquad \forall \mathbf{a}, \mathbf{b}, \mathbf{c} \in \mathbf{R}^3 \tag{3.9.15}$$

to note that

$$\sum_{i=1}^{N} m_i\boldsymbol{\omega} \wedge \left(\boldsymbol{\kappa}^{(i)} + (P_i - G)\right) \cdot \overset{*}{\boldsymbol{\kappa}}^{(i)} = \boldsymbol{\omega} \cdot \left(\sum_{i=1}^{N} m_i\left(\boldsymbol{\kappa}^{(i)} + (P_i - G)\right) \wedge \overset{*}{\boldsymbol{\kappa}}^{(i)}\right), \tag{3.9.16}$$

and one can remark that the quantity within brackets in the right-hand side of Eq. (3.9.16) is the angular momentum $\mathbf{K}_G^{(\mathrm{in})}$ "relative" to the frame $(G; \mathbf{i}_1, \mathbf{i}_2, \mathbf{i}_3)$ (also called the "internal angular momentum") and, furthermore, by Eq. (3.9.5) written componentwise in $(G; \mathbf{i}_1, \mathbf{i}_2, \mathbf{i}_3)$, by the $\overset{*}{\boldsymbol{\kappa}}^{(i)}$ definition and by the time independence of the components of $(P_i - G)$ in

$(G; \mathbf{i}_1, \mathbf{i}_2, \mathbf{i}_3)$, it follows that $\sum_{i=1}^{N} m_i (P_i - G) \wedge \dot{\tilde{\boldsymbol{\kappa}}}^{(i)} \equiv \mathbf{0}$,[11] so that

$$\mathbf{K}_G^{(\text{in})} = \sum_{i=1}^{N} m_i \big(\boldsymbol{\kappa}^{(i)} + (P_i - G) \big) \wedge \dot{\tilde{\boldsymbol{\kappa}}}^{(i)} \equiv \sum_{i=1}^{N} m_i \boldsymbol{\kappa}^{(i)} \wedge \dot{\tilde{\boldsymbol{\kappa}}}^{(i)}. \quad (3.9.17)$$

It is therefore true and, as we shall see, important that $\mathbf{K}_G^{(\text{in})} \equiv \mathbf{0}$ if $\boldsymbol{\kappa}^{(1)} = \cdots = \boldsymbol{\kappa}^{(N)} = \mathbf{0}$, i.e., if the system is, at the time considered, on Σ.

We shall write the second term of the right-hand side of Eq. (3.9.14) as

$$\begin{aligned} &\frac{1}{2} \sum_{i=1}^{N} m_i \Big(\omega \wedge \big(\boldsymbol{\kappa}^{(i)} + (P_i - G) \big) \Big)^2 \\ &\quad = \frac{1}{2} \omega \cdot \left[\sum_{i=1}^{N} m_i \big(\boldsymbol{\kappa}^{(i)} + (P_i - G) \big) \wedge \Big(\omega \wedge \big(\boldsymbol{\kappa}^{(i)} + (P_i - G) \big) \Big) \right] \quad (3.9.18) \\ &\quad = \frac{1}{2} (\omega \cdot \mathbf{I}(\omega)), \end{aligned}$$

where Eq. (3.9.15) has been used, and we set

$$I(\omega) = \sum_{i=1}^{N} m_i \big(\boldsymbol{\kappa}^{(i)} + (P_i - G) \big) \wedge \Big(\omega \wedge \big(\boldsymbol{\kappa}^{(i)} + (P_i - G) \big) \Big). \quad (3.9.19)$$

If we then define

$$T_G = \frac{1}{2} \left(\sum_{i=1}^{N} m_i \right) \dot{\mathbf{x}}_G^2, \quad (3.9.20)$$

"baricenter's kinetic energy",

$$T^{(\text{in})} = \frac{1}{2} \sum_{i=1}^{N} m_i (\dot{\tilde{\boldsymbol{\kappa}}}^{(i)})^2, \quad (3.9.21)$$

"internal kinetic energy",

$$T_C = \mathbf{K}_G^{(\text{in})} \cdot \omega, \quad (3.9.22)$$

"complementary" or "Coriolis' kinetic energy",

$$T_{\text{rot}} = \tfrac{1}{2} \omega \cdot I(\omega), \quad (3.9.23)$$

[11] Let $s = 1, 2, 3$; then Eq. (3.9.5) gives

$$\begin{aligned} 0 &= \frac{d}{dt} \left(\sum_{i=1}^{N} m_i (P_i - G) \wedge \boldsymbol{\kappa}^{(i)} \right)_s = \frac{d}{dt} \sum_{i=1}^{N} m_i \sum_{l', l''}^{1,3} (P_i - G)_{l'} \kappa_{l''}^{(i)} (\mathbf{i}_{l'} \wedge \mathbf{i}_{l''})_s \\ &\equiv \sum_{i=1}^{N} m_i \sum_{l', l''=1}^{3} (P_i - G)_{l'} \dot{\kappa}_{l''}^{(i)} (\mathbf{i}_{l'} \wedge \mathbf{i}_{l''})_s \\ &\equiv \left(\sum_{i=1}^{N} m_i (P_i - G) \wedge \dot{\tilde{\boldsymbol{\kappa}}}^{(i)} \right)_s \end{aligned}$$

since $(\mathbf{i}_{l'} \wedge \mathbf{i}_{l''})_s$ is either 0 or ± 1, $\forall t$, i.e., it has zero t-derivative.

"rotational kinetic energy", we have just shown that

$$T = T_G + T^{(\mathrm{in})} + T_{\mathrm{rot}} + T_C . \tag{3.9.24}$$

When $\boldsymbol{\omega} = \mathbf{0}$, this relation is called "König's theorem".

From Eq. (3.9.24), it is easy to conclude that the coordinate system defined by $\boldsymbol{\Xi}$ near $\boldsymbol{\xi}_0$ is well adapted and orthogonal on Σ.

In fact, one can note that at a point $\boldsymbol{\xi} \in \Sigma$, the coordinates $\boldsymbol{\kappa}^{(1)}, \ldots, \boldsymbol{\kappa}^{(N)}$, hence, the $\boldsymbol{\beta}_V$'s, vanish (i.e., the $\boldsymbol{\beta}$ coordinates are adapted to Σ). Furthermore, if the motion $t \to \mathbf{x}(t)$ happens to occupy the position $\boldsymbol{\xi} \in \Sigma$ at a certain time t_0, we have $T_C(t_0) = 0$, by Eqs. (3.9.17) and (3.9.22).

At the same time instant $T^{(\mathrm{in})}(t_0)$, as is easy to realize from the determination of $\boldsymbol{\kappa}^{(1)}, \boldsymbol{\kappa}^{(2)}, \boldsymbol{\kappa}^{(3)}$ via Eqs. (3.9.4) and (3.9.5), is a quadratic form in $\dot{\mathbf{b}}_V(t_0)$ with coefficients *only* depending upon the structure of Σ via the coordinates $(P_i - G)_l$, $l = 1, 2, 3$, $i = 1, \ldots, N$, which are given constants ($\boldsymbol{\xi}$ independent, hence, $\boldsymbol{\beta}_V$ independent).

Finally, $T_G(t_0)$ is a quadratic form in $\mathbf{b}_G(t_0)$ with constant coefficients, while $T_{\mathrm{rot}}(t_0)$ is a quadratic form in the components of $\boldsymbol{\omega}$ in $(G; \mathbf{i}_1, \mathbf{i}_2, \mathbf{i}_3)$ with coefficients depending only on the structure of Σ via the (constant) coordinates $(P_i - G)_l$, $l = 1, 2, 3$; $i = 1, \ldots, N$ [see, for more details, Eq. (3.9.29)]. Hence, since the components of $\boldsymbol{\omega}$ in $(G; \mathbf{i}_1, \mathbf{i}_2, \mathbf{i}_3)$ are, by Eq. (3.9.3),

$$\begin{aligned} \omega_1 &= \dot{\theta} \cos\psi + \dot{\varphi} \sin\theta \sin\psi, \\ \omega_2 &= -\dot{\theta} \sin\psi + \dot{\varphi} \sin\theta \cos\psi, \\ \omega_3 &= \dot{\varphi} \cos\theta + \dot{\psi}, \end{aligned} \tag{3.9.25}$$

it follows that the rotation kinetic energy is a quadratic form in $\dot{\mathbf{b}}_{\mathrm{rot}}(t_0) = (\dot{\theta}, \dot{\varphi}, \dot{\psi})_{t = t_0}$ with coefficients solely dependent on θ, φ, ψ [by Eq. (3.9.25)].

Hence, the quadratic forms defining T on Σ do not contain any mixed terms like $(\dot{b}_V)_i (\dot{b}_{\mathrm{rot}})_j$ or $(\dot{b}_V)_i (\dot{b}_G)_j$; therefore, the coordinate system is orthogonal on Σ (see Definition 12, §3.7, p. 178). It is also well adapted by the above observed constancy of the coefficients of the quadratic form in $\dot{\mathbf{b}}_V(t_0)$ expressing $T^{(\mathrm{in})}(t_0)$.

From the definition of W, it is clear that W depends only upon $\boldsymbol{\kappa}^{(1)}, \ldots, \boldsymbol{\kappa}^{(N)}$ through their components in $(G; \mathbf{i}_1, \mathbf{i}_2, \mathbf{i}_3)$, i.e., only upon $\boldsymbol{\beta}_V$ [in fact, as already remarked, such components can be reconstructed from the $\boldsymbol{\beta}_V$'s via Eqs. (3.9.4) and (3.9.5) and depend only on the $\boldsymbol{\beta}_V$'s and do not depend on $(\boldsymbol{\beta}_{\mathrm{rot}}, \boldsymbol{\beta}_G)$].

This concludes the perfection proof for the constraint model (Σ, W, λ) in the rigid case considered above. mbe

Observation. By deducing Eq. (3.9.24), we have explicitly shown that the kinetic energy of a rigid body, i.e., of a motion of a system of N point masses in $\mathbf{R}^3$ constrained to keep the mutual distances fixed, can be expressed in terms of six coordinates and their derivatives. If such coordinates are the baricenter's coordinates $\mathbf{x}_G$ and the three Euler angles (θ, φ, ψ)

of a co-moving frame of reference $(G; \mathbf{i}_1, \mathbf{i}_2, \mathbf{i}_3)$ with respect to a fixed frame $(0; \mathbf{i}, \mathbf{j}, \mathbf{k})$ and if $\omega_1, \omega_2, \omega_3$ are the components in $(G; \mathbf{i}_1, \mathbf{i}_2, \mathbf{i}_3)$ of the angular velocity $\boldsymbol{\omega}$ [see Eqs. (3.9.12), (3.9.3), and (3.9.25), then there exists a 3×3 matrix $I = (I_{ij})_{i,j=1,2,3}$ such that

$$T = \frac{1}{2} M \dot{\mathbf{x}}_G^2 + \frac{1}{2} \sum_{i,j=1}^{3} I_{ij} \omega_i \omega_j , \tag{3.9.26}$$

where $M = \Sigma_{i=1}^{N} m_i$ and T denotes the system's kinetic energy.

In fact, Eq. (3.9.26) follows from Eq. (3.9.24), since in this case $\boldsymbol{\kappa}^{(i)} \equiv \mathbf{0}$ (since the motion is rigid) and $T^{(\mathrm{in})} \equiv 0 \equiv T_C$, and from Eq. (3.9.18) showing

$$\begin{aligned} T_{\mathrm{rot}} &= \frac{1}{2} \boldsymbol{\omega} \cdot I(\boldsymbol{\omega}) = \frac{1}{2} \sum_{i=1}^{N} m_i \big[(P_i - G) \wedge (\boldsymbol{\omega} \wedge (P_i - G)) \big] \cdot \boldsymbol{\omega} \\ &= \frac{1}{2} \sum_{i=1}^{N} m_i (\boldsymbol{\omega} \wedge (P_i - G))^2 \end{aligned} \tag{3.9.27}$$

by Eq. (3.9.15); then Eq. (3.9.27) permits us to obtain Eq. (3.9.26) as follows.

If θ_i is the angle between $\boldsymbol{\omega}$ and $(P_i - G)$:

$$\begin{aligned} (\boldsymbol{\omega} \wedge (P_i - G))^2 &= (\boldsymbol{\omega})^2 (P_i - G)^2 (\sin \theta_i)^2 \\ &= (\boldsymbol{\omega})^2 (P_i - G)^2 \big(1 - (\cos \theta_i)^2\big) \\ &= (\boldsymbol{\omega})^2 (P_i - G)^2 \left[1 - \frac{(\boldsymbol{\omega} \cdot (P_i - G))^2}{(\boldsymbol{\omega})^2 (P_i - G)^2} \right] \\ &= (\boldsymbol{\omega})^2 (P_i - G)^2 - (\boldsymbol{\omega} \cdot (P_i - G))^2 \\ &= \left(\sum_{l=1}^{3} \omega_l^2 \right) \left(\sum_{l'=1}^{3} (P_i - G)_{l'}^2 \right) \\ &\quad - \sum_{l,l'=1}^{3} \omega_l \omega_{l'} (P_i - G)_l (P_i - G)_{l'} \\ &\equiv \sum_{l,l'=1}^{3} \omega_l \omega_{l'} \left(\left(\sum_{\bar{l}=1}^{3} (P_i - G)_{\bar{l}}^2 \right) \delta_{ll'} - (P_i - G)_l (P_i - G)_{l'} \right); \end{aligned} \tag{3.9.28}$$

hence,

$$I_{ll'} = \sum_{i=1}^{N} m_i \left\{ \left(\sum_{\bar{l}=1}^{3} (P_i - G)_{\bar{l}}^2 \right) \delta_{ll'} - (P_i - G)_l (P_i - G)_{l'} \right\}, \tag{3.9.29}$$

which are constants, $\forall l, l' = 1,2,3$, characteristic of the rigid body because such are the components $(P_i - G)_l$, $i = 1, \ldots, N$, $l = 1,2,3$, of the vectors $(P_i - G)$ in the co-moving frame $(G; \mathbf{i}_1, \mathbf{i}_2, \mathbf{i}_3)$.

We shall come back to Eqs. (3.9.26) and (3.9.29), deducing them independently of the constraint theory, to help the readers who have not paid attention to the proofs of this section.

Exercises and Problems for §3.9

1. Suppose that the reference system $(G; \mathbf{i}_1, \mathbf{i}_2, \mathbf{i}_3)$, with origin at the baricenter of a system of point masses, has a purely translational motion in the reference system $(0; \mathbf{i}, \mathbf{j}, \mathbf{k})$. Show that the system's kinetic energy is $T = T_G + T^{(\text{in})}$.

2. Let $t \to \omega(t)$, $t \in \mathbf{R}_+$, be the angular velocity for the triplet of orthogonal unit vectors $\mathbf{i}_1, \mathbf{i}_2, \mathbf{i}_3$ moving in the reference frame $(0; \mathbf{i}, \mathbf{j}, \mathbf{k})$. Let $t \to \mathbf{y}(t)$, $t \in \mathbf{R}_+$, be a $\mathbf{R}^3$-valued function and write it as $\mathbf{y}(t) = \Sigma_{l=1}^3 y_l(t) \mathbf{i}_l(t)$. Define

$$\mathbf{y}(t) = \frac{d\mathbf{y}}{dt} \quad \text{and} \quad \mathring{\mathbf{y}} = \sum_{l=1}^{3} \dot{y}_l(t) \mathbf{i}_l(t).$$

Show that $\dot{\mathbf{y}} = \mathring{\mathbf{y}} + \omega \wedge \mathbf{y}$ by using $d\mathbf{i}_l/dt = \omega \wedge i_l$. Show also that $\dot{\omega} \equiv \mathring{\omega}$.

3. Compute the components of ω, Eq. (3.9.12), in $(0; \mathbf{i}, \mathbf{j}, \mathbf{k})$ in terms of the Euler angles and their derivatives.

4. Evaluate the matrix I_{jl} for the rigid system in Fig. 3.2(a), assuming $m_0 = m_1 = m_2 = m_3 = 1$ or $m_0 = 1$, $m_1 = 2$, $m_2 = 3$, $m_3 = 4$, taking for the moving frame the one with axes parallel to those in Fig. 3.2(a) and origin in G (Hint: If the direct computation looks cumbersome, replace G by 0 using Problems 5 and 6 below.)

5. Consider a rigid system constrained to have one of its points fixed at the origin of the fixed frame of reference $(0; \mathbf{i}, \mathbf{j}, \mathbf{k})$. Show that if $(0; \mathbf{i}_1, \mathbf{i}_2, \mathbf{i}_3)$ is a co-moving frame, the kinetic energy can be expressed as

$$T = \frac{1}{2} \sum_{l=1}^{3} J_{ll'} \omega_l \omega_{l'},$$

where $J_{ll'}$ are constants depending only on the body's structure.

6. In the context of Problem 5, consider the cases when $0 = G$ and when $0 \neq G$, calling $I_{ll'}$ or $J_{ll'}$ the matrix expressing the kinetic energy in the frames $(0; \mathbf{i}_1, \mathbf{i}_2, \mathbf{i}_3)$ or $(G; \mathbf{i}_1, \mathbf{i}_2, \mathbf{i}_3)$. Show that if $M = \Sigma_{i=1}^N m_i$:

$$J_{ll'} = I_{ll'} + M\big[(G-0)^2 \delta_{ll'} - (G-0)_l (G-0)_{l'}\big].$$

§3.10. General Considerations on the Theory of Constraints

The approximate constraint theory, in the analysis of Proposition 13, §3.8, and Proposition 14, §3.9, still contains some unsatisfactory aspects that we wish to mention.

In applications in which a certain model (Σ, W, λ) of approximately ideal constraint is a good model, the rigidity parameter λ has a well-defined value $\lambda < +\infty$ which is fixed and, therefore, cannot tend to $+\infty$.

Therefore, the problem arises of how to estimate, in terms of λ, the error encountered when approximating the "real motions" $t \to \mathbf{x}_\lambda(t)$ with their limits as $\lambda \to +\infty$ (which are described by the equations of motion relative to ideal constraints, if (Σ, W, λ) is an approximately ideal constraint, i.e., by "simple" equations).

The theory of §3.8, if one carefully looks at the formulas derived in the proof, also provides some estimates of the errors made in the mentioned approximation.

However, it is sufficient to simply look at the calculations made there to realize that if N is a number of the order of magnitude of a few dozens (not to speak of the cases when it is on the order of Avogadro's number, as is sometimes the case), such estimates become ridiculously rough for reasonable values of λ and reasonable models of W.

As usual, the problem of finding good error estimates is a problem that should not be posed in too great generality but should be discussed in connection with precise and concrete questions of a physical nature concerning the behavior of physical entities which, in each case, appear as interesting. Even so, it remains a very difficult question and is a typical problem in statistical mechanics. Except in a few simple cases, it is an essentially open problem from a mathematical viewpoint.

Physicists and engineers have elaborated theories, mathematically nonrigorous consequences of the dynamics of point masses, which allow them to evaluate the errors involved in the perfect constraint approximation in a reasonable way, often experimentally correct.[12] However, it is often only through recourse to experiments that one is able to understand whether a certain constraint can or cannot be approximated by an ideal one.

It is good for the student to keep the above considerations in mind while solving the standard book-made problems concerning the constrained motions in order to appreciate their often purely didactic and abstract nature.

The above discussion, which we will not continue, gives an idea of the depth of notion of ideal constraint, and it can perhaps be useful to understand why long and learned discussions on the argument often take place. So many and so diverse are these arguments that they may leave those who realize their existence for the first time quite surprised.

Other problems naturally arise in the theory of the holonomous constraints. Some of them are:

(i) When an approximate constraint (Σ, W, λ) is not perfect, how can the motion be described in the limit $\lambda \to +\infty$? Is it possible, as the

[12] For instance, elasticity theory has, among other theories, this scope. Of course, elasticity theory can be set up as a mathematically rigorous theory in itself: what is nonrigorous is the connection between elasticity theory and the above microscopic theory of constraints. In other words, elasticity theory is itself a mathematical model which in this case "models" another mathematical model: even such things can happen!

considerations in §3.6 seem to suggest, to treat the constraint in this limit as ideal in the sense of §3.5, modifying the potential energy of the active forces, possibly as a function of the initial datum? See the example of §3.6, following Definition 10.

(ii) In case (i), how can we find the active forces? And how can we estimate the errors involved in the approximation $\lambda = +\infty$?

(iii) How can we treat the case when the constraint model (Σ, W, λ) is ideal but the system moves under the influence of a force which is the sum of the constraint's force, with potential energy λW, and a force law, in $C^\infty(\mathbf{R}^{Nd})$,

$$\mathbf{F}^{(a)} = \mathbf{F}^{(a)}(\boldsymbol{\xi}^{(1)}, \ldots, \boldsymbol{\xi}^{(N)}) \tag{3.10.1}$$

which is not necessarily conservative?

(iv) This is the same as (iii), replacing the constraint model by an approximate conservative model which is not approximately ideal.

(v) This is the same as (i) and (ii) in the situation described by (iii).

The preceding problems are not easy and are open problems to some extent (in the sense that there do not seem to be in the literature any interesting general propositions about them) except problem (iii) which is essentially completely solved by the following proposition, proved exactly in the same way as the analogous Proposition 13, §3.8, p. 187:

15 Proposition. *Let (Σ, W, λ) be a model for an ideal approximate bilateral s-codimensional constraint for N point masses in $\mathbf{R}^d$, with masses $m_1, \ldots, m_N > 0$.*

Consider an initial datum $(\boldsymbol{\eta}_0, \boldsymbol{\xi}_0) \in \mathbf{R}^{2Nd}$ such that $\boldsymbol{\xi}_0 \in \Sigma$. Let $t \to \mathbf{x}(t)$ be the motion that follows this initial datum and develops under the influence of the field of conservative forces with potential energy λW and of a field $\mathbf{F}^{(a)} \in C^\infty_{\lim}(\mathbf{R}^{Nd})$ of uniformly bounded forces, not necessarily conservative.

Then the limit

$$\lim_{\lambda \to +\infty} \mathbf{x}_\lambda(t) = \mathbf{x}(t) \tag{3.10.2}$$

exists for every $t \in \mathbf{R}$ and it is a motion constrained to Σ with initial datum

$$\mathbf{x}(0) = \boldsymbol{\xi}_0, \qquad \dot{\mathbf{x}}(0) = \boldsymbol{\eta}_0^\Sigma \tag{3.10.3}$$

[see Eqs. (3.7.24) and (3.7.26)]. Suppose that for $t \in [t_1, t_2]$ the motion $\mathbf{x}$ dwells in a neighborhood U where a system $(U, \boldsymbol{\Xi})$ of local regular coordinates adapted to Σ is established. Then $\mathbf{x}$ is described in the basis Ω for $(U, \boldsymbol{\Xi})$ by a motion $t \to \mathbf{b}(t)$ verifying:

$$b_1(t) = b_2(t) = \cdots = b_s(t) = 0, \tag{3.10.4}$$

$$\frac{d}{dt}\left(\frac{\partial T}{\partial \alpha_i}(\dot{\mathbf{b}}(t), \mathbf{b}(t))\right) - \left(\frac{\partial T}{\partial \beta_i}\right)(\dot{\mathbf{b}}(t), \mathbf{b}(t)) = \Phi_i(\mathbf{b}(t)), \tag{3.10.5}$$

$\forall i = s+1, \ldots, Nd$, *where*

$$T(\boldsymbol{\alpha}, \boldsymbol{\beta}) = \frac{1}{2} \sum_{i,j=1}^{Nd} g_{ij}(\boldsymbol{\beta})\alpha_i\alpha_j, \qquad (\boldsymbol{\alpha}, \boldsymbol{\beta}) \in \mathbf{R}^{Nd} \times \Omega \qquad (3.10.6)$$

if g is the kinetic matrix associated with the system $(U, \boldsymbol{\Xi})$ and

$$\Phi_i(\boldsymbol{\beta}) = \sum_{k=1}^{N} \mathbf{F}^{(a)(k)}(\boldsymbol{\Xi}(\boldsymbol{\beta})) \cdot \frac{\partial \boldsymbol{\Xi}^{(k)}}{\partial \beta_i}(\boldsymbol{\beta}) \qquad (3.10.7)$$

Observation. The functions (3.10.7) on Ω are called the "force components" of the force $\mathbf{F}^{(a)} = (\mathbf{F}^{(a)(1)}, \ldots, \mathbf{F}^{(a)N})$ in the reference system $(U, \boldsymbol{\Xi})$. The proof of Proposition 15 is a repetition of that of Proposition 13.

We conclude this section with a final comment on the theorems of §3.8 and §3.10.

The condition $\boldsymbol{\xi}_0 \in \Sigma$ appears to be somewhat unnatural, and one would like to change it to "$\boldsymbol{\xi}_0$ close enough to Σ". However, the problem is what is meant by "close enough"?

It is quite clear that the closeness notion should be λ dependent: in fact, we shall call $\boldsymbol{\xi}_0$ close to Σ only if the energy $\lambda W(\boldsymbol{\xi}_0)$ is not too large; i.e., if the initial deviation out of Σ does not involve "too large constraint forces" or "too large elastic deformation energy".

It is then clear that it will be possible to try to prove propositions analogous to Proposition 13 or Proposition 15 by replacing the hypothesis $\boldsymbol{\xi}_0 \in \Sigma$ with the hypothesis that the position of the initial datum is a function of $\lambda, \boldsymbol{\xi}_0(\lambda)$, such that the limit $\lim_{\lambda \to +\infty} \boldsymbol{\xi}_0(\lambda) = \boldsymbol{\xi}_0 \in \Sigma$. In this case, $\lambda W(\boldsymbol{\xi}_0(\lambda))/\lambda \to_{\lambda \to +\infty} 0$ (i.e., the initial "constraint's deformation energy" is not too large, being of lower order with respect to λ, which is the order of the energy of a λ-independent deformation).

The proof of the analogues of Propositions 13 and 15 would be identical under these more general assumptions: this could be realized via a detailed examination of their proofs.

Historical Note: The idea that the constrained systems, ideal or not, could be thought of as limiting cases of nonconstrained systems subject to strong forces is naturally ancient. However, to the best of this author's knowledge, it has been written down in the form of a precise theorem to be interpreted as a proof of the least-action principle in V. I. Arnold, *Methodes Mathematiques de la Mécanique Classique* (see References), pp. 80–82. Here, Arnold expresses this idea and shows how the least-action principle can be deduced through Proposition 13, §3.8, p. 187. This is, in my opinion, the most interesting and deepest of the "proofs" of the least-action principle (and, hence, of the virtual-work principle). There exist other proofs, sometimes very ingenious, which, however, are never more than pseudo-proofs in the sense well described by E. Mach (see References; e.g., in Chapter III, §5.6).

§3.11. The Differential Equations of Hamilton and Lagrange. Analytical Mechanics

Before beginning the study of concrete mechanical problems, it is convenient to deduce from what has already been seen some abstract mathematical structure naturally arising in the context of constraint theory and the least-action principle.

14 Definition. Let $U \subset \mathbf{R}^l \times \mathbf{R}^l \times \mathbf{R}$ be an open set and let $\mathfrak{L} \in C^\infty(U)$ be a real-valued function. $\mathfrak{L}$ will be called a "regular Lagrangian function" on U if the map $\boldsymbol{\Xi}$ transforming the point $(\boldsymbol{\alpha}, \mathbf{q}, t) \in U$ into the point

$$(\boldsymbol{\pi}, \mathbf{q}, t) = \boldsymbol{\Xi}(\boldsymbol{\alpha}, \mathbf{q}, t) \in \mathbf{R}^l \times \mathbf{R}^l \times \mathbf{R} \tag{3.11.1}$$

with

$$\pi_i = \frac{\partial \mathfrak{L}}{\partial \alpha_i}(\boldsymbol{\alpha}, \mathbf{q}, t) \tag{3.11.2}$$

maps the neighborhood U into a neighborhood $V \subset \mathbf{R}^l \times \mathbf{R}^l \times \mathbf{R}$, $V = \boldsymbol{\Xi}(U)$, invertibly and with a nonvanishing Jacobian ("nonsingularly").[13]

If $\mathfrak{L}$ is a regular Lagrangian on U, the equations for the motion $t \to (\dot{\mathbf{q}}(t), \mathbf{q}(t), t)$ in U,

$$\frac{d}{dt}\left(\frac{\partial \mathfrak{L}}{\partial \alpha_i}(\dot{\mathbf{q}}(t), \mathbf{q}(t), t)\right) = \frac{\partial \mathfrak{L}}{\partial q_i}(\dot{\mathbf{q}}(t), \mathbf{q}(t), t), \tag{3.11.3}$$

are called the "Lagrangian differential equations" for the Lagrangian $\mathfrak{L}$.

Since the map (3.11.1) does not really involve $\mathbf{q}$, the above definition makes sense without change if U is an open subset of $\mathbf{R}^l \times \mathbf{T}^l \times \mathbf{R}$ or of $\mathbf{R}^l \times (\mathbf{T}^{l_1} \times \mathbf{R}^{l_2}) \times \mathbf{R}$, with $l_1 + l_2 = l$, provided $C^\infty(U)$ is understood in the natural sense following Definition 13, p. 101, §2.21.

In these cases, V will have to be a subset of $\mathbf{R}^l \times \mathbf{T}^l \times \mathbf{R}$ or of $\mathbf{R}^l \times (\mathbf{T}^{l'_1} \times \mathbf{R}^{l'_2}) \times \mathbf{R}$, $l'_1 + l'_2 = l$, and the points on the tori are to be thought of as described in "angular coordinates", see Definition 12, p. 100, §2.21.

Observation.

(1) The usefulness of the clumsy-looking extension appearing in the second part of the definition can be understood by noting that, for instance, a point mass, with mass $m > 0$, bound to a vertically placed circle with radius R by an ideal constraint and subject to gravity has, if φ is the natural angular coordinate on the circle thought of as $\mathbf{T}^1$, a Lagrangian description in the sense of Definition 14 in terms of $\mathfrak{L}(\alpha, \varphi, t) = \frac{1}{2} m\alpha^2 + mgR\cos\varphi$. In this case, $U = \mathbf{R} \times \mathbf{T}^1 \times \mathbf{R}$ and Eq. (3.11.3) becomes the pendulum equation (g being gravitational acceleration).

[13] The Jacobian determinant coincides with the determinant of the matrix $J_{ij} = (\partial^2 \mathfrak{L}/\partial\alpha_i \partial\alpha_j)(\boldsymbol{\alpha}, \mathbf{q}, t)$, $i, j = 1, \ldots, l$, as is easily seen.

Similarly, a free particle ideally bound to a circle will be described on $\mathbf{R} \times \mathbf{T}^1 \times \mathbf{R}$ by $\mathfrak{L}_0(\alpha, \varphi, t) = \frac{1}{2} m\alpha^2$.

Hence, when the surface Σ generated by an ideal constraint is topologically a torus, we have the possibility of using "global angular coordinates" without having to cover Σ with several local systems of regular coordinates to describe the motions on Σ.

(2) When $\mathfrak{L}$ does not depend explicitly on time, i.e., $\mathfrak{L}(\boldsymbol{\alpha}, \boldsymbol{\beta}, t) \equiv \bar{\mathfrak{L}}(\boldsymbol{\alpha}, \boldsymbol{\beta})$, $\forall(\boldsymbol{\alpha}, \boldsymbol{\beta}, t) \in U$, for some $\bar{\mathfrak{L}}$, we say that $\mathfrak{L}$ is "time independent" and we shall write it without the variable t.

The following proposition holds.

16 Proposition. *Let $\mathfrak{L}$ be a regular Lagrangian on an open subset $U \subset \mathbf{R}^l \times \mathbf{R}^l \times \mathbf{R}$ (or $\mathbf{R}^l \times (\mathbf{R}^{l_1} \times \mathbf{T}^{l_2}) \times \mathbf{R}$, $l_1 + l_2 = l$, $l_1 \geqslant 0$), and let $t \to (\dot{\mathbf{q}}(t), \mathbf{q}(t), t) \in U$ be a motion defined for $t \in [t_1, t_2]$, verifying Eq.* (3.11.3). *Setting*

$$(\boldsymbol{\alpha}(\mathbf{p}, \mathbf{q}, t), \mathbf{q}, t) = \boldsymbol{\Xi}^{-1}(\mathbf{p}, \mathbf{q}, t), \tag{3.11.4}$$

$$H(\mathbf{p}, \mathbf{q}, t) = \sum_{i=1}^{l} p_i \alpha_i(\mathbf{p}, \mathbf{q}, t) - \mathfrak{L}(\boldsymbol{\alpha}(\mathbf{p}, \mathbf{q}, t), \mathbf{q}, t), \tag{3.11.5}$$

for $(\mathbf{p}, \mathbf{q}, t) \in V$ *(see Definition* 14), *the motion in V, image of the preceding motion in U via Eq.* (3.11.1), $t \to (\mathbf{p}(t), \mathbf{q}(t), t) = \boldsymbol{\Xi}(\dot{\mathbf{q}}(t), \mathbf{q}(t), t)$, *verifies the equations*:

$$\dot{p}_i = -\frac{\partial H}{\partial q_i}(\mathbf{p}(t), \mathbf{q}(t), t), \qquad i = 1, \ldots, l \tag{3.11.6}$$

$$\dot{q}_i = \frac{\partial H}{\partial p_i}(\mathbf{p}(t), \mathbf{q}(t), t), \qquad i = 1, \ldots, l. \tag{3.11.7}$$

Observation. Note that Eqs. (3.11.6) and (3.11.7) are equations to which the local existence, uniqueness, and regularity theorems for differential equations can be immediately applied; this is not the case for Eq. (3.11.3), where the highest derivatives do not necessarily appear with constant coefficients: see also the final part of the proof of §3.8 p. 196, to realize that this is really an inconvenience.

Proof. We only discuss the case $U \subset \mathbf{R}^l \times \mathbf{R}^l \times \mathbf{R}$, leaving the other two cases $((U \subset \mathbf{R}^l \times \mathbf{T}^l \times \mathbf{R}$ or $U \subset \mathbf{R}^l \times (\mathbf{R}^{l_1} \times \mathbf{T}^{l_2}) \times \mathbf{R})$, $l_1 + l_2 = l)$ as easy exercises. In any case, the proof is a trivial algebraic check. Equation (3.11.3) can be written by Eq. (3.11.2) as

$$\frac{d}{dt} p_i(t) = \frac{\partial \mathfrak{L}}{\partial q_i}(\dot{\mathbf{q}}(t), \mathbf{q}(t), t), \qquad i = 1, \ldots, l, \tag{3.11.8}$$

but, by Eq. (3.11.5), $\forall i = 1, \ldots, l$,

$$\frac{\partial H}{\partial q_i} = -\frac{\partial \mathfrak{L}}{\partial q_i} - \sum_{j=1}^{l} \frac{\partial \mathfrak{L}}{\partial \alpha_j} \frac{\partial \alpha_j}{\partial q_i} + \sum_{j=1}^{l} p_j \frac{\partial \alpha_j}{\partial q_i}, \tag{3.11.9}$$

and by Eqs. (3.11.2) and (3.11.4), implying $p_j \equiv (\partial \mathcal{L}/\partial \alpha_j)(\boldsymbol{\alpha}(\mathbf{p},\mathbf{q},t),\mathbf{q},t)$, the two sums cancel and Eqs. (3.11.8) and (3.11.9) become Eq. (3.11.6).

Furthermore, by Eqs. (3.11.5) and (3.11.2),

$$\frac{\partial H}{\partial p_i} = \alpha_i + \sum_{j=1}^{l} p_j \frac{\partial \alpha_j}{\partial p_i} - \sum_{j=1}^{l} \frac{\partial \mathcal{L}}{\partial \alpha_j}\frac{\partial \alpha_j}{\partial p_i} = \alpha_i = \dot{q}_i, \tag{3.11.10}$$

i.e., Eq. (3.11.7) follows. mbe

The above proposition suggests a definition.

15 Definition. Let V be an open set in $\mathbf{R}^l \times \mathbf{R}^l \times \mathbf{R}$ (or $\mathbf{R}^l \times \mathbf{T}^l \times \mathbf{R}$ or $\mathbf{R}^l \times (\mathbf{R}^{l_1} \times \mathbf{T}^{l_2}) \times \mathbf{R}$, $l_1 + l_2 = l$, $l_2 > 0$)[14] and let H be a real-valued $C^\infty(V)$ function. H will be said to be a "regular Hamiltonian function" on V if the map $\boldsymbol{\Psi}$ transforming the point $(\boldsymbol{\pi}, \boldsymbol{\beta}, t) \in V$ into the point

$$\boldsymbol{\Psi}(\boldsymbol{\pi}, \boldsymbol{\beta}, t) = (\boldsymbol{\alpha}, \boldsymbol{\beta}, t) \tag{3.11.11}$$

with

$$\alpha_i = \frac{\partial H}{\partial \pi_i}(\boldsymbol{\pi}, \boldsymbol{\beta}, t), \qquad i = 1, \ldots, l, \tag{3.11.12}$$

maps V in a neighborhood $U \subset \mathbf{R}^l \times \mathbf{R}^l \times \mathbf{R}$ (or $\mathbf{R}^l \times \mathbf{T}^l \times \mathbf{R}$ or $\mathbf{R}^l \times (\mathbf{R}^{l_1} \times \mathbf{T}^{l_2}) \times \mathbf{R}$, respectively), $U = \boldsymbol{\Psi}(V)$, invertibly and nonsingularly.[15]

If H is a regular Hamiltonian on V, the equations for the motion $t \to (\mathbf{p}(t), \mathbf{q}(t), t)$ in V:

$$\dot{p}_i(t) = -\frac{\partial H}{\partial \beta_i}(\mathbf{p}(t), \mathbf{q}(t), t), \qquad i = 1, \ldots, l \tag{3.11.13}$$

$$\dot{q}_i(t) = \frac{\partial H}{\partial \pi_i}(\mathbf{p}(t), \mathbf{q}(t), t), \qquad i = 1, \ldots, l \tag{3.11.14}$$

are called the "Hamiltonian differential equations" for the Hamiltonian H.

A proposition similar to Proposition 16 holds.

17 Proposition. *Let H be a regular Hamiltonian function on $V \subset \mathbf{R}^l \times \mathbf{R}^l \times \mathbf{R}$ (or $V \subset \mathbf{R}^l \times \mathbf{T}^l \times \mathbf{R}$ or $V \subset \mathbf{R}^l \times (\mathbf{R}^{l_1} \times \mathbf{T}^{l_2}) \times \mathbf{R}$, $l_1 + l_2 = l$, $l_2 > 0$). Let $t \to (\mathbf{p}(t), \mathbf{q}(t), t)$ be a motion in V defined for $t \in [t_1, t_2]$ and verifying Eqs.* (3.11.13) *and* (3.11.14). *Setting*

$$(\boldsymbol{\pi}(\boldsymbol{\alpha}, \boldsymbol{\beta}, t), \boldsymbol{\beta}, t) = \boldsymbol{\Psi}^{-1}(\boldsymbol{\alpha}, \boldsymbol{\beta}, t) \tag{3.11.15}$$

for $(\boldsymbol{\alpha}, \boldsymbol{\beta}, t) \in U$ *(see definition* 15*) and*

$$\mathcal{L}(\boldsymbol{\alpha}, \boldsymbol{\beta}, t) = \sum_{j=1}^{l} \pi_j(\boldsymbol{\alpha}, \boldsymbol{\beta}, t)\alpha_j - H(\boldsymbol{\pi}(\boldsymbol{\alpha}, \boldsymbol{\beta}, t), \boldsymbol{\beta}, t), \tag{3.11.16}$$

[14] As in Definition 14, this definition makes sense without change if U is an open subset of $\mathbf{R}^l \times \mathbf{T}^l \times \mathbf{R}$ or of $\mathbf{R}^l \times (\mathbf{T}^{l_1} \times \mathbf{R}^{l_2}) \times \mathbf{R}$, $l_1 + l_2 = l$ (see Definition 14 and Observation (1) to Definition 14).

[15] i.e., with nonvanishing Jacobian determinant. Such a Jacobian determinant is easily seen to be the determinant of the matrix $J_{ij} = (\partial^2 H/\partial \pi_i \partial \pi_j)$, $i, j = 1, \ldots, l$.

the motion in U, $t \to (\boldsymbol{a}(t), \boldsymbol{q}(t), t) = \Psi^{-1}(\boldsymbol{p}(t), \boldsymbol{q}(t), t)$, $t \in [t_1, t_2]$, *verifies the equations*:

$$\dot{\mathbf{q}}(t) = \mathbf{a}(t), \qquad i = 1, \ldots, l \tag{3.11.17}$$

$$\frac{d}{dt}\left(\frac{\partial L}{\partial \alpha_i}(\dot{\mathbf{q}}(t), \mathbf{q}(t), t)\right) = \frac{\partial L}{\partial \beta_i}(\dot{\mathbf{q}}(t), \mathbf{q}(t), t). \qquad i = 1, \ldots, l \tag{3.11.18}$$

Proof. The proof is basically identical to that of Proposition 16.

Observations.

(1) Propositions 16 and 17 show that "a system of Lagrangian differential equations, regular on U, is equivalent to a system of Hamiltonian differential equations, regular on V, and vice versa". The sets U and V are related by the relations:

(i) V is the image of U via the map

$$\Xi : (\boldsymbol{\alpha}, \boldsymbol{\beta}, t) \to (\boldsymbol{\pi}, \boldsymbol{\beta}, t) = \left(\frac{\partial \mathcal{L}}{\partial \boldsymbol{\alpha}}(\boldsymbol{\alpha}, \boldsymbol{\beta}, t), \boldsymbol{\beta}, t\right), \tag{3.11.19}$$

where $\mathcal{L}$ is the Lagrangian function on U.

(ii) U is the image of V via the map

$$\Psi : (\boldsymbol{\pi}, \boldsymbol{\beta}, t) \to (\boldsymbol{\alpha}, \boldsymbol{\beta}, t) = \left(\frac{\partial H}{\partial \boldsymbol{\pi}}(\boldsymbol{\pi}, \boldsymbol{\beta}, t), \boldsymbol{\beta}, t\right), \tag{3.11.20}$$

where H is the Hamiltonian function on V corresponding to $\mathcal{L}$.

(iii) $\mathcal{L}$ and the corresponding H are related by

$$H(\boldsymbol{\pi}, \boldsymbol{\beta}, t) = \sum_{i=1}^{l} \pi_i \alpha_i(\boldsymbol{\pi}, \boldsymbol{\beta}, t) - \mathcal{L}(\boldsymbol{\alpha}(\boldsymbol{\pi}, \boldsymbol{\beta}, t), \boldsymbol{\beta}, t), \tag{3.11.21}$$

$$\mathcal{L}(\boldsymbol{\alpha}, \boldsymbol{\beta}, t) = \sum_{i=1}^{l} \pi_i(\boldsymbol{\alpha}, \boldsymbol{\beta}, t)\alpha_i - H(\boldsymbol{\pi}(\boldsymbol{\alpha}, \boldsymbol{\beta}, t), \boldsymbol{\alpha}, t). \tag{3.11.22}$$

(2) In the applications met so far, $\mathcal{L}(\boldsymbol{\alpha}, \boldsymbol{\beta}, t)$ has always had the form

$$\mathcal{L}(\boldsymbol{\alpha}, \boldsymbol{\beta}, t) = \frac{1}{2} \sum_{i,j=1}^{l} g_{ij}(\boldsymbol{\beta})\alpha_i\alpha_j - V(\boldsymbol{\beta}), \tag{3.11.23}$$

where g is a positive definite matrix. $\mathcal{L}$ is usually defined in neighborhoods $U = \mathbf{R}^l \times U_0 \times \mathbf{R}$, $U_0 \subset \mathbf{R}^l$ (or $U_0 \subset \mathbf{T}^l$ or $U_0 \subset \mathbf{R}^{l_1} \times \mathbf{T}^{l_2}$, $l_1 + l_2 = l$, $l_2 > 0$) open. It is easy to check that Eq. (3.11.23) is regular on U because Eq. (3.11.19) [or Eq. (3.11.2)] becomes

$$\pi_i = \sum_{j=1}^{l} g_{ij}(\boldsymbol{\beta})\alpha_j, \qquad i = 1, \ldots, l, \tag{3.11.24}$$

which is obviously invertible and nonsingular if thought of as defining [see Eq. (3.11.19)] a map of U onto $V = \mathbf{R}^l \times U_0 \times \mathbf{R}$: this is so by virtue of Proposition 11, §3.7, p. 183, on the kinetic matrices (implying $\det g(\boldsymbol{\beta}) \neq 0$).

The Hamiltonian function associated with Eq. (3.11.23) is, by Eqs. (3.11.24) and (3.11.21),

$$H(\boldsymbol{\pi}, \boldsymbol{\beta}, t) = \frac{1}{2} \sum_{i,j=1}^{l} \left(g(\boldsymbol{\beta})^{-1} \right)_{ij} \pi_i \pi_j + V(\boldsymbol{\beta}), \tag{3.11.25}$$

where $g(\boldsymbol{\beta})^{-1}$ is the inverse matrix to $g(\boldsymbol{\beta})$.

(3) Note that in the case of the Lagrangian (3.11.23), Eq. (3.11.2) [i.e., Eq. (3.11.24)] is simply the condition expressing that the gradient of the function of $\boldsymbol{\alpha} \in \mathbf{R}^l$,

$$\boldsymbol{\alpha} \to \boldsymbol{\pi} \cdot \boldsymbol{\alpha} - \mathcal{L}(\boldsymbol{\alpha}, \boldsymbol{\beta}, t) \tag{3.11.26}$$

vanishes. A simple discussion shows that for such a value of $\boldsymbol{\alpha}$, Eq. (3.11.26) actually reaches its only absolute maximum [Note that in the case considered here, Eq. (3.11.26) is a quadratic form in $\boldsymbol{\alpha}$ plus a linear form in $\boldsymbol{\alpha}$.] So

$$H(\boldsymbol{\pi}, \boldsymbol{\beta}, t) = \max_{\boldsymbol{\alpha} \in \mathbf{R}^l} (\boldsymbol{\pi} \cdot \boldsymbol{\alpha} - \mathcal{L}(\boldsymbol{\alpha}, \boldsymbol{\beta}, t)), \tag{3.11.27}$$

when $\mathcal{L}$ is given by Eq. (3.11.23) or, more generally, whenever the function of Eq. (3.11.26) has only one stationarity point in $\boldsymbol{\alpha}$ which is a maximum (exercise).

Similarly,

$$\mathcal{L}(\boldsymbol{\alpha}, \boldsymbol{\beta}, t) = \max_{\boldsymbol{\pi} \in \mathbf{R}^l} (\boldsymbol{\pi} \cdot \boldsymbol{\alpha} - H(\boldsymbol{\pi}, \boldsymbol{\alpha}, t)) \tag{3.11.28}$$

if H is given by Eq. (3.11.25) or, more generally, whenever the function of $\boldsymbol{\pi}$ inside the parenthesis on the right-hand side has only one stationarity point in $\boldsymbol{\alpha}$ which is a maximum.

Equations (3.11.27) and (3.11.28) are often called "Legendre's duality" or "Legendre's transformations" on $\mathcal{L}$ or H, respectively.

(4) Definitions 14 and 15 and Propositions 16 and 17 assume a simpler form if one is interested in Lagrangian or Hamiltonian functions not explicitly depending on time and defined on sets U or V of the form $\hat{U} \times J$ or $\hat{V} \times J$ with $J = \{\text{open interval in } \mathbf{R}\}$ and $\hat{U}, \hat{V} \subset \mathbf{R}^{2l}$ or $\mathbf{R}^l \times \mathbf{T}^l$ or $\mathbf{R}^l \times (\mathbf{R}^{l_1} \times \mathbf{T}^{l_2})$, $l_1 + l_2 = l$, open sets.

In such cases, the t parameter can be eliminated from the definition of the sets U, V (replacing them by $\hat{U}, \hat{V}$) and of the maps $\boldsymbol{\Xi}, \boldsymbol{\Psi}$ in Definitions 14 and 15, and $\mathcal{L}$ or H will be functions in $C^\infty(\hat{U})$ or on $C^\infty(\hat{V})$.

We shall call $\mathcal{L}$ or H "time-independent" Lagrangian or Hamiltonian functions and they generate autonomous Lagrangian or Hamiltonian equations via Eqs. (3.11.3), (3.11.6), and (3.11.7).

When $\hat{V} = \mathbf{R}^l \times U_0$, the space $\hat{V}$ is usually called the "phase space" if it is regarded as the initial data space for some time-independent Hamiltonian equations: this name is often used even when $\hat{V}$ is just an open set (not necessarily of the form $\mathbf{R}^l \times U_0$). Similarly, when $\hat{U} = \mathbf{R}^l \times U_0$, the space $\hat{U}$ is called the "data space" if it is regarded as the initial data space for a time-independent Lagrangian equation.

The formal wording of the above concepts is straightforward and will be left to the reader. We shall freely refer to time-independent Lagrangian or Hamiltonian functions and equations on the data space or the phase space.

It is interesting to note the following abstract version of the energy conservation theorem.

18 Proposition. *Consider a system of Hamiltonian equations in a neighborhood* $U = \mathbf{R}^l \times U_0 \times \mathbf{R}$, $U_0 \subset \mathbf{R}^l$ *(or* $U_0 \subset \mathbf{T}^l$ *or* $U_0 \subset \mathbf{R}^{l_1} \times \mathbf{T}^{l_2}$, $l_1 + l_2 = l$*) and let* $(\boldsymbol{\pi}, \boldsymbol{\beta}, t) \to H(\boldsymbol{\pi}, \boldsymbol{\beta}, t)$ *be the (regular) Hamiltonian function.*

If $t \to (\mathbf{p}(t), \mathbf{q}(t), t) \in U$, $t \in [t_1, t_2]$, *is a motion verifying in U the Hamiltonian equations, then*

$$\frac{d}{dt}(H(\mathbf{p}(t), \mathbf{q}(t), t)) = \frac{\partial H}{\partial t}(\mathbf{p}(t), \mathbf{q}(t), t). \tag{3.11.29}$$

Hence, if H is time independent, i.e., $H(\boldsymbol{\pi}, \boldsymbol{\beta}, t) \equiv h(\boldsymbol{\pi}, \boldsymbol{\beta})$ *for some* $h \in C^\infty(\mathbf{R}^l \times U_0)$, *Eq.* (3.11.29) *implies the existence of a constant E depending on the motion under investigation such that*

$$h(\mathbf{p}(t), \mathbf{q}(t)) = E, \qquad t \in [t_1, t_2]. \tag{3.11.30}$$

Observations.

(1) In the cases met so far, the Lagrange function had the form of Eq. (3.11.23) and $\frac{1}{2}\sum_{i,j=1}^{l} g_{ij}(\mathbf{q}(t))\dot{q}_i(t)\dot{q}_j(t)$ had the interpretation of kinetic energy $T(t)$ of the motion, while $V(\mathbf{q}(t))$ had the interpretation of potential energy $V(t)$. Furthermore, the relation between $\mathbf{p}(t)$ and $\mathbf{q}(t)$ was [Eq. (3.11.24)]:

$$\mathbf{p}(t) = g(\mathbf{q}(t))\dot{\mathbf{q}}(t). \tag{3.11.31}$$

Then, by Eq. (3.11.31),

$$\frac{1}{2}\sum_{i,j=1}^{l}\left(g(\mathbf{q}(t))^{-1}\right)_{ij} p_i(t)p_j(t) = \frac{1}{2}\sum_{i,j=1}^{l} g(\mathbf{q}(t))_{ij}\dot{q}_i(t)\dot{q}_j(t) \equiv T(t). \tag{3.11.32}$$

Hence Eq. (3.11.30) becomes

$$T(t) + V(t) = E. \tag{3.11.33}$$

(2) When a system of N point masses without constraints, with the Lagrangian function

$$\mathcal{L}(\boldsymbol{\alpha}, \boldsymbol{\beta}) = \frac{1}{2}\sum_{i=1}^{l} m_i\alpha_i^2 - V(\boldsymbol{\beta}) \tag{3.11.34}$$

is considered, we see that $\boldsymbol{\pi} = (\boldsymbol{\pi}^{(1)}, \ldots, \boldsymbol{\pi}^{(N)})$ with $\boldsymbol{\pi}^{(i)} = m_i\boldsymbol{\alpha}^{(i)}$, $i = 1, \ldots, N$, so that if $t \to \mathbf{x}(t)$, $t \in [t_1, t_2]$, is a system's motion:

$$p_i(t) = m_i\dot{x}_i(t), \qquad i = 1, \ldots, l, \tag{3.11.35}$$

which explains the name "generalized momenta" given to the variables π_i in general.

Here let us mention that the variables π_i are also called the "conjugated momenta" with respect to β_i, $i = 1, \ldots, l$, and the $2l$ variables $(\boldsymbol{\pi}, \boldsymbol{\beta})$ are called "canonical" variables in the phase space of a Hamiltonian equation.

The word conjugation is used here because of the obvious symmetric role played by the $\mathbf{p}$ and $\mathbf{q}$ variables in the Hamiltonian equations. This symmetry could be used to build even more abstract structures associated with the theory of mechanical equations of motion for conservative systems; however, we do not need them here.

PROOF. In fact,

$$\begin{aligned}\frac{d}{dt}H(\mathbf{p}(t),\mathbf{q}(t),t) &= \frac{\partial H}{\partial t}(\mathbf{p}(t),\mathbf{q}(t),t) + \sum_{i=1}^{l}\left(\frac{\partial H}{\partial \pi_i}\dot{p} + \frac{\partial H}{\partial \beta_i}\dot{q}_i\right)\\ &= \frac{\partial H}{\partial t}(\mathbf{p}(t),\mathbf{q}(t),t) + \sum_{i=1}^{l}(\dot{q}_i\dot{p}_i - \dot{p}_i\dot{q}_i) \qquad (3.11.36)\\ &\equiv \frac{\partial H}{\partial t}(\mathbf{p}(t),\mathbf{q}(t),t).\end{aligned}$$

mbe

Another consequence, already mentioned in Problem 10, §2.24, p. 138, of the symmetry of Hamiltonian equations is the following:

19 Proposition. *Let $V = \mathbf{R}^l \times U_0$ with U_0 open subset of $\mathbf{R}^l$ (or $\mathbf{R}^{l_1} \times \mathbf{T}^{l_2}$, $l_1 + l_2 = l$). Let $h \in C^\infty(\mathbf{R}^l \times U_0)$ be a time-independent regular Hamiltonian function.*[16]

Call $S_t(\boldsymbol{\pi}, \boldsymbol{\beta})$ the point into which the initial datum $(\boldsymbol{\pi}, \boldsymbol{\beta})$ evolves through the equations:

$$\begin{aligned}\dot{\mathbf{p}} &= -\frac{\partial h}{\partial \boldsymbol{\beta}}(\mathbf{p},\mathbf{q}),\\ \dot{\mathbf{q}} &= \frac{\partial h}{\partial \boldsymbol{\pi}}(\mathbf{p},\mathbf{q}).\end{aligned} \qquad (3.11.37)$$

Suppose that for $\tau \in [0,t]$, the data $(\boldsymbol{\pi}, \boldsymbol{\beta}) \in A \subset U$ are such that $S_\tau(\boldsymbol{\pi}, \boldsymbol{\beta}) \subset U$, i.e., *$S_\tau A \subset U$ if $\tau \in [0,t]$ or, in other words, the evolution of the points in A takes place inside U for all $\tau \in [0,t]$, and suppose that A is measurable; then*

$$\text{volume } S_t A = \int_{S_t A} d\mathbf{p}\, d\mathbf{q} = \text{volume } A \qquad (3.11.38)$$

Observation. This is read by saying "the Hamiltonian flow preserves the phase space volume" and it is called the "Liouville theorem".

PROOF. This is an immediate consequence of the fact that the Hamiltonian equations have zero divergence:

$$\sum_{i=1}^{l} -\frac{\partial^2 h}{\partial \pi_i \partial \beta_i} + \sum_{i=1}^{l}\frac{\partial^2 h}{\partial \beta_i \partial \pi_i} \equiv 0$$

(see the hint to Problem 10, §2.24, where the argument is given in detail).

[16] i.e., see Observation (4) to Proposition 17; the function $H(\boldsymbol{\pi}, \boldsymbol{\beta}, t) \equiv h(\boldsymbol{\pi}, \boldsymbol{\beta})$ is a regular Hamiltonian on $\tilde{V} = V \times \mathbf{R}$ in the sense of the Definition 15, p. 214.

A corollary to the above proposition is the following.

20 Proposition. *Given the same assumptions as in Proposition* 19, *suppose, also, that the set of the* $(\boldsymbol{\pi}, \boldsymbol{\beta})$ *such that* $h(\boldsymbol{\pi}, \boldsymbol{\beta}) < E$ *is a set* Ω_E *whose closure in* $\mathbf{R}^l \times \mathbf{R}^l$ (*or* $\mathbf{R}^l \times \mathbf{T}^l$ *or* $\mathbf{R}^l \times (\mathbf{R}^{l_1} \times \mathbf{T}^{l_2})$, $l_1 + l = l$) *is contained in* V *and is bounded.*

Then given any $(\boldsymbol{\pi}_0, \boldsymbol{\beta}_0) \in \Omega_E$ *and* $t_0 > 0$ *and a neighborhood* $W \subset \Omega_E$ *of* $(\boldsymbol{\pi}_0, \boldsymbol{\beta}_0)$, *there exists* $t > t_0$ *such that* $S_t W \cap W \neq \emptyset$.

Observation. So "if the energy E surface is bounded" close to every point "inside" it, there is another point coming "as close" after a given time: this is the famous Poincaré recursion theorem. If our systems contains $N(\sim 10^{24})$ points enclosed in a box (modeled by a potential tending very quickly to $+\infty$ outside the box) and if it is initially in a configuration in which all the points are confined to the left half of the box (say), then as close to it as we wish there is a configuration which evolves so that, waiting "long enough", we shall be surprised to see that all the particles will again occupy the left half of the box. This nice paradox ("Zermelo's paradox") gave some problems to Boltzmann.

PROOF. The proof is a very simple consequence of Proposition 19 and is described in greater generality (for divergenceless differential equations) in Problem 11, §2.24, p. 139 (see hint).

In connection with the Hamiltonian equations, the notion of "canonical transformation" plays an important role. A transformation of coordinates is canonical when it leaves the structure of the Hamiltonian equations unchanged. Such a notion has remarkable importance in the algorithms used in the theory of perturbations, which we shall introduce in Chapter 5.

16 Definition. Let V be an open set in $\mathbf{R}^l \times \mathbf{R}^l \times \mathbf{R}$ (or in $\mathbf{R}^l \times \mathbf{T}^l \times \mathbf{R}$ or $\mathbf{R}^l \times (\mathbf{R}^{l_1} \times \mathbf{T}^{l_2}) \times \mathbf{R}$, $l_1 + l_2 = l$) and let H be a regular Hamiltonian function on V (see Definition 15).

Suppose that on V a C^∞ map is defined such that:

(i) The image of $(\mathbf{p}, \mathbf{q}, t) \in V$ has the form $(\boldsymbol{\pi}, \boldsymbol{\kappa}, t) = C(\mathbf{p}, \mathbf{q}, t)$, i.e., C is an "isochronous map" since it does not affect t.

(ii) The map C maps V onto $W = C(V)$, which is an open subset of $\mathbf{R}^l \times \mathbf{R}^l \times \mathbf{R}$ (or $\mathbf{R}^l \times \mathbf{T}^l \times \mathbf{R}$ or $\mathbf{R}^l \times (\mathbf{R}^{l'_1} \times \mathbf{T}^{l'_2}) \times \mathbf{R}$, $l'_1 + l'_2 = l$), and it is invertible and nonsingular,[17] i.e., C is a regular change of coordinates on V.

(iii) There is a real-valued function $H' \in C^\infty(W)$ such that if $t \to (\mathbf{p}(t), \mathbf{q}(t), t) \in V$, $t \in [t_1, t_2]$, is any motion in V verifying the Hamiltonian equations with Hamiltonian H, then $t \to (\boldsymbol{\pi}(t), \boldsymbol{\kappa}(t), t) = C(\mathbf{p}(t), \mathbf{q}(t), t) \in W$, $t \in [t_1, t_2]$, verifies the Hamiltonian equations relative to H' and vice versa.

[17] i.e., its Jacobian determinant does not vanish. Hence, C^{-1} has the same properties by the implicit function theorem.

One says that C is a "canonical transformation of V in W with respect to the pair of conjugate Hamiltonians H and H'".

Observation. In general, if a map C is canonical for the pair H, H', it will not be canonical for the pair $(\overline{H}, H'')$ no matter how H'' is chosen, if $\overline{H} \neq H$ (for an example, see Problem 38, at the end of this section).

It is therefore tempting to call "completely canonical" a map C between V and W such that for *any* choice of a Hamiltonian function H on V, one can find a conjugated Hamiltonian function H' on W in *some* standard way (Levi-Civita).

We shall make the notion of "complete canonicity" precise only in the simple case of "time-independent" canonical transformations.

17 Definition. Let $V = \hat{V} \times \mathbf{R}$ be an open subset of $\mathbf{R}^l \times \mathbf{R}^l \times \mathbf{R}$ (or of $\mathbf{R}^l \times \mathbf{T}^l \times \mathbf{R}$ or of $\mathbf{R}^l \times (\mathbf{R}^{l_1} \times \mathbf{T}^{l_2}) \times \mathbf{R}$, $l_1 + l_2 = l$) and let C have the form $C(\mathbf{p}, \mathbf{q}, t) = (\hat{C}(\mathbf{p}, \mathbf{q}), t)$ with $\hat{C}$ being a regular change of coordinates between $\hat{V}$ and its image $\hat{W}$ ($\hat{W} \subset \mathbf{R}^l \times \mathbf{R}^l$ or $\hat{W} \subset \mathbf{R}^l \times \mathbf{T}^l$ or $\hat{W} \subset \mathbf{R}^l \times (\mathbf{R}^{l'_1} \times \mathbf{T}^{l'_2})$, $l'_1 + l'_2 = l$).

We shall say that $\hat{C}$ is a "completely canonical time-independent" or, simply, "completely canonical" transformation if C is a transformation which conjugates canonically every regular Hamiltonian function H on V with

$$H'(\boldsymbol{\pi}, \boldsymbol{\kappa}, t) = H\big(\hat{C}^{-1}(\boldsymbol{\pi}, \boldsymbol{\kappa}), t\big), \qquad \forall (\boldsymbol{\pi}, \boldsymbol{\kappa}) \in \hat{W}. \tag{3.11.39}$$

Observation. In other words, a time-independent completely canonical transformation is one with the property that any Hamiltonian function is conjugated to itself computed in the new coordinates.

The following proposition provides a very general method of construction of canonical transformations and of completely canonical transformations.

21 Proposition. *Let H be a regular Hamiltonian function on the open set V in $\mathbf{R}^l \times \mathbf{R}^l \times \mathbf{R}$ (or in $\mathbf{R}^l \times \mathbf{T}^l \times \mathbf{R}$ or $\mathbf{R}^l \times (\mathbf{R}^{l_1} \times \mathbf{T}^{l_2}) \times \mathbf{R}$, $l_1 + l_2 = l$).*

Let $F \in C^\infty(\mathbf{R}^{2l+1})$ be a function denoted

$$(\mathbf{q}, \boldsymbol{\kappa}, t) \to F(\mathbf{q}, \boldsymbol{\kappa}, t) \in \mathbf{R}. \tag{3.11.40}$$

For $i = 1, \ldots, l$, set

$$p_i = \frac{\partial F}{\partial q_i}(\mathbf{q}, \boldsymbol{\kappa}, t), \qquad \pi_i = -\frac{\partial F}{\partial \kappa_i}(\mathbf{q}, \boldsymbol{\kappa}, t) \tag{3.11.41}$$

and assume that Eq. (3.11.41) *establishes a one-to-one map C_F between $(\mathbf{p}, \mathbf{q}, t) \in V$ and $(\boldsymbol{\pi}, \boldsymbol{\kappa}, t) = C_F(\mathbf{p}, \mathbf{q}, t) \in W$. Suppose that C_F is a regular change of coordinates*[18] *between V and W ($W \subset \mathbf{R}^l \times \mathbf{R}^l \times \mathbf{R}$ or $\mathbf{R}^l \times \mathbf{T}^l \times$*

[18] i.e., it is one-to-one and with nonvanishing Jacobian determinant.

$\mathbf{R}$ or $\mathbf{R}^l \times (\mathbf{R}^{l_1'} \times \mathbf{T}^{l_2'}) \times \mathbf{R}$, $l_1' + l_2' = l$). Then if we define $(\mathbf{p}(\boldsymbol{\pi}, \boldsymbol{\kappa}, t), \mathbf{q}(\boldsymbol{\pi}, \boldsymbol{\kappa}, t), t) \equiv C_F^{-1}(\boldsymbol{\pi}, \boldsymbol{\kappa}, t)$ and

$$H'(\boldsymbol{\pi}, \boldsymbol{\kappa}, t) = H(\mathbf{p}(\boldsymbol{\pi}, \boldsymbol{\kappa}, t), \mathbf{q}(\boldsymbol{\pi}, \boldsymbol{\kappa}, t), t) + \frac{\partial F}{\partial t}(\mathbf{q}(\boldsymbol{\pi}, \boldsymbol{\kappa}, t), \boldsymbol{\kappa}, t), \tag{3.11.42}$$

the map C_F is a canonical transformation of V onto W with respect to H and H'.

Observations.

(1) Note that F is required to be in $C^\infty(\mathbf{R}^{2l+1})$ even when V is in $\mathbf{R}^l \times \mathbf{T}^l \times \mathbf{R}$ or in $\mathbf{R}^l \times (\mathbf{R}^{l_1} \times \mathbf{T}^{l_2}) \times \mathbf{R}$, $l_1 + l_2 = l$. Recall that the points on a torus are, in the present contexts, always thought of as described in "flat or angular coordinates" (i.e., by thinking of the torus $\mathbf{T}^l$ as obtained by identifying mod 2π the points of $\mathbf{R}^l$ (see Definition 14, p. 212).

(2) From the proof of Proposition 21, it will follow that other coordinate transformations analogous to Eq. (3.11.41) are canonical: for instance, from a function $\Phi \in C^\infty(\mathbf{R}^{2l+1})$,

$$(\mathbf{q}, \boldsymbol{\pi}, t) \to \Phi(\mathbf{q}, \boldsymbol{\pi}, t) \in \mathbf{R}, \tag{3.11.43}$$

one builds a canonical transformation[19] C_Φ by setting, $\forall i = 1, 2, \ldots, l$,

$$p_i = \frac{\partial \Phi}{\partial q_i}(\mathbf{q}, \boldsymbol{\pi}, t), \qquad \kappa_i = \frac{\partial \Phi}{\partial \pi_i}(\mathbf{q}, \boldsymbol{\pi}, t), \tag{3.11.44}$$

$$H'(\boldsymbol{\pi}, \boldsymbol{\kappa}, t) = H(\mathbf{p}(\boldsymbol{\pi}, \boldsymbol{\kappa}, t), \mathbf{q}(\boldsymbol{\pi}, \boldsymbol{\kappa}, t), t) + \frac{\partial \Phi}{\partial t}(q(\pi, \kappa, t), \pi, t), \tag{3.11.45}$$

where we denote $(\mathbf{p}(\boldsymbol{\pi}, \boldsymbol{\kappa}, t), \mathbf{q}(\boldsymbol{\pi}, \boldsymbol{\kappa}, t), t) = C_\Phi(\boldsymbol{\pi}, \boldsymbol{\kappa}, t)$.

Similarly, with analogous notations, if $\Psi \in C^\infty(\mathbf{R}^{2l+1})$,

$$(\mathbf{p}, \boldsymbol{\kappa}, t) \to \Psi(\mathbf{p}, \boldsymbol{\kappa}, t) \in \mathbf{R}, \tag{3.11.46}$$

one defines a canonical transformation[19] C_Ψ by setting, $\forall i = 1, \ldots, l$:

$$q_i = -\frac{\partial \Psi}{\partial p_i}(\mathbf{p}, \boldsymbol{\kappa}, t), \qquad \pi_i = -\frac{\partial \Psi}{\partial \kappa_i}(\mathbf{p}, \boldsymbol{\kappa}, t), \tag{3.11.47}$$

$$H' = H + \frac{\partial \Psi}{\partial t},$$

and if $R \in C^\infty(\mathbf{R}^{2l+1})$,

$$(\mathbf{p}, \boldsymbol{\pi}, t) \to R(\mathbf{p}, \boldsymbol{\pi}, t) \in \mathbf{R} \tag{3.11.48}$$

defines a canonical transformation[19] C_R by setting, $\forall i = 1, \ldots, l$:

$$q_i = -\frac{\partial R}{\partial p_i}(\mathbf{p}, \boldsymbol{\pi}, t), \qquad \kappa_i = \frac{\partial R}{\partial \pi_i}(\mathbf{p}, \boldsymbol{\pi}, t) \tag{3.11.49}$$

$$H' = H + \frac{\partial R}{\partial t}.$$

[19] between regions where the regularity, invertibility and nonsingularity requirements for the maps C_Φ (or, see below, C_Ψ, C_R) similar to those put on C_F are verified. Such regions V, W may be very small or even nonexistent: in the last cases no canonical transformation is really associated with F, Φ, Ψ, R.

(3) However, it will appear that the class of canonical transformations built starting from F as described by Proposition 20 is not essentially less ample than that obtained by adding to it the canonical transformations associated with the functions Φ, Ψ, and R as described in the preceding observation.

With some natural exceptions, to every F it is possible to associate a Φ, a Ψ, and an R producing the same canonical transformation.

(4) If F is time independent, then C_F defines a completely canonical (time-independent) map.

(5) F is in general called a "generating function" of C_F. So one calls also the functions Φ, Ψ, R above.

PROOF. The proof by direct check is of course possible. However, if performed straightforwardly, it quickly becomes quite intricate.

It is certainly more convenient to proceed in the following elegant fashion, which also exhibits a new form of the least-action principle: the "Hamilton's principle".

Let $\mathfrak{M}^V = \mathfrak{M}_{t_1 t_2}(\mathbf{p}_1, \mathbf{q}_1 t_1; \mathbf{p}_2, \mathbf{q}_2 t_2; V) = \{$set of the motions in V having the form $t \to \mathbf{m}(t) = (\mathbf{p}(t), \mathbf{q}(t), t) \in V$, $t \in [t_1, t_2]$, and such that $\mathbf{p}(t_1) = \mathbf{p}_1$, $\mathbf{q}(t_1) = \mathbf{q}_1$, $\mathbf{p}(t_2) = \mathbf{p}_2$, $\mathbf{q}(t_2) = \mathbf{q}_2\}$ ("synchronous motions in V"). Consider the function on $\mathfrak{M}^V$:

$$S(\mathbf{m}) = \int_{t_1}^{t_2} \left(\sum_{i=1}^{l} p_i(t) \dot{q}_i(t) - H(\mathbf{p}(t), \mathbf{q}(t), t) \right) dt \tag{3.11.50}$$

and, with the methods of §2.2 and §3.4 by now familiar, it is easy to see that the stationarity condition for S on $\mathbf{m}$ in $\mathfrak{M}^V$ is simply that the motion $\mathbf{m}$ verifies the Hamiltonian equations in V with Hamiltonian function H [which are, essentially, the Euler–Lagrange equations for the action of Eq. (3.11.50)].

Now let $t \to \boldsymbol{\mu}(t) = (\boldsymbol{\pi}(t), \boldsymbol{\kappa}(t), t) = C_F(\mathbf{p}(t), \mathbf{q}(t), t)$, $t \in [t_1, t_2]$ be the image motion of a motion $\mathbf{m} \in \mathfrak{M}^V$: it is a motion in

$$\mathfrak{M}^W = C_F(\mathfrak{M}^V) = \mathfrak{M}_{t_1 t_2}(C_F(\mathbf{p}_1, \mathbf{q}_1, t_1), C_F(\mathbf{p}_2, \mathbf{q}_2, t_2); W).$$

If $\boldsymbol{\mu}$ verifies the Hamiltonian equations for some Hamiltonian H' on W in $\mathfrak{M}^W$, it must make the action

$$\Sigma(\boldsymbol{\mu}) = \int_{t_1}^{t_2} \left\{ \sum_{i=1}^{l} \pi_i(t) \dot{\kappa}_i(t) - H'(\boldsymbol{\pi}(t), \boldsymbol{\kappa}(t), t) \right\} dt \tag{3.11.51}$$

stationary. A *sufficient* condition for this to occur is that

$$S(\mathbf{m}) = \Sigma(C_F(\mathbf{m})) + \text{constant}, \qquad \forall \mathbf{m} \in \mathfrak{M}^V, \tag{3.11.52}$$

of course. Equation (3.11.52) is certainly verified if the differential form on V:

$$\sum_{i=1}^{l} p_i \, dq_i - H(\mathbf{p}, \mathbf{q}, t) \, dt \tag{3.11.53}$$

and the differential form

$$\sum_{i=1}^{l} \pi_i \, d\kappa_i - H'(\boldsymbol{\pi}, \boldsymbol{\kappa}, t) \, dt \tag{3.11.54}$$

are transformed into each other by the transformation C_F up to a total differential.

This condition can be imposed by requiring the existence of a function G on W such that

$$\sum_{i=1}^{l} p_i \, dq_i - H(\mathbf{p}, \mathbf{q}, t) \, dt = \sum_{i=1}^{l} \pi_i \, d\kappa_i - H'(\boldsymbol{\pi}, \boldsymbol{\kappa}, t) \, dt + dG, \tag{3.11.55}$$

where $(\mathbf{p}, \mathbf{q}, t)$ are to be thought of as functions of $(\boldsymbol{\pi}, \kappa, t)$ via the transformation C_F.

To use Eq. (3.11.55), it is more convenient to think of G as a function of $(\mathbf{q}, \boldsymbol{\kappa}, t)$ instead of $(\boldsymbol{\pi}, \boldsymbol{\kappa}, t)$ via Eq. (3.11.41); i.e., set $\tilde{G}(\mathbf{q}, \boldsymbol{\kappa}, t) = G((\partial F/\partial \mathbf{q})(\mathbf{q}, \boldsymbol{\kappa}, t), \boldsymbol{\kappa}, t)$. Then it follows from Eq. (3.11.55) that

$$d\tilde{G} = \sum_{i=1}^{l} p_i \, dq_i - \sum_{i=1}^{l} \pi_i \, d\kappa_i - (H - H') \, dt, \tag{3.11.56}$$

so we realize that Eq. (3.11.56) holds if and only if there is a function $\tilde{G}$ which is such that, $\forall i = 1, \ldots, l$,

$$p_i = \frac{\partial \tilde{G}}{\partial q_i}, \qquad \pi_i = -\frac{\partial \tilde{G}}{\partial \kappa_i}, \qquad H - H' = \frac{\partial \tilde{G}}{\partial t}, \tag{3.11.57}$$

thinking the coefficients of the right-hand side differentials in Eq. (3.11.56) as functions of $\mathbf{q}, \boldsymbol{\kappa}, t$, via Eq. (3.11.41). Such relations are obviously satisfied by the function F, setting $\tilde{G} \equiv F$. mbe

Observations. Subtracting the differential $d(\sum_{i=1}^{l} p_i q_i)$ from both sides of Eq. (3.11.56) and thinking of

$$\Psi = F - \sum_{i=1}^{l} p_i q_i \tag{3.11.58}$$

as a function of $\mathbf{p}, \boldsymbol{\kappa}, t$ via Eq. (3.11.41)[20] one finds that the transformation C_F may also be thought of as C_Ψ described by Eq. (3.11.47).

Similarly, setting

$$\Phi = F + \sum_{i=1}^{l} \pi_i \kappa_i \tag{3.11.59}$$

and thinking of Φ as a function of $(\mathbf{q}, \boldsymbol{\pi}, t)$ via Eq. (3.11.41),[20] one finds that C_F may also be thought of as C_Φ described by Eq. (3.11.44).

Finally, setting

$$R = F + \sum_{i=1}^{l} \pi_i \kappa_i - \sum_{i=1}^{l} p_i q_i \tag{3.11.60}$$

[20] assuming that the necessary inversions can actually be made.

and thinking of R as a function of $(\mathbf{p}, \boldsymbol{\pi}, t)$ via Eq. (3.11.41),[20] one finds that the C_F may also be thought of as C_R described by Eq. (3.11.49).

In the problems at the end of this section, we show that the inversions mentioned above (see footnote 20) can be performed at least in small regions under the respective conditions that the matrices $\partial^2 F/\partial\kappa_i\partial\kappa_j$, $\partial^2 F/\partial\kappa_i\partial q_j$, $\partial^2 F/\partial q_i\partial q_j$ have nonvanishing determinants.

This somewhat clarifies Observation (3) to Proposition 21. A complete clarification arises from the analysis of Problems (6)–(11) at the end of this section. The reader should try to think of these observations again after looking at the problems.

A simple corollary to the proof of Proposition 21 is the following.

22 Proposition. *Let* $(\boldsymbol{\pi}, \boldsymbol{\kappa}) \to C(\boldsymbol{\pi}, \boldsymbol{\kappa})$ *be a nonsingular invertible* C^∞ *map of the open set* $V \subset \mathbf{R}^{2l}$ *or* $\mathbf{R}^l \times (\mathbf{R}^{l_1} \times \mathbf{T}^{l_2})$, $l_1 + l_2 = l$, *onto* $W \subset \mathbf{R}^{2l}$ *or* $\mathbf{R}^l \times (\mathbf{R}^{l'_1} \times \mathbf{T}^{l'_2})$, $l'_1 + l'_2 = l$. *Write* C *explicitly as*

$$\mathbf{p} = \mathbf{P}(\boldsymbol{\pi}, \boldsymbol{\kappa}), \qquad \mathbf{q} = \mathbf{Q}(\boldsymbol{\pi}, \boldsymbol{\kappa}) \tag{3.11.61}$$

and suppose that the differential form on V,

$$\boldsymbol{\pi} \cdot d\boldsymbol{\kappa} - \mathbf{p} \cdot d\mathbf{q} \equiv \sum_{i=1}^{l} (\pi_i\, d\kappa_i - p_i\, dq_i) \tag{3.11.62}$$

is exact, i.e., *setting*

$$\begin{aligned} X_i &= \mathbf{P}(\boldsymbol{\pi}, \boldsymbol{\kappa}) \cdot \frac{\partial \mathbf{Q}}{\partial \pi_i}(\boldsymbol{\pi}, \boldsymbol{\kappa}), \\ Y_i &= \mathbf{P}(\boldsymbol{\pi}, \boldsymbol{\kappa}) \cdot \frac{\partial \mathbf{Q}}{\partial \kappa_i}(\boldsymbol{\pi}, \boldsymbol{\kappa}) - \pi_i, \end{aligned} \tag{3.11.63}$$

assume that, $\forall i = 1, \ldots, l$,

$$\frac{\partial X_i}{\partial \pi_j} = \frac{\partial X_j}{\partial \pi_i}, \qquad \frac{\partial X_i}{\partial \kappa_j} = \frac{\partial Y_j}{\partial \pi_i}, \qquad \frac{\partial Y_i}{\partial \kappa_j} = \frac{\partial Y_j}{\partial \kappa_i}. \tag{3.11.64}$$

Then C *is a completely canonical time-independent map.*

In particular, if $\boldsymbol{\pi} \cdot d\boldsymbol{\kappa} - \mathbf{p} \cdot d\mathbf{q} = 0$, *the map* C *is completely canonical: it is called "homogeneous" in the variables* $(\boldsymbol{\kappa}, \mathbf{q})$.

Similar results hold if $\mathbf{p} \cdot d\mathbf{q} + \boldsymbol{\kappa} \cdot d\boldsymbol{\pi}$, *or* $-\mathbf{q} \cdot d\mathbf{p} + \boldsymbol{\kappa} \cdot d\boldsymbol{\pi}$, *or* $-\mathbf{q} \cdot d\mathbf{p} - \boldsymbol{\pi} \cdot d\boldsymbol{\kappa}$ *are exact differentials: one similarly defines the homogeneous canonical maps with respect to* $(\mathbf{q}, \boldsymbol{\pi})$, *or* $(\mathbf{p}, \boldsymbol{\pi})$, *or* $(\mathbf{p}, \boldsymbol{\kappa})$.

Observations.

(1) If C is as above and homogeneous in $(\boldsymbol{\kappa}, \mathbf{q})$ variables, then it cannot be generated by a generating function $F(\boldsymbol{\kappa}, \mathbf{q})$ as in Eq. (3.11.41). The vanishing of the differential in Eq. (3.11.62) and the Eqs. (3.11.41), (3.11.42) written as

$$dF = \boldsymbol{\pi} \cdot d\mathbf{k} - \mathbf{p} \cdot d\mathbf{q} + (H' - H)\, dt \tag{3.11.65}$$

imply that $dF = (H' - H)\,dt$, i.e., $H' - H$, is a function of t only and so is

F as well, so that Eq. (3.11.41) gives $\boldsymbol{\pi} = \mathbf{0}$, $\mathbf{p} = \mathbf{0}$ which is obviously not usable to define an invertible map between $\mathbf{q}, \mathbf{p}$ and $\boldsymbol{\kappa}, \boldsymbol{\pi}$.

(2) If C is homogeneous as in Observation (1), it might be generated by functions $\Phi(\boldsymbol{\pi}, \mathbf{q})$ or $\Psi(\boldsymbol{\kappa}, \mathbf{p})$ or $R(\boldsymbol{\pi}, \mathbf{p})$: for instance, the map $p = a\pi$, $q = a^{-1}\kappa$ is homogeneous in (q, κ) variables (as $p\,dq = \pi\,d\kappa$) and it is generated by $\Psi(p, \kappa) = a^{-1}p\kappa$!

(3) A very interesting homogeneous canonical mapping is met in the theory of the motion of a rigid body (see Problems to §4.11).

PROOF. If $\boldsymbol{\pi} \cdot d\boldsymbol{\kappa} - \mathbf{p} \cdot d\mathbf{q}$ is an exact differential, one sees, by going through the proof of Proposition 11, that Eq. (3.11.56) can be satisfied by choosing $H = H'$. mbe

Observations.

(1) From the proof of Proposition 22 and from Eqs. (3.11.41), (3.11.44), (3.11.47), and (3.11.49), we see that a sufficient condition in order that any Hamiltonian H on V is conjugated to a Hamiltonian H' on W given by

$$H'(\boldsymbol{\pi}, \boldsymbol{\kappa}, t) = H\big(C^{-1}(\boldsymbol{\pi}, \boldsymbol{\kappa}, t)\big) \tag{3.11.66}$$

is that the transformation C mapping V onto W be generated by a time-independent function F or Φ or Ψ or R or be homogeneous in the sense of Proposition 22.

(2) The interest in canonical transformations consists of the fact that sometimes it is possible to solve the Hamiltonian equations by finding a canonical transformation transforming the system of Hamiltonian equations into a conjugate system with "trivial" Hamiltonian H', i.e., trivially soluble (e.g., $H' \equiv 0$ or $H'(\boldsymbol{\pi}, \boldsymbol{\kappa}) = h(\boldsymbol{\kappa})$ which yield trivial Hamiltonian equations, indeed).

A concrete method to look for such a transformation ("Hamilton–Jacobi method") consists of trying to find, using Proposition 21, a function F defined in a suitable neighborhood $\Omega \subset \mathbf{R}^{2l+1}$ such that, $\forall(\mathbf{q}, \boldsymbol{\kappa}, t)$,

$$H' = H\left(\frac{\partial F}{\partial \mathbf{q}}(\mathbf{q}, \boldsymbol{\kappa}, t), \mathbf{q}, t\right) + \frac{\partial F}{\partial t}(\mathbf{q}, \boldsymbol{\kappa}, t) = 0 \tag{3.11.67}$$

or, for some h,

$$H' = H\left(\frac{\partial F}{\partial \mathbf{q}}(\mathbf{q}, \boldsymbol{\kappa}, t), \mathbf{q}, t\right) + \frac{\partial F}{\partial t}(\mathbf{q}, \boldsymbol{\kappa}, t) = h(\boldsymbol{\kappa}). \tag{3.11.68}$$

Equations (3.11.67) and (3.11.68) are to be considered as equations in which $\boldsymbol{\kappa}$ is a parameter and, therefore, as partial differential equations for a function $(\mathbf{q}, t) \to f(\mathbf{q}, t)$:

$$H\left(\frac{\partial f}{\partial \mathbf{q}}(\mathbf{q}, t), \mathbf{q}, t\right) + \frac{\partial f}{\partial t}(\mathbf{q}, t) = 0 \tag{3.11.69}$$

or

$$H\left(\frac{\partial f}{\partial \mathbf{q}}(\mathbf{q}, t), \mathbf{q}, t\right) + \frac{\partial f}{\partial t}(\mathbf{q}, t) = \text{constant} \tag{3.11.70}$$

("Hamilton–Jacobi" equations). We wish to find solutions to Eq. (3.11.69) or Eq. (3.11.70) which depend on l parameters $\boldsymbol{\kappa} = (\kappa_1, \ldots, \kappa_l)$.

If we were able to find such a family, i.e., if we were able to find a C^∞ solution F of Eq. (3.11.69) or Eq. (3.11.70) depending on $(\mathbf{q}, \boldsymbol{\kappa}, t) \in \Omega$ $= \{$some open set in $\mathbf{R}^{2l+1}\}$, we could consider the transformation (3.11.41) and hope that it defines a canonical map C_F of some open set $\tilde{V} \subset V$ into a set W: the transformation C_F would then transform the Hamiltonian equations associated with H into trivial Hamiltonian equations in W, with Hamiltonian function 0 or $h(\boldsymbol{\kappa})$.

However, it is obvious that the difficulty of solving Eqs. (3.11.67) and (3.11.68) in the above sense is equivalent to or harder than solving the original Hamiltonian equations, and one should not think of Eq. (3.11.67) or Eq. (3.11.68) as a miraculous equation.

The usefulness of the above discussion on Hamilton–Jacobi equations consists of the possibility of finding approximation algorithms to the solutions to Eq. (3.11.67) or Eq. (3.11.68) and, therefore, to the original Hamiltonian equations, which are essentially different from the general recursive method seen in §2.3, valid for solving the most general first-order differential equations.

The methods devised to construct recursively successive approximations to Eq. (3.11.67) or Eq. (3.11.68) are methods in which the particular structure of the Hamiltonian equations is explicitly used. It is therefore not too surprising that they reveal themselves to be quite appropriate to the analysis of such equations and provide better approximations for a given amount of formal work done.

The reader can convince himself of the truth of the above statement only by seeing some concrete problems studied on the basis of approximation algorithms to the solutions of the Hamilton–Jacobi equations. The best known and most celebrated of these methods or some of its variants can be found in the theory of the motion of heavenly bodies and, more generally, in the stability theory of the motion of conservative systems. An important example will be illustrated in §5.9–§5.12. Some "trivial" examples can be found in the upcoming problems.

Exercises, Problems, and Complements for §3.11

1. Construct the canonical transformation with generating function $f(q, \kappa) = (m/2)\,\omega q^2 tg\kappa$, $q \in \mathbf{R}$, $\kappa \in \mathbf{T}^1$, and note that the above transformation simplifies the Hamiltonian $H(p,q) = (p^2/2m) + (\omega^2 m/2)q^2$. Find the harmonic oscillator's motion with the help of this transformation.

2. Consider a one-dimensional mechanical system consisting of a point mass with mass m subject to a force with potential energy $V \in C^\infty(\mathbf{R})$. Assume that $V(0) = 0$, $V'(q) \neq 0$ if $q \neq 0$, $V(q) \to_{|q| \to \infty} + \infty$.

Consider the canonical transformation $(p, q) \to (E, \tau)$ with generating function

$$f(E, q) = \int_0^q \sqrt{2m(E - V(q'))}\, dq'$$

near a point $(\bar{p},\bar{q})$ where $(\bar{p}^2/2m)+V(\bar{q})>0$. Write it explicitly, finding the Hamiltonian in the new coordinates and the physical interpretation of the τ coordinate and the E coordinate. (Hint: Do not try to "compute" the integral, but rather perform the necessary differentiations on the integral and then use the formulae for the one-dimensional motions found in §2.7.)

3. Interpret f defined in Problem 2 as a solution of the equation $(1/2m)[(\partial f)^2/\partial q]+V(q)=E$ and interpret this as a one-parameter solution of the Hamilton–Jacobi equation for the mechanical system in Problem 2, in the sense of Eq. (3.11.67), of the form $f(E,q)-Et$ (or, in the sense of Eq. (3.11.68), of the form $f(E,q)$ with $h(\kappa)=E$).

4. In the context of Problem 2, define, for $E>0$,

$$\omega(E)=\pi\Big/\int_{q_-(E)}^{q_+(E)}\left(\frac{2}{m}(E-V(q'))\right)^{1/2}dq',$$

where $q_\pm(E)$ are the roots of $E-V(q)=0$. For $E>0$, let

$$a(E)=\int_{E_0}^{E}\frac{1}{\omega(E')}\,dE'$$

and let $A\to e(A)$ be its inverse function (such that $e(a(E))\equiv E$). Consider the canonical transformation $(p,q)\to(A,\varphi)$ with generating function

$$S(A,q)=\int_{q_0}^{q}\sqrt{2m(e(A)-V(q'))}\,dq'.$$

near some $(p,q)\neq(0,0)$.

Compute the new Hamiltonian and show that the canonical transformation may be extended to a canonical transformation of $\mathbf{R}^2/\{0,0\}$ into $(0,\int_{E_0}^{+\infty}(dE'/\omega(E')))\times\mathbf{T}^1$. (Hint: Let $\varphi=(\partial S/\partial A)(A,q)$ mod 2π, $p=(\partial S/\partial q)(A,q)$ and show that this is a C^∞ map between the indicated sets.)

5. Show that the transformation in Problem 4 is a natural generalization of the Cartesian-polar coordinates in the plane (Hint: Consider the special case $(p^2+q^2)/2$, where it gives exactly the Cartesian-polar coordinates. Draw the curves $A=$ const and compare them with the circles.)

The angle defined in Problem 4 is called the "average anomaly" and, therefore, the time evolution of the average anomaly is always a uniform rotation.

6. Let A,C be two $l\times l$ symmetric matrices and let B be a $l\times l$ matrix. On $\mathbf{R}^{2l}$, define

$$F(\mathbf{q},\boldsymbol{\kappa},t)=\tfrac{1}{2}A\mathbf{q}\cdot\mathbf{q}+\tfrac{1}{2}C\boldsymbol{\kappa}\cdot\boldsymbol{\kappa}+B\boldsymbol{\kappa}\cdot\mathbf{q}.$$

Show that if $\det B\neq 0$, C_F is well defined as a completely canonical map between $\mathbf{R}^{2l+1}$ and itself. Show that its jacobian determinant is 1, at least in the case $l=1$ (the case $l>1$ is discussed in §3.12). (Hint: First deal in detail with the case $l=1$ when A, B, C are simply numbers.)

7. In the context of Problem 6, show that $\det B\neq 0$ is a necessary and sufficient condition for C_F to be defined. Hence, $F(q,\kappa,t)=(\kappa^2+q^2/2)$ does not define a canonical transformation

8. Let F be as in Problem 6. Construct explicitly the other generating functions for the canonical map [Eqs. (3.11.58), (3.11.59), and (3.11.60)] and check that, via Eqs. (3.11.44), (3.11.46), and (3.11.49), they all generate the same completely canonical transformation if $\det A,\det B,\det C\neq 0$. Check that all the inversions mentioned in connection with the quoted formulae can actually be performed, in the present situation.

9.* Let F be as in Proposition 21. Let $(\mathbf{p}_0, \mathbf{q}_0, t_0)$, $(\boldsymbol{\pi}_0, \boldsymbol{\kappa}_0, t_0)$ be two points related by Eq. (3.11.41). Define the $l \times l$ matrices

$$A_{ij} = \frac{\partial^2 F}{\partial q_i \partial q_j}(\mathbf{q}_0, \boldsymbol{\kappa}_0, t_0), \qquad B_{ij} = \frac{\partial^2 F}{\partial \kappa_i \partial q_i}(\mathbf{q}_0, \boldsymbol{\kappa}_0, t_0), \qquad C_{ij} = \frac{\partial^2 F}{\partial \kappa_i \partial \kappa_j}(\mathbf{q}_0, \boldsymbol{\kappa}_0, t_0).$$

Show that if $\det B \neq 0$, then the map C_F is defined in a neighborhood of $(\mathbf{p}_0, \mathbf{q}_0, t_0)$. (Hint: Use Problem 6 and apply the implicit function theorem to take into account that F no longer has constant second derivatives as in the cases of Problems 6 and 8.)

10.* Is it possible that C_F exists near $(\mathbf{p}_0, \mathbf{q}_0, t_0)$ in the context of Problem 9 when $\det B = 0$? (Answer: No; hence, $F(q, \kappa, t) = f(q) + g(\kappa)$ does not define a canonical transformation. Check this directly.)

11. Show that the invertibility properties of the matrices A, B, C mentioned in connection with the quotation of Eqs. (3.11.58), (3.11.59), and (3.11.60) in Problem 8 are necessary, in general, in order to be able to express C_F as C_Φ, C_Ψ, or C_R. (Hint: Consider for $l = 1$, $F(q, \kappa) = q\kappa$: this is a case where $A = C = 0$ and the inversion cannot be realized. In this case, it is impossible to generate the corresponding canonical transformation with a function $\Psi(p, \kappa, t)$, since the transformation is easily checked to be homogeneous with respect to (κ, p) as $q\,dp + \pi\,d\kappa = 0$. See Proposition 22, p. 224, and the subsequent Observation (1). Similar considerations hold for $\kappa^2 + \kappa q$, as $q\,dp + \pi\,dk = -\kappa\,d\kappa$, which is equally impossible for reasons similar to those used in Observation (1) to Proposition 22.)

12. Consider $\mathbf{x} \in \mathfrak{M}_{t_1,t_2}(\boldsymbol{\xi}_1, \boldsymbol{\xi}_2)$ and $\mathbf{y} \in \mathcal{V}_{\mathbf{x}}$. Call the variation $\mathbf{y}$ "nontrivial" if $t \to \mathbf{z}(t) = (\partial \mathbf{y}/\partial \epsilon)(t, 0)$, $t \in [t_1, t_2]$, is such that $\mathbf{z} \not\equiv \mathbf{0}$. Define $\mathbf{x}$ to be a "strict local minimum" for the action A relative to $\mathfrak{M}$ if for every variation $y \in \mathcal{V}_{\mathbf{x}}(\mathfrak{M})$ which is nontrivial, there exists $\eta_y > 0$ such that $A(y_\epsilon) > A(\mathbf{x})$, $\forall \epsilon \neq 0$, $|\epsilon| < \eta_y$, or if $\mathcal{V}_{\mathbf{x}}(\mathfrak{M})$ only contains trivial variations. Examine the proof of Proposition 38, §2.24, p. 133, to show that in the statements of Proposition 38, §2.24, Proposition 6, §3.3, p. 153, Proposition 8, §3.5, p. 163, one can replace the words "local minimum" by "strict local minimum" (Hint: Just look at the proof of Proposition 38 and Eq. (2.24.33).)

13. Let $t \to \mathbf{x}(t)$, $t \in [0, T]$, be a motion verifying the equations of motion associated with the Lagrangian (3.11.23) and taking place in the set $U_0 \subset \mathbf{R}^l$ where Eq. (3.11.23) is considered. Let E be the energy of the motion $\mathbf{x}$ [see Eq. (3.11.33)]. Consider the motion $\mathbf{x}$ for $t \in [t_1, t_2] \subset [0, T]$ and fix t_1: so $\mathbf{x} \in \mathfrak{M}_{t_1,t_2}(\mathbf{x}(t_1), \mathbf{x}(t_2); E)$ $=$ {space of the motions in $\mathfrak{M}_{t_1,t_2}(\mathbf{x}(t_1), \mathbf{x}(t_2))$ taking place in U_0 and with energy E}.

Show that $\mathbf{x}$ makes stationary and (if t_2 is close enough to t_1) strictly locally minimal (see Problem 12) the action

$$\tilde{A}(\mathbf{x}) = \int_{t_1}^{t_2} T(t)\,dt$$

in $\mathfrak{M}_{t_1,t_2}(\mathbf{x}(t_1), \mathbf{x}(t_2); E)$. (Hint: Simply note that if A is stationary or strictly locally minimal on $\mathbf{x}$ in $\mathfrak{M}_{t_1,t_2}(\mathbf{x}(t_1), \mathbf{x}(t_2))$, it is such in any $\mathfrak{M} \subset \mathfrak{M}_{t_1,t_2}(\mathbf{x}(t_1), \mathbf{x}(t_2))$. Then observe that

$$A(\mathbf{x}') = \int_{t_1}^{t_2} (T(t) - V(t))\,dt = 2\tilde{\mathbf{A}}(\mathbf{x}') - E(t_2 - t_1)$$

if $\mathbf{x}' \in \mathfrak{M}_{t_1,t_2}(\mathbf{x}(t_1), \mathbf{x}(t_2); E)$, as $T(t) + V(t) \equiv E$.)

14. Show through examples that it is possible that the set $\mathfrak{M}_{t_1,t_2}(\mathbf{x}(t_1), \mathbf{x}(t_2); E)$ considered in Problem 13) contains finitely many points (hence, $\mathcal{V}_{\mathbf{x}}(\mathfrak{M}_{t_1,t_2}(\mathbf{x}(t_1), \mathbf{x}(t_2); E)$ only contains trivial variations). Nevertheless, even in such cases the statement of Problem 12 is not an empty one: for instance, deduce from Problem 12 that the free motion in $\mathbf{R}^d$ takes places along straight lines. (Hint: Let $t \to \mathbf{x}(t)$ be a free motion in $\mathbf{R}^d$, then $T(t) = \frac{1}{2}\dot{\mathbf{x}}(t)^2$ and $V(t) = 0$. If $t \to \mathbf{x}(t)$ were not a straight line, show that $\mathcal{V}_{\mathbf{x}}(\mathfrak{M}_{t_1,t_2}(\mathbf{x}(t_1), \mathbf{x}(t_2); E))$ would not consist only of trivial variations: however, $\tilde{A}(\mathbf{x}') \equiv E(t_2 - t_2)$ for all $\mathbf{x}' \in \mathfrak{M}_{t_1,t_2}(\mathbf{x}(t_1), \mathbf{x}(t_2); E)$ and, therefore, $\mathbf{x}$ could not be a strict local minimum!)

15. On $\mathbf{R}^{Nd}$, consider the metric associated with the scalar product of Eq. (3.7.1) and let dl be the line element of the curve in $\mathbf{R}^{Nd}$ with equations $t \to \mathbf{x}(t)$, $t \in [t_1, t_2]$, which is a motion of energy E of N point masses, with masses $m_1, \ldots, m_N > 0$, under the influence of a force with potential energy $V \in C^\infty(\mathbf{R}^{Nd})$. Show that

$$A(\mathbf{x}) = \int_{\mathbf{x}_i}^{\mathbf{x}_2} \sqrt{2(E - V(\boldsymbol{\xi}(l)))}\, dl - E(t_2 - t_1) \qquad \text{if} \quad dl = \sqrt{2T(t)}\, dt,$$

where $A(\mathbf{x}) = \int_{t_1}^{t_2}(T(t) - V(t))\, dt$ and $l \to \boldsymbol{\xi}(l)$ is the description of the trajectory of $\mathbf{x}$ in terms of the curvilinear abscissa l on it and the integral $\int_{\mathbf{x}_1}^{\mathbf{x}_2}$ is the curvilinear integral on the trajectory.

16. Consider N point masses in $\mathbf{R}^d$, with masses $m_1, \ldots, m_N$. Assume that such a system is subject to an active force with potential energy $V^{(a)}$ and to an ideal holonomous constraint to a regular l-dimensional surface $\Sigma \subset \mathbf{R}^{Nd}$.

On Σ, consider two points $\boldsymbol{\xi}_1, \boldsymbol{\xi}_2$ and the set of the C^∞ curves, $\mathfrak{M}_{0,1}(\boldsymbol{\xi}_1, \boldsymbol{\xi}_2 \mid \Sigma)$, on Σ joining $\boldsymbol{\xi}_1$ and $\boldsymbol{\xi}_2$ parameterized by some parameter varying between 0 and 1. Given $E \in \mathbf{R}$, define on $\mathfrak{M}_{0,1}(\boldsymbol{\xi}_1, \boldsymbol{\xi}_2 \mid \Sigma)$ the function defined by the curvilinear integral on the curve $\hat{\mathbf{x}} \in \mathfrak{M}_{0,1}(\boldsymbol{\xi}_1, \boldsymbol{\xi}_2 \mid \Sigma)$ as

$$S(\hat{\mathbf{x}}) = (\hat{\mathbf{x}}) \int_{\boldsymbol{\xi}_1}^{\boldsymbol{\xi}_2} \sqrt{(E - V(\boldsymbol{\xi}))}\, ds,$$

where ds is the line element on Σ, measured with the kinetic energy metric $ds^2 = \sum_{i=1}^N m_i (d\mathbf{x}^{(i)})^2$, Eq. (3.7.1).

Show that the least-action principle implies that S is stationary on the curve $\hat{\mathbf{x}}$ if and only if $\hat{\mathbf{x}}$ is a trajectory of a motion with energy E leaving $\boldsymbol{\xi}_1$ and reaching $\boldsymbol{\xi}_2$, ("Maupertuis' principle"). (Hint: Consider a local system of local coordinates near Σ permitting representation of the points of $\Sigma \cap U$ through some parametric equations $\boldsymbol{\xi} = \mathbf{x}(a)$, $\mathbf{a} = (a_l, \ldots, a_1) \in \Omega \subset \mathbf{R}^l$. Suppose, for simplicity, that $\hat{\mathbf{x}}(t) \subset U \cap \Sigma$, $\forall t \in [t_1, t_2]$. Assume $\mathfrak{L}$ to be a Lagrangian of the form of Eq. (3.11.23) describing the system in these coordinates.

Write the stationarity conditions of S in $\mathfrak{M}_{0,1}(\boldsymbol{\xi}_1, \boldsymbol{\xi}_2 \mid \Sigma)$ for the curve $\hat{\mathbf{x}}$ with parametric equations $\tau \to \hat{\mathbf{a}}(\tau)$, $\tau \in [0, 1]$, in the chosen coordinates. Then, in the resulting Euler–Lagrange equations, perform the change of coordinates $\tau \leftrightarrow t$:

$$t = \int_0^\tau \sqrt{\frac{\sum_{i,j=1}^l g_{ij}(\mathbf{a}(\theta)) a_i'(\theta) a_j'(\theta)}{2(E - V(\mathbf{X}(\mathbf{a}(\theta))))}}\, d\theta,$$

where the prime denotes differentiation with respect to τ or θ. One finds that the motion $t \to \mathbf{X}(\mathbf{a}(\tau(t)))$ has energy E and verifies the Lagrangian equations for $\mathfrak{L}$.)

17. In the context of Problem 16, show that the Maupertuis' principle can be interpreted as saying that the motions developing on Σ with energy E take place along the geodesics of Σ with respect to the metric on Σ:

$$dh = \sqrt{2(E - V(\xi))}\, ds,$$

where ds is the kinetic energy metric on Σ. (We recall that by definition, a curve on Σ is called a geodesic for a given line element on Σ if it makes stationary the distance between any two of its points measured along the curve itself using the given line element.)

In other words, if we call the distance between two points $\xi_1, \xi_2 \in \Sigma$, measured with the line element dh along a given curve on Σ, with the name "mechanical path with energy E on Σ", we can say that the "motions with energy E on Σ take place along trajectories making stationary the mechanical path with energy E".

As usual, it is possible to show that the mechanical systems of the type considered here have the property that a trajectory of any of their motions with energy E, taking place on Σ, not only makes the mechanical path stationary but actually strictly minimizes it on short enough segments.

18.* By the assumptions of Problem 17, let $s \to \hat{\mathbf{x}}(s)$, $s \in [\bar{s}_1, s]$, be a geodesic segment on Σ for the line element dh. Suppose that $E - V(\hat{x}(s)) > 0$, $\forall s \in [s_1, \bar{s}]$. Show that there is $\bar{\bar{s}} > s_1$ such that if $s_2 \in [s_1, \bar{\bar{s}}]$, the curve $s \to \hat{\mathbf{x}}(s)$, $s \in [s_1, s_2]$, makes strictly locally minimal the mechanical path with energy E between $\hat{\mathbf{x}}(s_1)$ and $\hat{\mathbf{x}}(s_2)$.

19. A point mass, with mass $m > 0$, is bound to a surface $\Sigma \subset \mathbf{R}^3$ by an ideal constraint and it is subject to no other forces. Show that as a consequence of the Maupertuis principle, Problems 16–18, the point mass runs on Σ in such a way that if two points on its trajectory are close enough, then the trajectory itself is the one minimizing the distance on Σ between the two points, i.e., the trajectory is the shortest path on Σ joining the two points, the distance being measured in the ordinary $\mathbf{R}^3$ sense ("geodesics' or Fermat's principle"). (Hint: Note that dh and ds are now proportional, and use Problem 18.)

20. Consider the line segment $(dx^2 + dy^2)/y^2$ defined on the half-plane $y > 0$. Determine its geodesics by thinking of them via the mechanical interpretation, permitted by Problem 16, which allows us to regard them as the zero energy motions of the mechanical system with Lagrangian $\mathfrak{L} = \frac{1}{2}(\dot{x}^2 + \dot{y}^2) + \frac{1}{2}1/y^2$.

21. Calling the geodesics of the Problem 20 "straight lines for the geometry defined by the line element ds", check the truth or the falsity of the following statements:

(i) Given two points in the half-plane $y > 0$, there is one and only one straight line through them.

(ii) Two points in the $y > 0$ region are joined by just one straight line segment (if a straight line segment is defined as a connected closed subset of a straight line).

(iii) Given a point, and a straight line not containing it, there exists just one straight line containing the point and "parallel" to the first straight line (i.e., without common points with it).

22. Same as Problems 20 and 21 for the geometries associated with the following line elements:

(i) $ds^2 = (x^2 + y^2)(dx^2 + dy^2)$, $\quad (x, y) \in \mathbf{R}^2 \backslash \{\mathbf{0}\}$;

(ii) $ds^2 = (1 - x^2 - y^2)^\alpha (dx^2 + dy^2)$, $\quad (x, y) \in \mathbf{R}^2$, $x^2 + y^2 < 1$, $\alpha \in \mathbf{R}$;

(iii) $ds^2 = \dfrac{dx^2 + dy^2}{\sqrt{x^2 + y^2}}$, $\quad (x, y) \in \mathbf{R}^2 \backslash \{\mathbf{0}\}$.

23. Same as Problems 20 and 21 for the geometry defined on the surface of a sphere by the line element induced by the Euclidean distance of R^3; i.e., $ds^2 = d\theta^2 + \sin^2\theta\, d\varphi^2$ in polar coordinates.

24.* Consider the geometry defined in the half-plane $y > 0$ by the line element of Problem 20. Define a "triangle" as a figure formed by the three points pairwise connected by geodesic segments. Given a triangle, denote α, β, γ the three angles relative to its three vertices (defined as the angles between the tangents to the two geodesic segments meeting at the various vertices). The quantity $\alpha + \beta + \gamma - \pi$ is called the "geodesic defect": show that it is < 0. Show that the same quantity computed in the analogous situation for the sphere's geometry of Problem 23 is > 0.

25. A light ray moves in a plane with refraction index

$$n(x, y) = \sqrt{1 - \epsilon y^2} \qquad \text{if} \quad |y| < 1.$$

Using Fermat's principle, show that the ray proceeds along a sinusoidal path, if it is assumed that the ray starts at the origin with an initial direction close to the horizontal. Recall, for this purpose, that Fermat's principle says that the rays follow a path that makes stationary the "optical path" between any two trajectory's points in the set of the paths joining them. The optical path, in a medium with index of refraction $n(x, y)$, associated with the curve $\hat{\mathbf{x}} \in \mathfrak{M}_{0,1}(\xi_1, \xi_2)$, is

$$(\hat{\mathbf{x}}) \int_{\xi_1}^{\xi_2} n(x, y)\, ds, \qquad ds = \sqrt{dx^2 + dy^2}\ .$$

(Hint: Interpret the above problem as a mechanical problem via Problems 16 and 17; then it becomes very simple.)

Hence, via Maupertuis' principle, the problem of the determination of a light path can be interpreted as a purely mechanical problem.

26. Solve the problems at the end of §18, §20, §21, §24, §32, §39, §44 in Landau and Lifsciz, "Mécanique", see references.

27. Perform the de Legendre transformation on the Lagrangian $\mathfrak{L} = \sqrt{\dot{x}^2 + \dot{y}^2}$ and explain why one gets strange results.

28. Consider the function on $\mathbf{R}^{2l}$: $\mathfrak{L}(\dot{\mathbf{q}}, \mathbf{q}) = \frac{1}{2} A\dot{\mathbf{q}} \cdot \dot{\mathbf{q}} + \frac{1}{2} C\mathbf{q} \cdot \mathbf{q} + B\dot{\mathbf{q}} \cdot \mathbf{q}$ where A, C are $l \times l$ symmetric matrices and B is an $l \times l$ matrix. Under which conditions on A, B, C is $\mathfrak{L}$ a regular Lagrangian on $\mathbf{R}^{2l}$? In these cases, write the corresponding Hamiltonian function. Similarly, consider the function on $\mathbf{R}^{2l}$: $H(\mathbf{p}, \mathbf{q}) = \frac{1}{2} A\mathbf{p} \cdot \mathbf{p} + \frac{1}{2} C\mathbf{q} \cdot \mathbf{q} + B\mathbf{p} \cdot \mathbf{q}$ and find the conditions for H to be a regular Hamiltonian and write the corresponding Lagrangian.

29. In the cases when the Lagrangian in Problem 28 is regular, write the energy conservation theorem, Proposition 18, §3.11, in terms of $\dot{\mathbf{q}}$ and $\mathbf{q}$. (Hint: $H = \mathbf{p} \cdot \dot{\mathbf{q}} - \mathfrak{L}(\dot{\mathbf{q}}, \mathbf{q})$, and then express $\mathbf{p}$ in terms of $\mathbf{q}, \dot{\mathbf{q}}$ and use Proposition 18.)

30. Show that the time-independent completely canonical linear transformations on $\mathbf{R}^{2l}$ form a group $\mathcal{S}_l$ under the natural composition law.

31. The set $\mathcal{G}$ of the linear completely canonical transformations of $\mathbf{R}^2$ with generating functions $\Phi(\pi, q) = \frac{1}{2} a\pi^2 + \frac{1}{2} cq^2 + b\pi q$, $b \neq 0$, which we denote (a, c, b), does not form a subgroup of $\mathcal{S}_1$. Prove this by finding the composition law of (a, c, b) and (a', c', b'). Show that $(a, c, b) \cdot (a', c', b') \in \mathcal{G}$ if and only if $a'c \neq 1$. (Hint: The composition law is: if $\delta = ac - b^2$, $\delta' = a'c' - \delta'^2$,

$$(a, c, b) \cdot (a', c', b') = \left(\frac{a - a'\delta}{1 - a'c}, \frac{-c' + c\delta'}{1 - a'c}, \frac{bb'}{1 - a'c} \right).)$$

32. Same as Problem 31 for the class $\mathcal{G}'$ of the canonical transformations generated by functions $F(\kappa, q) = (a/2)q^2 + (c/2), \kappa^2 + bq\kappa$, $b \neq 0$. (Hint: The composition law is now, for suitable δ, δ':

$$(a,c,b)\cdot(a',c',b') = \left(\frac{aa'-\delta'}{a+c'}, \frac{cc'-\delta}{a+c'}, \frac{-bb'}{a+c'}\right).)$$

33. Find a generating function for the transformation $(\boldsymbol{\pi}, \boldsymbol{\kappa}) \to (\mathbf{p}, \mathbf{q})$ defined by $\mathbf{p} = R\boldsymbol{\pi}$, $\mathbf{q} = (R^T)^{-1}\boldsymbol{\kappa}$, where R is a nonsingular $l \times l$ matrix: this transformation is completely canonical. (Hint: Look for a generating function like $\Phi(\boldsymbol{\pi}, \mathbf{q}) = B\boldsymbol{\pi} \cdot \mathbf{q}$ with B being an $l \times l$ matrix.)

34. Let $\boldsymbol{\kappa} \to \mathbf{q} = \mathbf{f}(\boldsymbol{\kappa})$ be an invertible nonsingular transformation of $\mathbf{R}^l$ onto itself. Find out how to define $\mathbf{p} = \mathbf{F}(\boldsymbol{\pi}, \boldsymbol{\kappa})$ so that the map $(\boldsymbol{\pi}, \boldsymbol{\kappa}) \to (\mathbf{p}, \mathbf{q}) = (\mathbf{F}(\boldsymbol{\pi}, \boldsymbol{\kappa}), \mathbf{f}(\boldsymbol{\kappa}))$ will be completely canonical. (Answer: If $R_{ij}(\boldsymbol{\kappa}) = (\partial f_i / \partial \kappa_j)(\boldsymbol{\kappa})$, then $\mathbf{p} = ((R(\boldsymbol{\kappa}))^T)^{-l}\boldsymbol{\pi})$.

35. Let $f \in C^\infty(\mathbf{R}^l)$ be multiperiodic with periods 2π. Is the function $\Phi(\mathbf{A}', \boldsymbol{\varphi}) = \mathbf{A}' \cdot \boldsymbol{\varphi} + f(\boldsymbol{\varphi})$ a generating function of a canonical map of $\mathbf{R}^l \times \mathbf{T}^l$ onto $\mathbf{R}^l \times \mathbf{T}^l$? Find a sufficient condition.

36. Let $(\mathbf{A}', \boldsymbol{\varphi}) \to f(\mathbf{A}', \boldsymbol{\varphi})$ be a $C^\infty(\mathbf{R}^{2l})$ function multiperiodic with periods 2π in the $\boldsymbol{\varphi}$'s. Suppose that the transformation

$$\boldsymbol{\varphi}' = \boldsymbol{\varphi} + \frac{\partial f}{\partial \mathbf{A}'}(\mathbf{A}', \boldsymbol{\varphi}) \bmod 2\pi$$

establishes a nonsingular invertible map of $\mathbf{T}^l$ onto itself for each $\mathbf{A}' \in \mathbf{R}^l$. Suppose, also, that the transformation

$$\mathbf{A} = \mathbf{A}' + \frac{\partial f}{\partial \boldsymbol{\varphi}}(\mathbf{A}', \boldsymbol{\varphi})$$

establishes a nonsingular invertible map of $\mathbf{R}^l$ onto itself for each $\boldsymbol{\varphi} \in \mathbf{T}^l$.

Show that the function $\Phi(\mathbf{A}', \boldsymbol{\varphi}) = \mathbf{A}' \cdot \boldsymbol{\varphi} + f(\mathbf{A}', \boldsymbol{\varphi})$ generates a completely canonical map of $\mathbf{R}^l \times \mathbf{T}^l$ onto itself.

37. Find a "local version" of Problem 36 when $\mathbf{R}^l \times \mathbf{T}^l$ is replaced by $V \times \mathbf{T}^l$, $V \subset \mathbf{R}^l$ open.

38. Consider the maps $(p,q) \to C(p,q) = (A, \varphi)$ and $(p,q) \to \tilde{C}(p,q) = (B, \varphi)$ of $\mathbf{R}^2/\{\mathbf{0}\} \leftrightarrow \mathbf{R}_+ \times \mathbf{T}^1$ defined by

$$\varphi = \text{polar angle of } (p,q),$$

$$A = \frac{p^2+q^2}{2}, \qquad B = \sqrt{A}\,.$$

Show that while C is completely canonical, the map $\tilde{C}$ is such that the Hamiltonians $H = \frac{1}{2}(p^2 + q^2)$ and $H' = B$ are canonically conjugated by it, but the Hamiltonian $H = \frac{1}{2}p^2$ has no canonically conjugated Hamiltonian with respect to $\tilde{C}$. (Hint: C is studied in Problems 1 and 2. Show that a general measure-preserving flow on $\mathbf{R}^2/\{\mathbf{0}\}$ is not mapped by $\tilde{C}$ into a measure-preserving flow on $\mathbf{R}_+ \times \mathbf{T}^1$: the evolution associated with $H = \frac{1}{2}p^2$ is actually mapped by $\tilde{C}$ into a non-measure-preserving one. So the image flow cannot be a Hamiltonian flow since the latter would, instead, preserve the measure by the Liouville theorem, Proposition 19, in the case of a time-dependent Hamiltonian (or by an extension of Proposition 19 in the time-dependent case; see Problem 39).)

39. Extend Proposition 19, §3.11, to the case of time-dependent Hamiltonian equations. (Hint: Replace the semigroup property $S_t S_{t'} = S_{t+t'}$ used in the proof of Proposition 19 (see Problem 10, §2.24) by the more general relation $S(t,t') \cdot S(t',t_0) = S(t,t_0)$, $t > t' > t_0$, where $S(t,t')$ denotes the solution map of the nonautonomous Hamiltonian equations when the initial data are assigned at t'. The proof proceeds unchanged.)

40. Let $(\mathbf{q},t) \to S(\mathbf{q},t)$ be defined and C^∞ on a set $U \times J$, $U \subset \mathbf{R}^l$, $J \subset \mathbf{R}$, both open and connected. Let H be a regular Hamiltonian on $V = \mathbf{R}^l \times U \times J$ and suppose that S is a solution to the Hamilton–Jacobi equation

$$H\left(\frac{\partial S}{\partial \mathbf{q}}(\mathbf{q},t),\mathbf{q},t\right) + \frac{\partial S}{\partial t}(\mathbf{q},t) = 0.$$

Consider the differential equation for $t \to \mathbf{q}(t)$,

$$\dot{\mathbf{q}} = H\left(\frac{\partial S}{\partial \mathbf{q}}(\mathbf{q},t),\mathbf{q},t\right), \qquad \mathbf{q}(t_0) = \mathbf{q}_0$$

and suppose that for all $(\mathbf{q}_0,t_0) \in U \times J$, one can solve it for t near t_0 by $t \to \mathbf{q}(t)$. Show that setting

$$\mathbf{p}(t) = \frac{\partial S}{\partial \mathbf{q}}(\mathbf{q}(t),t),$$

the functions $t \to (\mathbf{p}(t),\mathbf{q}(t))$ are solutions to the Hamiltonian equations verifying the initial data

$$\mathbf{q}(t_0) = \mathbf{q}_0, \qquad \mathbf{p}(t_0) = \frac{\partial S}{\partial \mathbf{q}}(\mathbf{q}_0,t_0);$$

i.e., "every solution to the Hamilton–Jacobi equation provides a bundle of solutions to the Hamiltonian equation". (Hint: Check it directly by substitution.)

§3.12. Completely Canonical Transformations: Their Structure

Among the canonical transformations, it is clear that the completely canonical transformations are very simple and interesting [see Eq. (3.11.39)].

It is therefore somewhat important to find some general results about the structure of such transformations.

Let $V \subset \mathbf{R}^l \times \mathbf{R}^l$ or $\mathbf{R}^l \times \mathbf{T}^l$ or $\mathbf{R}^l \times (\mathbf{R}^{l_1} \times \mathbf{T}^{l_2})$, $l_1 + l_2 = l$, be an open set which is regarded as the phase space of the Hamiltonian systems of differential equations with regular Hamiltonian functions $H \in C^\infty(V)$ (see Observation (4) to Proposition 17, p. 216.)

18 Definition. Let V, W be open sets as above and let C be an invertible nonsingular[21] C^∞ map between V and W. Denote C as

$$\begin{aligned} \mathbf{p} &= \mathbf{P}(\boldsymbol{\pi},\boldsymbol{\kappa}),\\ \mathbf{q} &= \mathbf{Q}(\boldsymbol{\pi},\boldsymbol{\kappa}). \end{aligned} \tag{3.12.1}$$

[21] i.e., with nonvanishing Jacobian determinant.

Let $(\mathbf{p}_0, \mathbf{q}_0) = C(\boldsymbol{\pi}_0, \boldsymbol{\kappa}_0)$, $(\mathbf{p}_0, \mathbf{q}_0) \in V$, $(\boldsymbol{\pi}_0, \boldsymbol{\kappa}_0) \in W$, be two C-corresponding points.

We define the "linearized C map near $(\boldsymbol{\pi}_0, \boldsymbol{\kappa}_0)$" as the map of $\mathbf{R}^{2l} \times \mathbf{R}^{2l}$:

$$\begin{aligned} \mathbf{p} &= \mathbf{p}_0 + A(\boldsymbol{\pi} - \boldsymbol{\pi}_0) + B(\boldsymbol{\kappa} - \boldsymbol{\kappa}_0), \\ \mathbf{q} &= \mathbf{q}_0 + C(\boldsymbol{\pi} - \boldsymbol{\pi}_0) + D(\boldsymbol{\kappa} - \boldsymbol{\kappa}_0), \end{aligned} \tag{3.12.2}$$

where A, B, C and D are four $l \times l$ matrices:

$$\begin{aligned} A_{ij} &= \frac{\partial P_i}{\partial \pi_j}(\boldsymbol{\pi}_0, \boldsymbol{\kappa}_0), \qquad & B_{ij} &= \frac{\partial P_i}{\partial \kappa_j}(\boldsymbol{\pi}_0, \boldsymbol{\kappa}_0), \\ C_{ij} &= \frac{\partial Q_i}{\partial \pi_j}(\boldsymbol{\pi}_0, \boldsymbol{\kappa}_0), \qquad & D_{ij} &= \frac{\partial Q_i}{\partial \kappa_j}(\boldsymbol{\pi}_0, \boldsymbol{\kappa}_0), \end{aligned} \tag{3.12.3}$$

$i, j = 1, \ldots, l$, and in Eq. (3.12.2), $\boldsymbol{\pi}_0, \boldsymbol{\kappa}_0, \mathbf{p}_0, \mathbf{q}_0$ are regarded as elements of $\mathbf{R}^l$ (even though $\boldsymbol{\kappa}_0, \mathbf{q}_0$ might be in $\mathbf{T}^l$ or $\mathbf{R}^{l_1} \times \mathbf{T}^{l_2}$).[22]

The $2l \times 2l$ matrix L,

$$L = \begin{pmatrix} A & B \\ C & D \end{pmatrix}, \tag{3.12.4}$$

is the Jacobian matrix of the map C and, therefore, $\det L \neq 0$, $\forall (\boldsymbol{\pi}_0, \boldsymbol{\kappa}_0) \subset W$.

The main structure theorem for the completely canonical maps transforming V onto W and time independent is as follows.

23 Proposition. *A necessary and sufficient condition for the complete canonicity of a map C of the type considered in the Definition* 18, *Eq.* (3.12.1), *is that the map obtained by linearizing C at $(\boldsymbol{\pi}_0, \boldsymbol{\kappa}_0) \in W$ is a completely canonical map of $\mathbf{R}^{2l}$ onto $\mathbf{R}^{2l}$, $\forall (\boldsymbol{\pi}_0, \boldsymbol{\kappa}_0) \in W$.*

This is the case if and only if the inverse matrix to the matrix (3.12.4) *is*

$$L^{-1} = \begin{pmatrix} D^T & -B^T \\ -C^T & A^T \end{pmatrix}, \tag{3.12.5}$$

where the superscript T denotes the transposition of the matrix.

Observations.

(1) In other words, C is completely canonical in W if and only if its linearization around any point in W is completely canonical.

(2) Hence, complete canonicity is a "purely local" property of a map: this explains why the completely canonical maps are sometimes called "contact transformations" (although it does not explain why they are often called "symplectic").

[22] We recall that on $\mathbf{T}^l$ we use the flat coordinates: the ambiguity mod 2π of some of the coordinates of $\boldsymbol{\kappa}_0$ or $\mathbf{q}_0$ is arbitrarily solved here and it is irrelevant in the following.

Proof. Let $H \in C^\infty(V)$ and $H'(\boldsymbol{\pi}, \boldsymbol{\kappa}) = H(C(\boldsymbol{\pi}, \boldsymbol{\kappa})) = H(\mathbf{P}(\boldsymbol{\pi}, \boldsymbol{\kappa}), \mathbf{Q}(\boldsymbol{\pi}, \boldsymbol{\kappa}))$. The Hamiltonian equations in V are

$$\dot{\mathbf{p}} = -\frac{\partial H}{\partial \mathbf{q}}(\mathbf{p}, \mathbf{q}),$$
$$\dot{\mathbf{q}} = \frac{\partial H}{\partial \mathbf{p}}(\mathbf{p}, \mathbf{q}) \tag{3.12.6}$$

and if C is completely canonical, they must be equivalent to the equations

$$\dot{\boldsymbol{\pi}} = -\frac{\partial H'}{\partial \boldsymbol{\kappa}}(\boldsymbol{\pi}, \boldsymbol{\kappa}),$$
$$\dot{\boldsymbol{\kappa}} = \frac{\partial H'}{\partial \boldsymbol{\pi}}(\boldsymbol{\pi}, \boldsymbol{\kappa}), \tag{3.12.7}$$

i.e., if $t \to (\boldsymbol{\pi}(t), \boldsymbol{\kappa}(t))$ solves Eq. (3.12.7), then $t \to C(\boldsymbol{\pi}(t), \boldsymbol{\kappa}(t)) \equiv (\mathbf{P}(\boldsymbol{\pi}(t), \boldsymbol{\kappa}(t)), \mathbf{Q}(\boldsymbol{\pi}(t), \boldsymbol{\kappa}(t)) \equiv (\mathbf{p}(t), \mathbf{q}(t))$ has to solve Eq. (3.12.6).

By differentiating $p_i(t) = P_i(\boldsymbol{\pi}(t), \boldsymbol{\kappa}(t))$, $q_i(t) = Q_i(\boldsymbol{\pi}(t), \boldsymbol{\kappa}(t))$, with respect to t, one finds

$$\dot{p}_i = \sum_{k=1}^{l} A_{ik}(\boldsymbol{\pi}, \boldsymbol{\kappa})\dot{\pi}_k + \sum_{k=1}^{l} B_{ik}(\boldsymbol{\pi}, \boldsymbol{\kappa})\dot{\kappa}_k = \sum_{k=1}^{l}\left(-A_{ik}\frac{\partial H'}{\partial \kappa_k} + B_{ik}\frac{\partial H'}{\partial \pi_k}\right),$$
$$\tag{3.12.8}$$
$$\dot{q}_i = \sum_{k=1}^{l} C_{ik}(\boldsymbol{\pi}, \boldsymbol{\kappa})\dot{\pi}_k + \sum_{k=1}^{l} D_{ik}(\boldsymbol{\pi}, \boldsymbol{\kappa})\dot{\kappa}_k = \sum_{k=1}^{l}\left(-C_{ik}\frac{\partial H'}{\partial \kappa_k} + D_{ik}\frac{\partial H'}{\partial \pi_k}\right)$$

for $i = 1, \ldots, l$, where, of course, the derivatives of H' have to be computed in $(\boldsymbol{\pi}, \boldsymbol{\kappa})$ and in the first step we have dropped the explicit t dependence of the $\boldsymbol{\pi}, \boldsymbol{\kappa}, \dot{\boldsymbol{\pi}}, \dot{\boldsymbol{\kappa}}$ functions and, in the second step, we have dropped even the explicit dependence on the $(\boldsymbol{\pi}, \boldsymbol{\kappa})$ variables, to simplify the notations.

Using the expression of H' in terms of H, we find from Eq. (3.12.8), $\forall i = 1, \ldots, l$:

$$\dot{p}_i = \sum_{k,s}\left\{\left(-A_{ik}\left(\frac{\partial H}{\partial p_s}B_{sk} + \frac{\partial H}{\partial q_s}D_{sk}\right) + B_{ik}\left(\frac{\partial H}{\partial p_s}A_{sk} + \frac{\partial H}{\partial q_s}C_{sk}\right)\right)\right\}$$
$$\tag{3.12.9}$$
$$\dot{q}_i = \sum_{k,s}\left\{\left(-C_{ik}\left(\frac{\partial H}{\partial p_s}B_{sk} + \frac{\partial H}{\partial q_s}D_{sk}\right) + D_{ik}\left(\frac{\partial H}{\partial p_s}A_{sk} + \frac{\partial H}{\partial q_s}C_{sk}\right)\right)\right\},$$

where, of course, the derivatives of H have to be computed in $(\mathbf{P}(\boldsymbol{\pi}, \boldsymbol{\kappa}), \mathbf{Q}(\boldsymbol{\pi}, \boldsymbol{\kappa}))$, and the matrices A, B, C, and D have to be computed in $(\boldsymbol{\pi}, \boldsymbol{\kappa})$.

Equation (3.12.9) can be more compactly written with matrix-product notations:

$$\begin{pmatrix} \dot{\mathbf{p}} \\ \dot{\mathbf{q}} \end{pmatrix} = \begin{pmatrix} (AB^T - BA^T) & (-AD^T + BC^T) \\ (-DA^T + CB^T) & (-CD^T + DC^T) \end{pmatrix} \begin{pmatrix} -\dfrac{\partial H}{\partial \mathbf{p}} \\ \dfrac{\partial H}{\partial \mathbf{q}} \end{pmatrix}. \tag{3.12.10}$$

We now impose that Eq. (3.12.10) reduces to Eq. (3.12.6), $\forall H \in C^{\infty}(V)$. Of course, since the vector in the right-hand side of Eq. (3.12.10) can be made arbitrary by varying H, if the point $(\mathbf{p}, \mathbf{q})$ where the derivatives are evaluated is kept fixed, it follows that

$$\begin{aligned} AB^T - BA^T &= 0, \\ CD^T - DC^T &= 0, \\ -AD^T + BC^T &= -I, \end{aligned} \tag{3.12.11}$$

where $I = (l \times l \text{ identity matrix})$. Note that

$$\begin{bmatrix} (AB^T - BA^T) & (-AD^T + BC^T) \\ (-DA^T + CB^T) & (-CD^T + DC^T) \end{bmatrix} = \begin{bmatrix} A & B \\ C & D \end{bmatrix} \begin{bmatrix} B^T & -D^T \\ -A^T & C^T \end{bmatrix}; \tag{3.12.12}$$

hence, Eq. (3.12.11) can be written as

$$\begin{pmatrix} A & B \\ C & D \end{pmatrix} \begin{pmatrix} B^T & -D^T \\ -A^T & C^T \end{pmatrix} = \begin{pmatrix} 0 & -I \\ -I & 0 \end{pmatrix} \tag{3.12.13}$$

or, multiplying both sides on the right by $\left(\begin{smallmatrix} 0 & -I \\ -I & 0 \end{smallmatrix}\right)$:

$$\begin{pmatrix} A & B \\ C & D \end{pmatrix} \begin{pmatrix} D^T & -B^T \\ -C^T & A^T \end{pmatrix} = \begin{pmatrix} I & 0 \\ 0 & I \end{pmatrix} \tag{3.12.14}$$

which implies Eq. (3.12.5).

Vice versa, if Eq. (3.12.5) holds everywhere in W, the above equalities can be run backwards.

If H is explicitly time dependent, its conjugacy via C with H' defined by $H'(\boldsymbol{\pi}, \boldsymbol{\kappa}, t) = H(C(\boldsymbol{\pi}, \boldsymbol{\kappa}), t)$ follows in an identical fashion. mbe

24 Proposition. *The Jacobian determinant of any completely canonical transformation is* ± 1.

Proof. Equation (3.12.14) can be written

$$\begin{pmatrix} A & B \\ C & D \end{pmatrix} \begin{pmatrix} 0 & -I \\ I & 0 \end{pmatrix} \begin{pmatrix} A^T & C^T \\ B^T & D^T \end{pmatrix} \begin{pmatrix} 0 & -I \\ I & 0 \end{pmatrix} = -\mathbf{1}, \tag{3.12.15}$$

where $\mathbf{1}$ denotes the $2l \times 2l$ identity matrix; i.e.,

$$L \begin{pmatrix} 0 & -I \\ I & 0 \end{pmatrix} L^T \begin{pmatrix} 0 & -I \\ I & 0 \end{pmatrix} = -\mathbf{1}. \tag{3.12.16}$$

Hence, taking the determinant of both sides and remarking that the matrix $E = \left(\begin{smallmatrix} 0 & -I \\ I & 0 \end{smallmatrix}\right)$ has determinant $\det E = +1$, it follows that

$$(\det L)^2 = 1. \tag{3.12.17}$$

mbe

It could be shown that, actually, $\det L = +1$.

The conditions (3.12.15) or (3.12.11) for complete canonicity, equivalent to Eq. (3.12.5), can be expressed in terms of the following notion of "Poisson bracket" of two observables.

19 Definition. Let V be an open subset of $\mathbf{R}^l \times \mathbf{R}^l$ or $\mathbf{R}^l \times \mathbf{T}^l$ or $\mathbf{R}^l \times (\mathbf{R}^{l_1} \times \mathbf{T}^{l_2})$, $l_1 + l_2 = l$, regarded as the phase space for the Hamiltonian equations in V.

Let $F, G \in C^\infty(V)$ be two "observables". One defines the "Poisson bracket" $\{F, G\} \in C^\infty(V)$ of F and G as

$$\{F,G\}(\mathbf{p},\mathbf{q}) = \sum_{s=1}^{l} \left(\frac{\partial F}{\partial p_s}(\mathbf{p},\mathbf{q}) \frac{\partial G}{\partial q_s} - \frac{\partial F}{\partial q_s}(\mathbf{p},\mathbf{q}) \frac{\partial G}{\partial p_s}(\mathbf{p},\mathbf{q}) \right). \quad (3.12.18)$$

Observations.

(1) Clearly, $\forall i, j = 1, \ldots, l$,

$$\{p_i, q_j\} = \delta_{ij}, \qquad \{p_i, p_j\} = 0, \qquad \{q_i, q_j\} = 0. \quad (3.12.19)$$

Also, if $\varphi_1, \ldots, \varphi_r$ are C^∞ functions on $\mathbf{R}^n$ and $F_1, \ldots, F_n \in C^\infty(V)$, and if one defines

$$\Phi_j(\mathbf{p},\mathbf{q}) = \varphi_j(F_1(\mathbf{p},\mathbf{q}), \ldots, F_n(\mathbf{p},\mathbf{q})), \quad (3.12.20)$$

one finds:

$$\{\Phi_i, \Phi_j\} = \sum_{h,k=1}^{n} \frac{\partial \varphi_i}{\partial F_k} \frac{\partial \varphi_j}{\partial F_k} \{F_k, F_h\}. \quad (3.12.21)$$

(2) Sometimes the definition (3.12.18) is defined with the opposite sign: this is totally irrelevant despite claims to the contrary.

(3) Equations (3.12.19) are also called the "canonical commutation" relations.

The notion of Poisson bracket is remarkable in that it allows the following corollary to the Proposition 23, p. 234.

25 Corollary. *A necessary and sufficient condition for the complete canonicity of an invertible nonsingular map C between V and W (as in Definition 17, p. 220, above) is that the functions defining it,* $\mathbf{P}(\boldsymbol{\pi},\boldsymbol{\kappa}), \mathbf{Q}(\boldsymbol{\pi},\boldsymbol{\kappa})$, *have the property,* $\forall(\boldsymbol{\pi},\boldsymbol{\kappa}) \in W$, $\forall i, j = 1, \ldots, l$:

$$\{P_i, Q_j\} = \delta_{ij}, \qquad \{P_i, P_j\} = 0, \qquad \{Q_i, Q_j\} = 0. \quad (3.12.22)$$

Observations.

(1) So C is completely canonical if and only if it "preserves the canonical commutation relations".

(2) If C preserves the canonical commutation relations, it follows that it preserves the Poisson brackets of any pair of observables: this means that if $F, G \in C^\infty(V)$ and if we define

$$F_C(\boldsymbol{\pi},\boldsymbol{\kappa}) = F(C(\boldsymbol{\pi},\boldsymbol{\kappa})), \qquad G_C(\boldsymbol{\pi},\boldsymbol{\kappa}) = G(C(\boldsymbol{\pi},\boldsymbol{\kappa})), \quad (3.12.23)$$

then, as is easily checked by Eqs. (3.12.21) and (3.12.22):

$$\{F,G\}(\mathbf{p},\mathbf{q}) = \{F_C, G_C\}(\boldsymbol{\pi},\boldsymbol{\kappa}) \quad \text{if} \quad (\mathbf{p},\mathbf{q}) = C(\boldsymbol{\pi},\boldsymbol{\kappa}). \quad (3.12.24)$$

So C is completely canonical if and only if it "preserves the commutation relations of any pair of observables".

PROOF. Explicitly write Eq. (3.12.22) in terms of the derivatives of Eq. (3.12.3): one finds that they become Eq. (3.12.11), i.e., Eq. (3.12.5). mbe

In §3.11 we saw that a class of completely canonical transformations can be built from a generating function. One can wonder how general this construction is.

26 Proposition. *Let C be a completely canonical map between V and W as in Definition 17, p. 220. Given two corresponding points $(\mathbf{p}_0, \mathbf{q}_0) \in V$, $(\boldsymbol{\pi}_0, \boldsymbol{\kappa}_0) \in W$, $(\mathbf{p}_0, \mathbf{q}_0) = C(\boldsymbol{\pi}_0, \boldsymbol{\kappa}_0)$, consider the matrices (3.12.3).*

The map C can be generated in a neighborhood of $(\mathbf{p}_0, \mathbf{q}_0)$, $(\boldsymbol{\pi}_0, \boldsymbol{\kappa}_0)$ by a generating function, as in Proposition 21, p. 220, and in the observations following it, having the form:

(i) $F(\mathbf{q}, \boldsymbol{\kappa})$ if $\det C \neq 0$;
(ii) $\Phi(\mathbf{p}, \boldsymbol{\kappa})$ if $\det A \neq 0$;
(iii) $\Psi(\boldsymbol{\pi}, \mathbf{q})$ if $\det D \neq 0$; (3.12.25)
(iv) $R(\mathbf{p}, \boldsymbol{\pi})$ if $\det B \neq 0$.

Observations.

(1) There exist completely canonical transformations for which $\det A = \det B = \det C = \det D = 0$.

For instance, the map of $\mathbf{R}^2 \times \mathbf{R}^2 \leftrightarrow \mathbf{R}^2 \times \mathbf{R}^2$:

$$(p_1, p_2; q_1, q_2) \leftrightarrow (p_1, -q_2; q_1, p_2). \tag{3.12.26}$$

This canonical transformation cannot be generated by a generating function of the above types.

(2) Obviously, if C is completely canonical, defined on $V \subset \mathbf{R}^{2l}$, it must have a Jacobian matrix L with nonvanishing determinant, (see Corollary 24, p. 236.) Hence, there must be a choice of indices $i_1, \ldots, i_s$, $j_1, \ldots, j_{l-s}$, pairwise distinct such that

$$\det \frac{\partial(p_{i_1}, \ldots, p_{i_s}, q_{j_1}, \ldots, q_{j_{l-s}})}{\partial(\pi_1, \ldots, \pi_l)} \neq 0. \tag{3.12.27}$$

This means, as it can be understood with a little thought, that C can be locally constructed by composing a canonical transformation of the type:

$$(p_1, \ldots, p_l; q_1, \ldots, q_l)$$
$$\leftrightarrow (p_{i_1}, \ldots, p_{i_s}, -q_{j_1}, \ldots, -q_{j_{l-s}}; q_{i_1}, \ldots, q_{i_l}, p_{j_1}, \ldots, p_{j_{l-s}}) \tag{3.12.28}$$

[like Eq. (3.12.26)] with a completely canonical transformation generated by a function $\Phi(\mathbf{p}, \boldsymbol{\kappa})$.

(3) So any completely canonical transformation is, near a point, a composition of a trivial ("permutation type") completely canonical trans-

formation and a completely canonical transformation generated by a generating function (Arnold).

PROOF. Suppose, for instance, $\det D \neq 0$. Then it is possible to invert the equation $\mathbf{q} = \mathbf{Q}(\boldsymbol{\pi}, \boldsymbol{\kappa})$ to express

$$\boldsymbol{\kappa} = \mathbf{G}(\boldsymbol{\pi}, \mathbf{q}) \tag{3.12.29}$$

for $\mathbf{q}, \boldsymbol{\pi}, \boldsymbol{\kappa}$ near $\mathbf{q}_0, \boldsymbol{\pi}_0, \boldsymbol{\kappa}_0$ using the implicit function theorem (see Appendix G).

Then we can write

$$\begin{aligned} \mathbf{p} &= \mathbf{P}(\boldsymbol{\pi}, \mathbf{G}(\boldsymbol{\pi}, \mathbf{q})) \equiv \mathbf{F}(\boldsymbol{\pi}, \mathbf{q}), \\ \boldsymbol{\kappa} &= \mathbf{G}(\boldsymbol{\pi}, \mathbf{p}). \end{aligned} \tag{3.12.30}$$

We have to show the existence of $\Psi(\boldsymbol{\pi}, \mathbf{q})$ such that

$$\begin{aligned} \frac{\partial \Psi}{\partial \mathbf{q}}(\boldsymbol{\pi}, \mathbf{q}) &= \mathbf{F}(\boldsymbol{\pi}, \mathbf{q}), \\ \frac{\partial \Psi}{\partial \boldsymbol{\pi}}(\boldsymbol{\pi}, \mathbf{q}) &= \mathbf{G}(\boldsymbol{\pi}, \mathbf{q}) \end{aligned} \tag{3.12.31}$$

defined near $\boldsymbol{\pi}_0, \mathbf{q}_0$.

This means that we must show the validity of the integrability conditions:

$$\frac{\partial F_i}{\partial q_j} = \frac{\partial F_j}{\partial q_i}, \qquad \frac{\partial F_i}{\partial \pi_j} = \frac{\partial G_j}{\partial q_i} \qquad \frac{\partial G_i}{\partial \pi_j} = \frac{\partial G_j}{\partial \pi_i}. \tag{3.12.32}$$

But, by differentiating the first of Eqs. (3.12.30), we see that

$$\begin{aligned} \frac{\partial F_i}{\partial \pi_j} &= A_{ij} + \sum_{k=1}^{l} B_{ik} \frac{\partial G_k}{\partial \pi_j}, \\ \frac{\partial F_i}{\partial q_j} &= \sum_{k=1}^{l} B_{ik} \frac{\partial G_k}{\partial q_j}, \qquad \forall i, j = 1, \ldots, l, \end{aligned} \tag{3.12.33}$$

with the obvious choice of arguments of these functions; e.g., $\boldsymbol{\pi} = \boldsymbol{\pi}_0$, $\mathbf{q} = \mathbf{q}_0$.

By differentiating the identity $\boldsymbol{\kappa} \equiv \mathbf{G}(\boldsymbol{\pi}, \mathbf{Q}(\boldsymbol{\pi}, \boldsymbol{\kappa}))$, we find

$$\frac{\partial G_j}{\partial q_j} = (D^{-1})_{ij}, \tag{3.12.34}$$

$$\frac{\partial G_i}{\partial \pi_j} + \sum_{s=1}^{l} \frac{\partial G_i}{\partial q_s} C_{sj} = 0, \qquad \forall i, j = 1, \ldots, l.$$

More concisely, we rewrite Eqs. (3.12.33) and (3.12.34) as

$$\begin{aligned} \frac{\partial F}{\partial \boldsymbol{\pi}} &= A - BD^{-1}C, \qquad & \frac{\partial F}{\partial \mathbf{q}} &= BD^{-1}, \\ \frac{\partial G}{\partial \mathbf{q}} &= D^{-1}, & \frac{\partial G}{\partial \boldsymbol{\pi}} &= -D^{-1}C, \end{aligned} \tag{3.12.35}$$

and the conditions (3.12.32) become

$$A - BD^{-1}C = (D^{-1})^T,$$

$$BD^{-1} = (BD^{-1}), \quad \text{i.e.,} \quad BD^{-1} = (D^{-1})^T B^T, \quad \text{i.e.,} \quad D^T B = BD^T,$$

$$D^{-1}C = (D^{-1}C)^T, \quad \text{i.e.,} \quad D^{-1}C = C^T(D^{-1})^T \quad \text{i.e.,} \quad CD^T = DC^T. \tag{3.12.36}$$

So we simply need to check that Eqs. (3.12.36) are a disguised form of the complete canonicity conditions (3.12.11) and the proof will be complete.

In fact, the third of Eqs. (3.12.36) is implied by the second of Eqs. (3.12.11). Furthermore, using the first and the transposition of the third of Eqs. (3.12.11) we see that[23]

$$AB^T = BA^T = BD^{-1}DA^T = BD^{-1}(1 + CB^T) = BD^{-1} + B(D^{-1}C)B^T \tag{3.12.37}$$

which shows that BD^{-1} is symmetric since such is AB^T and $B(D^{-1}C)B^T$ (having already seen that $D^{-1}C$ is symmetric). So the second of Eqs. (3.12.36) also holds. Finally the first of Eq. (3.12.36) means

$$AD^T - BD^{-1}CD^T = I \tag{3.12.38}$$

which, since $CD^T = DC^T$ by the Eq. (3.12.11), shows that the second equality in Eqs. (3.12.38) simply means that $AD^T - BC^T = I$ which is true since it is the first of Eqs. (3.12.11). mbe

Problems and Complements for §3.12

1. Let C be a map of W onto W' and suppose that there is $\Phi \in C^\infty(G(C))$, $G(C) \equiv \{\mathbf{p}, \mathbf{q}, \mathbf{p}', \mathbf{q}' \,|\, (\mathbf{p}, \mathbf{q}, \mathbf{p}', \mathbf{q}') \in \hat{W} \times \hat{W}', (\mathbf{p}, \mathbf{q}) = C(\mathbf{p}', \mathbf{q}')\}$, such that

$$\mathbf{p} \cdot d\mathbf{q} = \mathbf{p}' \cdot d\mathbf{q}' + d\Phi.$$

Show that C is a time independent completely canonical map and that it is also "action preserving" in the sense that, if λ is a closed curve in W' and $C\lambda$ is its C-image in W, it is

$$\int_{C\lambda} \mathbf{p} \cdot d\mathbf{q} = \int_\lambda \mathbf{p}' \cdot d\mathbf{q}'$$

2. Consider in $\mathbf{R}^2$ the annulus $D = \{(q_1, q_2) \,|\, \alpha < q_1^2 + q_2^2 < \beta\}$, $\alpha, \beta > 0$, and let $f(q_1, q_2)dq_1 + g(q_1, q_2)dq_2$ be an exact but nonintegrable differential form on D. Define the map

$$C(p_1, p_2, q_1, q_2) = (p'_1, p'_2, q'_1, q'_2) \equiv (p_1 + f(q_1, q_2), p_2 + g(q_1, q_2), q_1, q_2).$$

Show that it is completely canonical (time independent, of course). (Hint: Note that $p'_1 dq'_1 + p'_2 dq'_2 = p_1 dq_1 + p_2 dq_2 + f(q_1, q_2)dq_1 + g(q_1, q_2)dq_2$ and recall that every ex-

[23] We use the fact that if M is a symmetric $l \times l$ matrix and E is an arbitrary $l \times l$ matrix, then $E^T M E$ is a symmetric matrix (exercise).

act form is locally integrable and that the complete canonicity is a local property and use problem 1.)

3. Show that not all completely canonical maps are action preserving in the sense of problem 1. (Hint: Consider the map in Problem 2 and choose λ to be the curve $p_1 = p_2 = 0$, $q_1^2 + q_2^2 = (\alpha + \beta)/2$.)

4.* Show that the existence of $\Phi \in C^\infty(G(C))$ verifying the property introduced in Problem 1 is a necessary and sufficient condition in order that C be an action preserving time independent completely canonical map of W onto W'.

5. Show that the map $C(p_x, p_y, x, y) = (p_1, p_2, q_1, q_2)$ between $\mathbf{R}^2 \times (\mathbf{R} \times \mathbf{R}_+)/\{$set of points with $p_x = 0\}$ and $\mathbf{R}^4/\{$set of points with $p_1q_1 + p_2q_2 < 0$ or with $p_1 < 0\}$ defined by the following relation:

$$p \equiv p_x + ip_y = \frac{i}{2}(p_1 + ip_2)^2, \qquad i = \sqrt{-1}\,,$$

$$q \equiv x + iy = \frac{p_2 + iq_1}{p_1 - iq_2}$$

is completely canonical (time independent). (Hint: Check that it is one-to-one and $p_x dx + p_y dy \equiv \operatorname{Re} \overline{p\,dq} = p_1 dq_1 + p_2 dq_2 + \frac{1}{2} d(p_1q_1 + p_2q_2)$.)

6. Let $(p', q') = C(p, q)$ be a map from $W \subset \mathbf{R}^2$ onto $W' \subset \mathbf{R}^2$. Show that a necessary and sufficient condition for being completely canonical is that it is orientation and area preserving (recall that a map is "orientation preserving" if its jacobian matrix L has positive determinant). (Hint: Note that the matrices A, B, C, D are numbers and therefore (3.12.5) holds if and only if $\det L = 1$.)

7. Extend the notion of completely canonical time independent map by observing that (3.11.52) could be replaced and extended as $S(\mathbf{m}) = \lambda\Sigma(\boldsymbol{\mu}) +$ constant. Discuss the case $\lambda = -1$ and prove a proposition like Proposition 23. Find the physical meaning of λ.

8.* Consider the Hamiltonian on $\mathbf{R}^2 \times (\mathbf{R} \times \mathbf{R}_+)$: $H_0(p_x, p_y, x, y) = y^2(p_x^2 + p_y^2)/2$, ("Hamiltonian for the geodesic motion for the geometry $ds^2 = (dx^2 + dy^2)/y^2$ on $\mathbf{R} \times \mathbf{R}_+$", see Problems 19–24, p. 230, and show that the canonical map in Problem 5 transforms H_0 into $(p_1q_1 + p_2q_2)^2/8$. Write and solve the Hamilton's equations in the new coordinates.

9.* In the context of Problem 8 consider the canonical map $p_1' = (p_1 + q_1)/\sqrt{2}$, $q_1' = (p_1 - q_1)/\sqrt{2}$, $p_2' = (p_2 + q_2)/\sqrt{2}$, $q_2' = (p_2 - q_2)/\sqrt{2}$ and show that H is transformed by it into $((p_1')^2 - (q_1')^2 + (p_2')^2 - (q_2')^2)^2/2$. Interpret this as saying that the geodesic motions of H_0 taking place at a given energy E can be thought of as describing the motions of two independent hyperbolic oscillators (i.e. two particles in a negative quadratic potential). How does this picture change as E varies?

10.* Show that the map $(p_1, p_2, q_1, q_2) \to (p_x, p_y, x, y)$ defined in Problem 5 is two-to-one from $\tilde{G} = \mathbf{R}^4/\{$set of points for which $p_1q_1 + p_2q_2 < 0\}$ onto $G' = \mathbf{R}^2 \times \mathbf{R} \times \mathbf{R}_+/\{$set of points for which $p_x = p_y = 0\}$. If however the "opposite" points (p_1, p_2, q_1, q_2) and $(-p_1, -p_2, -q_1, -q_2)$ are identified, the map becomes one-to-one. Then observe that (p_1, p_2, q_1, q_2) may be regarded as coordinates (modulo the sign) for the points of the set $G = \{\tilde{G}$ with opposite points identified$\}$ in the same sense as a point $\varphi \in \mathbf{R}^l$ can be regarded as a coordinate (modulo 2π) for a point in T^l.

Using this remark extend the notion of time independent completely canonical maps to cover the case when W instead of being a subset of $V \times (\mathbf{T}^{l_1} \times \mathbf{R}^{l_2})$ is a subset of G and show that the map under consideration is completely canonical, in this new sense, as a map between G and G'.

11.* Try to extend the notion of completely canonical time independent map to maps of arbitrary open surfaces of dimension $2l$ by abstracting the essential properties of the examples discussed in definition 17, p. 220, and in Problem 10 where the $2l$-dimensional surfaces are very special, i.e. they are, respectively, of the form $V \times \mathbf{T}^{l_1} \times \mathbf{R}^{l_2}$, $l_1 + l_2 = l$, $V \subset \mathbf{R}^l$, or the set $\tilde{G}$ with opposite points identified.

12. Let $\hat{W}$ be the phase for a regular time independent Hamiltonian function H, see Observation (4), p. 216. Let $T > 0$, $(\mathbf{p}, \mathbf{q}) \in W$, and suppose that the solution $S_t(p,q)$ to the Hamiltonian equations with initial datum $(\mathbf{p}, \mathbf{q})$ stays in W for all $t \in [0, T]$: $S_t(\mathbf{p}, \mathbf{q}) \in W$. Define $F_t(\mathbf{p}, \mathbf{q}) \equiv F(S_t(\mathbf{pq}))$ and show that

$$\frac{dF_t(\mathbf{p}, \mathbf{q})}{dt} = \{H, F_t\}(\mathbf{p}, \mathbf{q}),$$

i.e., "the time derivative of an observable F is given by its Poisson bracket with the Hamiltonian"

13. In the context of (12) show that $\{H, F_t\}(\mathbf{p}, \mathbf{q}) \equiv \{H, F\}(S_t(\mathbf{p}, \mathbf{q}))$: since in Physics the operation of associating with $F \in C^\infty(W)$ the function $\mathfrak{L}F = \{H, F\}$ is called the "Liouville operator's action" this can be read: the "Liouville operator commutes with the time evolution".

14. Let E, F, G be in $C^\infty(W)$, where W is the phase space for a regular time independent Hamiltonian H. Show that

$$\{E, \{F, G\}\} + \{F, \{G, E\}\} + \{G, \{E, F\}\} \equiv 0, \qquad \{E, F\} \equiv -\{F, E\}$$
$$\{E, FG\} = \{E, F\}G + \{E, G\}F$$

These relations are called respectively "the Jacobi identity", the "antisymmetry" and the "derivation property" of the Poisson bracket.

15. Show that, in the context of Problem 12, the relations ("Liouville's equations")

$$\frac{dF(S_t(\mathbf{p}, \mathbf{q}))}{dt} = \{H, F\}(S_t(\mathbf{p}, \mathbf{q})) = \mathfrak{L}F(S_t(\mathbf{p}, \mathbf{q}))$$

imply, if valid for all $F \in C^\infty(W)$, for all $(p,q) \in W$ and for t small (depending possibly on $(\mathbf{p}, \mathbf{q})$), that $t \to s_t(\mathbf{p}, \mathbf{q})$ verifies the Hamilton's equations, ("equivalence between the Hamilton's equations and the Liouville's equations").

Other problems on canonical maps can be found at the end of §4.9–4.12 and §5.10 and 5.12.

Concluding Comments to Chapter 3

(1) We have described by the word "action" certain quantities which, in fact, do not motivate such a nice name [see Eqs. (3.3.4), etc.].

Actually, in contemporary literature, the convention of calling Eq. (3.3.4) "action of a motion", or "least action principle" the corresponding variational principle, prevails.

This is perhaps historically incorrect: the action was introduced by Maupertuis when he formulated the variational principle bearing his name,

Problem 16 p. 229.[24] The numerical value of the quantity that Maupertuis called "action of a motion", computed on the real motion developing under the influence of given conservative forces and ideal constraints, is related, in a very simple way, to the value of the action of Eq. (3.3.4) computed on the real motion (see Problem 15, §3.11 p. 229). The same occurs for the numerical value of other quantities also sometimes called "action", see, for instance, Eq. (3.11.50). These simple relations explain why there is so much confusion in the names. However, it should be stressed that among the various notions of action there are simple relations only if we compare the numerical values that they have on the real motions: it would not make sense to ask if there is a simple relation between the values taken on the varied motions (mainly because in the different variational principles, the motions are described and parameterized differently and, therefore, one cannot compare them).

(2) It is interesting to quote Maupertuis in connection with his definition of action, afterwards interpreted by Euler as in Problem 16, p. 229 (quoted from E. Mach, Chapter III, §2.8):

> We must explain what is meant by quantity of action. When a body is moved from one point to another, a certain action is necessary. This action depends upon the velocity of body, upon the space it covers, but it is neither the velocity nor the space separately considered. The greater the body's velocity and the longer the path that it covers, the greater the action; the action is proportional to the sum of the spaces, each multiplied by the speed with which the bodies cover them. It is the quantity of action, the true expenditure of Nature, which she administers with as much economy as possible in the movement of light.

The last line refers to Maupertuis' application of his principle to the propagation of light. The other lines are a nice way of saying

$$A = \int_{\xi_1}^{\xi_2} \mathbf{v} \cdot d\mathbf{q} = \frac{1}{m} \int_{\xi_1}^{\xi_2} \mathbf{p} \cdot d\mathbf{q},$$

and the condition of stationarity of A on a motion $t \to (\mathbf{p}(t), \mathbf{q}(t))$ of given energy E can be shown to be equivalent to the stationarity of the quantity in Problem 16, §3.11 (a further problem for the reader).

For a comment on Maupertuis' definition, see the angry pages of E. Mach (Chapter III, §8.4).

(3) To understand the historical development of the various principles, one can consult Mach, where they are critically discussed, paying due attention to history. In Mach's book (Chapter IV, §2), one also finds an interesting comment on the "theological, animistic and mystical points of view in mechanics" (see, also, Observation (2), p. 164).

[24] The original formulation was, in fact, quite obscure and it was later clarified by Euler (see E. Mach Ch. III).

(4) Some very concrete exercises for this chapter can be found in the book by L. Metcherskij, *Recueil de problèmes de Mécanique Rationelle, Mir*, Moscow, 1973. For §3.1 and §3.3, see:

Chapter 3 §10, §11, §12;
Chapter 4 §21, §22, §23;
Chapter 9 §26, §27, §28, §29, §30, §31, §32, §33;
Chapter 10 §4, §35, §36, §38.

For §3.3, §3.4, §3.5, and §3.8, see:

Chapter 4, §13, §14;
Chapter 5, §15, §16, §17, §18;
Chapter 10, §37, §39, §43;
Chapter 11, §46, §47, §48;
Chapter 6, §19, §20.

For §3.11 and §3.12, see:

Chapter 11 §49.

One can also consult the book B. Finzi, P. Udeschini, *Esercizi di Meccanica Razionale*, Tamburini, Milano, 1974.

For §3.1 and §3.2, see: Chapters 6 and 11.
For §3.3, §3.4, §3.5, and §3.8, see Chapters 2, 3, 5, 7, 8, 10, 12, 13, 14, 17, 18, and 21.

CHAPTER 4

Special Mechanical Systems

§4.1. Systems of Linear Oscillators

In this chapter we adhere systematically to the convention of denoting and writing the Lagrangian functions that we shall meet as $\mathcal{L}(\dot{\mathbf{x}}, \mathbf{x}, t)$ or $\mathcal{L}(\dot{\mathbf{x}}, \mathbf{x})$ or $\mathcal{L}(\dot{\mathbf{q}}, \mathbf{q}, t)$, rather than as functions of generic variables $\boldsymbol{\alpha}$, $\boldsymbol{\beta}$, t: the notation is obviously improper since in such cases the variables $\dot{\mathbf{x}}$ and $\mathbf{x}$ are not Cartesian coordinates but local (or toral) coordinates, and often the mechanical systems will be described directly in local coordinates omitting the obvious but tedious discussion necessary when the local coordinates are not global (i.e., they are not globally equivalent to Cartesian coordinates).

A typical example of this situation is when one says that a point mass ideally bound to remain on the surface of the unit sphere is described by a Lagrangian function given, in polar coordinates, by

$$\mathcal{L}(\dot{\theta}, \dot{\varphi}, \theta, \varphi, t) = \tfrac{1}{2} m\big(\dot{\theta}^2 + (\sin\theta)^2 \dot{\varphi}^2\big). \tag{4.1.1}$$

After a little practice and thought, this notational convention, very common in literature, will appear natural and should not give rise to any confusion.

Hence, a system of linear oscillators with l degrees of freedom is the mechanical system defined by

$$\mathcal{L}(\dot{\mathbf{x}}, \mathbf{x}, t) = \frac{1}{2} \sum_{i,j}^{1,l} g_{ij} \dot{x}_i \dot{x}_j - \frac{1}{2} \sum_{i,j}^{1,l} v_{ij} x_i x_j, \tag{4.1.2}$$

where $G = (g_{ij})_{i,\, j=1,\ldots,l}$, $V = (v_{ij})_{i,\, j=1,\ldots,l}$ are two $l \times l$ symmetric positive-definite matrices (see Appendix F).

The Lagrangian equations corresponding to Eq. (4.1.2) are

$$\sum_{j=1}^{l} g_{ij}\ddot{x}_j = -\sum_{j=1}^{l} v_{ij}x_j, \qquad i = 1, \ldots, l. \tag{4.1.3}$$

They can be treated in full generality and their theory is summarized by the following proposition stating that essentially Eq. (4.1.3) is equivalent, through a "simple" transformation, to l equations of the type:

$$\ddot{y}_i = -\omega_i^2 y_i, \qquad i = 1, \ldots, l. \tag{4.1.4}$$

1 Proposition. *The most general solution of Eq.* (4.1.2) *for* $t \in \mathbf{R}$ *can be written in terms of* l *arbitrary non-negative constants* $\mathbf{A} = (A_1, \ldots, A_l)$ *and of* l *angles* $\boldsymbol{\varphi} = (\varphi_1, \ldots, \varphi_l)$ *as*

$$\mathbf{x}(t) = \sum_{i=1}^{l} \sqrt{\frac{2A_i}{\omega_i}}\, \boldsymbol{\eta}^{(i)} \cos(\omega_i t + \varphi_i), \tag{4.1.5}$$

where $\omega_1, \ldots, \omega_l$ *are the* l *positive solutions of the* lth *order equation for* ω^2:

$$\det(-\omega^2 G + V) = 0 \tag{4.1.6}$$

and the vectors $\boldsymbol{\eta}^{(1)}, \ldots, \boldsymbol{\eta}^{(l)}$ *verify the equation*:

$$-\omega_i^2 G\boldsymbol{\eta}^{(i)} + V\boldsymbol{\eta}^{(i)} = 0, \qquad i = 1, \ldots, l \tag{4.1.7}$$

and they can be chosen so that

$$(G\boldsymbol{\eta}^{(i)}) \cdot \boldsymbol{\eta}^{(j)} = \delta_{ij}, \qquad i, j = 1, \ldots, l. \tag{4.1.8}$$

Observations.

(1) In Eq. (4.1.5), one could of course write A_i instead of $\sqrt{2A_i/\omega_i}$: however, the square root is more convenient since in this way the map $(\dot{\mathbf{x}}(0), \mathbf{x}(0)) \to (\mathbf{A}, \boldsymbol{\varphi})$ can be related to a canonical transformation [see Exercises for §4.1 and Observation (3) to Corollary 3, p. 249].

(2) Therefore, Eq. (4.1.3) admits periodic solutions like

$$\sqrt{\frac{2A}{\omega}}\, \boldsymbol{\eta} \cos(\omega t + \varphi). \tag{4.1.9}$$

Such oscillations are called "normal vibration modes" or "normal motions". The preceding proposition tells us that there exist l independent normal modes, orthogonal in the sense of Eq. (4.1.8) and that every oscillation is a "superposition" of normal modes.

To underline the interest of the orthogonality of the normal oscillation modes, let us deduce from Proposition 1, and before its proof, the following corollary.

2 Corollary. *The energy of the oscillation* [*Eq.* (4.1.5)] *is*

$$E = \sum_{i=1}^{l} \omega_i A_i , \tag{4.1.10}$$

i.e., *it is the sum of the energies of each normal mode component.*

PROOF. The energy is [see §3.11, Observation (1), p. 217]

$$E = \frac{1}{2} \sum_{i,j=1}^{l} g_{ij}\dot{x}_i\dot{x}_j + \frac{1}{2} \sum_{i,j=1}^{l} v_{ij}x_i x_j , \tag{4.1.11}$$

which can be written in vector form as $E = \frac{1}{2}(G\dot{\mathbf{x}}) \cdot \dot{\mathbf{x}} + \frac{1}{2}(V\mathbf{x}) \cdot \mathbf{x}$ or, explicitly, from Eq. (4.1.5):

$$E = \frac{1}{2} \sum_{i,j=1}^{l} \left\{ \sqrt{\frac{4A_iA_j}{\omega_i\omega_j}}\, \omega_i\omega_j \sin(\omega_i t + \varphi_i)\sin(\omega_j t + \varphi_j) \cdot (\boldsymbol{\eta}^{(i)}, G\boldsymbol{\eta}^{(j)}) + \sqrt{\frac{4A_iA_j}{\omega_i\omega_j}}\, \cos(\omega_i t + \varphi_i)\cos(\omega_j t + \varphi_j) \cdot (\boldsymbol{\eta}^{(i)}, V\boldsymbol{\eta}^{(j)}) \right\} \tag{4.1.12}$$

and using Eq. (4.1.7), we can replace $\boldsymbol{\eta}^{(i)} \cdot V\boldsymbol{\eta}^{(j)}$ with $\omega_j^2(\boldsymbol{\eta}^{(i)} \cdot (G\boldsymbol{\eta}^{(j)}))$, and using Eq. (4.1.8) plus trigonometry, one realizes that Eq. (4.1.12) becomes Eq. (4.1.10). mbe

PROOF OF PROPOSITION 1. Assume the existence of $\omega_1, \ldots, \omega_l$, the l positive roots of Eq. (4.1.16), and of l linearly independent vectors $\boldsymbol{\eta}^{(1)}, \ldots, \boldsymbol{\eta}^{(l)}$ verifying Eq. (4.1.7). Then by direct substitution of Eq. (4.1.5) into Eq. (4.1.3), one sees that the function (4.1.5) verifies, $\forall(A_1, \ldots, A_l) \in \mathbf{R}_+^l$, $\forall(\varphi_1, \ldots, \varphi_l) \in \mathbf{T}^l$, the equations (4.1.3).

It is also easy to see that given $(\boldsymbol{\eta}, \boldsymbol{\xi}) \in \mathbf{R}^{2l}$ arbitrarily, it is possible to determine $\mathbf{A} \in \mathbf{R}_+^l$, $\boldsymbol{\varphi} \in \mathbf{T}^l$ so that Eq. (4.1.5) verifies the datum $\mathbf{x}(0) = \boldsymbol{\xi}$, $\dot{\mathbf{x}}(0) = \boldsymbol{\eta}$ for $t = 0$.

In fact, the conditions

$$\begin{aligned} \boldsymbol{\xi} &= \sum_{j=1}^{l} \sqrt{\frac{2A_j}{\omega_j}}\, \boldsymbol{\eta}^{(j)} \cos\varphi_j , \\ \boldsymbol{\eta} &= \sum_{j=1}^{l} -\sqrt{2\omega_j A_j}\, \omega_j \boldsymbol{\eta}^{(j)} \sin\varphi_j \end{aligned} \tag{4.1.13}$$

imply, by scalar multiplication of both sides of Eq. (4.1.13) by $G\boldsymbol{\eta}^{(i)}$, $i = 1, \ldots, l$:

$$\begin{aligned} (G\boldsymbol{\eta}^{(i)}) \cdot \boldsymbol{\xi} &= \sqrt{\frac{2A_i}{\omega_i}}\, \cos\varphi_i , \\ (G\boldsymbol{\eta}^{(i)}) \cdot \boldsymbol{\eta} &= -\sqrt{2\omega_i A_i}\, \omega_i \sin\varphi_i , \qquad i = 1, \ldots, l, \end{aligned} \tag{4.1.14}$$

by Eq. (4.1.8). Equation (4.1.14) determines A_i and φ_i as:

$$(\sqrt{A_i}, \varphi_i) = \{\text{polar coordinates of the point with Cartesian coordinates } (\sqrt{\omega_i/2}\, G\boldsymbol{\eta}^{(i)} \cdot \boldsymbol{\xi}, (1/\sqrt{2\omega_i}) G\boldsymbol{\eta}^{(i)} \cdot \boldsymbol{\eta}) \in \mathbf{R}^2\}. \tag{4.1.15}$$

Vice versa, if (A_i, φ_i) verifies Eq. (4.1.14), it is easy to see that, since $\boldsymbol{\eta}^{(1)}, \ldots, \boldsymbol{\eta}^{(l)}$ are l linearly independent vectors in $\mathbf{R}^l$ (by assumption) and, therefore, form a basis in $\mathbf{R}^l$, Eq. (4.1.13) necessarily follows.

By virtue of the existence and uniqueness theorems of differential equations, Eq. (4.1.3) is the most general C^∞ solution to Eq. (4.1.3), $t \in \mathbf{R}$.

It remains to show the actual existence of l linearly independent vectors $\boldsymbol{\eta}^{(1)}, \ldots, \boldsymbol{\eta}^{(l)}$ and of l numbers $\omega_1, \ldots, \omega_l > 0$. This is a well-known proposition of algebra (see Appendix F). mbe

It will turn out to be useful to stress a simple corollary of Proposition 1. For this purpose, we recall the definition of the l-dimensional torus $\mathbf{T}^l$ obtained by identifying the opposite sides of the square $[0, 2\pi]^l$, see Definitions 12 and 13, pp. 100 and 101, and that of a function in $C^\infty(\mathbf{T}^l)$ and let us give the following definition.

1 Definition. Given $\boldsymbol{\theta} = (\theta_1, \ldots, \theta_l) \in \mathbf{R}^l$, the transformation of $\mathbf{T}^l$ into itself,

$$\boldsymbol{\varphi} = (\varphi_1, \ldots, \varphi_l) \to (\boldsymbol{\varphi} + \boldsymbol{\theta}) \equiv (\varphi_1 + \theta_1, \ldots, \varphi_l + \theta_l) \bmod 2\pi, \tag{4.1.16}$$

will be called a "rotation of $\mathbf{T}^l$ with parameters $\boldsymbol{\theta} = (\theta_1, \ldots, \theta_l) \in \mathbf{R}^l$". The group $(S_t)_{t \in \mathbf{R}}$ of transformations of $\mathbf{T}^l$ into itself defined by

$$S_t \boldsymbol{\varphi} = S_t(\varphi_1, \ldots, \varphi_l) = (\varphi_1 + \omega_1 t, \ldots, \varphi_l + \omega_l t) \bmod 2\pi \tag{4.1.17}$$

will be called the "flow on $\mathbf{T}^l$ generated by the rotation of $\mathbf{T}^l$ with speed $\boldsymbol{\omega}$" or the "quasi-periodic flow on $\mathbf{T}^l$ with pulsation $\boldsymbol{\omega}$".

The following is then a corollary to Proposition 1.

3 Corollary. *It is possible to establish a correspondence between all the initial data* $(\boldsymbol{\eta}, \boldsymbol{\xi}) \in \mathbf{R}^{2l}$ *for Eq.* (4.1.3) *and the set of the points* $(\mathbf{A}, \boldsymbol{\varphi}) \in \mathbf{R}^l_+ \times \mathbf{T}^l$ *via Eq.* (4.1.15).

The correspondence is one to one, nonsingular, and of class C^∞ *between* $(0, +\infty) \times \mathbf{T}^l$ *and its image in* $\mathbf{R}^{2l}$.

In $(\mathbf{A}, \boldsymbol{\varphi})$ *coordinates, the motion of Eq.* (4.1.5) *is simply*

$$t \to (\mathbf{A}, \boldsymbol{\varphi} + \boldsymbol{\omega} t), \tag{4.1.18}$$

i.e., *it is a quasi-periodic flow on the torus* $\{\mathbf{A}\} \times \mathbf{T}^l$.

Observations.

(1) Corollary 3 and Eq. (4.1.18) say that the motion of l harmonic oscillators "consists of quasi-periodic motions taking place on a family of l-dimensional tori parameterized by l parameters $\mathbf{A}$".

If one discards the data for which some of the normal modes are at rest (i.e., those for which some of the A's vanish), one can also say that the initial data space can be thought of as "foliated" by an l-dimensional family of l-dimensional tori.

(2) The parameter A_i is called the "action of the ith normal mode". If one describes the system in $(\mathbf{A}, \boldsymbol{\varphi})$ coordinates in the region where $\mathbf{A} \in (0, +\infty)^l$, it is clear that it can be regarded as a Hamiltonian system on $(0, +\infty)^l \times \mathbf{T}^l$ with Hamiltonian

$$h(\mathbf{A}, \boldsymbol{\varphi}) = \sum_{i=1}^{l} \omega_i A_i = \boldsymbol{\omega} \cdot \mathbf{A} \tag{4.1.19}$$

which leads immediately to Eq. (4.1.18).

(3) Observation (2) leads us to think that if the original system with Lagrangian (4.1.2) is described in the Hamiltonian form by the Hamiltonian

$$H(\mathbf{p}, \mathbf{x}) = \tfrac{1}{2} G^{-1}\mathbf{p} \cdot \mathbf{p} + \tfrac{1}{2} V\mathbf{x} \cdot \mathbf{x} \tag{4.1.20}$$

[see Eq. (3.11.25)], the map $(\mathbf{A}, \boldsymbol{\varphi}) \leftrightarrow (\mathbf{p}, \mathbf{x})$ between $(0, +\infty)^l \times \mathbf{T}^l$ and the part of phase space where all the normal modes are excited is a completely canonical transformation: this is in fact true and it is the reason for writing Eq. (4.1.5) with $\sqrt{2A_i/\omega_i}$ instead of the simpler A_i (see exercises).

Exercises for §4.1

1. Using Problems 1, 2, and 33, §3.11, show that the map $(\mathbf{p}, \mathbf{q}) \leftrightarrow (\pi, \kappa)$ with $\pi = p/\sqrt{\omega m}$, $\kappa = q\sqrt{\omega m}$, and $(\pi, \kappa) \to ((\pi^2 + \kappa^2)/2, \varphi) \equiv (A, \varphi)$ with $\varphi = \{$polar angular coordinate of $(\kappa, \pi) \in \mathbf{R}^2\}$ are completely canonical maps. Show that performing such transformations successively, one builds a completely canonical transformation changing $H = p^2/2m + (\omega^2 m/2)q^2$ into $H = \omega A$.

2. Let $H(\mathbf{p}, \mathbf{q}) = \frac{1}{2} G^{-1}\mathbf{p} \cdot \mathbf{p} + \frac{1}{2} V\mathbf{q} \cdot \mathbf{q}$ with G, V being two positive-definite matrices, $l \times l$. By Problem 33 of §3.11, the map $(\mathbf{p}, \mathbf{q}) \leftrightarrow (\boldsymbol{\pi}, \boldsymbol{\kappa})$ defined by $\mathbf{p} = \sqrt{G}\,\boldsymbol{\pi}$, $\mathbf{q} = \sqrt{G^{-1}}\,\boldsymbol{\kappa}$ (see Appendix F for the definition and the existence of the matrix $\sqrt{G}$ such that $\sqrt{G}^{\,2} = G$) is completely canonical. Show that it transforms H into $\frac{1}{2}\boldsymbol{\pi} \cdot \boldsymbol{\pi} + \frac{1}{2}\tilde{V}\boldsymbol{\kappa} \cdot \boldsymbol{\kappa}$ with $\tilde{V} = \sqrt{G^{-1}}\, V \sqrt{G^{-1}}$.

Let R be an orthogonal matrix (see Appendix E), transforming $\tilde{V}$ into a diagonal matrix Ω with diagonal elements $\omega_1^2, \ldots, \omega_l^2$, i.e., $R^T \tilde{V} R = \Omega$ (see Appendix F for the existence of R). Show that the further completely canonical change of coordinates $\boldsymbol{\pi} = R^T \hat{\boldsymbol{\pi}}$, $\boldsymbol{\kappa} = (R^T)^{-1}\hat{\boldsymbol{\kappa}} \equiv R\hat{\boldsymbol{\kappa}}$ changes H into

$$\frac{1}{2}\hat{\boldsymbol{\pi}}^2 + \frac{1}{2}\Omega\hat{\boldsymbol{\kappa}} \cdot \hat{\boldsymbol{\kappa}} \equiv \frac{1}{2}\sum_{i=1}^{l}(\hat{\pi}_i^2 + \omega_i^2 \hat{\kappa}_i^2).$$

Then, for each i, by further applying the maps in Exercise 1, the Hamiltonian is changed in $\Sigma_{i=1}^l \omega_i A_i$: also prove this fact.

3. Check that the variables $(\mathbf{A}, \boldsymbol{\varphi})$ constructed in Exercise 2 are the same as those appearing in Proposition 1.

§4.2. Irrational Rotations on l-Dimensional Tori

In §4.1 we showed how one can describe naturally the motion of a harmonic oscillator as a quasi-periodic flow on $\mathbf{T}^l$ of the form

$$S_t\varphi = (\varphi + t\omega) = (\varphi_1 + \omega_1 t, \dots, \varphi_l + \omega_l t) \bmod 2\pi. \tag{4.2.1}$$

Let us therefore analyze a few properties of the quasi-periodic flows.

2 Definition. One says that the flow of Eq. (4.2.1) is "irrational" if the numbers $(\omega_1, \dots, \omega_l) \in \mathbf{R}^l$ are "rationally independent", i.e., if the relation

$$\mathbf{n} \cdot \omega = \sum_{i=1}^{l} n_i \omega_i = 0, \qquad n_1, \dots, n_l, \text{ integers}, \tag{4.2.2}$$

implies $n_1 = \cdots = n_l = 0$.

Then we arrive at the following proposition.

4 Proposition. *Let $(S_t)_{t \in \mathbf{R}}$ be a quasi-periodic flow defined on $\mathbf{T}^l$ by Eq. (4.2.1) with $\omega \in \mathbf{R}^l$. If $\omega \in \mathbf{T}^l$ and $t_0 \in \mathbf{R}$, the trajectory*

$$\Omega(t_0) = \{\varphi' \mid \varphi' = S_t\varphi, \text{ for some } t \geqslant t_0\} \tag{4.2.3}$$

is dense on $\mathbf{T}^l$ if and only if the flow is irrational.

Observation. It would be possible to provide a direct proof of Proposition 4 along the lines of the analogous Proposition 27, p. 94, §2.20, in the case $l = 2$.

However, we prefer to give an alternative proof based on the Fourier series and on the following proposition which is interesting in and of itself.

5 Proposition. *Let $f \in C^\infty(\mathbf{T}^l)$ and let $(S_t)_{t \in \mathbf{R}_+}$ be a flow of the type of Eq. (4.2.1) on $\mathbf{T}^l$ which is irrational. Then, $\forall \varphi \in \mathbf{T}^l$, the average value*

$$\bar{f}(\varphi) = \lim_{T \to +\infty} \frac{1}{T} \int_0^T f(S_t\varphi)\, dt \tag{4.2.4}$$

exists and is φ-independent and equal to

$$\bar{f} = \frac{1}{(2\pi)^l} \int_{T^l} f(\varphi_1', \dots, \varphi_l')\, d\varphi_1' \dots d\varphi_l' \equiv \int_{T^l} f(\varphi') \frac{d\varphi'}{(2\pi)^l}. \tag{4.2.5}$$

PROOF. Since $f \in C^\infty(\mathbf{T}^l)$, f may be represented as

$$f(\varphi) = \sum_{n_1, \dots, n_l}^{-\infty, +\infty} \hat{f}_{n_1 \dots n_l} e^{i\sum_{j=1}^{l} n_j\varphi_j} \equiv \sum_{\mathbf{n} \in \mathbf{Z}^l} \hat{f}_{\mathbf{n}} e^{i\mathbf{n} \cdot \varphi}, \tag{4.2.6}$$

where $(\hat{f}_{\mathbf{n}})_{\mathbf{n} \in \mathbf{Z}^l}$ are the Fourier harmonics of f (see Proposition 28, §2.21, p. 103), and they decrease faster than any power in $|\mathbf{n}|$ as $|\mathbf{n}| \to \infty$.

Furthermore, the right-hand side of Eq. (4.2.5) is just f_0 [see Eq. (2.21.13)]. Then

$$\frac{1}{T}\int_0^T f(\varphi+\omega t)\,dt=\sum_{\mathbf{n}\in\mathbf{Z}^l}\hat{f}_{\mathbf{n}}e^{i\mathbf{n}\cdot\varphi}\left\{\frac{1}{T}\int_0^T e^{it\,\mathbf{n}\cdot\omega}\,dt\right\} \tag{4.2.7}$$

and the series in Eq. (4.2.7) is majorized by the convergent series

$$\sum_{\mathbf{n}\in\mathbf{Z}^l}|\hat{f}_{\mathbf{n}}|<+\infty \tag{4.2.8}$$

because the number in curly brackets in Eq. (4.2.7) clearly has a modulus not exceeding 1, being an average of numbers of modulus 1. Then we can take the limit in Eq. (4.2.7), as $T\to+\infty$, term by term.

But the integral in the right-hand side of Eq. (4.2.7) is

$$\frac{1}{T}\,\frac{e^{iT\mathbf{n}\cdot\omega}-1}{i\mathbf{n}\cdot\omega}\xrightarrow[T\to\infty]{}0,\qquad \text{if}\quad \mathbf{n}\cdot\omega\neq 0, \tag{4.2.9}$$

while it is 1 if $\mathbf{n}\cdot\omega=0$. However, $\mathbf{n}\cdot\omega=0$ only for $\mathbf{n}=\mathbf{0}$ and all the terms in Eq. (4.2.7) vanish except that with $\mathbf{n}=\mathbf{0}$, as $T\to+\infty$, and Eq. (4.2.5) is proved. mbe

Note that Proposition 5 is also an immediate consequence of Proposition 30, §2.21, p. 105. The same method of proof of Proposition 5 could be used to prove the following proposition which we describe before proving Proposition 4.

6 Proposition. *With the same hypothesis as that of Proposition* 5, *let* $T\in\mathbf{R}$, $T\neq 0$, *and consider the limit*

$$\lim_{N\to\infty}\frac{1}{N}\sum_{h=0}^{N-1}f(S_{hT}\varphi). \tag{4.2.10}$$

Such a limit exists and is given by Eq. (4.2.5) *if the* $(l+1)$ *numbers* $\omega,\omega_1,\ldots,\omega_l$ $(\omega\equiv 2\pi T^{-1})$ *are rationally independent.*

Observations.

(1) Proposition 6 is the generalization to the $l>1$ case of the Observations (5) and (6), p. 111. The proof is left to the reader as an exercise on the proof of Proposition 5.

(2) A simple analysis of the proof of the Propositions 5 and 6 allows us to conclude that the limits of Eqs. (4.2.4) and (4.2.10) exist in general, but they will not generally be φ independent unless $\omega_1,\ldots,\omega_l$ (or $\omega_1,\ldots,\omega_l$, $2\pi T^{-1}$) are rationally independent.

An immediate corollary to Proposition 5 is the following proof of Proposition 4.

PROOF OF PROPOSITION 4. Assume that S_t is an irrational flow. Let $\varphi_0\in\mathbf{T}^l$ and let $\chi\in C^\infty(\mathbf{T}^l)$ be a non-negative function having the value 1 in φ_0 and zero outside a small sphere $\sigma_\epsilon \subset \mathbf{T}^l$ with center φ_0 and radius ϵ (in the metric of $\mathbf{T}^l$ [see Eq. (2.21.5), p. 101.]

Let us apply Proposition 5 to χ. We see that the average value of $t \to \chi(S_t\varphi)$ cannot approach zero, $\forall \varphi \in \mathbf{T}^l$. Hence, for every t_0, there must be $t > t_0$ such that $\chi(S_t\varphi) > 0$, i.e., $S_t\varphi$ is closer to φ_0 than ϵ. This means that $\Omega(t_0)$ is dense. Vice versa, if there exist integers, $\bar{n}_1, \ldots, \bar{n}_l$, not all equal to zero, such that $\bar{\mathbf{n}} \cdot \omega = 0$, the function on $\mathbf{T}^l$ defined by

$$\varphi \to \cos(\bar{n}_1\varphi_1 + \cdots + \bar{n}_l\varphi_l) = \cos \bar{\mathbf{n}} \cdot \varphi \tag{4.2.11}$$

is not constant on $\mathbf{T}^l$, as is easy to see, but is constant on the trajectory $t \to S_t(\varphi)$, $t \in \mathbf{R}_+$, for all $\varphi \in \mathbf{T}^l$ (since $\bar{\mathbf{n}} \cdot \omega = 0$). Therefore, for instance, the origin's trajectory cannot approach too closely any point φ such that $\cos \varphi \cdot \bar{\mathbf{n}} < 1$ and vice versa. So $\Omega(t_0)$ is not dense. mbe

In the same way in which Proposition 5 implies Proposition 4, one sees that Proposition 6 implies the following corollary.

7 Corollary. *With the same hypothesis as that of Proposition* 4, *let* $\tau > 0$. *The denumerable subset of* $\mathbf{T}^l$,

$$\Omega_\tau(t_0) = \{\psi \mid \exists h \text{ integer}, h\tau > t_0, \psi = S_{h\tau}\varphi\} \tag{4.2.12}$$

is dense in $\mathbf{T}^l$ *if and only if the* $l+1$ *numbers* $\omega, \omega_1, \ldots, \omega_l$, $\omega = 2\pi\tau^{-1}$, *are rationally independent.*

PROOF. Exercise.

§4.3. Ordered Systems of Oscillators. Phenomenological Discussion and Heuristic Formulation of the Model of the Perfect Elastic Body (String, Film, and Solid)

In applications, serious difficulties may be met in the use of the general theory of §4.1, and §4.2. Such use, in fact, presupposes the actual possibility of constructing the proper pulsations $\omega_1, \ldots, \omega_l$ and the respective eigenvectors $\eta^{(1)}, \ldots, \eta^{(l)}$: we just saw that their construction passes through the solution of an lth-degree algebraic equation, Eq. (4.1.6), and of l linear systems of l equations, Eq. (4.1.7).

However, it is also true that in important applications, the matrices G and V of §4.1 are not arbitrary, but rather they have special properties sometimes permitting the explicit solution of the normal modes' construction.

In §4.3–4.6, we examine some of the most interesting cases, while this section is devoted to the precise mathematical formulation of the models that we wish to consider.

Let $\mathbf{Z}_a^d$ be the d-dimensional lattice of the points $\xi \in \mathbf{R}^d$ with coordinates which are integer multiples of $a > 0$:

$$\xi = (n_1a, n_2a, \ldots, n_da), \qquad n_1, \ldots, n_d \text{ integers.} \tag{4.3.1}$$

We shall imagine that in every site $\xi \in \mathbf{Z}_a^d$, a mass m oscillates bound by ideal constraints to move on a straight line through ξ and orthogonal to $\mathbf{R}^d$.

Furthermore, suppose that if y_ξ is the elongation with respect to ξ of the oscillator in ξ then:

(i) Every oscillator is subject to a restoring elastic force with potential energy

$$\frac{K}{2} y_\xi^2. \tag{4.3.2}$$

(ii) Every oscillator is subject to an external force with potential energy

$$mg(\xi) y_\xi, \tag{4.3.3}$$

where $g \in C^\infty(\mathbf{R}^d)$ ("weight").

(iii) Between the oscillators adjacent in $\mathbf{Z}_a^d$, an elastic force acts whose potential energy is

$$\tfrac{1}{2} K' \left[(y_\xi - y_{\xi'})^2 + a^2 \right], \tag{4.3.4}$$

where $|\xi' - \xi| = a$ and the term in square brackets represents the square of the elongation of a spring between the two oscillators.

(iv) An ideal constraint forcing all the oscillators outside an open connected bounded region Ω, with boundary $\partial\Omega$ which is a C^∞-regular surface, to have zero elongation. We set $\Omega_a = \Omega \cap \mathbf{Z}_a^d$.

We shall only consider the cases $d = 1$ or $d = 2$, the $d = 3$ case being a not too interesting model of an elastic solid since it can only "vibrate in one direction".

The situation in the $d = 1$ case is pictured in Fig. 4.1, and the $d = 2$ case is pictured in Fig. 4.2.

Analytically, the system under the analysis seems to be described by the Lagrangian function:

$$\begin{aligned} \mathfrak{L}_0 = \frac{1}{2} \sum_{\xi \in \Omega_a} m \dot{y}_\xi^2 - \frac{K}{2} \sum_{\xi \in \Omega_a} y_\xi^2 - m \sum_{\xi \in \Omega_a} g(\xi) y_\xi \\ - \frac{K'}{2} \sum_{\xi \in \Omega_a} \sum_{\mathbf{e}} \frac{1}{\nu(\mathbf{e}, \xi)} (y_\xi - y_{\xi + a\mathbf{e}})^2, \end{aligned} \tag{4.3.5}$$

where $\Sigma_{\mathbf{e}}$ denotes the sum over the $2d$ unit vectors directed as the axes of $\mathbf{Z}_a^d : \mathbf{e} = \mathbf{e}_1, -\mathbf{e}_1, \mathbf{e}_2, -\mathbf{e}_2, \ldots, \mathbf{e}_d, -\mathbf{e}_d$ if $\mathbf{e}_1, \ldots, \mathbf{e}_d$ are the d unit vectors associated with $\mathbf{Z}_a^d$, and, to avoid double counting, $\nu(\mathbf{e}, \xi) = 2$ if $\xi, \xi + a\mathbf{e} \in \Omega_a$, $\nu(\mathbf{e}, \xi) = 1$ otherwise.

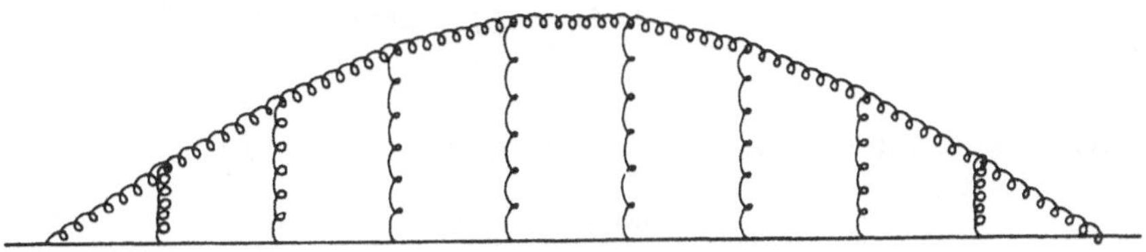

Figure 4.1.

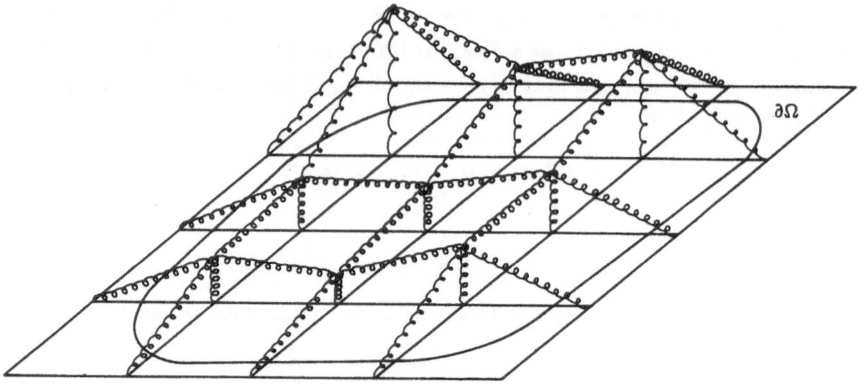

Figure 4.2.

In the last sum in the right-hand side of Eq. (4.3.5), we have neglected writing the term a^2 appearing in Eq. (4.3.4) since it produces an additive constant to $\mathcal{L}_0$ (dynamically irrelevant).

In Eq. (4.3.5) there appear terms y_ξ with $\xi \notin \Omega_a$ (in fact, if ξ is close to $\partial\Omega$, it can happen that $\xi + a\mathbf{e} \notin \Omega_a$). Such terms, conforming to (iv), must be interpreted by setting $y_\xi = 0$.

From a physical viewpoint, the interest of the mechanical system (4.3.5) lies in the fact suggested by the above pictures that if a is very small, it can be considered as a discrete model for an elastic string or film (if $d = 1$ or $d = 2$).

We can imagine that for small a, every "regular" initial datum $(\dot{y}_\xi, y_\xi)_{\xi \in \Omega_a}$, i.e., every datum having the form

$$y_\xi = u(\xi), \qquad \dot{y}_\xi = v(\xi), \qquad \xi \in \Omega_a, \tag{4.3.6}$$

where u and $v \in C^\infty(\mathbf{R}^d)$, evolves remaining approximately regular, thus simulating the motion of a string or film. In order for this to occur, it is, however, clear that the parameters m, K, K' must be suitably chosen as functions of a: their choice, which we adopt in the following, is motivated by a heuristic discussion.

(a) The mass m of each oscillator must have the form

$$m = \mu a^d, \qquad \mu > 0, \tag{4.3.7}$$

since each oscillator should intuitively correspond to a small piece of the body with dimension a: the body will then have density μ.

(b) The constants K, K' have to be determined so as to produce forces proportional to a^d on the oscillator in ξ; otherwise their effects would vanish in the $a \to 0$ limit (if $\ll a^d$) or they would produce infinite accelerations (if $\gg a^d$).

Hence, since the force associated with K is $-Ky$, it must be:

$$K = \sigma a^d, \qquad \sigma > 0. \tag{4.3.8}$$

The force exerted by the two oscillators in $\xi - a\mathbf{e}$ and $\xi + a\mathbf{e}$ on the oscillator in ξ is

$$-K'\left[(y_\xi - y_{\xi+ae_1}) + (y_\xi - y_{\xi-ae_1})\right], \tag{4.3.9}$$

and if y_ξ can be assimilated to $u(\xi)$, $u \in C^\infty(\mathbf{R}^d)$, we can compute Eq. (4.3.9) using the Taylor–Lagrange expansion to second order as

$$y_\xi - y_{\xi\pm ae_i} \cong u(\xi) - u(\xi \pm ae_i) = \mp a(\partial_i u)(\xi) - \frac{a^2}{2}(\partial_i^2 u)(\xi) + O(a^3), \tag{4.3.10}$$

where $\partial_i u, \partial_i^2 u$ are short notations for $\partial u/\partial\xi_i, \partial^2 u/\partial\xi_i^2$. Then Eq. (4.3.9) becomes

$$K'a^2(\partial_i^2 u)(\xi) + O(a^3) \tag{4.3.11}$$

which allows us to infer that we must set

$$K'a^2 = \tau a^d, \qquad \tau > 0. \tag{4.3.12}$$

With the above choices of K, m, K', Eq. (4.3.5) becomes

$$\begin{aligned}\mathfrak{L}_0^{(a)} &= \frac{\mu}{2}a^d \sum_{\xi\in\Omega_a} \dot y_\xi^2 - \frac{\sigma}{2}a^d \sum_{\xi\in\Omega_a} y_\xi^2 - \mu a^d \sum_{\xi\in\Omega_a} g(\xi) y_\xi \\ &\quad - \frac{\tau}{2}a^d \sum_{\xi\in\Omega_a}\sum_{\mathbf{e}} \frac{1}{\nu(\mathbf{e},\xi)} \frac{(y_\xi - y_{\xi+ae})^2}{a^2}.\end{aligned} \tag{4.3.13}$$

This model is not yet completely correct from a physical point of view. The heuristic discussion so far presented has been dealt with by supposing that ξ was far from $\partial\Omega$: if ξ is adjacent to $\partial\Omega$, it is not quite clear what is meant by y_ξ being regular since the functions u, v approximating it in Eq. (4.3.6) cannot be a independent, as supposed. A look at Fig. 4.3 suffices to realize this. The points outside Ω and adjacent to it have a rather erratic structure and, quite delicately, are a dependent.

Though this point may superficially appear irrelevant, it in fact has some importance at least as far as the correct formulation of the meaning of "regular datum y_ξ, $\dot y_\xi$" is concerned.

In the $d = 1$ case, the difficulty can be simply avoided by supposing that a is chosen always so that $\partial\Omega$ (which now consists of two points) is always on $\mathbf{Z}_a^1$: in this case, therefore, we shall actually do so and we shall assume that the system (4.3.13), with the above restriction on the "allowed values" of a, is a vibrating string model.

In the $d = 2$ case, it is obviously not possible to circumvent so easily the difficulty and, to understand what to do, let us again refer to some heuristic physical considerations.

When one imagines an elastic homogeneous film oscillating with a fixed boundary $\partial\Omega$, one probably has in mind the following situation: one deposits an elastic homogeneous film on a plane and then "glues" the film on the plane at $\partial\Omega$ and, *afterwards*, lets it oscillate and studies or watches the oscillations.

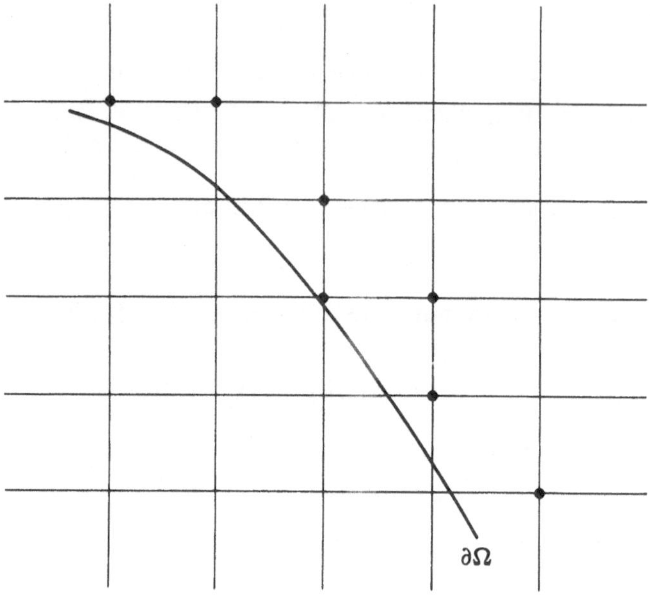

Figure 4.3.

When the surface is described, as in our case, by linked oscillators, the corresponding procedure is that of setting the oscillators in their equilibrium positions on $\mathbf{Z}_a^d$ and then pinching (with "glue" or "nails") the springs connecting the points $\boldsymbol{\xi} \in \Omega$ to the points $\boldsymbol{\xi}' = \boldsymbol{\xi} + a\mathbf{e} \notin \Omega$ at the point where the segment $\overline{\boldsymbol{\xi}\boldsymbol{\xi}'}$ crosses $\partial\Omega$. Once this is done, the system is allowed to oscillate.

On the boundary of Ω, the situation drawn in Fig. 4.4. is produced. This means that the elastic constant binding $y_{\boldsymbol{\xi}}$ to $\partial\Omega$ is different from K' contrary to what, instead, is hypothesized in $\mathcal{L}_0^{(a)}$, Eq. (4.3.13). In fact, $y_{\boldsymbol{\xi}}$ is pulled from $\partial\Omega$ by a spring with elastic constant

$$\tilde{K} = K' \frac{a}{\epsilon} \tag{4.3.14}$$

because the elastic constant of a piece of spring with elastic constant K' obtained by pinching it at a distance ϵ when the spring is elongated by a is manifestly given by Eq. (4.3.14).

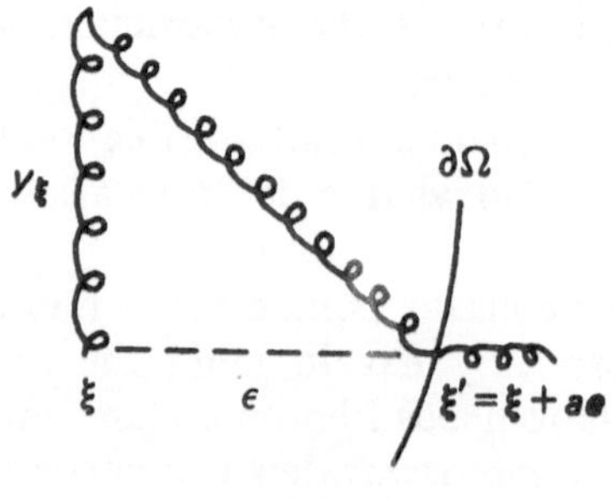

Figure 4.4.

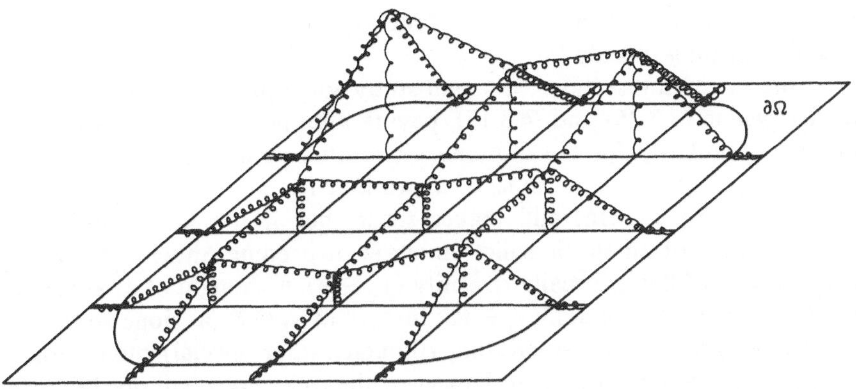

Figure 4.5. Illustration of the system of oscillators corresponding to Eq. (4.3.16).

Then, for $\xi \in \Omega_a$, we set

$$\epsilon_a(\xi, \mathbf{e}) = a \qquad \text{if} \quad \xi + a\mathbf{e} \in \Omega_a$$
$$\epsilon_a(\xi, \mathbf{e}) = \left\{\text{distance between } \xi \text{ and } \partial\Omega \cap \overline{\xi(\xi + a\mathbf{e})}\right\} \qquad \text{otherwise} \tag{4.3.15}$$

and the above considerations are summarized in the following Lagrangian function which will be supposed to be our discrete model of the elastic string or film (see Fig. 4.5), discarding the simpler but more naive model of Eq. (4.3.13):

$$\begin{aligned}\mathfrak{L}^{(a)} = {} & \frac{\mu}{2} a^d \sum_{\xi \in \Omega_a} \dot{y}_\xi^2 - \mu a^d \sum_{\xi \in \Omega_a} g(\xi) y_\xi - \frac{\sigma}{2} a^d \sum_{\xi \in \Omega_a} y_\xi^2 \\ & - \frac{\tau}{2} a^{d-2} \sum_{\xi \in \Omega_a} \sum_{\mathbf{e}} \frac{1}{\nu(\mathbf{e}, \xi)} \frac{a}{\epsilon_a(\xi, \mathbf{e})} (y_\xi - y_{\xi + a\mathbf{e}})^2.\end{aligned} \tag{4.3.16}$$

Here the values of $y_{\xi'}$ when $\xi' \notin \Omega$, present in Eq. (4.3.16) if ξ is close to $\partial\Omega$ and $\xi + \epsilon_a(\xi, \mathbf{e}) = \xi' \in \partial\Omega$, have to be thought of as vanishing. Or, more generally, we may fix the film or string at preassigned elongations on $\partial\Omega$, described by a function $h \in C^\infty(\partial\Omega)$[1]. In this case, the values of $y_{\xi'}$ for the above ξ''s are to be thought of as given by

$$y_{\xi'} = h(\xi'). \tag{4.3.17}$$

It is clear that Eq. (4.3.16) differs from Eq. (4.3.13) only because of the terms for which ξ is adjacent to $\partial\Omega$.

It is also clear that the critique to Eq. (4.3.13) raised above can no longer be applied. For instance, if $h \equiv 0$, the initial datum $y_\xi, \dot{y}_\xi$, very naturally, can be called "regular" if, $\forall \xi \in \Omega_a$,

$$y_\xi = u(\xi), \qquad \dot{y}_\xi = v(\xi) \tag{4.3.18}$$

[1] A function f defined on a regular surface $\Sigma \subset \mathbf{R}^d$ is said to be in $C^\infty(\Sigma)$ if in any local system (U, Ξ) of regular coordinates, its restriction to $\Sigma \cap U$ is a C^∞ function of the coordinates of the points of $U \cap \Sigma$ in (U, Ξ) (see Definition 10, §3.6, p. 171).

and $u, v \in C_0^\infty(\bar{\Omega}) = \{$set of the C^∞ functions defined in a neighborhood of $\bar{\Omega}$ and vanishing outside $\bar{\Omega}\}$.

In the upcoming sections, we shall study some properties of the motions of the system (4.3.16) and (4.3.17), paying attention to the problem of the motion's regularity for the motions with initial conditions (4.3.18) and to their interpretability as motions of a string or film.

If $d = 3$, Eq. (4.3.16) still makes sense, but it not longer provides a natural model of an elastic solid. However, it becomes much more natural if y_ξ, instead of being a scalar quantity ($y_\xi \in \mathbf{R}$), is thought of as a vector in $\mathbf{R}^3$. In this case, by thinking $y_\xi \in \mathbf{R}^3$, instead of $y_\xi \in \mathbf{R}$ (as done so far), Eq. (4.3.16) would yield an interesting (though rather special) model for the elastic deformations of a solid. However, the case $d = 3$ will not be further examined.

§4.4. Oscillator Chains and the Vibrating String

Consider the Lagrangian function of Eqs. (4.3.16) and (4.3.17), supposing $\Omega = [0, L]$ and a such that $L/a = N$ is an integer.

Therefore, this function describes a system of $N + 1$ oscillators, the first and the last of which are fixed at given heights. The Lagrangian of Eqs. (4.3.16) and (4.3.17) becomes

$$\mathfrak{L} = \frac{\mu}{2} a \sum_{i=1}^{N-1} \dot{y}_{ia}^2 + \mu a \sum_{i=1}^{N-1} g(ia) y_{ia} - \frac{\sigma}{2} a \sum_{i=1}^{N} y_{ia}^2 - \frac{\tau}{2} a \sum_{i=0}^{N-1} \frac{(y_{ia} - y_{ia+a})^2}{a^2}, \tag{4.4.1}$$

$$y_0 = h_0, \qquad y_L = h_L, \qquad g \in C^\infty(R). \tag{4.4.2}$$

The equations of motion for Eqs. (4.4.1) and (4.4.2) become

$$\mu a \ddot{y}_{ia} = \mu a g(ia) - \sigma a y_{ia} - \frac{\tau}{a}(2y_{ia} - y_{ia+a} - y_{ia-a}) \tag{4.4.3}$$

for $i = 1, \ldots, N-1$, where $y_0 = h_0$, $y_L = h_L$, $\mu > 0$, $\sigma \geqslant 0$, $\tau > 0$.

This is a system of linear nonhomogeneous differential equations which, as usual, we shall study by writing its solutions as sums of a particular solution and of a solution of the homogeneous equation, which is obtained by setting $g = 0$ and $h_0 = h_L = 0$.

Let us first study the homogeneous equation. The results of the following analysis are summarized by Proposition 9 at the end of this section.

In the homogeneous case, Eq. (4.4.3) correspond to the Lagrangian equations (4.4.1) and (4.4.2) with $g = 0$, $h_0 = h_L = 0$. This is a system of oscillators of the type considered in §4.1, with

$$G_{ij} = \mu a \delta_{ij}, \qquad i, j = 1, \ldots, N-1, \tag{4.4.4}$$

$$V_{ij} = \sigma a \delta_{ij} + \frac{\tau}{a}(2\delta_{ij} - \delta_{ij+1} - \delta_{ij-1}), \qquad i, j = 1, \ldots, N-1, \tag{4.4.5}$$

This can be checked immediately by noting that if $\gamma = (\gamma_i)_{i=1,\ldots,N-1}$, one finds (setting $\gamma_0 = \gamma_N = 0$) that Eq. (4.4.5) yields

$$\sum_{i,j=1}^{N-1} V_{ij}\gamma_i\gamma_j = a\sigma \sum_{j=1}^{N-1} \gamma_j^2 + \frac{\tau}{a} \sum_{j=0}^{N-1} \left((\gamma_i - \gamma_{i+1})^2\right) \tag{4.4.6}$$

Let us now try to solve the system $\omega^2 G\boldsymbol{\eta} - \mathbf{V}\boldsymbol{\eta} = \mathbf{0}$ [see Eqs. (4.1.6) and (4.1.7)]; such a system has the explicit form

$$-\mu\omega^2\eta_{ja} = -\sigma\eta_{ja} - \tau\frac{(2\eta_{ja} - \eta_{ja+a} - \eta_{ja-a})}{a^2}, \tag{4.4.7}$$

where $j = 1, \ldots, N-1$ and $\eta_0 = \eta_L = 0$.

Observing the clear analogy between this equation and the linear differential equation $-\mu\omega^2\eta = -\sigma\eta - \tau\eta''$, let us look for solutions of Eq. (4.4.7) having the form

$$\eta_{ja} = \sum_\rho \beta_\rho e^{\alpha_\rho ja}, \qquad \beta_\rho, \alpha_\rho \in \mathbf{C}, \tag{4.4.8}$$

where ρ is a summation index.

In order that $\exp\alpha_\rho ja$ is a solution of Eq. (4.4.7) for $j = 2, \ldots, N-2$, it must be [by substitution of Eq. (4.4.8) into Eq. (4.4.7), $j = 2, \ldots, N-2$]:

$$(-\mu\omega^2 + \sigma) = -\frac{2\tau}{a^2}\left(1 - \frac{e^{\alpha_\rho a} + e^{-\alpha_\rho a}}{2}\right). \tag{4.4.9}$$

If ω^2 is such that this equation for α_ρ has a solution α, then $-\alpha$ is also a solution. Hence, we shall try to solve Eq. (4.4.7) with $\boldsymbol{\eta}$ given by

$$\eta_{ja} = \beta_+ e^{\alpha ja} + \beta_- e^{-\alpha ja}, \qquad j = 1, \ldots, N-1, \tag{4.4.10}$$

where α and ω are related by Eq. (4.4.9).

The only equations of the system of Eq. (4.4.7) that Eq. (4.4.10) still may fail to verify are the first and last. If $\boldsymbol{\eta}$ has the form of Eq. (4.4.10), such equations become equations for $\beta_\pm$:

$$\sum_{\rho=\pm} \left((-\mu\omega^2 + \sigma) + \frac{\tau}{a^2}(2 - e^{-\rho\alpha a})\right)\beta_\rho e^{\rho\alpha(N-1)a} = 0 \tag{4.4.11}$$

corresponding to Eq. (4.4.7) with $i = (N-1)$, or for $i = 1$,

$$\sum_{\rho=\pm} \left((-\mu\omega^2 + \sigma) + \frac{\tau}{a^2}(2 - e^{\rho\alpha a})\right)\beta_\rho e^{\rho\alpha a} = 0 \tag{4.4.12}$$

which, by using Eq. (4.4.9), become, respectively,

$$\sum_{\rho=\pm} \frac{\tau}{a^2} e^{\rho\alpha Na}\beta_\rho = 0, \qquad \sum_{\rho=\pm} \frac{\tau}{a^2}\beta_\rho = 0. \tag{4.4.13}$$

In order that these two homogeneous equations in the two unknowns β_+ and β_- have a nontrivial solution, it is necessary that the determinant of the coefficients vanishes, i.e., it must be

$$\exp 2\alpha Na = 1 \tag{4.4.14}$$

and, in this case, $\beta_+ = -\beta_-$.

Hence, $(i = \sqrt{-1})$:

$$\alpha = i\frac{\pi}{Na}h, \qquad h = 0, 1, \ldots, N-1, \ldots \tag{4.4.15}$$

to which correspond the solutions [see Eq. (4.4.10)]

$$\eta_{ja}^{(h)} = \beta \sin\frac{\pi}{N}hj, \qquad h = 0, 1, \ldots, N-1 \tag{4.4.16}$$

with the respective eigenvalues ω_h^2 given by Eq. (4.4.9):

$$\omega_h^2 = \frac{\sigma}{\mu} + \frac{\tau}{\mu}\,\frac{2\left[1 - \cos(\pi h/Na)a\right]}{a^2}. \tag{4.4.17}$$

The $N-1$ solutions (4.4.16) are linearly independent vectors: they are, in fact, orthogonal, as is easy to check. This also follows from the general theory of Appendix F since $\omega_{h+1}^2 > \omega_h^2$, $h = 1, \ldots, N-2$, but the direct check is somewhat instructive. Let, in fact, $1 \leqslant h, h' \leqslant N-1$; then[2]

$$\begin{aligned}
\sum_{j=1}^{N-1} \eta_{ja}^{(h)}\eta_{ja}^{(h')} &\equiv \beta^2 \sum_{j=0}^{N-1} \sin\frac{\pi h}{N}j \sin\frac{\pi h'}{N}j \\
&= \frac{\beta^2}{2}\sum_{j=0}^{N-1}\left(\cos\frac{\pi(h-h')j}{N} - \cos\frac{\pi(h+h')j}{N}\right) \\
&= \frac{\beta^2}{2}\sum_{j=0}^{N-1} \mathrm{Re}\left(e^{i[\pi(h-h')/N]j} - e^{i[\pi(h+h')/N]j}\right) \\
&= \frac{\beta^2}{2}\mathrm{Re}\left(\frac{e^{i\pi[(h-h')} - 1}{e^{i\pi[(h-h')/N]} - 1} - \frac{e^{i\pi(h+h')} - 1}{e^{i\pi[(h+h')/N]} - 1}\right)
\end{aligned}$$

which, if $h = h'$, has to be interpreted as $\beta^2 N/2$ and, if $h \neq h'$, is zero since $e^{i\pi(h-h')} = e^{i\pi(h+h')} = \pm 1$ and $\mathrm{Re}(e^{i\alpha} - 1)^{-1} \equiv -\frac{1}{2}$, $\forall \alpha \in \mathbf{R}$. Therefore,

$$\boldsymbol{\eta}^{(h)} \cdot \boldsymbol{\eta}^{(h')} = \beta^2 \frac{N}{2}\delta_{hh'}. \tag{4.4.18}$$

Hence, using the results of §4.1, the most general motion of the $N-1$ oscillators described by Eqs. (4.4.1) and (4.4.2) with $h_0 = h_L = 0$ and $g = 0$ is, $\forall j = 1, \ldots, N-1$,

$$y_{ja}(t) = \sum_{h=1}^{N-1} A_h \sqrt{\frac{2}{N}}\left(\sin\frac{\pi h}{Na}ja\right)\cos(\omega_h t + \varphi h), \qquad h, h' = 1, \ldots, N-1, \tag{4.4.19}$$

where $\omega_h > 0$ is given by Eq. (4.4.17) and $A_h \geqslant 0$, $\varphi_h \in [0, 2\pi]$ are arbitrary constants.

Let us now go back to Eq. (4.4.3) to look for a particular solution to it. Obviously, the simplest particular solution is, if existing, a stationary one,

[2] Since $\cos\varphi = \mathrm{Re}(e^{i\varphi})$.

$\bar{\mathbf{y}}(t) = \mathbf{c}$, i.e., a solution of the system

$$\sigma c_{ja} + \tau \frac{(2c_{ja} - c_{ja+a} - c_{ja-a})}{a^2} = \mu g(ja) + \frac{\tau}{a^2}(\delta_{jN-1}h_L + \delta_{j1}h_a), \qquad j = 1, \ldots, N-1, \tag{4.4.20}$$

where $c_0 = c_L = 0$. These equations immediately follow from Eq. (4.4.3) in which the terms with the time derivatives have been eliminated and the inhomogeneous terms depending on g and h have been brought to the right-hand side.

Call $\boldsymbol{\gamma}$ the vector $\boldsymbol{\gamma} = (\gamma_{ia})_{i=1,\ldots,N-1}$ defined by the right-hand side of Eq. (4.4.20). Recalling the definition of V, Eq. (4.4.5), we see that Eq. (4.4.20) can be written as

$$a^{-1}V\mathbf{c} = \boldsymbol{\gamma}. \tag{4.4.21}$$

This equation has one and only one solution because V, by Eq. (4.4.6), is positive definite (so $\det V > 0$) if $\sigma \geqslant 0$, $\tau > 0$ and its solution $\mathbf{c}$ is the particular solution to Eq. (4.4.3) that we were looking for, and, in fact, it is the only stationary solution to Eq. (4.4.3).

It is even possible to find a useful expression for $\mathbf{c}$.

If in Eq. (4.4.16) we choose $\beta = \sqrt{2/N}$, we see that Eq. (4.4.18) says that $\boldsymbol{\eta}^{(1)}, \ldots, \boldsymbol{\eta}^{(N-1)}$ are $(N-1)$ vectors with $N-1$ components forming an orthonormal basis in $\mathbf{R}^{N-1}$.

Furthermore, these vectors are such that, by construction, [see, also, Eq. (4.4.7)]

$$a^{-1}V\boldsymbol{\eta}^{(h)} = \mu\omega_h^2\boldsymbol{\eta}^{(h)}. \tag{4.4.22}$$

Hence, we can write

$$\boldsymbol{\gamma} = \sum_{k=1}^{N-1} \hat{\gamma}(k)\boldsymbol{\eta}^{(k)}, \tag{4.4.23}$$

$$\mathbf{c} = \sum_{k=1}^{N-1} \hat{c}(k)\boldsymbol{\eta}^{(k)}, \tag{4.4.24}$$

where the $\hat{c}(k)$ are unknown and, setting $Na = L$,

$$\hat{\gamma}(k) = (\boldsymbol{\eta}^{(k)} \cdot \boldsymbol{\gamma}) = \sqrt{\frac{2}{N}}\left\{\left(\sum_{j=1}^{N-1} \mu g(ja) \sin\frac{\pi k}{L} ja\right) + \frac{\tau}{a^2}\left(h_0 \sin\frac{\pi k}{L} a + h_L \sin\frac{\pi k}{L}(N-1)a\right)\right\}. \tag{4.4.25}$$

Using Eq. (4.4.22), Eq. (4.4.21) becomes

$$\hat{c}(k) = \mu^{-1}\omega_k^{-2}\hat{\gamma}(k) \tag{4.4.26}$$

and provides an explicit expression for the components of $\mathbf{c}$ on the "natural basis" $\boldsymbol{\eta}^{(1)}, \ldots, \boldsymbol{\eta}^{(N-1)}$.

Before stating a proposition summarizing all of the above remarks, it is useful to give a very interesting definition allowing a suggestive interpretation of Eq. (4.4.21).

3 Definition. Let $\Omega = [0, L]$, $L/a = N =$ integer. One defines the "finite-differences Laplace operator relative to $\mathbf{Z}_a$" as the $(N-1) \times (N+1)$ matrix D associating the vector $((D\boldsymbol{\delta})_{ja})_{j=1}^{N-1}$ to the vector $\boldsymbol{\delta} = (\delta_{ja})_{j=0}^{N} \in \mathbf{R}^{N+1}$ so that[3]

$$(D\boldsymbol{\delta})_{ja} = \frac{\delta_{ja+a} - 2\delta_{ja} + \delta_{ja-a}}{a^2}, \qquad j = 1, \ldots, N-1. \tag{4.4.27}$$

In this notation, Eq. (4.4.21) can be written as

$$\begin{aligned} (\sigma \mathbf{c} - \tau D\mathbf{c})_\xi &= \mu g(\xi), \qquad && \xi \in \Omega_a \backslash \partial\Omega, \\ c_\xi &= h_\xi, && \xi \in \partial\Omega. \end{aligned} \tag{4.4.28}$$

4 Definition. Equation (4.4.28) for the vector $\mathbf{c}$ will be called, for $\sigma \geqslant 0$, $\tau > 0$, a "discrete nonhomogeneous Dirichlet problem for the region Ω on $\mathbf{Z}_a^1$ with boundary datum h, interior datum μg".

The already remarked existence and uniqueness of the solutions of Eq. (4.4.21) can be phrased as follows.

8 Proposition. *Equation* (4.4.28) *admits one and only one solution for arbitrarily given boundary and interior data and for all* $\sigma \geqslant 0$, $\tau > 0$.

We conclude this section by summarizing its results.

9 Proposition. *The motions associated with Eqs.* (4.4.1) *and* (4.4.2) *have the form*

$$y_\xi^{(a)}(t) = c_\xi^{(a)} + \sum_{h=1}^{N-1} A_h \sqrt{\frac{2}{N}} \left(\sin \frac{\pi h}{L} \xi\right) \cos(\omega_h t + \varphi_h) \tag{4.4.29}$$

for $\xi \in \Omega_a$ *with*

$$\omega_h = \sqrt{\frac{\sigma}{\mu} + \frac{\tau}{\mu} \frac{2[1 - \cos(\pi h/L)a]}{a^2}}, \tag{4.4.30}$$

[3] The matrix elements of D are: $D_{ij} = -(2/a^2)\delta_{ij} + [(\delta_{ij+1} + \delta_{ij-1})/a^2]$, $i = 1, \ldots, N-1$, $j = 0, \ldots, N$.

and the vector $\mathbf{c}^{(a)} = (c_\xi)_{\xi \in \Omega_a}$ *is the solution to the Dirichlet problem* (4.4.28) *with boundary datum* h *and interior datum* g. *The vector* $\mathbf{c}$ *is given by*

$$c_\xi^{(a)} = \sum_{k=1}^{N-1} \frac{\sin(\pi/L)k\xi}{\omega_k^2} \left\{ \left(\frac{2}{N} \sum_{\xi' \in \Omega_a} g(\xi') \sin \frac{\pi k}{L} \xi' \right) + \frac{2}{N} \frac{\tau}{a^2} \left(h_0 \sin \frac{\pi k}{L} a + h_L \sin \frac{\pi k}{L} a(N-1) \right) \right\}. \tag{4.4.31}$$

Observation. The normal modes have a remarkable "spatial structure", i.e., a remarkable ξ dependence. They are in fact interpolated by sinusoidal functions with "two nodes", i.e., two zeros, at the "extremes of the string", 0 and L, and in the hth normal mode such a function has exactly $(h-1)$ other nodes in $[0, L]$. This is a complete description of the "wave-form" of the modes.

§4.5. The Vibrating String as a Limiting Case of a Chain of Oscillators. The Case of Vanishing g and h. Wave Equation

The motivation for the choice of the Lagrangian (4.4.1) and (4.4.2) lies in the request that the mechanical system described by it be a good model for the oscillations of an elastic string.

In this section we shall prove some propositions showing in a mathematically precise sense how this property is actually realized in the models of Eqs. (4.4.1) and (4.4.2) when g and h vanish. We shall suppose $\sigma \geqslant 0$, $\tau > 0$.

To get an idea of what we should try to prove, we first observe that Eq. (4.4.3) has a formal limit given by

$$\mu \frac{\partial^2 y_\xi}{\partial t^2} = \mu g_\xi - \sigma y_\xi - \tau \frac{\partial^2 y_\xi}{\partial \xi^2}, \tag{4.5.1}$$

$$y_0 = h_0, \qquad y_\xi = h_L, \tag{4.5.2}$$

as $a \to 0$, while Eq. (4.4.28) for the "center" of the oscillations becomes, still formally,

$$\begin{aligned} \sigma c_\xi - \frac{d^2}{d\xi^2} c_\xi &= \mu g(\xi), \qquad \xi \in [0, L], \\ c_0 = h_0, \qquad & \qquad c_L = h_L. \end{aligned} \tag{4.5.3}$$

We now discuss the following proposition.

10 Proposition. *Let $t \to \mathbf{y}^{(a)}(t)$, $t \in \mathbf{R}$, be the solution of Eq.* (4.4.3) *with $g = h = 0$, $\sigma \geqslant 0$, $\tau > 0$, $\mu > 0$, following the initial datum*

$$y_{ja}^{(a)}(0) = u_0(ja), \qquad j = 1, \ldots, N-1, \tag{4.5.4}$$

$$\dot{y}_{ja}^{(a)}(0) = v_0(ja), \qquad j = 1, \ldots, N-1, \tag{4.5.5}$$

where $u_0, v_0 \in C_0^\infty((0, L)) \equiv \{$functions in $C^\infty([0, L])$ vanishing in a neighborhood of 0 and $L\}$.

Then, $\forall t \in \mathbf{R}$, $\forall x \in [0, L]$, the limit

$$\lim_{\substack{a \to 0 \\ \xi \to x}} y_\xi^{(a)}(t) = w(x, t) \tag{4.5.6}$$

exists and defines a C^∞ function on $[0, L] \times \mathbf{R}$, *verifying the equations:*

$$\mu \frac{\partial^2 w}{\partial t^2} - \tau \frac{\partial^2 w}{\partial x^2} + \sigma w = 0, \qquad \forall (x, t) \in [0, L] \times \mathbf{R}, \tag{4.5.7}$$

$$w(x, 0) = u_0(x), \qquad \forall x \in [0, L], \tag{4.5.8}$$

$$\frac{\partial w}{\partial t}(x, 0) = v_0(x), \qquad \forall x \in [0, L], \tag{4.5.9}$$

$$w(0, t) = 0 = w(L, t), \qquad \forall t \in \mathbf{R}. \tag{4.5.10}$$

Equations (4.5.7)–(4.5.10) *admit one and only one C^∞ solution: this solution is explicitly given by Eq.* (4.5.19) *below.*

Observations.

(1) It is clear that this proposition makes precise the fact that a "regular" initial datum evolves through Eq. (4.4.3) into a "regular configuration". Furthermore, it explains why Eq. (4.5.7) is called the "wave equation" describing the oscillations of a string with density μ, tension τ, and restoring constant σ. In the case $\sigma = 0$, Eq. (4.5.7) is the "D'Alembert wave equation" for the vibrating string oscillating under the only action of its tension τ.

(2) The derivation of the wave equation presented here and its theory, as expressed by Proposition 10, starting from the theory of harmonic oscillators, is a celebrated theorem of Lagrange.

(3) Another explicit solution to Eqs. (4.5.7)–(4.5.10) can be found in Problem 11, p. 270, (see, also, §4.7).

Proof. Let us write Eq. (4.4.29) as

$$y_\xi^{(a)}(t) = \sum_{h=1}^{N-1} \left\{ \tilde{A}_h \sqrt{\frac{2}{N}} \sin \frac{\pi h}{L} \xi \cdot \cos \omega_h t + \tilde{B}_h \sqrt{\frac{2}{N}} \sin \frac{\pi h}{L} \xi \cdot \sin \omega_h t \right\}, \tag{4.5.11}$$

where $\xi = ia$, $i = 1, \ldots, N-1$ and let us determine $\tilde{A}_h$ and $\tilde{B}_h$ by imposing the initial data.

Consider the initial data of Eqs. (4.5.4) and (4.5.5) as $(N-1)$-component vectors and express them as linear combinations, with suitable coefficients, of the vectors $\boldsymbol{\eta}^{(1)}, \dots, \boldsymbol{\eta}^{(N-1)}$ with components $(\boldsymbol{\eta}^{(h)})_i = \sqrt{2/N}\sin(\pi i h a/L)$, which (as seen in §4.4) form an orthogonal basis in $\mathbf{R}^{N-1}$ [see Eqs. (4.4.16) and (4.4.18)]:

$$u_0(\xi) = \sum_{h=1}^{N-1} \hat{u}_0(h)\sqrt{\frac{2}{N}}\sin\frac{\pi h}{L}\xi, \qquad \xi = ia, \quad i = 1, \dots, N-1,$$
$$v_0(\xi) = \sum_{h=1}^{N-1} \hat{v}_0(h)\sqrt{\frac{2}{N}}\sin\frac{\pi h}{L}\xi, \qquad \xi = ia, \quad i = 1, \dots, N-1. \tag{4.5.12}$$

After Eq. (4.5.12), it becomes immediate to impose the initial data of Eqs. (4.5.4) and (4.5.5) to Eq. (4.5.11):

$$\tilde{A}_h = \hat{u}_0(h), \qquad \tilde{B}_h = \hat{v}_0(h)\omega_h^{-1}. \tag{4.5.13}$$

Since, on the other hand, $\hat{u}_0(h)$ and $\hat{v}_0(h)$ can be obtained by scalar multiplication of the vectors of Eqs. (4.5.4) and (4.5.5) by $\boldsymbol{\eta}^{(h)}$, Eq. (4.5.13) yields

$$\sqrt{\frac{2}{N}}\,\tilde{A}_h = \frac{2}{N}\sum_{\xi}\left(\sin\frac{\pi h}{L}\xi\right)u_0(\xi), \tag{4.5.14}$$

$$\sqrt{\frac{2}{N}}\,\tilde{B}_h = \frac{1}{\omega_h}\,\frac{2}{N}\sum_{\xi}\left(\sin\frac{\pi h}{L}\xi\right)v_0(\xi) \tag{4.5.15}$$

and Σ_ξ runs over $\xi = ia$, $i = 1, \dots, N-1$.

Then, by the assumptions on u_0 and v_0, one sees that Eqs. (4.5.14) and (4.5.15) contain summations over ξ which, after being multiplied by a, are the Riemann sums for the integrals between 0 and L of the functions $x \to (\sin \pi h x/L)u_0(x)$ and $x \to (\sin \pi h x/L)v_0(x)$, $x \in [0, L]$. Hence,

$$\lim_{a\to 0}\sqrt{\frac{2}{N}}\,\tilde{A}_h = \frac{2}{L}\int_0^L u_0(x)\left(\sin\frac{\pi h}{L}x\right)dx, \tag{4.5.16}$$

$$\lim_{a\to 0}\sqrt{\frac{2}{N}}\,\tilde{B}_h = \frac{1}{\bar{\omega}(h)}\,\frac{2}{L}\int_0^L v_0(x)\left(\sin\frac{\pi h}{L}x\right)dx, \tag{4.5.17}$$

where, for $h = 1, 2, \dots$ [see Eq. (4.4.30)],

$$\bar{\omega}(h) = \lim_{a\to 0}\omega_h = \sqrt{\frac{\sigma}{\mu} + \frac{\tau}{\mu}\left(\frac{\pi h}{L}\right)^2}\,. \tag{4.5.18}$$

Hence, we see that the sum (4.5.11), thought of as a series in h (with vanishing terms for $h \geqslant N$), converges term by term, as $a \to 0$ and $\xi \to x \in [0, L]$, to the series

$$w(x,t) = \sum_{h=1}^{\infty}\sin\frac{\pi h}{L}x\left\{\left(\frac{2}{L}\int_0^L u_0(x')\sin\frac{\pi h}{L}x'\,dx'\right)\cos\bar{\omega}(h)t + \left(\frac{2}{L}\int_0^L v_0(x')\sin\frac{\pi h}{L}x'\,dx'\right)\frac{\sin\bar{\omega}(h)t}{\bar{\omega}(h)}\right\}. \tag{4.5.19}$$

We now show that the series in Eq. (4.5.19) is uniformly convergent in t and x and defines a function w verifying Eqs. (4.5.7)–(4.5.10). This will mean that a function w verifying Eqs. (4.5.7)–(4.5.10) does exist. Then we shall prove Eq. (4.5.6), and the proof will finally be concluded by proving the uniqueness of the solution to Eqs. (4.5.7)–(4.5.10).

All of the above deductions are based on the following lemma, a corollary to the Fourier theorem, proved in Appendix I.

11 Lemma. *Let $\bar{C}^\infty([0,L])$ be the set of the $C^\infty([0,L])$ real functions vanishing together with all their even derivatives in the points 0 and L.*

Then setting

$$\bar{u}_k = \frac{2}{L}\int_0^L u(x')\sin\frac{\pi k}{L}x'\,dx' \tag{4.5.20}$$

$\forall u \in \bar{C}^\infty([0,L])$, *it follows that*:

(i) $\forall \alpha > 0$, $\exists C_\alpha$ *such that*

$$|\bar{u}_k| \leqslant C_\alpha(1+k^\alpha)^{-1}, \qquad \forall k = 1,2,\ldots. \tag{4.5.21}$$

(ii) $u(x) = \Sigma_{k=1}^\infty \bar{u}_k \sin(\pi k x/L)$. (4.5.22)

(iii) *Equation* (4.5.22) *can be differentiated term by term an arbitrary number of times, giving rise to uniformly convergent series.*

(iv) *Every function of the form of Eq.* (4.5.22) *with $\bar{u}_k$ verifying Eq.* (4.5.21) *is in $\bar{C}^\infty([0,L])$.*

Observation. Clearly $\bar{C}^\infty([0,L]) \supset C_0^\infty((0,L))$.

The proof of Proposition 10 can be continued as follows.

The uniform convergence in t and x of Eq. (4.5.19), as well as the admissibliity of its term-by-term differentiations, follow from (i), Eq. (4.5.21). We call w the sum of the series (4.5.19): it clearly verifies Eq. (4.5.7) because every term of Eq. (4.5.19) does [see Eq. (4.5.18) and do a direct check].

Equation (4.5.10) is clearly true since $\sin(\pi h x/L)$ vanishes in 0 and in L, for all integers h.

Equations (4.5.8) and (4.5.9) can be checked by computing $w(x,0)$, $(\partial w/\partial t)(x,0)$, from Eq. (4.5.19), using (ii) of Lemma 11.

It remains to prove Eq. (4.5.6) and uniqueness. Since Eq. (4.5.11), thought of as a series in h by setting $\tilde{A}_h, \tilde{B}_h \equiv 0$ for $h \geqslant N$, converges term by term to the function in Eq. (4.5.19), we simply have to show that the series (4.5.11) is uniformly convergent in a and ξ (or, what amounts to the same, in N and ξ). To obtain this, it suffices to show that given $\alpha > 0$ there exists C'_α such that, $\forall \alpha > 0$,

$$\sqrt{\frac{2}{N}}\,|\tilde{A}_h| \leqslant C'_\alpha(1+h^\alpha)^{-1}, \qquad h = 1,2,\ldots, \tag{4.5.23}$$

$$\sqrt{\frac{2}{N}}\,|\tilde{B}_h| \leqslant C'_\alpha(1+h^\alpha)^{-1}, \qquad h = 1,2,\ldots, \tag{4.5.24}$$

having set $\tilde{A}_h = \tilde{B}_h = 0$ for $h = N, N+1, \ldots$.

Let us, for instance, prove Eq. (4.5.23). From Eqs. (4.5.14) and (4.5.22), one obtains, $\forall h = 1, \ldots, N-1$,

$$\begin{aligned}\sqrt{\frac{2}{N}}\,\tilde{A}_h &= \frac{2}{N} \sum_{\substack{\xi = ia \\ i=1,\ldots,N-1}} \sin\frac{\pi h}{L}\xi\left(\sum_{k=1}^{\infty} \bar{u}_{0k}\sin\frac{\pi k}{L}\xi\right) \\ &= \sum_{k=1}^{\infty} \bar{u}_{0k}\left(\frac{2}{N} \sum_{\substack{\xi = ia \\ i=1,\ldots,N-1}} \sin\frac{\pi h}{L}\xi\sin\frac{\pi k}{L}\xi\right)\end{aligned} \tag{4.5.25}$$

and it is clear by Eqs. (4.4.16) and (4.4.18) that for $h = 1, \ldots, N-1$ and k arbitrary (even for $k \geqslant N$),

$$\begin{aligned}\frac{2}{N} \sum_{\substack{\xi = ia \\ i=1,\ldots,N-1}} \sin\frac{\pi h}{L}\xi\sin\frac{\pi k}{L}\xi &= \delta_{k,h} - \delta_{k,2N-h} + \delta_{k,h+2N} - \cdots \\ &= \sum_{p=0}^{\infty} \delta_{k,h+2pN} - \sum_{p=1}^{\infty} \delta_{k,2pN-h}\,.\end{aligned} \tag{4.5.26}$$

Hence, by Eq. (4.5.21), $\forall \alpha > 0$, $\forall h = 1, \ldots, N-1$,

$$\begin{aligned}\left|\sqrt{\frac{2}{N}}\,\tilde{A}_h\right| &= |\bar{u}_{0h} - \bar{u}_{02N-h} + \ldots| \leqslant \sum_{p=0}^{\infty} |\bar{u}_{0h+p}| \leqslant \sum_{p=0}^{\infty} \frac{C_\alpha}{(1+(h+p)^\alpha)} \\ &\leqslant \sum_{p=0}^{\infty} \frac{C_\alpha}{\sqrt{(1+h^\alpha)}}\,\frac{1}{\sqrt{(1+p^\alpha)}} = \frac{C_\alpha}{\sqrt{1+h^\alpha}}\left(\sum_{p=0}^{\infty}\frac{1}{\sqrt{1+p^\alpha}}\right),\end{aligned} \tag{4.5.27}$$

implying Eq. (4.5.23) by the arbitrariness of α and because $\tilde{A}_h \equiv 0$ for $h \geqslant N$.

To show uniqueness, it is enough to show that if $w^0 \in C^\infty([0, L] \times \mathbf{R})$ and verifies Eqs. (4.5.7)–(4.5.10), with $u_0 = v_0 = 0$, then $w^0 \equiv 0$.

The idea of the proof is based on energy conservation. Equations (4.5.7)–(4.5.10) should "keep memory" of the fact that they are a formal limit of Eq. (4.4.3) and it should be possible to define, for every motion w verifying them, a function which is constant as t varies and which can be obtained as the limit $a \to 0$ of the energy expression for Eq. (4.4.3). If $y_0 = y_L = 0$, the energy of the motions of Eq. (4.4.3) is [see Eq. (4.3.13)]

$$\begin{aligned}E^{(a)} &= \frac{a\mu}{2} \sum_{\substack{\xi = ja \\ j=1,\ldots,N-1}} \dot{y}_\xi^2 + \frac{a\sigma}{2} \sum_{\substack{\xi = ja \\ j=1,\ldots,N-1}} y_\xi^2 \\ &\quad + \frac{a\tau}{2} \sum_{\substack{\xi = ja \\ j=0,\ldots,N-1}} \frac{(y_\xi - y_{\xi+a})^2}{a^2},\end{aligned} \tag{4.5.28}$$

formally becoming

$$E(w,t) = \frac{\mu}{2}\int_0^L\left(\frac{\partial w}{\partial t}\right)^2 dx + \frac{\sigma}{2}\int_0^L w^2\,dx + \frac{\tau}{2}\int_0^L\left(\frac{\partial w}{\partial x}\right)^2 dx. \tag{4.5.29}$$

as $a \to 0$.

If we show that the solutions of Eqs. (4.5.7)–(4.5.10) in $C^{\infty}([0,L]\times\mathbf{R})$ are such that $E(w,t)$ remains constant as t varies, uniqueness is proved. In fact, if $w(x,0)=0$ and $(\partial w/\partial t)(x,0)=0$, then $E(0)=0$, on the other hand, it is clear that $E(w,t)=0 \Rightarrow w(t,x)=0$, $\forall x\in[0,L]$, if $\sigma \geqslant 0$, $\tau>0$. But, the difference between two solutions of Eqs. (4.5.7)–(4.5.10) is a solution with $u_0=v_0=0$ with zero energy: hence, it vanishes identically.

To show the constancy of Eq. (4.5.29), one proceeds as follows:

$$\frac{d}{dt}E(w,t)=\mu\int_0^L \frac{\partial w}{\partial t}\frac{\partial^2 w}{\partial t^2}\,dx+\sigma\int_0^L w\frac{\partial w}{\partial t}\,dx+\tau\int_0^L \frac{\partial w}{\partial x}\frac{\partial}{\partial x}\frac{\partial w}{\partial t}\,dx. \tag{4.5.30}$$

Then we integrate the last term in the right-hand side by parts using $(\partial w/\partial t)(0,t)=(\partial w/\partial t)(L,t)\equiv 0$, by Eq. (4.5.10). Collecting the integrals into a single integral and taking Eq. (4.5.7) into account, one finds

$$\frac{dE}{dt}=\int_0^L \frac{\partial w}{\partial t}\left(\mu\frac{\partial^2 w}{\partial t^2}+\sigma w-\tau\frac{\partial^2 w}{\partial x^2}\right)dx=0 \tag{4.5.31}$$

and the proof is complete. mbe

Observations.

(1) From the proof, one can see that the condition $u_0, v_0 \in C_0^{\infty}((0,L))$ has only been used to apply the Lemma 11 through the observation that $C_0^{\infty}((0,L))\subset \overline{C}^{\infty}([0,L])$.

It is then clear that Proposition 10 can be strengthened by replacing the assumption $u_0, v_0\in C_0^{\infty}((0,L))$ with the assumption $u_0, v_0 \in \overline{C}^{\infty}([0,L])$ and by substituting Eq. (4.5.10) with

$$w(\cdot,t) \text{ and } \frac{\partial w}{\partial t}(\cdot,t)\in \overline{C}^{\infty}([0,L]), \qquad \forall t\in\mathbf{R}. \tag{4.5.32}$$

(where $\cdot$ denotes a dummy variable; in this case, $x\in[0,L]$).

In this way the existence and uniqueness theorem for the waves equations (4.5.7)–(4.5.9) and (4.5.32) with initial datum $u_0, v_0\in\overline{C}^{\infty}([0,L])$ is more satisfactory because the initial regularity condition is not modified as t evolves. In fact, from the above proof it is not possible to conclude (and it is generally false) that when the initial configuration u_0, v_0 is built with elements of $C_0^{\infty}((0,L))$, then also the evolved configuration at time t, $w(x,t)$, $(\partial w/\partial t)(x,t)$ consists of elements in $C_0^{\infty}((0,L))$ (i.e., the initial regularity is generally not preserved).

(2) One may think that $u_0, v_0\in\overline{C}^{\infty}([0,L])$ is still not optimal and that, perhaps, the optimal condition could be $u_0, v_0\in C^{\infty}([0,L])$ plus $u_0(0)=u_0(L)=0$, $v_0(0)=v_0(L)=0$. By counterexamples, it can be shown easily that this is not the case (see exercises). To further extend the set of the initial configurations, one has to give up the C^{∞} smoothness.

Exercises and Problems for §4.5

1. Consider the wave equation for $(x, t) \in \mathbf{R}^2$:

$$\frac{\partial^2 w}{\partial t^2} - c^2 \frac{\partial^2 w}{\partial x^2} = 0.$$

Given $u, v \in C^\infty(\mathbf{R})$, show that

$$w(x,t) = \frac{u(x+ct) + u(x-ct)}{2} + \int_{x-ct}^{x+ct} v(\xi) \frac{d\xi}{2c}$$

is a C^∞ solution verifying the initial datum (u, v).

2. In the context of Problem 1, suppose that $\int_{-\infty}^{+\infty}(v^2(x) + c^2u^2(x))\,dx < +\infty$. Show that w is the only $C^\infty(\mathbf{R}^2)$ solution "with finite energy E" and datum (u, v), where

$$E = \int_{-\infty}^{+\infty}\left(\left(\frac{\partial w}{\partial t}\right)^2 + c^2\left(\frac{\partial w}{\partial x}\right)^2\right)dx.$$

(Hint: Repeat the energy conservation argument at the end of the proof of Proposition 10.)

3. Find the relations between u and v, in the context of Problem 1, necessary to guarantee that w is a "purely progressive" or "purely regressive" wave, i.e., $w(x,t) = a(x - ct)$ or $w(x,t) = b(x + ct)$.

4. Let $u \in C_0^\infty((0, +\infty))$ and suppose that $u(x) = 0$, unless $x \in (a, b)$, $0 < a < b < +\infty$, and $u(x) > 0$ for $x \in (a, b)$. Let $v(x) = c(du/dx)(x)$. Show that, up to a time $t_0 > 0$, the solution w of the equation $\partial^2 w/\partial t^2 - c^2(\partial^2 w/\partial x^2) = 0$ with initial data (u, v) is such that $w(x, t) \in C_0^\infty((0, +\infty))$ for $t < t_0$. Show that for $t = t_0$, $w(\cdot, t) \in \overline{C}^\infty((0 + \infty))$. (Hint: Use Problem 3 by noting that up to $t_0 = a/c$ the solution is $w(x,t) = u(x + ct)$.)

5. Consider the wave equation on $[0, 1]$, $\partial^2 w/\partial t^2 - c^2(\partial^2 w/\partial x^2) = 0$, with the initial data $v_0 = c(du_0/dx)$, $u_0(x) = x^{2n}e^{-(1/2-x)^{-2}}$ for $0 < x < \frac{1}{2}$, $u_0(x) \equiv 0$ for $|x| \geqslant \frac{1}{2}$. Letting $n \geqslant 1$, show that up to $t_0 = 1/2c$, the function $w(x, t) = u_0(x - ct)$ if $0 \leqslant x - ct \leqslant \frac{1}{2}$ or $w(x, t) = 0$, otherwise, is a $C^{(2n)}$ solution following a $C^\infty([0, 1])$ datum. Infer that the conditions $u_0, v_0 \in \overline{C}^\infty([0, L])$ in Proposition 10 cannot be replaced by the more general ones of the Observation (2), p. 268. (Hint: Show by the same energy conservation argument at the end of the proof of Proposition 10 that there is uniqueness for the $C^{(2)}$ solutions of the wave equation, etc.)

6. Is the condition $\tau > 0$ in Proposition 10 essential? If yes, give a physical interpretation of the reason.

7. A solution to the equation $\partial^2 w/\partial t^2 - c^2(\partial^2 w/\partial x^2) + m^2 w = 0$ having the form $\exp i(kx \pm et)$ is called a "plane wave" solution. Its real and imaginary parts are called "real plane waves" solutions. Find the plane wave solutions to the above equation.

8. Find the energy per unit length of a real plane wave solution to the equation in Problem 7. (Hint: $E = \lim_{L\to\infty}(1/2L)\int_{-L}^{L}\{(\partial w/\partial t)^2 + c^2(\partial w/\partial x)^2 + m^2 w^2\}$ $dx \ldots$).

9. Formulate and prove Proposition 10 in the case when the segment $[0, L]$ is replaced by a closed circle, i.e., the oscillators in Fig. 4.1 are ideally bound to the set of equispaced lines orthogonal to a circle with radius R, obviously without fixed extreme oscillators ("periodic boundary conditions"). Show that Eqs. (4.5.7)–(4.5.9) remain the same while Eq. (4.5.10) is replaced by $u_0, v_0 \in C^\infty(\mathbf{T}^1(2\pi R)) = C^\infty$-periodic functions with period $2\pi R$. (Hint: The ordinary Fourier theorem replaces Lemma 11 in the proof (which actually looks easier).)

10. In the context of Problem 1, call $V_0(x) = \int_0^x v_0(\xi)\, d\xi$. Show that to compute w at the point (x, t), it is enough to know the data u_0, V_0 at the points $x \pm ct$ ("propagation along characteristic lines").

11. Consider the wave equations (4.5.7)–(4.5.10). Define $\bar{u}_0, \bar{v}_0$ as

$$\bar{u}_0(x) = u_0(x), \qquad \text{if} \quad 0 \leqslant x \leqslant L,$$
$$\bar{u}_0(L + x) = -u_0(L - x), \qquad \text{if} \quad L \leqslant L + x \leqslant 2L,$$

and

$$\bar{u}_0(x) = u_0(x - 2kL), \qquad \text{if} \quad x - 2kL \in [0, 2L].$$

Likewise, define $\bar{v}_0$.

Show that $\bar{u}_0, \bar{v}_0$ are $C^\infty(\mathbf{R})$ functions if and only if $u_0, v_0 \in \overline{C}^\infty([0, L])$.

Let $\overline{V}_0(x) \equiv \int_0^x \bar{v}_0(x)\, dx$. Show that the solution to Eqs. (4.5.7)–(4.5.10) can be written

$$w(x, t) = \frac{\bar{u}_0(x - ct) + \bar{u}_0(x + ct)}{2} + \frac{\overline{V}_0(x - ct) + \overline{V}_0(x + ct)}{2}$$

(see, also, §4.7). Find a statement analogous to the one in Problem 10 in terms of $\bar{u}_0, \overline{V}_0$.

§4.6. Vibrating String: General Case. The Dirichlet Problem in $[0, L]$

Having in mind the results of §4.4, it is convenient to study preliminarily what happens to the stationary solution $\mathbf{c}^{(a)}$ [see Eqs. (4.4.29) and (4.4.31)] in the limit $a \to 0$, $\xi \to x$.

The heuristic considerations at the beginning of §4.5 suggest the following proposition.

12 Proposition. *The stationary solution $\mathbf{c}^{(a)}$ of the oscillator-chain equations* (4.4.1) *and* (4.4.2) *given by Eq.* (4.4.31) *is such that the limit*

$$c(x) = \lim_{\substack{a \to 0 \\ \xi \to x}} c_\xi^{(a)} \tag{4.6.1}$$

exists for $x \in [0, L]$ and defines a function $c \in C^\infty([0, L])$ such that

$$\sigma c - \tau \frac{d^2 c}{dx^2} = \mu g, \qquad x \in [0, L], \tag{4.6.2}$$

$$c(0) = h_0, \qquad c(L) = h_L. \tag{4.6.3}$$

Proof. Define

$$c_\xi^{(a)1} = \sum_{k=1}^{N-1} \left(\sin \frac{\pi k}{L} \xi\right) \frac{1}{\omega_k^2} \left(\frac{2}{N} \sum_{\xi'} g(\xi') \sin \frac{\pi k}{L} \xi' \right), \tag{4.6.4}$$

$$c_\xi^{(a)2} = \sum_{k=1}^{N-1} \left(\sin \frac{\pi k}{L} \xi\right) \frac{1}{\mu \omega_k^2} \frac{\tau}{a^2} \frac{2}{N} \left(\sin \frac{\pi a}{L} k\right)\left(h_0 - (-1)^k h_L\right) \tag{4.6.5}$$

for $\xi = ia$, $i = 0, 1, \ldots, N$, $N = L/a$, and by Eq. (4.4.31),

$$\mathbf{c}^{(a)} = \mathbf{c}^{(a)1} + \mathbf{c}^{(a)2} \tag{4.6.6}$$

and $\mathbf{c}^{(a)1}$ solves Eq. (4.4.28) for $h = 0$, while $\mathbf{c}^{(a)2}$ solves it for $g = 0$.

We shall separately show the existence of the limits:

$$\lim_{\substack{a \to 0 \\ \xi \to x}} c_\xi^{(a)1} = c^{(1)}(x), \tag{4.6.7}$$

$$\lim_{\substack{a \to 0 \\ \xi \to x}} c_\xi^{(a)2} = c^{(2)}(x), \tag{4.6.8}$$

and that, furthermore, they define two $C^\infty([0, L])$ functions verifying Eqs. (4.6.2) and (4.6.3) with $h = 0$ or $g = 0$, respectively.

Let us first study Eq. (4.6.7) using Eq. (4.6.4) as a starting point. If we think of Eq. (4.6.4) as a series in k with all the terms with $k \geqslant N$ vanishing, we see that such a series converges term by term, when $\xi \to x$, $a \to 0$ to the series

$$c^{(1)}(x) = \sum_{k=1}^{\infty} \left(\sin \frac{\pi k}{L} x\right) \frac{1}{\bar{\omega}(k)^2} \left(\frac{2}{L} \int_0^L g(x') \sin \frac{\pi k}{L} x' \, dx' \right), \tag{4.6.9}$$

where $\bar{\omega}(k)^2 = \sigma/\mu + \tau(\pi k/L)^2/\mu = \lim_{a \to 0} \omega_k^2$ is given by Eq. (4.5.18).

If $g \in \bar{C}^\infty([0, L])$, we could infer from the Lemma 11, p. 266, Eq. (4.5.21), that the above series is a uniformly convergent series, differentiable indefinitely term by term. It would then be clear that $c^{(1)}$ verifies Eqs. (4.6.2) and (4.6.3) with $h = 0$ since

$$\sigma c^{(1)} - \tau \frac{d^2 c^{(1)}}{dx^2} = \sum_{k=1}^{\infty} \left(\sin \frac{\pi k}{L} x\right) \mu \left(\frac{2}{L} \int_0^L g(x') \sin \frac{\pi k}{L} x' \, dx' \right), \tag{4.6.10}$$

and by Lemma 11 the right-hand side is just μg.

It would also be easy to prove the validity of Eq. (4.6.7) with $c^{(1)}$ defined by Eq. (4.6.9). One should repeat, word by word, the §4.5 proof where the convergence of $y_\xi^{(a)}(t)$ to its "term-by-term limit", Eq. (4.5.19), is discussed.

In the present case, however, $g \in C^\infty([0, L])$ but not necessarily $g \in \bar{C}^\infty([0, L])$, and the proof of Eq. (4.6.7), of the convergence of Eq. (4.6.9), and of the $C^\infty([0, L])$ nature of $c^{(1)}$ is more delicate.

Technically, such a problem must be present and it takes place because the series (4.6.10) *cannot* converge too well to $g(x)$: if, in fact, it did converge absolutely and if it had g as its sum, it would follow that $g(0) = g(L) = 0$, for instance, which may be false for a given g.

This phenomenon always appears, whenever one tries to approximate a function g with functions (in our case $\sin(\pi kx/L)$) with properties too different from those of g (for instance, $g(0) \neq 0$ in general, but all the approximating functions vanish in 0!).

The upcoming discussion is interesting because it illustrates how it is sometimes possible to bypass the obstacle just met: it is in fact a type of problem that often occurrs in mathematical analysis.

We shall first show that the series in Eq. (4.6.9) converges to some function $c^{(1)}$ on $[0, L]$, continuous and once differentiable term by term. Then we shall show that Eq. (4.6.9) also verifies Eq. (4.6.7).

Finally, and this will be the most interesting part, we shall show that Eq. (4.6.9) verifies the Dirichlet problem, Eq. (4.6.2); and this will imply, by the regularity theorem, Proposition 1, p. 13, that, actually, $c^{(1)} \in C^\infty([0, L])$, although, of course, it may be that $c^{(1)} \notin \bar{C}^\infty([0, L])$.

To show that the series (4.6.9) is convergent and once differentiable term by term, we can observe that, setting $g' = dg/dx$:

$$\begin{aligned}\bar{g}_k &= \frac{2}{L}\int_0^L g(x')\sin\frac{\pi k}{L}x'\,dx' \\ &= \left[\frac{-1}{\pi k/L}\,\frac{2}{L}\,g(x')\cos\frac{\pi k}{L}x'\right]_0^L + \frac{1}{\pi k/L}\,\frac{2}{L}\int_0^L g'(x')\cos\frac{\pi k}{L}x'\,dx' \\ &= \frac{1}{\pi k}\,\frac{2}{L}\left[g(0) - (-1)^k g(L)\right] + \frac{2}{\pi k}\int_0^L g'(x')\cos\frac{\pi k}{L}x'\,dx', \\ &\qquad k = 1, 2, \ldots . \end{aligned} \tag{4.6.11}$$

This implies, if $M_{g'} = \max_{x\in[0,L]}|g'(x)|$:

$$|\bar{g}_k| \leq \frac{2}{L}\,\frac{|g(0)| + |g(L)| + LM_{g'}}{\pi k} \tag{4.6.12}$$

which means that the series (4.6.9) is uniformly convergent together with its derivative series, since $\bar{\omega}(k)^2$ diverges as k^2 for $k \to \infty$, in fact, such series are respectively majorized by the convergent series [see Eq. (4.6.12)]:

$$\sum_{k=1}^\infty \frac{|\bar{g}_k|}{\bar{\omega}(k)^2} \quad \text{and} \quad \sum_{k=1}^\infty \frac{|\bar{g}_k|}{\bar{\omega}(k)^2}\,\frac{\pi k}{L}\,. \tag{4.6.13}$$

Hence, by the series differentiation theorems, Eq. (4.6.9) converges and its derivative can be computed by series differentiation and is a continuous function (as a sum of a uniformly convergent series of continuous functions).

We now show that Eq. (4.6.9) verifies Eq. (4.6.7). Since, as already observed, the term-by-term limit of Eq. (4.6.4), thought of as a series in k, is Eq. (4.6.9), it will suffice to show that such a term-by-term limit is actually correct. In other words, it will suffice to show that Eq. (4.6.4), thought of as a series in k with all the terms with $k \geqslant N$ vanishing, is uniformly convergent with respect to a and ξ.

We shall show this by dominating the series (4.6.4) by the series

$$\sum_{k=1}^{N-1} \frac{1}{\omega_k^2} 2M_g, \qquad \text{if} \quad M_g = \max_{x \in [0,L]} |g(x)|, \tag{4.6.14}$$

where the terms with $k \geqslant N$ are thought to be zero.

Recalling the form of ω_k^2, see Eq. (4.4.17), and using the inequality

$$2\frac{(1-\cos\varphi)}{\varphi^2} \geqslant \frac{4}{\pi^2} \qquad \text{if} \quad \varphi \in [0, \pi], \tag{4.6.15}$$

we see that if $0 \leqslant (\pi k/L)a \leqslant \pi$:

$$\omega_k^{-2} = \left[\frac{\sigma}{\mu} + \frac{\tau}{\mu}\frac{2\left(1-\cos\frac{\pi k}{L}a\right)}{a^2}\right]^{-1} \leqslant \left(\frac{\sigma}{\mu} + \frac{\tau}{\mu}\frac{4}{L^2}k^2\right)^{-1}. \tag{4.6.16}$$

Hence, Eq. (4.6.14) is a series which is majorized by the series in which ω_k^{-2} is replaced by the right-hand side of Eq. (4.6.16), and this last series is dominated by

$$\sum_{k=1}^{\infty} 2M_g\left(\frac{\sigma}{\mu} + \frac{\tau}{\mu}\frac{4k^2}{L^2}\right)^{-1} < +\infty, \tag{4.6.17}$$

having removed only in this last step the restriction $k \leqslant N-1$. This proves that Eq. (4.6.4) is uniformly convergent with respect to the parameters a, ξ, N and, hence, Eq. (4.6.7) follows.

We now must show that Eq. (4.6.9) is a $C^{\infty}([0,L])$ function verifying Eqs. (4.6.2) and (4.6.3) with $h_0 = 0$, $h_L = 0$. Equation (4.6.3) is obvious since Eq. (4.6.9) has been proved to converge (and all its terms vanish for $x = 0$ or $x = L$). To prove Eq. (4.6.2), we use the fact that, as already remarked, it would be obvious if $g \in \bar{C}^{\infty}([0,L])$.

Given $\epsilon > 0$, let $g_\epsilon \in C_0^{\infty}((0,L)) \subset \bar{C}^{\infty}([0,L])$ be a function such that:

(i) $g_\epsilon(x) = g(x)$ if $\epsilon \leqslant x \leqslant L - \epsilon$. (4.6.18)

(ii) $(1/L)\int_0^L |g_\epsilon(x) - g(x)|\,dx < \epsilon$. (4.6.19)

(iii) The derivative g'_ϵ of g_ϵ, see Fig. 4.6, is such that

$$\int_0^L |g'_\epsilon(x)|\,dx \leqslant \int_0^L |g'(x)|\,dx + 2M_g. \tag{4.6.20}$$

We leave as an exercise based on Appendix C, the proof that such a function indeed exists (note that (iii) expresses that g_ϵ can be chosen to go from zero to $g(\epsilon)$ or from $g(L-\epsilon)$ to 0 without oscillating too much, i.e., with a derivative changing sign once at most without growing too large).

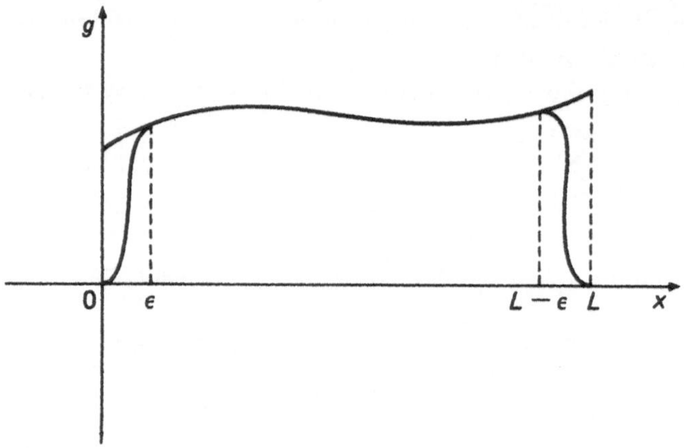

Figure 4.6.

Then define

$$\bar{g}_{\epsilon,k} = \frac{2}{L}\int_0^L g_\epsilon(x')\sin\frac{\pi k}{L}x'\,dx',$$
$$\bar{c}^{(1)\epsilon}(x) = \sum_{k=1}^{\infty}\left(\sin\frac{\pi k}{L}x\right)\frac{1}{\bar{\omega}(k)^2}\,\bar{g}_{\epsilon,k}, \tag{4.6.21}$$

and, since $g_\epsilon \in \bar{C}^\infty([0,L])$, we already mentioned that

$$\sigma\bar{c}^{(1)\epsilon} - \tau\frac{d^2\bar{c}^{(1)\epsilon}}{dx^2} = \mu g_\epsilon, \qquad \bar{c}^{(1)\epsilon}(0) = \bar{c}^{(1)\epsilon}(L) = 0 \tag{4.6.22}$$

which implies

$$\bar{c}^{(1)\epsilon}(x) \equiv \int_0^x \frac{d\bar{c}^{(1)\epsilon}}{dx'}(x')\,dx' \equiv \int_0^x dx'\left[\frac{d\bar{c}^{(1)\epsilon}}{dx'}(0) + \int_0^{x'}\frac{d^2\bar{c}^{(1)\epsilon}}{dx''^2}(x'')\,dx''\right]$$
$$= x\frac{d\bar{c}^{(1)\epsilon}}{dx}(0) + \int_0^x dx'\int_0^{x'} dx''\left[\frac{\sigma\bar{c}^{(1)\epsilon}(x'') - \mu g_\epsilon(x'')}{\tau}\right]. \tag{4.6.23}$$

If we show that uniformly in $x \in [0,L]$:

$$c^{(1)}(x) = \lim_{\epsilon\to 0}\bar{c}^{(1)\epsilon}(x), \qquad \frac{dc^{(1)}}{dx}(x) = \lim_{\epsilon\to 0}\frac{d\bar{c}^{(1)\epsilon}}{dx}(x), \tag{4.6.24}$$

we shall be able to take the limit in Eq. (4.6.23) and obtain

$$c^{(1)}(x) = x\frac{dc^{(1)}}{dx}(0) + \int_0^x dx'\int_0^{x'} dx''\left[\frac{\sigma c^{(1)}(x'') - \mu g(x'')}{\tau}\right], \tag{4.6.25}$$

implying by assumed continuity of g and by the above proved continuity of $c^{(1)}$ that $c^{(1)}$ is twice differentiable and by twofold differentiation of Eq. (4.6.25) that it verifies Eq. (4.6.2).

The regularity theorem of §2.2, Proposition 1, p. 13, will then permit us to deduce from the fact that $c^{(1)}$ is twice differentiable with continuous derivatives and verifies Eq. (4.6.2) that $c^{(1)}$ is in $C^{\infty}([0,L])$.[4]

Therefore, it remains to prove that the limits of Eq. (4.6.24) are correct and uniform in $x \in [0,L]$.

We already know that $c^{(1)}$ and its first derivative are given by the series (4.6.9) and by the sum of its term-by-term derivative. Such series are also the limits, term-by-term, of the series in Eq. (4.6.21) and of its derivative series because by Eqs. (4.6.19) and (4.6.21):

$$|\bar{g}_{\epsilon,k} - \bar{g}_k| < 2\epsilon, \qquad \forall k > 0. \tag{4.6.26}$$

Hence, the proof of Eq. (4.6.24) is again a problem of exchanging a limit with a series summation.

The necessary uniformity of the limit and the convergence of the series follow from the identity:

$$\begin{aligned} \bar{g}_{\epsilon,k} &= \frac{2}{L}\int_0^L g_\epsilon(x)\Big(\sin\frac{\pi k}{L}x\Big)\,dx = \frac{1}{\pi k/L}\,\frac{2}{L}\int_0^L g'_\epsilon(x)\Big(\cos\frac{\pi k}{L}x\Big)\,dx \\ &= \frac{1}{\pi k/L}\,\frac{2}{L}\int_\epsilon^{L-\epsilon} g'(x)\Big(\cos\frac{\pi k}{L}x\Big)\,dx \\ &\quad + \frac{1}{\pi k/L}\,\frac{2}{L}\int_{x\notin[\epsilon,L-\epsilon]} g'_\epsilon(x)\Big(\cos\frac{\pi k}{L}x\Big)\,dx. \end{aligned} \tag{4.6.27}$$

Hence, by Eq. (4.6.20),

$$\begin{aligned} |\bar{g}_{\epsilon,k}| &\leqslant \frac{1}{\pi k}2LM_{g'} + \frac{2}{\pi k}\int_{x\notin[\epsilon,L-\epsilon]}|g'_\epsilon(x)|\,dx \\ &\leqslant \frac{1}{\pi k}\Big(2LM_{g'} + 2\int_0^L|g'_\epsilon(x)|\,dx\Big) \leqslant \frac{1}{\pi k}(4LM_{g'} + 4M_g) \end{aligned} \tag{4.6.28}$$

and, therefore, the series (4.6.21) and its derivative series are dominated by the series (ϵ independent and convergent):

$$\sum_{k=1}^{\infty}\frac{[4LM_{g'} + 4M_g]}{\pi k\bar{\omega}(k)^2} \quad \text{and} \quad \sum_{k=1}^{\infty}\frac{\pi}{L}\,\frac{[4LM_{g'} + 4M_g]}{\pi\bar{\omega}(k)^2}, \tag{4.6.29}$$

proving their uniform convergence and, hence, Eq. (4.6.24).

To conclude the proof of Proposition 12, we still have to treat $\mathbf{c}^{(a)2}$ defined by Eq. (4.6.5) or by being the unique solution to the equations [see Eq. (4.4.28)]:

$$\begin{aligned} &(\sigma\mathbf{c}^{(a)2} - \tau D\mathbf{c}^{(a)2})_\xi = 0, \qquad \xi = ja, \;\; j = 1, \ldots, N-1, \\ &c_0^{(a)2} = h_0, \qquad\qquad c_L^{(a)2} = h_L. \end{aligned} \tag{4.6.30}$$

[4] This also follows directly from Eq. (4.6.2) since it shows that the second derivative of $c^{(1)}$ is continuously differentiable because such are g and $c^{(1)}$, etc.

Suppose, first, that $\sigma > 0$. The expression (4.6.5) is not too helpful for investigating the limit $a \to 0$, $\xi \to x$. We therefore look for an alternative representation for $\mathbf{c}^{(a)2}$ in analogy with the theory of linear differential equations.

We look for a solution of Eq. (4.6.30) having the form

$$c_{ja}^{(a)} = \beta_0 e^{-\lambda ja} + \beta_1 e^{-\lambda(L-ja)}, \qquad j = 0, \ldots, N, \tag{4.6.31}$$

where in the second term we use (instead of an arbitrary constant factor β) the constant factor $\beta_1 e^{-\lambda L}$, still arbitrary because such is β_1 but yielding a more symmetric expression (in which 0 and L "play the same role").

The parameters β_0, β_1, are to be determined so that Eq. (4.6.30) is verified.

Equation (4.6.30) will hold for $j = 2, \ldots, N-2$ if

$$\sigma + 2\frac{\tau}{a^2}\left(1 - \frac{e^{\lambda a} + e^{-\lambda a}}{2}\right) = 0 \tag{4.6.32}$$

which, via a simple discussion, is shown to admit a unique positive solution λ such that

$$\lim_{a \to 0} \lambda = \sqrt{\frac{\sigma}{\tau}} \equiv \lambda_0 \tag{4.6.33}$$

Furthermore, Eq. (4.6.30) for $j = 1$ or $N-1$ says, by taking Eq. (4.6.32) into account,

$$\begin{aligned} \beta_0 + \beta_1 e^{-\lambda L} &= h_0 & & \beta_0 = \frac{h_0 - h_L e^{-\lambda L}}{1 - e^{-2\lambda L}} \\ & & \Rightarrow & \\ \beta_0 e^{-\lambda L} + \beta_1 &= h_L & & \beta_1 = \frac{h_L - h_0 e^{-\lambda L}}{1 - e^{-2\lambda L}}. \end{aligned} \tag{4.6.34}$$

From Eqs. (4.6.31), (4.6.33), and (4.6.34), it is now immediate to take the limit $a \to 0$, $ja \to x$. One finds

$$c^{(2)}(x) = \lim_{\substack{a \to 0 \\ ja \to x}} c_{ja}^{(a)2} = \frac{h_0 - h_L e^{-\lambda_0 L}}{1 - e^{-2\lambda_0 L}} e^{-\lambda_0 x} + \frac{h_L - h_0 e^{-\lambda_0 L}}{1 - e^{-2\lambda_0 L}} e^{-\lambda_0 (L-x)} \tag{4.6.35}$$

which is immediately checked to verify Eqs. (4.6.2) and (4.6.3) with $g = 0$. The case $\sigma = 0$ is analogously treated by replacing Eq. (4.6.31) with

$$c_{ja}^{(a)2} = \beta_0 + \beta_1 ja, \tag{4.6.36}$$

and one eventually finds

$$c^{(2)}(x) = h_0 + \frac{x}{L}(h_L - h_0), \tag{4.6.37}$$

and Proposition 12 is completely proved. mbe

It is useful to collect all the results of this and the preceding section into a single statement.

13 Corollary. *Let $t \to \mathbf{y}^{(a)}(t)$ be a motion verifying Eq.* (4.4.3) *with initial data*

$$y_\xi^{(a)}(0) = c_\xi^{(a)}(0) + u_0(\xi), \qquad \dot{y}_\xi^{(a)}(0) = v_0(\xi), \tag{4.6.38}$$

where $u_0, v_0 \in \overline{C}^\infty([0,L])$ and $\mathbf{c}^{(a)}$ is a solution to the discrete Dirichlet problem, Eq. (4.4.28). *Then the limit*

$$\lim_{\substack{a\to 0\\ \xi\to x}} y_\xi^{(a)}(t) = c(x) + \overline{w}(x,t) \tag{4.6.39}$$

exists and $c \in C^\infty([0,L])$ is the solution to the "Dirichlet problem"

$$\sigma c - \tau \frac{d^2c}{dx^2} = \mu g, \qquad c(0) = h_0, \quad c(L) = h_L, \tag{4.6.40}$$

while $\overline{w} \in C^\infty([0,L] \times \mathbf{R})$ verifies the wave equations (4.5.7)–(4.5.10) *and $\overline{w}(\cdot,t) \in \overline{C}^\infty([0,L])$, $\forall t \in \mathbf{R}$.*

§4.7. Elastic Film. The Dirichlet Problem in $\Omega \subset \mathbf{R}^2$ and General Considerations on the Waves

The theory of the oscillations of an elastic film is considerably more complex and interesting than that of the elastic string of §4.3–§4.6. The results, however, are very similar. We shall not enter into the details of a theory that would lead us quite far from our program of analysis of the simplest mechanical systems.

We only give some terminology and formulate for illustrative purposes some easy propositions.

We shall then conclude our introduction to wave theory by defining the wave propagation velocity, studying it in the simple case of the elastic string subject only to tension forces ($\sigma = 0$, $h = 0$, $g = 0$).

5 Definition. Let $\Omega \subset \mathbf{R}^2$ be a bounded open connected region with a boundary $\partial\Omega$ which is a regular surface (see Definition 10, p. 171). Let $\Omega_a = \Omega \cap \mathbf{Z}_a^2$; $\partial\Omega_a = \{$set of points of $\partial\Omega$ lying on the intersections between $\partial\Omega$ and the bonds of the lattice $\mathbf{Z}_a^2\}$.

One defines the discrete Laplace operator on Ω relative to $\mathbf{Z}_a^2$ as the linear transformation D associating to every vector $\boldsymbol{\delta} = (\delta_\xi)_{\xi \in \Omega_a \cup \partial\Omega_a}$ the vector $((D\boldsymbol{\delta})_\xi)_{\xi \in \Omega_a}$ defined by

$$(D\boldsymbol{\delta})_\xi = -\sum_{\mathbf{e}} \frac{a}{\epsilon_a(\xi,\mathbf{e})} \frac{(\delta_\xi - \delta_{\xi+\epsilon_a(\xi,\mathbf{e})\mathbf{e}})}{a^2}, \qquad \xi \in \Omega_a, \tag{4.7.1}$$

where $\mathbf{e} = \pm\mathbf{e}_1, \pm\mathbf{e}_2$ ($\mathbf{e}_1$ and $\mathbf{e}_2$ being the two unit vectors parallel to the axes of $\mathbf{Z}_a^2$) and, for $\xi \in \Omega_a$:

$$\epsilon_a(\xi,\mathbf{e}) = \{\text{distance between } \xi \text{ and its nearest neighbor in } \Omega_a \cup \partial\Omega_a \text{ in the direction } \mathbf{e}\}. \tag{4.7.2}$$

The "$\mathbf{Z}_a^2$-discretized" Dirichlet problem in Ω with interior data $\mathbf{g} = (g_\xi)_{\xi\in\Omega_a}$ and boundary data $\mathbf{h} = (h_\xi)_{\xi\in\partial\Omega_a}$ are the equations

$$\sigma\delta_\xi - \tau(D\delta)_\xi = g_\xi, \qquad \xi\in\Omega_a, \tag{4.7.3}$$

$$\delta_\xi = h_\xi, \qquad \xi\in\partial\Omega_a. \tag{4.7.4}$$

Using the invertibility of positive-definite matrices one checks the following proposition along the same pattern of the proof of Proposition 8, §4.4, p. 262.

14 Proposition. *If $\sigma \geqslant 0$, $\tau > 0$, the Dirichlet problem* [*Eqs.* (4.7.3) *and* (4.7.4)] *always admits one and only one solution for any given boundary and interior data.*

Again, in the same way as in §4.4 and §4.5, one may check the following proposition.

15 Proposition. *Given $g\in C^\infty(\mathbf{R}^2)$, $h\in C^\infty(\partial\Omega)$,*[5] *consider the mechanical system with Lagrangian function* [*see Eqs.* (4.3.13) *and* (4.3.5)]

$$\begin{aligned}\mathcal{L} = \frac{\mu}{2}a^2\sum_{\xi\in\Omega_a}\dot{y}_\xi^2 + \mu a^2\sum_{\xi\in\Omega_a} g(\xi)y_\xi - \frac{\sigma}{2}a^2\sum_{\xi\in\Omega_a} y_\xi^2 \\ - \frac{\tau}{2}a^2\sum_{\xi\in\Omega_a}\sum_{\mathbf{e}}\frac{a}{\epsilon_a(\xi,\mathbf{e})}\,\frac{1}{\nu(\mathbf{e},\xi)}\,\frac{(y_\xi - y_{\xi+\epsilon_a(\xi,\mathbf{e})\mathbf{e}})^2}{a^2},\end{aligned} \tag{4.7.5}$$

$$y_{\xi'} = h(\xi'), \qquad \forall\xi'\in\partial\Omega. \tag{4.7.6}$$

This mechanical system has one and only one equilibrium configuration $\mathbf{y} = \mathbf{c}^{(a)}$. It is described by the solution $\mathbf{c}^{(a)}$ of the $\mathbf{Z}_a^2$-discretized Dirichlet problem with interior data $(\mu g(\xi))_{\xi\in\Omega_a}$ and boundary data $(h(\xi))_{\xi\in\partial\Omega_a}$.

Observation. More generally, if one is not interested in the limit $a\to 0$, the conditions $g\in C^\infty(\mathbf{R}^2)$, $h\in C^\infty(\partial\Omega)$ can be replaced by

$$g = (g_\xi)_{\xi\in\Omega_a} \quad\text{and}\quad h = (h_\xi)_{\xi\in\partial\Omega_a}.$$

Difficulties arise when one wishes to study the $a\to 0$ limit. Basically, one can say that the difficulties are due to the impossibility of providing the eigenvalues $\omega_1^2, \omega_2^2, \ldots$ and the respective eigenvectors $\boldsymbol{\eta}^{(1)}, \boldsymbol{\eta}^{(2)}, \ldots,$ describing the normal modes of the system of Eqs. (4.7.5) and (4.7.6), in a very explicit way, as in the case $d = 1$. Hence, the theory has to be developed in a somewhat more abstract way.

An example of a result that *should* be possible to obtain is as follows.

16 Proposition. *The stationary solution $\mathbf{c}^{(a)}$ of the equations for the mechanical system of Eqs.* (4.7.5) *and* (4.7.6) *with $g\in C^\infty(\mathbf{R}^2)$, h*

[5]See footnote 1.

$\in C^\infty(\partial\Omega)$ is such that the limit

$$\lim_{\substack{\xi\to\mathbf{x}\\ a\to 0}} c_\xi^{(a)} = c(\mathbf{x}), \qquad \mathbf{x}\in\overline{\Omega}, \tag{4.7.7}$$

exists and defines a function $c\in C^\infty(\overline{\Omega})$ such that

$$\sigma c(\mathbf{x}) - \tau\Delta c(\mathbf{x}) = \mu g(\mathbf{x}), \qquad \mathbf{x}\in\Omega, \tag{4.7.8}$$

$$c(\mathbf{x}) = h(\mathbf{x}), \qquad \mathbf{x}\in\partial\Omega, \tag{4.7.9}$$

where $\Delta f(\mathbf{x}) = \Sigma_{i=1}^2(\partial^2 f/\partial x_i^2)(\mathbf{x})$, $\forall f\in C^\infty(\overline{\Omega})$. Furthermore, Eqs. (4.7.8) and (4.7.9) have a unique solution in $C^\infty(\overline{\Omega})$.

The motions $t\to\mathbf{y}^{(a)}(t)$, $t\in\mathbf{R}$, of the above mechanical system, fulfilling the initial conditions

$$y_\xi^{(a)}(0) = c_\xi^{(a)} + u_0(\xi), \qquad \xi\in\Omega_a, \tag{4.7.10}$$

$$\dot{y}_\xi^{(a)}(0) = v_0(\xi), \qquad \xi\in\Omega_a \tag{4.7.11}$$

with $u_0, v_0\in C_0^\infty(\Omega)$ are such that the limit

$$\lim_{\substack{a\to 0\\ \xi\to\mathbf{x}}} y_\xi^{(a)}(t) = w(\mathbf{x},t), \qquad \mathbf{x},t\in\overline{\Omega}\times\mathbf{R} \tag{4.7.12}$$

exists and defines a $C^\infty(\overline{\Omega}\times\mathbf{R})$ function.

Furthermore, setting

$$w(\mathbf{x},t) = c(\mathbf{x}) + w(\mathbf{x},t), \tag{4.7.13}$$

it is

$$\sigma\overline{w}(\mathbf{x},t) - \tau\Delta\overline{w}(\mathbf{x},t) + \mu\frac{\partial^2\overline{w}}{\partial t^2}(\mathbf{x},t) = 0, \tag{4.7.14}$$

$$\overline{w}(\mathbf{x},0) = u_0(\mathbf{x}), \qquad \frac{\partial\overline{w}}{\partial t}(\mathbf{x},0) = v_0(\mathbf{x}), \tag{4.7.15}$$

$$\overline{w}(\mathbf{x},t) = 0, \qquad \frac{\partial\overline{w}}{\partial t}(\mathbf{x},t) = 0, \qquad \forall\mathbf{x}\in\partial\Omega, \quad \forall t\in\mathbf{R}. \tag{4.7.16}$$

Finally, there is a family of functions $S^{(h)}\in C^\infty(\overline{\Omega})$, $h=1,2,\ldots,$ vanishing on $\partial\overline{\Omega}$ and a sequence $\overline{\omega}(h), h=1,2,\ldots,$ of positive numbers such that

$$\overline{w}(\mathbf{x},t) = \sum_{h=1}^{\infty}\left[\hat{u}(h)S^{(h)}(\mathbf{x})(\cos\overline{\omega}(h)t) + \frac{\hat{v}(h)}{\overline{\omega}(h)}S^{(h)}(\mathbf{x})(\sin\overline{\omega}(h)t)\right], \tag{4.7.17}$$

where

$$\hat{u}(h) = \int_{\overline{\Omega}} S^{(h)}(\mathbf{x})u_0(\mathbf{x})\,d\mathbf{x}, \qquad \hat{v}(h) = \int_{\overline{\Omega}} S^{(h)}(\mathbf{x})v_0(\mathbf{x})\,d\mathbf{x} \tag{4.7.18}$$

and the series (4.7.17) converges, $\forall\mathbf{x}\in\overline{\Omega}$, $\forall t\in\mathbf{R}$.

Observations.

(1) The analogy between the vibrating string and the vibrating film would then be essentially complete. However, this author does not know if there is a proof of Proposition 16 (admitting its truth) in the above generality.

(2) There is a case in which an obvious variation of the above proposition holds and its proof is very simple. It is the case in which Ω is a torus (i.e., Ω is a "bicyle tire") and $\sigma > 0$. Mathematically, this is the system associated with the Lagrangian that follows; let $N = L/a =$ integer, $Q_L = [0, L-a] \times [0, L-a]$:

$$\mathcal{L}_{\text{per}} = \frac{\mu}{2} a^2 \sum_{\xi \in Q_L \cap \mathbf{Z}_a^2} \dot{y}_\xi^2 + \mu a^2 \sum_{\xi \in Q_L \cap \mathbf{Z}_a^2} g(\xi) Y_\xi$$
$$- \frac{\sigma}{2} a^2 \sum_{\xi \in Q_L \cap \mathbf{Z}_a^2} y_\xi^2 - \frac{\tau}{2} a^2 \sum_{\xi \in Q_L \cap \mathbf{Z}_a^2} \sum_{\mathbf{e}} \frac{(y_\xi - y_{\xi + a\mathbf{e}})^2}{2a^2} \tag{4.7.19}$$

and in the last sum the points which do not belong to $Q_L \cap \mathbf{Z}_a^2$ and which correspond to the points adjacent to the boundary ∂Q_L have to be identified with the points on ∂Q_L opposite to them.

In other words, $Q_L \cap \mathbf{Z}_a^2$ is thought of as a "discrete torus" and the film looses its boundary, becoming a tube.

The theory of Eq. (4.7.19) is identical to that of the vibrating string. Actually it is technically even easier (and analogous to Problem 9, §4.5, on the vibrating string). The role played by the functions $\sqrt{2/N}\,[\sin(\pi kja/N)]$ in the vibrating-string case is now played by

$$S^{(h_1,h_2)}(\xi) = \frac{1}{N} \exp \frac{2\pi i}{L}(j_1 a h_1 + j_2 a h_2) \qquad \text{if} \quad \xi = (j_1 a, j_2 a). \tag{4.7.20}$$

The ω_h^2 is now replaced by

$$\omega_{h_1 h_2}^2 = \frac{\sigma}{\mu} + \frac{\tau}{\mu} 2 \left[\frac{1 - \cos(2\pi h_1 a/L)}{a^2} + \frac{1 - \cos(2\pi h_2 a/L)}{a^2} \right], \tag{4.7.21}$$

while the role of Lemma 11, §4.5, is simply played by the two-dimensional Fourier theorem.

The detailed development of the theory of the motion of Eq. (4.7.19) (and of the analogous one-dimensional system, Problem 11, §4.5) is a very useful exercise. The reader will however realize that the assumption $\sigma > 0$ cannot, in the case of such periodic boundary conditions, be replaced by $\sigma \geqslant 0$. What is the physical meaning of this?

To conclude our analysis of the ordered systems of oscillators, we define and study concisely the notion of velocity of wave propagation.

6 Definition. Let Ω be an open region with regular boundary $\partial\Omega, \Omega \subset \mathbf{R}^d, d = 1$ or $d = 2$.

Consider the wave equation in Ω for $w \in C^\infty(\overline{\Omega} \times \mathbf{R})$:

$$\mu \frac{\partial^2 w}{\partial t^2} - \tau \Delta w + \sigma w = 0, \qquad (\mathbf{x}, t) \in \Omega \times \mathbf{R}. \tag{4.7.22}$$

$$w(\mathbf{x}, t) = 0 = \frac{\partial w}{\partial t}(\mathbf{x}, t), \qquad \mathbf{x} \in \partial\Omega \tag{4.7.23}$$

with initial data

$$w(\mathbf{x}, 0) = u_0(\mathbf{x}), \tag{4.7.24}$$

$$\frac{\partial w}{\partial t}(\mathbf{x}, 0) = v_0(\mathbf{x}) \tag{4.7.25}$$

and suppose that $u_0, v_0 \in C_0^\infty(\Omega)$ and vanish outside a neighborhood with radius ϵ around $\mathbf{x}_0 \in \Omega$.

Given $\mathbf{x}_1 \in \Omega$, $\mathbf{x}_1 \neq \mathbf{x}_0$, let

$$t_\epsilon(\mathbf{x}_0, \mathbf{x}_1) = \inf_{u_0 v_0} \{\text{inf of the set of the values } t \text{ for which there is } t' < t,\ t' > 0, \text{ when } w(\mathbf{x}_1, t') \neq 0\}. \tag{4.7.26}$$

Obviously, $t_\epsilon(\mathbf{x}_0, \mathbf{x}_1) \geqslant t_{\epsilon'}(\mathbf{x}_0, \mathbf{x}_1)$ if $\epsilon' > \epsilon$, and

$$t(\mathbf{x}_0, \mathbf{x}_1) = \sup_{\epsilon > 0} t_\epsilon(\mathbf{x}_0, \mathbf{x}_1) \tag{4.7.27}$$

is the "minimum time" needed for a perturbation of the equilibrium, (i.e., flat), string, or film, initially located around $\mathbf{x}_0$ to "reach" $\mathbf{x}_1$.

The "wave velocity" of the waves described by Eqs. (4.7.22) and (4.7.23) is naturally defined as

$$C = \sup_{\mathbf{x}_1 \neq \mathbf{x}_0} \frac{|\mathbf{x}_1 - \mathbf{x}_0|}{t(\mathbf{x}_0, \mathbf{x}_1)}. \tag{4.7.28}$$

Observation. In the $d = 2$ case, we did not prove existence and uniqueness theorems for Eqs. (4.7.22)–(4.7.25), while for $d = 1$ we did. However, if we set $t_\epsilon(\mathbf{x}_0, \mathbf{x}_1) = +\infty$ if for every (u_0, v_0) there is no solution to Eqs. (4.7.22)–(4.7.25) and if, in case of nonunique solutions, we take into account all the solutions in the infimum in Eq. (4.7.26), the above definition also makes sense for $d = 2$.

In any case, this is not a real problem since existence and uniqueness for Eqs. (4.7.22)–(4.7.25) for $u_0, v_0 \in C_0^\infty(\Omega)$ can be proved in a satisfactory sense.

Let us prove the following proposition for $\sigma = 0$:

17 Proposition. *Let $d = 1$, $\Omega = (0, L)$. The wave propagation velocity of the waves described by Eqs.* (4.7.22) *and* (4.7.23) *with $\sigma \geqslant 0$, $\tau > 0$, $\mu > 0$ is*

$$C = \sqrt{\frac{\tau}{\mu}} \tag{4.7.29}$$

independent of the value of σ.

PROOF (Case $\sigma = 0$ only). From Eq. (4.5.19), we derive by trigonometry:

$$w(x,t) = \sum_{h=1}^{\infty}\left(\sin\frac{\pi h}{L}x\right)\left[\hat{u}_0(h)\left(\cos\frac{\pi}{L}\sqrt{\frac{\tau}{\mu}}\,ht\right) + \frac{\hat{v}_0(h)}{\sqrt{\tau/\mu}\,\pi h/L}\left(\sin\frac{\pi}{L}\sqrt{\frac{\tau}{\mu}}\,ht\right)\right]$$

$$= \frac{1}{2}\sum_{h=1}^{\infty}\hat{u}_0(h)\left[\sin\frac{\pi h}{L}\left(x+\sqrt{\frac{\tau}{\mu}}\,t\right) + \sin\frac{\pi h}{L}\left(x-\sqrt{\frac{\tau}{\mu}}\,t\right)\right]$$
$$+ \frac{1}{2}\sum_{h=1}^{\infty}\frac{\hat{\imath}_0(h)}{\sqrt{(\tau/\mu)}\,\pi h/L}$$
$$\times\left[\cos\frac{\pi h}{L}\left(x+\sqrt{\frac{\tau}{\mu}}\,t\right) - \cos\frac{\pi h}{L}\left(x-\sqrt{\frac{\tau}{\mu}}\,t\right)\right] \tag{4.7.30}$$

since $\bar{\omega}(h)^2 = (\tau/\mu)(\pi h/L)^2$. Then, let $\forall x \in \mathbf{R}$,

$$u_0^*(x) = \sum_{h=1}^{\infty}\hat{u}_0(h)\sin\frac{\pi h}{L}x, \qquad v_0^*(x) = \sum_{h=1}^{\infty}\hat{v}_0(h)\sin\frac{\pi h}{L}x \tag{4.7.31}$$

and, by the Lemma 11, p. 266, plus the periodicity and parity properties of the sine:

(i) $u_0^*(x) = u_0(x)$, $v_0^*(x) = v_0(x)$ $\forall x \in [0, L]$.
(ii) $u_0^*,(L+x) = -u_0(L-x)$, $v_0^*(L+x) = -v_0(L-x)$, $\forall x \in [0, L]$. (4.7.32)
(iii) u_0^*, v_0^* are periodic C^∞ functions with period $2L$, i.e., u_0^*, v_0^* are obtained from u_0, v_0, by first reflecting them about L and then by periodic continuation of the function on $[0, 2L]$ thus constructed.

If u_0 has support in a neighborhood with radius ϵ around x_0, one finds that u_0^* is described in Fig. 4.7.

Equation (4.7.30) can be written in terms of u_0^*, v_0^* as

$$w(x,t) = \frac{1}{2}\left[u_0^*\left(x+\sqrt{\frac{\tau}{\mu}}\,t\right) + u_0^*\left(x-\sqrt{\frac{\tau}{\mu}}\,t\right)\right]$$
$$-\frac{1}{2}\frac{1}{\sqrt{\tau/\mu}}\int_{x-\sqrt{\tau/\mu}\,t}^{x+\sqrt{\tau/\mu}\,t} v_0^*(\xi)\,d\xi \tag{4.7.33}$$

[see, also, Problem 11, §4.5, for an alternative proof of Eq. (4.7.33)].

Then, for instance, one sees from the picture that in order that $u_0^*(x_1 - \sqrt{\tau/\mu}\,t) \neq 0$, the point $x_1 - \sqrt{\tau/\mu}\,t$ has to fall inside some of the intervals

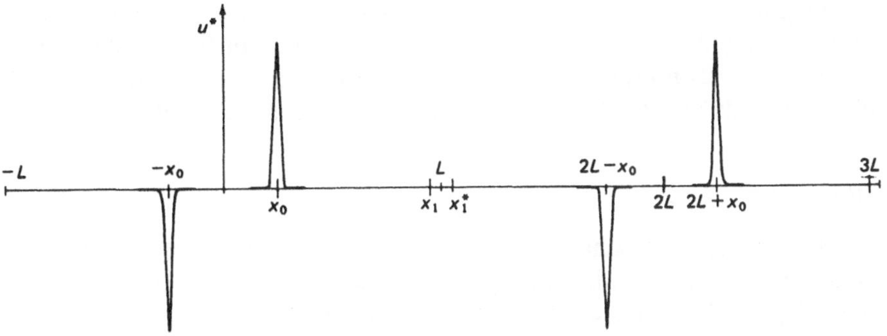

Figure 4.7. Example of graph of u_0^*.

where u_0^* is not zero; hence, t must be such that

$$\frac{|x_1 - x_0| + \epsilon}{\sqrt{\tau/\mu}} \geqslant t \geqslant \frac{|x_1 - x_0| - \epsilon}{\sqrt{\tau/\mu}} \tag{4.7.34}$$

and a similar bound on t can be found likewise, discussing the second and the third terms in Eq. (4.7.33). This clearly implies Eq. (4.7.29). mbe

§4.8. Anharmonic Oscillators. Small Oscillations and Integrable Systems

Consider an l-degrees-of-freedom system with Lagrangian function

$$\mathcal{L}(\dot{\boldsymbol{\beta}}, \boldsymbol{\beta}) = \sum_{i,j=1}^{l} \frac{1}{2} g_{ij}(\boldsymbol{\beta})\dot{\beta}_i\dot{\beta}_j - V(\boldsymbol{\beta}), \tag{4.8.1}$$

where g is a given $C^\infty(\mathbf{R}^l)$ positive-definite $l \times l$ matrix and $V \in C^\infty(\mathbf{R}^l)$ is a given potential energy function.

Assume that V has a second-order miminum in $\boldsymbol{\beta}_0 \in \mathbf{R}^l$; i.e., $\partial V = \mathbf{0}$ in $\boldsymbol{\beta}_0$ and that the matrix

$$I_{ij} = \frac{\partial^2 V}{\partial \beta_i \partial \beta_j}(\boldsymbol{\beta}_0), \qquad i, j = 1, \ldots, l, \tag{4.8.2}$$

is positive definite. Then $\boldsymbol{\beta}_0$ is an equilibrium point.

7 Definition. The "small oscillations" near $\boldsymbol{\beta}_0$ of the system described by Eq. (4.8.1) with V verifying Eq. (4.8.2) are the motions of the mechanical system with Lagrangian function

$$\mathcal{L}_{\text{small}}(\dot{\boldsymbol{\beta}}, \boldsymbol{\beta}, t) = \frac{1}{2} \sum_{i,j=1}^{l} g_{ij}(\boldsymbol{\beta}_0)\dot{\beta}_i\dot{\beta}_j - \frac{1}{2} \sum_{i,j=1}^{l} I_{ij}(\beta_i - \beta_{0i})(\beta_j - \beta_{0j}), \tag{4.8.3}$$

where $\boldsymbol{\beta} = (\beta_1, \ldots, \beta_l)$, $\boldsymbol{\beta}_0 = (\beta_{01}, \ldots, \beta_{0l})$. The normal modes' pulsations of Eq. (4.8.3) are called the "proper pulsations" of Eq. (4.8.1) near $\boldsymbol{\beta}_0$; their reciprocals multiplied by 2π are the "proper periods". The reciprocals of the periods are the "proper frequencies" of the small oscillations.

Observations.

(1) Therefore, the small oscillations are the motions of the Lagrangian system obtained by replacing the matrix g with its value at the equilibrium point $\boldsymbol{\beta}_0$ and by replacing the potential energy V by its Taylor expansion about $\boldsymbol{\beta}_0$ truncated to second order:

$$V(\boldsymbol{\beta}) = V(\boldsymbol{\beta}_0) + \frac{1}{2} \sum_{i,j=1}^{l} I_{ij}(\beta_i - \beta_{0i})(\beta_j - \beta_{0j}) + \cdots \tag{4.8.4}$$

and in Eq. (4.8.3), $V(\boldsymbol{\beta}_0)$ does not appear since it does not affect the associated notions.

(2) It is clear that on the basis of the above definitions, the "small oscillations" are not necessarily motions with small amplitude. However, one can expect or hope that if the energy of a motion described by Eq. (4.8.1) is just slightly above $V(\boldsymbol{\beta}_0)$ (hence, the motion takes place in the vicinity of $\boldsymbol{\beta}_0$ when it is initially there), then the motion of Eq. (4.8.3) with the same initial data approximates well the exact motion.

Since the small oscillations are, by definition, harmonic motions, hence "simple motions", one understands the interest in the following question: in what sense do the small oscillations approximate the real motions of Eq. (4.8.1) near $\boldsymbol{\beta}_0$?

In Chapter 2 we met and essentially solved this problem for systems with one degree of freedom. The generalization to systems with $l > 1$ degrees of freedom is, however, surprisingly difficult and interesting. In Chapter 5 we shall discuss some of its aspects. For the moment we shall only provide a definition of a class of systems behaving "as if they were linear oscillators" and we shall continue by discussing a few remarkable examples of such systems warning the reader, however, that it should not be hoped that Definition 10 to follow is a definition covering many cases.

8 Definition. Consider a system of N point masses in $\mathbf{R}^d$ subject to ideal bilateral constraints with l degrees of freedom and to a conservative force.

We assume that the equations of motion are normal in the future as well as in the past; i.e., they admit a global solution $t \to S_t(\dot{\mathbf{x}}, \mathbf{x})$ for every initial datum $(\dot{\mathbf{x}}, \mathbf{x})$ compatible with the constraints. We shall call "space of the initial data" the set $\mathbb{S} \subset \mathbf{R}^{2Nd}$ of all the pairs $(\dot{\mathbf{x}}, \mathbf{x})$, where $\mathbf{x}$ is a constraint-compatible configuration and $\dot{\mathbf{x}}$ is a constraint-compatible velocity.

We define on $\mathbb{S}$ the "time evolution flow", $(S_t)_{t \in \mathbf{R}}$, as the group of transformations mapping $(\dot{\mathbf{x}}, \mathbf{x})$ into $S_t(\dot{\mathbf{x}}, \mathbf{x}) =$ (datum into which $(\dot{\mathbf{x}}, \mathbf{x})$ evolves in the time t according to the system's equations of motion).

Observations

(1) This generalizes the initial data space introduced in §2.22 to constrained systems.

(2) $\mathbb{S}$ will be considered to be a surface in $\mathbf{R}^{2Nd}$. The geometric structure of $\mathbb{S}$ is very simple.

18 Proposition. *The surface $\mathbb{S}$ of the preceding definition is a regular surface in $\mathbf{R}^{2Nd}$.*

PROOF. By Definition 10, §3.6, p. 171, given $(\dot{\mathbf{x}}_0, \mathbf{x}_0) \in \mathbb{S}$, we have to find a neighborhood W of $(\dot{\mathbf{x}}_0, \mathbf{x}_0)$ on which it is possible to establish a local system of regular coordinates adapted to the surface $\mathbb{S}$.

Let U be a neighborhood of $\mathbf{x}_0$ on which it is possible to establish a local system of regular coordinates $\boldsymbol{\xi} = \boldsymbol{\Xi}(\boldsymbol{\beta})$, with basis Ω, adapted to the surface Σ in $\mathbf{R}^{Nd}$ defined by the constraint. The set U exists by the very definition of an l-degrees-of-freedom holonomous constraint.

In this coordinate system, the possible velocity vectors $\dot{\boldsymbol{\beta}}$ for the system which are compatible with the constraints are those such that

$$\dot{\beta}_1 = \dot{\beta}_2 = \cdots = \dot{\beta}_{Nd-l} = 0, \tag{4.8.5}$$

while the possible position vectors $\boldsymbol{\beta}$ are those for which

$$\beta_1 = \beta_2 = \cdots = \beta_{Nd-l} = 0, \qquad (0, \ldots, 0, \beta_{Nd-l+1}, \ldots, \beta_{Nd}) \in \Omega. \tag{4.8.6}$$

Hence, the correspondence between $\mathbf{R}^{Nd} \times \Omega$ and $\mathbf{R}^{2Nd}$ described by

$$\begin{aligned} \dot{\mathbf{x}}^{(i)} &= \sum_{k=1}^{Nd} \dot{\beta}_k \frac{\partial \boldsymbol{\Xi}^{(i)}}{\partial \beta_k}(\boldsymbol{\beta}), \\ \mathbf{x}^{(i)} &= \boldsymbol{\Xi}(\boldsymbol{\beta}) \end{aligned} \tag{4.8.7}$$

establishes on the image $W \subset \mathbf{R}^{2Nd}$ of $\mathbf{R}^{Nd} \times \Omega$ a coordinate system near $(\dot{\mathbf{x}}_0, \mathbf{x}_0) \in W$ adapted to $\mathbb{S}$ with basis $\mathbf{R}^{Nd} \times \Omega$, and it is easily checked that the Jacobian determinant of this coordinate change at the point with coordinates $(\dot{\boldsymbol{\beta}}, \boldsymbol{\beta})$ is the square of the Jacobian determinant in $\boldsymbol{\beta}$ of the transformation $\boldsymbol{\Xi}$. By the regularity assumption, on the coordinate system $(U, \boldsymbol{\Xi})$, such a determinant does not vanish. mbe

Observations.

(1) The above proof shows that setting

$$\begin{aligned} \dot{\boldsymbol{\kappa}} &= (\dot{\beta}_{Nd-l+1}, \ldots, \dot{\beta}_{Nd}) \equiv (\dot{\kappa}_1, \dot{\kappa}_2, \ldots, \dot{\kappa}_l), \\ \boldsymbol{\kappa} &= (\beta_{Nd-l+1}, \ldots, \beta_{Nd}) \equiv (\kappa_1, \kappa_2, \ldots, \kappa_l), \end{aligned} \tag{4.8.8}$$

Eqs. (4.8.7) establish a coordinate system, $(\dot{\boldsymbol{\kappa}}, \boldsymbol{\kappa})$, for the points of $W \cap \mathbb{S}$, where W is the image via Eqs. (4.8.7) of $\mathbf{R}^{Nd} \times \Omega$. Furthermore, as $(\dot{\mathbf{x}}, \mathbf{x})$

varies in $W \cap \mathbb{S}$, the point $(\dot{\boldsymbol{\kappa}}, \boldsymbol{\kappa})$ varies in $\mathbf{R}^l \times V$ where V is an open convex set in $\mathbf{R}^d$ (as $V = \Omega \cap \{\text{plane } \beta_1 = \cdots = \beta_{Nd-l} = 0\}$, and Ω is convex).

One refers to this remark by saying that the data space $\mathbb{S}$ of a system with l degrees of freedom locally has the structure $\mathbf{R}^l \times V$ with $V \subset \mathbf{R}^{l}$".

For this reason, and with an abuse of notation very useful and widely used, one often denotes the points of $\mathbb{S}$ as $(\dot{\boldsymbol{\kappa}}, \boldsymbol{\kappa})$, where $(\dot{\boldsymbol{\kappa}}, \boldsymbol{\kappa})$ are local regular coordinates, in a neighborhood W of a point in $\mathbb{S}$ such that $W \cap \mathbb{S}$ has the structure $\mathbf{R}^l \times V$ (which have to be deduced from the context and which often are really local (i.e., nonglobal) coordinates)

Coherently, the Lagrangian of the constrained system is described as a function $\mathcal{L}(\dot{\boldsymbol{\kappa}}, \boldsymbol{\kappa})$ of $(\dot{\boldsymbol{\kappa}}, \boldsymbol{\kappa})$.

(2) Since $\mathbb{S}$ is a regular surface, it makes sense to define the open sets on $\mathbb{S}$ and the space $C^\infty(\mathbb{S})$. A set $E \subset \mathbb{S}$ is open on $\mathbb{S}$ if it is the intersection of an open set in $\mathbf{R}^{Nd}$ with $\mathbb{S}$. A function f is in $C^\infty(\mathbb{S})$ if its restriction to a neighborhood U, on which it is possible to set up a local system of regular coordinates transforming U into $\mathbf{R}^l \times V$, has the property that, if thought of as a function of the local coordinates $(\dot{\boldsymbol{\kappa}}, \boldsymbol{\kappa})$, it is a $C^\infty(\mathbf{R}^l \times V)$ function.

9 Definition. Let $\mathbb{S}$ be the initial data space for a system of N point masses with l degrees of freedom subject to ideal holonomous constraints and to conservative forces.

Let $A \in C^\infty(\mathbb{S})$ be a real-valued function on $\mathbb{S}$. We shall say that A is a "prime integral" for the system's motions $t \to S_t(\dot{\mathbf{x}}, \mathbf{x}) \equiv (\dot{\mathbf{x}}(t), \mathbf{x}(t))$, $t \in \mathbf{R}$, if

$$A(\dot{\mathbf{x}}(t), \mathbf{x}(t)) = \text{constant} \tag{4.8.9}$$

for all $(\dot{\mathbf{x}}, \mathbf{x}) \in \mathbb{S}$.

Examples

(1) The energy

$$E(\dot{\mathbf{x}}, \mathbf{x}) = \frac{1}{2} \sum_{i=1}^{N} m_i (\dot{\mathbf{x}}^{(i)})^2 + V^{(a)}(\mathbf{x}^{(1)}, \ldots, \mathbf{x}^{(N)}) \tag{4.8.10}$$

is a typical example of a prime integral. Often it is the only prime integral admitted by the system's motions.

(2) If the system is isolated, i.e., subject to zero external forces, the d components of the linear momentum

$$\mathbf{Q}(\dot{\mathbf{x}}, \mathbf{x}) = \sum_{i=1}^{N} m_i \dot{\mathbf{x}}^{(i)} \tag{4.8.11}$$

are also prime integrals when the third law of dynamics holds. In the same situation, the angular momentum components also give rise to prime integrals.

In light of the above definition, we can reinterpret Corollary 3, §4.1, p. 248, as saying the following.

19 Proposition. *A system of l harmonic oscillators with Lagrangian function* (4.1.2) *admits l prime integrals $A_1, \ldots, A_l$ given by Eq.* (4.1.5). *Furthermore, it is possible to parametrize the initial data space $\mathbb{S}$ through the values $A_1, \ldots, A_l$ and a point $\boldsymbol{\varphi} \in \mathbf{T}^l$, $\boldsymbol{\varphi} = (\varphi_1, \ldots, \varphi_l)$, so that $\mathbb{S}$ can be thought of as the product $[0, +\infty)^l \times \mathbf{T}^l$, and the motion $t \to (\dot{\mathbf{x}}(t), \mathbf{x}(t))$, $t \in \mathbf{R}_+$, of the system is described in these coordinates as*

$$(A_1, \ldots, A_l; \varphi_1, \ldots, \varphi_l) \to (A_1, \ldots, A_l; \varphi_1 + \omega_1 t, \ldots, \varphi_l + \omega_l t), \tag{4.8.12}$$

where $\omega_1, \ldots, \omega_l$ are positive constants.

Finally, the correspondence $(\mathbf{A}, \boldsymbol{\varphi}) \to (\dot{\mathbf{x}}, \mathbf{x})$ is a C^∞ invertible nonsingular[6] *correspondence between $(0, +\infty)^l \times \mathbf{T}^l$ and the subset of $\mathbf{R}^{2l}$ which is its image.*

Proposition 19 suggests the following definition.

10 Definition. Let $\mathbb{S}$ be the initial data space for a system with l degrees of freedom subject to ideal constraints and to conservative active forces.

We shall say that the system is "integrable" on the open region $W \subset \mathbb{S}$ if on W it is possible to define l prime integrals. $\mathbf{A} = (A_1, \ldots, A_l)$ and l $\mathbf{T}^l$-valued $C^\infty(W)$ functions $\boldsymbol{\varphi} = (\varphi_1, \ldots, \varphi_l)$ such that:

(1) The image of W under the map

$$(\dot{\mathbf{x}}, \mathbf{x}) \to I(\dot{\mathbf{x}}, \mathbf{x}) = (\mathbf{A}, \boldsymbol{\varphi}) \tag{4.8.13}$$

has the form $V \times \mathbf{T}^l$, where V is an open set in $\mathbf{R}^l$ and the correspondence I between W and $V \times \mathbf{T}^l$ is an invertible nonsingular (i.e., with nonvanishing Jacobian determinant) correspondence.

(2) There are l real C^∞ functions on V, $\mathbf{A} \to \boldsymbol{\omega}(\mathbf{A}) = (\omega_1(\mathbf{A}), \ldots, \omega_l(\mathbf{A})) \in \mathbf{R}^l$ such that if $t \to \mathbf{x}(t)$ is a system's motion with initial data $(\dot{\mathbf{x}}(0), \mathbf{x}(0)) \in W$, then, $\forall t \in \mathbf{R}_+$, $(\dot{\mathbf{x}}(t), \mathbf{x}(t)) \in W$ and

$$\begin{aligned} I(\dot{\mathbf{x}}(t), \mathbf{x}(t)) &= (\mathbf{A}(\dot{\mathbf{x}}(0), \mathbf{x}(0)), \boldsymbol{\varphi}(\dot{\mathbf{x}}(0), \mathbf{x}(0)) + \boldsymbol{\omega}(\mathbf{A}(\dot{\mathbf{x}}(0)), \mathbf{x}(0))t) \\ &\equiv (\mathbf{A}(\dot{\mathbf{x}}(0), \mathbf{x}(0)), \boldsymbol{\varphi}(\dot{\mathbf{x}}(0), \mathbf{x}(0)) + \boldsymbol{\omega}(\mathbf{A}_0)t) \qquad (4.8.14) \\ &\equiv (\mathbf{A}_0, \boldsymbol{\varphi}_0 + \boldsymbol{\omega}(\mathbf{A}_0)t), \end{aligned}$$

where $\boldsymbol{\varphi} \to \boldsymbol{\varphi} + \boldsymbol{\omega}(\mathbf{A}_0)t$ denotes the quasi-periodic flow on $\mathbf{T}^l$ with speed $\boldsymbol{\omega}(\mathbf{A}_0)$, see Definition 1, p. 248, and $\mathbf{A}_0 = \mathbf{A}(\dot{\mathbf{x}}(0), \mathbf{x}(0))$, $\boldsymbol{\varphi}_0 = (\dot{\mathbf{x}}(0), \mathbf{x}(0))$.

The numbers $\omega_i(\mathbf{A})$, $T_i(\mathbf{A}) = 2\pi/\omega_i(\mathbf{A})$, $\nu_i(\mathbf{A}) = T_i(\mathbf{A})^{-1}$, $i = 1, \ldots, l$, are, respectively, called the pulsations, the periods, and the frequencies of the motions in W with amplitudes $\mathbf{A}$.

Observations.

(1) In the case of a system of harmonic oscillators, there are various choices of W for which the system is integrable on W: the most natural one

[6] See Definition 13 and related observations, p. 101, for the meaning of the derivatives.

takes W to be the set in $\mathbb{S}$ whose image under the map of Eq. (4.1.15) is $(0, +\infty)^l \times \mathbf{T}^l$ (i.e., the set of data having all the normal modes excited: $A_i > 0$ for $i = 1, \ldots, l$).

(2) One can interpret Eqs. (4.8.13) and (4.8.14) as saying that the data space W of an integrable system is "foliated by an l-parameter family of l-dimensional invariant tori". The parameters are the values of the l prime integrals. The torus with parameters $\mathbf{A} \in V$ is the set $I(\{\mathbf{A}\} \times \mathbf{T}^l)$ image of $\{\mathbf{A}\} \times \mathbf{T}^l$ under the "integration map" I.

(3) In the case of harmonic oscillators, $\boldsymbol{\omega}(\mathbf{A})$ is $\mathbf{A}$ independent: "isochrony of the harmonic oscillations". As seen in the case $l = 1$, §2.10, it is obvious that this should be a very special property of the harmonic oscillators. Therefore, it is better not to introduce it into the definition of integrable system, to avoid giving a too restrictive definition.

(4) In the context of the theory of small oscillations, the above definition seems especially designed to formulate the conjecture that in a small enough neighborhood W of an equilibrium position $(\mathbf{0}, \boldsymbol{\beta}_0)$ for a mechanical system described by a Lagrangian (4.8.1) verifying Eq. (4.8.2), the system is integrable.

Such a conjecture, true if $l = 1$, is generally false if $l > 1$; i.e., there may be motions which stay indefinitely close to an equilibrium point and, nevertheless, move in a fashion substantially different from a quasi-periodic motion. However, a conjecture similar to this one is true. We shall discuss this matter in Chapter 5, §5.9–§5.12.

(5) To establish the integrability of a system with l degrees of freedom, one usually proceeds to show that it is possible to describe the motions which develop in W in terms of $2l$ parameters $(\mathbf{A}, \boldsymbol{\varphi}) \in V \times \mathbf{T}^l$ and of N C^∞ functions on $V \times \mathbf{T}^l$, $\boldsymbol{\Phi}^{(1)}, \ldots, \boldsymbol{\Phi}^{(N)}$, such that if $\mathbf{x}$ is a motion, one has, $\forall i = 1, 2, \ldots, N$,

$$\mathbf{x}^{(i)}(t) = \boldsymbol{\Phi}^{(i)}(A_1, \ldots, A_l, \varphi_1 + \omega_1(\mathbf{A})t, \ldots, \varphi_l + \omega_l(\mathbf{A})t), \quad (4.8.15)$$

where $\omega_1(\mathbf{A}), \ldots, \omega_l(\mathbf{A})$ are l C^∞ functions of $\mathbf{A} \in V$.

Successively, one proceeds to check the invertibility, regularity, and nonsingularity of the map $(\dot{\mathbf{x}}(0), \mathbf{x}(0)) \leftrightarrow (\mathbf{A}, \boldsymbol{\varphi})$. This check is usually an easy matter and without direct interest once Eq. (4.8.15) has been established for all the motions in W. Actually, the true analytic difficulty that is met in the intregrability proofs lies in the proof of the validity of a consequence of Eq. (4.8.15): precisely, in checking that all the motions in W are "quasi-periodic" in the sense that their coordinates depend quasi-periodically on time (see §2.21 for the notion of quasi-periodicity). See, however, Problem 20 to §4.15.

Therefore, in the upcoming sections, we shall often stop our analysis of integrability when we find that the system's motions taking place in a given W are quasi-periodic, without entering into the sometimes long analysis necessary to prove the invertibility and smoothness properties required by integrability.

The following extension of Definition 10 is natural in the context of the concepts of analytical mechanics of §3.11 and §3.12.

11 Definition. Let $\mathfrak{L} \in C^\infty(W)$, $W \subset \mathbf{R}^{2l}$ or $W \subset \mathbf{R}^l \times \mathbf{T}^l$ or $W \subset \mathbf{R}^l \times (\mathbf{R}^{l_1} \times \mathbf{T}^{l_2})$, $l_1 + l_2 = l$, be a time-independent regular Lagrangian on W (see Definition 14, §3.11, p. 212).

We say that $\mathfrak{L}$ is integrable on the data space W if there is an integrating map I transforming W into $V \times \mathbf{T}^l$ enjoying the properties (1) and (2) of Definition 10, where the motion $t \to \mathbf{x}(t)$ is now a solution to the Lagrangian equations relative to $\mathfrak{L}$.

Similarly, if $H \in C^\infty(\tilde{W})$, $\tilde{W} \subset \mathbf{R}^{2l}$ or $\tilde{W} \subset \mathbf{R}^l \times \mathbf{T}^l$ or $\tilde{W} \subset \mathbf{R}^l \times (\mathbf{R}^{l_1} \times \mathbf{T}^{l_2})$, $l_1 + l_2 = l$, is a regular time-independent Hamiltonian function on the phase space $\tilde{W}$, we say that H is integrable on $\tilde{W}$ if the corresponding Lagrangian function $\mathfrak{L}$ is integrable on the data space subset $W = \Xi^{-1}(\tilde{W})$, Ξ being the map inducing the Legendre transformation between H and $\mathfrak{L}$ (see §3.11).

In this case, if I is the integrating map for $\mathfrak{L}$, the map

$$\tilde{I}(\mathbf{p}, \mathbf{q}) = I(\Xi^{-1}(\mathbf{p}, \mathbf{q})) \tag{4.8.16}$$

maps $\tilde{W}$ onto $V \times T^l$ and it is called an "integrating map" for H.

If $\tilde{I}$ is a completely canonical map of $\tilde{W}$ onto $V \times \mathbf{T}^l$ for a suitable, we say that H is "canonically integrable" on the phase space $\tilde{W}$.

If H is analytic[7] on $\tilde{W}$ and $\tilde{I}$ is also analytic, we say that H is "analytically integrable on $\tilde{W}$".

If H is analytic and if $\tilde{I}$ is an analytic completely canonical map of $\tilde{W}$ onto $V \times \mathbf{T}^l$, we shall say that H is "canonically analytically integrable" on $\tilde{W}$.

Observations.

(1) If $H \in C^\infty(W)$ is canonically integrable and if $\tilde{I}$ is a completely canonical map integrating H, then

$$H(\tilde{I}^{-1}(\mathbf{A}, \boldsymbol{\varphi})) \equiv h(\mathbf{A}) = (\boldsymbol{\varphi}\text{-independent function}) \tag{4.8.17}$$

and $\boldsymbol{\omega}(\mathbf{A}) = [(\partial h/\partial A_1)(\mathbf{A}), \ldots, (\partial h/\partial A_l)(\mathbf{A})]$:

$$\boldsymbol{\omega}(\mathbf{A}) = \frac{\partial h}{\partial \mathbf{A}}(\mathbf{A}) \tag{4.8.18}$$

and in this case the variables $(\mathbf{A}, \boldsymbol{\varphi})$ are called "action-angle" variables and are canonical variables.

(2) It turns out that all the systems that we shall consider in the upcoming sections are analytically canonically integrable on vast regions of

[7] Analytic means "having convergent Taylor series" near every point of the definition's domain. For a formal definition, see Definitions 13, 14 and 15, §4.13, p. 333–334.

phase space. However, this will not always be explicitly checked and it will be left to the reader, in the problems, to draw this conclusion from the properties discussed in the text.

(3) In an obvious way, one could also define the notion of a Lagrangian analytically integrable on some set W in the data space. The corresponding Hamiltonian system would then be analytically integrable on the corresponding phase-space subset $\tilde{W}$ and vice versa.

Problems for §4.8

1. Given $\boldsymbol{\omega} \in \mathbf{R}^l$ and $g \in C^\infty(\mathbf{T}^l)$, suppose that $\forall \boldsymbol{\nu} \in \mathbf{Z}^l, \boldsymbol{\nu} \neq \mathbf{0}, |\boldsymbol{\omega} \cdot \boldsymbol{\nu}|^{-1} < C|\boldsymbol{\nu}|^\alpha$, for some $C > 0$, $\alpha > 0$. Show that the system on $\mathbf{R}^l \times \mathbf{T}^l$ with Hamiltonian $H(\mathbf{A}, \boldsymbol{\varphi}) = \mathbf{A} \cdot \boldsymbol{\omega} + g(\boldsymbol{\varphi})$ is integrable and find an expression for l prime integrals. Show that this is an isochronous system. (Hint: Write the equations of motion and solve the one for the A's by developing g into a Fourier series $g(\boldsymbol{\varphi}) = \Sigma_{\boldsymbol{\nu} \in \mathbf{Z}^l} g_{\boldsymbol{\nu}} e^{i\boldsymbol{\nu} \cdot \boldsymbol{\varphi}}$ before integration. The prime integrals can be chosen

$$\mathbf{B} = \mathbf{A} + \sum_{\substack{\boldsymbol{\nu} \neq 0 \\ \boldsymbol{\nu} \in \mathbf{Z}^l}} \boldsymbol{\nu} g_{\boldsymbol{\nu}} \frac{e^{i\boldsymbol{\nu} \cdot \boldsymbol{\varphi}}}{\boldsymbol{\nu} \cdot \boldsymbol{\omega}}$$

and the condition on $\boldsymbol{\omega}$ is required to insure the convergence of the series.) Note that the equations of motion can be solved explicitly for general $\boldsymbol{\omega}$.

2. In the context of Problem 1, show that if g is a trigonometric polynomial (i.e., it has finitely many nonvanishing Fourier coefficients), then the results of Problem 1 hold under the sole assumption that the components of $\boldsymbol{\omega}$ are rationally independent.

3. In the context of Problem 1, suppose that there is $\boldsymbol{\nu}_0 \in \mathbf{Z}^l, \boldsymbol{\nu}_0 \neq \mathbf{0}$, such that $\boldsymbol{\omega} \cdot \boldsymbol{\nu}_0 = 0$. Show that the Hamiltonian system with Hamiltonian $H(\mathbf{A}, \boldsymbol{\varphi}) = \mathbf{A} \cdot \boldsymbol{\omega} + \varepsilon \cos(\boldsymbol{\nu} \cdot \boldsymbol{\varphi})$ is not integrable. (Hint: Show that its motions are not quasi-periodic.)

4. In the context of Problem 1, suppose that $|\boldsymbol{\omega} \cdot \boldsymbol{\nu}|^{-1} < C|\boldsymbol{\nu}|^\alpha$, $\forall \boldsymbol{\nu} \neq \mathbf{0}$ and $\boldsymbol{\nu} \notin N_0$, where N_0 is a subset of $\mathbf{Z}^l$. Suppose also that $g \in C^\infty(\mathbf{T}^l)$ is such that $\hat{g}_{\boldsymbol{\nu}} \equiv 0$, $\forall \boldsymbol{\nu} \in N_0$. Show that the Hamiltonian $H(\mathbf{A}, \boldsymbol{\varphi}) = \mathbf{A} \cdot \boldsymbol{\varphi} + g(\boldsymbol{\varphi})$ is integrable on $\mathbf{R}^l \times \mathbf{T}^l$.

5. Show that the integrability in Problems 1, 2, and 4 is analytical and canonical (Hint: $\mathbf{A}' \cdot \boldsymbol{\varphi} + \Sigma_{\boldsymbol{\nu} \neq 0} \hat{g}_{\boldsymbol{\nu}} (e^{i\boldsymbol{\nu} \cdot \boldsymbol{\varphi}})/(-i\boldsymbol{\omega} \cdot \boldsymbol{\nu}) = \Phi(\mathbf{A}', \boldsymbol{\varphi})$ is a generating function for the integrating map.)

§4.9. Integrable Systems. Central Motions with Nonvanishing Areal Velocity. The Two-Body Problem

The best-known integrable mechanical system consists of two point masses with masses $m_1, m_2 > 0$ interacting through a conservative force with potential energy $\overline{V}$ depending only on the distance between the two points:

$$\overline{V}(\boldsymbol{\xi}_1, \boldsymbol{\xi}_2) = V(|\boldsymbol{\xi}_1 - \boldsymbol{\xi}_2|); \tag{4.9.1}$$

and we shall assume that the function $\rho \to V(\rho)$ is defined for $\rho > 0$ and that it is a C^∞ function such that

$$\lim_{\rho \to 0} \rho^2 V(\rho) = 0, \tag{4.9.2a}$$

$$\inf_{\rho > \epsilon} V(\rho) = -V_\epsilon > 0, \qquad \forall \epsilon > 0. \tag{4.9.2b}$$

Note that $V(0)$ is undefined, and this means that we shall only consider motions $t \to (\mathbf{x}_1(t), \mathbf{x}_2(t))$, $t \in \mathbf{R}_+$, such that $|(\mathbf{x}_2(t) - \mathbf{x}_1(t)| > 0$, $t \in \mathbf{R}_+$. This restriction will be imposed via the condition of nonvanishing areal velocity.

If $t \to (\mathbf{x}_1(t), \mathbf{x}_2(t))$, $t \in \mathbf{R}_+$, is a motion of the system, the two points will move so that the total linear and angular momentum will be conserved. In fact, the force generated by Eq. (4.9.1) is easily seen to verify the third law of dynamics, so that the cardinal equations hold and imply the above-mentioned conservation laws.

Hence, the center of mass G moves in a uniform rectilinear fashion and, possibly by changing the reference system, we may suppose that G coincides with the origin 0 of the reference system $(0; \mathbf{i}, \mathbf{j}, \mathbf{k})$ in which we study the motion.

In this situation, the motion $t \to (\mathbf{x}_1(t), \mathbf{x}_2(t)$, $t \in \mathbf{R}_+$, will be such that

$$m_1 \mathbf{x}_1(t) = -m_2 \mathbf{x}_2(t), \qquad \forall t \in \mathbf{R}_+, \tag{4.9.3}$$

and to determine the positions of the two points it will suffice to give the vector $\boldsymbol{\rho}(t) = \mathbf{x}_2(t) - \mathbf{x}_1(t)$:

$$\mathbf{x}_1(t) = -\frac{m_2}{m_1 + m_2} \boldsymbol{\rho}(t), \qquad \mathbf{x}_2(t) = \frac{m_1}{m_1 + m_2} \boldsymbol{\rho}(t). \tag{4.9.4}$$

Since the angular momentum with respect to 0 is a constant vector $\mathbf{K}$, we may assume, without loss of generality, that $\mathbf{K}$ is parallel to $\mathbf{k}$:

$$\mathbf{K} = \tilde{A}\mathbf{k}. \tag{4.9.5}$$

We shall only consider motions for which $\tilde{A} > 0$. We shall see that $\tilde{A}$ is proportional to the areal velocity.

From Eqs. (4.9.3)–(4.9.5), it follows that

$$\mathbf{K} = \tilde{A}\mathbf{k} = m_1 \mathbf{x}_1 \wedge \dot{\mathbf{x}}_1 + m_2 \mathbf{x}_2 \wedge \dot{\mathbf{x}}_2 = m_1 \mathbf{x}_1 \wedge (\dot{\mathbf{x}}_1 - \dot{\mathbf{x}}_2) = -\frac{m_1 m_2}{m_1 + m_2} \boldsymbol{\rho} \wedge \dot{\boldsymbol{\rho}}, \tag{4.9.6}$$

i.e., $\boldsymbol{\rho}$ and $\dot{\boldsymbol{\rho}}$ must both lie in the plane $(\mathbf{i}, \mathbf{j})$. Therefore, the motion $t \to \boldsymbol{\rho}(t)$, $t \in \mathbf{R}_+$, takes place on the plane $(\mathbf{i}, \mathbf{j})$. Recalling the considerations in §3.4 about the constraints, we can find the equations of motion by parameterizing the motion by the polar coordinates (ρ, θ) of $\boldsymbol{\rho}$ in the plane $(\mathbf{i}, \mathbf{j})$ and then writing the Lagrangian equations for the Lagrangian

$$\mathfrak{L} = \tfrac{1}{2} m_1 (\dot{\mathbf{x}}_1)^2 + \tfrac{1}{2} m_2 (\dot{\mathbf{x}}_2)^2 - V(|\mathbf{x}_2 - \mathbf{x}_1|) \tag{4.9.7}$$

computed on the motions parameterized as above.

For such motions, Eq. (4.9.7) becomes

$$\begin{aligned}\mathcal{L}(\dot\rho,\dot\theta,\rho,\theta) &= \frac{1}{2}m_1\left(\frac{m_2}{m_1+m_2}\right)^2\dot{\boldsymbol\rho}^2 + \frac{1}{2}m_2\left(\frac{m_1}{m_1+m_2}\right)^2\dot{\boldsymbol\rho}^2 - V(\rho)\\ &= \frac{1}{2}\frac{m_1m_2}{m_1+m_2}\dot{\boldsymbol\rho}^2 - V(\rho) \qquad (4.9.8)\\ &= \frac{1}{2}\left(\frac{m_1m_2}{m_1+m_2}\right)(\dot\rho^2+\rho^2\dot\theta^2) - V(\rho),\end{aligned}$$

where the well-known formula expressing the square velocity $\dot{\boldsymbol\rho}^2$ as $\dot\rho^2 + \rho^2\dot\theta^2$ in polar coordinates has been used together with Eq. (4.9.4).

Equation (4.9.8) yields the following.

20 Proposition. *The theory of the motion of two point masses, with masses $m_1, m_2 > 0$, under the action of a mutual central conservative force with potential energy given by Eq.* (4.9.1) *is equivalent to the theory of the motion of a single point mass with mass m:*

$$m = \frac{m_1m_2}{m_1+m_2} \qquad (4.9.9)$$

moving on a plane under the action of a conservative force centrally acting on the mass, from a point 0 *in the plane, with the same potential energy V.*

The motions described by the Lagrangian function (4.9.8) and such that $\tilde A \neq 0$ are called the "central motions".

21 Proposition. *The motions of the mechanical system described by Eq.* (4.9.8) *admit two prime integrals:*

$$E = \tfrac{1}{2}m(\dot\rho^2+\rho^2\dot\theta^2) + V(\rho), \qquad (4.9.10)$$

$$A = \rho^2\dot\theta, \qquad (4.9.11)$$

and, if $A \neq 0$, they indefinitely stay away from the origin at a distance greater than some time-independent positive quantity (A and E dependent).

Proof. Equation (4.9.10) is the total energy and, by Eq. (4.9.6), $\tilde A = m\rho^2\dot\theta$ is the angular momentum along the z axis. Hence, E and A are both prime integrals. Note that $\rho^2\dot\theta$ is twice the "areal velocity", i.e., twice the area spanned by $\boldsymbol\rho$ per unit time.

By substituting Eq. (4.9.11) into Eq. (4.9.10) we see that

$$E = \frac{1}{2}m\left(\dot\rho^2 + \frac{A^2}{\rho^2}\right) + V(\rho) \qquad (4.9.12)$$

and Eq. (4.9.2a) implies the existence of $\rho_0 > 0$ such that $E - mA^2/2\rho^2 - V(\rho) < 0$ for $\rho < \rho_0$. So $\rho(t) \geqslant \rho_0$, $\forall t \in \mathbf{R}_+$. mbe

Let $t \to (\rho(t), \theta(t))$, $t \in \mathbf{R}_+$, be a motion associated with Eq. (4.9.8) with $A > 0$. We can write the equation of motion for ρ by considering the

Lagrangian equation relative to Eq. (4.9.8) and corresponding to the coordinate ρ:

$$m\ddot{\rho} = m\rho\dot{\theta}^2 - \frac{\partial V}{\partial \rho}(\rho) \tag{4.9.13}$$

which, by Eq. (4.9.11), becomes

$$m\ddot{\rho} = m\frac{A^2}{\rho^3} - \frac{\partial V}{\partial \rho}(\rho) \equiv -\frac{\partial V_A}{\partial \rho}(\rho), \tag{4.9.14}$$

where

$$V_A(\rho) = \frac{mA^2}{2\rho^2} + V(\rho), \tag{4.9.15}$$

showing that the ρ coordinate evolves in time as the abscissa of a point mass, with mass m, on a line subject to a conservative force with potential energy V_A.

Since the motion, by Proposition 21, is such that $\rho(t) \geqslant \rho_0 > 0$, we can ignore the singularities of V and V_A in $\rho = 0$ and we can also ignore the constraint $\rho \geqslant 0$ due to the fact that ρ is the polar radial coordinate, and we can apply the theory of Chapter 2 for the conservative C^∞ forces acting upon one-dimensional systems.

22 Proposition. *Let $\rho \to V(\rho)$, $\rho > 0$, be a $C^\infty((0, +\infty))$ function verifying Eq.* (4.9.2). *Let W be the open set, in the data space of the system described by Eq.* (4.9.8), *consisting of the data with E and A in Eqs.* (4.9.10) *and* (4.9.11) *such that*

(i) $A > 0$. (4.9.16)

(ii) *The equation*

$$V_A(\rho) = \frac{mA^2}{2\rho^2} + V(\rho) = E \tag{4.9.17}$$

admits just two solutions $\rho_-(E,A), \rho_+(E,A)$ such that $\rho_+(E,A) > \rho_-(E,A)$ and $(\partial V_A/\partial\rho)(\rho) \neq 0$ for $\rho = \rho_\pm(E,A)$.

Then the system is integrable in a neighborhood of every point in W and has two periods, see Definition 10, §4.8, *p.* 287, *given by*

$$T_1(E,A) = 2\int_{\rho_-}^{\rho_+}\left(\frac{2}{m}(E - V_A(\rho))\right)^{-1/2} d\rho, \tag{4.9.18}$$

$$T_2(E,A) = \frac{2\pi}{A}\,\frac{\int_{\rho_-}^{\rho_+}((2/m)(E - V_A(\rho)))^{-1/2}\,d\rho}{\int_{\rho_-}^{\rho_+}\rho^{-2}((2/m)(E - V_A(\rho)))^{-1/2}\,d\rho}, \tag{4.9.19}$$

where $\rho_+ = \rho_+(E,A) = \rho_- = \rho_-(E,A)$.

Proof. Let $(\dot{\rho}_0, \dot{\theta}_0, \rho_0, \theta_0) \in W$ be an initial datum with energy E and areal velocity $A/2$.

In the course of the proof we shall state that some functions are C^∞, leaving the proof to the reader.

Consider the solution of Eq. (4.9.14),

$$t \to R(t, E, A), \qquad t \in \mathbf{R}_+ \tag{4.9.20}$$

with initial datum

$$R(0, E, A) = \rho_-(E, A), \qquad \dot{R}(0, E, A) = 0. \tag{4.9.21}$$

By the theory of one-dimensional motions, §2.7, the function R is a C^∞ function periodic in t with period

$$T_1(E, A) = 2\int_{\rho_-}^{\rho_+} \frac{d\rho}{\sqrt{(2/m)(E - V_A(\rho))}}, \tag{4.9.22}$$

where $\rho_- = \rho_-(E, A)$, $\rho_+ = \rho_+(E, A)$.

If $t_0(\rho_0, \dot{\rho}_0)$ is the shortest time such that

$$R(t_0, E, A) = \rho_0, \qquad \dot{R}(t_0, E, A) = \dot{\rho}_0, \tag{4.9.23}$$

necessarily existing by our assumptions on W, it follows that

$$\rho(t) = R(t + t_0(\rho_0, \dot{\rho}_0), E, A), \qquad t \in R_+ . \tag{4.9.24}$$

To complete the analysis of the motion, we still have to determine $\theta(t)$. Using Eq. (4.9.11):

$$\theta(t) = \theta_0 + \int_0^t \frac{A}{R(t' + t_0(\rho_0, \dot{\rho}_0), E, A)^2}\, dt'; \tag{4.9.25}$$

and observe that the integrand function in Eq. (4.9.25) is a C^∞ periodic function of t' with the period of Eq. (4.9.22), since such is R and also $R \geqslant \rho_-(E, A) > 0$. Then by the Fourier theorem, if $T_1 \equiv T_1(E, A)$,

$$AR(t, E, A)^{-2} = \sum_{k \in \mathbf{Z}} \chi_k(A, E) e^{(2\pi i/T_1)kt}, \tag{4.9.26}$$

where $(\chi_k)_{k \in \mathbf{Z}}$ are the Fourier coefficients of AR^{-2}. They vanish as $k \to \infty$ faster than any power in k.

Inserting Eq. (4.9.26) into Eq. (4.9.25), it appears that

$$\theta(t) = \theta_0 + \chi_0(A, E)t + \sum_{\substack{k \in \mathbf{Z} \\ k \neq 0}} \chi_k(A, E) \frac{e^{(2\pi i/T_1)kt} - 1}{2\pi i k/T_1} e^{(2\pi i/T_1)kt_0(\rho_0, \dot{\rho}_0)} \tag{4.9.27}$$

which we shall write as

$$\theta(t) = \theta_0 + \chi_0(A, E)t + S(t + t_0(\rho_0, \dot{\rho}_0), E, A) - S(t_0(\rho_0, \dot{\rho}_0), E, A), \tag{4.9.28}$$

where

$$S(t, E, A) = \sum_{\substack{k \in \mathbf{Z} \\ k \neq 0}} \chi_k(A, E) \frac{e^{2\pi i kt/T_1(E,A)}}{2\pi i k/T_1(E, A)} \tag{4.9.29}$$

is a C^∞ function, periodic with period $T_1 \equiv T_1(E, A)$.

It is then clear that the coordinates of $\rho(t)$ have the form of Eq. (4.8.15). For instance, if $\rho(t) = (\rho_1(t), \rho_2(t)) \in \mathbf{R}^2$:

$$\begin{aligned}\rho_1(t) = \rho(t)\cos\theta(t) &= R(t+t_0)(\cos(\theta_0 + \chi_0 t + S(t+t_0) - S(t_0))\\ &= R(t+t_0)\{\cos(\theta_0+\chi_0 t)\cos(S(t+t_0) - S(t_0))\\ &\quad - \sin(\theta_0+\chi_0 t)\sin(S(t+t_0) - S(t_0))\},\end{aligned} \tag{4.9.30}$$

where the dependence on the $E, A, \rho_0, \dot\rho_0$ variables has not been explicitly written. By Observation 4 to Definition 10, p. 288, this shows the integrability of the system and that the two periods are $T_1(E,A)$ and $T_2(E,A) = 2\pi\chi_0(A,E)^{-1}$.

It is also easy to find explicitly the integrating transformation I: the prime integrals are E and A, the angles $(\varphi_1, \varphi_2) \in \mathbf{T}^2$ are, for instance, by Eqs. (4.9.24) and (4.9.28),

$$\varphi_1(\dot\rho_0, \dot\theta_0, \rho_0, \theta_0) = \frac{2\pi}{T_1(E,A)} t_0(\rho_0, \dot\rho_0) \qquad (\text{"average anomaly"}), \tag{4.9.31}$$

$$\varphi_2(\dot\rho_0, \dot\theta_0, \rho_0, \theta_0) = \theta_0 - S(t_0(\rho_0, \dot\rho_0), E, A) \qquad (\text{"average longitude"}), \tag{4.9.32}$$

and the respective periods are, as already mentioned, $T_1(E,A)$ and $T_2(E,A)$ [see Eq. (4.9.28)].

We do not check explicitly the regularity and the invertibility of the transformation I on suitable neighborhoods of the trajectory starting in $(\dot\rho_0, \rho_0, \dot\theta_0, \theta_0)$.

It remains to check Eq. (4.9.19). Again we do not write explicitly the E and A dependence in the functions $\rho_-(E,A)$, $\rho_+(E,A)$, $\chi_0(A,E)$, $T_1(E,A)$, $R(t,E,A)$, $S(T,E,A)$. By the Fourier theorem,

$$\chi_0 = T_1^{-1}\int_0^{T_1} AR(t)^{-2}\,dt = 2T_1^{-1}\int_0^{T_1/2} AR(t)^{-2}\,dt \tag{4.9.33}$$

because $R(t)$ behaves specularly when t varies from 0 to $T_1/2$ or from $T_1/2$ to T_1 (when R varies between ρ_- and ρ_+ or between ρ_+ and ρ_-). But for $t \in [0, T_1/2]$,

$$t = \int_{\rho_-}^{R(t)} \left(\frac{2}{m}(E - V_A(\rho))\right)^{-1/2} d\rho \tag{4.9.34}$$

by Eqs. (4.9.15) and (4.9.20). Hence, by changing variables "$t \to R$", via Eq. (4.9.34),

$$dt = \left(\frac{2}{m}(E - V_A(R))\right)^{-1/2} dR, \tag{4.9.35}$$

it follows from Eq. (4.9.33) that

$$\chi_0 = (2A/T_1)\int_{\rho_-}^{\rho_+} R^{-2}\frac{2}{m}(E - V_A(R))^{-1/2}\,dR \tag{4.9.36}$$

which implies Eq. (4.9.19), as $T_2 = 2\pi\chi_0^{-1}$. mbe

Observation. If we regard Eq. (4.9.8) as defining a three-dimensional problem with Lagrangian

$$\mathcal{L}(\dot{\rho}, \rho_-) = \tfrac{1}{2} m \dot{\rho}^2 - V(\rho), \tag{4.9.37}$$

it follows, of course, that under the same assumptions as in Proposition 22, the system is integrable. Now the prime integrals will be E, A and the angle of inclination i of the orbit's plane with the reference $(\mathbf{i}, \mathbf{j})$ plane. The third angle will be the longitude in the $(\mathbf{i}, \mathbf{j})$ plane, counted from the $\mathbf{i}$ axis (say), of the intersection of the orbit's plane with the $(\mathbf{i}, \mathbf{j})$ plane ("nodes line").

However, the third angle thus defined remains constant over time. This means that the system's pulsations in these coordinates will be $\omega_1 = 2\pi T_1^{-1}$, $\omega_2 = 2\pi T_2^{-1}$, $\omega_3 = 0$.

Problems for §4.9

1. Let $m = 1$ and consider the motions associated with the Lagrangian (4.9.8) under the assumptions of Proposition 22. Following the idea of Problem 4, p. 227 and substituting L for A in that problem, define

$$L \equiv \lambda(E, A) = \int_{\rho_-(E,A)}^{\rho_+(E,A)} (2(E - V_A(\rho)))^{1/2} \frac{d\rho}{\pi}.$$

Suppose that this relation between L, E, A can be inverted with respect to E, for E, A in some open set V, in the form $E = \epsilon(L, A)$ so that

$$L \equiv \lambda(\epsilon(L, A), A)$$

with ϵ of class C^∞. Show that if $E = \epsilon(L, A)$,

$$2\pi T_1(E, A)^{-1} = \frac{\partial \epsilon}{\partial L}(L, A),$$

$$2\pi T_2(E, A)^{-1} = \frac{\partial \epsilon}{\partial A}(L, A).$$

(Hint: Note that $1 \equiv (\partial\lambda/\partial E)(\partial\epsilon/\partial L)$ and $0 \equiv (\partial\lambda/\partial E)(\partial E/\partial A) + \partial\lambda/\partial A$ and then use Eqs. (4.9.18) and (4.9.19) noting that the derivatives with respect to the integration extremes vanish as, by the definition of ρ_-, ρ_+, the integrand vanishes at the extremes.)

2. In the context of Problem 1 write the Hamiltonian function corresponding to Eq. (4.9.8) as, $(m = 1)$:

$$H(p_\rho, p_\theta, \rho, \theta) = \frac{1}{2}\left(p_\rho^2 + \frac{p_\theta^2}{\rho^2}\right) + V(\rho)$$

and note that the function

$$\tilde{S}(L, A, \rho, \theta) = A\theta + \int_{\rho_-(\epsilon(L,A),A)}^{\rho} (2(\epsilon(L, A) - V_A(\rho')))^{1/2} d\rho'$$

solves the Hamilton–Jacobi equation

$$H\left(\frac{\partial \tilde{S}}{\partial \rho}, \frac{\partial \tilde{S}}{\partial \theta}, \rho, \theta\right) = \epsilon(L, A)$$

From this fact, infer that S generates a change of coordinates (completely canonical):

$$(p_\rho, p_\theta, \rho, \theta) \leftrightarrow (L, A, l, g),$$

where l and g are the angular variables defined in Eqs. (4.9.31) and (4.9.32) in terms of the data:

$$l = \frac{2\pi}{T_1(\epsilon(L,A),A)}\, t_0, \qquad g = \theta - S(t_0, E, A),$$

and the Hamiltonian in the new variables is simply $\epsilon(L,A)$.

3. In the context of Problem 2, if one defines

$$\hat{S}(E,A,\rho,\theta) = A\theta + \int_{\rho_-(E,A)}^{\rho} (2(E - V_A(\rho')))^{-1/2}\, d\rho',$$

then one finds a two-parameters local solution to the Hamiltonian–Jacobi equation

$$H\left(\frac{\partial \hat{S}}{\partial \rho}, \frac{\partial \hat{S}}{\partial \theta}, \rho, \theta\right) = E$$

(the parameters being E and A).

Then $\hat{S}$ generates a completely canonical transformation $(p_\rho, p_\theta, \rho, \theta) \leftrightarrow (E, A, \tau, \alpha)$ in which α is a constant angle and τ varies linearly over time.

Show that these new coordinates *cannot* be extended to a well-defined system of coordinates in the vinicity of a full trajectory of the motion if this trajectory corresponds to a quasi-periodic motion with two periods having irrational ratio. Note that this is not the case for the other coordinate transformation of Problem 2.

4. In the context of Problem 3, the change of coordinates introduced there can be extended to a well-defined system of coordinates in the vicinity of a full trajectory for which $\rho_+(E,A) = +\infty$ (i.e., in the vicinity of an unbounded trajectory) if $\limsup_{\rho \to +\infty} V_A(\rho) < E$. In this case, the pair of variables (E, τ) are called "energy-time" coordinates. Why?

5. Solve Problems 1 and 2 for arbitrary $m > 0$.

§4.10. Kepler's Marvelous Laws

Leva dunque, lettore, a l'alte ruote
Meco la vista, dritto a quella parte
Dove l'un moto e l'altro si percuote;
E lí comincia a vagheggiar ne l'arte
Di quel maestro che dentro a sé l'ama
Tanto che mai da lei l'occhio non parte,
Vedi come da indi si dirama
L' oblico cerchio che i pianeti porta
Per sodisfare al mondo che li chiama.[8]

The main result of §4.9, expressed by Proposition 22, is that the motion of a point mass in a central force field under some hypotheses on the initial data

[8] In basic English:
Look up now, reader, to the high wheels
together with me, straight there
where several motions hit each other.
And there begin to wonder about the art
Of that master who inside himself moves them with his love
so much that he never drops his eyes away.
Look up how the oblique circle bearing the planets develops there
to satisfy the world that calls them.
(Dante, *Paradiso*, Canto X)

is a quasi-periodic motion with two periods given by Eqs. (4.9.18) and (4.9.19) depending upon the energy E and the areal velocity $A/2$.

By contemplating Eqs. (4.9.18) and (4.9.19), it is easy to convince oneself that in general $T_1(E,A)$ and $T_2(E,A)$ are "independent". Hence, unless

$$\frac{T_1(E,A)}{T_2(E,A)} = \text{rational number,} \tag{4.10.1}$$

which is "exceptional" when E and A vary, the motion is actually quasi-periodic and not periodic.

Note, however, that the set of the space points where Eq. (4.10.1) holds will generally be dense in the region W where the motion is integrable. As an exercise, the reader may show the truth of this statement near a point of W where the E and A values are such that the Jacobian determinant of the map $(E,A) \leftrightarrow (T_1(E,A), T_2(E,A))$ does not vanish.

However, there are two exceptional and marvelous cases.

The first, already implicitly studied in §4.1, is the harmonic oscillator bound to O by a force with potential

$$V(\rho) = \frac{k}{2}\rho^2, \tag{4.10.2}$$

leading to

$$2T_1 \equiv T_2 = 2\pi\sqrt{\frac{k}{m}} \equiv T, \tag{4.10.3}$$

and the orbits are ellipses centered in O. Equation (4.10.3) could be proved by computing the integrals of Eqs. (4.9.18) and (4.9.19) (which is a long but straightforward calculation). However, the reader should try to find a simple argument leading to Eq. (4.10.3) without any explicit calculations beyond the ones already done in §4.9.

The other case corresponds to

$$V(\rho) = -m\frac{g}{\rho}. \tag{4.10.4}$$

This is the case of the so-called "Newtonian two-body problem" or "Kepler's problem". If $E < 0$, the motion is periodic and $T_1 = T_2$, although T_1 and T_2 now actually depend on A and E, and the orbits are ellipses with focus in O.

We treat this problem in some detail by proving the following proposition.

23 Proposition. *The motions with energy $E < 0$ and areal velocity $A/2 \neq 0$ are periodic and the integrals of Eqs.* (4.9.18) *and* (4.9.19) *coincide, $\forall E < 0$, $\forall A > 0$. Furthermore:*

(i) *the trajectories $t \to \rho_-(t)$, $t \in \mathbf{R}_+$, are ellipses with focus in O;*
(ii) *such ellipses are run with the constant areal velocity $A/2$;*

(iii) *the ratio between the square of the revolution period T and the cube of the length of the ellipse's major axis is a constant solely depending on g.*

Finally, if ρ_+ and ρ_- are the focal distances of the ellipse on which a given motion takes place:

$$\rho_+ + \rho_- = \frac{mg}{-E}, \tag{4.10.5}$$

$$\rho_+ \rho_- = \frac{mA^2}{-2E}, \tag{4.10.6}$$

$$T = \frac{\pi}{\sqrt{2g}} (\rho_+ + \rho_-)^{3/2}. \tag{4.10.7}$$

Observations.

(1) (i), (ii), and (iii) are Kepler's laws. Starting from them, Newton realized that if one wanted to describe a planet's motion by a second-order differential equation $m\ddot{\boldsymbol{\rho}} = \mathbf{F}(\boldsymbol{\rho})$, the only possibility was that $\mathbf{F}(\boldsymbol{\rho}) = -mg\, \rho^{-2}(\boldsymbol{\rho}/\rho)$. This led him to assume, by symmetry, $g = kM$, $M =$ Sun's mass, i.e., $V(\rho) = -kmM\rho^{-1}$ which is the universal law of gravitation.

Of course, he also assumed that (i), (ii), and (iii) would describe the motion laws of an arbitrary body revolving around the Sun, whatever its initial position and speed.

Newton's argument is interesting and different in spirit from the one based on the analytic theory of differential equations. It is based on some beautiful geometric considerations relying on the theory of conic sections: it can be found in the first book of the *Principia* (see References).

(2) One could also easily study the $E > 0$ or $E = 0$ motions: they are not periodic motions and the trajectories become a hyperbola's wing or a parabola, respectively. This is a simple exercise along the lines of the upcoming proof and it will be left to the reader.

(3) The heavenly bodies have finite extension. Hence, if a satellite revolves circularly around a primary body (planet or Sun), turning always the same face to it, a situation in apparent contradiction to Kepler's laws is produced. In fact, if the satellite is thought of as decomposed into small point masses, the points on one face rotate on an orbit with radius smaller than the orbit of the points of the opposite face. Hence, if one could neglect the mutual interactions between the points of the body, they would have to have a different rotation period around the main body (by the Kepler's third law) and the satellite would disintegrate over time. This means that if the above catastrophic event does not occur, the body must be subject to some internal stresses ("tidal stresses") which cannot be stronger "than the body's material resistance" (otherwise, the satellite could not exist). So Kepler's laws and the gravitation law provide a mechanism for explaining Saturn's rings and why, in general, satellites stay quite far away from a planet (see problems at the end of this Section).

PROOF. With the notation of §4.9, we set $V_A(\rho) = m(A^2/2\rho^2 - g/\rho)$ and we see that the energy and angular momentum conservation laws become

$$\frac{1}{2}\left(\dot{\rho}^2 + \frac{A^2}{\rho^2}\right) - \frac{g}{\rho} = \frac{E}{m} \tag{4.10.8}$$

or

$$\dot{\rho}^2 = 2\left(\frac{E - V_A(\rho)}{m}\right), \tag{4.10.9}$$

$\rho^2\dot{\theta} = A$. The graph of V_A is illustrated in Fig. 4.8. Hence, if $E < 0$, $E > -mg^2/2A^2$, the roots of the equation $V_A(\rho) = E$ are $\rho_-(E,A), \rho_+(E,A)$ and they can be explicitly found by solving a second-degree equation in the unknown $1/\rho$. By the factorization properties of polynomials, we shall write the polynomial in $1/\rho$ given by $E/m - V_A(\rho)/m$ in terms of its roots:

$$\frac{E}{m} - \frac{A^2}{2\rho^2} + \frac{g}{\rho} = \frac{A^2}{2}\left(\frac{1}{\rho_-} - \frac{1}{\rho}\right)\left(\frac{1}{\rho} - \frac{1}{\rho_+}\right). \tag{4.10.10}$$

The radii ρ_- and ρ_+ are, as we shall shortly see, the focal distances of the ellipse on which the motion develops. They obviously verify Eqs.

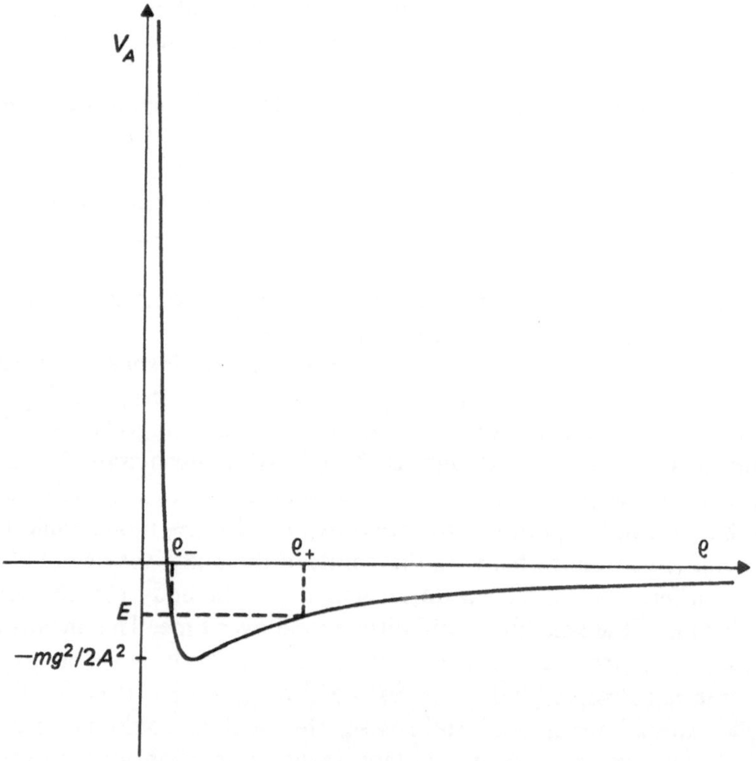

Figure 4.8.

(4.10.5) and (4.10.6) since

$$\rho_+^{-1} + \rho_-^{-1} = 2gA^{-2}, \qquad \rho_+^{-1}\rho_-^{-1} = 2E/mA^2.$$

Use Eq. (4.10.10) to rewrite Eqs. (4.10.8) and (4.10.9) as

$$\dot{\rho} = \pm A\sqrt{\left(\frac{1}{\rho_-} - \frac{1}{\rho}\right)\left(\frac{1}{\rho} - \frac{1}{\rho_+}\right)}, \tag{4.10.11}$$

$$\dot{\theta} = \frac{A}{\rho^2}. \tag{4.10.12}$$

Suppose that for $t = 0$, it is $\rho(0) = \rho_-, \theta(0) = \pi$. Since the motion of ρ is periodic, being a solution to Eq. (4.9.14), and oscillates between ρ_- and ρ_+, this hypothesis does not affect the generality.

Then in Eq. (4.10.11) the + sign holds for $t \in [0, T/2]$ if T is the period of the ρ motion [Eq. (4.9.18)]:

$$T = 2\int_{\rho_-(E,A)}^{\rho_+(E,A)} \frac{d\rho}{\sqrt{(2/m)(E - V_A(\rho))}}. \tag{4.10.13}$$

Hence, for $t \in [0, T/2]$, observing that Eq. (4.10.12) implies that θ is a strictly increasing function of t and, therefore, allows us to think of ρ as a function of θ instead of t, we can say that Eq. (4.10.11) divided by Eq. (4.10.12) yields

$$\frac{d\rho}{d\theta} = \rho^2\sqrt{\left(\frac{1}{\rho_-} - \frac{1}{\rho}\right)\left(\frac{1}{\rho} - \frac{1}{\rho_+}\right)} \tag{4.10.14}$$

which for $\rho_- < \rho < \rho_+$ implies

$$\theta - \pi = \int_{\rho_-}^{\rho} \frac{d\rho'}{(\rho')^2\sqrt{(1/\rho_- - 1/\rho')(1/\rho' - 1/\rho_+)}}. \tag{4.10.15}$$

This is an elementary integral. Changing the variable as $y = \rho^{-1}$, after some algebra, one finds

$$\frac{1}{\rho} = \frac{1}{2}\left(\frac{1}{\rho_+} + \frac{1}{\rho_-}\right) + \frac{1}{2}\left(\frac{1}{\rho_-} - \frac{1}{\rho_+}\right)\cos(\theta - \pi), \tag{4.10.16}$$

showing that when θ reaches 2π, ρ reaches ρ_+.

The study of the trajectory for $t \in [T/2, T]$ proceeds likewise, changing the choice of sign in Eq. (4.10.11), and one finds that the trajectory still verifies Eq. (4.10.16), and at time T when ρ takes the value ρ_-, the angle θ takes the value 3π. This means that after time T has elapsed, not only ρ but also θ take on the initial value (of course θ has to be measured mod 2π). Hence, the trajectory is closed because $\dot{\theta}$ and $\dot{\rho}$ also take on again the initial values, by Eqs. (4.10.11) and (4.10.12) (i.e., $\dot{\rho} = 0, \dot{\theta} = A\rho_-^{-2}$), and because of the autonomy of equations of motion.

Equation (4.10.16), as is well known from elementary geometry, is the polar coordinate equation of an ellipse with focus at the origin, focal distances ρ_- and ρ_+, and major axis along the x axis (and "perihelion" on the negative x axis).

To compute the period of the motion, it suffices to calculate the integral of Eq. (4.10.13), elementary after the substitution $y = \rho^{-1}$. However, this calculation can be avoided by recalling that the ellipse is run with constant areal velocity $A/2$ and, hence, T can be obtained by dividing the area of the ellipse of Eq. (4.10.16) by $A/2$.

This area is

$$\pi \frac{\rho_+ + \rho_-}{2} \sqrt{\rho_+ \rho_-}, \tag{4.10.17}$$

since the semi-axes of an ellipse with focal distances ρ_+ and ρ_- are $(\rho_+ + \rho_-)/2$ and $\sqrt{\rho_+ \rho_-}$. Hence,

$$T = \pi \frac{\rho_+ + \rho_-}{2} \sqrt{\rho_+ \rho_-} \frac{2}{A} = \frac{\pi}{\sqrt{2g}} (\rho_+ + \rho_-)^{3/2} \tag{4.10.18}$$

by Eqs. (4.10.5) and (4.10.6). mbe

Exercises and Problems for §4.10

Use the tables in Appendix P for the numerical values. Problems 1 through 9 are inspired by Berry, see references.

1. Let T' be a heavenly body identical to the Earth. Could a satellite T'' identical to the Earth (i.e., a twin) be eternally eclipsed by T' while they revolve around the Sun S on a circular orbit in a one-year period? Compute the $T'T''$ distance as well as the ST' distance, comparing the percentage difference between ST' and the actual average distance between the Sun and the Earth.

2. Could a point mass M have two homogeneous rigid gravitational satellites with radius $\delta/2$ and mass μ whose surfaces touch at a point at distance ρ from M? Find the necessary relations among ρ, δ, μ, M assuming $\delta \ll \rho$ to first order in δ/ρ. Compute the force τ ("disruptive force") due to the spheres' contact. (Answer: $\delta < \rho\sqrt[3]{\frac{2}{3}\,\mu/M}$; $\tau = (k\mu^2/\delta^2)(1 - \frac{3}{2}(\mu/M)(\delta/\rho)^3) > 0$, k being the gravitational constant. Suppose that the force τ cannot be negative, i.e., that the two bodies can only push each other.)

3. Same as Problem 2, but assuming that the body with mass M is a homogeneous sphere with readius R and that both the planet and the satellites have the same density σ: $M = (4\pi/3)\sigma R^3$, $\mu = (4\pi/3)\sigma(\delta/2)^3$. Show that to first order in δ/ρ there is no condition on δ but only a condition on the ratio between R and ρ. (Answer: $\tau > 0 \Leftrightarrow 1 - \frac{3}{2}8(R/\rho)^3 > 0 \Leftrightarrow \rho > 2.29R$.)

4. Use Problem 3 to show that a heuristic estimate for the minimum distance of a planet to the Sun's center is $\rho \sim 2.29R$ if R is the radius of the Sun. Compute ρ in km and compare it to Mercury's orbital radius, schematizing the Sun as a sphere with radius equal to its optically apparent radius.

5. Same as Problem 4 to estimate at what distance from the Earth can one find the closest satellite with the same density ($\sim 2.29 \times 6.3 \times 10^3$ km). Why can the artificial satellites gravitate much closer? (See Exercise 6.)

6. Assume that a satellite to a planet is made of rock with density σ, cohesion force per unit surface γ, and with diameter δ. Using Problem 3, find a heuristic estimate of how large must δ be in order that the satellite cannot gravitate at distance ρ from the planet (supposed to have the same density) if ρ is in the forbidden band ($\rho < 2.29R$). (Hint: Compute the tidal force τ of Exercise 3 and compare it with the cohesion force $\pi(\delta/2)^2\gamma$: if $[k(4\pi/3)\sigma(\delta/2)^3]\delta^{-2}(\frac{3}{2}8(R/\rho)^3 - 1) > \pi(\delta/2)^2\gamma$, the tidal force prevails over the cohesion force and the body breaks up.)

7. Let $\sigma = 5.5$ g/cm^3, $\gamma = 100$ kg$_w$/cm^2, $k = 6.67 \times 10^{-8}$ cm^3/g sec^3, $\rho = 7.0 \times 10^3$ km, $R = 6.33 \times 10^3$ km. How big should a rocky satellite be in order to apply the instability argument of Problem 4? Same for $\rho = 2R$. (1 kg$_w$ = weight of a mass of 1 kg at the Earth's surface.)

8. At what distance from Saturn can one find its closest satellite? Compare it with Mima's distance.

9. Assuming that Saturn's rings consist of rocky satellites with a cohesion modulus γ like that of Exercise 7 and a density equal to that of Saturn (3 g/cm^3), heuristically estimate how big can the rings' stones be as a function of the radius r of the ring. Compare their maximum diameter with the observed width of the rings ($\sim$2–20 km).

10. Solve explicitly Problems 1, 2, and 4 in §4.9 in the case of Kepler's problem, explicitly computing L and $\epsilon(L,A)$. (Answer: $\epsilon(L,A) = -k^2m^3/2(L+mA)^2$ if $V(\rho) = -km/\rho$.)

11. Given a Kepler motion in $\mathbf{R}^3$ with energy E, set $a = (\rho_+ + \rho_-)/2$, $e = (\rho_+ - \rho_-)/(\rho_+ + \rho_-)$ = (eccentricity of the ellipse with focal distances ρ_- and ρ_+), and set

$$L = m\sqrt{k}\,\sqrt{a} \equiv \frac{m^{3/2}k}{\sqrt{-2E}}, \qquad G = L\sqrt{1-e^2} \equiv mA.$$

Applying Problems 1, 2, and 5 of §4.9, consider the canonical transformation $(p_\rho, p_\theta, \rho, \theta) \leftrightarrow (L, G, l, g)$ associated with the generating function

$$\tilde{S}(L,G,\rho,\theta) = \theta G + \int_{\rho_-}^{\rho} \left(2m(\epsilon(L) - V_{G/m}(\rho))\right)^{1/2} d\rho,$$

where $\epsilon(L) = -k^2m^3/2L^2 = E$ and ρ_+ and ρ_- depend on L, G being equal to $\rho_+(E,A), \rho_-(E,A)$, i.e., consider the map I generated by

$$p_\rho = \frac{\partial \tilde{S}}{\partial \rho}, \qquad l = \frac{\partial \tilde{S}}{\partial L},$$

$$p_\theta = \frac{\partial \tilde{S}}{\partial \theta}, \qquad g = \frac{\partial \tilde{S}}{\partial G}.$$

Applying Problems 1, 2, and 5 of § 4.9 and Problem 10 above, show that the above completely canonical transformation I can be extended to the entire set of initial data such that $G > 0$, $E_G \equiv -m^3k^2/2G^2 < E < 0$ and that the image of this set of data via I has the form $V \times \mathbf{T}^2$, where $V = \{(L,G) \mid G > 0,\ E_G < -k^2m^3/2G^2 < 0\} \equiv \{(L,G) \mid G > 0,\ L > G\}$, and check that l, g are "angles".

12. Show that the physical interpretation of the angle g canonically conjugate to G in Problem 11 is that of the longitude of the major semiaxis of the ellipse, while the angle l conjugated to L, "average anomaly", is $l=(2\pi t/T)$, where t is the time necessary to reach the initial point of the orbit starting, say, at time zero from the "perihelion", i.e., from the extreme point on the major axis closest to the center of force.

13. Consider a point attracted to the origin by a gravitational force. Suppose that its energy is negative so that it moves on an ellipse and let (L, G, l, g) be its Keplerian coordinates (see Problems 11 and 12). Let $(p_\rho, p_\theta, \rho, \theta)$ be the corresponding "natural canonical coordinates" (see Problem 2, §4.9) and let β be the polar angle formed by the position vector with the major semiaxis of the ellipse on which the motion develops following the initial data $(p_\rho, p_\theta, \rho, \theta)$. Call a, b, and e the major semiaxis, the minor semiaxis, and the eccentricity of the ellipse, respectively, and write its equation as

$$\rho=\frac{p}{1-e\cos\beta},\qquad p=\frac{b^2}{a}=\frac{\rho_+\rho_-}{\frac{1}{2}(\rho_+ +\rho_-)}$$

[see Eq. (4.10.16)] and define ξ as

$$\rho=a(1+e\cos\xi).$$

Find the relation expressing ρ, θ, β in terms of the Keplerian variables (L, G, l, g). Show that

$$l=\sqrt{1-e^2}^{\,3}\int_0^\beta\frac{d\beta'}{(1-e\cos\beta')^2}=\beta+2e\sin\beta+\frac{3}{4}e^2\sin 2\beta+\cdots,$$

$$\beta=l-2e\sin l+\tfrac{5}{4}e^2\sin 2l+\cdots,$$

$$l=\xi+e\sin\xi,$$

$$\xi=l-e\sin l+\frac{e^2}{2}\sin 2l+\cdots,$$

$$\theta=g+\beta,$$

$$\rho=p(1-e\cos\beta)^{-1}=a(1+e\cos\xi).$$

(Hint: Use Eq. (4.10.12) and $\dot{l}=2\pi/T$ to find that

$$\frac{d\beta}{dl}=(1-e^2)^{-3/2}(1-e\cos\beta)^2$$

(noting that β is the analogue of the angle θ of §4.10) and having used all the relations between ρ_+, ρ_-, A, T in Eqs. (4.10.5)–(4.10.7). Use Eq. (4.10.11) to see analogously that

$$\frac{d\rho}{dl}=\frac{a}{\rho}\sqrt{a^2e^2-(\rho-a)^2}\,.$$

Then integrate the first equation to express l in terms of β and the second equation to express l in terms of ξ after changing variables as $\rho=a(1+e\cos\xi)$.

To prove the expansion for l as an eccentricity series, consider the integral expression of l in terms of β found above and expand the function $(1-e^2)^{3/2}(1-e\cos\beta')^{-2}$ in powers of e before integrating and, then, integrate term by term.)

14. Using Problem 13, express the Cartesian coordinates of the point's position in terms of (L, G, l, g), proving that

$$x=a(1+e\cos\xi)\cos(g+\beta),\qquad y=a(1+e\cos\xi)\sin(g+\beta),$$

or

$$x = p(1 - e\cos\beta)^{-1}\cos(g+\beta), \qquad y = p(1 - e\cos\beta)^{-1}\sin(g+\beta),$$

where ξ, β have to be expressed in terms of (L, G, l, g) via the formulae of Problem 13.

15. Using Problems 13 and 14, show that the Cartesian coordinates can be expressed correctly in terms of (L, G, l, g) up to second order in the eccentricity e as

$$x = a\left[\cos(g+l) + eA_x(g,l) + e^2B_x(g,l)\right] + O(e^3),$$
$$y = a\left[\sin(g+l) + eA_y(g,l) + e^2B_y(g,l)\right] + O(e^3),$$

where

$$A_x = \cos g + \sin l \sin(g+l), \qquad A_y = \sin g - \sin l\cos(g+l),$$
$$B_x = \sin g \sin l - \tfrac{3}{4}\sin(g+l)\sin 2l,$$
$$B_y = -\cos g \sin l + \tfrac{3}{4}\cos(g+l)\sin 2l,$$

which can also be written in complex form, and calling ϵ the eccentricity to avoid confusion with $e = 2.71\ldots$,

$$x + iy = ae^{i(g+l)}\left[1 + \epsilon(e^{-il} - i\sin l) + \epsilon^2\left(-i(\sin l)e^{-il} + \tfrac{3}{4}i\sin 2l\right)\right].$$

16. Consider the problem analogous to Problems 10 and 11 in the case of the Kepler motion in $\mathbf{R}^3$ and look for a completely canonical transformation between the natural polar coordinates $(p_\rho, p_\varphi, p_\theta, \rho, \varphi, \theta)$ in terms of which the system's Hamiltonian is

$$(2m)^{-1}\left(p_\rho^2 + (\rho\sin\theta)^{-2}p_\varphi^2 + \rho^{-2}p_\theta^2\right) - km/\rho,$$

corresponding to the Lagrangian

$$\frac{m}{2}\left(\dot\rho^2 + (\rho^2(\sin\theta)^2\dot\varphi^2 + \rho^2\dot\theta^2\right) + km/\rho$$

and the coordinates $(L, G, \Theta, l, g, \tau)$, where L, G, Θ are defined in terms of the energy E of the areal velocity A and of the orbit's planar inclination i with respect to the z axis by

$$L = \frac{m^{3/2}k}{\sqrt{-2E}}, \qquad G = mA, \qquad \Theta = G\cos i$$

and l, g, τ are their canonically conjugated variables which will turn out to be $l = \{$average anomaly in the ellipse's plane$\}$, $g = \{$longitude of the major semiaxis of the ellipse in the ellipse's plane measured, say, from the nodal line, i.e., from the line of intersection of the ellipse's plane and the $(\mathbf{i}, \mathbf{j})$ plane of the reference system $(O; \mathbf{i}, \mathbf{j}, \mathbf{k})$ to which the motion is referred$\}$, $\tau = \{$longitude of the nodal line of the ellipse's plane in the plane $(\mathbf{i}, \mathbf{j})$ measured, say, from $\mathbf{i}\}$ = "angle of ascension".

Show that, if $\epsilon(L) = m^3k^2/-2L^2$, the above transformation is completely canonical and is generated by the solution of the Hamilton–Jacobi equation

$$\frac{1}{2m}\left(\left(\frac{\partial S}{\partial\rho}\right)^2 + \frac{1}{\rho^2\sin^2\theta}\left(\frac{\partial S}{\partial\varphi}\right)^2 + \frac{1}{\rho^2}\left(\frac{\partial S}{\partial\theta}\right)^2\right) - \frac{km}{\rho} = \epsilon(L),$$

parametrized by L, G, Θ and having the form

$$S(\rho,\theta,\varphi; L, G, \Theta) = -\left(\frac{\pi}{2} - \varphi\right)\Theta + \tilde\sigma_{G,\Theta}(\theta) + \tilde\sigma_{G,L}(\rho) - \frac{\pi}{2}G$$

with

$$\left(\frac{d\sigma_{G,\Theta}}{d\theta}\right)^2 = G^2 - \frac{\Theta^2}{(\sin\theta)^2},$$

$$\left(\frac{d\tilde{\sigma}_{G,L}}{d\rho}\right)^2 = 2m(\epsilon(L) - V_{G/m}(\rho)) = 2m(-m^3k^2/2L - G^2/2m^2\rho^2 + km/\rho).$$

(Hint: The angle θ varies between $\theta_- = \pi/2 - i$ and $\theta_+ = i + \pi/2$, assuming to have chosen the axis normal to the ellipse's plane oriented so that $i < \pi/2$. The ρ variable varies between ρ_- and ρ_+. $\theta_-, \theta_+, \rho_-, \rho_+$ can be computed from L, G, Θ. Write

$$\tilde{\sigma}_{G,\Theta}(\theta) = \int_{\theta_-}^{\theta}\left(G^2 - \frac{\Theta^2}{(\sin\theta')^2}\right)^{1/2} d\theta',$$

$$\tilde{\sigma}_{G,L}(\rho) = \int_{\rho_-}^{\rho}\left(2m(\epsilon(L) - V_{G/m}(\rho'))\right)^{1/2} d\rho'$$

and note that the variables τ, g, l are defined by

$$\tau = -\frac{\pi}{2} + \varphi + \frac{\partial\tilde{\sigma}_{G,\Theta}}{\partial\Theta}(\theta),$$

$$g = \frac{\partial\tilde{\sigma}_{G,\Theta}}{\partial G}(\theta) + \frac{\partial\tilde{\sigma}_{G,L}}{\partial G}(\rho) - \frac{\pi}{2},$$

$$l = \frac{\partial\tilde{\sigma}_{G,L}}{\partial L}(\rho).$$

In the new variables, the Hamiltonian becomes $\epsilon(L)$ so that the Keplerian evolution is, in these variables, $\tau =$ constant, $g =$ constant. So we can compute τ and g by choosing special phase-space points on the orbit. To find the meaning of τ, consider the time when the point occupies the "highest position", $\theta = \theta_-$. Show that $(\partial\tilde{\sigma}_{G,\Theta}/\partial\Theta)(\theta_-) = 0$, noting that the argument of the integral for $\tilde{\sigma}_{G,\Theta}$ vanishes for $\theta = \theta_\pm$. Hence, $\tau = -\pi/2 + \varphi$ when $\theta = \theta_-$. Geometrically, this means that τ is the angle formed with the x axis by the line in the xy plane orthogonal to the projection on the xy plane of the normal to the orbit's plane, i.e., the nodal line.

Similarly, to find the meaning of g, consider the time when $\rho = \rho_-$ (i.e., the point is at the perihelion). Now, $(\partial\tilde{\sigma}_{G,L}/\partial G)(\rho_-) = 0$ and

$$g = \int_{\theta_-}^{\theta_0} \frac{G}{\left(G^2 - \Theta^2/(\sin\theta)^2\right)^{1/2}} d\theta + \frac{\pi}{2},$$

where θ_0 is the polar angle corresponding to the perihelion position. This relation can be interpreted as saying that $g - \pi/2$ is the time necessary for a point moving according to the equation

$$\dot{\theta}^2 = 1 - \frac{\Theta^2}{G^2}\frac{1}{(\sin\theta)^2} = 1 - \frac{(\cos i)^2}{(\sin\theta)^2}$$

to go from θ_- to θ_0. On the other hand, it is easy to see that the above equation also describes the θ variation over time in a circular uniform motion on the unit circle in the plane of the ellipse (inclined by i) with unit speed. So $g - \pi/2$ is the angle between the major semiaxis of the ellipse and the intersection between the ellipse's plane and the plane containing the normal to the ellipse and the z axis ("azimuthal plane" of the normal). Since the angle between the latter line and the nodal line is

$\pi/2$, it follows that g has the desired interpretation. The angle l has the same expression found for the planar case (see Problem 11). Hence, it has the same interpretation of average anomaly.)

17. Express the Cartesian coordinates of the position corresponding to $(L, G, \Theta, l, g, \tau)$ of Problem 16. (Hint: Use the results of Problem 15 directly.)

§4.11. Integrable Systems. Solid with a Fixed Point

This is another interesting and famous integrable system.

Consider N point masses, with masses $m_1, \ldots, m_N > 0$, subject to an ideal constraint imposing that the system be rigid and have a fixed point 0. Suppose $N \geqslant 3$ and that the points are not aligned.

We shall describe the motions in a reference frame $(O; \bar{\mathbf{i}}, \bar{\mathbf{j}}, \bar{\mathbf{k}})$, conventionally called "fixed", and we shall fix a "comoving" frame $(O; \mathbf{i}_1, \mathbf{i}_2, \mathbf{i}_3)$ with axes suitably chosen.

To determine the position of the body, it will suffice to give the position of the reference frame $(O; \mathbf{i}_1, \mathbf{i}_2, \mathbf{i}_3)$ since, in this system of coordinates, the ith point has constant coordinates by the rigidity constraint. We shall use the Euler angles $(\bar{\theta}, \bar{\varphi}, \bar{\psi})$ to define $(O; \mathbf{i}_1, \mathbf{i}_2, \mathbf{i}_3)$; they are defined in §3.9, Fig. 3.3 (see Fig. 4.9):

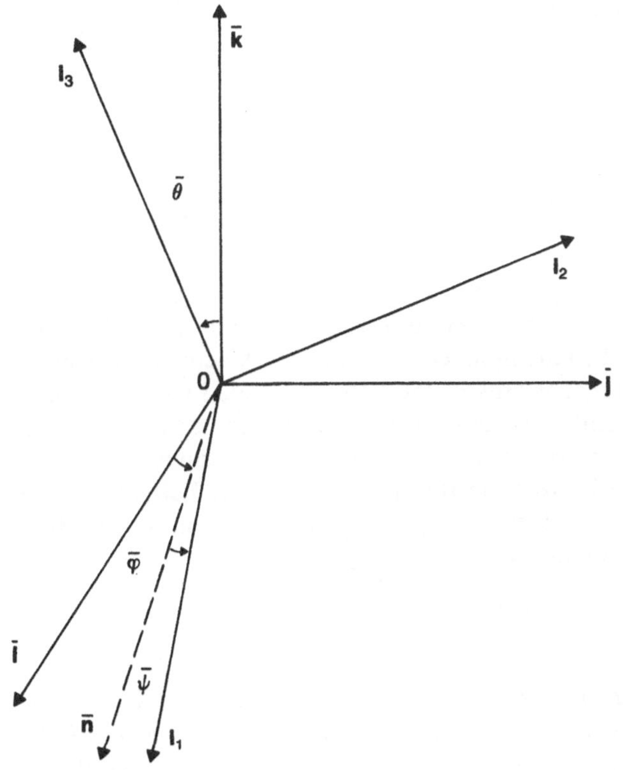

Figure 4.9.

The kinetic energy can be expressed in terms of the angular velocity ω of $(O; \mathbf{i}_1, \mathbf{i}_2, \mathbf{i}_3)$ with respect to $(O; \bar{\mathbf{i}}, \bar{\mathbf{j}}, \bar{\mathbf{k}})$, [see Eqs. (3.9.11) and (3.9.12)]:

$$\omega = \dot{\bar{\theta}}\bar{\mathbf{n}} + \dot{\bar{\varphi}}\bar{\mathbf{k}} + \dot{\bar{\psi}}\mathbf{i}_3 . \tag{4.11.1}$$

In fact, the velocity of the ith point can simply be written as

$$\dot{\mathbf{x}}^{(i)} = \omega \wedge (P_i - O) \tag{4.11.2}$$

(see footnote 10, p. 203, last formula).

Therefore,

$$\begin{aligned} T &= \frac{1}{2}\sum_{i=1}^{N} m_i (\dot{\mathbf{x}}^{(i)})^2 = \frac{1}{2}\sum_{i=1}^{N} m_i (\omega \wedge (P_i - O)) \cdot (\omega \wedge (P_i - O)) \\ &= \frac{1}{2}\omega \cdot \sum_{i=1}^{N} (P_i - O) \wedge (\omega \wedge (P_i - O)) = \frac{1}{2}\omega \cdot I\omega, \end{aligned} \tag{4.11.3}$$

by vector calculus, see Eq. (3.9.15), where

$$I\omega = \sum_{i=1}^{N} (P_i - O) \wedge (\omega \wedge (P_i - O)) = \mathbf{K}_O . \tag{4.11.4}$$

The components of $I\omega$ in the co-moving frame $(O; \mathbf{i}_1, \mathbf{i}_2, \mathbf{i}_3)$ have, by Eq. (4.11.4), the form

$$(I\omega)_\alpha = \sum_{\beta=1}^{3} I_{\alpha\beta}\omega_\beta \qquad \alpha = 1, 2, 3 \tag{4.11.5}$$

and from Eq. (4.11.4) it is easy to check that

$$I_{\alpha\beta} = \sum_{i=1}^{N} m_i \left[(P_i - O)^2 \delta_{\alpha\beta} - (P_i - O)_\alpha (P_i - O)_\beta \right], \tag{4.11.6}$$

for instance, using the identity $\mathbf{a} \wedge (\mathbf{b} \wedge \mathbf{c}) = (\mathbf{a} \cdot \mathbf{c})\mathbf{b} - (\mathbf{a} \cdot \mathbf{b})\mathbf{c}$. Since the components of $(P_i - O)$ in $(O; \mathbf{i}_1, \mathbf{i}_2, \mathbf{i}_3)$ are constants, by the rigidity constraint, the nine numbers of Eq. (4.11.5), actually six since $I_{\alpha\beta} \equiv I_{\beta\alpha}$, are characteristic constants of the body associated with the frame $(O; \mathbf{i}_1, \mathbf{i}_2, \mathbf{i}_3)$.

At this point, it is convenient to choose the co-moving frame so that the matrix I ("inertia matrix") is as simple as possible.

Note that by rotating the $\mathbf{i}_1, \mathbf{i}_2, \mathbf{i}_3$ axes to $\mathbf{i}'_1, \mathbf{i}'_2, \mathbf{i}'_3$, the coordinates of the vectors $(P_i - O)$ become $(P_i - O)'_\alpha$, $\alpha = 1, 2, 3$, in the new frame, related to the old coordinates by

$$(P_i - O)_\alpha = \sum_{\beta=1}^{3} R_{\alpha\beta}(P_i - O)'_\beta \tag{4.11.7}$$

and R is an orthogonal matrix $RR^T = R^TR = 1$ $(R_{\alpha\beta} = \mathbf{i}_\alpha \cdot \mathbf{i}'_\beta)$. And, vice versa, any orthogonal matrix corresponds to some frame $(O; \mathbf{i}'_1, \mathbf{i}'_2, \mathbf{i}'_3)$ so that Eq. (4.11.7) gives the relation of the change of coordinates.

Therefore, the inertia matrix depends on the co-moving frame and changes to I', in $(O; \mathbf{i}'_1, \mathbf{i}'_2, \mathbf{i}'_3)$, related to I by

$$I = RI'R^T \tag{4.11.8}$$

by Eqs. (4.11.7) and (4.11.6), in matrix notations.

Then we can choose R so that I' becomes

$$I' = \begin{bmatrix} I_1 & 0 & 0 \\ 0 & I_2 & 0 \\ 0 & 0 & I_3 \end{bmatrix}, \tag{4.11.9}$$

where $0 < I_1 \leqslant I_2 \leqslant I_3$. Such an R exists because I is a symmetric positive-definite matrix[9] and every such matrix can be "diagonalized" by an orthogonal transformation (see Appendix F).

We use the above observation to suppose that the choice of the co-moving frame $(O; \mathbf{i}_1, \mathbf{i}_2, \mathbf{i}_3)$, is, since the beginning, such that I takes the form of Eq. (4.11.9).

With this choice of the co-moving axes, the kinetic energy and the angular momentum become [see Eqs. (4.11.3), (4.11.4), and (4.11.5)]

$$T = \tfrac{1}{2}\left(I_1\omega_1^2 + I_2\omega_2^2 + I_3\omega_3^2\right), \tag{4.11.10}$$

$$\mathbf{K}_O = I_1\omega_1\mathbf{i}_1 + I_2\omega_2\mathbf{i}_2 + I_3\omega_3\mathbf{i}_3 . \tag{4.11.11}$$

To write the lagrangian function describing the solid's motion, with O fixed and subject to no force other than that of the ideal constraints of fixed O and of rigidity, it will be enough to express the kinetic energy in terms of the Euler angles (Fig. 4.9) and of their time derivatives, through Eqs. (4.11.1) and (4.11.10). The components of ω become explicitly

$$\omega_1 = \dot{\bar\theta}\cos\bar\psi + \dot{\bar\varphi}\sin\bar\theta\sin\bar\psi \tag{4.11.12}$$

$$\omega_2 = \dot{\bar\theta}\sin\bar\psi + \dot{\bar\varphi}\sin\bar\theta\cos\bar\psi \tag{4.11.13}$$

$$\omega_3 = \dot{\bar\varphi}\cos\bar\theta + \dot{\bar\psi}. \tag{4.11.14}$$

by Eqs. (3.9.3) and (4.11.1). The result is not particularly illuminating in the general case and we write it only in the "gyroscope case" when, say, $I_1 = I_2 = I$.

One finds

$$\mathfrak{L} = \tfrac{1}{2}I\left(\dot{\bar\theta}^2 + (\sin\bar\theta)^2\dot{\bar\varphi}^2\right) + \tfrac{1}{2}I_3\left(\dot{\bar\varphi}\cos\bar\theta + \dot{\bar\psi}\right)^2. \tag{4.11.15}$$

Before treating the general case, let us study the system described by Eq.

[9] Since $\frac{1}{2}I\omega\cdot\omega = \{\text{body's kinetic energy}\} > 0$, $\forall\omega \in \mathbf{R}^3$, and it can vanish only if $\omega = \mathbf{0}$ because the points are assumed to be not aligned, see Eq. (4.11.2).

(4.11.15), i.e., the gyroscope. In this case, the results are easier and particularly suggestive.

As is often the case, it is not convenient to write down only the Lagrange equations for Eq. (4.11.15) and discuss them. It is better to combine them with other information which can be obtained by general conservation principles (of energy and angular momentum, in the present case). Such information, although implicitly present in the Lagrange equations, is not very obvious there.

Since $\mathbf{K}_O$ is a constant of the motion, given a motion $t \to (\bar{\theta}(t), \bar{\varphi}(t), \bar{\psi}(t))$ with initial datum $(\dot{\bar{\theta}}_0, \dot{\bar{\varphi}}_0, \dot{\bar{\psi}}_0, \bar{\theta}_0, \bar{\varphi}_0, \bar{\psi}_0)$, we can suppose without affecting generality that $\mathbf{K}_O$ is parallel to some fixed axis $\mathbf{k}$:

$$\mathbf{K}_O = A\mathbf{k}, \qquad A > 0. \tag{4.11.16}$$

(The $A = 0$ case corresponds to a motionless solid which remains so forever, of course.)

Let $(O; \mathbf{i}, \mathbf{j}, \mathbf{k})$ be a reference frame with z axis oriented as $\mathbf{k}$ and choose $\mathbf{i}$ on the intersection between the $(\mathbf{i}, \mathbf{j})$ plane and the $(\bar{\mathbf{i}}, \bar{\mathbf{j}})$ plane. We suppose that such planes do not concide (otherwise, we change $(\bar{\mathbf{i}}, \bar{\mathbf{j}}, \bar{\mathbf{k}})$).

We shall discuss the motion in this new fixed frame $(O; \mathbf{i}, \mathbf{j}, \mathbf{k})$ calling (θ, φ, ψ) the Euler angles of $(O; \mathbf{i}_1, \mathbf{i}_2, \mathbf{i}_3)$ with respect to the frame $(O; \mathbf{i}, \mathbf{j}, \mathbf{k})$.

The components of $\mathbf{K}_O = A\mathbf{k}$ in the co-moving frame are expressed, see Eq. (3.9.3), in terms of the new Euler angles as

$$\begin{aligned}(\mathbf{K}_O)_3 &= A\cos\theta,\\ (\mathbf{K}_O)_1 &= A\sin\theta\cos\psi,\\ (\mathbf{K}_O)_2 &= A\sin\theta\sin\psi,\end{aligned} \tag{4.11.17}$$

and by relations like Eqs. (4.11.12)–(4.11.14), written with the new angles, the angular momentum conservation gives the following relations:

$$A\cos\theta = I_3\omega_3 = I_3(\dot{\varphi}\cos\theta + \dot{\psi}), \tag{4.11.18}$$

$$A\sin\theta\cos\psi = I\omega_2, \tag{4.11.19}$$

$$A\sin\theta\sin\psi = I\omega_1 \tag{4.11.20}$$

which are three differential equations for the three unknowns θ, φ, ψ and A is a constant [ω_1, ω_2 is also expressed in terms of the angles (θ, φ, ψ) and of their derivatives by relations like Eqs. (4.11.12) and (4.11.13)].

Instead of discussing the above equations, which, in principle, should be sufficient to determine the motion, we shall combine them with some of the Lagrangian equations associated with Eq. (4.11.15), written in the new θ, φ, ψ variables (i.e., without the overbars).

Since Eq. (4.11.15) does not explicitly depend upon φ, ψ, one deduces two conservation laws from Eq. (4.11.15) by writing the Lagrange equations corresponding to the variables φ, ψ:

$$\frac{d}{dt} I_3(\dot{\varphi}\cos\theta + \dot{\psi}) = 0, \tag{4.11.21}$$

corresponding to ψ and

$$\frac{d}{dt}\left(I(\sin\theta)^2\dot{\varphi}+I_3(\dot{\varphi}\cos\theta+\dot{\psi})\cos\theta\right)=0, \tag{4.11.22}$$

corresponding to φ.

Equations (4.11.18)–(4.11.22) form a redundant system of equations permitting us to easily determine the functions $(\theta(t),\varphi(t),\psi(t))$ in terms of the initial data.

In fact, Eq. (4.11.21) implies that $(\dot{\varphi}\cos\theta+\dot{\psi})$ is constant as t varies; hence, Eq. (4.11.18) implies that $\cos\theta$ is constant, i.e.,

$$\theta(t)=\theta_0. \tag{4.11.23}$$

(Hence, in the particular reference system which we recall was chosen after the particular motion had been selected, $\dot{\theta}=0$, hence, $\dot{\theta}_0=0$.)

Using Eqs. (4.11.23) and (4.11.21) in Eq. (4.11.22), we see that $\ddot{\varphi}=0$, i.e.,

$$\varphi(t)=\varphi_0+\dot{\varphi}_0 t. \tag{4.11.24}$$

Then the constancy of $\dot{\varphi}$ and of θ and Eq. (4.11.21) imply that $\dot{\psi}$ is also a constant:

$$\psi(t)=\psi_0+\dot{\psi}_0 t. \tag{4.11.25}$$

Hence, Eqs. (4.11.23)–(4.11.25) provide a full description of the motion in the chosen coordinates (which, we stress once more, is a reference frame depending on the motion itself since it has the z axis parallel to the constant angular momentum). Clearly, it appears that the motion expressed in the Cartesian coordinates is quasi-periodic with periods

$$T_1=\frac{2\pi}{\dot{\varphi}_0},\qquad T_2=\frac{2\pi}{\dot{\psi}_0}. \tag{4.11.26}$$

Hence, the motion is quasi-periodic but generally not periodic, although the set of the initial data for which $\dot{\varphi}_0/\dot{\psi}_0$ is rational is a dense set of data lying on periodic orbits.

By Observation (5) to Definition 10, p. 288, the above system should be integrable in the sense of Definition 10, p. 287, on vast regions of the data space.

Let us study the general case, assuming $0<I_1<I_2<I_3$ and using a method quite different from the preceding one, (inspired from Landau–Lifschitz, *Mécanique*, see references).

As before, given a motion, the angular momentum is a constant together with the kinetic energy. This implies

$$I_1\omega_1^2+I_2\omega_2^2+I_3\omega_3^2=2E=\text{const}, \tag{4.11.27}$$

$$I_1^2\omega_1^2+I_2^2\omega_2^2+I_3^2\omega_3^2=A^2=\text{const} \tag{4.11.28}$$

which permit us to deduce two of the three components of $\boldsymbol{\omega}$ in terms of the third:

$$\omega_1 = \pm\sqrt{\frac{(2EI_3 - A^2) - (I_3 - I_2)I_2\omega_2^2}{I_1(I_3 - I_1)}}, \tag{4.11.29}$$

$$\omega_3 = \pm\sqrt{\frac{(A^2 - 2EI_1) - (I_2 - I_1)I_2\omega_2^2}{I_3(I_3 - I_1)}}. \tag{4.11.30}$$

To find an equation allowing the determination of ω_2, one can observe that Eq. (4.11.28) contains less information than the constancy of the angular momentum as a vector.

In fact, the angular momentum conservation means [recalling Eq. (3.9.12)]

$$\begin{aligned}\mathbf{0} = \frac{d\mathbf{K}_O}{dt} &= \frac{d}{dt}(I_1\omega_1\mathbf{i}_1 + I_2\omega_2\mathbf{i}_2 + I_3\omega_3\mathbf{i}_3)\\ &= I_1\dot{\omega}_1\mathbf{i}_1 + I_2\dot{\omega}_2\mathbf{i}_2 + I_3\dot{\omega}_3\mathbf{i}_3 + I_1\omega_1\frac{d\mathbf{i}_1}{dt} + I_2\omega_2\frac{d\mathbf{i}_2}{dt} + I_3\omega_3\frac{d\mathbf{i}_3}{dt}\\ &= I_1\dot{\omega}_1\mathbf{i}_1 + I_2\dot{\omega}_2\mathbf{i}_2 + I_3\dot{\omega}_3\mathbf{i}_3 + I_1\omega_1\boldsymbol{\omega}\wedge\mathbf{i}_1 + I_2\omega_2\boldsymbol{\omega}\wedge\mathbf{i}_2 + I_2\omega_3\boldsymbol{\omega}\wedge\mathbf{i}_3,\end{aligned} \tag{4.11.31}$$

which, written in components on $(O; \mathbf{i}_1, \mathbf{i}_2, \mathbf{i}_3)$, is

$$I_1\dot{\omega}_1 = (I_2 - I_3)\omega_2\omega_3, \tag{4.11.32}$$

$$I_2\dot{\omega}_2 = (I_3 - I_1)\omega_1\omega_3, \tag{4.11.33}$$

$$I_3\dot{\omega}_3 = (I_1 - I_2)\omega_1\omega_2. \tag{4.11.34}$$

These very beautiful equations are the "Euler equations" for the solid's motion.

Equation (4.11.33) together with Eqs. (4.11.29) and (4.11.30) give the equation for ω_2:

$$\dot{\omega}_2 = \pm\sqrt{\frac{\{(2EI_3 - A^2) - (I_3 - I_2)I_2\omega_2^2\}\{(A^2 - 2EI_1) - (I_2 - I_1)I_2\omega_2^2\}}{I_1I_2^2I_3}} \tag{4.11.35}$$

and the discussion of the choice of sign in Eq. (4.11.35) leads to the usual result: initially, $\dot{\omega}_2$ has some sign which is kept until it vanishes, then the sign changes until the next time $\dot{\omega}_2$ vanishes, etc., alternating[10] (see §2.7).

Hence, recalling §2.7, Eq. (4.11.35) tells us that ω_2 varies over time as the abscissa of a point mass with mass 2, total energy 0, moving under the

[10] As in the one-dimensional conservative problems, if $\dot{\omega}_2$ vanishes initially the choice of sign for $t > 0$ and small can be inferred from the initial value of ω_2, (see §2.7).

action of a conservative force with potential energy:

$$V_{E,A}(x) = -\frac{\{(2EI_3 - A^2) - (I_3 - I_2)I_2x^2\}\{(A^2 - 2EI_1) - (I_2 - I_1)I_2x^2\}}{I_1I_2^2I_3}. \tag{4.11.36}$$

Therefore, $t \to \omega_2(t)$ is a C^∞-periodic function of t oscillating between two extreme values $\alpha_+(E,A)$, $\alpha_-(E,A)$ which are the extremes of the smaller of the two intervals $(-a_1, a_1), (-a_3, a_3)$ with $a_j =$ roots of $V_{E,A}(x) = 0$, $a_j > 0, j = 1, 3$:

$$a_1(E,A) = \sqrt{\frac{2EI_3 - A^2}{I_2(I_3 - I_1)}}, \qquad a_3(E,A) = \sqrt{\frac{A^2 - 2EI_1}{I_1(I_2 - I_1)}}, \tag{4.11.37}$$

provided

$$a_1(E,A) \neq a_3(E,A); \tag{4.11.38}$$

otherwise, the equation $V_{E,A} = 0$ has only two solutions, $\pm a$, and $V'_{E,A}$ vanishes there so that the motion, by the analysis of §2.7, will be aperiodic.

The period of $t \to \omega_2(t)$ is

$$T_1(E,A) = 2\int_{\alpha_-(E,A)}^{\alpha_+(E,A)} \frac{dx}{\sqrt{-V_{E,A}(x)}}, \tag{4.11.39}$$

and a better expression for $\omega_2(t)$ can be obtained by defining

$$t \to \Omega(t, E, A), \qquad t \in \mathbf{R}, \tag{4.11.40}$$

to be the solution of $2\ddot{\Omega} = -(\partial V_{E,A}/\partial\omega_2)(\Omega)$, hence, of Eq. (4.11.35), with initial datum

$$\Omega(0, E, A) = \alpha_-(E,A), \qquad \dot{\Omega}(0, E, A) = 0. \tag{4.11.41}$$

Then

$$\omega_2(t) = \Omega(t + t_0(\omega_2(0), \dot{\omega}_2(0)), E, A), \tag{4.11.42}$$

where $t_0(\omega_2(0), \dot{\omega}_2(0))$ is the minimum time necessary in order that the solution (4.11.40) "reaches" the datum $(\dot{\omega}_2(0), \omega_2(0))$. Furthermore, for $0 \leqslant t \leqslant T_1(E,A)/2$, it is

$$t = \int_{\alpha_-(E,A)}^{\Omega(t,E,A)} \frac{dx}{\sqrt{-V_{E,A}(x)}}. \tag{4.11.43}$$

To find the motion $t \to (\theta(t), \varphi(t), \psi(t))$, we have to go back to the equations expressing the conservation of angular momentum and its identity with $A\mathbf{k}$, assuming, again, to have chosen a reference frame $(O; \mathbf{i}, \mathbf{j}, \mathbf{k})$ with $\mathbf{k}$ and $\mathbf{K}_O$ parallel and $\mathbf{i}$ along the node line of the planes $(\mathbf{i}, \mathbf{j})$ and $(\bar{\mathbf{i}}, \bar{\mathbf{j}})$, see

Eqs. (4.11.18)–(4.11.20). Now

$$\begin{aligned} I_3\omega_3 &= A\cos\theta, \\ I_2\omega_2 &= A\sin\theta\sin\psi, \\ I_1\omega_1 &= A\sin\theta\cos\psi \end{aligned} \tag{4.11.44}$$

tell us that

$$\theta(t) = \arccos\frac{I_3\omega_3(t)}{A}, \tag{4.11.45}$$

$$\psi(t) = \operatorname{arctg}\frac{I_2\omega_2(t)}{I_1\omega_1(t)}, \tag{4.11.46}$$

where the determination of the arctangent has to be chosen so that $t \to \psi(t)$ is continuous.

From Eqs. (4.11.12) and (4.11.13) written without overbars (i.e., for the Euler angles of $(O; \mathbf{i}_1, \mathbf{i}_2, \mathbf{i}_3)$ with respect to $(O; \mathbf{i}, \mathbf{j}, \mathbf{k})$), we deduce $\dot\varphi$:

$$\dot\varphi = \frac{\omega_1\sin\psi + \omega_2\cos\psi}{\sin\theta} = A\frac{I_1\omega_1^2 + I_2\omega_2^2}{I_1^2\omega_1^2 + I_2^2\omega_2^2}, \tag{4.11.47}$$

where the second equality follows from Eqs. (4.11.19), (4.11.20), and (4.11.18) (recalling that we are supposing $\mathbf{K}_O$ parallel to $\mathbf{k}$). Let

$$\Phi(t, E, A) = A\frac{I_1\Omega_1(t, E, A)^2 + I_2\Omega_2(t, E, A)^2}{I_1^2\Omega_1(t, E, A)^2 + I_2^2\Omega_2(t, E, A)^2}, \tag{4.11.48}$$

where Ω_1 is connected with Ω as ω_1 with ω_2 in Eq. (4.11.29) (note that the sign's ambiguity has no relevance here).

Then Eq. (4.11.47) becomes

$$\dot\varphi = \Phi(t + t_0(\omega_2(0), \dot\omega_2(0)), E, A). \tag{4.11.49}$$

Using the periodicity with period Eq. (4.11.39) of $t \to \Phi(t, E, A)$ and calling $(\chi_n(E, A))_{n \in \mathbf{Z}}$ the Fourier coefficients of this function, if one writes

$$\Phi(t, E, A) = \sum_{n=-\infty}^{+\infty} \chi_n(E, A) e^{(2\pi i/T_1(E,A))t}, \tag{4.11.50}$$

one finds, by integrating Eq. (4.11.49),

$$\begin{aligned} \varphi(t) = \varphi_0 + \chi_0(E, A)t &+ S(t + t_0(\omega_2(0), \dot\omega_2(0)), E, A) \\ &- S(t_0(\omega_2(0), \dot\omega_2(0)), E, A), \end{aligned} \tag{4.11.51}$$

where

$$S(t, E, A) = \sum_{\substack{n=-\infty \\ n \neq 0}}^{+\infty} \chi_n(E, A)\frac{e^{2\pi i n t/T_1(E,A)}}{2\pi i n/T_1(E, A)} \tag{4.11.52}$$

which is a C^∞ function periodic in t with period $T_1(E, A)$.

Equations (4.11.45), (4.11.40), (4.11.42), (4.11.51), (4.11.46), (4.11.29), and (4.11.30) give a complete description of the motion under investigation.

The analogy of the above results with those of the two-body problem lead to the formulation of the following proposition.

24 Proposition. *The motion of a solid with a fixed point and inertia moments $0 < I_1 < I_2 < I_3$ is integrable in the sense of Definition* 10, §4.8, *p.* 287, *in a family of regions covering the region W of the data space where $A \neq 0$, $a_3(E,A) \neq a_1(E,A)$* [*see Eq.* (4.11.38)], *and in such cases the motion is quasi-periodic with two periods*:

$$T_1(E,A) = 2\int_{\alpha_-(E,A)}^{\alpha_+(E,A)} \frac{dx}{\sqrt{-V_{E,A}(x)}}, \tag{4.11.53}$$

$$T_2(E,A) = \frac{2\pi}{\chi_0(E,A)} = \frac{2\pi}{A}\left[\int_{\alpha_-(E,A)}^{\alpha_+(E,A)} \frac{dx}{\sqrt{-V_{E,A}(x)}}\right]$$
$$\times\left[\int_{\alpha_-(E,A)}^{\alpha_+(E,A)} \frac{dx}{\sqrt{-V_{E,A}(x)}}\left[\frac{(2E-A^2)-(I_2-I_1)I_2x^2}{I_1(2E-A^2)-I_2I_3(I_2-I_1)x^2}\right]\right]^{-1} \tag{4.11.54}$$

and $\alpha_\pm(E,A)$ are the two positive roots of the smallest modulus of $V_{E,A}(x)=0$, with $V_{E,A}$ being defined in Eq. (4.11.36).

Similar (*and simpler*) *results hold if $I_1 = I_2 \neq I_3$, $I_1 \neq I_2 = I_3$, $I_1 = I_2 = I_3$.*

Observations.

(1) The proof of Proposition 24 is essentially a different way of stating what has already been discussed above. The analysis of this section (as well as that on the central forces) is a classical proof. Somehow, it seems unsatisfactory because it looks like "magic", with its use of redundant equations chosen, without apparent a priori logic, to reach the goal of finding explicit expressions for the motions. However, with further thought, it appears quite simple and, in particular, no need of the theory of elliptic functions emerges (despite this frequent claim).

(2) However, there is a deeper critique of the above deductions. It is not at all clear that the systems are canonically integrable in the sense of Definition 11, §4.8, p. 289. This becomes very serious when one tries to study by the Hamilton–Jacobi theory the perturbations provoked by small conservative forces on the above simple motions. The reader will realize this problem more clearly in §5.10–§5.12, where the theory of the Hamiltonian perturbations based on the Hamilton–Jacobi equations is developed.

In the problems to §4.9 and §4.10 we have shown, however, how to deduce for the central motions complete canonical integrability from the integrability proof. Likewise, in the problems of this section, we show how to deduce canonical integrability of the solid's motion from parts of the proof of the above proposition. The derivation is simple and nice, not so much because it leads very quickly to the quadrature formulae (4.11.42), (4.11.47), (4.11.46), and (4.11.52), but mainly because it achieves the proof of canonical integrability at the same time. This integrability property had always been discussed either abstractly or quite obscurely until recently when the "De Prit canonical transformation" was introduced.

PROOF. We discuss the proof in some further details because it is useful to illustrate Observation (5) to Definition 10, p. 288.

Let $(O;\bar{\mathbf{i}},\bar{\mathbf{j}},\bar{\mathbf{k}})$ be the original fixed frame and let $(O,\mathbf{i},\mathbf{j},\mathbf{k})$ be the "adapted" fixed frame chosen, once a particular motion is given, with the $\mathbf{k}$ axis parallel to the angular momentum. Suppose that $\mathbf{i}$ is parallel to the node of the planes $(\bar{\mathbf{i}},\bar{\mathbf{j}})$ and $(\mathbf{i},\mathbf{j})$ (i.e., to their intersection).

To determine the initial datum in the $I_1 = I_2$ case, we use the following coordinates:

(1) the angle γ between $\bar{\mathbf{i}}$ and $\mathbf{i}$;
(2) the angle δ between $\mathbf{K}_O$ and $\bar{\mathbf{k}}$;
(3) the Euler angles φ,ψ of $(O;\mathbf{i}_1,\mathbf{i}_2,\mathbf{i}_3)$ in $(O;\mathbf{i},\mathbf{j},\mathbf{k})$;
(4) the angular velocity variables $\dot{\varphi}$ and $\dot{\psi}$.

From the preceding analysis, it follows that the motion of the system has three prime integrals $(\delta,\dot{\varphi},\dot{\psi})$ and, given them, it is described by the points $(\gamma,\varphi,\psi)\in\mathbf{T}^3$, and the time evolution on $\mathbf{T}^3$ is described by quasi-periodic flow with pulsations

$$\sigma_1(\delta,\dot{\varphi},\dot{\psi}) = 0, \qquad \sigma_2(\delta,\dot{\varphi}\dot{\psi}) = \dot{\varphi}, \qquad \sigma_3(\delta,\dot{\varphi},\dot{\psi}) = \dot{\psi}, \tag{4.11.55}$$

having denoted them with σ instead of ω to avoid confusion with the above angular velocity components.

The integrating map is thus $I(\dot{\bar{\theta}},\dot{\bar{\varphi}},\dot{\bar{\psi}},\bar{\theta},\bar{\varphi},\bar{\psi})\leftrightarrow(\delta,\dot{\varphi},\dot{\psi},\gamma,\varphi,\psi)$. It should still be checked that this map is C^∞ nonsingular and invertible on a suitable family of neighborhoods W' which, as one uses the arbitrariness of the choice of $(O;\bar{\mathbf{i}},\bar{\mathbf{j}},\bar{\mathbf{k}})$, cover W. We do not enter into this analysis.

In the general case $(I_1 < I_2 < I_3)$, we replace the variables $(\delta,\dot{\varphi},\dot{\psi})$, which are no longer conserved with the exception of δ, with (δ,E,A), and we also replace the angles (γ,φ,ψ), which no longer rotate uniformly with the exception of γ which is constant, with $(\gamma,\tilde{\varphi},\tilde{\psi})$, where

$$\begin{aligned}\tilde{\psi} &= \frac{2\pi}{T_1(E,A)}\,t_0(\omega_2(0),\dot{\omega}_2(0)),\\ \tilde{\varphi} &= \varphi - S\big(t_0(\omega_2(0),\dot{\omega}_2(0)),E,A\big)\end{aligned} \tag{4.11.56}$$

[see Eqs. (4.11.51), (4.11.52), and (4.11.42)].

By the discussion preceding Proposition 24, it appears that $\gamma,\tilde{\psi},\tilde{\varphi}$ are angles rotating with pulsations

$$\sigma_1(A,E,\delta) = 0, \qquad \sigma_2(A,E,\delta) = \frac{2\pi}{T_1(E,A)}, \qquad \sigma_3(A,E,\delta) = \frac{2\pi}{T_2(E,A)}. \tag{4.11.57}$$

This follows after some contemplation of Eqs. (4.11.42) and (4.11.51).

Again we do not enter into the analysis of the regularity and invertibility of the integration map I: $(\dot{\bar{\theta}},\dot{\bar{\varphi}},\dot{\bar{\psi}},\bar{\theta},\bar{\varphi},\bar{\psi})\leftrightarrow(\delta,E,A,\gamma,\tilde{\varphi},\tilde{\psi})$.

Note that the coordinates chosen in the general case do not reduce to those of the symmetric case ($I_1 = I_2$) when $I_2 \to I_1$.

However, there is great arbitrariness in defining the prime integrals because any function of δ, E, A is still a prime integral, and it is possible to find two other prime integrals Φ, Ψ becoming $\dot{\varphi}$ and $\dot{\psi}$ in the $I_1 = I_2$ case.

In fact, let

$$\Phi = \frac{1}{T_1(E,A)}\int_0^{T_1(E,A)} A\,\frac{I_1\Omega_1(t)^2 + I_2\Omega(t)^2}{I_1^2\Omega_1(t)^2 + I_2\Omega(t)^2}\,dt, \tag{4.11.58}$$

where $\Omega(t)$ is defined in Eq. (4.11.30) and $\Omega_1(t)$ is related to Ω by Eq. (4.11.29) with $\Omega(t)$ replacing ω_2 [and, likewise, we could define $\Omega_3(t)$ by Eq. (4.11.30)].

Note that Φ is the average value of $\dot{\varphi}$ along a period [since $\dot{\varphi}$ is periodic with period $T_1(E,A)$, see Eq. (4.11.47)].

Analogously, from Eq. (4.11.12), (4.11.13) written without overbars, one can find an expression of $\dot{\psi}$ in terms of $\omega_1,\omega_2,\omega_3$:

$$\dot{\psi} = \frac{(A^2 - 2EI_3)\omega_3}{I_1^2\omega_1^2 + I_2^2\omega_2^2}\,. \tag{4.11.59}$$

So $\dot{\psi}$ is a periodic function with period $T_1(E,A)$ and we can define the prime integral

$$\Psi = \frac{1}{T_1(E,A)}\int_0^{T_1(E,A)} \frac{(A^2 - 2EI_3)\Omega_3(t)}{I_1^2\Omega_1(t)^1 + I_2^2\Omega(t)^2}\,dt, \tag{4.11.60}$$

where the ambiguity of the sign in the definition of Ω_3 has to be resolved now by remarking that ω_3 from Eq. (4.11.30) never vanishes if $A \neq 0$ and, therefore, it has a constant sign which we attribute also to Ω_3.

It could also be possible to change $\tilde{\psi}$ to a variable reducing to ψ when $I_1 \to I_2$. However, we shall not do this.

It remains to check Eq. (4.11.54); $T_2 = 2\pi/\chi_0$:

$$\begin{aligned}\chi_0(E,A) &= \frac{1}{T_1(E,A)}\int_0^{T_1(E,A)}\Phi(t,E,A)\,dt\\ &= \frac{2}{T_1(E,A)}\int_0^{T_1(E,A)/2}\Phi(t,E,A)\,dt\end{aligned} \tag{4.11.61}$$

and changing variable $t \to \Omega(t,E,A)$, one has [see Eq. (4.11.43)]

$$dt = \frac{d\Omega}{\sqrt{-V_{E,A}(\Omega)}}\,. \tag{4.11.62}$$

Hence, recalling that Ω_1 can be expressed in terms of Ω, we can express the integral on the right-hand side of Eq. (4.11.61) as an integral over the variable Ω, via Eqs. (4.11.48) and (4.11.29) and, after some algebra, Eq. (4.11.54) follows. mbe

Problems and Complements for §4.11

1. Let $\bar{\mathfrak{L}}$ be the Lagrangian function describing the motion of a rigid body in a fixed frame $(O; \bar{\mathbf{i}}, \bar{\mathbf{j}}, \bar{\mathbf{k}})$ in Euler angle coordinates

$$\bar{\mathfrak{L}} = \tfrac{1}{2} I_1 \left(\dot{\bar{\theta}} \cos \bar{\psi} + \dot{\bar{\varphi}} \sin \bar{\theta} \sin \bar{\psi} \right)^2$$
$$+ \tfrac{1}{2} I_2 \left(-\dot{\bar{\theta}} \sin \bar{\psi} + \dot{\bar{\varphi}} \sin \bar{\theta} \cos \bar{\psi} \right)^2 + \tfrac{1}{2} I_3 \left(\dot{\bar{\varphi}} \cos \bar{\theta} + \dot{\bar{\psi}} \right)_0^2 .$$

Compute the canonical variables $p_{\bar{\theta}}, p_{\bar{\varphi}}, p_{\bar{\psi}}$ associated with $\bar{\theta}, \bar{\varphi}, \bar{\psi}$ via $\bar{\mathfrak{L}}$. Show that if $\mathbf{K}_O$ is the angular momentum of the solid with respect to the fixed point O, and if $\bar{\mathbf{n}}$ is the node line unit vector, then

$$p_{\theta} = \mathbf{K}_O \cdot \bar{\mathbf{n}}, \qquad p_{\bar{\varphi}} = \mathbf{K}_O \cdot \bar{\mathbf{k}}, \qquad p_{\bar{\psi}} = \mathbf{K}_O \cdot \mathbf{i}_3 .$$

(Hint: Just apply the definition of p [Eq. (3.11.1)] and use Eqs. (4.11.11) and (3.9.3).)

2. Call $A = |\mathbf{K}_O|$, $K_z = p_{\bar{\varphi}}$, $L = \mathbf{K}_O \cdot \mathbf{i}_3 = p_{\bar{\psi}}$ and let (γ, φ, ψ) be the angles considered on p. 316 (see Fig. 4.10). $(K_z, A, L, \gamma, \varphi, \psi)$ are the "De Prit variables" (from De Prit, see references).

$$\bar{\mathbf{n}} = (\bar{\mathbf{i}}, \bar{\mathbf{j}}) \cap (\mathbf{i}_1, \mathbf{i}_2)$$
$$\mathbf{n} = (\mathbf{i}_1, \mathbf{i}_2) \cap (\mathbf{i}, \mathbf{j})$$
$$\mathbf{m} = (\bar{\mathbf{i}}, \bar{\mathbf{j}}) \cap (\mathbf{i}, \mathbf{j}) \equiv \mathbf{i}$$

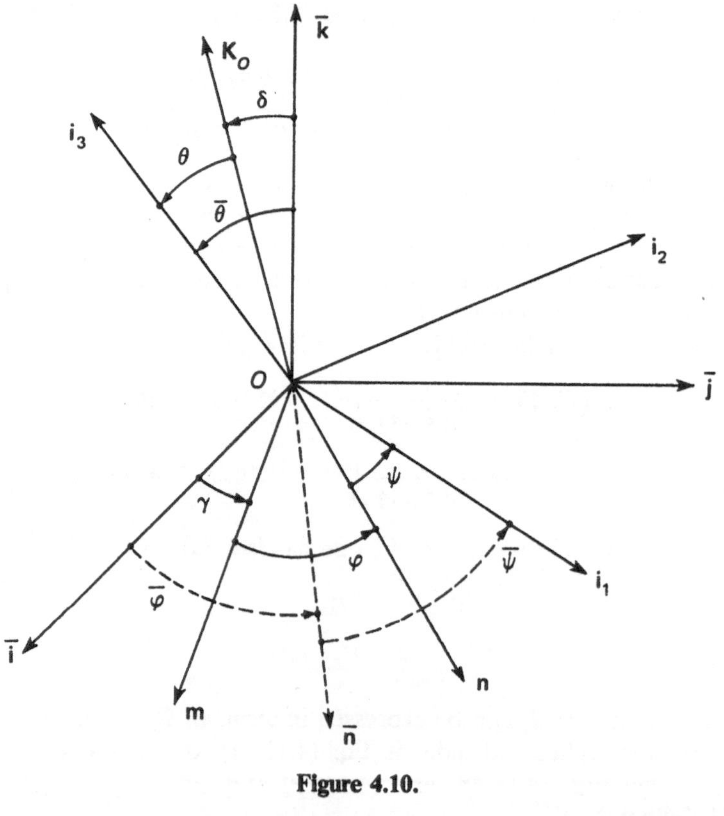

Figure 4.10.

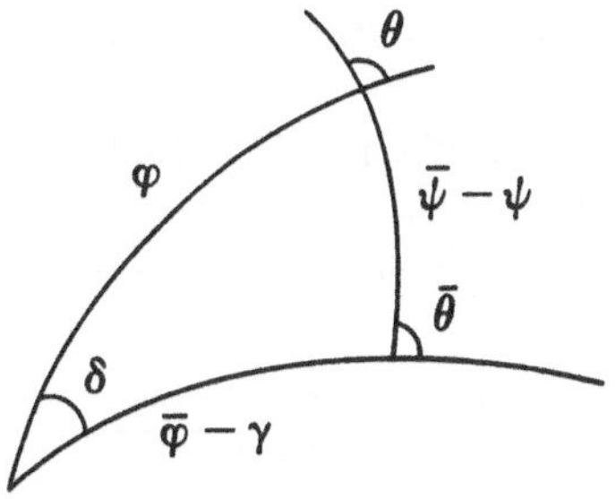

Figure 4.11.

Show that given $(p_{\bar{\theta}}, p_{\bar{\varphi}}, p_{\bar{\psi}}, \bar{\theta}, \bar{\varphi}, \bar{\psi})$, the De Prit variables are determined and vice versa. (Hint: Note that $p = A\cos\delta = K_z$, $p_{\bar{\psi}} = A\cos\theta = L$, $p_{\bar{\theta}} = a(\sin\theta)\sin(\psi - \bar{\psi}))$ and note that the angles $\varphi, \theta, \bar{\psi}-\psi, \bar{\varphi}-\gamma, \delta, \bar{\theta}$ can be arranged in a spherical triangle (Fig. 4.11). Theorefore, given the De Prit variables, one computes $p_{\bar{\varphi}} = K_z$, then $\cos\delta = K_z/A$, then $p_{\bar{\psi}} = L$, then $\cos\theta = L/A$. Hence, at this point, one knows the elements φ, θ, δ of the spherical triangle in Fig. 4.11 and by solving it one computes, by spherical trigonometry, the three other elements, i.e., $\bar{\psi}-\psi, \bar{\varphi}-\gamma$, $\bar{\theta}$, and since γ, ψ are known, one gets $\bar{\psi}$,

3. Consider the spherical triangle of Fig. 4.12. Recalling the basic spherical trigonometry relations:

$$\cos A = \cos B\cos C + \sin B\sin C\cos\alpha,$$

$$\frac{\sin A}{\sin\alpha} = \frac{\sin B}{\sin\beta} = \frac{\sin C}{\sin\gamma},$$

show that

$$dA = \cos\beta\, dC + \cos\gamma\, dB + \sin\beta\sin C\, d\alpha.$$

(Hint: First vary α only, using the first of the above relations and then use the second to simplify the results. Then vary C and B successively and sum the results.)

4. Show that the map $(p_{\bar{\theta}}, p_{\bar{\varphi}}, p_{\bar{\psi}}, \bar{\theta}, \bar{\varphi}, \bar{\psi}) \leftrightarrow (K_z, A, L, \gamma, \varphi, \psi)$ has the property ("De Prit's theorem"):

$$K_z\, d\gamma + A\, d\varphi + L\, d\psi = p_{\bar{\theta}}\, d\bar{\theta} + p_{\bar{\varphi}}\, d\bar{\varphi} + p_{\bar{\psi}}\, d\bar{\psi}.$$

(Hint: Using Problems 2 and 3, show that $d\varphi = \cos\theta\, d(\bar{\psi}-\psi) + \cos\delta\, d(\bar{\varphi}-\gamma) - \sin\theta\sin(\bar{\psi}-\psi)\, d\bar{\theta}$; then substitute into the left-hand side using $K_z = p_{\bar{\varphi}}$, $L = p_{\bar{\psi}}$ and $A\sin\theta\sin(\bar{\psi}-\psi) = p_{\bar{\theta}}$).

5. The map $(p_{\bar{\theta}}, p_{\bar{\varphi}}, p_{\bar{\psi}}, \bar{\theta}, \bar{\varphi}, \bar{\psi}) \leftrightarrow (K_z, A, L, \gamma, \varphi, \psi)$, defined in Problems 2 and 4 maps six variables into six others without any reference to a rigid body. Interpret Problem 4 as saying that this map is a completely canonical map homogeneous in

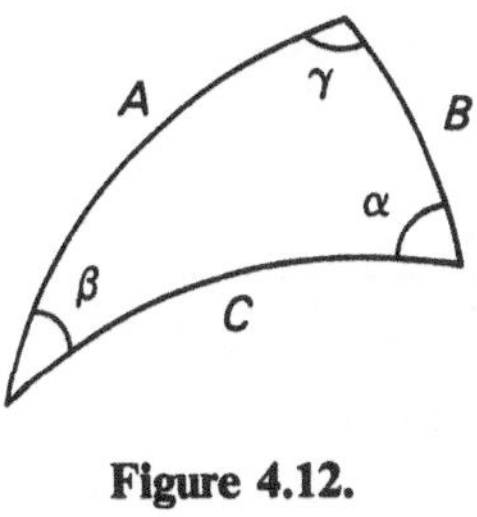

Figure 4.12.

the variables $(\gamma, \varphi, \psi, \bar{\theta}, \bar{\varphi}, \bar{\psi})$ in the sense of Proposition 22, §3.11, p. 224. (Hint: Apply Proposition 22, §3.11.)

6. Compute the rigid body's Hamiltonian $\tilde{H}$ in De Prit variables, remarking that, by Problem 5, it must simply be the kinetic energy expressed in these variables (see the general properties of the completely canonical transformations, §3.12), and show that

$$\tilde{H}(K_z, A, L, \gamma, \varphi, \psi) = \frac{1}{2}\frac{L^2}{I_3} + \frac{1}{2}\left(\frac{(\cos\psi)^2}{I_1} + \frac{(\sin\psi)^2}{I_2}\right)(A^2 - L^2).$$

Deduce the Hamilton equations of the motion and check that they are identical to Eqs. (4.11.47) and (4.11.59). Use this Hamiltonian formulation to rederive directly the integrability of the motions of a solid with a fixed point. (Hint: Write the equations of motion and integrate by quadratures.)

7. Using the Hamiltonian in Problem 6, show that the solid with a fixed point gives rise to canonically integrable motions (see Definition 11, §4.8, p. 289). (Hint: Since the map $(p_{\bar{\theta}}, p_{\bar{\varphi}}, p_{\bar{\psi}}, \bar{\theta}, \bar{\varphi}, \bar{\psi}) \leftrightarrow (K_z, A, L, \gamma, \varphi, \psi)$ is completely canonical, it is enough to show that the Hamiltonian motions generated by the Hamiltonian in Problem 6 are canonically integrable.

Define the canonical transformation with generating function

$$\Phi(K_z', A', M, \gamma, \varphi, \psi) = K_z'\gamma + A'\varphi + S(A', M, \psi)$$

with S chosen so that Φ solves the Hamilton–Jacobi equation for $\tilde{H}$:

$$\frac{1}{2}\left(\frac{1}{I_3} - \frac{(\cos\psi)^2}{I_1}\right) - \frac{(\sin\psi)^2}{I_2}\left(\frac{\partial S}{\partial\psi}\right)^2 + \frac{1}{2}A'^2\left(\frac{(\cos\psi)^2}{I_1} + \frac{(\sin\psi)^2}{I_2}\right) = e(A', M),$$

where the function $e(A', M)$ is naturally chosen so that the function S does generate a canonical transformation on the Hamiltonian $\tilde{H}$, regarded as a function of L, ψ only (parametrized by $A' \equiv A$), bringing it to action angle variables (M, μ). By Problem 5, §3.11, this means that the function $e(A', M)$ has to be chosen so that

$$\frac{\partial e(A', M)}{\partial M} = \omega(A', E),$$

where $\omega(A', E)$ is the pulsation of the motion (of this one-dimensional system parameterized by A') with energy $E = e(A', M)$. Since the equation of motion for ψ in this auxiliary one-dimensional system is

$$\dot{\psi} = -\frac{\partial\tilde{H}}{\partial L}(A', L, \psi),$$

the pulsation will be such that

$$\frac{2\pi}{\omega(A', E)} = \int_0^{2\pi}\frac{d\psi}{\dot{\psi}(t)} = \int_0^{2\pi}\frac{d\psi}{-(\partial\tilde{H}/\partial L)(A', L, \psi)},$$

where L has to be fixed so that $\tilde{H}(A', L, \psi) = E$; i.e., L has to be taken as a function $L(E, A', \psi)$:

$$L = L(E, A', \psi) = \left(2\frac{E - A'^2[(\cos\psi)^2/I_2 + (\sin\psi)^2/I_3]}{1/I_3 - (\cos\psi)^2/I_2 - (\sin\psi)^2/I_3}\right)^{1/2}$$

which permits us to compute $\omega(A', E)$.

The function $e(A',M)$ can be computed in terms of its inverse $m(A',E)$ (such that $m(A',e(A',M)) \equiv M$), since $\partial m/\partial E)(A',E)$ must be

$$\left(\frac{\partial e}{\partial M}(A',M)\right)^{-1} = \frac{1}{\omega(A',E)}.$$

So, for instance, e can be defined by inverting the relation:

$$m(A',E) = \int_{E_0(A')}^{E} \frac{dE'}{\omega(A',E')},$$

where $E_0(A') = \min_{L,\psi} \tilde{H}(A',L,\psi)$.

Coming back to S, we see that

$$S(A',M,\psi) = \int_0^{\psi} L(e(A',M),A',\psi')\,d\psi'$$

is the explicit solution of the Hamilton–Jacobi equation (recall the expression of L).

The above Φ-generated canonical transformation leaves A, K_z, γ unchanged and changes L to M, φ to some new φ' and ψ to some new μ with

$$\varphi' = \varphi + \frac{\partial S}{\partial A}(A,M,\psi),$$

$$\mu = \frac{\partial S}{\partial M}(A,M,\psi)$$

and transforms the $\tilde{H}$ into $e(A,M)$.

The above transformation is "globally" defined because one can show that

$$S(A',M,2\pi) \equiv 2\pi M.$$

In fact, $(\partial S/\partial M)(A',M,2\pi) \equiv 2\pi$ (since this can be checked directly by differentiating the integral giving S and by comparing it to the integral for computing $\omega(A,E)$ explicitly and then using $\partial e/\partial M = 1/\omega(A,E)$; so $S(A',M,2\pi) = 2\pi M + g(A')$ for some function $g(A')$. But $M = 0$ means $E = E_0(A')$; hence, $L = 0$; hence, $S = 0$; hence, $g(A') \equiv 0$. This means that when (θ,φ) vary on $\mathbf{T}^2$, (φ', μ) also vary on $\mathbf{T}^2$.)

§4.12. Integrable Systems. Geodesic Motion on the Surface of an Ellipsoid and Other Systems

In general, given a closed regular surface $\Sigma \subset \mathbf{R}^d$, the "geodesic motions" are those which a point mass with mass 1 can undergo on Σ when it is ideally bound to Σ and subject to no other active forces.

The Lagrangian of such motions is, of course,

$$\mathfrak{L} = \tfrac{1}{2}\dot{\mathbf{x}}^2 \tag{4.12.1}$$

and energy conservation, §3.5, implies that

$$T = \tfrac{1}{2}\dot{\mathbf{x}}^2 = \text{constant} \tag{4.12.2}$$

on the considered motions. For instance, the set of the motions with initial speed of modulus 1 consists entirely of motions in which the speed has modulus 1 at all times.

The "geodesic flow" on Σ is the flow on $\mathbb{S}$ = {data space for the motions on Σ} ≡ {set of pairs $(\boldsymbol{\eta}, \mathbf{x})$ with $\mathbf{x} \in \Sigma$ and $\boldsymbol{\eta}$ compatible with the constraint, i.e., tangent to Σ}[11] which to every point $(\boldsymbol{\eta}, \mathbf{x}) \in \mathbb{S}$ associates $S_t(\boldsymbol{\eta}, \mathbf{x}) \in \mathbb{S}$, the configuration into which the datum $(\boldsymbol{\eta}, \mathbf{x})$ evolves in time t under the only influence of an ideal constraint to Σ.

Clearly, since $|\dot{\mathbf{x}}|$ = constant, there is an a priori bound on the distance that a point can travel in a given time and, therefore, the geodesic flow is well defined, $\forall t \in \mathbf{R}$.

The conservation of speed has an interesting consequence: the action of a geodesic motion $t \to \mathbf{x}(t)$ computed between t_1 and t_2 can be expressed in terms of the curvilinear abscissas on the trajectory $\mathfrak{T}$ on which $\mathbf{x}$ moves. If $V = |\dot{\mathbf{x}}|$,

$$\begin{aligned} A_{t_1,t_2} &= \{\text{action of } \mathbf{x} \text{ between } t_1 \text{ and } t_2\} \\ &= \frac{V^2}{2}(t_2 - t_1) = \frac{V}{2}(s_2 - s_1). \end{aligned} \tag{4.12.3}$$

By the least-action principle, Proposition 8, §3.5, p. 163, we know that the motion $t \to \mathbf{x}(t)$ makes the action locally minimal in sufficiently small time intervals.

From this, it follows that the trajectory $\mathfrak{T}$, as a curve in $\mathbf{R}^d$, makes the distance between $\mathbf{x}(t_1)$ and $\mathbf{x}(t_2)$ measured along Σ locally minimal if t_2 is close enough to t_1 ("Maupertuis' principle", see problems to §3.11). In fact, given $\mathbf{x}_1 = \mathbf{x}(t_1) \in \mathfrak{T}$, suppose that for $|t_2 - t_1| < \epsilon$, the action $A_{t_1,t_2}(\mathbf{y})$ is minimal on $\mathbf{x}$ as $\mathbf{y}$ varies in $\mathfrak{M}_{t_1,t_2}(\mathbf{x}(t_1), \mathbf{x}(t_2); \Sigma)$ = {motions on Σ defined for $t \in [t_1, t_2]$ and leading from $\mathbf{x}(t_1)$ to $\mathbf{x}(t_2)$}. If there existed a curve $\mathcal{C}_{1,2}$ connecting $\mathbf{x}(t_1)$ with $\mathbf{x}(t_2)$, lying on Σ and shorter than $(s_2 - s_1)$ = {length of the part of $\mathfrak{T}$ between $\mathbf{x}(t_1)$ and $\mathbf{x}(t_2)$}, then one could run it with uniform speed starting from $\mathbf{x}(t_1)$ at time t_1 so as to reach $\mathbf{x}(t_2)$ at time t_2. Such a motion $\mathbf{x}_{\mathcal{C}_{1,2}} \in \mathfrak{M}_{t_1,t_2}(\mathbf{x}(t_1), \mathbf{x}(t_2); \Sigma)$ would have an action

$$\begin{aligned} A_{t_1,t_2}(\mathbf{x}_{\mathcal{C}_{1,2}}) &= \frac{1}{2}\left(\frac{|\mathcal{C}_{1,2}|}{(t_2 - t_1)}\right)^2 (t_2 - t_1) = \frac{1}{2}\,\frac{|\mathcal{C}_{1,2}|^2}{t_2 - t_1} \\ &< \frac{1}{2}\,\frac{(s_2 - s_1)^2}{t_2 - t_1} = \frac{1}{2}V^2(t_2 - t_1) = \frac{1}{2}V(s_2 - s_1) \end{aligned} \tag{4.12.4}$$

as $|\mathcal{C}_{1,2}| < s_2 - s_1$ = {length of $\mathfrak{T}$}. This contradicts the minimality of A_{t_1,t_2} on $\mathbf{x}$.

The curves on a surface Σ which make minimal the distance between the points that they connect, provided such points are close enough, are called "geodesics" on Σ, and this explains the name given to the motions with Lagrangian (4.12.1) on Σ.

The simplest nontrivial example of a geodesic motion is the motion on

[11] In fancy language, call this the "tangent fiber bundle" to Σ.

the surface of the sphere in $\mathbf{R}^3$. It is easy to see that the possible trajectories of this motion are great circles. It is possible to interpret this statement in terms of the integrability of the geodesic motion on the sphere's surface in the sense of Definition 10, §4.8, p. 287. In this case, the motions are all periodic [see Observation (5), p. 288)].

A less simple example is the motion on the ellipsoid's surface. We shall only treat the case of the ellipsoid of revolution. However, the motion on an arbitrary ellipsoid is also integrable (see problems at the end of this section for a glimpse of the theory).

In the case of an ellipsoid of revolution, we choose as z axis the symmetry axis of the ellipsoid $\mathfrak{S}$ and determine the position on $\mathfrak{S}$ of a point through the two coordinates (θ, φ) as

$$\begin{aligned} x &= a \sin\theta \cos\varphi, \\ y &= a \sin\theta \sin\varphi, \\ z &= b \cos\theta, \end{aligned} \tag{4.12.5}$$

where a and b are the principal semi-axes of the ellipsoid.

The Lagrangian (4.12.1) of the geodesic motion on $\mathfrak{S}$ can be immediately written by Eq. (4.1.25) as

$$\mathfrak{L}(\dot\theta, \dot\varphi, \theta, \varphi) = \frac{1}{2}\left[\left(b^2(\sin\theta)^2 + a^2(\cos\theta)^2\right)\dot\theta^2 + a^2\dot\varphi^2(\sin\theta)^2\right]. \tag{4.12.6}$$

Hence, the equations of motion are

$$\frac{d}{dt} a^2\dot\varphi(\sin\theta)^2 = 0, \tag{4.12.7}$$

$$\frac{d}{dt}\left(b^2(\sin\theta)^2 + a^2(\cos\theta)^2\right)\dot\theta = \frac{\partial\mathfrak{L}}{\partial\theta}(\dot\theta, \dot\varphi, \theta, \varphi). \tag{4.12.8}$$

However, it is convenient to discuss only Eq. (4.12.7), combining it with the energy conservation principle:

$$\tfrac{1}{2}\left[\left(b^2(\sin\theta)^2 + a^2(\cos\theta)^2\right)\dot\theta^2 + a^2(\sin\theta)^2\dot\varphi^2\right] = E. \tag{4.12.9}$$

Equations (4.12.9) and (4.12.7), which we use to define the prime integral $A = \dot\varphi(\sin\theta)^2$, yield

$$\dot\theta = \pm\sqrt{\frac{\left(2E(\sin\theta)^2 - a^2A^2\right)}{(\sin\theta)^2\left(b^2(\sin\theta)^2 + a^2(\cos\theta)^2\right)}} \equiv \pm\sqrt{-V_{E,A}(\theta)} \tag{4.12.10}$$

which by the usual argument tells us that $t \to \theta(t)$ is a periodic function with period:

$$T_1(E, A) = 2\int_{\theta_-(E,A)}^{\theta_+(E,A)} \frac{d\theta}{\sqrt{-V_{E,A}(\theta)}}, \tag{4.12.11}$$

where $\theta_-(E,A)$ and $\theta_+(E,A)$ are the two solutions of $V_{E,A}=0$ of the form $\theta_\pm(E,A)=\pi/2\pm\theta_0(E,A)$ or $\theta_\pm(E,A)=-\pi/2\pm\theta_0(E,A)$ and $\theta_0=\arcsin(aA/\sqrt{2E})$.

Furthermore, θ verifies the equation

$$\ddot{\theta}=-\frac{\partial V_{E,A}}{\partial\theta}(\theta) \tag{4.12.12}$$

and, therefore, it is a C^∞ function of t (see §2.7) and can be expressed in terms of the solution $t\to R(t,E,A)$ of Eq. (4.12.12) with initial data $R(0,E,A)=\theta_-(E,A)$, $\dot{R}(0,E,A)=0$. Such a function is defined, recalling §2.7, by

$$t=\int_{\theta_-(E,A)}^{R(t,E,A)}\frac{d\theta}{\sqrt{-V_{E,A}(\theta)}} \tag{4.12.13}$$

for $0\leqslant t<T_1(E,A)/2$ and it is continued naturally for $T_1(E,A)/2<t<T_1(E,A)$. Furthermore, $\theta(t)$ is given by

$$\theta(t)=R\big(t+t_0(\theta_0,\dot{\theta}_0),E,A\big), \tag{4.12.14}$$

where $t_0(\theta(0),\dot{\theta}(0))=\{$first time when the motion $t\to R(t,E,A)$ reaches $(\dot{\theta}(0),\theta(0))\}$, with $\theta_0=\theta(0)$, $\dot{\theta}_0=\dot{\theta}(0)$.

As one sees, the analysis of this problem by "quadratures" is entirely analogous to the ones seen in § 4.9–§4.11. As on those occasions, the motion $t\to\varphi(t)$ can be deduced from Eq. (4.12.7) by quadrature:

$$\varphi(t)=\varphi_0+\int_0^t\frac{A}{(\sin\theta(\tau))^2}\,d\tau \tag{4.12.15}$$

that can be treated, as already seen in §4.9–§4.11, by noting that

$$\frac{A}{(\sin R(t,E,A))^2}=\sum_{k=-\infty}^{+\infty}\chi_k(A,E)e^{2\pi ikt/T_1(E,A)} \tag{4.12.16}$$

by the periodicity of R and by the Fourier theorem, because $t\to A/(\sin R(t,E,A))^2$ is a $T_1(E,A)$-periodic C^∞ function with Fourier coefficients $(\chi_k(A,E))_{k\in\mathbf{Z}}$. Setting

$$S(t,E,A)=\sum_{\substack{k=-\infty\\k\neq0}}^{+\infty}\chi_k(A,E)\frac{e^{2\pi ikt/T_1(E,A)}}{2\pi ik/T_1(E,A)}, \tag{4.12.17}$$

the quadrature (4.12.15) yields

$$\varphi(t)=\varphi_0+\chi_0(A,E)t-S\big(t_0(\theta_0,\dot{\theta}_0),E,A\big)+S\big(t+t_0(\theta_0,\dot{\theta}_0),E,A\big). \tag{4.12.18}$$

Hence, from Eqs. (4.12.14) and (4.12.18), we can conclude that all the motions are quasi-periodic with periods $T_1(E,A)$ given by Eq. (4.12.11) and $T_2(E,A) = 2\pi\chi_0(A,E)^{-1}$:

$$T_2(E,A) = 2\pi\chi_0(A,E)^{-1} = 2\pi\left(\frac{1}{T_1(E,A)}\int_0^{T_1(E,A)}\frac{A\,dt}{(\sin R(t,E,A))^2}\right)^{-1}$$

$$= 2\pi\left[\int_{\theta_-(E,A)}^{\theta_+(E,A)}\frac{d\theta}{\sqrt{-V_{E,A}(\theta)}}\right. \tag{4.12.19}$$

$$\left.\times\left[\int_{\theta_-(E,A)}^{\theta_+(E,A)}\frac{A}{(\sin\theta)^2}\frac{d\theta}{\sqrt{-V_{E,A}(\theta)}}\right]^{-1}\right]$$

after changing variables $\theta = R(t,E,A)$ along the lines already seen in Proposition 24, §4.11, p. 315 and Proposition 22, §4.9, p. 293.

It could be checked that as E, A vary the two periods $T_1(E,A)$, $T_2(E,A)$ will generally have an irrational ratio.

The above analysis basically achieves the proof of the following proposition, (if one disregards the checks of regularity and invertibility in suitably large regions W' in the data space of the map $I(\dot\theta(0),\dot\varphi(0),\theta(0),\theta(0)) = (E, A, \alpha, \beta)$ with $\alpha = (2\pi/T_1(E,A))t_0(\theta(0),\dot\theta(0))$, $\beta = \varphi(0) - S(t_0(\theta(0),\dot\theta(0)), E,A))$:

25 Proposition. *The set W of the data for the geodesic motions on an ellipsoid of revolution E and such that $E \neq 0$, $A \neq 0$ can be covered by sets $W' \subset W$ on which the geodesic motions are integrable in the sense of Definition* 10, §4.8, *p.* 287. *Such motions are quasi-periodic with periods $T(E,A), T_2(E,A)$, given by Eqs.* (4.12.11) *and* (4.12.19).

If the ellipsoid's semi-axes are different, the motion is generally quasi-periodic and nonperiodic.

Observations.

(1) The discussion preceding Proposition 25 is very general and could be repeated with essentially no change to cover very general classes of surfaces of revolution like those parametrically described by equations like

$$\begin{aligned} z &= f(\theta), \\ x &= g(\theta)\cos\varphi, \qquad (\theta,\varphi)\in[0,2\pi]\times[0,2\pi] \\ y &= g(\theta)\sin\varphi, \end{aligned} \tag{4.12.20}$$

with $f, g, \in C^\infty(\mathbf{T}^1)$ such that the curve in $\mathbf{R}^2$ with parametric equations $\xi = g(\theta)$, $\eta = f(\theta)$, $\theta\in[0,2\pi]$ is a simple closed curve symmetric under reflection around the η axis.

Other surfaces covered by the above method are those with parametric equations

$$\begin{aligned} z &= a(\varphi), \\ x &= b(\varphi)\cos\psi, \qquad (\varphi,\psi)\in[0,2\pi]\times[0,2\pi] \\ y &= b(\varphi)\sin\psi, \end{aligned} \tag{4.12.21}$$

with $f, g \in C^\infty(\mathbf{T}^1)$ such that the curve with parametric equations $\xi = g(\theta)$, $\eta = f(\theta)$ is a simple closed curve contained in the half-plane $\xi > 0$.

We leave it to the reader to check the above statements as an exercise on the quadrature method.

(2) Surfaces like Eq. (4.12.20) generalize the ellipsoid of revolution while those like Eq. (4.12.21) generalize the "torus of revolution":

$$\begin{aligned} x &= (a + b\cos\varphi)\cos\psi, \\ y &= (a + b\cos\varphi)\sin\psi, \\ z &= b\sin\varphi, \end{aligned} \tag{4.12.22}$$

$a, b > 0$, $a > b$.

We conclude this list of remarkable integrable systems by citing a few other systems integrable on suitable regions W.

(1) A point mass on an ellipsoid of revolution, with symmetry axis along the major axis of the revolving ellipse, subject to a force with potential energy

$$V(\mathbf{x}) = g|\mathbf{x} - \mathbf{f}_1|^{-1}|\mathbf{x} - \mathbf{f}_2|^{-1}, \tag{4.12.23}$$

where $\mathbf{f}_1, \mathbf{f}_2$ are the foci of the ellipse generating the ellipsoid.

This system can be easily integrated by the quadrature method of §4.9–§4.11, and one obtains similar results in elliptic coordinates defined in terms of Cartesian coordinates (x, y, z) of $\mathbf{x}$ as

$$\begin{aligned} \sqrt{x^2+y^2} &= \sigma\sqrt{(\xi^2-1)(1-\eta^2)}\,, \\ z &= \sigma\xi\eta, \\ \text{azimuth of } \mathbf{x} &= \varphi, \end{aligned} \tag{4.12.24}$$

where $\xi \in [1, +\infty]$, $\eta \in [-1, 1]$, $\varphi \in [0, 2\pi]$ and the parameter σ has to be chosen so that the considered ellipsoid is a $\xi = \text{constant}$ surface. Such surfaces are the ellipsoids

$$\frac{z^2}{\sigma^2\xi^2} + \frac{x^2+y^2}{\sigma^2(\xi^2-1)} = 1. \tag{4.12.25}$$

(2) A point mass on the surface of a sphere with potential energy in polar coordinates:

$$U(\theta,\varphi) = b(\theta) + \frac{c(\varphi)}{(\sin\theta)^2} \tag{4.12.26}$$

with b, c C^∞-periodic functions with period 2π. This system is easily integrated by quadratures by writing its Lagrangian function in polar coordinates and discussing the Lagrange equations.

(3) A solid body with a symmetry axis fixed at a point 0 of this axis, which we call $\mathbf{i}_3$, different from the center of mass G and subject to ideal constraints plus the weight, i.e., a force $m_i\mathbf{g}$ on the ith point (equivalent by the observations in §3.2, Observation (5), p. 149, to a force $M\mathbf{g}$ applied to G as far as the force momentum calculation is concerned, $M = \Sigma_i m_i$).

This system is also integrable by quadratures. One proceeds as in §3.11. One chooses the fixed reference frame $(O;\mathbf{i},\mathbf{j},\mathbf{k})$ with $\mathbf{k}$ axis antiparallel to $\mathbf{g}$, and then one writes the Lagrangian function in terms of the Euler angles. The Lagrange equations can then be combined with the conservation laws, for energy and for the $\mathbf{k}$ component of the angular momentum, to quickly reduce the problem to that of the analysis of one-dimensional systems, i.e., to the quadratures. See, also, problems at the end of this section and problems to §3.5 in Landau and Lifshitz, *Mécanique* (see References).

(4) Two more difficult classical integrable systems are the geodesic motions on the surface of a nonsymmetric ellipsoid (see problems at the end of this section for an introduction to this theory) and the motions of a heavy rigid body with a fixed point O, with the baricenter in the 1-2 plane, say on the 1 axis at distance a from O, and with inertia moments $I_1 = I_2 = 2I_3 = 2G$.

Such systems can be shown to be integrable by quadratures (as discovered by Jacobi and Kowaleskaia, respectively, see exercises).

(5) Other systems are that of N point masses on the line $\mathbf{R}$ with Lagrangian functions

$$\frac{m}{2}\sum_{i=1}^{N}\dot{x}_i^2 - g\sum_{i=1}^{N}\exp\alpha(x_i - x_{i+1}), \tag{4.12.27}$$

or

$$\frac{m}{2}\sum_{i=1}^{N}\dot{x}_1^2 - g^2\sum_{i<j}^{1,N}\frac{1}{(x_i - x_j)^2} - \frac{\omega^2}{2}\sum_{i=1}^{N}x_i^2, \tag{4.12.28}$$

called the "Toda" or "Calogero" systems, respectively. These were discovered very recently and are also integrable. Some variants of such systems with the same properties are also known.

(6) Obviously, there are other integrable systems: it suffices to perform an arbitrary change of coordinates in the Lagrangian functions which we have just examined to obtain integrable Lagrangian functions.

However, only very "few" other systems are known that have the integrability property and that are "interesting", i.e., not obtained by trivial changes of coordinates from those so far listed. Some can be found among the problems for §4.12.

Finally, we remark that all the integrable systems of §4.9–§4.12 could be shown to be not only integrable in the sense of Definition 10, §4.8, p. 287, but also analytically and canonically integrable in the sense of Definition 11, §4.8, p. 289 in large regions of the phase space. In the problems of §4.10–4.12, the main steps towards such a proof are given.

Exercises and Problems for §4.12

1. Integrate explicitly by quadratures the systems mentioned in the points (1), (2), and (3) of the list of integrable systems in §4.12. By the Hamilton–Jacobi method, show their canonical integrability.

2. Integrate the heavy gyroscope system (3) p. 327, by using the De Prit variables (see problems to §4.11). First show that the Hamiltonian (i.e., the energy) can be written in the De Prit variables as

$$H(K_z, A, L, \gamma, \varphi, \psi)$$
$$= \frac{A^2 - L^2}{2I} + \frac{L^2}{2I_3} + \mu\left[\frac{K_z L}{A^2} - \left(1 - \frac{K_z^2}{A^2}\right)^{1/2}\left(1 - \frac{L^2}{A^2}\right)^{1/2}\cos\psi\right],$$

where $\mu = Mgd$, $M =$ total mass, $g =$ gravity constant, $I \equiv I_1 = I_2$ and I_3 are the moments of inertia. Show also the canonical integrability of this system.

3. Consider the "Kowaleskaya gyroscope", see p. 327, and show that its Lagrangian is

$$\mathfrak{L} = G\dot{\theta}^2 + G(\sin\theta)^2\dot{\varphi}^2 + \frac{G}{2}(\dot{\psi} + \dot{\varphi}\cos\theta)^2 + Mga\sin\theta\cos\psi$$

and explicitly write the Lagrange equations relative to the θ, ψ, φ variables.

4. In the context of Exercise 3, eliminate the $\dot{\psi}$ between the Lagrange equations relative to φ and ψ and add the resulting equation to the equation relative to the θ variable multiplied by $+i$ or $-i$, successively. Show that the two equations that result imply

$$\frac{1}{U}\frac{dU}{dt} = -\frac{1}{V}\frac{dV}{dt} = \text{"same function"},$$

where ($i = \sqrt{-1}$):

$$U = (\dot{\varphi}\sin\theta + i\dot{\theta})^2 + Mgae^{-i\psi}\sin\theta = \bar{V},$$

so $UV \equiv |U|^2 =$ constant and UV is a prime integral.[12]

5. In the context of Problems 3 and 4, show that the Kowaleskaya gyroscope is integrable by quadratures on vast regions of phase space.

6. Consider the geodesic motion on the surface of the ellipsoid $\mathfrak{E} : x^2/a + y^2/b + z^2/c = 1$, $a < b < c$. Introduce the local coordinate system ("Jacobi's system") described by

$$x = \sqrt{a\frac{(u-a)(v-a)}{(b-a)(c-a)}},$$

$$y = \sqrt{b\frac{(u-b)(v-b)}{(c-b)(a-b)}}, \qquad (u,v) \in [b,c] \times [a,b] \text{ or } [a,b] \times [b,c]$$

$$z = \sqrt{c\frac{(u-c)(v-c)}{(a-c)(b-c)}},$$

[12] See Wittaker (in References), Chap. VI.

Defining for $\lambda \in \mathbf{R}$,

$$A(\lambda) = \frac{1}{4} \frac{\lambda}{(a-\lambda)(b-\lambda)(c-\lambda)},$$

show that the kinetic energy is given by

$$T(\dot{u}, \dot{v}, u, v) = \tfrac{1}{2}(u-v)\big(A(u)\dot{u}^2 - A(v)\dot{v}^2\big).$$

Applying the Hamilton–Jacobi method to the Lagrangian system with Lagrangian $T(\dot{u}, \dot{v}, u, v)$, show that the geodesic motion on the ellipsoid admits a second prime integral:

$$M(\dot{u}, \dot{v}, u, v) = (u-v)\big(vA(u)\dot{u}^2 - uA(v)\dot{v}^2\big).$$

(Hint: Write the Hamilton–Jacobi equation in (u, v) variables after finding the Hamiltonian function in (u, v) and in their canonically conjugate momenta p_u, p_v:

$$\left(\frac{\partial f}{\partial t}\right) + \frac{1}{2}\left[\left(\frac{\partial f}{\partial u}\right)^2 \frac{1}{H(u,v)} + \left(\frac{\partial f}{\partial v}\right)^2 \frac{1}{G(u,v)}\right] = 0,$$

where

$$H(u,v) = (u-v)A(u),$$
$$G(u,v) = (v-u)A(v),$$

and look for solutions of the form

$$f(u,v,t) = -\frac{E}{2}t + \psi(u,v), \qquad \psi(u,v) = \alpha(u) + \beta(v).$$

The equation becomes

$$A(v)\left(\frac{\partial \psi}{\partial u}\right)^2 - A(u)\left(\frac{\partial \psi}{\partial v}\right)^2 = (u-v)(A(u)A(v))E,$$

admitting a family of solutions parametrized by E and a new arbitrary parameter a:

$$\psi(u, v \mid a, E) = \int_{u_0}^{u} \sqrt{EA(u')(u'+a)}\, du' + \int_{v_0}^{v} \sqrt{EA(v')(v'+a)}\, dv'.$$

Now, applying the canonical transformation G generated by $-(E/2)t + \psi(u, v, a, E)$, deduce that the trajectories of the motion (geodesics on $\mathcal{E}$) are given by the equation $d\psi/da = c =$ constant, i.e., $F_a(u) + F_a(v) = 2c$ if $F_a(u)$ is a primitive function to $\sqrt{A(m)/(u+a)}$. This also implies that a is a prime integral. Writing the canonical transformation G, it is possible to express a in terms of μ, v, p_μ, p_v or $u, v, \dot{u}, \dot{v}$. The computation gives $a = -M(\dot{u}, \dot{v}, u, v)/T(\dot{u}, \dot{v}, u, v) \equiv -M/E$; so M is a prime integral.) (see Courant-Hilbert in the reference list.)

7. Consider the system ("atom in electric field")

$$H(\mathbf{p}, \mathbf{q}) = \frac{\mathbf{p}^2}{2m} - \frac{g}{|\mathbf{q}|} + Fx$$

$\mathbf{p} = (p_x, p_y, p_z)$, $\mathbf{q} = (x, y, z)$ and study it in "squared parabolic coordinates"

$$x = \tfrac{1}{2}(u^2 - v^2)$$
$$y = uv\cos\varphi$$
$$z = uv\sin\varphi$$

and show by the method of problem 6 (i.e., by the Hamilton–Jacobi method) that this system has three prime integrals and that it can be integrated by quadratures (from Sommerfeld, see references).

8. Consider the Hamiltonian ("ionized hydrogen molecule")

$$H(\mathbf{p},\mathbf{q})=\frac{\mathbf{p}^2}{2m}-\frac{g}{|\mathbf{q}-\mathbf{f}_1|}-\frac{g}{|\mathbf{q}-\mathbf{f}_2|}$$

with $\mathbf{p}=(p_x,p_y,p_z)\in\mathbf{R}^3$, $\mathbf{q}=(x,y,z)\in\mathbf{R}^3$, $\mathbf{f}_1,\mathbf{f}_2=\mathbf{R}^3$ and study it in elliptical coordinates (see (4.12.24) and (4.12.25)) and show by the methods of problems 6 and 7 that it has three prime integrals and that it can be integrated by quadratures. Find canonical action angle variables (from Sommerfeld, see references).

§4.13. Some Integrability Criteria. Introduction: Geometric Considerations and Preliminary Definitions

Considering the "rarity" of the mechanical systems known as integrable one wonders whether it is possible to easily recognize, a priori, the nonintegrability of a mechanical system.

For instance, the integrability on a region W of the data space $\mathbb{S}$ implies the existence of l-"independent" prime integrals. Therefore, a way of showing nonintegrability might be that of showing the nonexistence of as many prime integrals as the number of degrees of freedom.

In any concrete case, however, it is very difficult to decide whether or not a system possesses prime integrals (other than the total energy and its functions). The Poincaré proof of nonintegrability, in a sense stricter than the above, of the motion of three heavenly bodies is based on showing the nonexistence of enough prime integrals (also defined in a more stringent way). It is still a famous proof (see H. Poincaré: *Mécanique Celeste*, vol. I, ch. VI).

Hence, it is useful to try to identify other special properties of the integrable systems to use them as necessary conditions for integrability or to formulate sufficient nonintegrability conditions.

In the following sections we go through an analysis that will allow us to classify the motions of the integrable systems as "simple and ordered" motions and those of the nonintegrable ones as "complex and disordered".

Coming back into the atmosphere of §2.21 and §4.8, we see that the notions of observable, average values, etc. introduced there have a natural extension to the systems with several degrees of freedom.

We consider an l-degrees-of-freedom system described by a Lagrangian function

$$\mathfrak{L}(\dot{\mathbf{x}},\mathbf{x})+\text{ideal constraints} \tag{4.13.1}$$

regular in the sense of § 3.11, Definition 14, and generating Hamiltonian equations admitting global solutions, in the future and in the past, for all the constraint-compatible initial data.

As usual, we denote $\mathbb{S}$ the data space for the system of Eq. (4.13.1). By Proposition 18, p. 285, it is a regular surface in $\mathbf{R}^d\times\mathbf{R}^d$, where d is the dimension of the unconstrained system, usually $d = 3N$, $N = \{$number of the system's points$\}$, see Definiton 9, §4.8, p. 286.

12 Definition. The elements of $C^\infty(\mathbb{S})$[13] will be called the "observables" of the mechanical system of Eq. (4.13.1).

Given an increasing sequence $\mathbf{t} = t_0, t_1 \ldots$ such that $t_i \to_{i\to+\infty} +\infty$ and given $f \in C^\infty(\mathbb{S})$, we shall call the "t-history of f" on the motion of $(\dot{\mathbf{x}}, \mathbf{x}) \in \mathbb{S}$ the sequence

$$\left(f(S_{t_i}(\dot{\mathbf{x}}, \mathbf{x}))\right)_{i=0}^{\infty}. \tag{4.13.2}$$

It is the sequence of the results of the successive observations of the values of f on the motion starting at $(\dot{\mathbf{x}}, \mathbf{x})$ at times $t_0, t_1, \ldots$. We shorten the notation by simply referring to the "$(f, \mathbf{t})$ history of $(\dot{\mathbf{x}}, \mathbf{x})$".

If $f \in C^\infty(\mathbb{S})$ and $\mathbf{t}$ is a sequence like

$$t_i = it_1, \qquad i = 0, 1, \ldots; \qquad t_1 > 0 \tag{4.13.3}$$

and if $(\dot{\mathbf{x}}, \mathbf{x}) \in W \subset \mathbb{S}$, where W is a region on which the mechanical system of Eq. (4.13.1) is integrable, then the $(f, \mathbf{t})$ history of $(\dot{\mathbf{x}}, \mathbf{x})$ is far from being an "arbitrary" sequence of numbers. Proposition 6, §4.2, p. 251, allows us to state, for instance, the following obvious reformulation of its contents.

26 Proposition. *If in $W \subset \mathbb{S}$ the system of Eq.* (4.13.1) *is integrable and $f \in C^\infty(\mathbb{S})$ is an observable and if* $\mathbf{t}$ *is as in Eq.* (4.13.3), *the* $(f, \mathbf{t})$ *histories of the points* $(\dot{\mathbf{x}}, \mathbf{x}) \in W$ *have a well-defined average value* i.e., *the following limit exists*:

$$\bar{f}(\dot{\mathbf{x}}, \mathbf{x}) = \lim_{N\to\infty} \frac{1}{N} \sum_{j=0}^{N-1} f(S_{jt_1}(\dot{\mathbf{x}}, \mathbf{x})). \tag{4.13.4}$$

Furthermore, if $(\mathbf{A}, \boldsymbol{\varphi}) = I(\dot{\mathbf{x}}, \mathbf{x})$ *is the integrating transformation mapping W onto $V \times \mathbf{T}^l$, see Definition* 10, §4.8, *p.* 287, *and if the* $(l+1)$ *numbers* $(\omega_1(\mathbf{A}), \ldots, \omega_l(\mathbf{A}), \sigma)$, *with* $\sigma = 2\pi/t_1$, *are rationally independent, then*

$$\bar{f}(\dot{\mathbf{x}}, \mathbf{x}) = (2\pi)^{-l} \int_{T^l} F_f(\mathbf{A}, \boldsymbol{\varphi}')\, d\boldsymbol{\varphi}', \tag{4.13.5}$$

having set

$$F_f(\mathbf{A}, \boldsymbol{\varphi}) = f\left(I^{-1}(\mathbf{A}, \boldsymbol{\varphi})\right). \tag{4.13.6}$$

Observations.

(1) Obviously, F_f is the observable f in the new $(\mathbf{A}, \boldsymbol{\varphi})$ coordinates.

(2) Hence, the nonintegrability of the system of Eq. (4.13.1) in W can be proved "just" by exhibiting a single point $(\dot{\mathbf{x}}, \mathbf{x}) \in W$ and a single observ-

[13] See Observation (2) to Definition 7, p. 294.

able f whose $(f, \mathbf{t})$ history on $(\dot{\mathbf{x}}, \mathbf{x})$ does not have a well-defined average value.

(3) However, this criterion is very difficult to apply in practice: the $(f, \mathbf{t})$ histories are very hard to analyze in concrete interesting cases and "usually" they admit an average value even in nonintegrable systems.

The following proposition provides a more geometric integrability criterion different in spirit from the one above.

27 Proposition. *If in $W \subset \mathfrak{S}$ the system of Eq.* (4.13.1) *is integrable, the closure of every trajectory of a point $(\dot{\mathbf{x}}, \mathbf{x}) \in W$ is a set $\bar{\mathfrak{T}}$ which can be mapped continuously and in a one-to-one way onto a torus $\mathbf{T}^s$ with $1 \leqslant s \leqslant l$, if l is the number of degrees of freedom of the system.*

Observations.

(1) Proposition 27 is also essentially a way of rephrasing some properties of the integrability of Definition 10, §4.8, p. 287. In fact, the motions of an integrable system take place on invariant tori of dimension l run quasi-periodically. If $(\omega_1, \ldots, \omega_l)$ are the pulsations of a given motion and are rationally independent, then the trajectory fills densely a set homeomorphic[14] to $\mathbf{T}^l$, see Proposition 4, p. 250, §4.2. In general, if s is the number of elements of a maximal subset of $\{\omega_1, \ldots, \omega_l\}$ consisting of rationally independent numbers, then $\bar{\mathfrak{T}}$ will be homeomorphic to $\mathbf{T}^s$. The proof of this fact is left to the reader and is essentially described in the hints to the problems for §4.14.

(2) So to prove nonintegrability in W, it suffices to find "just" one $(\dot{\mathbf{x}}, \mathbf{x}) \in W$ whose trajectory has a closure which is not homeomorphic to a smooth surface or, more particularly, to an s-dimensional torus, $s \leqslant l$.

(3) The geometric structure of a trajectory of a point mass bound to a surface Σ can be found by the Maupertuis principle by noting that the trajectory of an energy-E motion is a geodesic for the metric $dh = \sqrt{2(E - V(\xi)}\, ds$ on Σ, where ds is the line element on Σ and W is the potential energy of the active forces.

In geometry, some criteria for the existence of dense geodesics on a bounded surface with some metric are known (e.g., if the metrics' curvature is everywhere negative, there are dense geodesics). So with the help of Proposition 27 and of the Maupertuis principle, some examples of nonintegrable systems can be easily built.

To obtain deeper insight into integrable systems, it is convenient to restrict attention to the "analytically integrable" systems.

They are connected with some interesting geometrical notions which we have to illustrate before continuing the analysis.

[14] i.e., a set which is a one-to-one bicontinuous image of $\mathbf{T}^l$.

To help the reader avoid getting lost in the labyrinth of the geometric concepts that follow, it is better to state our aim at the beginning. Basically we wish to define sets $G \subset \mathbf{R}^d \times \mathbf{T}^d$ with "piecewise analytic boundary" (see the following definition of analiticity). Such sets have the remarkable property that not only are they measurable in the Riemann sense, but also that their intersection with planar surfaces are measurable with respect to the Riemann measure on the surface. This is a property which might not hold for sets with C^∞ boundary (see Problems).

We shall need, in an essential way, the above simple property and its invariance with respect to some changes of coordinates. There are several ways of constructing families of sets and classes of coordinate changes with this property. However, none of them seem describable in few words, although this fact might seem incredible. It will be an amusing puzzle for the reader to try to find (possibly giving up analiticity) some alternative definitions which would allow us to retain the substance of §4.14 and §4.15.

13 Definition. If $\Omega \subset \mathbf{R}^d$ is open and $f \in C^\infty(\Omega)$, we say that f is "analytic" on Ω if $\forall \xi_0 \in \Omega$, $\exists \epsilon(\xi_0) > 0$ such that f can be developed $\forall |\xi - \xi_0| < \epsilon(\xi_0)$, as

$$f(\xi) = \sum_{k_1, \dots, k_d}^{0,\infty} \frac{\partial^{k_1 + \cdots k_d} f}{\partial \xi_1^{k_1} \dots \partial \xi_d^{k_d}}(\xi_0) \frac{(\xi_1 - (\xi_0)_1)^{k_1}}{k_1!} \cdots \frac{(\xi_d - (\xi_0)_d)^{k_d}}{k_d!} \tag{4.13.7}$$

and

$$\sum_{k_1, \dots, k_d}^{0,+\infty} \left| \frac{\partial^{k_1 + \cdots + k_d} f}{\partial \xi_1^{k_1} \dots \partial \xi_d^{k_d}}(\xi_0) \right| \frac{\epsilon(\xi_0)^{k_1 + \cdots + k_d}}{k_1! \dots k_d!} < +\infty. \tag{4.13.8}$$

If $\mathbf{k} = (k_1, \dots, k_d) \in \mathbf{Z}_+^d$, we shall sometimes use the notations: $\mathbf{k}! = k_1! \dots k_d!$, $(\xi - \xi_0)^{\mathbf{k}} = (\xi_1 - (\xi_0)_1)^{k_1} \dots (\xi_d - (\xi_0)_d)^{k_d}$ and $\partial^{\mathbf{k}} f(\xi)$ for

$$\frac{\partial^{k_1 + \cdots + k_d} f}{\partial \xi_1^{k_1} \dots \partial \xi_d^{k_d}}(\xi),$$

so that Eq. (4.13.7) will be rewritten as

$$f(\xi) = \sum_{\mathbf{k} \in \mathbf{Z}_+^d} \frac{\partial^{\mathbf{k}} f(\xi_0)}{\mathbf{k}!} (\xi - \xi_0)^{\mathbf{k}}, \tag{4.13.9}$$

which is far better.

A function on Ω with values in $\mathbf{R}^l$ is called analytic if its components are analytic.

In the following, we will also need the obvious extension of the notion of analytic function f to the case when Ω is an open subset in $\mathbf{R}^d \times \mathbf{T}^{d'}$ and its values are in $\mathbf{R}^l \times \mathbf{T}^{l'}$, $d + d' > 0$, $l + l' > 0$, $d, d', l, l' \geqslant 0$.

First, observe that a real function on $\Omega \subset \mathbf{R}^d \times \mathbf{T}^{d'}$ can be "canonically extended" to a function $\tilde{f}$ on a set $\tilde{\Omega} \subset \mathbf{R}^d \times \mathbf{R}^{d'}$ by setting

$$\begin{aligned} \tilde{\Omega} = \big\{ &\text{set of pairs } (\xi, \eta) \in \mathbf{R}^d \times \mathbf{R}^{d'} \text{ such that} \\ &(\xi, \eta \bmod 2\pi) \equiv (\xi, \varphi) \in \Omega \big\}, \qquad (4.13.10) \\ \tilde{f}(\xi, \eta) &= f(\xi, \eta \bmod 2\pi) \equiv f(\xi, \varphi). \end{aligned}$$

With the above convention (4.13.10) and with the above restrictions on $d, d', l, l,'$, we state the following definition (for some examples see Exercises).

14 Definition. A function on $\Omega \subset \mathbf{R}^d$ taking values in $\mathbf{R}^l \times \mathbf{T}^{l'}$ and associating to $\xi \in \Omega$ the value $(\mathbf{x}, \varphi) \in \mathbf{R}^l \times \mathbf{T}^{l'}$ will be called "analytic" on Ω if, $\forall \xi_0 \in \Omega$, there is a function F, "representative of f", defined in the vicinity of ξ_0, taking values in $\mathbf{R}^l \times \mathbf{R}^{l'}$ and analytic, such that

$$\text{if } F(\xi) = (\mathbf{x}, \eta) \quad \text{then} \quad f(\xi) = (\mathbf{x}, \varphi) \quad \text{with } \varphi = \eta \bmod 2\pi \qquad (4.13.11)$$

for all ξ near ξ_0.

A function f on an open set $\Omega \subset \mathbf{R}^d \times \mathbf{T}^{d'}$ taking values in $\mathbf{R}^l \times \mathbf{T}^{l'}$ will be called analytic on Ω if its canonical extension $\tilde{f}$ to $\tilde{\Omega}$ is analytic.

The derivatives of f will obviously be defined as the "derivatives of the canonical extension of a representative" and they will be denoted by the usual symbols.

Observation. If some of the integers l, l', d, d' vanish, we interpret $\mathbf{R}^l \times \mathbf{T}^{l'}$ or $\mathbf{R}^d \times \mathbf{T}^{d'}$ in the obvious way: $\mathbf{R}^0 \times \mathbf{T}^p \equiv \mathbf{T}^p$, $\mathbf{R}^p \times \mathbf{T}^0 \equiv \mathbf{R}^p$, $\forall p > 0$.

Together with the notion of analytic function, we need the notion of analytic coordinates.

15 Definition. Let $U \subset \mathbf{R}^d \times \mathbf{T}^{d'}$ be open and let Ξ be an $\mathbf{R}^d \times \mathbf{T}^{d'}$-valued analytic function defined on an open set $\Omega \subset \mathbf{R}^d \times \mathbf{R}^{d'}$ such that:

(i) Ξ is invertible as a map between U and Ω;
(ii) the Jacobian determinant of Ξ never vanishes on Ω ("Ξ is nonsingular");[15]
(iii) Ξ and Ξ^{-1} are analytic in Ω and U, respectively.

Then we say that (U, Ξ) is an analytic system of local coordinates on U.

If $U \subset \mathbf{R}^d \times \mathbf{T}^{d'}$, $V \subset \mathbf{R}^{\bar{d}} \times \mathbf{T}^{\bar{d}'}$ are open sets and $d + d' = \bar{d} + \bar{d}'$, and if Ξ is an analytic function on U taking values in V and establishing between U and V a one-to-one nonsingular correspondence with analytic inverse, then we shall say that Ξ is an analytic correspondence between U and V.

Observation. Some among $d, d', \bar{d}, \bar{d}'$, may vanish: see the Observation to Definition 14.

[15] Naturally, the Jacobian determinant of Ξ in ξ_0 is the Jacobian determinant of a representative of Ξ near ξ_0 (see Definition 14, above).

It is now possible to establish the definition of an analytic surface. The reader should try to make drawings to see the various geometrical objects discussed in the following definitions and observations.

16 Definition. A regular surface $\Sigma \subset \mathbf{R}^d \times \mathbf{T}^{d'}$ is said to be "locally analytic" in an open set $U \subset \mathbf{R}^d \times \mathbf{T}^{d'}$ if there is a family of local analytic systems of coordinates $(U_\alpha, \Xi_\alpha)_{\alpha \in A}$ with bases $(\Omega_\alpha)_{\alpha \in A}$ such that:

(i) the points of $\Sigma \cap U_\alpha$ are those which in (U_α, Ξ_α) have coordinates $\beta_1 = \cdots = \beta_{d+d'-l} = 0$, where l is the "dimension" of Σ, i.e., (U_α, Ξ_α) are adapted to Σ;
(ii) as α varies in A, the sets U_α cover $\Sigma \cap U$ and A is a finite set of indices.

If Σ is a locally analytic surface in U and f is an $\mathbf{R}^d \times \mathbf{T}^{d'}$-valued function on Σ, we shall say that "f is analytic on Σ" if it is the restriction to Σ of an analytic function on an open set $\tilde{U} \supset \Sigma \cap U$.

If $\Sigma \subset U$ is a closed set and if Σ is a locally analytic surface in U, we shall say that Σ is an "analytic surface" (this notion is U independent).

Observations.

(1) If some of the $d, d', \bar{d}, \bar{d}'$ vanish see the Observation to Definition 14.

(2) Examples are discussed in the problems and exercises at the end of this section.

Finally, we define the "analytically regular sets".

17 Definition. A closed set $G \subset \mathbf{R}^d \times \mathbf{T}^{d'}$ will be called "locally analytic" in the open set $U \subset \mathbf{R}^d \times \mathbf{T}^{d'}$ if ∂G is a surface locally analytic in U.

If G is locally analytic in U and $G \subset U$, then G will be called "an analytic set" (this notion is U independent).

A closed set $G \subset \mathbf{R}^d \times \mathbf{T}^{d'}$ will be called "analytically regular" if there is an open set $U \supset G$ and a family of sets locally analytic in U through which, via a finite number of union and intersection operations, one can build G.

Observations.

(1) If d or d' vanish, see comment (1) to Definition 14.

(2) Any analytic surface is an analytic set (since either $\partial\Sigma \equiv \Sigma$ or $\Sigma \equiv \mathbf{R}^d \times \mathbf{T}^{d'}$).

(3) If Ξ is an analytic transformation of $U \subset \mathbf{R}^d \times \mathbf{T}^{d'}$ onto $V \subset \mathbf{R}^d \times \mathbf{T}^{d'}$ and if $G \subset U$ is an analytically regular set, then $\Xi(G) \subset V$ is also analytically regular, i.e., the above notion is invariant under analytic maps. This follows from the fact that composing analytic functions, one obtains analytic functions.[16]

[16] A proof can be found in the calculus books. The reader can attempt a proof starting with the $l = 1$ case.

(4) If Σ is a surface locally analytic in U and if $G \subset U$ is an analytically regular set also, $G \cap \Sigma$ is an analytically regular set: this is the "invariance under the intersection operations" of the analytic regularity.

(5) Let $\bar{d} \geqslant d$, $\bar{d}' \geqslant d'$ and let $\mathbf{R}^d \times \mathbf{T}^{d'}$ be regarded as a subset of $\mathbf{R}^{\bar{d}} \times \mathbf{T}^{\bar{d}'}$ by identifying it as the subset of $\mathbf{R}^{\bar{d}} \times \mathbf{T}^{\bar{d}'}$ consisting of the points $(\bar{\mathbf{x}}, \bar{\boldsymbol{\varphi}})$ such that $\bar{\mathbf{x}} = (\mathbf{x}, \mathbf{0})$, $\bar{\boldsymbol{\varphi}} = (\boldsymbol{\varphi}, \mathbf{0})$ with $(\mathbf{x}, \boldsymbol{\varphi}) \in \mathbf{R}^d \times \mathbf{T}^{d'}$ and the $\mathbf{0}$'s denote the origins in $\mathbf{R}^{\bar{d}-d}$ and $\mathbf{R}^{\bar{d}'-d'}$, respectively. Then $\mathbf{R}^d \times \mathbf{T}^{d'}$ is an analytic surface in $\mathbf{R}^{\bar{d}} \times \mathbf{T}^{\bar{d}'}$.

If $G \subset \mathbf{R}^d \times \mathbf{T}^{d'}$ is analytically regular, then its "extension $\hat{G}$ to $\mathbf{R}^{\bar{d}} \times \mathbf{T}^{\bar{d}'}$", $\hat{G} = \{(\mathbf{x}, \boldsymbol{\varphi}) \,|\, (\bar{\mathbf{x}}, \bar{\boldsymbol{\varphi}}) \in \mathbf{R}^{\bar{d}} \times \mathbf{T}^{\bar{d}'}, \bar{\mathbf{x}} = (\mathbf{x}, \mathbf{y}), \bar{\boldsymbol{\varphi}} = (\boldsymbol{\varphi}, \boldsymbol{\psi}) \text{ with } (\mathbf{x}, \boldsymbol{\varphi}) \in G\}$ is analytically regular in $\mathbf{R}^d \times \mathbf{T}^{d'}$.

(6) On every regular (or locally analytic) surface $\Sigma \subset \mathbf{R}^d \times \mathbf{T}^{d'}$, one can define the "area measure": if $(U, \boldsymbol{\Xi})$ is a regular (or analytic) system of local coordinates adapted to Σ with basis Ω, there is a regular (or analytic) function σ on Ω such that for $E \subset \Sigma \cap U$:

$$\begin{aligned} \text{area } E &= \int_{\boldsymbol{\Xi}^{-1}(E)} \sigma(0, \ldots, 0, \beta_{d+d'-l+1}, \ldots, \beta_{d+d'}) \\ &\quad \times d\beta_{d+d'-l+1}, \ldots, d\beta_{d+d'} \end{aligned} \tag{4.13.13}$$

provided $\boldsymbol{\Xi}^{-1}(E)$ is measurable in the Riemanian sense (in this case, one says that E is measurable with respect to the area measure).

If $\Sigma \subset \mathbf{R}^d$ is a regular surface and $(U, \boldsymbol{\Xi})$ is a well-adapted orthogonal and of Fermi type system of local coordinates (in the sense of Definition 12, §3.7, p. 178 and Proposition 12, p. 184) with respect to the scalar product $\boldsymbol{\eta} \cdot \boldsymbol{\chi}$ on $\mathbf{R}^d$, one has

$$\sigma(0, \ldots, 0, \beta_{d+d'-l+1}, \ldots, \beta_{d+d'}) = \sqrt{\gamma^l} \tag{4.13.14}$$

essentially by (a very reasonable) definition; σ in the other coordinate systems is computed by ordinary coordinate transformations.

The simplicity of Eq. (4.13.14) provides a further illustration of the notion of "well-adapted orthogonal" systems of coordinates of §3.7.

(7) One may think that it is possible to define something like "C^∞-regular" sets by simply replacing the word analytic by C^∞ everywhere above. However, the property in Observation (4), for instance, would not hold. See exercises to §4.13.

The problem of the (Riemann) measurability of sets is not always trivial and the interest in the above digression on the definition of analytically regular sets rests mainly on the validity of the following proposition.

28 Proposition. *Let $\Sigma \subset \mathbf{R}^d \times \mathbf{T}^{d'}$ be a surface locally analytic in the open set U and let E be the analytically regular set contained in U.*

The set $E \cap \Sigma$ is then measurable with respect to the area measure on Σ, i.e., given $\epsilon > 0$, there exist two functions χ^+ and χ^- of class C^∞ on Σ,

$0 \leqslant \chi^- \leqslant \chi^+ \leqslant 1$, *such that if* χ_E *is the characteristic function of* $E \cap \Sigma$:

$$\text{(i)}\ \chi^-(\xi) \leqslant \chi_E(\xi) \leqslant \chi^+(\xi), \qquad \forall \xi \in \Sigma \cap E, \tag{4.13.15}$$
$$\text{(ii)}\ \int_{E\cap\Sigma} (\chi^+(\xi) - \chi^-(\xi))\, d\sigma_\xi < \epsilon, \tag{4.13.16}$$

where the integral denotes the surface integral on Σ, *see Observation* (6) *above.*

We do not describe the proof of this proposition. Although it is not particularly difficult, it would require a preliminary analysis of the structure of the analytic surfaces and their intersections which, being marginal for us, would lead us too far away from our problem of discussing the integrability criteria for mechanical systems. Therefore, the reader should consult geometry textbooks.

Exercises and Problems for §4.13

1. Show that the function on $\mathbf{T}^1 : \varphi \to \cos\varphi$ is analytic.

2. Show that the $\mathbf{T}^1$-valued function $x \to x \bmod 2\pi$ is analytic.

3. If $f \in C^\infty(\mathbf{T}^1)$ and if its Fourier coefficients can be bounded as $|\hat{f}_k| < Fc^k$, $c < 1$, $F > 0$, then f is analytic on $\mathbf{T}^1$. Prove this statement.

4. Generalize Problem 3 to the case of a function on $\mathbf{T}^l$.

5. Show that a "surface" of $\mathbf{R}$ relatively closed in $U \subset \mathbf{R}$ and locally analytic in U(U open) is, inside U, a union of at most denumerably many points without accumulation points in U or coincides with U. Show that a bounded analytically regular set in $\mathbf{R}$ is a union of finitely many points and intervals.

6. Show that straight lines, planes, half-lines, half-planes, and half-spaces are analytic sets in $\mathbf{R}^2$ and $\mathbf{R}^3$.

7. Show that triangles, polygons, disks and their boundaries are analytically regular in $\mathbf{R}^2$ and in $\mathbf{R}^3$.

8. Show that the regular solids, the spheres, the diedra, the triedra, etc., and their boundaries are analytically regular in $\mathbf{R}^3$.

9. Show that the disk and the ellipse, or the sphere and the ellipsoid, and their boundaries are analytic sets in $\mathbf{R}^2$ or $\mathbf{R}^3$, respectively.

10. Show that a disk in $\mathbf{R}^3$ is not an analytic set although it is analytically regular.

11. Let $x_1, x_2 \ldots$ be a numeration of the rational numbers in $[0, 1]$. For every x_k, consider the open interval with length 2^{-1-k} and center x_k. Show that the union of such intervals is an open set dense in $[0, 1]$ with external measure (in the Riemannian sense) $\geqslant 1$ and with the internal measure $< \frac{1}{2}$. Call this union A.

12. Let $g \in C^\infty(\mathbf{R})$ be a positive function on $(-\frac{1}{2}, \frac{1}{2})$ and zero elsewhere. Set

$$f(x) = \sum_{k=1}^{\infty} k!^{-1} g(2^{k+1}(x - x_k))$$

and show that f is positive on A (see Problem 11) and zero outside. Show also that f is in $C^\infty(\mathbf{R})$.

13. Show that the set $D_\infty = \{(x, y) \mid 0 \leqslant x \leqslant 1, f(x) \leqslant y < +\infty\} \subset \mathbf{R}^2$, with f as defined in Problem 12, has a piecewise C^∞ boundary. Show that the intersection between D_∞ and the x axis is not measurable in the Riemannian sense.

14. Let $v \in C^\infty(\mathbf{R}^1)$ and consider the surface in $\mathbf{R}^2$ with equations $z = v(x)$, $x \in \mathbf{R}$. Show that the "area" of a line element over dx is, according to Eq. (4.13.14), $d\sigma = \{1 + [(dv/dx)(x)]^2\}^{1/2}\, dx$

15. Let $v \in C^\infty(\mathbf{R}^2)$ and consider the surface in $\mathbf{R}^3$ with equations $z = v(x, y)$, $(x, y) \in \mathbf{R}^2$. Show that the area of a surface element over $dx\,dy$ is, according to Eq. (4.13.14),

$$d\sigma = \left(1 + \left(\frac{\partial v}{\partial x}(x, y)\right)^2 + \left(\frac{\partial v}{\partial y}(x, y)\right)^2\right)^{1/2} dx\, dy.$$

§4.14. Analytically Integrable Systems. Frequency of Visits and Ergodicity

In §4.8, Definition 11, p. 289, we introduced the notion of "analytically integrable" Hamiltonian systems defined on an open set $W \subset \mathbf{R}^l \times \mathbf{R}^l$ or $\mathbf{R}^l \times \mathbf{T}^l$ or $\mathbf{R}^l \times \mathbf{R}^{l_1} \times \mathbf{T}^{l_2}$, $l_1 + l_2 = l$, in phase space.

The interest in analytically integrable systems is twofold: essentially all concrete integrable systems so far met were analytically integrable (and this could be verified with some labor); furthermore, if $\mathbf{t} = (it_1)_{i=0}^\infty$, the $(f, \mathbf{t})$ histories of the points in the integrability region W of phase space have a well-defined average value for all the $f \in C^\infty(W)$, and also for many other more singular functions f, for instance for the characteristic functions of the analytically regular sets.

To illustrate this remarkable property, it is convenient to introduce the following notions.

18 Definition. Let W be a subset of the phase space ($\subset \mathbf{R}^l \times \mathbf{R}^l$, or $\mathbf{R}^l \times \mathbf{T}^l$ or $\mathbf{R}^l \times \mathbf{R}^{l_1} \times \mathbf{T}^{l_2}$, $l_1 + l_2 = l$) of an analytic time-independent Hamiltonian system. Suppose that the system is analytically integrable on W.

Let $\mathcal{E} = (E_0, E_1, \ldots, E_p)$ be a family of subsets of W such that:

(i) $\cup_{i=0}^p E_i = W$, $E_i \cap E_j = \emptyset$ if $i \neq j$, i.e., $\mathcal{E}$ is a "partition of W";
(ii) $E_1, \ldots, E_p$ are analytically regular;
(iii) $d(E_i, E_j) > 0$ if $i \neq j$, $i, j = 1, \ldots, p$.

Obviously, $E_0 = W \setminus \cup_{j=1}^p E_j$ is an open set.

The partition $\mathcal{E}$ will be called an "analytically regular partition" of W.

We shall denote χ_{E_i} the characteristic function of the sets E_i, $i = 0, 1, \ldots, p$, and we set

$$f_{\mathcal{E}}(\xi) = \sum_{j=0}^{p} j\chi_{E_j}(\xi), \qquad \xi \in W \tag{4.14.1}$$

and, finally, we call "$(\mathfrak{S}, \mathbf{t})$ history of $(\mathbf{p}, \mathbf{q}) \in W$" the $(f_{\mathfrak{S}}, \mathbf{t})$ history of $(\mathbf{p}, \mathbf{q})$ when $\mathbf{t} = (t_i)_{i=0}$ is a divergent monotonic sequence.

Observation. The $(\mathfrak{S}, \mathbf{t})$ history is a sequence of integers between 0 and p: the kth element of this sequence simply indicates into which set among those of $\mathfrak{S}$ the point $S_{t_k}(\boldsymbol{\xi})$ falls, if $t \to S_t(\boldsymbol{\xi})$, $t \geqslant 0$, is the solution to the Hamiltonian equations with initial datum $\boldsymbol{\xi} = (\mathbf{p}, \mathbf{q})$.

The following proposition is very remarkable.

29 Proposition. *Given an analytic Hamiltonian system analytically integrable in the subset W of phase space, the limit*

$$\lim_{N \to \infty} \frac{1}{N} \sum_{j=0}^{N-1} \chi_E(S_{jt_1}(\boldsymbol{\xi})) \tag{4.14.2}$$

exists, $\forall t_1 > 0$ *and for all analytically regular subsets E of W.*

This limit will be called naturally the "frequency of visit to E by the motion starting in $\boldsymbol{\xi}$*" with respect to the sequence of observation times* $\mathbf{t} = (it_1)_{i=0}^{\infty}, t_1 > 0$.

PROOF. The image $I(E) \subset V \times \mathbf{T}^l$ of E via the analytic integrating transformation I, see Definition 11, §4.8, p. 289, will still be analytically regular, see observation (3) to Definition 17, p. 335. Then it is clear, since for $I(\boldsymbol{\xi}) = (\mathbf{A}, \boldsymbol{\varphi})$,

$$I(S_t\boldsymbol{\xi}) = (\mathbf{A}, \boldsymbol{\varphi}, +\boldsymbol{\omega}(\mathbf{A})t), \tag{4.14.3}$$

that the problem of the proof of the above proposition is "reduced" to the one contemplated in the following proposition.

30 Proposition. *Let* $\boldsymbol{\omega} = (\omega_1, \ldots, \omega_1)$ *be an l-tuple of real numbers and let* $E \subset \mathbf{T}^l$ *be an analytically regular subset of* $\mathbf{T}^l$. *If* $\mathbf{t} = (it_1)_{i=0}^{\infty}$, $t_1 > 0$, *the frequency of visits*

$$\nu_E(\boldsymbol{\varphi}) = \lim_{N \to \infty} \frac{1}{N} \sum_{j=0}^{N-1} \chi_E(\boldsymbol{\varphi} + jt_1\boldsymbol{\omega}) \tag{4.14.4}$$

exists, $\forall \boldsymbol{\varphi} \in \mathbf{T}^l$.

Furthermore, if the numbers $\omega_1, \ldots, \omega_l$ *and* $\sigma = 2\pi/t_1$ *are rationally independent it turns out that*

$$\nu_E(\boldsymbol{\varphi}) = \frac{1}{(2\pi)^l} \int_E d\boldsymbol{\varphi}'. \tag{4.14.5}$$

PROOF. We shall only treat the simple case when $\omega_1, \ldots, \omega_l, \sigma$ are rationally independent, because it is easy. The general case can be reduced to this one with some patient though interesting work which we leave to the reader, referring, as a guide, to the sequence of problems at the end of this section.

The idea of the proof is to use the Riemann measurability of E (consequence of its analytic regularity, see Proposition 28) to find two $C^{\infty}(\mathbf{T}^l)$ functions χ^- and χ^+ verifying Eqs. (4.13.15) and (4.13.16) to infer that $\nu_E(\varphi)$, if existing, must be between the averages of χ^- and χ^+ which, in turn, exist and differ at most by ϵ because of Eq. (4.13.16) and Proposition 6, §4.2, Eq. (4.2.10), p. 251.

Then the arbitrariness of ϵ implies the actual existence of $\nu_E(\varphi)$ and the fact that it is between the averages $(2\pi)^{-l}\int_{T^l}\chi^{\pm}(\varphi')d^l\varphi'$. Again the arbitrariness of ϵ and Eqs. (4.13.16) and (4.13.15) imply Eq. (4.14.5). mbe

Observations.

(1) The reader will note that the analytic regularity of E is used in the above proof only to infer the Riemann measurability of E. However, if $\omega_1, \ldots, \omega_1, \sigma$ were not rationally independent, the analytic regularity should again be used to prove the reducibility of the general case to the rationally independent one. This is the reason why the Riemann measurability is not in itself a general sufficient condition for the existence of the limit of Eq. (4.14.4). See problems at the end of this section.

(2) Of course if $\omega_1, \ldots, \omega_l, \sigma$ are rationally independent, the Riemann measurability of E suffices, alone, to deduce Eqs. (4.14.5) and (4.14.6) as it appears clear from the above proof.

So every motion of a Hamiltonian system analytically integrable in W visits an analytically regular set E with a well-defined frequency of visit. One can wonder about the frequency of joint visits to two given analytically regular sets E and E'. The remarkable fact is that they are, on the average, "independent". The frequency of a visit to E followed j time units later by a visit to E' is, on the average over j, equal to the product of the frequency of visit to E and of that of E': $\forall \xi \in W, \forall t_1 > 0$,

$$\lim_{N\to\infty} \frac{1}{N} \sum_{j=0}^{N-1} \nu_{E\cap S_{jt_1}(E')}(\xi) = \nu_E(\xi)\nu_{E'}(\xi). \tag{4.14.6}$$

In other words, the visit to E by a given motion does not put any restrictions on the possibility of a visit to E' j time units later, at least on the average on j.

This is the content of the following proposition.

31 Proposition. *In the assumptions of Proposition* 29, *let* $E, E' \subset W$ *be two analytically regular sets. Then the property* (4.14.6) *holds for all* $\xi \in W$, $\forall t_1 > 0$.

Observation. This proposition is a corollary of Proposition 32 on the quasi-periodic motions on $\mathbf{T}^l$ in the same way in which Proposition 29 appears to be a corollary of Proposition 30.

32 Proposition. *Let $E, E' \subset \mathbf{T}^l$ be two analytically regular sets and let $\omega \in \mathbf{R}^l$, $t_1 > 0$. Denote $E' + t\omega$ the set of points $\varphi' + t\omega \bmod 2\pi$ as φ' varies in E' ($E' + t\omega$ is the set into which E' evolves in time t under the quasi-periodic flow on $\mathbf{T}^l$ with pulsations ω). If $\nu_E(\varphi)$ is the frequency of visits of the points $\varphi + jt_1\omega$, $j = 0, 1, \ldots,$ to E, it is, $\forall \varphi \in \mathbf{T}^l$,*

$$\lim_{N\to\infty} \frac{1}{N} \sum_{j=0}^{N-1} \nu_{E\cap(E'+jt_1\omega)}(\varphi) = \nu_E(\varphi)\nu_{E'}(\varphi). \tag{4.14.7}$$

Observations.

(1) When $\omega_1, \ldots, \omega_l$ and $\sigma = 2\pi/t_1$ are rationally independent, $\nu_E(\varphi)$ is the measure of E, see Eq. (4.14.5). Hence, Eq. (4.14.7) means that in this case the fraction of E occupied by images of points of E' is a fraction of E equal, on the average, to the measure of E'. In other words, $E' + jt_1\omega$ is uniformly scattered in $\mathbf{T}^l$, on the average. This holds for $\forall E'$ analytically regular.

(2) By considering the case $l = 1$ and taking E and E' to be two small intervals, one sees that the limit of $\nu_{E\cap(E'+jt_1\omega)}(\varphi)$ as $j \to \infty$ does not exist in general: even in the case of rational independency of $\omega_1, \ldots, \omega_l, 2\pi/t_1$ the average over j in Eq. (4.14.7) is essential. Therefore, even though on the average $E' + jt_1\omega$ is uniformly scattered in $\mathbf{T}^l$, it is not true that for large times j this set is uniformly scattered. This is due manifestly to the fact that the rotations of the torus are "rigid" transformations and they do not "mix" the points of $\mathbf{T}^l$.

PROOF. As in the case of Proposition 29, let us only treat the simple case when $\omega_1, \omega_2, \ldots, \omega_l, \sigma = 2\pi/t_1$ are rationally independent. The general case can be treated by solving the last of the problems at the end of this section.

Proceeding as in the proof of Proposition 30 and using the Riemann measurability [see Eq. (4.13.16)] of the sets E, E', the problem of proving Eq. (4.14.7) is reduced to that of proving $\forall f, g \in C^\infty(\mathbf{T}^l)$:

$$\lim_{N\to\infty} \frac{1}{N} \sum_{j=0}^{N-1} \overline{f(\varphi)g(\varphi + jt_1\omega)} = \overline{f(\varphi)}\ \overline{g(\varphi)}, \tag{4.14.8}$$

where the bar over a function of φ denotes the average:

$$\overline{f(\varphi)} = \lim_{N\to\infty} \frac{1}{N} \sum_{j=0}^{N-1} f(\varphi + jt_1\omega). \tag{4.14.9}$$

Note that Eq. (4.14.8) would directly become Eq. (4.14.7) if one could take $f = \chi_E$, $g = \chi_{E'}$.

To prove this proposition, Eq. (4.14.8) shall be applied to the functions χ^+, χ'^+ which, according to Proposition 28, approximate $\chi_E, \chi_{E'}$ from above and to the functions χ^-, χ'^- which approximate $\chi_E, \chi_{E'}$, from below, following the approximation idea of the proof of Proposition 30.

Let us now check Eq. (4.14.8). By the simplifying assumption of rational independence, see Proposition 6, §4.2, p. 251,

$$\begin{aligned} \overline{f(\varphi)} &= \frac{1}{(2\pi)^l}\int_{T^l} f(\varphi')\,d\varphi' = \hat{f}_0, \\ \overline{g(\varphi)} &= \frac{1}{(2\pi)^l}\int_{T^l} g(\varphi')\,d\varphi' = g_0, \end{aligned} \tag{4.14.10}$$

where $\hat{f}_{\mathbf{n}}, \hat{g}_{\mathbf{n}}, \mathbf{n} \in \mathbf{Z}^l$, are the Fourier coefficients of f and g, respectively. Furthermore, since

$$\begin{aligned} f(\varphi)\,g(\varphi + jt_1\omega) &= \sum_{\mathbf{n}}\sum_{\mathbf{n}'} \hat{f}_{\mathbf{n}}\hat{g}_{\mathbf{n}'}e^{i(\mathbf{n}\cdot\varphi + \mathbf{n}'\cdot\varphi + jt_1\mathbf{n}'\cdot\omega)} \\ &= \sum_{\mathbf{m}} e^{i\mathbf{m}\cdot\varphi}\Big(\sum_{\mathbf{n}+\mathbf{n}'=\mathbf{m}} \hat{f}_{\mathbf{n}}\hat{g}_{\mathbf{n}'}e^{ijt_1\mathbf{n}'\cdot\omega}\Big), \end{aligned} \tag{4.14.11}$$

one finds, still by Proposition 6, §4.2,

$$\begin{aligned} \overline{f(\varphi)g(\varphi + jt_1\omega)} &= \{\text{Fourier coefficient of order } \mathbf{0} \\ &\qquad \text{of the function in Eqs. (4.14.11)}\} \\ &= \sum_{\mathbf{n}} \hat{f}_{\mathbf{n}}\hat{g}_{-\mathbf{n}}e^{-ijt_1\mathbf{n}\cdot\omega}. \end{aligned} \tag{4.14.12}$$

Then

$$\begin{aligned} N^{-1}\sum_{j=0}^{N-1} f(\varphi)g(\varphi + jt_1\omega) &= \sum_{\mathbf{n}} \hat{f}_{\mathbf{n}}\hat{g}_{-\mathbf{n}}N^{-1}\sum_{j=0}^{N-1} e^{-ijt_1\omega\cdot\mathbf{n}} \\ &= \hat{f}_0\hat{g}_0 + \sum_{\mathbf{n}\neq\mathbf{0}} \hat{f}_{\mathbf{n}}\hat{g}_{-\mathbf{n}}\frac{1}{N}\,\frac{e^{-it_1\omega\cdot\mathbf{n}N} - 1}{e^{-it_1\omega\cdot\mathbf{n}} - 1}, \end{aligned} \tag{4.14.13}$$

and, by the usual argument of passage to the limit under the series sign, it follows that the limit as $N \to \infty$ of Eq. (4.14.13) is just $\hat{f}_0\hat{g}_0$ which shows, recalling (4.14.10), the validity of Eq. (4.14.8) and, hence, the above proposition's validity (in the special case treated here). mbe

From the above propositions, we can deduce some simple consequences.

Let $\mathfrak{E} = (E_0, E_1, \ldots, E_s)$ be a partition of the phase space W of an analytically integrable Hamiltonian system. Suppose that $\mathfrak{E}$ is analytically regular in W in the sense of Definition 18, p. 338.

Given $\mathbf{t} = (it_1)_{i=0}^{\infty}$, the partition $\mathfrak{E}$ and $k \geq 0$, $0 \leq j_1 < j_2 < \cdots < j_k$, we shall define

$$E\begin{pmatrix} j_1 \ldots j_k \\ \alpha_1 \ldots \alpha_k \end{pmatrix} = S_{-j_1t_1}(E_{\alpha_1}) \cap S_{-j_2t_1}(E_{\alpha_2}) \cap \cdots \cap S_{-j_kt_1}(E_{\alpha_k}). \tag{4.14.14}$$

This is the set of the points $\xi \in W$ such that

$$S_{j_1t_1}\xi \in E_{\alpha_1}, S_{j_2t_1}\xi \in E_{\alpha_2}, \ldots, S_{j_kt_1}\xi \in E_{\alpha_k}. \tag{4.14.15}$$

From Eq. (4.14.15) and from the fact that $\mathfrak{E}$ is a partition of W, it is obvious that

$$E\begin{pmatrix} j_1 \cdots j_k \\ \alpha_1 \cdots \alpha_k \end{pmatrix} \cap E\begin{pmatrix} j_1 \cdots j_k \\ \beta_1 \cdots \beta_k \end{pmatrix} = \emptyset \tag{4.14.16}$$

unless $\alpha_1 = \beta_1, \alpha_2 = \beta_2, \ldots, \alpha_k = \beta_k$. Also,

$$\bigcup_{\alpha_1, \ldots, \alpha_k}^{0,s} E\begin{pmatrix} j_1 \cdots j_k \\ \alpha_1 \cdots \alpha_k \end{pmatrix} = W, \tag{4.14.17}$$

$$\bigcup_{\alpha}^{0,s} E\begin{pmatrix} j_1 \cdots j_{p-1} j_p j_{p+1} \cdots j_k \\ \alpha_1 \cdots \alpha_{p-1} \alpha\, \alpha_{p+1} \cdots \alpha_k \end{pmatrix} = E\begin{pmatrix} j_1 \cdots j_{p-1} j_{p+1} \cdots j_k \\ \alpha_1 \cdots \alpha_{p-1} \alpha_{p+1} \cdots \alpha_k \end{pmatrix}. \tag{4.14.18}$$

It is also clear that if $\alpha_1 \neq 0, \alpha_2 \neq 0, \ldots, \alpha_k \neq 0$, the set $E\binom{j_1, \ldots, j_k}{\alpha_1, \ldots, \alpha_k}$ is analytically regular because the time evolution transformations $(S_t)_{t \in \mathbf{R}}$ are analytic (being such after the analytic change of coordinates I, [see Eq. (4.8.14)] which integrates the system,[17] and because the analytic image of an analytically regular set is still analytically regular [see Observation (3) to Definition 17, p. 335).

The sets $E\binom{j_1 \cdots j_k}{\alpha_1 \cdots \alpha_k}$ have a simple physical meaning.

We imagine that the partition $\mathfrak{E}$ models an actual observation of some physical quantity. The results of the observations, read on a dial, give a finite number of results $1, 2, \ldots, s$ or 0 ("off the dial"). Since the results of physical measurements can always be numbered from 1 to some s, this is a very general model.

Thus the phase space W is divided by collecting together all the physical configurations $\xi \in W$ that produce the same result for the value of the physical quantity described by Eq. (4.14.1) in this model.

Given a sequence of observation times $0, t_1, t_2, \ldots, t_j = jt_1$, we can decide to record the results of the observations made at times $j_1 t_1$, $j_2 t_1, \ldots, j_k t_1$. We see that the possible outcomes of such observations are $(s+1)^k$ k-tuples $(\alpha_1, \ldots, \alpha_k)$ and we can partition W into $(s+1)^k$ sets of the form $E\binom{j_1 \cdots j_k}{\alpha_1 \cdots \alpha_k}$, collecting the points falling in E_{α_1} at time $j_1 t_1$, in E_{α_2} at time $j_2 t_1, \ldots,$ in E_{α_k} at time $j_k t_1$.

In terms of the above mathematical notions, it is possible to formulate an interesting proposition whose physical meaning can easily be gathered from the just discussed interpretation.

33 Proposition. *Let W be the phase space of a time-independent analytically integrable Hamiltonian system. Let $t_1 > 0$ and let $\mathfrak{E} = (E_0, \ldots, E_s)$ be an analytically regular partition of W. Then $\forall \xi \in W$, the frequencies of*

[17] Here we use a well-known fact that when composing analytic functions, one obtains analytic functions. The reader can attempt a proof of this starting with the $l = 1$ case.

visits to the sets of the form of Eq. (4.14.14) *by the motion starting at* ξ *exist.*

Denoting such frequencies

$$\mathbf{p}\left(\begin{matrix} j_1 \cdots j_k \\ \alpha_1 \ldots \alpha_k \end{matrix}\middle|\xi\right) \equiv \nu_{E\left(\begin{smallmatrix} j_1 \cdots j_k \\ \alpha_1 \ldots \alpha_k \end{smallmatrix}\right)}(\xi), \tag{4.14.19}$$

it also follows that:

(i) $\forall k \in \mathbf{Z}_+$, $\forall 0 \leqslant j_1 < j_2 < \cdots < j_k$ *integers*, $\forall \alpha_1, \ldots, \alpha_k$ *in* $\{0, 1, \ldots, s\}$:

$$\mathbf{p}\left(\begin{matrix} j_1 \cdots j_k \\ \alpha_1 \ldots \alpha_k \end{matrix}\middle|\xi\right) \geqslant 0; \tag{4.14.20}$$

(ii) $\forall k \in \mathbf{Z}_+$, $\forall 0 \leqslant j_1 < j_2 < \cdots < j_k$ *integers*, $\forall \alpha_1, \ldots, \alpha_k$ *in* $\{0, 1, \ldots, s\}$, $\forall q = 1, 2, \ldots, k$:

$$\sum_{\alpha=0}^{s} \mathbf{p}\left(\begin{matrix} j_1 \cdots j_{s-1} j_s j_{s+1} \cdots j_k \\ \alpha_1 \ldots \alpha_{s-1} \alpha\, \alpha_{s+1} \ldots \alpha_k \end{matrix}\middle|\xi\right) = \mathbf{p}\left(\begin{matrix} j_1 \cdots j_{s-1} j_{s+1} \cdots j_k \\ \alpha_1 \ldots \alpha_{s-1} \alpha_{s+1} \ldots \alpha_k \end{matrix}\middle|\xi\right); \tag{4.14.21}$$

(iii) $\forall k \in \mathbf{Z}_+$, $\forall 0 \leqslant j_1 < \cdots < j_k$ *integers*:

$$\sum_{\alpha_1 \ldots \alpha_k}^{0,s} \mathbf{p}\left(\begin{matrix} j_1 \cdots j_k \\ \alpha_1 \ldots \alpha_k \end{matrix}\middle|\xi\right) \equiv 1; \tag{4.14.22}$$

(iv) $\forall k, h \in Z_+$, $\forall 0 \leqslant j_1 < \cdots < j_k$, $0 \leqslant i_1 < \cdots < i_h$ *integers and* $\alpha_1, \ldots, \alpha_k, \beta_1, \ldots, \beta_h$ *in* $\{0, \ldots, s\}$:

$$\begin{aligned} \lim_{N\to\infty} \frac{1}{N} \sum_{l=0}^{N-1} \mathbf{p}\left(\begin{matrix} j_1 \cdots j_k, i_1 + l \ldots i_h + l \\ \alpha_1 \ldots \alpha_k, \;\beta_1 \ldots \;\beta_h \end{matrix}\middle|\xi\right) \\ = \mathbf{p}\left(\begin{matrix} j_1 \cdots j_k \\ \alpha_1 \ldots \alpha_k \end{matrix}\middle|\xi\right) \mathbf{p}\left(\begin{matrix} i_1 \cdots i_h \\ \beta_1 \ldots \beta_h \end{matrix}\middle|\xi\right). \end{aligned} \tag{4.14.23}$$

Properties (*i*), (*ii*), (*iii*), *and* (*iv*) *will be referred to, respectively, as "positivity", "compatibility", "normalization", and "ergodicity" properties of the motion "generated by* ξ *and observed on* $\mathfrak{S}$".

Observation. It will appear that the above proposition is just a fancy statement of the results already obtained. However, it is very useful because it allows us to introduce some qualitative notions which are very natural and important.

PROOF. First suppose the existence of the frequencies of Eq. (4.14.19). Then (i) is obvious, while (ii) and (iii) follow from Eqs. (4.14.16)–(4.14.18) and Eqs. (4.14.16) and (4.14.17), respectively.

So it remains to prove the existence of the frequencies and (iv).

The existence of the frequencies for the sets $E\binom{j_1 \cdots j_k}{\alpha_1 \ldots \alpha_k}$ with $\alpha_1 \neq 0$, $\alpha_2 \neq 0, \ldots, \alpha_k \neq 0$ follows from their analytic regularity stated after their definition and from Proposition 29. It remains, therefore, to examine the cases when some among $\alpha_1, \ldots, \alpha_k$ vanish.

This can be easily accomplished inductively by using Eqs. (4.14.16)–(4.14.18). If $k = 1$, it is clear from Eq. (4.14.17) and from the definition of frequency that $\mathbf{p}(\begin{smallmatrix} j_1 \\ 0 \end{smallmatrix} | \boldsymbol{\xi})$ exists and is actually

$$\mathbf{p}\left(\begin{matrix} j_1 \\ 0 \end{matrix}\middle|\boldsymbol{\xi}\right) = 1 - \sum_{\alpha=1}^{s} \mathbf{p}\left(\begin{matrix} j_1 \\ \alpha \end{matrix}\middle|\boldsymbol{\xi}\right). \tag{4.14.24}$$

In fact, in general, if E is visited with well defined frequency ν, its complement is visited with frequency equal to $(1 - \nu)$. If $k = 2$, by the same arguments, we deduce that for $\alpha > 0$,

$$\mathbf{p}\left(\begin{matrix} j_1 j_2 \\ 0\ \alpha \end{matrix}\middle|\boldsymbol{\xi}\right) = \mathbf{p}\left(\begin{matrix} j_2 \\ \alpha \end{matrix}\middle|\boldsymbol{\xi}\right) - \sum_{\alpha'=1}^{s} \mathbf{p}\left(\begin{matrix} j_1 j_2 \\ \alpha' \alpha \end{matrix}\middle|\boldsymbol{\xi}\right). \tag{4.14.25}$$

Hence the frequency

$$\mathbf{p}\left(\begin{matrix} j_1 j_2 \\ 0\ 0 \end{matrix}\middle|\boldsymbol{\xi}\right) = \mathbf{p}\left(\begin{matrix} j_1 \\ 0 \end{matrix}\middle|\boldsymbol{\xi}\right) - \sum_{\alpha=1}^{s} \mathbf{p}\left(\begin{matrix} j_1 j_2 \\ 0\ \alpha \end{matrix}\middle|\boldsymbol{\xi}\right) \tag{4.14.26}$$

exists for the same reasons, etc., inductively.

Finally, (iv) follows, by Proposition 31, immediately when $\alpha_1 \neq 0$, $\alpha_2 \neq 0, \ldots, \alpha_k \neq 0$, $\beta_1 \neq 0, \ldots, \beta_h \neq 0$, because $E\binom{j_1 \cdots j_k}{\alpha_1 \ldots \alpha_k}$ and $E\binom{i_1 \cdots i_h}{\beta_1 \ldots \beta_h}$ are analytically regular and Eq. (4.14.23) is just a transcription in other symbols of Eq. (4.14.6). However, it is clear that the general case when some of the α_i's or β_i's may be zero is treated in the same way as that used to show the existence of the frequencies of visit, see Eqs. (4.14.24)–(4.14.26). mbe

It is useful to reinterpret Proposition 33 as follows.

Given $\boldsymbol{\xi} \in W$, consider the $(\mathcal{E}, \mathbf{t})$ history of $\boldsymbol{\xi}$, see Definition 18, §4.14, p. 338. It is the sequence of $\mathbf{a} = (a_0, a_1, \ldots), a_i = 0, 1, \ldots, s$ such that

$$S_{it_1}(\boldsymbol{\xi}) \in E_{a_i}, \qquad i = 0, 1, \ldots . \tag{4.14.27}$$

The frequencies of Eq. (4.14.19) can be "computed" starting from the history $\mathbf{a}$ as

$$\mathbf{p}\left(\begin{matrix} j_1 \cdots j_k \\ \alpha_1 \ldots \alpha_k \end{matrix}\middle|\boldsymbol{\xi}\right) = \lim_{N\to\infty} \frac{1}{N} n_N\left(\begin{matrix} j_1 \cdots j_k \\ \alpha_1 \ldots \alpha_k \end{matrix}\middle|\mathbf{a}\right), \tag{4.14.28}$$

where $n_N(\ldots)$ is the number of values of h, integer and smaller than N, such that

$$S_{(h+j_1)t_1}(\boldsymbol{\xi}) \in E_{\alpha_1}, \ldots, S_{(h+j_k)t_1}(\boldsymbol{\xi}) \in E_{\alpha_k} \tag{4.14.29}$$

[see Eqs. (4.14.15) and (4.14.19)], i.e., it is the number of times when

$$a_{h+j_1} = \alpha_1, \ldots, a_{h+j_k} = \alpha_k \tag{4.14.30}$$

occur simultaneously, with h integer in $[0, N)$.

In other words, Eq. (4.14.28) says that $\mathbf{p}\binom{j_1 \cdots j_k}{\alpha_1 \ldots \alpha_k}\,|\,\xi)$ is the frequency of appearance of the "string $\alpha_1, \ldots, \alpha_k$ at sites following each other at successive distances $j_2 - j_1, j_3 - j_2, \ldots, j_k - j_{k-1}$" in the history $\mathbf{a}$ of ξ.

It is then natural to state the following general definition.

19 Definition. Let $\mathbf{a} = (a_i)_{i=0}^{\infty}$ be a sequence, $a_i = 0, 1, \ldots, s$, $\forall i \in \mathbf{Z}_+$. Given $k > 0$, $0 \leqslant j_1 < j_2 < \cdots < j_k$ integers and $\alpha_1, \ldots, \alpha_k$ in $\{0, \ldots, s\}$ we say that a "string homologous to $\binom{j_1 \cdots j_k}{\alpha_1 \ldots \alpha_k}$" is "realized in $\mathbf{a}$ at the hth site" if

$$a_{h+j_1} = \alpha_1, a_{h+j_2} = \alpha_2, \ldots, a_{h+jk} = \alpha_k . \tag{4.14.31}$$

The frequency of realization in $\mathbf{a}$ of strings homologous to $\binom{j_1 \cdots j_k}{\alpha_1 \ldots \alpha_k}$ will be defined in terms of the quantity

$$\mathbf{p}_N\left(\begin{matrix} j_1 \cdots j_k \\ \alpha_1 \ldots \alpha_k \end{matrix}\middle|\,\mathbf{a}\right) = N^{-1}\left\{\begin{matrix}\text{number of times in which a string} \\ \text{homologous to } \begin{pmatrix} j_1 \cdots j_k \\ \alpha_1 \ldots \alpha_k \end{pmatrix} \text{ is realized in} \\ \mathbf{a} \text{ at sites } h \text{ between 0 and } N\end{matrix}\right\} \tag{4.14.32}$$

We shall set

$$\mathbf{p}\left(\begin{matrix} j_1 \cdots j_k \\ \alpha_1 \ldots \alpha_k \end{matrix}\middle|\,\mathbf{a}\right) = \lim_{N\to\infty} \mathbf{p}_N\left(\begin{matrix} j_1 \cdots j_k \\ \alpha_1 \ldots \alpha_k \end{matrix}\middle|\,\mathbf{a}\right) \tag{4.14.33}$$

whenever the limit exists.

We shall say that a sequence $\mathbf{a}$ is "ergodic" if:

(i) it has well-defined frequencies of appearance for all the strings of symbols, i.e., the limits (4.14.33) exist for all choices of the indices;

(ii) there are at least two distinct symbols α, β occurring with positive frequency in $\mathbf{a}$:

$$\mathbf{p}\left(\begin{matrix} 0 \\ \alpha \end{matrix}\middle|\,a\right) > 0, \qquad \mathbf{p}\left(\begin{matrix} 0 \\ \beta \end{matrix}\middle|\,\mathbf{a}\right) > 0; \tag{4.14.34}$$

(iii) for all choices of indices,

$$\begin{aligned} &\lim_{N\to\infty} \frac{1}{N} \sum_{l=0}^{N} \mathbf{p}\left(\begin{matrix} j_1 \cdots j_k i_1 + l \ldots i_h + l \\ \alpha_1 \ldots \alpha_k \;\; \beta_1 \ldots \quad \beta_h \end{matrix}\middle|\,\mathbf{a}\right) \\ &= \mathbf{p}\left(\begin{matrix} j_1 \cdots j_k \\ \alpha_1 \ldots \alpha_k \end{matrix}\middle|\,\mathbf{a}\right)\mathbf{p}\left(\begin{matrix} i_j \cdots i_h \\ \beta_1 \ldots \beta_h \end{matrix}\middle|\,\mathbf{a}\right). \end{aligned} \tag{4.14.35}$$

As $k, j_1, \ldots, j_k, \alpha_1, \ldots, \alpha_k$ vary, the family of numbers (4.14.33) will be called the "distribution of $\mathbf{a}$".

If $\mathbf{a}$ only verifies (i) [or (i) and (ii)], it will be called a "sequence with well-defined frequencies" (respectively, a "sequence with nontrivial frequencies") of the occurrence of the symbols.

Finally, we shall say that an ergodic sequence is "mixing" if for all the choices of indices,

$$\lim_{l\to\infty} \mathbf{p}\left(\begin{matrix} j_1 \dots j_k i_l + l, \dots, i_h + l \\ \alpha_1 \dots \alpha_i \ \beta_1 \dots \qquad \beta_h \end{matrix} \middle| \mathbf{a} \right) = \mathbf{p}\left(\begin{matrix} j_1 \dots j_k \\ \alpha_1 \dots \alpha_k \end{matrix} \middle| \mathbf{a} \right) \mathbf{p}\left(\begin{matrix} i_1 \dots i_h \\ \beta_1 \dots \beta_h \end{matrix} \middle| \mathbf{a} \right) \tag{4.14.36}$$

which is obviously stronger than Eq. (4.14.35).

Observations.

(1) From the definition given by Eq. (4.14.33) of the distribution of $\mathbf{a}$ as a family of frequencies of certain events, it immediately follows that such numbers verify Eqs. (4.14.20), (4.14.21), and (4.14.22) with $\mathbf{a}$ replacing $\boldsymbol{\xi}$.

(2) Using the language of probability theory (see §2.23), we can say that to any sequence $\mathbf{a}$ with well-defined frequencies of occurrence of the symbols it is possible to associate a family $(\mathcal{E}_k, \mathbf{p}_k)_{k=1}^{\infty}$ of probability distributions as follows. $\mathcal{E}_k$ will be the set of $(s+1)^k$ events, which we can denote $\boldsymbol{\alpha} = (\alpha_0, \dots, \alpha_{k-1}), \alpha_i = 0, 1, \dots, s$, whose probability is $\mathbf{p}(\begin{smallmatrix} 0 \dots & k-1 \\ \alpha_1 \dots & \alpha_{k-1} \end{smallmatrix}) | \mathbf{a})$. By Eq. (4.14.33), this probability coincides, by definition, with the frequency of occurrence in $\mathbf{a}$ of strings homologous to $(\begin{smallmatrix} 0 \dots & k-1 \\ \alpha_0 \dots & \alpha_{k-1} \end{smallmatrix})$.

For this reason, the sequence $(\mathcal{E}_k, \mathbf{p}_k)_{k=1}$ is also called the "probability distribution of the symbols of $\mathbf{a}$" and $\mathbf{p}(\begin{smallmatrix} 0 \dots & k-1 \\ \alpha_0 \dots & \alpha_{k-1} \end{smallmatrix} | \mathbf{a})$ is called the "probability of the string $\boldsymbol{\alpha} = (\alpha_0, \dots, \alpha_{k-1})$ in $\mathbf{a}$".

Proposition 33 can be reinterpreted in terms of the above definition as follows.

34 Proposition. *By the assumptions of the preceding proposition, denote for $\boldsymbol{\xi} \in W$ the $(\mathcal{E}, \mathbf{t})$ history of $\boldsymbol{\xi}$ as $\mathbf{a}(\boldsymbol{\xi})$. Then, if Eq. (4.14.34) holds, $\mathbf{a}(\boldsymbol{\xi})$ is an ergodic nonmixing sequence.*

Observation. The only statement not already contained in Proposition 33 is the one concerning mixing.

PROOF. By the assumed analytic integrability of the system, we can imagine that $\mathbf{a} = \mathbf{a}(\boldsymbol{\xi})$ is the $(\mathcal{E}, \mathbf{t})$ history of a point $\boldsymbol{\varphi} \in \mathbf{T}^l$ with respect to an analytically regular partition $\mathcal{E} = (E_0, \dots, E_p)$ of $\mathbf{T}^l$ and to the transformations $(S_t)_{t\in\mathbf{R}}$ of $\mathbf{T}^l$ given by

$$S_t\boldsymbol{\varphi} = \boldsymbol{\varphi} + \boldsymbol{\omega} t \bmod 2\pi. \tag{4.14.37}$$

For simplicity, we shall only deal with the case when $\omega_1, \dots, \omega_l$, $\sigma = 2\pi/t_1$ are rationally independent and when it is also assumed that there are two sets E_α, E_β in $\mathcal{E}$, such that $\mathbf{p}(\begin{smallmatrix} 0 \\ \alpha \end{smallmatrix} | \mathbf{a}) > 0$, $\mathbf{p}(\begin{smallmatrix} 0 \\ \beta \end{smallmatrix} | \mathbf{a}) > 0$, having a diameter so small that there is a point $\varphi_0 \in E_\alpha$ at a distance from E_β greater than twice the diameter of E_β.

These are serious restrictions. However, the general case can be reduced to the above, as it will become apparent after having gone through the problems at the end of this section.

The rational independence assumption of $\omega_1, \ldots, \omega_l, \sigma$ and the analytic regularity of $\mathfrak{E}$ imply that

$$\mathbf{p}\begin{pmatrix} 0 & j \\ \gamma & \gamma' \end{pmatrix}\bigg|\mathbf{a}\bigg) = \frac{1}{(2\pi)^l}\int_{E_\gamma \cap S_{-jt_1}(E_{\gamma'})} d\varphi, \qquad \forall \gamma, \gamma' \tag{4.14.38}$$

[see Proposition 30, p. 339, Eq. (4.14.5)].

If $\mathbf{a}(\xi)$ were mixing, by Eq. (4.14.36), one would also have

$$\lim_{j\to\infty} \mathbf{p}\left(\begin{matrix} 0 & j \\ \alpha & \beta \end{matrix}\bigg|\mathbf{a}\right) = \mathbf{p}\left(\begin{matrix} 0 \\ \alpha \end{matrix}\bigg|\mathbf{a}\right)\mathbf{p}\left(\begin{matrix} 0 \\ \beta \end{matrix}\bigg|\mathbf{a}\right) > 0. \tag{4.14.39}$$

However this would mean that for j large enough, it should be

$$\mathbf{p}\left(\begin{matrix} 0 & j \\ \alpha & \beta \end{matrix}\bigg|\mathbf{a}\right) \equiv \int_{E_\alpha \cap S_{-jt_1}(E_\beta)} \frac{d\varphi}{(2\pi)^l} > 0. \tag{4.14.40}$$

Hence, $E_\alpha \cap S_{-jt_1}E_\beta \neq \emptyset$ eventually. But, by the rational independence of $\omega_1, \ldots, \omega_l, \sigma$, the trajectory $\tilde{\varphi} - jt_1\omega$, $j \geqslant j_0$, of any point $\tilde{\varphi}$ chosen in E_β is dense in $\mathbf{T}^l$, $\forall j_0$ (see §4.2). Therefore, given $\varphi_0 \in E_\alpha$, there must exist infinitely many values of $j \geqslant 0$ such that the distance of $\tilde{\varphi} - jt_1\omega$ from φ_0 is less than the diameter of E_β. It is clear that for such values of j, it must be that $E_\alpha \cap S_{-jt_1}E_\beta = \emptyset$ since these torus rotations do not deform the sets but they only translate them, and φ_0 is chosen so that $d(\varphi_0, E_\beta) > \{$twice the diameter of $E_\beta\}$.

mbe

Exercises and Problems for §4.14

Solve the following connected sequence of problems for $l = 2$ first, drawing graphical representations of the various maps and transformations. The notations are those of §4.14. The aim is to solve problem 8 below.

1. Let $\omega_1, \ldots, \omega_l$ be rationally dependent and not all zero. Show that there exists $\bar{l} < l$ rationally independent numbers $\hat{\omega}_1, \ldots, \hat{\omega}_{\bar{l}}$ and an $l \times \bar{l}$ matrix J with integer coefficients and such that $\omega = J\hat{\omega}$, i.e., $\omega_j = \sum_{k=1}^{\bar{l}} J_{jk}\hat{\omega}_k$, $j = 1, \ldots, l$.

2. In $\mathbf{R}^l$ consider the plane $\pi_l = J\mathbf{R}^{\bar{l}} = \{\mathbf{x} \mid x_j = \sum_{k=1}^{\bar{l}} J_{jk}y_k, \mathbf{y} \in \mathbf{R}^{\bar{l}}\}$ and the plane $\pi_l^\perp$ orthogonal to it. Show that there exists an $l \times (l - \bar{l})$ matrix $J^\perp$ with integer coefficients such that $\pi_l^\perp = J^\perp \mathbf{R}^{l-\bar{l}}$.

3. Define the map $(J \times J^\perp)_T$ of $\mathbf{T}^{\bar{l}} \times \mathbf{T}^{l-\bar{l}}$ onto $\mathbf{T}^l$, $\forall (\boldsymbol{\theta}, \boldsymbol{\nu}) \in \mathbf{T}^{\bar{l}} \times \mathbf{T}^{l-\bar{l}}$, as:

$$(J \times J^\perp)_T(\boldsymbol{\theta}, \boldsymbol{\nu}) = (J\boldsymbol{\theta} + J^\perp\boldsymbol{\nu}) \bmod 2\pi.$$

If one defines

$$\hat{E} = (J \times J^\perp)_T^{-1} E$$

for $E \subset \mathbf{T}^l$, and if E is analytically regular in $\mathbf{T}^l$, show that $\hat{E}$ is such in $\mathbf{T}^{\bar{l}} \times \mathbf{T}^{l-\bar{l}} \equiv \mathbf{T}^l$. (Hint: Note that $(J \times J^{\perp})_T$, regarded as a matrix denoted $(J \times J^{\perp})$, linearly maps $\mathbf{R}^l$ onto $\mathbf{R}^l$; hence, $\det(J \times J^{\perp}) \neq 0$. Hence, $(J \times J^{\perp})^{-1}E$ is analytically regular in $\mathbf{R}^l$ and $\hat{E}$ is obtained by considering $(J \times J^{\perp})^{-1}E$, after reducing mod 2π, the coordinates of its points, as a subset of the torus $\mathbf{T}^{\bar{l}} \times \mathbf{T}^{l-\bar{l}} \equiv \mathbf{T}^l$.)

4. If $\varphi_0 \in \mathbf{T}^l$ and $\varphi_0 = (J \times J^{\perp})_T(\boldsymbol{\theta}_0, \boldsymbol{\nu}_0)$, show that the frequency of visits to E of the trajectory of φ_0 under the transformation $\varphi_0 \to \varphi_0 + t\omega$ coincides with the frequency of visit to $\hat{E}$ of the trajectory of $(\boldsymbol{\theta}_0, \boldsymbol{\nu}_0)$ under the transformation $(\boldsymbol{\theta}_0, \boldsymbol{\nu}_0) \to (\boldsymbol{\theta}_0 + \hat{\omega}t, \boldsymbol{\nu}_0)$ (Hint: Note that $\varphi_0 + \omega t = (J \times J^{\perp})_T(\boldsymbol{\theta}_0 + \hat{\omega}t, \boldsymbol{\nu}_0)$ by the construction of J.)

5. Let $\hat{E}(\boldsymbol{\nu}_0) = \hat{E} \cap \{(\boldsymbol{\theta}, \boldsymbol{\nu}) \mid (\boldsymbol{\theta}, \boldsymbol{\nu}) \in \mathbf{T}^{\bar{l}} \times \mathbf{T}^{l-\bar{l}},\ \boldsymbol{\nu} = \boldsymbol{\nu}_0\}$, then the frequency of visits to E of the trajectory of φ_0 for the transformations $\varphi_0 \to \varphi_0 + \omega t$ coincides with the frequency of visits to $\hat{\hat{E}}(\boldsymbol{\nu}_0) = \{\boldsymbol{\theta} \mid \boldsymbol{\theta} \in \mathbf{T}^l,\ (\boldsymbol{\theta}, \boldsymbol{\nu}_0) \in \hat{E}(\boldsymbol{\nu}_0)\} \subset \mathbf{T}^{\bar{l}}$ by the trajectory of $\boldsymbol{\theta}_0$ under the transformation $\boldsymbol{\theta}_0 \to \boldsymbol{\theta}_0 + \hat{\omega}t$. Furthermore, if E is analytically regular in $\mathbf{T}^l$, then $\hat{\hat{E}}(\boldsymbol{\nu}_0)$ is such in $\mathbf{T}^{\bar{l}}$ (Hint: Interpret $\hat{\hat{E}}(\boldsymbol{\nu}_0)$ as the intersection of $\hat{E}(\boldsymbol{\nu}_0)$ with a "plane".)

6. If $\omega_1, \ldots, \omega_l$ are rationally independent but $\omega_1, \ldots, \omega_l$, $\sigma = 2\pi/t_1$ are not rationally independent, there are l integers $m_1, \ldots, m_l$ and $q > 0$, integer too, such that

$$\sigma = (\mathbf{m} \cdot \omega)q^{-1}.$$

The problem of the determination of the frequency of visits to $E \subset \mathbf{T}^l$ by the trajectory of $\varphi \in \mathbf{T}^l$ under the map $\varphi \to \varphi + \omega 2\pi j/\sigma$, $j = 0, 1, \ldots,$ is equivalent (via a suitable change of coordinates) to the analogous problem when the relation between σ and ω is simply $\sigma = (m_1/q)\omega_1$. (Hint: The transformation is analogous to that described in Problems 2 and 3 above. It is the transformation associated, in the same way as above, to the matrix J of the transformation

$$\omega_1 = \omega_1' - \sum_{i=2}^{l} m_i \omega_i',$$
$$\omega_j = \omega_j' m_1, \qquad j = 2, \ldots, l.)$$

7. Consider the trajectory of $\varphi_0 \in \mathbf{T}^l$ under the transformations $\varphi_0 \to \varphi_0 + (2\pi/\sigma)\omega j$ with $\sigma = (m/q)\omega_1, m, q$ integers, and assume that $\omega_1, \ldots, \omega_l$ are rationally independent.

Think of $\mathbf{T}^l$ as $\mathbf{T}^1 \times \mathbf{T}^{l-1}$ and, if $(\varphi, \psi) \in \mathbf{T}^1 \times \mathbf{T}^{l-1}$, show that the map under analysis can be written as $(\varphi, \psi) \to (\varphi + (2\pi q/m)j, \psi + \omega' j)$, where $\omega' = (\omega_2', \ldots, \omega_l')$ are $l-1$ rationally independent numbers which together with 2π form a set of l rationally independent numbers.

If $E \subset \mathbf{T}^l$ is analytically regular, show that the frequency of visit to E exists and depends only on φ. (Hint: Note that

$$\frac{1}{Mm} \sum_{j=0}^{Mm-1} \chi_E\left(\varphi_0 + \frac{2\pi q}{m} j, \psi_0 + \omega' j\right)$$
$$= \frac{1}{Mm} \sum_{k=0}^{m-1} \sum_{p=0}^{M-1} \chi_E\left(\varphi_0 + \frac{2\pi q}{m}(k + mp), (\psi_0 + k\omega') + mp\omega'\right)$$
$$= \frac{1}{m} \sum_{k=0}^{m-1} \left(\frac{1}{M} \sum_{p=0}^{M-1} \chi_E\left(\varphi_0 + \frac{2\pi q}{m} k, (\psi_0 + k\omega') + mp\omega'\right)\right),$$

and, letting $\varphi_k = \varphi_0 + (2\pi q/m)k, \psi_k = \psi_0 + k\omega'$, this can be rewritten

$$\frac{1}{m}\sum_{k=0}^{m-1}\left(\frac{1}{M}\sum_{p=0}^{M}\chi_E(\varphi_k, \psi_k + mp\omega')\right) = \frac{1}{m}\sum_{k=0}^{m-1}\left(\frac{1}{M}\sum_{p=0}^{M}\chi_{E_k(\varphi_0)}(\psi_k + mp\omega')\right),$$

where $E_k(\varphi_0) = \{\psi \mid \psi \in \mathbf{T}^{l-1}, (\varphi_0 + (2\pi q/m)k, \psi) \in E\}$ is still analytically regular for $k = 0, \ldots, m-1$. Hence, the frequence of visit to E exists because $m\omega'$ has rationally independent components and it is given by

$$\frac{1}{m}\sum_{k=0}^{m-1}\int_{E_k(\varphi_0)}\frac{d\varphi'}{(2\pi)^{l-1}}\,.\Bigg)$$

8. On the basis of the above problems, deduce the proofs of Propositions 30 and 31 in the general case from their validity in the rationally independent cases.

§4.15. Analytic Integrability Criteria. Complexity of the Motions and Entropy

Summarizing what has been discussed in the preceding sections, we have obtained the following criteria of nonanalytic integrability, on a phase-space subset W, for an analytic time-independent Hamiltonian system:

(i) if in W there is one ξ whose $(\mathfrak{S}, \mathbf{t})$ history on an analytically regular partition $\mathfrak{S}$ of W contains some strings without well-defined frequency of occurrence;
(ii) if in W there is one ξ whose trajectory $\mathfrak{T}$ has a closure $\bar{\mathfrak{T}}$ that cannot be mapped bicontinuously on a torus $\mathbf{T}^s, s \leqslant l, l$ being the number of degrees of freedom;
(iii) if in W there is one ξ whose $(\mathfrak{S}, \mathbf{t})$ history on an analytically regular partition $\mathfrak{S}$ of W has nontrivial frequency distributions but is not ergodic;
(iv) if in W there is one ξ whose $(\mathfrak{S}, \mathbf{t})$ history on an analytically regular partition $\mathfrak{S}$ of W is "too ergodic", i.e., mixing.

We conclude this review of nonintegrability criteria by examining another very interesting property of the anaytically integrable systems: we shall show that the motions of such systems have a "small complexity". This leads to another nonintegrability criterion, see (v), p. 356.

To obtain such a result, we must first try to give a quantitative meaning to the notion of "complexity" of the motions associated with points moving on a regular (analytic) surface under the action of a family ("semigroup") $(S_t)_{t \in \mathbf{R}_+}$ of C^∞ (analytic) transformations.

A natural way to evaluate the complexity of a motion is to count the number of different strings of history appearing in the $(\mathfrak{S}, \mathbf{t})$ history of the motion on an analytically regular partition.

20 Definition. Let $\mathbf{a}$ be a sequence $\mathbf{a} = (a_i)_{i=0}^{\infty}, a_i \in \{0, \ldots, s\}$. Assume that $\mathbf{a}$ has well-defined frequencies of symbol appearances (see Definition 19, §4.14, p. 346).

The "number of strings of symbols of length k appearing in $\mathbf{a}$" is defined as

$$N_{\text{abs}}(\mathbf{a}, k) = \Big\{ \text{number of choices of } (\alpha_0, \ldots, \alpha_{k-1}) \in \{0, \ldots, s\}^k \text{ such that } \mathbf{p}\left(\begin{matrix} 0 \ldots & k-1 \\ \alpha_0 \ldots & \alpha_{k-1} \end{matrix} \middle| \mathbf{a}\right) > 0 \Big\}, \tag{4.15.1}$$

where, we recall, $\mathbf{p}(\begin{smallmatrix} 0 \ldots & k-1 \\ \alpha_0 \ldots & \alpha_{k-1} \end{smallmatrix} | \mathbf{a})$ denotes the frequency of appearance in $\mathbf{a}$ of a string homologous to $(\begin{smallmatrix} 0 \ldots & k-1 \\ \alpha_0 \ldots & \alpha_{k-1} \end{smallmatrix})$.

Clearly $N_{\text{abs}}(\mathbf{a}, k) \leqslant (s+1)^k$. We shall set[18]

$$S_{\text{abs}}(\mathbf{a}) = \lim_{k \to +\infty} k^{-1} \log N_{\text{abs}}(\mathbf{a}, k) \tag{4.15.2}$$

which we call the "absolute complexity" of the sequence $\mathbf{a}$.

Observations.

(1) The number in Eq. (4.15.2) can give an idea of how complex the sequence $\mathbf{a}$ might be. However, it is clear that $S_{\text{abs}}(\mathbf{a})$ is a rather rough measure of the complexity of $\mathbf{a}$: in its evaluation, in fact, one puts on the same footing strings occurring in $\mathbf{a}$ with a frequency of occurrence much smaller than that of other strings or, by the Observation (2), to Definition 19, p. 346, with a "probability" much smaller than that of others.

(2) The existence of the limit of Eq. (4.15.8) is easy to prove and very instructive (see Problem 21 below).

The following more sophisticated definition takes into account the possibility that some strings may be present in $\mathbf{a}$ with extremely small probability and gives them less importance.

21 Definition. Let $\mathbf{a} = (a_i)_{i \in \mathbf{Z}_+}, a_i = 0, 1, \ldots, p$, be a sequence with well-defined frequencies of symbol occurrence as in Definition 20.

Given $\epsilon > 0$, consider all the possible subsets $\mathcal{C}_\epsilon$ of the set of the k-tuples $\alpha_0, \ldots, \alpha_{k-1}, \alpha_i = 0, \ldots, s$, such that

$$\sum_{(\alpha_0, \ldots, \alpha_{k-1})} \mathbf{p}\left(\begin{matrix} 0 \ldots & k-1 \\ \alpha_0 \ldots & \alpha_{k-1} \end{matrix} \middle| \mathbf{a}\right) < \epsilon. \tag{4.15.3}$$

These are the sets $\mathcal{C}_\epsilon$ of "k strings" (strings of length k) whose total frequency of occurrence is smaller than $\epsilon > 0$.

[18] The limit always exists (see Problem 21, p. 362).

Let

$$N(\mathbf{a},k,\epsilon) = \{\text{infimum, on the choices of } \mathcal{C}_\epsilon \text{, of the number of } k\text{-tuples outside } \mathcal{C}_\epsilon\}, \tag{4.15.4}$$

and let

$$S(\mathbf{a},\epsilon) = \limsup_{k\to+\infty} k^{-1}\log N(a,k,\epsilon), \tag{4.15.5}$$

$$S(\mathbf{a}) = \lim_{\epsilon\to 0} S(\mathbf{a},\epsilon). \tag{4.15.6}$$

This last quantity will be called the "entropy" of **a** and it can also be regarded as a measure of the complexity of **a**.

Observations.

(1) This is a measure of complexity more interesting than Eq. (4.15.2). Through Eq. (4.15.4) and the two limits in Eqs. (4.15.5) and (4.15.6), in some way, one discards from the number of strings of **a** those which appear with a very small frequency (see, also, Proposition 37 to follow).

(2) Obviously,

$$0 \leqslant S(\mathbf{a}) \leqslant S_{\text{abs}}(\mathbf{a}) \leqslant \log(s+1), \tag{4.15.7}$$

and one can note that the two numbers given in Eqs. (4.15.2) and (4.15.6) can be thought of as obtained by permuting the following two limits:

$$S_{\text{abs}}(\mathbf{a}) = \lim_{k\to\infty}\lim_{\epsilon\to 0}\frac{1}{k}\log N(\mathbf{a},k,\epsilon), \tag{4.15.8}$$

$$S(\mathbf{a}) = \lim_{\epsilon\to 0}\lim_{k\to\infty}\frac{1}{k}\log N(\mathbf{a},k,\epsilon) \tag{4.15.9}$$

if all the above limits exist.

(3) The term entropy given to Eq. (4.15.6) is due to the analogy of this definition with the Boltzmann's glorious idea on the proportionality between the entropy of the state of a system, in the thermodynamic sense of the word, and the number of ways of realizing the same macroscopic state by equivalent microscopic states.

This analogy is evident if one is not biased by the various limit steps taken in Eqs. (4.15.5), (4.15.6), (4.15.8), and (4.15.9), and at first one ignores them.

The following proposition holds.

35 Proposition. *Consider a Hamiltonian system analytically integrable on the phase-space subset W. Let $\mathfrak{S} = (E_0, \ldots, E_s)$ be an analytically regular partition of W. Let $t_1 > 0$, $\mathbf{t} = (it_1)_{i=0}^{\infty}$.*
For all $\xi \in W$, denote $\mathbf{a}(\xi)$ the $(\mathfrak{S}, t)$ history of ξ. Then

$$S(\mathbf{a}(\xi)) = 0, \qquad \forall \xi \in W. \tag{4.15.10}$$

Observation. As already seen in the propositions of §4.14, the statement of this proposition is an immediate consequence of an analogous proposition concerning the torus rotations. In this case, the proposition is the following.

36 Proposition. *Let $\omega \in \mathbf{R}^l$ and let $(S_t)_{t\in\mathbf{R}}$ be the quasi-periodic flow on $\mathbf{T}^l$ with pulsations ω (i.e., $S_t\varphi = \varphi + \omega t$). Consider the transformations $(S_{jt_1})_{j=0}^{\infty}$, $t_1 > 0$, and let $\mathcal{E} = (E_0, \ldots, E_s)$ be an analytically regular partition $\mathbf{T}^l$ into $(s+1)$ sets. The $(\mathcal{E}, \mathbf{t})$ history of $\varphi \in \mathbf{T}^l$, denoted by $\mathbf{a}(\varphi)$, is such that*

$$S(\mathbf{a}(\varphi)) = 0, \qquad \forall \varphi \in \mathbf{T}^l. \tag{4.15.11}$$

Observations.

(1) The argument presented in the proof below essentially gives the proof of a more general theorem of great importance in the theory of entropy ("Kousnirenko's theorem").

(2) Actually, one could prove a stronger result, namely,

$$S_{\text{abs}}(a(\varphi)) = 0, \qquad \forall \varphi \in \mathbf{T}^l. \tag{4.15.12}$$

However, in the course of the proof, we show Eq. (4.15.12) only in the $l = 1$ case. The argument could be adapted to prove Eq. (4.15.12) in general. However, for $l > 1$, we prefer to explain an alternative proof of the weaker result of Eq. (4.15.11) since the method of this proof is in itself interesting and, as mentioned in Observation (1), contains the germs of interesting extensions.

(3) Equations (4.15.10) and (4.15.11) have an interesting monotonicity property: if $\mathcal{E}'$ is a partition finer than $\mathcal{E}$ in the sense that every set in $\mathcal{E}$ can be thought of as a union of sets in $\mathcal{E}'$, then the absolute complexity (and the entropy) of $\mathbf{a}'(\varphi)$ is not smaller than that of $\mathbf{a}(\varphi)$. This reflects the intuitively clear fact that by increasing the precision of the measurements, the motion can only look more complicated since more of its features may become manifest.

PROOF. As mentioned in Observation (2), we shall consider separately the cases $l = 1$ and $l > 1$. We only treat the case when $\omega_1, \ldots, \omega_l, 2\pi/t_1$ are rationally independent. The problems of §4.14 show that the general case can be reduced to this special one.

Case $l = 1$. To fix the ideas, suppose $s = 1$ and $E_0 = (\lambda, 2\pi), E_1 = [0, \lambda]$, $\lambda \in (0, 2\pi)$. Consider the images of the points 0 and λ for the maps $\varphi \to \varphi + jt_1\omega_1, j = 0, \ldots, k-1$. There are at most $2(k+1)$ points (and at least 2) dividing the interval $[0, 2\pi]$ in $2(k+1)$, at most, consecutive intervals $J_1, J_2, \ldots$. It is clear that all the points internal to some such interval have the same $(\mathcal{E}, \mathbf{t})$ history in the first k sites of their history.

To the $2(k+1)$, at most, histories of the points internal to the above intervals, we can add the $2(k+1)$ histories, at most, of their extreme points.

We thus obtain all the possible strings of the history with length k that can appear in the $(\mathscr{E},\mathbf{t})$ history of a point $\varphi \in \mathbf{T}^1$. Hence,

$$N_{\text{abs}}(\mathbf{a}(\varphi),k) \leqslant 4(k+1) \tag{4.15.13}$$

and Eq. (4.15.12) follows from the definition given by Eq. (4.15.2).

Case $l > 1$. The entire proof will be based on the possibility of estimating the volume $|E|$ of a set E in terms of the area $|\partial E|$ of its boundary ∂E. If $E \subset \mathbf{R}^l$ is a bounded set, its volume $|E|$ cannot exceed the volume of the sphere with surface area equal to the surface area $|\partial E|$ of E ("isoperimetric inequality"). So an inequality of the type

$$|E| < C_l |\partial E|^{l/(l-1)} \tag{4.15.14}$$

holds C_l being a suitable E-independent constant. However, on $\mathbf{T}^l$, such an inequality is false for sets which "wrap around $\mathbf{T}^l$" (e.g., if $E = \mathbf{T}^l$, $|E| = (2\pi)^l$, $|\partial E| = 0$ as $\partial E = \varnothing$); but, of course, it is still true for sets with small enough diameter.

To apply isoperimetric inequalities in $\mathbf{T}^l$, it is therefore useful to think of $\mathbf{T}^l$ as the union of many small sets. We shall regard $\mathbf{T}^l$ as a union of 2^l cubes with side π parameterized by an index $\boldsymbol{\sigma}$:

$$C_{\boldsymbol{\sigma}} = \{\boldsymbol{\varphi} \mid \boldsymbol{\varphi} \in \mathbf{T}^l, \pi\sigma_i \leqslant \varphi_i \leqslant \pi(\sigma_i + 1),\ i = 1, \dots, l\}, \tag{4.15.15}$$

where each σ_i takes the value 0 or 1. We call Σ the set of the 2^l $\boldsymbol{\sigma}$'s.

Given $(\alpha_1, \dots, \alpha_{k-1}) \in \{0, \dots, s\}^k$ and $(\sigma_0, \dots, \sigma_{k-1}) \in \Sigma^k$, we can consider the sets

$$\begin{aligned} E\begin{pmatrix} 0 \dots k-1 \\ \alpha_1 \dots \alpha_{k-1} \end{pmatrix} &= E_{\alpha_0} \cap S_{-t_1} E_{\alpha_1} \cap \cdots \cap S_{-(k-1)t_1} E_{\alpha_{k-1}}, \\ B\begin{pmatrix} 0 \dots k-1 \\ \sigma_1 \dots \sigma_{k-1} \end{pmatrix} &= C_{\sigma_0} \cap S_{-t_1} C_{\sigma_1} \cap \cdots \cap S_{-(k-1)t_1} E_{\sigma_{k-1}}. \end{aligned} \tag{4.15.16}$$

Since the rotations of the torus are "rigid transformations", i.e., they do not change the form and volume of the sets that they transform, it will be possible to infer that the sum of the surfaces of the sets $E \cap B$, with E, B like Eq. (4.15.16) with the same value of k, is such that

$$\sum_{\substack{\alpha_0 \dots \alpha_{k-1} \\ \sigma_0 \dots \sigma_{k-1}}} \left| \partial\left(E\begin{pmatrix} 0 \dots k-1 \\ \alpha_0 \dots \alpha_{k-1} \end{pmatrix} \cap B\begin{pmatrix} 0 \dots k-1 \\ \sigma_0 \dots \sigma_{k-1} \end{pmatrix} \right) \right| \leqslant 2(k+1)L, \tag{4.15.17}$$

where $L = \sum_{j=0}^{s} |\partial E_j| + 2^l(2l\pi^{l-1})$. This simple relation follows from the geometric observation that

$$\begin{aligned} &\bigcup_{\substack{\alpha_0 \dots \alpha_{k-1} \\ \sigma_0 \dots \sigma_{k-1}}} \partial\left(E\begin{pmatrix} 0 \dots k-1 \\ \alpha_0 \dots \alpha_{k-1} \end{pmatrix} \cap B\begin{pmatrix} 0 \dots k-1 \\ \sigma_0 \dots \sigma_{k-1} \end{pmatrix} \right) \\ &\quad = \bigcup_{h=0}^{k-1} \left[S_{-ht_1}(\partial E_{\alpha_h}) \cup S_{-ht_1}(\partial C_{\sigma_h}) \right], \end{aligned} \tag{4.15.18}$$

and the right-hand-side points are counted twice in the left-hand side except for a subset of total area zero corresponding to the edges and corners of the sets

$$E\begin{pmatrix} 0 & \dots & k-1 \\ \alpha_0 & \dots & \alpha_{k-1} \end{pmatrix} \cap B\begin{pmatrix} 0 & \dots & k-1 \\ \sigma_0 & \dots & \sigma_{k-1} \end{pmatrix}.$$

We can now use Eq. (4.15.14) to bound

$$\begin{aligned}
\mathbf{p}\left(\begin{matrix} 0 \dots k-1 \\ \alpha_0 \dots \alpha_{k-1} \end{matrix}\middle|\mathbf{a}(\varphi)\right) &= (2\pi)^{-1}\int_{E\left(\begin{smallmatrix} 0 \dots k-1 \\ \alpha_0 \dots \alpha_{k-1}\end{smallmatrix}\right)} d\varphi = (2\pi)^{-1}\left|E\begin{pmatrix} 0 \dots k-1 \\ \alpha_0 \dots \alpha_{k-1} \end{pmatrix}\right| \\
&= (2\pi)^{-l} \sum_{\sigma_0 \dots \sigma_{k-1}} \left|E\begin{pmatrix} 0 \dots k-1 \\ \alpha_0 \dots \alpha_{k-1} \end{pmatrix} \cap B\begin{pmatrix} 0 \dots k-1 \\ \sigma_0 \dots \sigma_{k-1} \end{pmatrix}\right| \\
&\leqslant (2\pi)^{-l} C_l \sum_{\sigma_0 \dots \sigma_{k-1}} \left|\partial\left(E\begin{pmatrix} 0 \dots k-1 \\ \alpha_0 \dots \alpha_{k-1} \end{pmatrix} \cap B\begin{pmatrix} 0 \dots k-1 \\ \sigma_0 \dots \sigma_{k-1} \end{pmatrix}\right)\right|^{l/l-1} \\
&\leqslant (2\pi)^{-l} C_l \left(\sum_{\sigma_0 \dots \sigma_{k-1}} \left|\partial\left(E\begin{pmatrix} 0 \dots k-1 \\ \alpha_0 \dots \alpha_{k-1} \end{pmatrix} \cap B\begin{pmatrix} 0 \dots k-1 \\ \sigma_0 \dots \sigma_{k-1} \end{pmatrix}\right)\right|\right)^{l/l-1},
\end{aligned} \tag{4.15.19}$$

having used the rational independence of $(\omega_1, \dots, \omega_l, 2\pi/t_1)$ in the first step [applying Proposition 30, Eq. (4.14.5)], and in the last step the inequality $(\alpha+\beta)^x \geqslant \alpha^x + \beta^x, \forall x \geqslant 1, \forall \alpha, \beta > 0$, has also been used. The isoperimetric inequality has been used in the intermediate step.

Equations (4.15.19) and (4.15.17) will now be used to estimate the total frequency of the strings of length k in $\mathbf{a}(\varphi)$ having "small probability" and, "precisely" such that, given $\eta > 0$,

$$\mathbf{p}\left(\begin{matrix} 0 \dots k-1 \\ \alpha_0 \dots \alpha_{k-1} \end{matrix}\middle|\mathbf{a}(\varphi)\right) \leqslant e^{-k\eta}. \tag{4.15.20}$$

Recalling the ideas involved in the proof of the Chebysčev inequality, Proposition 34, p. 119, we find

$$\begin{aligned}
&\sum_{\substack{\alpha_0, \dots, \alpha_{k-1} \\ \mathbf{p}\left(\begin{smallmatrix} 0 \dots k-1 \\ \alpha_0 \dots \alpha_{k-1}\end{smallmatrix}\middle|\mathbf{a}(\varphi)\right) \leqslant e^{-\eta k}}} \mathbf{p}\left(\begin{matrix} 0 \dots k-1 \\ \alpha_0 \dots \alpha_{k-1} \end{matrix}\middle|\mathbf{a}(\varphi)\right) \\
&\leqslant \sum_{\alpha_0, \dots, \alpha_{k-1}} \left(\frac{e^{-\eta k}}{\mathbf{p}\left(\begin{smallmatrix} 0 \dots k-1 \\ \alpha_0 \dots \alpha_{k-1}\end{smallmatrix}\middle|\mathbf{a}(\varphi)\right)}\right)^{\gamma} \mathbf{p}\left(\begin{matrix} 0 \dots k-1 \\ \alpha_0 \dots \alpha_{k-1} \end{matrix}\middle|\mathbf{a}(\varphi)\right) \qquad (4.15.21) \\
&= \sum_{\alpha_0, \dots, \alpha_{k-1}} e^{-\gamma\eta k} \mathbf{p}\left(\begin{matrix} 0 \dots k-1 \\ \alpha_0 \dots \alpha_{k-1} \end{matrix}\middle|\mathbf{a}(\varphi)\right)^{1-\gamma}
\end{aligned}$$

no matter how $\gamma > 0$ is chosen.

Then let $\gamma = 1/l$, i.e., such that $(1-\gamma)l/(l-1) = 1$ and deduce from Eqs. (4.15.21), (4.15.19), and (4.15.17) that the total probability that Eq.

(4.15.20) holds is bounded by

$$e^{-\eta k/l}\left((2\pi)^{-l}C_l\right)^{1-\gamma}$$
$$\times \sum_{\alpha_0,\ldots,\alpha_{k-1}} \sum_{\sigma_0\ldots\sigma_{k-1}} \left|\partial\left(E\begin{pmatrix}0\ldots k-1\\ \alpha_0\ldots\alpha_{k-1}\end{pmatrix} \cap B\begin{pmatrix}0\ldots k-1\\ \sigma_0\ldots\sigma_{k-1}\end{pmatrix}\right)\right|$$
$$\leqslant \left((2\pi)^{1-\gamma}e^{-\eta k/l}2(k+1)L. \qquad (4.15.22)$$

Hence, given $\epsilon > 0$ and $\eta > 0$, Eq. (4.15.22) shows that if k is so large that the right-hand side of Eq. (4.15.20) is smaller than ϵ, we can find among the sets $\mathcal{C}_\epsilon$ appearing in Eq. (4.15.3) the set

$$\mathcal{C}_\epsilon(\eta) = \{\text{set of the } k\text{-tuples } \alpha_0\ldots\alpha_{k-1} \text{ verifying Eq. (4.15.20)}\}. \qquad (4.15.23)$$

Since [see Eq. (4.15.22) and Observation (1) to Definition 19, p. 346]

$$\sum_{\alpha_0\ldots\alpha_{k-1}} \mathbf{p}\left(\begin{matrix}0\ldots k-1\\ \alpha_0\ldots\alpha_{k-1}\end{matrix}\middle|\mathbf{a}(\varphi)\right) = 1,$$

it becomes clear that the complement of $\mathcal{C}_\epsilon(\eta)$ cannot contain more than $e^{\eta k}$ elements because it consists of sets with probability $\geqslant e^{-\eta k}$. One then finds that

$$N(\epsilon, \mathbf{a}(\varphi), k) \leqslant e^{\eta k} \qquad (4.15.24)$$

if k is large enough. Hence,

$$S(\mathbf{a}(\varphi), \epsilon) \leqslant \eta \qquad (4.15.25)$$

and Eq. (4.15.11) follows from the arbitrariness of η.

So far the analytic regularity of $\mathfrak{E}$ has been only used to deduce the first of Eqs. (4.15.19) which, as remarked elsewhere (see Observation (1), to Proposition 30, p. 339), follows simply from the Riemann measurability of $E_0, \ldots, E_s$. However, in the general case when $\omega_1, \ldots, \omega_l, 2\pi/t_1$ are not rationally independent, as assumed above, analytic regularity has to be used again to reduce the general case to the above-treated independent case. mbe

The above propositions provide a further nonintegrability criterion.

(v) If in W there is one ξ whose $(\mathfrak{E}, \mathbf{t})$ history on an analytically regular partition $\mathfrak{E}$ of W has positive entropy, then the system is not analytically integrable on W.

This criterion can be added to those listed at the beginning of §4.15, p. 350 and to the other criteria, also quite remarkable, that emerge from the problems at the end of this section, see problems 13–20.

We now quote, without proof, some results on entropy theory and nonintegrable systems showing that in fact the previously stated nonintegrability criteria (i), (iii), (iv), and (v) are not empty of content [(ii) has already been discussed in §4.13, Observation (3) Proposition 27, p. 332), i.e., the propositions below illustrate other properties of entropy (Proposition 37) or they show that there actually are systems whose nonintegrability could be decided on the basis of the above criteria (Proposition 38).

37 Proposition. *Let* $\mathbf{a} = (a_i)_{i \in Z_+}$, $a_i = 0, 1, \ldots, p-1$ *be an ergodic sequence.*

(i) *The entropy of* $\mathbf{a}$ *can be computed as*

$$S(\mathbf{a}) = \lim_{N\to\infty} -\frac{1}{N} \sum_{\alpha_0,\ldots,\alpha_{N-1}} \mathbf{p}\left(\begin{matrix} 0 \ldots N-1 \\ \alpha_0 \ldots \alpha_{N-1} \end{matrix}\middle|\boldsymbol{\alpha}\right) \log \mathbf{p}\left(\begin{matrix} 0 \ldots N-1 \\ \alpha_0 \ldots \alpha_{N-1} \end{matrix}\middle|\mathbf{a}\right). \tag{4.15.26}$$

(ii) *Given* $\epsilon > 0$, *there exists* N_ϵ *such that,* $\forall N \geqslant N_\epsilon$, *the* p^N *strings* $\left(\begin{smallmatrix} 0 \ldots N-1 \\ \alpha_0 \ldots \alpha_{N-1} \end{smallmatrix}\right)$ *of history with length N, a priori possible, can be divided into classes* $\mathcal{C}^1_\epsilon(N)$ *and* $\mathcal{C}^{\text{rare}}_\epsilon(N)$ *such that*

$$\sum_{\alpha_0,\ldots,\alpha_{N-1} \in \mathcal{C}^{\text{rare}}_\epsilon(N)} \mathbf{p}\left(\begin{matrix} 0 \ldots N-1 \\ \alpha_0 \ldots \alpha_{N-1} \end{matrix}\middle|\mathbf{a}\right) < \epsilon \tag{4.15.27}$$

and for every $\left(\begin{smallmatrix} 0 \ldots N-1 \\ \alpha_0 \ldots \alpha_{N-1} \end{smallmatrix}\right) \in \mathcal{C}^1_\epsilon(N)$:

$$e^{-(S(a)+\epsilon)N} \leqslant \mathbf{p}\left(\begin{matrix} 0 \ldots N-1 \\ \alpha_0 \ldots \alpha_{N-1} \end{matrix}\middle|\mathbf{a}\right) \leqslant e^{-(S(\mathbf{a})-\epsilon)N}. \tag{4.15.28}$$

(iii) *The number of elements in* $\mathcal{C}^1_\epsilon(N)$ *is such that*

$$e^{(S(\mathbf{a})-\epsilon)N} \leqslant |\mathcal{C}^1_\epsilon(N)| \leqslant e^{(S(\mathbf{a})+\epsilon)N}. \tag{4.15.29}$$

Observations.

(1) This is the "Shannon–McMillan theorem", see Khintchin, references.

(2) Equation (4.15.26) is very useful because it sometimes allows the explicit calculation of $S(\mathbf{a})$. The statement (ii) tells us that if N is large the number of strings of $\mathbf{a}$ that are "really important" is measured by $S(\mathbf{a})$. Furthermore, such strings have about the same probability of appearance, and their number is therefore estimated by Eq. (4.15.29).

In other words, one can think that in a rough (and weak) sense, see Eqs. (4.15.27) and (4.15.28), $\mathbf{a}$ consists of strings of large length each appearing equally probable (i.e., equally often) in $\mathbf{a}$.

If $\mathbf{a}$ is not ergodic, this last statement is not generally true: this is one of the reasons why the ergodic sequences are interesting.

The following proposition (Hopf–Asonov–Sinai theorem, see Arnold-Avez, references) gives an example of an analytic Hamiltonian system which is not analytically integrable.

38 Proposition. *Let* $\Sigma \subset \mathbf{R}^d$ *be an analytic surface, bounded and with negative curvature. The geodesic motion on* Σ *(i.e., the motion of a unit mass ideally constrained to* Σ*) is not analytically integrable because for every analytically regular partition* $\mathfrak{G}$ *of its phase space there exists a dense* *set of data whose $(\mathcal{E}, \mathbf{t})$ history, $\mathbf{t} = (jt_1)_{j=0}^{\infty}$, $t_1 > 0$, is mixing and also has positive entropy.*

These last two theorems are two important examples of "ergodic theory" problems. This is a young theory; nevertheless, it is already rich in interesting results and, even more, interesting open problems.

Exercises and Problems for §4.15

Can one build sequences of preassigned distribution? See Problems 1–12 below.

1. Find examples of sequences $\mathbf{a}$ of symbols $a_i = \pm 1$ with nondefinite frequencies.

2. Consider the sequence of symbols $a_i = \pm 1$:

$$a = (1, -1, 1, 1, -1, -1, 1, 1, 1, -1, -1, -1, \dots).$$

Show that it has well-defined frequencies and that $\mathbf{p}\left(\begin{smallmatrix}0\\1\end{smallmatrix}\middle|\mathbf{a}\right) = \frac{1}{2}$, $\mathbf{p}\left(\begin{smallmatrix}0 & j\\1 & 1\end{smallmatrix}\middle|\mathbf{a}\right) = \frac{1}{2}$.

3. Show that the sequence in Problem 2 is nonergodic. (Hint: Show that Eq. (4.14.35) is false for $j_1 = 0$, $i_1 = j$, $\alpha_1 = 1$, $\beta_1 = 1$.)

4. Find an example of a subset $A \subset \mathbf{T}^l$ such that, setting $E_0 = A$, $E_1 = \mathbf{T}^l / A$, there is in $\mathbf{T}^l$ a point p whose history on the partition $\mathcal{E} = (E_0, E_1)$ with respect to the rotation $\varphi \rightarrow (\varphi + \omega) \bmod 2\pi$, supposed irrational, does not have well-defined frequencies. (Hint: Let $\mathbf{a}$ be a sequence of 0's and 1's without well-defined frequencies, see Problem 2); then given p, let $A = \bigcup_{k, \alpha_k = 0} \{p + k\omega\}$.)

5. Using Proposition 28, §4.13, p. 336, and the method of proof of Proposition 30, §4.14, p. 339, show that, if $E \subset \mathbf{T}^l$ is Riemann measurable, then every point of $\mathbf{T}^l$ evolving under an irrational rotation transformation visits E with well-defined frequencies.

6. Let $\mathcal{E} = \{0, 1\}$, $p_0 = \frac{1}{2}$, $p_1 = \frac{1}{2}$, and consider the probability distribution $(\mathcal{E}, \mathbf{p})$, see §2.23, and $(\mathcal{E}, \mathbf{p})^N$, see Definition 20, §2.23, p. 118.

Let $A_N(0)$ be the subset in $\mathcal{E}^N$ consisting of the sequences $\alpha_0, \dots, \alpha_{N-1}$, $\alpha_j = 0, 1$, in which the symbol 0 appears with a frequency closer to $\frac{1}{2}$ than $N^{-1/8}$:

$$A_N(0) = \left\{ \alpha_0, \dots, \alpha_{N-1} \,\middle|\, \left| N^{-1} \left(\sum_{j=0}^{N-1} (1 - \alpha_j) \right) - \frac{1}{2} \right| < N^{-1/8} \right\}.$$

Show that the probability of $A_N(0)$ in $(\mathcal{E}, p)^N$ is such that

$$\mathbf{p}(A_N(0) > 1 - \frac{1}{8N^{3/4}} .$$

(Hint: Use the Chebysčev inequality, Proposition 34, §2.23, p. 119; see, also, Proposition 33, §2.23, p. 119.)

7. In the context of Problem 6, regard $A_N(0)$ as a subset $\tilde{A}_N(0)$ of the space of the infinite sequences $\mathbf{a} = (a_0, a_1, \dots)$ of 0's and 1's defined by $\mathbf{a} \in \tilde{A}_N(0) \leftrightarrow (a_0, \dots, a_{N-1}) \in A_N(0)$.

Show that the sets $\tilde{A}_{k^2}(0)$ have the finite intersection property, i.e., $\bigcap_{k=1}^{q} \tilde{A}_{k^2}(0) \neq \emptyset$, $\forall q > 1$. (Hint: Use Problem (6) to note that if $\tilde{A}_1, \tilde{A}_2, \tilde{A}_4, \dots, \tilde{A}_{k^2}$ are all regarded as subsets in $\mathcal{E}^{k^2}$ in a natural way, they have a probability in $\mathcal{E}^{k^2}$:

$$\mathbf{p}(A_{k^2}(0)) > 1 - 1/8k^{3/2}.$$

Hence, the complement of the intersection of any number of the A_{k^2}'s has a probability such that

$$\mathbf{p}((\cap A_{k^2}(0))^c) \leqslant \sum \mathbf{p}(A_{k^2}(0)^c) \leqslant \frac{1}{8} \sum_{k=1}^{\infty} \frac{1}{k^{3/2}} < 1$$

since $(\cap E_\alpha)^c \subset \cup E_\alpha^c$, in general. Hence, $\cap A_{k^2}(0)$ cannot be empty.)

8. Extend Problem 6 to show that for every given string $(\sigma_1, \ldots, \sigma_s)$ of 0's and 1's, the set $A_N(\sigma_1 \ldots \sigma_s) \subset (\mathcal{E}, \mathbf{p})^N$ consisting of the strings $\alpha = (\alpha_0, \ldots, \alpha_{N-1}) \in \mathcal{E}^N$ in which the string $(\sigma_1, \ldots, \sigma_s)$ appears somewhere, with a frequency differing from 2^{-s} by at most $N^{-1/8}$, is such that

$$\mathbf{p}(A_N(\sigma_1 \ldots \sigma_s)) > 1 - \frac{\epsilon_s}{N^{3/4}}$$

for some $\epsilon_s > 0$. (Hint: Proceed as in Problem 6, observing that

$$A_N(\sigma_1 \ldots \sigma_s) = \left\{ \alpha_0 \ldots \alpha_{N-1} \,\middle|\, \left| N^{-1} \times \left(\sum_{j=0}^{N-1} (\alpha_j - \sigma_1)^2 (\alpha_{j+1} - \sigma_2)^2 \cdots \times (\alpha_{j+s-1} - \sigma_s)^2 \right) - 2^{-s} \right| < N^{-1/8} \right\}.$$

9. Extend Problem 7 as follows: regard $A_N(\sigma_1 \ldots \sigma_s)$ as a subset $\tilde{A}_N(\sigma_1 \ldots \sigma_s)$ in the space of the infinite sequences $\mathbf{a}$ of 0's and 1's defined by $\mathbf{a} \in \tilde{A}_N(\sigma_1, \ldots, \sigma_s) \leftrightarrow (a_0 \ldots a_{N-1}) \in A_N(\sigma_1 \ldots \sigma_s)$.

Show that there is N_s, $s = 1, 2, \ldots,$ such that $\forall n, q \geqslant 1$:

$$B = \bigcap_{s=1}^{n} \bigcap_{\sigma_0, \ldots, \sigma_s}^{0,1} \bigcap_{k=1}^{q} \tilde{A}_{N_s + k^2}(\sigma_0 \ldots \sigma_s) \neq \emptyset.$$

(Hint: See the hint to Problem 7 to estimate the probability of the complement of B. One now finds the condition:

$$\sum_{k=1}^{\infty} \sum_{s=1}^{\infty} \frac{2^s \epsilon_s}{(N_{s+k^2})^{3/4}} < 1.$$

10. In the context of Problem 9, show that if $\bigcap_{n,q} B \neq \emptyset$ and $\mathbf{a} \in \bigcap_{n,q} B$, then $\mathbf{a}$ has well-defined frequencies and:

(i)

$$\mathbf{p}\left(\begin{matrix} 0 \ldots N-1 \\ \alpha_0 \ldots \alpha_{N-1} \end{matrix} \middle| \mathbf{a} \right) = 2^{-N} \qquad \forall N, \forall \alpha_0 \ldots \alpha_{N-1};$$

(ii) $\mathbf{a}$ is ergodic and mixing;

(iii) $S_{\text{abs}}(\mathbf{a}) = \log 2$, $S(\mathbf{a}) = \log 2$.

(Hint: For (ii), check the mixing directly; for (iii), apply, with patience, Definition 21, p. 351.)

11. Show that $\bigcap_{n,q} B$ in Problem 9 is nonempty. (Hint: Number the sets $A_{N_s+k^2}(\sigma_1 \ldots \sigma_s)$ from 1 to ∞ and denote them as $D_1, D_2, \ldots,$. Then, by Problem 9, $\bigcap_{j=1}^{m} D_j \text{n} \emptyset, \forall m$. Let $\mathbf{a}_m \in \bigcap_{j=1}^{m} D_j$. Since the sequences $\mathbf{a}_q$ have only two possible entries at each site, there must exist a subsequence $\mathbf{a}_{q_i}, q_i \to +\infty$, and a $\mathbf{a}_\infty$ such that $\mathbf{a}_{q_i}$ eventually coincides with $\mathbf{a}_\infty$ on any finite number of sites: $\mathbf{a}_\infty \in \bigcap_j D_j$.)

12. Extend Problems 6–11 to the case $\mathfrak{S} = \{0, 1\}, p_0 > 0, p_1 > 0, p_0 + p_1 = 1, p_0 \neq \frac{1}{2}$. Show that there are sequences of 0's and 1's such that $S_{\text{abs}}(\mathbf{a}) = \log 2$, $S(\mathbf{a}) = -p_0 \log p_0 - (1 - p_0)\log(1 - p_0) < S_{\text{abs}}(\mathbf{a})$.

Other necessary integrability criteria emerge from the following series of problems together with other remarkable properties of integrable systems.

13. Let $A_1, \ldots, A_l$ be l prime integrals for an l-degrees-of-freedom Hamiltonian system on $W \subset \mathbf{R}^{2l}$ or $W \subset \mathbf{R}^l \times \mathbf{T}^l$ or $W \subset \mathbf{R}^l \times (\mathbf{R}^{l_1} \times \mathbf{T}^{l_2})$, $l_1 + l_2 = l$, open. Call $\mathbf{A}(W)$ the set of the values of $(A_1, \ldots, A_l)$ on W: $\mathbf{A}(W) \subset \mathbf{R}^l$. Suppose that the equation

$$\mathbf{A}(\mathbf{p}, \mathbf{q}) = \mathbf{a}$$

can be inverted with nonzero Jacobian near $\mathbf{p}_0, \mathbf{q}_0, \mathbf{a}_0$ as $\mathbf{p} = \boldsymbol{\alpha}(\mathbf{a}, \mathbf{q})$ so that $\mathbf{A}(\boldsymbol{\alpha}(\mathbf{a}, \mathbf{q}), \mathbf{q}) \equiv \mathbf{a}$.

Define the $l \times l$ matrices:

$$M_{ij} = \frac{\partial A_i}{\partial p_j}, \qquad N_{ij} = \frac{\partial A_i}{\partial q_j}, \qquad T_{ij} = \frac{\partial \alpha_i}{\partial q_j}, \qquad R_{ij} = \frac{\partial \alpha_i}{\partial a_j}.$$

Study the "Hamilton–Jacobi" equations:

$$\mathbf{A}\left(\frac{\partial s}{\partial \mathbf{q}}, \mathbf{q}\right) = \mathbf{a}, \qquad \text{i.e.,} \quad \frac{\partial s}{\partial \mathbf{q}} = \boldsymbol{\alpha}(\mathbf{a}, \mathbf{q})$$

and find the conditions guaranteeing their solvability near $\mathbf{q}_0, \mathbf{a}_0$. Show that the conditions are $\{A_i, A_j\} = 0$, $\forall i, j = 1, \ldots, l$, i.e.,

$$\sum_{s=1}^{l} \left(\frac{\partial A_i}{\partial p_2} \frac{\partial A_j}{\partial q_s} - \frac{\partial A_i}{\partial q_s} \frac{\partial A_j}{\partial p_2} \right) = 0, \qquad i, j = 1, \ldots, l$$

(see, also, Definition 19, §3.12). (Hint: Clearly, one only needs that the differential form $\boldsymbol{\alpha} \cdot d\mathbf{q}$ be exact, i.e., $\partial \alpha_i / \partial q_j = \partial \alpha_j / \partial q_i$ or $T_{ij} = T_{ji}$. By the implicit function theorem and by the chain differentiation rule, it follows from $\mathbf{A}(\boldsymbol{\alpha}, (\mathbf{a}, \mathbf{q}), \mathbf{q}) \equiv \mathbf{a}$ that

$$\sum_{s=1}^{l} \frac{\partial A_i}{\partial p_s} \frac{\partial \alpha_s}{\partial a_j} = \delta_{ij}$$

and

$$\sum_{s=1}^{l} \left(\frac{\partial A_i}{\partial p_s} \frac{\partial \alpha_s}{\partial q_i} \right) + \frac{\partial A_i}{\partial q_j} = 0;$$

i.e., with the above notations, $NR = 1$ and $MT + N = 0$. So, since $T = -M^{-1}N$, the integrability condition becomes

$$M^{-1}N=(M^{-1}N)^T=N^T(M^{-1})^T \Leftrightarrow NM^T=MN^T$$

because $\det M \neq 0$. The last expression once written explicitly, yields the result ("Liouville's theorem").)

14. Show the following properties of the Poisson bracket, see Definition 19, §3.12:

$$\{F,G\}=-\{G,F\},$$
$$\{F,GL\}=\{F,G\}L+\{F,L\}G,$$
$$\{F,\{G,L\}\}+\{G,\{L,F\}\}+\{L,\{F,G\}\}=0.$$

Two observables on phase space F, G are said to be "in involution" if $\{F,G\}=0$.

15. In the context of Problems 13 and 14, suppose that $A_1, \ldots, A_l$ are l prime intregrals in involution. Consider the completely canonical transformation $\mathcal{C}$ generated by the function $(\mathbf{a},\mathbf{q}) \to s(\mathbf{a},\mathbf{q})$ in Problem 13 (via $\boldsymbol{\kappa}=(\partial s/\partial \mathbf{a})(\mathbf{a},\mathbf{q}), \mathbf{p}=(\partial s/\partial \mathbf{q})(\mathbf{a},\mathbf{q})$). Denote it $(\mathbf{a},\boldsymbol{\kappa})=\mathcal{C}(\mathbf{p},\mathbf{q})$. Show that $H(\mathcal{C}^{-1}(\mathbf{a},\boldsymbol{\kappa}))=h(\mathbf{a})$ is $\boldsymbol{\kappa}$ independent. (Hint: Since the A's are prime integrals and the map $(\mathbf{p},\mathbf{q})\leftrightarrow(\mathbf{a},\boldsymbol{\kappa})$ is completely canonical, it must be that

$$\dot{\mathbf{A}}=\frac{\partial H(\mathcal{C}^{-1}(\mathbf{A},\boldsymbol{\kappa}))}{\partial \boldsymbol{\kappa}}=\mathbf{0},$$

i.e., $H(\mathcal{C}^{-1}(\mathbf{A},\boldsymbol{\kappa}))$ is $\boldsymbol{\kappa}$ independent.)

16. Using the fact that the completely canonical transformations preserve the Poisson brackets, see Observation (2), p. 237, to Corollary 25, §3.12, show that a necessary condition for the canonical integrability of a Hamiltonian system on a region W of phase space is the existence in W of l independent prime integrals in involution.

17. Show that a necessary and sufficient condition in order that $A \in C^\infty(W)$ be a prime integral for a regular Hamiltonian system on W is that $\{A,H\}=0$, if H is the Hamiltonian function. More generally, if $S_t(\mathbf{p},\mathbf{q})$, $t\in J$, denotes a solution to the Hamilton equations in W and $F\in C^\infty(W)$, show that

$$\frac{d}{dt}F(S_s(\mathbf{p},\mathbf{q}))=\{H,F\}(S_t(\mathbf{p},\mathbf{q})), \qquad \forall t\in J.$$

Here $W\subset \mathbf{R}^{2l}$ or $\mathbf{R}^l\times\mathbf{T}^l$ or $\mathbf{R}^l\times(\mathbf{R}^{l_1}\times\mathbf{T}^{l_2})$, $l_1+l_2=l$, is open. (Hint: Just compute the derivative of F using the Hamilton equations to express $\mathbf{p},\mathbf{q}$ and the definition of the Poisson bracket.)

18. Let W be as in the above problems and let $H\in C^\infty(W)$ be a regular Hamiltonian function. Assume that H is integrable on W and let I be the integrating transformation I: $W\leftrightarrow V\times\mathbf{T}^l$, $V\subset\mathbf{R}^l$, let $(\mathbf{A},\boldsymbol{\varphi})=I(\mathbf{p},\mathbf{q})$ and denote $\boldsymbol{\omega}(\mathbf{A})$ the pulsations of the quasi-periodic motions on the torus $\{\mathbf{A}\}\times\mathbf{T}^l$. We say that the system is "nonisochronous" if the matrix $J_{ij}(\mathbf{A})=(\partial\omega_i/\partial A_j)(\mathbf{A})$ has a nonvanishing determinant.

Show that any prime integral $B\in C^\infty(W)$ for a nonisochronous integrable Hamiltonian system must be a function of $A_1,\ldots,A_l$ introduced above. (Hint: Let $B=b(\mathbf{A},\boldsymbol{\varphi})$ be a prime integral in the $(\mathbf{A},\boldsymbol{\varphi})$ variables. It must be $b(\mathbf{A},\boldsymbol{\varphi})\equiv b(\mathbf{A},\boldsymbol{\varphi}+\boldsymbol{\omega}(\mathbf{A})t)$, $\forall t$. If the components of $\boldsymbol{\omega}(\mathbf{A})$ are rationally independently the points $\boldsymbol{\varphi}+\boldsymbol{\omega}(\mathbf{A})t$, $t\in\mathbf{R}$, densely covers $\mathbf{T}^l$; hence, for such $\mathbf{A}$'s, B must depend only on $\mathbf{A}$ and not on $\boldsymbol{\varphi}$. However, if $\det J\neq 0$, the set of $\mathbf{A}$'s in V such that $\boldsymbol{\omega}(\mathbf{A})$ has rationally independent coordinates is dense in V (see Problems 9 and 15, §5.10, pp. 483 and 484). Hence, B must always depend only on $\mathbf{A}$.)

19. There is a theorem by Arnold concerning the case when W is an invariant open bounded set for a regular Hamiltonian flow generated by $H \in C^\infty(W)$ and on W one can define l independent prime integrals $\mathbf{A} = (A_1, \ldots, A_l)$ in involution (see Problem 14)), with $A_1 \equiv H$ and such that the sets $\mathbf{A}(\mathbf{p}, \mathbf{q}) = \mathbf{a}$ are, for $\mathbf{a} \in \mathbf{A}^{-1}(W)$, regular closed connected surfaces in W. Then H is integrable on W.

Are there systems integrable but not canonically integrable? (Answer: Only if $l = 2$, some partial results are known (Rüssmann, see reference.) The proof of Arnold's theorem can be found on page 269 of his book "Methodes Mathématiques . . . ", (see References).

20. Find an example of a Hamiltonian system whose motions are all quasi-periodic but which is not integrable. (Hint: consider two point masses free on a circle and on a line, respectively; let their positions be determined by $(\varphi_1, q_2) \in \mathbf{T}^1 \times \mathbf{R}$ if φ_1 is the angular position of the first particle and q_2 the position of the second. Let

$$H(p_1, p_2, \varphi_1, q_2) = \frac{p_1^2}{2}\,.\Big)$$

21. Let $k \to f(k)$ be a function defined for $k = 1, 2, \ldots,$ such that $0 \leqslant f(k)$, $f(k+h) \leqslant f(k) + f(h)$, for all $h, k = 1, 2, \ldots$. Show that

$$\lim_{k\to\infty} k^{-1} f(k) = \inf_k k^{-1} f(k)$$

and apply this result to prove the existence of the limit (4.15.2) by showing that $f(k) = \log N_{\text{abs}}(k, \mathbf{a})$ has the above properties. (Hint: let $\epsilon > 0$ and let k_ϵ be such that $s \leqslant k_\epsilon^{-1} f(k_\epsilon) \leqslant s + \epsilon$, where $s \equiv \inf_k k^{-1} f(k)$; write $k = hk_\epsilon + p$ with $h = 0, 1, \ldots$ and $p = 0, 1, \ldots, k_\epsilon - 1$ and note that $s \leqslant k^{-1} f(k) \leqslant (hk_\epsilon + p)^{-1}(hf(k_\epsilon) + f(p)) \to_{k\to\infty} k_\epsilon^{-1} f(k_\epsilon) < s + \epsilon$.)

CHAPTER 5

Stability Properties for Dissipative and Conservative Systems

§5.1. A Mathematical Model for the Illustration of Some Properties of the Dissipative Systems

In various possible senses, the stability properties of motions are more easily analyzed in systems moving in the presence of friction, as already noted in Chapter 2.

Therefore, we shall mainly concentrate our attention on such systems, studying some stability questions selected among others because they seem particularly significant for the generality of the methods used to treat them.

Similar questions will later be asked about conservative systems. However, the answers, when known, will be much harder to obtain.

The gyroscope is, in some sense, the prototype for systems with many degrees of freedom. In fact, general systems of linear oscillators trivially reduce to systems of independent one dimensional oscillators, as explained §4.1–4.4 in the conservative cases; this remains true even in the presence of linear friction.

On the other hand, the gyroscope with friction, or even some of its particular cases, already presents many of the possibilities and difficulties that can be met in more complex systems.

For this reason, in the upcoming sections, we shall illustrate the general theory through the treatment of a single example, described below and drawn from the gyroscope's theory, which will be used to motivate the successive steps of a theory and of a method of analysis which, as will become evident, is applicable to many other dissipative systems as well.

The example is given by Eqs. (5.1.18) and (5.1.19) and this section is devoted to their gyroscopic interpretation.

We consider a rigid body consisting of N masses, $m_1, m_2, \ldots, m_N > 0$, with a fixed point O (all the constraints being ideal) immersed in a viscous fluid opposing to the motion a frictional force at the ith point:

$$-\lambda m_i \dot{\mathbf{x}}^{(i)}. \tag{5.1.1}$$

The moment of the frictional force with respect to O is then given by

$$-\lambda \sum_{i=1}^{N} m_i (P_i - O) \wedge \dot{\mathbf{x}}^{(i)} = -\lambda \sum_{i=1}^{N} m_i (P_i - O) \wedge (\omega \wedge (P_i - O)) = -\lambda I \omega \tag{5.1.2}$$

with the notations of §4.11.

From the second cardinal equation,[1] it is easy to see that

$$I\dot{\omega} = -\lambda I \omega - \omega \wedge (I\omega), \tag{5.1.3}$$

where $\dot{\omega}$ is the vector whose components in a co-moving frame $(O; \mathbf{i}_1, \mathbf{i}_2, \mathbf{i}_3)$ are the derivatives of the homonymous components of ω in the same frame. Equation (5.1.3) extends Eq. (4.11.31) to the case when the moment of the external forces is $-\lambda I \omega$ instead of $\mathbf{0}$.

Assume that the co-moving frame has been fixed once and for all so that the inertia matrix I is diagonal, see Eq. (4.11.9), p. 309, with elements

$$I_1, I_2, I_3. \tag{5.1.4}$$

In order to obtain nontrivial motions, it will be convenient to imagine that the system is subject to the action of other forces having a moment $\mathbf{M}$ with respect to O. Otherwise, as is intuitively clear and as we shall shortly see, the system will just stop.

The simplest force laws are those with moment $\mathbf{M}$ having constant components on the axes of $(O; \mathbf{i}_1, \mathbf{i}_2, \mathbf{i}_3)$:

$$\mathbf{M} = R_1 \mathbf{i}_1 + R_2 \mathbf{i}_2 + R_3 \mathbf{i}_3, \tag{5.1.5}$$

or those with the moment's components in $(O; \mathbf{i}_1, \mathbf{i}_2, \mathbf{i}_3)$ dependent only upon the angular velocity

$$\mathbf{M}'(\omega) = R_1'(\omega)\mathbf{i}_1 + R_2'(\omega)\mathbf{i}_2 + R_3'(\omega)\mathbf{i}_3 \tag{5.1.6}$$

which can be imagined (as we shall see in more detail in an example) to be generated by some "inner mechanisms" regulating their action as a function of the body's motion.

In the presence of forces with moments of Eqs. (5.1.5) and (5.1.6) added to the friction forces, the equations of motion of the system would become

$$I\dot{\omega} = -\omega \wedge (I\omega) - \lambda I \omega + \mathbf{M} + \mathbf{M}'(\omega). \tag{5.1.7}$$

Even in the simplest situation, where we imagine

$$\mathbf{M} = R\mathbf{i}_3, \qquad \mathbf{M}'(\omega) = \{\text{linear function of } \omega\}, \tag{5.1.8}$$

[1] i.e., $\dot{K}_O = -\lambda I \omega$.

it could a priori happen that the differential equation (5.1.7) admits solutions $t \to S_t(\overline{\omega})$, with suitable initial datum $\overline{\omega}$, diverging as $t \to +\infty$.[2]

We wish to avoid having to deal with such phenomena, too idealized from a physical point of view, since it is clear that any real system "breaks down into pieces" if ω reaches too large a value, when the centrifugal forces exceed the materials' resistance. This is done by supposing that the friction coefficient λ has some extra dependence on ω. For instance,

$$\lambda(\omega) = (\lambda_1 + \lambda_2\omega^2) \quad \text{or} \quad \lambda(\omega) = (\lambda_1 + \lambda_2'\omega_1^2 + \lambda_2''\omega_2^2 + \lambda_2'''\omega_3^2) \tag{5.1.9}$$

which is a very special case of the more general and realistic friction model in which Eq. (5.1.1) is replaced by $-\lambda\mu_i(1 + L_i\dot{\mathbf{x}}^{(i)} \cdot \dot{\mathbf{x}}^{(i)})\dot{\mathbf{x}}^{(i)}$, λ, $\mu_i > 0$ and L_i is a 3×3 positive-definite matrix.

Summarizing the above discussion, the mechanical system whose properties we wish to analyze will be described by the equation

$$I\dot{\omega} = -\omega \wedge (I\omega) - \lambda(\omega)(I\omega) + \mathbf{M} + \mathbf{M}'(\omega), \tag{5.1.10}$$

where $\lambda(\omega)$ is given by Eq. (5.1.9) and $\mathbf{M}'(\omega)$ is a linear function of ω.

The above system is general enough to present a great variety of phenomena. For simplicity, we shall impose further restrictions, studying the following particular case of Eq. (5.1.10)

(i) The rigid body is a gyroscope:

$$I_1 = I_2 = I, \qquad I_3 = J. \tag{5.1.11}$$

(ii) $\mathbf{M} = R\mathbf{i}_3$, $R > 0$. (5.1.12)
(iii) $\mathbf{M}'(\omega) = \alpha_1\omega_1\mathbf{i}_1 + \alpha_2\omega_2\mathbf{i}_2$, $\alpha_1 = \alpha_2 = \alpha > 0$. (5.1.13)
(iv) $\lambda(\omega)$ is given by the first or the second of Eqs. (5.1.9).

It might be useful to have in mind a physical representation of the special system mathematically described by Eqs. (5.1.9)–(5.1.13).

One can think of the body as consisting of six points with mass m located at the points $\pm\rho\mathbf{i}_1$, $\pm\rho\mathbf{i}_2$, $\pm\rho'\mathbf{i}_3$. Then

$$\begin{aligned} I &= I_1 = I_2 = 2m(\rho^2 + \rho'^2), \\ J &= I_3 = 4m\rho^2. \end{aligned} \tag{5.1.14}$$

The force given by Eq. (5.1.12) can be imagined to be generated by small "jet motors" located at the four points of the $z_3 = 0$ plane, producing a thrust f identical at each site and perpendicular to the coordinate axis on which the site lies and parallel to the $z_3 = 0$ plane. The moment of such forces is

$$\mathbf{M} = 4\rho f\mathbf{i}_3 \tag{5.1.15}$$

like Eq. (5.1.12) with $R = 4\rho f$.

[2] However, in this case, the global existence of the solutions, assuming λ constant and M' linear in ω_1, ω_2, ω_3, can be insured by an a priori estimate, for $t \in \mathbf{R}_+$. Let $\boldsymbol{\Omega} = I\omega$ and multiply both sides of Eq. (5.1.7) scalarly by $\boldsymbol{\Omega}$. One easily finds $\frac{1}{2}(d\Omega^2/dt) \leqslant K\Omega^2 + K'$, for some K, $K' > 0$, which implies $(K\Omega(t)^2 + K') \leqslant (K\Omega(0)^2 + K')e^{2Kt}$, $\forall t \geqslant 0$.

The force given by Eq. (5.1.13) is generated by small jet motors located at the two points on the axis $\mathbf{i}_3$, exerting a thrust along $\mathbf{i}_1$ and $\mathbf{i}_2$, respectively, with intensities

$$f'\omega_2\mathbf{i}_1 \quad \text{and} \quad f'\omega_1\mathbf{i}_2 \tag{5.1.16}$$

and, therefore, their moment is

$$f'\rho'(\omega_1\mathbf{i}_1 + \omega_2\mathbf{i}_2) \tag{5.1.17}$$

like Eq. (5.1.13) with $\alpha = f'\rho'$.

The somewhat bizarre force given by Eq. (5.1.16) must be thought of as generated by jets producing a thrust proportional to the amount of air entering their mouths per unit time, which have to be supposed to be oriented as $\mathbf{i}_2$ and $\mathbf{i}_1$, respectively, and orthogonal to $\mathbf{i}_3$. The amount of air entering the jets' mouths per unit time is in this way proportional to ω_2 and ω_1, respectively.

Obviously, if $f' \neq 0$, the gyroscope will tend to increase its rotation speed around the axes $\mathbf{i}_1$, $\mathbf{i}_2$, but not indefinitely [just as long as the system reaches a rotation speed causing so strong a friction as to compensate for the force of the motor (as we shall see, this is what actually happens if $\lambda_2', \lambda_2'', \lambda_2''' > 0$)].

We conclude this section by explicitly writing Eq. (5.1.10) by components, given the assumptions of Eqs. (5.1.9)–(5.1.13):

$$\begin{aligned} \dot{\omega}_1 &= -(\lambda_1 + \lambda_2\omega^2)\omega_1 + \alpha\omega_1 - \omega_2\omega_3, \\ \dot{\omega}_2 &= -(\lambda_1 + \lambda_2\omega^2)\omega_2 + \alpha\omega_2 + \omega_1\omega_3, \qquad R, \alpha, \lambda_1, \lambda_2 > 0, \\ \dot{\omega}_3 &= -(\lambda_1 + \lambda_2\omega^2)\omega_3 + R, \end{aligned} \tag{5.1.18}$$

if the first of Eqs. (5.1.9) is assumed and if one sets $(J - I)/I = 1$, a case to which one can reduce by the change of variables $\omega_1' = \omega_i(J - I)/I$; α, R are real numbers which we assume positive, for definiteness.

If the second of Eqs. (5.1.9) is assumed, one instead finds

$$\begin{aligned} \dot{\omega}_1 &= -(\lambda_1 + \lambda_2'\omega_1^2 + \lambda_2''\omega_2^2 + \lambda_2'''\omega_3^2)\omega_1 + \alpha\omega_1 - \omega_2\omega_3, \\ \dot{\omega}_2 &= -(\lambda_1 + \lambda_2'\omega_1^2 + \lambda_2''\omega_2^2 + \lambda_2'''\omega_3^2)\omega_2 + \alpha\omega_2 + \omega_1\omega_3, \\ \dot{\omega}_3 &= -(\lambda_1 + \lambda_2'\omega_1^2 + \lambda_2''\omega_2^2 + \lambda_2'''\omega_3^2)\omega_3 + R, \qquad R, \alpha, \lambda_1, \lambda_2', \lambda_2'', \lambda_2'' > 0, \end{aligned} \tag{5.1.19}$$

with the same remarks as above. Call $\lambda_2 = \min(\lambda_2', \lambda_2'', \lambda_2''') > 0$.

Equation (5.1.18) has a symmetry, absent in Eq. (5.1.19), allowing us to eliminate one of the variables. In fact, if $\omega^2 = \omega_1^2 + \omega_2^2$, we find, by multiplying the first of Eqs. (5.1.18) by ω_1 and the second by ω_2 and adding them:

$$\begin{aligned} \frac{1}{2}\frac{d\omega^2}{dt} &= -(\lambda_1 + \lambda_2\omega^2 + \lambda_2\omega_3^2)\omega^2 + \alpha\omega^2, \\ \frac{d\omega_3}{dt} &= -(\lambda_1 + \lambda_2\omega^2 + \lambda_2\omega_3^2)\omega_3 + R, \qquad \lambda_1, \lambda_2, R, \alpha > 0 \end{aligned} \tag{5.1.20}$$

which is much simpler because it involves only two unknowns, ω^2 and ω_3.

§5.2 Stationary Motions for a Dissipative Gyroscope

We will first explicitly observe that Eqs. (5.1.18) and (5.1.19) admit global solutions in the future.

1 Proposition. *Equation* (5.1.19) *admits a solution* $t \to S_t(\omega_0)$, $t \in \mathbf{R}_+$, *for every initial datum* $\omega_0 \in \mathbf{R}^3$.

Furthermore, if $\lambda_2 = \min(\lambda_2', \lambda_2'', \lambda_2''')$ *and* $\Omega = (2\mathbf{R}/\lambda_2)^{1/3} + (2|\alpha - \lambda_1|/\lambda_2)^{1/2}$:

(i) $|S_t(\omega_0)| \leqslant |\omega_0| + \Omega, \ \forall t \geqslant 0$ (5.2.1)

(ii) $|S_t(\omega_0)| \leqslant 2\Omega, \ \forall t \geqslant (|\omega_0|^2 - 4\Omega^2)/2\lambda_2\Omega^4$. (5.2.2)

Observations.

(1) Equation (5.2.1) tells us that the trajectory of the motions of the ω's are bounded uniformly for $t \geqslant 0$.

(2) Equation (5.2.2) says that all the motions take place inside the sphere with radius 2Ω after a finite transient time (which may depend upon the initial datum).

PROOF. To show global existence, it suffices to show, on the basis of Definition 3 and Proposition 5, §2.5, p. 28, an a priori estimate, i.e., it suffices to show that if $t \to S_t(\omega_0)$ is a solution to Eq. (5.1.19) for $t \in [0, T]$ with datum ω_0, then it verifies the inequality (5.2.1), $\forall t \in [0, T]$.

This is a simple consequence of the structure of Eq. (5.1.19). In fact, let $\omega = S_t(\omega_0)$ and multiply each equation by ω_1, ω_2, ω_3, respectively, and add the results to find

$$\begin{aligned} \frac{d}{dt}\frac{1}{2}\omega^2 &= -\lambda(\omega)\omega^2 + \alpha(\omega_1^2 + \omega_2^2) + R\omega_3 \\ &\leqslant |\alpha - \lambda_1|\omega^2 - \lambda_2(\omega^2)^2 + R|\omega|. \end{aligned} \tag{5.2.3}$$

Hence, if the inequalities

$$\frac{\lambda_2}{2}(\omega^2)^2 > |\alpha - \lambda_1|\omega^2, \qquad \frac{\lambda_2}{2}(\omega^2)^2 > R|\omega| \tag{5.2.4}$$

hold, the right-hand side of Eq. (5.2.3) is negative.

Therefore, if initially $|\omega_0| > \Omega$ with

$$\Omega = \left(\frac{2R}{\lambda_2}\right)^{1/3} + \left(\frac{2|\alpha - \lambda_1|}{\lambda_2}\right)^{1/2}, \tag{5.2.5}$$

the quantity $|S_t(\omega_0)| \equiv |\omega|$ must diminish as t grows, at least until it becomes $\leqslant \Omega$. This implies both global existence and the estimate (5.2.1).

To find the estimate (5.2.2), note that $|\omega| \geqslant 2\Omega$ implies that the right-hand side of Eq. (5.2.3) is smaller than $-\lambda_2\Omega^4$. Hence, as long as $|S_t(\omega_0)| \geqslant 2\Omega$, one must have

$$|S_t(\omega_0)|^2 \leqslant |\omega_0|^2 - 2\lambda_2\Omega^4 t \tag{5.2.6}$$

which means that for $t \geqslant (|\omega_0|^2 - 4\Omega^2)/2\lambda_2\Omega^4$, it will be $|S_t(\omega_0)| \leqslant 2\Omega$. mbe

In general, the simplest information about the nature of the motions described by a differential equation can be obtained through the study of stationary solutions.

2 Proposition. *Equation* (5.1.19) *has*, $\forall R > 0$ *and* $\forall \alpha > 0$, *a unique stationary solution* $\hat{\omega}$. *This solution has* $\hat{\omega}_1 = \hat{\omega}_2 = 0$, *while* $\hat{\omega}_3$ *is the unique real solution to the equation*

$$-(\lambda_1 + \lambda_2'''\hat{\omega}_3^2)\hat{\omega}_3 + R = 0. \tag{5.2.7}$$

PROOF. Setting $\dot{\omega}_1 = \dot{\omega}_2 = 0$ in the first two of Eqs. (5.1.19) and imagining[3] that one knows $\lambda(\hat{\omega})$ and $\hat{\omega}_3$, one obtains two homogeneous linear equations for $\hat{\omega}_1, \hat{\omega}_2$ with determinant

$$(\alpha - \lambda(\hat{\omega}))^2 + \hat{\omega}_3^2 \tag{5.2.8}$$

which vanishes only for $\hat{\omega}_3 = 0$ and $\alpha = \lambda(\hat{\omega})$, but the third of Eqs. (5.1.19) does not admit a stationary solution with $\hat{\omega}_3 = 0$. Hence, Eq. (5.2.8) does not vanish and, therefore, $\hat{\omega}_1 = \hat{\omega}_2 = 0$, which in turn implies that $\hat{\omega}_3$ has to verify Eq. (5.2.7). This equation clearly admits just one solution by the strict monotonicity in $\hat{\omega}_3$ of the left-hand side. mbe

Naturally, one asks how does the actual motion of the gyroscope look if the angular velocity is $\hat{\omega}$?

3 Proposition. *The motion of the gyroscope corresponding to the stationary solution* $\hat{\omega}$ *of Eq.* (5.1.19) *is a rotation with constant angular velocity* $\hat{\omega}_3$ *around the axis* $\mathbf{i}_3$ *which remains fixed in space.*

PROOF. Let $t \to (\theta(t), \varphi(t), \psi(t))$ be a description of the motion in terms of the Euler angles (θ, φ, ψ) of $(O; \mathbf{i}, \mathbf{i}_2, \mathbf{i}_3)$ with respect to the fixed reference frame $(O, \mathbf{i}, \mathbf{j}, \mathbf{k})$.

From Eqs. (4.11.12), (4.11.13), and (4.11.14),[4] p. 309, one deduces the relationship between $\dot{\theta}(t)$, $\dot{\varphi}(t)$, $\dot{\psi}(t)$, and the vector $\omega(t)$. In general,

$$\dot{\theta} = \omega_1 \cos\psi - \omega_2 \sin\psi, \tag{5.2.9}$$

$$\dot{\varphi} = \frac{\omega_1 \sin\psi + \omega_2 \cos\psi}{\sin\theta}, \tag{5.2.10}$$

$$\dot{\psi} = \omega_3 - \frac{\cos\theta}{\sin\theta}(\omega_1 \sin\psi + \omega_2 \cos\psi). \tag{5.2.11}$$

Letting $\omega_1 = \omega_2 = 0$ and $\omega_3 = \hat{\omega}_3$, one deduces from Eq. (5.2.9) that θ is constant ($\dot{\theta} = 0$). We can suppose, without loss of generality, to have fixed $(O; \mathbf{i}, \mathbf{j}, \mathbf{k})$ so that $\theta(0) \neq 0$ or π.

The second equation, Eq. (5.2.10), implies $\dot{\varphi} = 0$. Hence, φ is a constant. Since θ and φ determine the position of $\mathbf{i}_3$ in $(O; \mathbf{i}, \mathbf{j}, \mathbf{k})$, it follows that $\mathbf{i}_3$ is fixed in $(O; \mathbf{i}, \mathbf{j}, \mathbf{k})$ and, therefore, the system rotates around $\mathbf{i}_3$ (fixed) with angular velocity given by $\dot{\psi} = \hat{\omega}_3$, by Eq. (5.2.11). mbe

[3] $\lambda(\omega)$ denotes $\lambda_1 + \lambda_2'\omega_1^2 + \lambda_2''\omega_2^2 + \lambda_2'''\omega_3^2$.

[4] Without the bars, since now there is no need of them.

We can now begin the study of nonstationary motions. If $\alpha < \lambda_1$, the motions are particularly simple.

4 Proposition. *If $\alpha < \lambda_1$, the solutions $t \to S_t(\omega)$ of Eq. (5.1.19) with initial datum ω verify*

$$|S_t(\omega) - \hat{\omega}| \leqslant c(|\omega|)e^{-(\lambda_1-\alpha)t}, \tag{5.2.12}$$

where $c(x)$ is a suitable increasing function of $x \in \mathbf{R}_+$.

The corresponding motion of the gyroscope asymptotically tends to become a uniform rotation with angular velocity $\hat{\omega}_3$ around the axis $\mathbf{i}_3$ which in turn tends to acquire a fixed position in $(O; \mathbf{i}, \mathbf{j}, \mathbf{k})$, the fixed reference frame.

More precisely, if $t \to (\theta(t), \varphi(t), \psi(t))$ is the description of the motion of the Euler angles, whose angular velocity is $\omega(t) = S_t(\omega)$, for $t \geqslant 0$, there exist constants $t_1 > 0$, $C_1 > 0$, $\bar{\theta}$, $\bar{\varphi}$, $\bar{\psi}$, depending on the initial data and such that

$$\begin{aligned} |\theta(t) - \bar{\theta}| &\leqslant C_1 e^{-t(\lambda_1-\alpha)}, \\ |\varphi(t) - \bar{\varphi}| &\leqslant C_1 (\sin \bar{\theta})^{-1} e^{-t(\lambda_1-\alpha)}, \\ |\psi(t) - \bar{\psi} - \hat{\omega}_3 t| &\leqslant C_1 (\sin \bar{\theta})^{-1} e^{-t(\lambda_1-\alpha)}. \end{aligned} \tag{5.2.13}$$

For instance, C_1 can be chosen as $C_1 = 4c(\Omega + |\omega|)/(\lambda_1 - \alpha)$, see Eq. (5.2.12).

PROOF. To begin with, it is easy to check that Eq. (5.2.12) implies Eq. (5.2.13). In fact, Eqs. (5.2.9) and (5.2.12) imply that $\dot{\theta}(t) \to_{t \to +\infty} 0$ exponentially. Hence, we can define

$$\bar{\theta} = \lim_{t \to +\infty} \theta(t) = \lim_{t \to +\infty} \left(\theta(0) + \int_0^t \dot{\theta}(\tau)\, d\tau \right) = \theta(0) + \int_0^{+\infty} \dot{\theta}(\tau)\, d\tau \tag{5.2.14}$$

because the integral converges.

Also,

$$|\theta(t) - \bar{\theta}| = \left| \int_t^{+\infty} \dot{\theta}(\tau)\, d\tau \right| \leqslant 2c(\Omega + |\omega|) \frac{e^{-(\lambda_1-\alpha)t}}{\lambda_1 - \alpha} \tag{5.2.15}$$

by Eqs. (5.2.9) and (5.2.12).

We can suppose, possibly by rotating the fixed frame, that $\bar{\theta} \neq 0$ and $\bar{\theta} \neq \pi$. Then Eq. (5.2.10) implies that $\dot{\varphi}$ tends to zero exponentially since, proceeding as above, we find

$$\bar{\varphi} = \varphi(0) + \int_0^{+\infty} \dot{\varphi}(\tau)\, d\tau, \tag{5.2.16}$$

$$|\varphi(t) - \bar{\varphi}| \leqslant \frac{2c(\Omega + |\omega|)}{\inf_{\tau \geqslant t} |\sin \theta(\tau)|} \frac{e^{-(\lambda_1-\alpha)t}}{(\lambda_1 - \alpha)} \tag{5.2.17}$$

which show the second of Eqs. (5.2.13).

Similarly Eqs. (5.2.12) and (5.2.11) imply that $\dot{\psi}$ approaches $\hat{\omega}_3$ exponentially, as $t \to +\infty$. Hence, setting

$$\bar{\psi} = \psi(0) + \int_0^{+\infty} (\dot{\psi}(\tau) - \hat{\omega}_3)\, d\tau, \tag{5.2.18}$$

one finds, by Eqs. (5.2.11) and (5.2.12),

$$|\psi(t) - \bar{\psi} - \overline{\omega}_3 t| = \left|\psi(0) + \int_0^t \dot{\psi}(\tau)\, d\tau - \bar{\psi} - \hat{\omega}_3 t\right| = \left|\int_t^{+\infty} (\dot{\psi}(\tau) - \hat{\omega}_3)\, d\tau\right|$$
$$\leqslant \frac{2c(\Omega + |\omega|)}{\inf_{\tau > t} |\sin\theta(\tau)|} \frac{e^{-(\lambda_1 - \alpha)t}}{(\lambda_1 - \alpha)}, \tag{5.2.19}$$

proving Eq. (5.2.13). Naturally, the time t_1 has to be chosen so that $\inf_{\tau > \bar{t}_1} |\sin\theta(\tau)| \geqslant |\sin\bar{\theta}|/2 > 0$, say.

Let us now prove Eq. (5.2.12).

From Eq. (5.1.19), multiplying the first equation by ω_1 and the second by ω_2 and adding the results, one finds

$$\frac{d}{dt}\frac{1}{2}(\omega_1^2 + \omega_2^2) \leqslant -(\lambda_1 - \alpha)(\omega_1^2 + \omega_2^2). \tag{5.2.20}$$

Hence,

$$\omega_1(t)^2 + \omega_2(t)^2 \leqslant (\omega_1(0)^2 + \omega_2(0)^2)e^{-2(\lambda_1 - \alpha)t}. \tag{5.2.21}$$

Furthermore, setting $z = \omega_3 - \hat{\omega}_3$, the third of the Eqs. (5.1.19) becomes

$$\begin{aligned} \dot{z} = \dot{\omega}_3 &= -\lambda_1 z - \lambda_2'''((\omega_3)^3 - (\hat{\omega}_3)^3) - (\lambda_2'(\omega_1)^2 + \lambda_2''(\omega_2)^2)\omega_3 \\ &= (-\lambda_1 - \lambda_2'(\omega_1)^2 - \lambda_2''(\omega_2)^2 - \lambda_2'''((\omega_3)^2 + \hat{\omega}_3\omega_3 + (\hat{\omega}_3)^2))z \\ &\quad - \hat{\omega}_3(\lambda_1'(\omega_1)^2 + \lambda_2''(\omega_2)^2). \end{aligned} \tag{5.2.22}$$

Hence, observing that the general solution to the equation

$$\dot{y} = f(t)y + g(t), \qquad t \geqslant 0, \tag{5.2.23}$$

is, $\forall f, g \in C^\infty(\mathbf{R})$,

$$y(t) = y(0)\exp\int_0^t f(\tau)\, d\tau + \int_0^t g(\tau)\left[\exp\int_\tau^t f(\theta)\, d\theta\right] d\tau, \tag{5.2.24}$$

one deduces, from Eq. (5.2.21),

$$\begin{aligned} z(t) = z(0)\exp &- \int_0^t (\lambda_1 + \lambda_2'(\omega_1)^2 + \lambda_2''(\omega_2)^2 + \lambda_2'''((\omega_3)^2 + \hat{\omega}_3\omega_3 + (\hat{\omega}_3)^2))\, d\tau \\ &- \int_0^t \hat{\omega}_3(\lambda_2'\omega_1^2 + \lambda_2''\omega_2^2) \\ &\times\left[\exp - \int_\tau^t (\lambda_1 + \lambda_2'''((\omega_3)^2 + \omega_3\hat{\omega}_3 + (\hat{\omega}_3)^2) \right. \\ &\left. + \lambda_2'(\omega_1)^2 + \lambda_2''(\omega_2)^2)\, d\theta\right] d\tau, \end{aligned} \tag{5.2.25}$$

and observing that the functions which multiply λ_2', λ_2'', λ_2''' are non-negative, we find, also using Eq. (5.2.21), that

$$|z(t)| \leqslant |z(0)|e^{-\lambda_1 t} + \bar{\lambda}_2|\hat{\omega}_3|\left(\omega_1(0)^2 + \omega_2(0)^2\right)\int_0^t e^{-2(\lambda_1-\alpha)\tau}e^{-\lambda_1(t-\tau)}\,d\tau$$
$$\leqslant e^{-(\lambda_1-\alpha)t}\left(|z(0)| + \bar{\lambda}_2|\hat{\omega}_3|\left(\omega_1(0)^2 + \omega_2(0)^2\right)(\lambda_1-\alpha)^{-1}\right)$$
$$\leqslant e^{-(\lambda_1-\alpha)t}\left(|\hat{\omega}_3| + |\omega_3(0)| + \frac{\bar{\lambda}_2}{\lambda_1-\alpha}|\hat{\omega}_3|\omega(0)^2\right),$$

where $\bar{\lambda}_2 = \max(\lambda_2', \lambda_2'')$.

Hence, Eq. (5.2.12) follows from Eqs. (5.2.26) and (5.2.21) with

$$c(x)^2 = \left(|\hat{\omega}_3| + x + \frac{\bar{\lambda}_2}{\lambda_1-\alpha}|\hat{\omega}_3|x^2\right)^2 + x^2. \tag{5.2.27}$$

mbe

The analysis of what happens for $\alpha > \lambda_1$ is much more interesting and involves quite a few general ideas which will be discussed in the upcoming sections.

We anticipate that the character of motion will change: for $\alpha \gg \lambda_1$, it will be described asymptotically for $t \to +\infty$ by a behavior very different from the one seen so far, where the gyroscope sets itself in a state of uniform rotation around the axis $\mathbf{i}_3$, fixed in space.

Exercises for §5.2

1. Suppose that $R = 0$ in Eq. (5.1.18). Show that for $\alpha < \lambda_1$ something analogous to the statement of Proposition 4 holds.

2. Same as Problem 1 for Eq. (5.1.19).

3. Show that for $\alpha > \lambda_1$, Eqs. (5.1.18) and (5.1.19) with $R = 0$ admit infinitely many stationary solutions, and find them.

4. Consider a gyroscope like the one considered in Eq. (5.1.14), but assume that the friction is linear, that the two little jets arranged along the $\mathbf{i}_2$ axis in $-\rho\mathbf{i}_2$ or $+\rho\mathbf{i}_2$ produce a thrust in the direction $\mathbf{i}_3$ equal to $f'\omega_1\mathbf{i}_1$, while the two jets on $\pm\rho'\mathbf{i}_3$ produce a constant thrust $R\mathbf{i}_1$. Show that the equations of motion become

$$\dot{\omega}_1 = -\lambda\omega_1 + \omega_2\omega_3 - \sigma\omega_3,$$
$$\dot{\omega}_2 = -\lambda\omega_2 - \omega_1\omega_3 + \alpha,$$
$$\dot{\omega}_3 = -\lambda\omega_3 + \sigma\omega_2$$

for suitably chosen α, σ and after a change of variables $\omega_i \to (J-I)\omega_i/I$. Find the stationary solutions for the above equation.

5. Set $\omega_3 = x$, $\omega_2 = z$, $\omega_1 = y$ and suppose that the friction is different for the different components of the angular velocity; i.e., suppose that the friction's moment is $(-\lambda_1\omega_1, -\lambda_2\omega_2, -\lambda_3\omega_3)$. Study the same problem as in Problem 4 with $\lambda_1 = 1$, $\lambda_2 = b$, $\lambda_3 = \sigma$, fixing $b = \frac{8}{3}$, $\sigma = 10$ ("Lorenz model").

6. Find whether an analogue of Proposition 4 holds for the equations in Problems 4 and 5 for some values of α.

7. Find the stationary solutions for the equations

$$\begin{aligned}
\dot\gamma_1 &= -2\gamma_1 + 4\gamma_2\gamma_3 + 4\gamma_4\gamma_5,\\
\dot\gamma_2 &= -9\gamma_2 + 3\gamma_1\gamma_3,\\
\dot\gamma_3 &= -5\gamma_3 - 7\gamma_1\gamma_2 + \alpha,\\
\dot\gamma_4 &= -5\gamma_4 - \gamma_1\gamma_5,\\
\dot\gamma_5 &= -\gamma_5 - 3\gamma_1\gamma_4.
\end{aligned}$$

Using the same method of the proof of Proposition 4, for α small, find a proof of the statement analogous to that appearing in Proposition 4, Eq. (5.2.12) ("five-mode approximation to the Navier–Stokes equations on $\mathbf{T}^2$").

8. Same as Problem 7 for the equations

$$\begin{aligned}
\dot\gamma_1 &= -2\gamma_1 + 4\sqrt{5}\,\gamma_2\gamma_3 + 4\sqrt{5}\,\gamma_4\gamma_5,\\
\dot\gamma_2 &= -9\gamma_2 + 3\sqrt{5}\,\gamma_1\gamma_3,\\
\dot\gamma_3 &= -5\gamma_3 - 7\sqrt{5}\,\gamma_1\gamma_2 + 9\gamma_1\gamma_7 + \alpha,\\
\dot\gamma_4 &= -5\gamma_4 - \sqrt{5}\,\gamma_1\gamma_5,\\
\dot\gamma_5 &= -\gamma_5 - 3\sqrt{5}\,\gamma_1\gamma_4 + 5\gamma_1\gamma_6,\\
\dot\gamma_6 &= -\gamma_6 + 5\gamma_1\gamma_5,\\
\dot\gamma_7 &= -5\gamma_7 + 9\gamma_1\gamma_3
\end{aligned}$$

("seven-mode truncation of the Navier–Stokes equations on $\mathbf{T}^2$").

§5.3. Attractors and Stability

For $\alpha \gg \lambda_1$ the motions of the model considered in §5.2 will exhibit a behavior qualitatively different from that seen for $\alpha < \lambda_1$. It is therefore convenient to introduce some notions well suited to discuss various results in suggestive and agile language.

The notions on stability and attractors that we shall introduce can be subjected to the same critiques already presented in Chapter 2 when we introduced similar notions; i.e., they should not be taken too seriously as absolute definitions. Usually everyone, motivated by their own scopes, ideas, and needs, introduce their own definitions and it makes no sense to insist on a standard nomenclature, as much as it makes no sense to agree once and for all on the choice of the units of measure of the various physical entities. Here we shall choose some significant definitions and not discuss alternative definitions, recalling that in applications the correct notions of stability and attractivity will be determined by the applications themselves.

In this and in the following sections, we shall consider autonomous differential equations in $\mathbf{R}^d$ of the form

$$\dot{\mathbf{x}} = \mathbf{f}(\mathbf{x}) \tag{5.3.1}$$

which we shall suppose to have bounded trajectories, see Definition 3, §2.5,

p. 28, i.e., such that the solution flow S_t to Eq. (5.3.1) has the property that there exists a function $\mu: \mathbf{R}_+ \to \mathbf{R}_+$ such that

$$|S_t(\mathbf{u})| \leqslant \mu(|\mathbf{u}|), \qquad \forall t \geqslant 0, \quad \forall \mathbf{u} \in \mathbf{R}^d. \tag{5.3.2}$$

Proposition 1, §5.2, p. 367, shows that Eq. (5.1.19) has this property with $\mu(|\mathbf{u}|) = |\mathbf{u}| + \Omega$.

The first interesting notion is that of a stable set.

1 Definition. Consider the flow S_t solving, for $t \geqslant 0$, a differential equation in $\mathbf{R}^d$, like Eq. (5.3.1), with bounded trajectories.

If $A \subset \mathbf{R}^d$, we denote $S_t(A)$ the set of the points $\mathbf{u}$ having the form $\mathbf{u} = S_t(\mathbf{w})$ for some $\mathbf{w} \in A$.

A set A will be called "invariant" for Eq. (5.3.1) [or for the motions of Eq. (5.3.1) or its trajectories] if

$$S_t(A) \subset A, \qquad \forall t \geqslant 0, \tag{5.3.3}$$

i.e., A is invariant if the trajectories originating in A develop entirely within A. If the inclusion in (5.3.3) holds also for $t < 0$, the set A will be called "bi-invariant".

An invariant set A will be called "stable" for the evolution described by Eq. (5.3.1) if every neighborhood U of A contains a neighborhood V of A such that

$$S_t(V) \subset U, \qquad \forall t \geqslant 0, \tag{5.3.4}$$

i.e., A is stable if the motions starting sufficiently close to A do not go too far from it.

Examples

(1) The equation of the harmonic oscillator,

$$\dot{x} = -y, \qquad \dot{y} = x, \tag{5.3.5}$$

is an equation in $\mathbf{R}^2$ such that every circle around the origin is invariant and stable.

(2) Proposition 1, §5.2, relative to the gyroscope equation (5.1.19) provides another example. Equation (5.2.1) says that the sphere with radius 2Ω is invariant. From Eq. (5.2.1), it also follows that it is stable.

Another notion, closely related to the above, is that of attractor.

2 Definition. A closed set $A \subset \mathbf{R}^d$, invariant for the evolution associated to Eq. (5.3.1), is called an "attractor" for the motions of Eq. (5.3.1) if there exists an open set $U \supset A$ such that

$$\lim_{t \to +\infty} d(S_t(\mathbf{u}), A) = 0, \qquad \forall \mathbf{u} \in U, \tag{5.3.6}$$

where $d(\mathbf{x}, A) = \{\text{distance of } \mathbf{x} \text{ from } A\}$ and the set U is said to be a "partial basin of attraction" for A.

The union of all the partial basins of attraction will be called the "attraction basin" of A and denoted as $B(A)$.

An attractor A is called minimal if it does not contain any proper subset which is also an attractor.

A partial basin of attraction U for an attractor A will be called "normal" if it is possible to associate with every $\mathbf{u} \in U$ at least one point $\pi(\mathbf{u}) \in A$ such that

$$\lim_{t\to+\infty} d(S_t(\mathbf{u}), S_t(\pi(\mathbf{u}))) = 0, \tag{5.3.7}$$

and the point $\pi(\mathbf{u})$ will be called a "projection" of $\mathbf{u}$ on A.

Examples and *Observations*

(1) The sphere with radius 2Ω, as well as that with radius Ω, are attractors for Eq. (5.1.19). The first statement follows from Eq. (5.2.2), while the second can easily be deduced from the remark following Eq. (5.2.5) by slightly improving it (exercise).

(2) For $\alpha < \lambda_1$, the point $\hat{\omega}$ is an attractor for Eq. (5.1.19) as is shown by Eq. (5.2.12). Its basin is all of $\mathbf{R}^3$, and it is a normal basin. Clearly, every basin of attraction for an attractor consisting of just one point is normal for it.

(3) The unit circle is an attractor for the solutions of the differential equation in $\mathbf{R}^2$:

$$\begin{aligned} \dot{x} &= -\frac{x}{2}(x^2+y^2-1), \\ \dot{y} &= -\frac{y}{2}(x^2+y^2-1). \end{aligned} \tag{5.3.8}$$

One can find this by multiplying the first of Eqs. (5.3.8) by x, the second by y, and adding the results

$$\frac{d}{dt}\frac{x^2+y^2}{2} = -(x^2+y^2-1)\frac{x^2+y^2}{2}, \tag{5.3.9}$$

which, setting $\rho = x^2+y^2$, becomes $\dot{\rho} = -\rho(\rho-1)$, implying

$$\frac{\rho(t)-1}{\rho(t)} = \frac{\rho(0)-1}{\rho(0)}e^{-t} \tag{5.3.10}$$

if $\rho(0) \neq 0$, hence $\lim_{t\to+\infty}\rho(t) = 1$ and the attraction basin for the unit circle consists of $\mathbf{R}^2/\{\mathbf{0}\}$. The basin is normal for the attractor since the point $(x, y) \neq \mathbf{0}$ has projection

$$\pi(x, y) = \left(\frac{x}{\sqrt{x^2+y^2}}, \frac{y}{\sqrt{x^2+y^2}}\right) \tag{5.3.11}$$

on it, and, in this case, the projection on the attractor is unique.

As an exercise one can look at the trajectories of Eqs. (5.3.8) and at the geometrical meaning of Eq. (5.3.11). The unit circle is an attractor consisting of fixed points; it is also minimal.

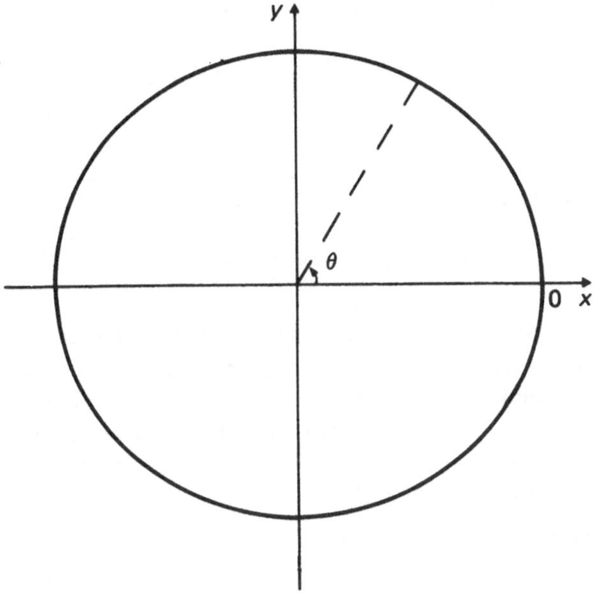

Figure 5.1.

(4) In general, it is not true that an attractor is a stable set.

To obtain some understanding of the mechanism (somewhat pathological, in fact) by which a point may be attractive without being stable, consider the unit circle S^1 in $\mathbf{R}^2$ and let $f \in C^\infty(S^1)$ be a function described as $\theta \to f(\theta)$, where $\theta \in [0, 2\pi]$ parameterizes a point on S^1. Suppose that $f(\theta) > 0$, $\forall \theta \in (0, 2\pi)$ and $f(0) = 0 = f(2\pi)$; then, by the Taylor expansion, one realizes that $1/f(\theta)$ is not summable to either the right or to the left of 0. Consider the equation

$$\dot{\theta} = f(\theta) \tag{5.3.12}$$

as an equation of motion of a point moving on S^1, interpreting the angle θ, in Fig. 5.1, as the point's position.

It appears immediately that since $f(0) = 0$, the point $\theta = 0$ is an equilibrium position for Eq. (5.3.12). But if $\theta_0 > 0$, then $S_t(\theta_0) = \theta(t)$ increases with t, because $f \geqslant 0$ and f vanishes only for $\theta = 0$ or $\theta = 2\pi$, and it takes an infinite amount of time to reach 2π. This is so because the time needed to reach 2π starting from $\theta_0 < 2\pi$ is $\int_{\theta_0}^{2\pi} d\theta / f(\theta) = +\infty$ since $f(\theta)^{-1}$ is not integrable.

However, in a finite time, $\theta(t)$ reaches any other position $\theta' \in (\theta_0, 2\pi)$ (as $\int_{\theta_0}^{\theta'} d\theta / f(\theta) < +\infty$). Hence,

$$\lim_{t \to +\infty} \theta(t) = 2\pi, \qquad \forall \theta_0 \in (0, 2\pi). \tag{5.3.13}$$

All the points on the circle evolve counterclockwise towards 2π, reaching it from the left, with the obvious exception of the points $\theta_0 = 0$ and $\theta_0 = 2\pi$.

Now let $\mathbf{f}$ be an $\mathbf{R}^2$-valued function in $C^\infty(\mathbf{R}^2)$ which in a circular annulus U around the unit circle has the value

$$\mathbf{f}(x, y) = \left(-yf(\theta) - \frac{x}{2}(x^2 + y^2 - 1), xf(\theta) - \frac{y}{2}(x^2 + y^2 - 1)\right) \quad (5.3.14)$$

if (r, θ) are the polar coordinates of (x, y).

The equation $\dot{\mathbf{x}} = \mathbf{f}(\mathbf{x})$ associated with Eq. (5.3.14) can be written in polar coordinates and for $(x, y) \in U$:

$$\dot{\theta} = f(\theta), \qquad \frac{d}{dt} r^2 = -r^2(r^2 - 1) \quad (5.3.15)$$

and the second relation shows that the set U is invariant and that the unit circle is an attractor. The first of Eqs. (5.3.15) shows that the point $\theta = 0$, $r = 1$ is a minimal attractor on the unit circle, which is unstable since arbitrarily close to it, there are points reaching it after going as far as ~ 2 away (e.g., the point $r = 1$, $\theta = \epsilon > 0$), i.e., after traveling a distance approximately equal to the circle's diameter.

(5) As the reader may guess, the problem of finding the basin $B(A)$ of attraction of an attractor is a difficult problem. Very often one is only able to determine some partial basins of attraction. The same remark applies to the determination of the minimal attractors.

In many applications, knowing partial domains of attraction or non-minimal attractors is often sufficient and one does not really need the knowledge of such "global properties" as the maximal basins or the minimal attractors.

(6) It is convenient not to require that a partial basin of attraction U for A be invariant. Clearly, this may rightly be considered a natural requirement; note, however, that $V = \bigcup_{t \geqslant 0} S_t(U)$ is an open invariant basin of attraction for A, i.e., any partial basin of attraction for A is contained inside an invariant partial basin of attraction. The total basin $B(A)$ is obviously invariant.

(7) If the differential equation (5.3.1) is normal in the past also, it is possible to construct $B(A)$ from a partial basin U for A as $B(A) = \bigcup_{t \in R} S_t(U)$.

The question of the normality of a basin U for an attractor A is obviously quite important. For simplicity, assume A bi-lateral.

Intuitively, the normality of U with respect to A depends on two factors: the speed of approach of the points of U to A and the speed of reciprocal separation of two points in A. One can expect that U is normal with respect to A if the speed of reciprocal separation of two points in A is much smaller than the speed of approach to A by the points of U.

To make precise this intuitive idea, let us introduce some new concepts.

3 Definition. Let U be a partial attraction basin for an attractor A for Eq. (5.3.1). We define the "attraction modulus" of A for U (or the "attraction speed") as the function

$$d_U(t) = \sup_{\substack{u \in U \\ \tau \geqslant t}} d(S_\tau(\mathbf{u}), A), \tag{5.3.16}$$

and $d_U(t)$ may be $+\infty$. Note that $d_U(t)$ decreases monotonically with t.

Together with this notion, it is convenient to introduce another notion measuring how quickly two points on A can separate from each other. Note that from the regularity theorem for differential equations, see §2.4, it follows that if Γ is a bounded closed set, the quantity

$$\sup_{\substack{\mathbf{x},\mathbf{x}' \in \Gamma \\ \mathbf{x} \neq \mathbf{x}'}} \frac{|S_t(\mathbf{x}) - S_t(\mathbf{x}')|}{|\mathbf{x} - \mathbf{x}'|} = m_t(\Gamma) \tag{5.3.17}$$

is finite for all $t \geqslant 0$ and bounded on every finite interval $[0, T]$, $T > 0$. It can be naturally called the "maximal expansion rate" for Eq. (5.3.1) relative to $t \in \mathbf{R}$ and to $\Gamma \subset \mathbf{R}^d$. To this notion, the following definition is related.

4 Definition. Let A be a bi-invariant attractor for Eq. (5.3.1) which is not a single point. The "uniform coefficient of maximal expansion" for Eq. (5.3.1) on A will be defined as the quantity

$$M_t(A) = \sup_{\substack{\mathbf{x} \neq \mathbf{x}' \in A \\ |\tau| \leqslant t}} \frac{|S_\tau(\mathbf{x}) - S_\tau(\mathbf{x}')|}{|\mathbf{x} - \mathbf{x}'|} = \sup_{|\tau| \leqslant t} m_\tau(A), \tag{5.3.18}$$

Note that $M_t(A)$ is monotonically increasing with t for $t \geqslant 0$.

Observations.

(1) Note that the assumptions of normality and of boundedness of trajectories on Eq. (5.3.1) made at the beginning of this section do not provide a guarantee of the existence of global solutions in the past for all initial data.[5] Hence, it is very important to stress that in Eq. (5.3.18) A is bi-invariant and negative times are also involved.

(2) Another important remark is that even if A is bounded so that $|S_\tau(\mathbf{x}) - S_\tau(\mathbf{x}')| \leqslant \{\text{diameter of } A\}$ for all τ, the function $M_t(A)$ can increase very rapidly with t. A simple though rather trivial example is the following. Let $f \in C^\infty(\mathbf{R})$ be such that

$$\begin{aligned} f(x) &= x && \text{if } |x| < \tfrac{1}{2}, \\ f(x) &= -x(x^2-1) && \text{if } |x| > 1. \end{aligned} \tag{5.3.19}$$

Then the interval $[-1, 1]$ is an attractor for the solutions of the differential equation $\dot{x} = f(x)$ and it is easy to see that

$$M_t([-1,1]) \geqslant e^t. \tag{5.3.20}$$

[5] For instance, the differential equation $\dot{x} = -x^3/2$ is normal in the future but not in the past; its solutions cannot be extended beyond $t_0 = -x(0)^{-2}$.

As an exercise, the reader can deduce Eq. (5.3.20) by considering the evolutions of the points $x_0 = 0$ and $x_1 = \epsilon$.

(3) By definition, $M_t(A) \geqslant 1$. When A is a single point, we shall set $M_t(A) \equiv 1$ by definition.

(4) If A is a periodic orbit with minimal period $T > 0$, then $M_t(A)$ is bounded in t. In fact, it is clear that $M_t(A) \leqslant M_T(A)$, $\forall t \geqslant 0$.

The following proposition makes quantitative the idea discussed above about the normality of an attraction basin U for an attractor A of Eq. (5.3.1). It provides a sufficient, though by no means necessary, condition for the normality of a basin.

5 Proposition. *Let A be a bounded bi-invariant attractor for Eq.* (5.3.1) *and let U be an attraction basin for A.*

Assume the existence of $C > 0$, $\epsilon > 0$ such that for all $t \geqslant 0$:

$$M_{t+1}(A)^2 d_U(t) < \frac{C}{(1+t)^{1+\epsilon}}. \tag{5.3.21}$$

Then U is normal for A.

If A is a periodic trajectory, it is normal if there is a $C_1 > 0$ such that

$$d_U(t) \leqslant \frac{C_1}{(1+t)^{1+\epsilon}}. \tag{5.3.22}$$

Observations.

(1) Note that the statement concerning the periodic orbits is a consequence of the general statement. In fact, if A is a periodic orbit, it is clear that $M_t(A)$ is bounded, see Observation (4), to Definition 4 above.

(2) Equation (5.3.21) implies the existence of a constant C_2 such that

$$m_\tau(A) \leqslant C_2, \qquad \forall \tau \in [-1, 1]. \tag{5.3.23}$$

It also implies

$$\begin{aligned} \text{diameter of } U &< (2d_U(0) + \text{diameter of } A) \\ &\leqslant (2C + \text{diameter of } A) < +\infty. \end{aligned} \tag{5.3.24}$$

PROOF. Let $t_n = n$, $n = 0, 1, \ldots$, and let $\mathbf{x} \in U$. Let $\mathbf{a}_n \in A$ be a point with minimal distance from $S_n(\mathbf{x})$, among the points of A.

The natural idea is that a projection $\pi(\mathbf{x})$ of $\mathbf{x}$ can be defined as

$$\pi(\mathbf{x}) = \lim_{n\to\infty} S_{-n}(\mathbf{a}_n). \tag{5.3.25}$$

To prove the existence of the above limit, let us compare $S_{-n}(\mathbf{a}_n)$ with $S_{-n-1}(\mathbf{a}_{n+1})$, assuming that A is not a single point (a case in which everything becomes trivial).

Let $\overline{U}$ be the closure of U, bounded by Eq. (5.3.24). Then, by the remark after Eq. (5.3.17), $\sup_{\tau\in[0,1]} m_\tau(\overline{U}) \leqslant \mu < +\infty$. So, by Eq. (5.3.21),

$$\begin{aligned}
&|S_{-n}(\mathbf{a}_n) - S_{-(n+1)}(\mathbf{a}_{n+1})| \\
&\quad \leqslant M_{n+1}(A)|S_1(\mathbf{a}_n) - \mathbf{a}_{n+1}| \\
&\quad \leqslant M_{n+1}(A)(|S_1(\mathbf{a}_n) - S_{n+1}(\mathbf{x})| + |S_{n+1}(\mathbf{x}) - \mathbf{a}_{n+1}|) \\
&\quad \leqslant M_{n+1}(A)(|S_1(\mathbf{a}_n) - S_1(S_n(\mathbf{x}))| + d_U(n+1)) \qquad (5.3.26)\\
&\quad \leqslant M_{n+1}(A)(m_1(U)d_U(n) + d_U(n+1)) \leqslant M_{n+1}(A)d_U(n)(1 + m_1(U)) \\
&\quad \leqslant \frac{C(1+m_1(U))}{(1+n)^{1+\epsilon}} M_{n+1}(A)^{-1}.
\end{aligned}$$

Hence, the series $\sum_{n=0}^{\infty}|S_{-n}(\mathbf{a}_n) - S_{-(n+1)}(\mathbf{a}_{n+1})|$ converges and, therefore, the limit of Eq. (5.3.25) exists. It also verifies

$$\begin{aligned}
|\pi(\mathbf{x}) - S_{-n}(\mathbf{a}_n)| &= \left|\pi(\mathbf{x}) - \mathbf{a}_0 - \sum_{k=1}^{n}(S_{-k}(\mathbf{a}_k) - S_{-(k-1)}(\mathbf{a}_{k-1}))\right| \\
&\leqslant CM_{n+1}(A)^{-1}(1+m_1(U)) \sum_{h=n+1}^{\infty} \frac{1}{h^{1+\epsilon}} \xrightarrow[n\to\infty]{} 0.
\end{aligned} \qquad (5.3.27)$$

We now compare $S_n(\pi(\mathbf{x}))$ with $S_n(\mathbf{x})$:

$$\begin{aligned}
|S_n(\pi(\mathbf{x})) - S_n(\mathbf{x})| &\leqslant |S_n(\pi(\mathbf{x})) - \mathbf{a}_n| + |\mathbf{a}_n - S_n(\mathbf{x})| \\
&\leqslant |S_n(\pi(\mathbf{x})) - S_n(S_{-n}(\mathbf{a}_n))| + d_U(n) \\
&\leqslant M_n(A)|\pi(\mathbf{x}) - S_{-n}(\mathbf{a}_n)| + d_U(n) \qquad (5.3.28)\\
&\leqslant C(1+m_1(U)) \sum_{h=n+1}^{\infty} \frac{1}{h^{1+\epsilon}} + d_U(n) \xrightarrow[n\to\infty]{} 0.
\end{aligned}$$

Finally, if $t = n + \tau$, $\tau \in (0, 1)$, is large enough,

$$\begin{aligned}
|S_t(\mathbf{x}) - S_t(\pi(\mathbf{x}))| &= |S_\tau S_n(\mathbf{x}) - S_\tau S_n(\pi(\mathbf{x}))| \\
&\leqslant m_\tau(U)|S_n(\mathbf{x}) - S_n(\pi(\mathbf{x}))| \leqslant \mu|S_n(\mathbf{x}) - S_n(\pi(\mathbf{x}))|
\end{aligned} \qquad (5.3.29)$$

Since $S_n(\mathbf{x}) \in U$ for n large enough.

Since the right-hand side of Eq. (5.3.29) approaches zero, by Eq. (5.3.28) the proposition is proved. mbe

Exercises and Problems for §5.3

1. Investigate the normality of some basins of partial attraction for the attractors associated with the equation $\dot{\mathbf{x}} = \mathbf{f}(\mathbf{x})$ in $\mathbf{R}^2$:

$$\mathbf{f}(x, y) = \left(-y(x^2+y^2-1) - \frac{x}{2}\psi(x^2+y^2), x(x^2+y^2-1) - \frac{y}{2}\psi(x^2+y^2)\right),$$

where $\psi \geqslant 0$ is a C^∞ function of its argument vanishing in 1 only. Show that the normality of the attractor is related to the convergence of the integral $\int (dr^2/r^2)((r^2-1)/\psi(r^2))$ near $r=1$.

2. An attractor may be minimal and nonconnected. Find an example. (Hint: Starting from Observation (4) to Definition 2, p. 375, improve the idea, i.e., take $f(\theta)$ vanishing not only in 0 and 2π, but also in π, positive elsewhere, so that the integral $\int d\theta/f(\theta)$ diverges near 0 and π.)

3. Consider a Hamiltonian system with l degrees of freedom, integrable on some region W of its phase space. Show that each of the tori covering W is an invariant set for the Hamiltonian flow. Each is stable, but none are attractive.

4. In the context of Problem 3, note that the invariant tori in W having pulsations ω with rational components are covered by periodic orbits. Show that none of these orbits are stable if the Jacobian matrix $J_{ij} = (\partial\omega_i/\partial A_j)(\mathbf{A})$ has a nonvanishing determinant in W. (Hint: As close as we wish to a given "rational" torus, there must be one with rationally independent components if $\det J \neq 0$ (use the implicit function theorem, or see Problem 15, §5.10, p. 484). Every point on such "irrational torus" evolves covering it densely, and this implies instability because .... On the contrary if $\partial\omega/\partial\mathbf{A} \equiv 0$, the periodic orbits, when existing, are stable.)

5. Let A_1, A_2 be two attractors with partial basins U_1, U_2, respectively. Show that $A_1 \cap A_2$ is an attractor with partial basin $U_1 \cap U_2$, if $A_1 \cap A_2 \neq \emptyset$.

6. Show that if the set of the attractors contained in a given bounded attractor A is finite then there is a minimal attractor in A.

7. Find an example of "an attractor without minimal attractors". (Hint: Let $f \in C^\infty(\mathbf{R})$ be everywhere positive for $x<0$ except at the points $x_j = -1/j$, $j = 1,2,\ldots$, where it vanishes (so that $\int dx/f(x)$ does not converge near any of the x_j). Suppose, also, that $f(x)<0$ for $x>0$. Then $\dot{x} = f(x)$ admits $[x_j, 0]$ as attractors; however, it has no minimal attractors because $\{0\}$ is not an attractor.)

8. Show that the bi-invariance assumption is essential in proposition 5. (Hint: Consider $\dot{x} = -x$, $A = [-1,1]$ and show that A is not normal.)

9. Show that every bounded attractor A contains a bi-invariant attractor $\tilde{A}$. (Hint: $\tilde{A} = \bigcap_{t>0} S_t(A)$.)

§5.4. The Stability Criterion of Lyapunov

Consider a differential equation, like Eq. (5.3.1), with bounded trajectories. A simple and useful criterion for the stability of one of its stationary solutions ("fixed points") is the following proposition ("Lyapunov's theorem").

6 Proposition. *Let* $\mathbf{x}_0$ *be an equilibrium point for Eq.* (5.3.1), $\dot{\mathbf{x}} = \mathbf{f}(\mathbf{x})$, *with* $\mathbf{f} \in C^\infty(\mathbf{R}^d)$:

$$\mathbf{f}(\mathbf{x}) = \left(f^{(1)}(\mathbf{x}), \ldots, f^{(d)}(\mathbf{x})\right) \tag{5.4.1}$$

and define the "stability matrix" (or "Lyapunov matrix")

$$L_{ij} = \frac{\partial}{\partial x_j} f^{(i)}(\mathbf{x}_0), \qquad i,j = 1,2,\ldots,d. \tag{5.4.2}$$

If the eigenvalues of L, i.e., the solutions of the dth degree equation in λ:

$$\det(L - \lambda) = 0 \tag{5.4.3}$$

(see Appendix E), have a negative real part, then $\mathbf{x}_0$ *is stable and is locally attractive with exponential speed.*[6]

If at least one of the eigenvalues has a positive real part, then $\mathbf{x}_0$ *is unstable.*

Observations.

(1) More precisely, if all the eigenvalues $\lambda_1, \ldots, \lambda_d$ of L have a negative real part, there exists $t_0 > 0$ ("halving time") and $\rho > 0$ such that for $|\mathbf{x}_0 - \mathbf{w}| < \rho$, one has

$$d(S_t(\mathbf{w}), \mathbf{x}) \leqslant 2 \cdot 2^{-t/t_0}|\mathbf{w}|, \qquad \forall t \geqslant t_0. \tag{5.4.4}$$

(2) The reason why the above proposition is true and very natural is made clear by the analysis of the "linear case", i.e., by the analysis of Eq. (5.3.1) with

$$f^{(i)}(\mathbf{x}) = \sum_{j=1}^{d} L_{ij} x_j = (L\mathbf{x})_j. \tag{5.4.5}$$

In this case, $\mathbf{x}_0 = \mathbf{0}$ is a stationary point for the equation; the equation itself can now be written as

$$\dot{\mathbf{x}} = L\mathbf{x}, \tag{5.4.6}$$

and its stability matrix is just L.

As seen in the problems of §2.2–§2.6, one can look for d linearly independent solutions of Eq. (5.4.6) having the form

$$\mathbf{x}(t) = e^{\lambda t} v. \tag{5.4.7}$$

Such a solution exists if there exists $\mathbf{v}$ such that

$$Lv = \lambda v. \tag{5.4.8}$$

If we assume that the dth-degree algebraic equation for λ, $\det(L - \lambda) = 0$, has d pairwise distinct roots $\lambda_1, \ldots, \lambda_d$ and if $\mathbf{v}^{(1)}, \ldots, \mathbf{v}^{(d)}$ are the associated eigenvectors of Eq. (5.4.8), it is well known that $\mathbf{v}^{(1)}, \ldots, \mathbf{v}^{(d)}$ are linearly independent (see Appendix E, p. 528.)

Then the function of $t \in \mathbf{R}$:

$$\mathbf{x}(t) = \sum_{j=1}^{d} \alpha_j e^{\lambda_j t} \mathbf{v}^{(j)} \tag{5.4.9}$$

is, for every choice of $\alpha_1, \ldots, \alpha_d \in \mathbf{C}$, a solution to Eq. (5.4.6).

By the linear independence of the vectors $\mathbf{v}^{(1)}, \ldots, \mathbf{v}^{(d)}$, it is clear that by suitably fixing the coefficients $\alpha_1, \ldots, \alpha_d$ one can impose that Eq. (5.4.9) verifies any preassigned initial condition. Hence, Eq. (5.4.9) is the most general solution of Eq. (5.4.6).

[6] i.e., if U is a small enough neighborhood of $\mathbf{x}_0$, it is a partial basin of attraction for $\mathbf{x}_0$ with an exponential speed of attraction, see Definition 3, p. 376.

If $\operatorname{Re}\lambda_i < 0$, $i = 1, \ldots, d$, it is clear that

$$|\mathbf{x}(t)| \leqslant e^{-\nu t} \sum_{j=1}^{d} |\alpha_j||\mathbf{v}^{(j)}|, \qquad \forall t \geqslant 0 \tag{5.4.10}$$

if $-\nu = \max_{i=1,\ldots,d} \operatorname{Re}\lambda_i < 0$; hence, the origin is an attractor with basin $\mathbf{R}^d$ itself. Every bounded sphere is attracted by the origin with exponential speed, by Eq. (5.4.10).

If, instead, $\operatorname{Re}\lambda_1 > 0$ and $\operatorname{Im}\lambda_1 \neq 0$, say, and if $\lambda_2 = \bar{\lambda}_1$, $\mathbf{v}^{(2)} = \overline{\mathbf{v}^{(1)}}$ (the bar denotes complex conjugation),[7] it is clear that by (5.4.9), the initial datum $\epsilon(\mathbf{v}^{(1)} + \overline{\mathbf{v}^{(1)}})$ evolves into

$$2\epsilon e^{(\operatorname{Re}\lambda_1)t} \operatorname{Re}(\mathbf{v}^{(1)} e^{i(\operatorname{Im}\lambda_1)t}). \tag{5.4.11}$$

Hence, arbitrarily close to the origin, there are points evolving indefinitely far away from the origin. Therefore, O not only does not attract, but it is unstable.

The following proof will reduce the nonlinear case to the linear one. If $\operatorname{Re}\lambda_i < 0$, $i = 1, 2, \ldots, d$, one shows that if a point is close enough to the origin, then the nonlinear terms of $\mathbf{f}$ can initially be neglected for the purposes of studying the equation of the motion and, by the preceding argument, the point starts approaching O. Therefore, the nonlinear terms become even less important and, more and more precisely, the system will move as if it were subject to a linear equation.

If $\operatorname{Re}\lambda_1 > 0$, on the contrary, O cannot be stable because the initial datum $\epsilon(\mathbf{v}^{(1)} + \overline{\mathbf{v}^{(1)}})$ moves away from the origin, if ϵ is small enough, at least as much as it is needed so that the equation's nonlinear terms become sizeable. Clearly, this suffices to exclude stability of the origin, even though it cannot permit the exclusion of its attractivity (since the point could go far from O in the $\mathbf{v}^{(1)}, \overline{\mathbf{v}^{(1)}}$ plane (roughly) and, then, under the influence of nonlinearity, it could come back towards O along a direction i where $\operatorname{Re}\lambda_i < 0$, except, of course, when $\operatorname{Re}\lambda_i > 0$, for all $i = 1, \ldots, d$.

The reader will recognize the above ideas in the following proof.

Proof. Let U_R be a sphere centered at the origin and with radius R. Assuming that $\operatorname{Re}\lambda_i < 0$, $i = 1, \ldots, d$, we must determine ρ_0 so that the evolution $t \to S_t(\mathbf{w})$ of an initial datum $\mathbf{w} \in U_{\rho_0}$ develops, $\forall t \geqslant 0$, in U_R: $S_t(\mathbf{w}) \in U_R$, $\forall t \geqslant 0$.

For simplicity, we treat only the case when $\lambda_1, \ldots, \lambda_d$ are pairwise distinct. The reader can think of the general case as a problem (basically, it is just an algebraic problem).

We proceed as in the theory of small oscillations of §2.14, Proposition 20, p. 67. We write Eq. (5.3.1), assuming, without loss of generality, that $\mathbf{x}_0 = \mathbf{0}$:

$$\dot{\mathbf{x}} = L\mathbf{x} + (\mathbf{f}(\mathbf{x}) - L\mathbf{x}) \equiv L\mathbf{x} + N(\mathbf{x}), \tag{5.4.12}$$

[7] Since L is a real matrix, its eigenvalues appear in complex-conjugate pairs or are real. Similarly, the eigenvectors can be chosen to be either real or appearing in complex-conjugate pairs corresponding to complex-conjugate eigenvalues.

where $\mathbf{N}$ is an $\mathbf{R}^d$-valued $C^\infty(\mathbf{R}^d)$ function with a second-order zero at the origin.

By Taylor's theorem, see Appendix B, given $R > 0$, there is a constant C_R such that

$$|N(\mathbf{x})| \leqslant C_R|\mathbf{x}|^2, \qquad \forall \mathbf{x} \in U_R. \tag{5.4.13}$$

We now consider Eq. (5.4.12) as a differential equation in which $\mathbf{N}(\mathbf{x}(t))$ is thought of as a known function of t, $\forall t \geqslant 0$. Then a particular solution of Eq. (5.4.12), thought of in this way, is

$$\mathbf{p}(t) = \int_0^t \sum_{i=1}^d e^{\lambda_i(t-\tau)}\alpha_i(N(\mathbf{x}(\tau)))\mathbf{v}^{(i)}\,d\tau, \tag{5.4.14}$$

where, in general, given $\mathbf{w} \in \mathbf{R}^d$ we shall set

$$\mathbf{w} \equiv \sum_{i=1}^d \alpha_i(\mathbf{w})\mathbf{v}^{(i)}. \tag{5.4.15}$$

Since $\mathbf{v}^{(1)}, \ldots, \mathbf{v}^{(d)}$ is a basis in $\mathbf{C}^d$, such a representation is possible and defines the coefficients $\alpha_i(\mathbf{w})$ (which, in general, may be complex even for real $\mathbf{w}$); and, furthermore, there is a constant A such that

$$\sum_{i=1}^d |\alpha_i(\mathbf{w})| \leqslant A|\mathbf{w}|. \tag{5.4.16}$$

We shall suppose to have chosen the vectors $\mathbf{v}^{(i)}$ so that $|\mathbf{v}^{(i)}| \equiv 1$, $i = 1, \ldots, d$, which implies that $A \geqslant 1$.

Then the solution to Eq. (5.4.12), $t \to \mathbf{x}(t)$, $t \geqslant 0$, with the initial datum $\mathbf{w}$ will be

$$S_t(\mathbf{w}) = \sum_{i=1}^d e^{\lambda_i t}\alpha_i(\mathbf{w})\mathbf{v}^{(i)} + \int_0^t \sum_{i=1}^d e^{\lambda_i(t-\tau)}\alpha_i(N(S_\tau(\mathbf{w})))\mathbf{v}^{(i)}\,d\tau. \tag{5.4.17}$$

We now use the boundedness assumption on the trajectories to infer the existence of $\mu(R) < +\infty$ such that $|S_t(\mathbf{w})| \leqslant \mu(R)$, $\forall t \geqslant 0$, $\forall \mathbf{w} \in U_R$. Then, setting, $\forall \rho \leqslant R$,

$$D(t) = \max_{\substack{0 \leqslant \tau \leqslant t \\ |\mathbf{w}| \leqslant \rho}} |S_\tau(\mathbf{w})|, \tag{5.4.18}$$

one deduces from Eqs. (5.4.17), (5.4.18), (5.4.16), and (5.4.13):

$$|S_t(\mathbf{w})| \leqslant e^{-\nu t}A|\mathbf{w}| + A\int_0^t e^{-\nu(t-\tau)}C_{\mu(R)}D(t)^2\,d\tau \leqslant A\rho + \frac{A}{\nu}C_{\mu(R)}D(t)^2, \tag{5.4.19}$$

where $\nu = \min_{i=1,\ldots,d}|\mathrm{Re}\,\lambda_i|$.

By the arbitrariness of t and by the monotonicity of D, as a function of t, Eq. (5.4.19) means that

$$D(t) \leqslant A\rho + \frac{A}{\nu}C_{\mu(R)}D(t)^2, \tag{5.4.20}$$

i.e., if $4A^2C_{\mu(R)}\gamma^{-1}\rho < 1$, it must either be that

$$D(t) \geqslant \frac{1+\sqrt{1-4A^2C_{\mu(R)}\nu^{-1}\rho}}{2AC_{\mu(R)}\nu^{-1}} \geqslant \frac{1}{2AC_{\mu(R)}\nu^{-1}} \tag{5.4.21}$$

or

$$D(t) \leqslant \frac{1-\sqrt{1-4A^2C_{\mu(R)}\nu^{-1}\rho}}{2AC_{\mu(R)}\nu^{-1}} \leqslant K\rho \tag{5.4.22}$$

if $K > 1$ is a suitably chosen constant (R dependent).

If $|\mathbf{w}| \leqslant \rho_0 < R$ and ρ_0 is chosen so that also

$$\rho_0 = \tfrac{1}{2}\left(2AC_{\mu(R)}\nu^{-1}\right)^{-1}, \tag{5.4.23}$$

we see that Eq. (5.4.22) must hold for all $t \geqslant 0$, by continuity, since for $t = 0$,

$$|\mathbf{w}| \equiv D(0) \leqslant \rho_0. \tag{5.4.24}$$

Hence for all $\mathbf{w} \in U_{\rho_0}$,

$$D(t) \leqslant K|\mathbf{w}| \tag{5.4.25}$$

which clearly implies that O is stable.

To show the attractivity of O, one can use the autonomy of the differential equation (5.4.12) or (5.3.1). If, in fact, we show that there is a time $t_0 > 0$ and a $\bar{\rho} < \rho_0$ [here choose ρ_0 as given by Eq. (5.4.23) with $R = 1$, say], such that

$$|S_t(\mathbf{w})| \leqslant \tfrac{1}{2}|\mathbf{w}|, \qquad \forall t \geqslant t_0, \quad \mathbf{w} \in U_{\bar{\rho}}, \tag{5.4.26}$$

then, by the autonomy of the differential equation,

$$|S_t(\mathbf{w})| \leqslant 2^{-n}|\mathbf{w}|, \qquad \forall t \geqslant nt_0, \quad |\mathbf{w}| \in U_{\bar{\rho}}, \tag{5.4.27}$$

as one sees by iterating Eq. (5.4.26). Hence, from Eq. (5.4.27),

$$|S_t(\mathbf{w})| \leqslant 2 \cdot 2^{-t/t_0}|\mathbf{w}|, \qquad \forall t \geqslant t_0 \tag{5.4.28}$$

which shows that the origin attracts the points of $U_{\bar{\rho}}$ with exponential speed.

It remains, then, to show Eq. (5.4.26).

The first of Eqs. (5.4.19), together with Eq. (5.4.25), implies

$$|S_t(\mathbf{w})| \leqslant e^{-\nu t}A|\mathbf{w}| + \frac{A}{\nu}C_{\mu(1)}K^2|\mathbf{w}|^2 \leqslant |\mathbf{w}|\left(e^{-\nu t}A + \frac{A}{\nu}C_{\mu(1)}K^2\bar{\rho}\right), \tag{5.4.29}$$

$\forall |\mathbf{w}| \leqslant \bar{\rho}$, with $\bar{\rho}$ arbitrary provided $\bar{\rho} \leqslant \rho_0$.

If $\bar{\rho}$ is chosen so small that $A\nu^{-1}C_{\mu(1)}K^2\bar{\rho} < \frac{1}{4}$, it follows that

$$|S_t(\mathbf{w})| \leqslant \left(e^{-\nu t}A + \tfrac{1}{4}\right)|\mathbf{w}| \tag{5.4.30}$$

and Eq. (5.4.26) follows by choosing t_0 so that $Ae^{-\nu t_0} = \frac{1}{4}$, i.e., $t_0 = \nu^{-1}\log 4A$.

The statement concerning the instability is left to the reader as a problem. mbe

Exercises and Problems for §5.4

1. Compute the Lyapunov matrix for the stationary points of the equation $\dot{x} = ax(1-x)$, $a \in \mathbf{R}$, and find for which values of a they are stable.

2. Consider the pendulum's differential equation on $\mathbf{R}^2$:

$$\dot{x} = y, \qquad \dot{y} = -g \sin x.$$

Find the stationary points and compute their Lyapunov matrices, identifying the unstable ones. Find the explicit value of the eigenvalues of the Lyapunov matrix relative to all the stationary points and find the stable ones. (Hint: Stability cannot, in this case, be decided on the basis of Lyapunov's criterion; use the conservation of energy, instead.)

3. Consider the Euler equations (4.11.32)–(4.11.34), p. 312. Assume $I_1 < I_2 < I_3$ and compute the Lyapunov matrix of the stationary solutions different from $\boldsymbol{\omega} = \mathbf{0}$. Show that the only other stationary solutions are uniform rotations around either the $\mathbf{i}_1$ axis, or the $\mathbf{i}_2$ axis, or around the $\mathbf{i}_3$ axis. Show that the only solutions of this type for which the Lyapunov criterion does not exclude the stability are those around the $\mathbf{i}_1$ axis or around the $\mathbf{i}_3$ axis.

4. Same as Problem 3, but with $I_1 = I_2 = I$, $I_3 = J$. Show that the Lyapunov criterion is useless to decide the stability of any of the stationary rotations.

5. Suppose that the differential equation in $\mathbf{R}^d$, $\dot{\mathbf{x}} = \mathbf{f}(\mathbf{x})$, admits a prime integral $A(\mathbf{x})$, i.e., a function $A \in C^\infty(\mathbf{R}^d)$ such that, $\forall \mathbf{x} \in \mathbf{R}^d$, $\forall t \geqslant 0$, $A(S_t(\mathbf{x})) = A(\mathbf{x})$. Suppose that A has a strict minimum at $\mathbf{x}_0 \in \mathbf{R}^d$. Show that $\mathbf{x}_0$ is a stable stationary point.

6. Use Problem 5 and the conservation of energy to discuss the stationary rotations of the frictionless gyroscope and their stability properties along the following lines. First find the De Prit variables of the uniform stationary rotations (see §4.11, p. 318 and p. 320) around the axis of inertia $\mathbf{i}_k$, $k = 1,2,3$. (Answer: K_2, A, A, γ, φ, $\psi + \omega t$). Then, using the De Prit Hamiltonian as a prime integral and Problem 5, show that the rotation around $\mathbf{i}_3$ is stable if $I_3 > I_1, I_2$. (Hint: Note that the De Prit Hamiltonian can be written as

$$H = \frac{A^2}{2I_3} + \frac{1}{2}\left(\frac{(\cos\psi)^2}{I_1} + \frac{(\sin\psi)^2}{I_2} - \frac{1}{I_3} \right)(A^2 - L^2)$$

which has a minimum when $A = L$ if and only if $I_2, I_1 < I_3$.)

7. If the differential equation $\dot{\mathbf{x}} = \mathbf{f}(\mathbf{x})$ on $\mathbf{R}^d$ is such that there exists a function $A \in C^\infty(U)$, $U \subset \mathbf{R}^d$, which is monotonically nonincreasing along the motions (i.e., $A(S_t(\mathbf{x})) \leqslant A(\mathbf{x})$, $\forall t \geqslant 0$, $\forall \mathbf{x} \in U$, as long as $S_\tau(\mathbf{x}) \in U$, $\forall \tau \in [0,t]$), we shall say that A is a "monotonic function" for the given differential equation. If A is monotonically decreasing, we call it a "Lyapunov function" for the differential equation.

Show that every point where a monotonic function (for a given differential equation) has a strict minimum is a stable fixed point.

8. In the context of Problem 7, and under the assumptions of Proposition 6, one can define the function

$$A(\mathbf{w}) = \int_0^{+\infty} |S_t(\mathbf{w})|^2 2^{t/2t_0}\, dt$$

for $|\mathbf{w}|$ small enough, say, $|\mathbf{w}| < \rho$.

Show that:

(i) A is well defined for all $|\mathbf{w}| < \bar{\rho}$ if $\bar{\rho}$ is chosen as in Eq. (5.4.29).
(ii) $A \in C^\infty(U_{\bar{\rho}})$, where $U_{\bar{\rho}} = \{|\mathbf{w}| \,|\mathbf{w}| < \bar{\rho}\}$.
(iii) A is a Lyapunov function in the sense of Problem 7.
(iv) $2^{t/2t_0} A(S_t(\mathbf{w}))$ is monotonic in $t \geqslant 0$, $\forall \mathbf{w} \in U_{\bar{\rho}}$.
(v) A has a strict minimum at $\mathbf{w} = \mathbf{0}$.

This is the "second Lyapunov theorem" (on the existence of a Lyapunov function whenever the stationary point has a stability matrix with eigenvectors having a negative real part).

9. Compute the function A of Problem 8 for the linear equation $\dot{\mathbf{x}} = L\mathbf{x}$. Assume that all the eigenvalues of L are pairwise distinct and have a negative real part. Show that A is a quadratic form in $\mathbf{w}$ which is positive definite. (Answer: If $\gamma_0 = (1/2t_0)\log 2 > (\log 2)\nu/2 \log 4A$, with A being the constant introduced in Eq. (5.4.16) and not to be confused with the quadratic form A that we wish to compute, it is

$$A(\mathbf{w}) = \sum_{i,j=1}^{d} \left(\lambda_i + \lambda_j + \gamma_0\right)^{-1} \overline{\alpha_i(\mathbf{w})}\, \alpha_j(\mathbf{w}),$$

where $\alpha_i(\mathbf{w})$ is defined as in Eq. (5.4.15).)

10. In the context of Problem 9, show that the ellipsoid $A(\mathbf{w}) = a > 0$ has in $\mathbf{w}$ an outer normal $\mathbf{n}(\mathbf{w})$ such that $\mathbf{n}(\mathbf{w}) \cdot L\mathbf{w} < 0$. (Hint: Note that $\mathbf{n}(\mathbf{w}) = \partial A(\mathbf{w})/|\partial A(\mathbf{w})|$, ∂ denoting the gradient; furthermore, by Problem 9 (iii), the derivative of $2^{t/2t_0} A(S_t(\mathbf{w}))$ is nonpositive:

$$\left(\frac{1}{2t_0}(\log 2)A + \frac{dA}{dt}\right) 2^{t/2t_0} \leqslant 0 \Rightarrow \frac{dA}{dt} \leqslant \frac{-\log 2}{2t_0} A,$$

so if $dA/dt = 0$, it must be that $A = 0$, i.e., $\mathbf{w} = \mathbf{0}$ because A is positive definite. However, $dA/dt = (\partial A)(\mathbf{w}) \cdot L\mathbf{w}$; hence, $\partial A(\mathbf{w}) \cdot L\mathbf{w} < 0$ if $\mathbf{w} \neq \mathbf{0}$.)

11. Show that the proof of Proposition 6 can be interpreted as saying that if $A_0(\mathbf{w})$ denotes the Lyapunov function of the linear differential equation $\dot{\mathbf{x}} = L\mathbf{x}$, see Problems 9 and 10, and if $A(\mathbf{w})$ is the Lyapunov function of the differential equation $\dot{\mathbf{x}} = \mathbf{f}(\mathbf{x})$ with the origin as a fixed point with Lyapunov matrix L, then

$$A(\mathbf{w}) = A_0(\mathbf{w}) + O(|\mathbf{w}|^3)$$

(assuming that the real part of the eigenvalues of L is negative).

12. Consider a one-parameter family of differential equations in $\mathbf{R}^d$: $\dot{\mathbf{x}} = \mathbf{f}(\mathbf{x}, \alpha)$, with $\mathbf{x}_0 = \mathbf{0}$ being a stationary point for all values of $\alpha \in (a, b) \subset \mathbf{R}$. Suppose that the origin's Lyapunov matrix $L(\alpha)$ has pairwise distinct eigenvalues, all with real part $\leqslant -\nu < 0$, $\forall \alpha \in (a, b)$. Let $\alpha_0 \in (a, b)$ and let A_{α_0} be the Lyapunov function of Problem 11, relative to the equation $\dot{\mathbf{x}} = \mathbf{f}(\mathbf{x}, \alpha_0)$.

Show the existence of $\delta > 0$, $\epsilon > 0$ such that the neighborhood $V_\delta = \{\mathbf{x} \,|\, A_{\alpha_0}(\mathbf{x}) < \delta\}$ has an outer normal $\mathbf{n}(\mathbf{x})$ such that, $\forall \mathbf{x} \in \partial V_\delta$, $\mathbf{n}(\mathbf{x}) \cdot \mathbf{f}(\mathbf{x}, \alpha) < 0$, $\forall \alpha \in [\alpha_0 - \epsilon, \alpha_0 + \epsilon]$. (Hint: First consider the linear case, then Problems 10 and 11).

13. From Problem 12, deduce that V_δ is invariant for $\dot{\mathbf{x}} = \mathbf{f}(\mathbf{x}, \alpha)$ for all $\alpha \in [\alpha_0 - \epsilon, \alpha_0 + \epsilon]$. (Hint: Suppose the contrary and proceed per absurdum.)

14. Consider a Hamiltonian differential equation in $\mathbf{R}^{2d}$ associated with the Hamiltonian function $H(\mathbf{p}, \mathbf{q}) = \mathbf{p}^2/2 + V(\mathbf{q})$. Let $(\mathbf{0}, \mathbf{q}_0)$ be an equilibrium point. Show that its Lyapunov matrix has eigenvalues that can be collected into pairs of opposite value, either both real or both purely imaginary. Furthermore, show that this implies that its stability cannot be settled on the basis of the Lyapunov criterion, while its instability can sometimes be settled on this basis. (Hint: Note that the Lyapunov matrix has the structure $L = \left(\begin{smallmatrix} A & -B \\ C & D \end{smallmatrix}\right)$ where A, B, C, D are the $d \times d$ matrices

$$A = 0, \qquad B_{ij} = \frac{\partial^2 V}{\partial q_i \partial q_j}(\mathbf{q}_0), \qquad C = 1, \qquad D = 0.$$

So if $L\left(\begin{smallmatrix} \mathbf{u} \\ \mathbf{v} \end{smallmatrix}\right) = \lambda\left(\begin{smallmatrix} \mathbf{u} \\ \mathbf{v} \end{smallmatrix}\right)$, with $\mathbf{u}, \mathbf{v} \in \mathbf{R}^d$, it must be that $\lambda\mathbf{u} + B\mathbf{v} = \mathbf{0}$, $\mathbf{u} = \lambda\mathbf{v}$ so that $-\lambda^2\mathbf{v} = B\mathbf{v}$. But B is symmetric so that its eigenvalues are real (see Appendix F), hence . . .).

15. Show that the Lyapunov matrix eigenvalues are invariant under regular changes of coordinates $\mathbf{y} = \boldsymbol{\sigma}(\mathbf{x})$. (Hint: If $\boldsymbol{\sigma}$ is defined in the vicinity of the stationary point $\mathbf{x}_0 \in \mathbf{R}^d$ for $\dot{\mathbf{x}} = \mathbf{f}(\mathbf{x})$ and if $J_{ij}(\mathbf{y}) = (\partial\sigma_i/\partial x_j)(\mathbf{x})$, for $\mathbf{y} = \sigma(\mathbf{x})$, is the Jacobian matrix of the nonsingular change of coordinates, (i.e., such that $\det J \neq 0$), then the differential equation becomes, in $\mathbf{y}$ coordinates, $\dot{\mathbf{y}} = J(\mathbf{y})\mathbf{f}(\boldsymbol{\sigma}^{-1}(\mathbf{y}))$, and this implies that the Lyapunov matrix at $\mathbf{y}_0 = \boldsymbol{\sigma}^{-1}(\mathbf{x}_0)$ is $L' = J(\mathbf{y}_0)LJ(\mathbf{y}_0)^{-1}$; hence, $\det(L' - \lambda) = \det(JLJ^{-1} - \lambda) = \det(J(L - \lambda)J^{-1}) = \det(L - \lambda)$.)

16. Let H be a Hamiltonian function describing in some local system of coordinates N point masses in $\mathbf{R}^d$ subject to conservative active forces and ideally constrained by a bilateral ideal constraint to a surface Σ (in the sense of Chapter 3).

Let $(\mathbf{0}, \mathbf{x}_0)$, $\mathbf{x}_0 \in \Sigma$, be a stationary point. Give arguments (or prove) that the Lyapunov matrix for the Hamiltonian equations corresponding to the given stationary point appear in pairs of opposite eigenvalues either both real or both purely imaginary. This is a refinement of Problem 15. (Hint: In a system of local regular coordinates around $\mathbf{x}_0$ and adapted to Σ, the system's Lagrangian takes the form [see Eq. (3.11.23), p. 215]: $\mathcal{L} = \frac{1}{2}\Sigma'_{i,j=1} g_{ij}(\boldsymbol{\beta})\dot{\beta}_i\dot{\beta}_j - V(\boldsymbol{\beta})$, with g being a C^∞ positive-definite matrix function and with V also of class C^∞. So the Hamiltonian is [see Eq. (3.11.25), p. 216] $H = \frac{1}{2}\Sigma'_{i,j=1} g_{ij}^{-1}(\boldsymbol{\beta})p_ip_j + V(\boldsymbol{\beta})$. Hence, the matrix L is $\left(\begin{smallmatrix} A & -B \\ C & D \end{smallmatrix}\right)$ with

$$A = 0, \qquad B_{ij} = \frac{\partial^2 V}{\partial \beta_i \partial \beta_j}(\boldsymbol{\beta}_0), \qquad C = G^{-1}, \qquad D = 0,$$

where $G_{ij} = g(\boldsymbol{\beta}_0)_{ij}$ and $\boldsymbol{\beta}_0$ is the point representing $\mathbf{x}_0$ in our system of coordinates. So if $L\left(\begin{smallmatrix} \mathbf{u} \\ \mathbf{v} \end{smallmatrix}\right) = \lambda\left(\begin{smallmatrix} \mathbf{u} \\ \mathbf{v} \end{smallmatrix}\right)$, $\mathbf{u}, \mathbf{v} \in \mathbf{R}^l$, this means $B\mathbf{v} + \lambda\mathbf{u} = \mathbf{0}$, $G^{-1}\mathbf{u} = \lambda\mathbf{v}$, i.e., $(B + \lambda^2 G)\mathbf{v} = \mathbf{0}$; hence, $0 = \det(B + \lambda^2 G) = \det(B + \lambda^2\sqrt{G}\sqrt{G}) = \det(\sqrt{G}(\sqrt{G^{-1}}B\sqrt{G^{-1}} + \lambda^2) \times \sqrt{G}) = (\det G)\det(\sqrt{G^{-1}}B\sqrt{G^{-1}} + \lambda^2)$, see Appendix F for the definition of the square root of a positive-definite matrix). So, since $\sqrt{G^{-1}}B\sqrt{G^{-1}}$ is a symmetric matrix (because G is such, see Appendix F), it follows that λ^2 is real, positive or negative, etc.).

17. Show that Proposition 6 holds if the hypothesis of bounded trajectories is weakened into that of normality or even into no assumption at all (in the latter case, show that global solutions exist for $t \geqslant 0$ for initial data close enough to $\mathbf{x}_0$. (Hint: Simply carefully examine the proof of Proposition 6.)

§5.5. Application to the Model of §5.1. The Notion of Vague Attractivity of a Stationary Point

In the case of Eq. (5.1.19), it is easy to compute the Lyapunov matrix relative to the stationary solution $\hat{\omega}$:

$$L = \begin{bmatrix} \alpha - \lambda_1 - \lambda_2'''\hat{\omega}_3^2 & -\hat{\omega}_3 & 0 \\ \hat{\omega}_3 & \alpha - \lambda_1 - \lambda_2'''\hat{\omega}_3^2 & 0 \\ 0 & 0 & -\lambda_1 - 3\lambda_2'''\hat{\omega}_3^2 \end{bmatrix}, \quad (5.5.1)$$

whose eigenvalues are

$$(\alpha - \lambda_1 - \lambda_2'''\hat{\omega}_3^2) \pm i\hat{\omega}_3, \; -\lambda_1 - 3\lambda_2'''\hat{\omega}_3^2. \quad (5.5.2)$$

Hence, we see that $\hat{\omega}$ is stable and attractive for some of its neighborhoods not only if $\alpha < \lambda_1$, as already seen in §5.2 and §5.3, but also for $\lambda_1 \leqslant \alpha < \lambda_1 + \lambda_2'''\hat{\omega}_3^2$, see Proposition 6, §5.4, p. 380. The attractivity of $\hat{\omega}$ in this interval of variability of α is exponential near $\hat{\omega}$:

$$|S_t(\omega) - \hat{\omega}| \leqslant 2 \cdot 2^{-t/t_0}|\omega - \hat{\omega}|. \quad (5.5.3)$$

if $|\omega - \hat{\omega}|$ is small enough; $t_0 > 0$ depends only on the matrix L [see §5.4, comment after Eq. (5.4.30)], and it can be estimated as inversely proportional to $(\lambda_1 + \lambda_2'''\hat{\omega}_3^2 - \alpha)$.

A discussion identical to the one developed in the case $\alpha < \lambda_1$ shows that Eq. (5.5.3) implies that every motion of the gyroscope associated with an evolution $t \to S_t(\omega)$ like Eq. (5.5.3), for the angular velocity of the co-moving frame, asymptotically tends to become a uniform rotation around the $\mathbf{i}_3$ axis fixed in space.

The difference between the cases $\alpha < \lambda_1$ and $\lambda_1 \leqslant \alpha < \lambda_1 + \lambda_2\hat{\omega}_3^2$ lies in the fact that now we can no longer guarantee that $\hat{\omega}$ is a "global attractor", i.e., with basin of attraction coinciding with $\mathbf{R}^3$. The criterion of Lyapunov has, in fact, only a local character, and thus it can only lead to the recognition of local stability, instability, or attractivity.

Of course, it is of interest to investigate whether or not the attraction basin for $\hat{\omega}$ is all of $\mathbf{R}^3$, and if not, it would be important to understand where the other attractors for the equation are located. However, this analysis could not be done using general results such as the Lyapunov criterion and we shall not discuss this point in further detail, contenting ourselves with the local information found so far. In any event, it has to be stressed that these kinds of problems are very difficult and very little understood in general.

The motion $\hat{\omega}$ is no longer stable for $\alpha > \lambda_1 + \lambda_2\hat{\omega}_3^2$, not even locally, by the second part of Proposition 6, §5.4.

We then inquire about what happens to a solution of Eq. (5.1.19) following an initial datum ω slightly different from $\hat{\omega}$ and for $\alpha > \alpha_c \equiv \lambda_1 + \lambda_2'''\hat{\omega}_3^2$, at least for small $\alpha - \alpha_c$.

The first question that one can ask is whether for α slightly larger than α_c,

$$\alpha_c = \lambda_1 + \lambda_2'''\hat{\omega}_3^2, \tag{5.5.4}$$

the motion of the data ω close to $\hat{\omega}$ departs very much from the motion $\hat{\omega}$. As we shall see, this question naturally leads to the following interesting notion of "vague attractivity".

5 Definition. Let $(\mathbf{x},\alpha)\to\mathbf{f}(\mathbf{x},\alpha)$ be an $\mathbf{R}^d$-valued $C^\infty(\mathbf{R}^d\times I)$ function with $I=$ open interval, such that the differential equations

$$\dot{\mathbf{x}} = \mathbf{f}(\mathbf{x},\alpha), \tag{5.5.6}$$

parametrized by $\alpha\in I$, have uniformly bounded trajectories[8] with respect to $\alpha \in I$ and, furthermore, admit a stationary solution $\mathbf{x}_0\in\mathbf{R}^d$ such that

$$\mathbf{f}(\mathbf{x}_0,\alpha)\equiv\mathbf{0}. \tag{5.5.7}$$

We shall say that $\mathbf{x}_0$ is "vaguely attractive" near $\alpha_c\in I$ if there is a neighborhood U of $\mathbf{x}_0$ such that for every $\delta>0$, one can find $t_\delta>0$, $\epsilon_\delta>0$, $\rho_\delta>0$ such that

$$\begin{aligned} S_t^{(\alpha)}U &\subset \Gamma(\delta), && \forall t \geqslant t_\delta, \quad \forall\alpha\in(\alpha_c-\epsilon_\delta,\alpha_c+\epsilon_\delta),\\ S_t^{(\alpha)}\Gamma(\delta) &\subset \Gamma(\rho_\delta), && \forall t\geqslant 0, \quad \forall\alpha\in(\alpha_c-\epsilon_\delta,\alpha_c+\epsilon_\delta) \end{aligned} \tag{5.5.8}$$

with $\rho_\delta\to_{\delta\to 0}0$. Here $S_t^{(\alpha)}$ is the solution flow for Eq. (5.5.5) and $\Gamma(\delta)$ = cube with side 2δ centered around $\mathbf{x}_0$.

Observations.

(1) In other words, $\mathbf{x}_0$ is vaguely attractive near α_c if there is a neighborhood U which is a basin of attraction for an attractor containing $\mathbf{x}_0$ and having a diameter smaller than any arbitrarily prefixed length $\delta>0$ for all α's close enough to α_c. Furthermore, this attractor, contained in $\Gamma(\delta)$, "uniformly attracts" the points of U and has a "weak stability" as expressed more precisely by the first and second of Eqs. (5.5.7), respectively.

Note that for $\alpha=\alpha_c$, the point $\mathbf{x}_0$ must be attractive for the points in U. In fact $\mathbf{x}_0$ is vaguely attractive for α near α_c if and only if it is stable and attractive for Eq. (5.5.6) with $\alpha\equiv\alpha_c$.

(2) One can also say that $\mathbf{x}_0$ is vaguely attractive near α_c if it is the attractor of a neighborhood U of $\mathbf{x}_0$, for $\alpha=\alpha_c$, while for α close to α_c, it still attracts the points of U not too close to $\mathbf{x}_0$. The "attractivity away from $\mathbf{x}_0$ is uniform in α" near α_c.

(3) If in α_c the Lyapunov matrix $L(\alpha)$ for Eq. (5.5.5) relative to $\mathbf{x}_0$ has

[8] If $S_t^{(\alpha)}$ denotes the flow generated by Eq. (5.5.5), this means that the bound on the trajectory of $\mathbf{x}\in\mathbf{R}^d$, $|S_t^{(\alpha)}(\mathbf{x})| < \mu(|\mathbf{x}|)$ holds and μ is continuous and is α independent.

eigenvalues with a negative real part, it follows from the arguments of the proof of Proposition 6, §5.4, that $\mathbf{x}_0$ is vaguely attractive near α_c. Actually, the set U can be taken such that for some $\epsilon_0 > 0$ it is $S_t^{(\alpha)}U \subset U$. (Basically this follows from the Problems 12 and 13, §5.4, p. 386.)

(4) Hence, the vague-attractivity notion is interesting only when $L(\alpha_c)$ has some eigenvalues with a vanishing real part.

(5) All the examples of vaguely attractive points that we will meet will have the property that U can be chosen to fulfill Eq. (5.5.8). It seems not impossible that the neighborhood U of vague attractivity could always be chosen to verify Eq. (5.5.8).

(6) The condition that $\Gamma(\delta)$ be a cube with side 2δ centered at $\mathbf{x}_0$ could be equivalently replaced by the requirement that $\Gamma(\delta)$ be a family of neighborhoods of $\mathbf{x}_0$ with diameter tending to zero with δ.

(7) The assumption that $\mathbf{x}_0$ should be α independent is only apparently more restrictive than the natural assumption of the existence of a stationary solution $\mathbf{x}^{(\alpha)}$ depending in a C^∞-regular way on α. With a change of coordinates, one can always reduce the stability theory of such a stationary point to that relative to the case when $\mathbf{x}^{(\alpha)} = \mathbf{0}$.

(8) If $\mathbf{x}_0$ is replaced in the above definition by an invariant set A and $\Gamma_A(\delta) = \{$set of the points at distance $\leqslant \delta$ from $A\}$, one defines the notion of a "vague attractor".

This notion could be extended to the case when A depends on α, though not as straightforwardly and as unambigously, as in the case $A = \{\mathbf{x}^{(\alpha)}\}$ discussed in Observation (7).

(9) Last but not least, it is not difficult to check that the vague attractivity of $\mathbf{x}_0$ is a notion invariant under changes of coordinates; it is also invariant under changes of the equation itself (i.e., of the function $\mathbf{f}(\mathbf{x}, \alpha)$), for $\mathbf{x}$ outside some neighborhood of $\mathbf{x}_0$. One says that the vague attractivity is an "intrinsic local property" of Eq. (5.5.5) near $\mathbf{x}_0$ and α_c.

These facts play an important role in the formulation of simple vague attractivity criteria.

Vague attractivity is a property that can often be inferred from the knowledge of the $\mathbf{x}$ derivatives of $\mathbf{f}(\mathbf{x}, \alpha_c)$ in $\mathbf{x}_0$ of order not exceeding 3.

To illustrate this important fact and to provide, in this way, some simple vague-attractivity criteria, it is convenient to introduce the notion of "normal form" of a differential equation near a stationary solution.

6 Definition. Let $\dot{\mathbf{x}} = \mathbf{f}(\mathbf{x}, \alpha)$, $\mathbf{f} \in C^\infty(\mathbf{R}^d \times \mathbf{R})$, be a differential equation in $\mathbf{R}^d$ with uniformly bounded trajectories (see footnote 8 to p. 389) and with $\mathbf{x}_0 = \mathbf{0}$ as a stationary point, for $\alpha \in I = (a, b)$.

Let $L(\alpha)$ be the stability matrix at $\mathbf{x}_0 = \mathbf{0}$ and suppose that $\lambda_1(\alpha)$, $\overline{\lambda_1(\alpha)}, \dots, \lambda_p(\alpha), \overline{\lambda_p(\alpha)}, \lambda'_1(\alpha), \dots, \lambda'_q(\alpha)$ are $2p + q$ of its eigenvalues, the first $2p$ being arranged into complex-conjugate nonreal pairs and the last q being real.

We say that the differential equation has, for $\alpha \in I$, a "normal form" with respect to the mentioned eigenvalues of $L(\alpha)$ if, writing the coordinates of $\mathbf{x}$ as $(\mathbf{x}^{(1)}, \mathbf{x}^{(2)}, \ldots, \mathbf{x}^{(p)}, y^{(1)}, \ldots, y^{(q)}, \mathbf{z})$ with $\mathbf{x}^{(j)} \in \mathbf{R}^2$, $j = 1, \ldots, p$, $y^{(j)} \in \mathbf{R}$, $j = 1, \ldots, q$, and $\mathbf{z} \in \mathbf{R}^{d-2p-q}$, the equation has the form, $\forall \alpha \in I$,

$$\begin{aligned}
\dot{x}_1^{(j)} &= (\operatorname{Re}\lambda_1(\alpha))x_1^{(j)} - (\operatorname{Im}\lambda_1(\alpha))x_2^{(j)} \\
&\quad + N_1^{(j)}(\mathbf{x}^{(1)}, \ldots, \mathbf{x}^{(p)}, y^{(1)}, \ldots, y^{(q)}\mathbf{z}, \alpha), \\
\dot{x}_2^{(j)} &= (\operatorname{Im}\lambda_1(\alpha))x_1^{(j)} - (\operatorname{Re}\lambda_1(\alpha))x_2^{(j)} \\
&\quad + N_2^{(j)}(\mathbf{x}^{(1)}, \ldots, \mathbf{x}^{(p)}, y^{(1)}, \ldots, y^{(q)}\mathbf{z}, \alpha), \\
\dot{y}^{(h)} &= \lambda_h'(\alpha)y^{(h)} + M^{(h)}(\mathbf{x}^{(1)}, \ldots, \mathbf{x}^{(p)}, y^{(i)}, \ldots, y^{(q)}, \mathbf{z}, \alpha), \\
\dot{\mathbf{z}} &= \tilde{L}(\alpha)\mathbf{z} + \tilde{\mathbf{P}}(\mathbf{x}^{(1)}, \ldots, \mathbf{x}^{(p)}, y^{(1)}, \ldots, y^{(q)}, \mathbf{z}, \boldsymbol{\alpha}),
\end{aligned} \tag{5.5.9}$$

$j = 1, \ldots, p$, $h = 1, \ldots, q$, $\tilde{L}(\alpha)$ being a $(d - 2p - q) \times (d - 2p - q)$ matrix with C^∞ entries (as functions of α), and $N_1^{(j)}$, $N_2^{(j)}$, $M^{(h)}$, $\tilde{\mathbf{P}}$ being C^∞ functions of their arguments with the extra property that $N_1^{(j)}$, $N_2^{(j)}$, $M^{(h)}$ have a zero of third order at the origin in the $\mathbf{x}$, $\mathbf{y}$, $\mathbf{z}$ variables, for all $\alpha \in I$, while $\tilde{\mathbf{P}}$ has a second-order zero, at least, at the origin (in the same variables).

Observation. If $p = 0$ or $q = 0$ or $d = 2p + q$, the above definition makes sense in an obvious way by deleting parts of Eq. (5.5.9).

As we shall see, the vague attractivity near α_c may be easily discussed once the equation is written in normal form with respect to the eigenvalues of $L(\alpha)$ whose real part vanishes for $\alpha = \alpha_c$.

In general, the equations that one wishes to study will not have normal form, but they may acquire such a form after a change of variables. This is as suitable for vague-attractivity analysis, by Observation (9), p. 390.

For instance, Eq. (5.1.19) does not have normal form near α_c with respect to the two complex eigenvalues of $L(\alpha)$.

Therefore, before discussing a vague-attractivity criterion, it is convenient to remark that there is a simple and rather weak sufficient condition for the existence of a system of coordinates where the equation $\dot{\mathbf{x}} = \mathbf{f}(\mathbf{x}, \alpha)$ assumes normal form.

7 Proposition. *Let $\dot{\mathbf{x}} = \mathbf{f}(\mathbf{x}, \alpha)$, $\mathbf{f} \in C^\infty(\mathbf{R}^d \times \mathbf{R})$, be a differential equation parameterized by α and with uniformly bounded trajectories as $\alpha \in I = (a, b)$. Suppose that $\mathbf{f}(\mathbf{0}, \alpha) \equiv \mathbf{0}$ and let $L(\alpha)$ be the stability matrix of $\mathbf{0}$.*

Suppose that for $\alpha \in I$, $L(\alpha)$ has d pairwise-distinct eigenvalues $\Lambda_1(\alpha), \ldots, \Lambda_d(\alpha)$, among which $2p$ are nonreal; write them as $\lambda_1(\alpha)$, $\overline{\lambda_1(\alpha)}, \ldots, \lambda_p(\alpha), \overline{\lambda_p(\alpha)}, \lambda_1'(\alpha), \ldots, \lambda_q'(\alpha)$, $(2p + q = d)$, and arrange them so that the functions $\alpha \to \Lambda_j(\alpha)$ are C^∞ functions of α, for $\alpha \in I$.[9]

(i) *There is a (global) coordinate system on $\mathbf{R}^d \times I$:$(\mathbf{R}^d \times I, \boldsymbol{\Xi})$, with basis $\mathbf{R}^d \times I$, denoted $(\mathbf{x}, \alpha) = \boldsymbol{\Xi}(\boldsymbol{\xi}^{(1)}, \ldots, \boldsymbol{\xi}^{(p)}, \eta^{(1)}, \ldots, \eta^{(q)}, \alpha')$ with*

[9] Since the eigenvalues are supposed to be pairwise distinct and they are the roots of a dth-order polynomial, this is possible and it follows from general results in algebra.

$\alpha' = \alpha$, $\boldsymbol{\xi}^{(j)} \in \mathbf{R}^2$, $\eta^{(h)} \in \mathbf{R}$, such that in the new coordinates, the equation takes the form, $(j = 1, \ldots, p; h = 1, \ldots, q)$,

$$\begin{aligned} \dot{\xi}_1^{(j)} &= (\operatorname{Re}\lambda_1(\alpha))\xi_1^{(j)} - (\operatorname{Im}\lambda_1(\alpha))\xi_2^{(j)} + F_1^{(j)}(\boldsymbol{\xi}^{(1)}, \ldots, \alpha), \\ \dot{\xi}_2^{(j)} &= (\operatorname{Im}\lambda_1(\alpha))\xi_1^{(j)} + (\operatorname{Re}\lambda_1(\alpha))\xi_2^{(j)} + F_2^{(j)}(\boldsymbol{\xi}^{(1)}, \ldots, \alpha), \\ \dot{\eta}^{(h)} &= \lambda_h'(\alpha)\eta^{(h)} + F^{(h)}(\boldsymbol{\xi}^{(1)}, \ldots, \alpha), \end{aligned} \tag{5.5.10}$$

where $F_1^{(j)}$, $F_2^{(j)}$, $F^{(h)}$ are in $C^\infty(\mathbf{R}^d \times I)$ and have a second-order zero at the origin in the variables $\boldsymbol{\xi}$, $\boldsymbol{\eta}$ for each $\alpha \in I$.

(ii) *If $\Lambda_k(\alpha_c) \neq \Lambda_h(\alpha_c) + \Lambda_l(\alpha_c)$, $k, h, l = 1, \ldots, d$, and if the equation $\dot{\mathbf{x}} = \mathbf{f}(\mathbf{x}, \alpha)$ has already the form of Eq. (5.5.10) for $\alpha \in I$, there is a coordinate system on a suitable neighborhood $U \times J$ of $(\mathbf{0}, \alpha_c)$, $(U \times J, \boldsymbol{\Xi})$, such that in the new coordinates, Eq. (5.5.10) takes normal form with respect to all the eigenvalues of $L(\alpha)$. Calling $(\boldsymbol{\beta}, \alpha')$ the new coordinates, the transformation $\boldsymbol{\Xi}$ can be chosen as*

$$\begin{aligned} \beta_j &= x_j - \sum_{k,l=1}^{d} S_{jkl}(\alpha) x_k x_l, \\ \alpha' &= \alpha \end{aligned} \tag{5.5.11}$$

with $S_{jkl} = S_{jlk} \in C^\infty(J)$, "quadratic change of coordinates"; its inverse will (therefore[10]) have the form

$$\begin{aligned} x_j &= \beta_j + \sum_{k,l=1}^{d} S_{jkl}(\alpha)\beta_k\beta_l + G_j(\boldsymbol{\beta}, \alpha) \\ \alpha &= \alpha', \end{aligned} \tag{5.5.12}$$

where $G_j \in C^\infty(\boldsymbol{\Xi}(U \times J))$ has a third-order zero at $\boldsymbol{\beta} = \mathbf{0}$, $\forall \alpha \in J$.

Observations.

(1) Note that defining, for $j = 1, \ldots, p$, $h = 1, \ldots, q$,

$$\begin{aligned} z^{(j)} &= \xi_1^{(j)} + i\xi_2^{(j)}, \qquad \bar{z}^{(j)} = \xi_1^{(j)} - i\xi_2^{(j)}, \\ N^{(j)}&(z^{(1)}, \bar{z}^{(1)}, \ldots, z^{(p)}, \bar{z}^{(p)}, \eta^{(1)}, \ldots, \eta^{(q)}, \alpha) \\ &= F_1^{(j)}(\boldsymbol{\xi}^{(1)}, \ldots) + iF_2^{(j)}(\boldsymbol{\xi}^{(1)}, \ldots), \\ M^{(h)}&(z^{(1)}, \bar{z}^{(1)}, \ldots, z^{(p)}, \bar{z}^{(p)}, \eta^{(1)}, \ldots, \eta^{(q)}, \alpha) \\ &= F^{(h)}(\boldsymbol{\xi}^{(1)}, \ldots), \end{aligned} \tag{5.5.14}$$

Eq. (5.5.10) assumes the more symmetric form

$$\begin{aligned} \dot{z}^{(j)} &= \lambda_j(\alpha) z^{(j)} + N^{(j)}(z^{(1)}, \bar{z}^{(1)}, \ldots, \alpha), \qquad j = 1, \ldots, p, \\ \dot{\eta}^{(h)} &= \lambda_h'(\alpha)\eta^{(h)} + M^{(h)}(z^{(1)}, \bar{z}^{(1)}, \ldots, \alpha), \qquad h = 1, \ldots, q \end{aligned}$$

which is sometimes very convenient.

[10] By the implicit function theorem (see Appendix G).

(2) If the eigenvalues $\lambda_1(\alpha_c)$, $\overline{\lambda_1(\alpha_c)}$ are nondegenerate and $\mathrm{Re}\,\lambda_1(\alpha_c)=0$, $\mathrm{Im}\,\lambda_1(\alpha_c)\neq 0$, it follows from (ii) that it will be possible to put the equation $\dot{\mathbf{x}}=\mathbf{f}(\mathbf{x},\alpha)$ into normal form with respect to $\lambda_1(\alpha)$, $\overline{\lambda_1(\alpha)}$ in the sense of Definition 6 above. This is obvious if $\Lambda_k(\alpha_c)\neq\Lambda_h(\alpha_c)+\Lambda_l(\alpha_c)$, $\forall k,h,l$, but it is also generally true as a consequence of (ii) (see below).

Suppose, in fact, that the equation already has the form of Eq. (5.5.10). We then perform the quadratic change of coordinates that would put into normal form, (with respect to all the eigenvalues), the equation obtained from Eq. (5.5.10) by replacing the eigenvalues $(\lambda_1(\alpha),\overline{\lambda_1(\alpha)},\ldots,\lambda_p(\alpha),\overline{\lambda_p(\alpha)},\lambda_1'(\alpha),\ldots,\lambda_q'(\alpha))$ by $(\tilde{\Lambda}_1(\alpha),\ldots,\tilde{\Lambda}_d(\alpha))=(\lambda_1(\alpha),\overline{\lambda_1(\alpha)},\lambda_2(\alpha)+\epsilon_2,\overline{\lambda_2(\alpha)}+\bar{\epsilon}_2,\ldots,\lambda_1'(\alpha)+\epsilon_1',\ldots)$, where $\epsilon_2,\ldots,\epsilon_p,\epsilon_1',\ldots,\epsilon_q'$ are chosen so that the condition $\tilde{\Lambda}_k(\alpha)+\tilde{\Lambda}_h(\alpha)\neq\tilde{\Lambda}_l(\alpha)$, $\forall k,h,l$, is fulfilled (and the ϵ_h' are real).

Taking into account the quadratic nature of the maps of Eqs. (5.5.11) and (5.5.12), it is clear that the original equation will take, in the new coordinates, normal form with respect to the only two eigenvalues which have not been modified, i.e., $\lambda_1(\alpha)$, $\overline{\lambda_1(\alpha)}$.

If the equation does not have the form of Eq. (5.5.10), but $\lambda_1(\alpha_c)$, $\overline{\lambda_1(\alpha_c)}$ are nondegenerate, one can apply a similar argument.

(3) From the proof, it appears that the normal-form coordinates (for all the eigenvalues or, via Observations (2) suitably adapted, for some of them) can sometimes be found even when the "nonresonance condition" on the eigenvalues [in (ii) above] is not fulfilled, provided the equation verifies additional properties. Such conditions can be explicitly stated by requiring that Eq. (5.5.22) below be solvable. In the problems at the end of this section, we give some examples of the explicit use of this remark (see Problems 16 and 17).

Proof. To find the $\xi^{(1)},\ldots,\xi^{(p)},\eta^{(1)},\ldots,\eta^{(q)}$ coordinates, we simply consider the eigenvectors $(\mathbf{w}^{(1)}(\alpha),\ldots,\mathbf{w}^{(d)}(\alpha))\equiv(\mathbf{v}^{(1)}(\alpha),\overline{\mathbf{v}^{(1)}(\alpha)},\ldots,\mathbf{v}^{(p)}(\alpha),\overline{\mathbf{v}^{(p)}(\alpha)},\mathbf{v}'^{(1)}(\alpha),\ldots,\mathbf{v}'^{(q)}(\alpha))$ of $L(\alpha)$ associated with the eigenvalues $(\lambda_1(\alpha),\lambda_1(\alpha),\ldots,\lambda_1'(\alpha),\ldots,\lambda_q'(\alpha))$, $\alpha\in I$, respectively. At fixed $\alpha\in I$, the above vectors are linearly independent, by the assumption of distinct eigenvalues.

We may and shall assume that the above eigenvectors are C^∞ functions of α. Since such vectors form a basis in $\mathbf{C}^d$, any $\mathbf{x}\in\mathbf{R}^d$ can be written as

$$\mathbf{x}=\sum_{j=1}^{p}\left[(\xi_1^{(j)}+i\xi_2^{(j)})\mathbf{v}^{(j)}(\alpha)+(\xi_1^{(j)}-i\xi_2^{(j)})\overline{\mathbf{v}^{(j)}(\alpha)}\right]$$
$$+\sum_{h=1}^{q}\eta^{(h)}\mathbf{v}'^{(h)}(\alpha). \tag{5.5.15}$$

Since the $\mathbf{v}^{(j)}(\alpha)$, $\mathbf{v}'^{(h)}(\alpha)$ are eigenvectors of $L(\alpha)$, it is immediate to check that in the $(\xi^{(1)},\ldots,\xi^{(p)},\eta^{(1)},\ldots,\eta^{(q)})$ coordinates, the equation $\dot{\mathbf{x}}=\mathbf{f}(\mathbf{x},\alpha)$ takes the form of Eq. (5.5.10).

To prove (ii), write Eq. (5.5.10) as

$$\dot{x}_j = f_j(\mathbf{x}, \alpha) = \sum_{k=1}^{d} L_{jk}(\alpha) x_k + \sum_{k,l=1}^{d} F_{jkl}(\mathbf{x}, \alpha) x_k x_l \tag{5.5.16}$$

with $F_{jkl} = F_{jlk} \in C^\infty(\mathbf{R}^d \times J)$, $j = 1, \ldots, d$.

Performing the change of coordinates in Eqs. (5.5.11) and (5.5.12), one easily finds that after some algebra Eq. (5.5.16) becomes, in the new coordinates $\hat{\beta}$,

$$\begin{aligned}\dot{\beta}_j = \dot{x}_j - 2\sum_{k,l=1}^{d} S_{jkl}(\alpha) x_k \dot{x}_l &= \sum_{k=1}^{d} L_{jk}(\alpha)\beta_k \\ &+ \sum_{h,l=1}^{d}\left\{\sum_{k=1}^{d} (L_{jk}(\alpha) S_{khl}(\alpha) - S_{jhk}(\alpha) L_{kl}(\alpha) - S_{jlk}(\alpha) L_{kh}(\alpha))\beta_h \beta_l\right\} \\ &+ F_{jhl}(\mathbf{0}, \alpha)\beta_h \beta_l + \overline{G}_j(\boldsymbol{\beta}, \alpha)\end{aligned} \tag{5.5.17}$$

where $\overline{G}_j$ has a third-order zero at $\boldsymbol{\beta} = \mathbf{0}$, $\forall \alpha \in J$.

Therefore, if we can find a solution $S_{jhk}(\alpha)$ to the linear system of $d[d(d+1)/2]$ equations in $d[d(d+1)/2]$ unknowns (recall that $S_{jhl} = S_{jlh}$, $F_{jhl} = F_{jlh}$) described for $j, h, k = 1, \ldots, d$, by

$$\sum_{k=1}^{d} (L_{jk}(\alpha) S_{khl}(\alpha) - S_{jhk}(\alpha) L_{kl}(\alpha) - S_{jlk}(\alpha) L_{kh}(\alpha)) + F_{jhl}(\mathbf{0}, \alpha) = 0, \tag{5.5.18}$$

and if the solution S_{jhk} depends on α in a C^∞ way for α near α_c, then Proposition 7 will have been proved.

Define a matrix $W(\alpha)$ in terms of the eigenvectors of $L(\alpha)$, $\mathbf{w}^{(1)}(\alpha), \ldots, \mathbf{w}^{(d)}(\alpha)$, as

$$W(\alpha)_{hk} = w_h^{(k)}(\alpha), \qquad h, k = 1, \ldots, d. \tag{5.5.19}$$

The linear independence of the eigenvectors $\mathbf{w}^{(j)}$ implies that $\det W(\alpha) \neq 0$, $\forall \alpha \in I$, so that $W(\alpha)^{-1}$ exists and is a C^∞-matrix function of $\alpha \in I$, and if a matrix $\Lambda(\alpha)$ is defined as $\Lambda(\alpha)_{hk} = \Lambda_h(\alpha)\delta_{hk}$, it is

$$L(\alpha) = W(\alpha)\Lambda(\alpha)W(\alpha)^{-1} \tag{5.5.20}$$

(see Appendix F for some details on this well-known relation between a matrix, its eigenvalues, and its eigenvectors).

Inserting Eq. (5.5.20) into Eq. (5.5.18) one finds

$$\begin{aligned}\sum_{k,s=1}^{d} \Big(W(\alpha)_{js}\Lambda_s(\alpha) W(\alpha)^{-1}_{sk} S_{khl}(\alpha) - S_{jhk}(\alpha) W(\alpha)_{ks}\Lambda_s(\alpha) W(\alpha)^{-1}_{sl} \\ - S_{jlk}(\alpha) W(\alpha)_{ks}\Lambda_s(\alpha) W(\alpha)^{-1}_{sh}\Big) + F_{jhl}(\mathbf{0}, \alpha) = 0\end{aligned}$$

and multiplying both sides by $W(\alpha)^{-1}_{rj} W(\alpha)_{lp} W(\alpha)_{hq}$, summing over j, h, l, and setting

$$\sigma_{spq}(\alpha) = \sum_{j,h,l} W(\alpha)^{-1}_{sk} S_{khl}(\alpha) W(\alpha)_{hq} W(\alpha)_{lp},$$
$$\varphi_{spq}(\alpha) = \sum_{j,h,l} W(\alpha)^{-1}_{sk} F_{khl}(\mathbf{0},\alpha) W(\alpha)_{hq} W(\alpha)_{lp}, \tag{5.5.21}$$

one finds that Eq. (5.5.18) becomes

$$\varphi_{spq} + (\Lambda_s(\alpha) - \Lambda_p(\alpha) - \Lambda_q(\alpha))\sigma_{spq}(\alpha) = 0 \tag{5.5.22}$$

which can certainly be solved uniquely for σ and via Eq. (5.5.21) yields a C^∞ solution to Eq. (5.5.18), $\forall \alpha \in J$. This solution is real because Eq. (5.5.18) is a linear equation with real coefficients and real known terms. mbe

Observation. Note that in the above proof, the determination of the change of coordinates leading to the form of Eq. (5.5.10) only involves the matrix $L(\alpha)$, i.e., the first $\mathbf{x}$ derivatives at the origin of $\mathbf{f}(\mathbf{x},\alpha)$. The definition of $S_{jkl}(\alpha)$, i.e., of the coordinates putting the equation into the normal form, only involves $L(\alpha)$ and $F_{jkl}(\mathbf{0},\alpha)$, i.e., the first and second derivatives of $\mathbf{f}(\mathbf{x},\alpha)$ at $\mathbf{0}$.

It is now possible to discuss a simple vague-attractivity criterion.

8 Proposition. *Let $\dot{\mathbf{x}} = \mathbf{f}(\mathbf{x},\alpha)$, $\mathbf{f} \in C^\infty(\mathbf{R}^d \times \mathbf{R})$, be a differential equation parameterized by α, with uniformly bounded trajectories as $\alpha \in I = (a,b)$, (see footnote 8) and such that $\mathbf{f}(\mathbf{0},\alpha) = \mathbf{0}$, $\forall \alpha \in I$.*

Suppose that for $\alpha = \alpha_c$, the stability matrix of the origin, $L(\alpha_c)$, has one pair of conjugate imaginary eigenvalues $\lambda_1(\alpha_c) = \overline{\lambda_2(\alpha_c)} \neq 0$, while all the other $d-2$ eigenvalues have negative real parts.

Also suppose that the equation has normal form with respect to λ_1, λ_2 near α_c, say for $\alpha \in I$, and write the differential equations for the first two components of $\mathbf{x}$, x_1 and x_2, as

$$\begin{aligned} \dot{x}_1 &= (\operatorname{Re}\lambda_1(\alpha))x_1 - (\operatorname{Im}\lambda_1(\alpha))x_2 + N_1(x_1, x_2, \mathbf{y}, \alpha),\\ \dot{x}_2 &= (\operatorname{Im}\lambda_1(\alpha))x_1 + (\operatorname{Re}\lambda_1(\alpha))x_2 + N_2(x_1, x_2, \mathbf{y}, \alpha) \end{aligned} \tag{5.5.23}$$

with $N_1, N_2 \in C^\infty(\mathbf{R}^d \times \mathbf{R})$ and having a third-order zero at the origin $x_1 = x_2 = 0$, $\mathbf{y} = \mathbf{0}$, for all $\alpha \in I$, having denoted $\mathbf{y}$ the last $d-2$ coordinates of $\mathbf{x}$.

If $x_1 + ix_2 = \rho e^{i\theta}$ define

$$\gamma_\alpha(\theta) = \lim_{\rho\to 0}\lim_{\mathbf{y}\to\mathbf{0}} \frac{x_1 N_1 + x_2 N_2}{(x_1^2 + x_2^2)^2} \tag{5.5.24}$$

for $\alpha \in I$. Then the origin is vaguely attractive near α_c if

$$\bar{\gamma} = \frac{1}{2\pi}\int_0^{2\pi} \gamma_{\alpha_c}(\theta)\,d\theta < 0, \tag{5.5.25}$$

while if $\bar{\gamma} > 0$, it is not a vague attractor.

The same conclusions can be drawn under the sole assumption that the differential equation takes the form of Eq. (5.5.23) without requiring that N_1 and N_2 be of third order, but only requiring the existence of the limit Eq. (5.5.24), i.e., only requiring that $x_1N_1 + x_2N_2$ be of fourth order.

Observations.

(1) As already remarked, the assumption on the normality of the equation with respect to $\lambda_1(\alpha)$, $\overline{\lambda_1(\alpha)}$ is not really restrictive if (as assumed above) $\operatorname{Im}\lambda_1(\alpha_c) \neq 0$ and if all the remaining eigenvalues have a negative real part. In fact, one can always change coordinates and put the equation in this form (see Observation (2), p. 393 to Proposition 7).

(2) The number $\bar{\gamma}$ can, in principle, be computed in any system of coordinates in terms of the derivatives of first order, second order, and third order of $\mathbf{f}(\mathbf{x}, \alpha_c)$ at $\mathbf{x} = \mathbf{0}$, with respect to the $\mathbf{x}$ coordinates. However, this calculation may be very long in practical cases, very very long indeed. For the computation of $\bar{\gamma}$, it is more practical to first reduce the equation to the form of Eq (5.5.23) using Observation (2) to Proposition 7, p. 393, and then to compute $\bar{\gamma}$ via Eq. (5.5.25).

(3) A similar criterion holds if the equation has one real eigenvalue $\lambda'(\alpha)$ vanishing at α_c, while all the others remain with negative real part near α_c if $\mathbf{x} = (x_1, \mathbf{y})$ and assuming

$$\dot{x}_1 = \lambda_1'(\alpha)x_1 + N_1(x_1, \mathbf{y}, \alpha) \tag{5.5.26}$$

with N_1 having a zero of third order at $x_1 = 0$, $\mathbf{y} = \mathbf{0}$, $\forall \alpha \in I$, then a vague-attractivity criterion is that $\bar{\gamma} = \lim_{x_1 \to 0} x_1 N_1(x_1, \mathbf{0}, \alpha_c)/x_1^4 < 0$.

However, the above normal-form assumption, i.e., the assumption that N should be of third order, is now restrictive. Sometimes it might be impossible to find coordinates in which the equation for x_1 takes the form of Eq. (5.5.26).

PROOF. For simplicity, we suppose that the only nonreal eigenvalues of $L(\alpha)$ are $\lambda(\alpha) \equiv \lambda_1(\alpha) = \sigma(\alpha) + i\mu(\alpha) = \overline{\lambda_2(\alpha)}$ and $\lambda_2(\alpha)$, for α near α_c; we also suppose that the other eigenvalues are pairwise distinct and $\mu(\alpha) > 0$.

Let $\nu > 0$, $a > 0$ be such that

$$\lambda_1'(\alpha), \ldots, \lambda_{d-2}'(\alpha) \leqslant -\nu < 0, \qquad \mu(\alpha) > \nu, \qquad \forall \alpha \in (\alpha_c - a, \alpha_c + a).$$

We may and shall suppose that the equation takes the form

$$\begin{aligned} \dot{x}_1 &= \sigma(\alpha)x_1 - \mu(\alpha)x_2 + N_1(x_1, x_2, \mathbf{y}, \alpha), \\ \dot{x}_2 &= \mu(\alpha)x_1 + \sigma(\alpha)x_2 + N_2(x_1, x_2, \mathbf{y}, \alpha), \\ \dot{y}_j &= \lambda_j'(\alpha)y_j + \tilde{N}_j(x_1, x_2, \mathbf{y}, \alpha), \qquad j = 1, \ldots, d-2 \end{aligned} \tag{5.5.27}$$

with N_1, N_2 having a third-order zero at $x_1 = 0$, $x_2 = 0$, $\mathbf{y} = \mathbf{0}$, $\forall \alpha \in (\alpha_c - a, \alpha_c + a)$, while $\tilde{N}_j$ has at least a second-order zero at the same point, $\forall \alpha \in (\alpha_c - a, \alpha_c + a)$, see Proposition 7 (i), p. 393.

By the Lagrange–Taylor theorem, see Appendix B, we can write, for $j = 1,2$,

$$N_j(x_1, x_2, \mathbf{y}, \alpha) = \sum_{h,k,l=1}^{2} \overline{N}_{jhkl}(\alpha) x_h x_k x_l + \sum_{h,k=1}^{2} \sum_{l=1}^{d-2} \overline{N}'_{jhkl}(\alpha) x_h x_k y_l + \sum_{h=1}^{2} \sum_{k,l=1}^{2} \overline{N}''_{jhkl}(\alpha) x_h y_k y_l + \sum_{h,k,l=1}^{2} \overline{N}_j'''{}_{hkl}(\alpha) y_h y_k y_l + \hat{N}_j(x_1, x_2, \mathbf{y}, \alpha), \tag{5.5.28}$$

where $\overline{N}$, $\overline{N}'$, $\overline{N}''$, $\overline{N}'''$ are C^∞ functions of $\alpha \in (\alpha_c - a, \alpha_c + a)$ and $\hat{N}$ is a C^∞ function of its arguments, for $\alpha \in (\alpha_c - a, \alpha_c + a)$, and it has a fourth-order zero at the origin in the $x_1, x_2, \mathbf{y}$ variables, $\forall \alpha \in (\alpha_c - a, \alpha_c + a)$.

Clearly, $\forall \alpha \in (\alpha_c - a, \alpha_c + a)$, if $x_1 + ix_2 = \rho e^{i\theta}$,

$$\gamma_\alpha(\theta) = \sum_{j,h,k,l=1}^{2} \overline{N}_{jhkl}(\alpha) \frac{x_j x_k x_h x_l}{(x_1^2 + x_2^2)^2} \cdot \tag{5.5.29}$$

To continue, first assume that $\gamma_\alpha(\theta) \equiv \bar{\gamma}_\alpha < 0$, $\forall \alpha \in (\alpha_c - a, \alpha_c + a)$, i.e., suppose that $\gamma_\alpha(\theta)$ is θ independent.

We shall later remove this severe restriction.

Multiply the Eqs. (5.5.27) by x_1, x_2, y_j, respectively, and sum the first two and, separately, the last $d-2$; one finds

$$\frac{1}{2}\frac{d\rho^2}{dt} = \sigma(\alpha)\rho^2 + \bar{\gamma}\rho^4 + D_4, \qquad \frac{1}{2}\frac{d\mathbf{y}^2}{dt} \leqslant -\nu \mathbf{y}^2 + D_3, \tag{5.5.30}$$

where D_3 and D_4 are C^∞ functions of $x_1, x_2, \mathbf{y}$ and of $\alpha \in (\alpha_c - a, \alpha_c + a)$ such that $\exists C_1, C_2 > 0$ which, for $x_1, x_2, \mathbf{y}$ near zero, verify

$$|D_4| \leqslant C_1(|\mathbf{y}| + \rho^2)(\rho + |\mathbf{y}|)^3; \qquad |D_3| \leqslant C_2(\rho + |\mathbf{y}|)^3. \tag{5.5.31}$$

Let $\beta, \delta_0 > 0$ be such that $\beta(1+\beta)^3 C_1 < |\bar{\gamma}|/2$, $\nu\beta^2\delta_0^2 > C_2(1+\beta)^3\delta_0^3/2$, and let δ_0 be so small that for all $\delta \leqslant \delta_0$ Eqs. (5.5.31) hold in

$$\Gamma(\delta) = \{x_1, x_2, \mathbf{y} \mid \rho < \delta, |\mathbf{y}| < \delta\beta\}. \tag{5.5.32}$$

Then we see that for $(x_1, x_2, \mathbf{y}) \in \partial\Gamma(\delta)$, $\delta \leqslant \delta_0$, it is

$$\begin{aligned} \frac{1}{2}\frac{d\rho^2}{dt} &\leqslant \sigma(\alpha)\delta^2 + \frac{\bar{\nu}}{2}\delta^4 \qquad \text{if} \quad \rho = \delta, \\ \frac{1}{2}\frac{d\mathbf{y}^2}{dt} &\leqslant -\frac{\nu}{2}\beta^2\delta^2 \qquad \text{if} \quad |\mathbf{y}| = \beta\delta. \end{aligned} \tag{5.5.33}$$

Hence, if α is very close to α_c, the right-hand sides of both of Eqs. (5.5.33) are negative. We use this to infer in a standard fashion that there is a function $\epsilon_\delta > 0$ such that $\forall \alpha \in (\alpha_c - \epsilon_\delta, \alpha_c + \epsilon_\delta)$, the set $\Gamma(\delta)$ is $S_t^{(\alpha)}$ invariant (where, as usual, the solution flow for our equation is denoted $S_t^{(\alpha)}$).

In fact, let ϵ_δ be a monotonically decreasing function of $\delta \in (0, \delta_0]$ such that $\sigma(\alpha) < |\bar{\gamma}|\delta^2/4$ for $\alpha \in (\alpha_c - \epsilon_\delta, \alpha_c + \epsilon_\delta)$. For such values of α, the right-hand sides of both of Eqs. (5.5.33) are negative. Then let $\mathbf{x} = (x_1, x_2, y) \in \Gamma(\delta)$ and let $\bar{t} = \{$first time > 0 such that $S_t^{(\alpha)}(\mathbf{x}) \notin \Gamma(\delta)\}$ and note that either the first or the second of Eqs. (5.5.33) (according to which side of $\partial\Gamma(\delta)$ is crossed) implies that $S_t^{(\alpha)}(\mathbf{x}) \notin \Gamma(\delta)$ for some earlier time $t < \bar{t}$ against the definition of $\bar{t}$. So $\bar{t} = +\infty$ and $\Gamma(\delta)$ is invariant for $S_t^{(\alpha)}$, $\forall t \geqslant 0$, $\forall \alpha \in (\alpha_c - \epsilon_\delta, \alpha_c + \epsilon_\delta)$, $\forall \delta \leqslant \delta_0$.

To prove vague attractivity, see Definition 5, p. 389 and Observation (8), p. 389, it seems natural to try to choose $U = \Gamma(\delta_0)$. Therefore, we ask the following question: given $\delta \leqslant \delta_0$ and $\alpha \in (\alpha_c - \epsilon_\delta, \alpha_\epsilon + \epsilon_\delta)$, can we find $t_\delta > 0$ such that $S_t^{(\alpha)}\Gamma(\delta_0) \subset \Gamma(\delta)$, $\forall t > t_\delta$?

Let $\mathbf{x} \in \Gamma(\delta_0)/\Gamma(\delta)$ and suppose that for t in some interval $[0, T]$, $S_t^{(\alpha)}(\mathbf{x}) \in \Gamma(\delta_0)/\Gamma(\delta)$. If we define $\delta(t)^2 = \max(\rho(t)^2, (y(t))^2/\beta^2)$, the point $S_t^{(\alpha)}(\mathbf{x})$ is in $\partial\Gamma(\delta(t))$ and Eq. (5.5.33), together with the assumption that $\forall t \in [0, T]$, $\delta \leqslant \delta(t) \leqslant \delta_0$, imply

$$\delta(t)^2 \leqslant \delta(0)^2 + 2T\max\left(\frac{\bar{\gamma}}{4}\delta^2, -\frac{\nu}{2}\delta^2\right) \leqslant \delta_0^2 - TM_\delta \qquad (5.5.34)$$

with $M_\delta > 0$.[11] Hence, if $t_\delta = (\delta_0^2 - \delta^2)/M_\delta$, it follows that $T < t_\delta$. Hence, $S_t^{(\alpha)}\Gamma(\delta_0) \subset \Gamma(\delta)$, $\forall t > t_\delta$, $\forall \alpha \in (\alpha_c - \epsilon_\delta, \alpha_c + \epsilon_\delta)$.

It is now clear that $\mathbf{x}_0 = \mathbf{0}$ is vaguely attractive. By Definition 5, p. 389 and its Observation (8), p. 389, one can take $U = \Gamma(\delta_0)$, $\rho_\delta = \delta$, $\forall \delta \leqslant \delta_0$, ϵ_δ, t_δ as above for all $\delta \leqslant \delta_0$ and Eq. (5.5.7) holds for $\delta \leqslant \delta_0$. If $\delta > \delta_0$, Eq. (5.5.7) is a trivial consequence of the uniform boundedness of the trajectories, supposed at the beginning.

So the proof of vague attractivity is complete as long as $\gamma_\alpha(\theta) \equiv \bar{\gamma}_\alpha$, $\forall \alpha \in (\alpha_c - a, \alpha_c + a)$, $a > 0$.

[11] Here we use a lemma on integration theory: if $a, b > 0$ are two C^∞ functions bounded below by a positive constant $\sigma > 0$ and if $d(t) = \max(a(t), b(t))$ and $c(t) = \dot{a}(t)$ for $a(t) > b(t)$, $c(t) = \dot{b}(t)$ for $a(t) < b(t)$, $c(t) = (\dot{a}(t) + \dot{b}(t))/2$ if $a(t) = b(t)$, then

$$d(t) = d(0) + \int_0^t c(\tau)d\tau \leqslant d(0) + t \sup_{0 \leqslant \tau \leqslant t} c(\tau).$$

This can be proved by noting that

$$d(t) = \lim_{N\to\infty} \left(a(t)^N + b(t)^N\right)^{1/N} = d(0) + \lim_{N\to\infty} \int_0^t \frac{d}{d\tau}\left(a(\tau)^N + b(\tau)^N\right)^{1/N} d\tau$$

$$= d(0) + \lim_{N\to\infty} \int_0^t \left(a(\tau)^N + b(\tau)^N\right)^{-1+1/N} \left(a(\tau)^{N-1}\dot{a}(\tau) + b(\tau)^{N-1}\dot{b}(\tau)\right) d\tau$$

and the function under the integration sign is uniformly bounded in N by $\max_{0 \leqslant \tau \leqslant t}(|\dot{a}(\tau)|, |\dot{b}(\tau)|)$ and it is pointwise convergent to $c(\tau)$. If $a(\tau) = b(\tau)$ has only a finite number of solutions τ, the possibility of taking the limit under the integral sign is easily proved. If $a(\tau) = b(\tau)$ has infinitely many roots one can find a simple approximation argument, recalling that a, b are bounded below by $\sigma > 0$, to infer $d(t) \leqslant d(0) + t\sup_{0 \leqslant \tau \leqslant t} c(\tau)$. Alternatively, one can apply the dominated convergence theorem of Lebesgue.

We must now remove this restriction.

This will be achieved by studying a coordinate change $(x_1, x_2, \mathbf{y}, \alpha) \to (\xi_1, \xi_2, \mathbf{y}', \alpha')$ with $\alpha \equiv \alpha'$, $\mathbf{y}' \equiv \mathbf{y}$, and

$$\begin{aligned} \xi_i &= x_i + \sum_{j,k=1}^{2} a_{ijk}(\alpha) x_j x_k , && i = 1,2, \\ x_i &= \xi_i - \sum_{j,k=1}^{2} a_{ijk}(\alpha) \xi_j \xi_k + H_i(\xi, \alpha), && i = 1,2, \end{aligned} \tag{5.5.35}$$

where H_i, a_{ijk} are C^∞ functions of their arguments and defined in the neighborhoods of $(\mathbf{0}, \alpha_c)$ of the form $V \times I$, $V \subset \mathbf{R}^d$ open.

We must show that $a_{ijk}(\alpha)$ can be so chosen that the two equations

$$\begin{aligned} \dot{x}_1 &= \sigma(\alpha) x_1 - \mu(\alpha) x_2 + \sum_{h,k=1}^{2} \overline{N}_{1hkl}(\alpha) x_h x_k x_l , \\ \dot{x}_2 &= \mu(\alpha) x_1 + \sigma(\alpha) x_2 + \sum_{h,k=1}^{2} \overline{N}_{2hkl}(\alpha) x_h x_k x_l \end{aligned} \tag{5.5.36}$$

[see Eqs. (5.5.28) and (5.5.29)] are changed into

$$\begin{aligned} \dot{\xi}_1 &= \sigma(\alpha)\xi_1 - \mu(\alpha)\xi_2 + \bar{\gamma}_\alpha \xi_1(\xi_1^2 + \xi_2^2) + \text{fourth-order terms}, \\ \dot{\xi}_2 &= \mu(\alpha)\xi_1 + \sigma(\alpha)\xi_2 + \bar{\gamma}_\alpha \xi_2(\xi_1^2 + \xi_2^2) + \text{fourth-order terms}. \end{aligned} \tag{5.5.37}$$

In fact, such a change of variables would manifestly change Eq. (5.5.27) into an equation of the same normal form but with $\gamma_\alpha(\theta) \equiv \bar{\gamma}_\alpha$.

The existence of such a change of coordinates is easier to discuss introducing the variables $z = x_1 + ix_2$, $\bar{z} = x_1 - ix_2$, $\lambda = \sigma + i\mu$, $\zeta = \xi_1 + i\xi_2$, $\bar{\zeta} = \xi_1 - i\xi_2$ and writing Eq. (5.5.36) as an equation for z, multiplying the second equation by i and adding it to the first (see Lanford, References):

$$\dot{z} = \lambda(\alpha) z + a_3(\alpha) z^3 + a_2(\alpha) z^2 \bar{z} + a_1(\alpha) z \bar{z}^2 + a_0(\alpha) \bar{z}^3, \tag{5.5.38}$$

where $a_0, \ldots, a_3$ are complex numbers that can be obtained from the N's by suitable linear combinations.

Similarly, we can write Eq. (5.5.35) in complex form:

$$\begin{aligned} \zeta &= z + A_3(\alpha) z^3 + A_2(\alpha) z^2 \bar{z} + A_1(\alpha) z \bar{z}^2 + A_0(\alpha) \bar{z}^3, \\ z &= \zeta - A_3(\alpha)\zeta^3 - A_2(\alpha)\zeta^2\bar{\zeta} - A_1(\alpha)\zeta\bar{\zeta}^2 - A_0(\alpha)\bar{\zeta}^3 + H(\zeta, \bar{\zeta}, \alpha) \end{aligned} \tag{5.5.39}$$

with H having a zero of fourth order in $|\zeta|$ as $\zeta \to 0$, $\forall \alpha \in (\alpha_c - a, \alpha_c + a)$.

Note that from Eqs. (5.5.38) and (5.5.39) it is also easy to find an expression for $\gamma_\alpha(\theta)$: if $z = \rho e^{i\theta}$, then

$$\begin{aligned} \gamma_\alpha(\theta) &= \rho^{-4} \mathrm{Re}\big(a_3(\alpha) \bar{z} z^3 + a_2(\alpha) z^2 \bar{z}^2 + a_1(\alpha) z \bar{z}^3 + a_0(\alpha) \bar{z}^4\big) \\ &= \mathrm{Re}\big(a_3(\alpha) e^{2i\theta} + a_2(\alpha) + a_1(\alpha) e^{-2i\theta} + a_0(\alpha) e^{-4i\theta}\big) \end{aligned} \tag{5.5.40}$$

which follows after some algebra, starting with the observation that $(x_1 N_1 + x_2 N_2) = \mathrm{Re}(\bar{z} N)$, if $N = N_1 + iN_2$ denotes the complex combination of the nonlinear terms in the right-hand side of Eq. (5.5.38). Hence,

$$\bar{\gamma}_\alpha \equiv \frac{1}{2\pi}\int_0^{2\pi}\gamma_\alpha(\theta)\,d\theta = \operatorname{Re} a_2(\alpha).$$

So our goal is to determine A_3, A_2, A_1, A_0 in Eq. (5.5.39) so that (5.5.38) in the ζ variables has a third-order term of the form $a_2(\alpha)\zeta^2\bar{\zeta}$.

A simple calculation shows that the equation for ζ is

$$\begin{aligned}\dot{\zeta} &= \dot{z} + 3A_3 z^2\dot{z} + 2A_2 z\bar{z}\dot{z} + A_2 z^2\dot{\bar{z}} + A_1\dot{z}\bar{z}^2 + 2A_1 z\bar{z}\dot{\bar{z}} + 3A_0\bar{z}^2\dot{\bar{z}}\\ &= \lambda\zeta + \zeta^3(a_3 + 2\lambda A_3) + \zeta^2\bar{\zeta}\big(a_2 + (\lambda+\bar{\lambda})A_2\big) + \zeta\bar{\zeta}^2\big(a_1 + 2\bar{\lambda}A_1\big)\\ &\quad + \bar{\zeta}^3\big(a_0 - (\lambda - 3\bar{\lambda})A_0\big) + \text{fourth-order terms.}\end{aligned} \tag{5.5.41}$$

hence, we take

$$A_3 = -\frac{a_3}{2\lambda}, \qquad A_2 = 0, \qquad A_1 = -\frac{a_1}{2\bar{\lambda}}, \qquad A_0 = \frac{a_0}{\lambda - 3\bar{\lambda}} \tag{5.5.42}$$

and, near α_c, the equation becomes

$$\dot{\zeta} = \lambda(\alpha)\zeta + a_2(\alpha)\bar{\zeta}\zeta^2 + \text{fourth-order terms} \tag{5.5.43}$$

whose $\gamma_\alpha(\theta)$ function is $\gamma_\alpha(\theta) \equiv \bar{\gamma}_\alpha = \operatorname{Re} a_2(\alpha)$.

The proof that if $\bar{\gamma} > 0$ the origin is not vaguely attractive is left as a problem for the reader.

mbe

It may be interesting to state explicitly some elementary invariance criteria for sets, which have been implicitly proved in the course of the above proof of Proposition 8.

9 Proposition. (i) *Let $U \subset \mathbf{R}^d$ be an open set with regular boundary ∂U. Then U is invariant for $\dot{\mathbf{x}} = \mathbf{f}(\mathbf{x})$, $\mathbf{f} \in C^\infty(\mathbf{R}^d)$, if*

$$\mathbf{f}(\mathbf{x}) \cdot \mathbf{n}(\mathbf{x}) < 0, \qquad \forall \mathbf{x} \in \partial U, \tag{5.5.44}$$

where $\mathbf{n}$ is the outer normal to ∂U in $\mathbf{x}$.

(ii) *Let $V \in C^1(\mathbf{R}^d)$ and let $U(\mu) = \{\mathbf{x} \mid \mathbf{x} \in \mathbf{R}^d, V(\mathbf{x}) < \mu\}$. If $\dot{\mathbf{x}} = \mathbf{f}(x)$, $\mathbf{f} \in C^\infty(\mathbf{R}^d)$, is a differential equation and*

$$\frac{\partial V}{\partial \mathbf{x}}(\mathbf{x}) \cdot \mathbf{f}(\mathbf{x}) < 0, \qquad \forall \mathbf{x} \in \partial U(\mu), \tag{5.5.45}$$

then the set $U(\mu)$ is invariant. Furthermore, if

$$\sup_{V(\mathbf{x}) \in (\mu_1, \mu_2)} \frac{\partial V}{\partial \mathbf{x}}(\mathbf{x}) \cdot \mathbf{f}(\mathbf{x}) = -C < 0 \tag{5.5.46}$$

and if $\mu' < \mu''$, $[\mu', \mu''] \subset (\mu_1, \mu_2)$, then

$$S_t U(\mu'') \subset U(\mu'), \qquad \forall t > \frac{\mu'' - \mu'}{C}. \tag{5.5.47}$$

Observations

(1) This proposition can be extended to the case when V is "piecewise C^∞" by replacing $(\partial V/\partial \mathbf{x})(\mathbf{x})$ with the set of the convex linear combina-

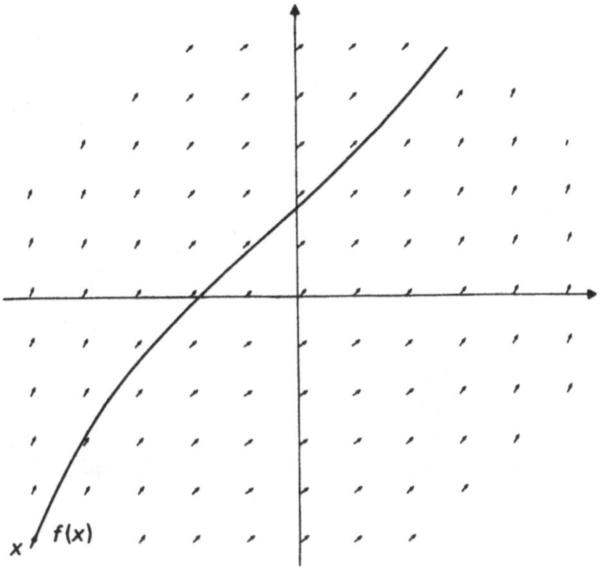

Figure 5.2.

tions of its extreme values (i.e., by a suitable bundle of vectors "pointing out of $U(V(\mathbf{x}))$" in $\mathbf{x}$). This is useful because sometimes V may have a square or a cylinder as its level surface, as was the case for $\Gamma(\delta)$ after Eq. (5.5.33), where $V(\mathbf{x}) = \max((x_1^2 + x_2^2), |\mathbf{y}|^2/\beta^2))$.

(2) The geometric interpretion of a differential equation is the following: At every $\mathbf{w} \in \mathbf{R}^d$, draw a vector $\mathbf{f}(\mathbf{w})$, i.e., think of $\mathbf{f}$ as a "vector field" over $\mathbf{R}^d$. A solution to $\dot{\mathbf{x}} = \mathbf{f}(\mathbf{x})$ is associated with a curve in $\mathbf{R}^d$ which at every point is tangent to the vector field at the same point. This curve is run at a speed which at every point is equal to the modulus of the field's vector at that point and has the same direction, see Fig. 5.2. The first statement of Proposition 9 is illustrated in Fig. 5.3. Observation (1) is illustrated in

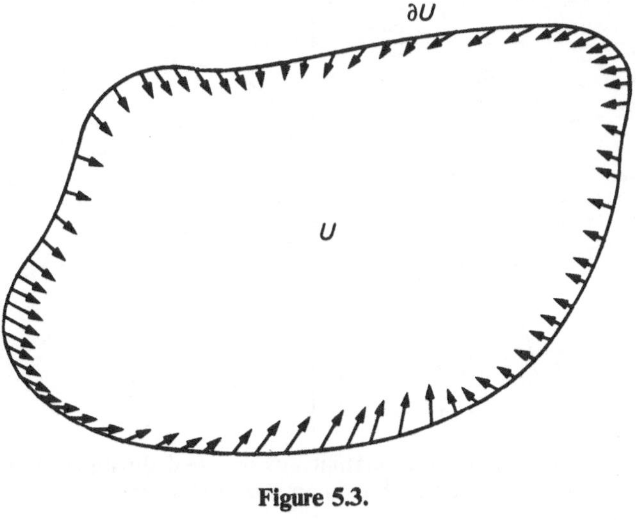

Figure 5.3.

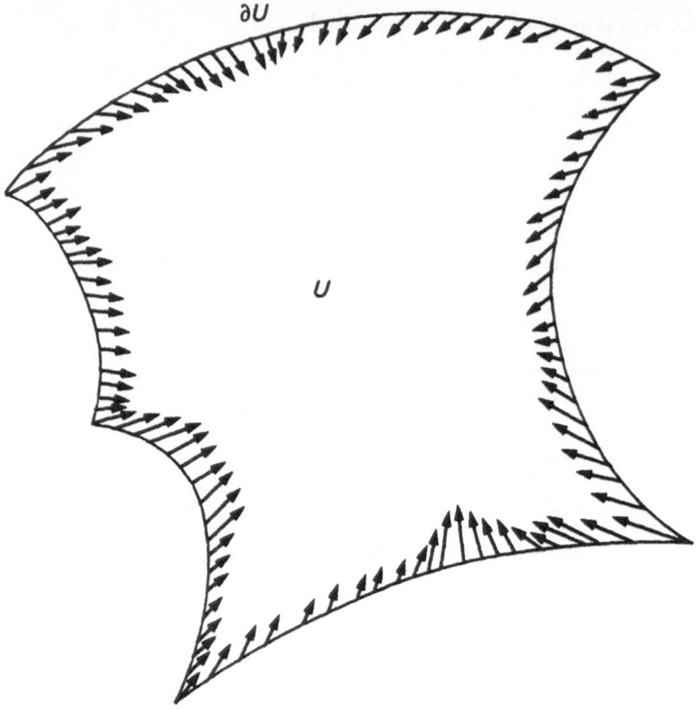

Figure 5.4.

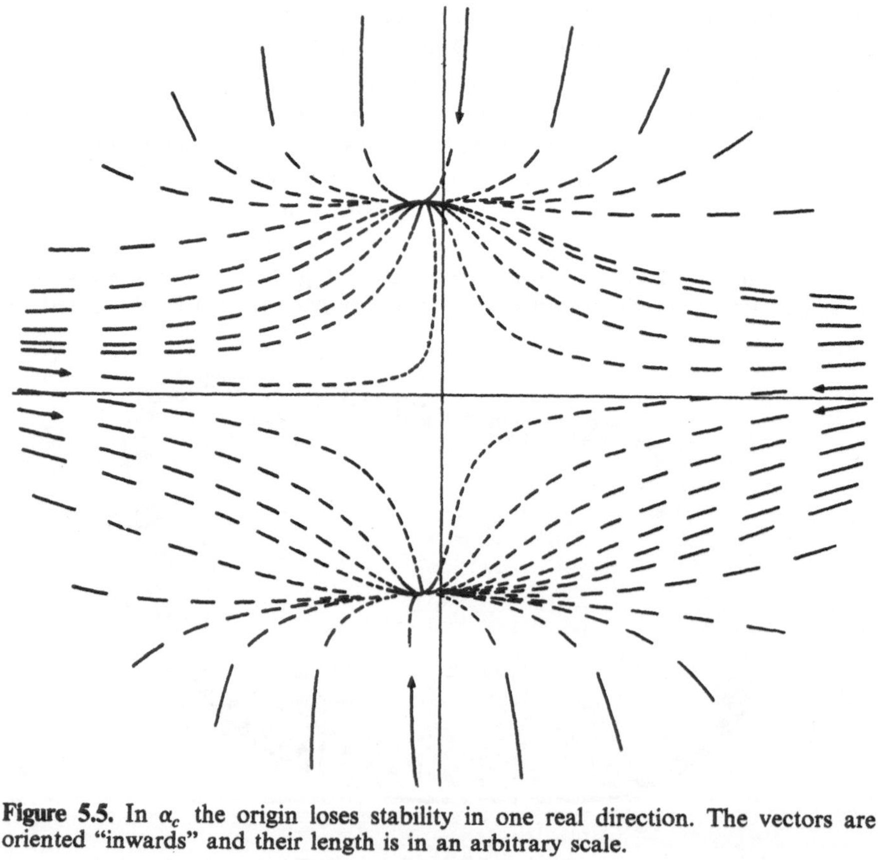

Figure 5.5. In α_c the origin loses stability in one real direction. The vectors are oriented "inwards" and their length is in an arbitrary scale.

Fig. 5.4. In connection with the above remarks, it is useful to see some pictures of a vaguely attractive point for an equation $\dot{\mathbf{x}} = \mathbf{f}(\mathbf{x}, \alpha)$. In Figs. 5.5 and 5.6, we draw the vector field for α slightly larger than α_c in a typical case.

The reader should try to understand such pictures by trying to draw them on the basis of the above information and comments on how a vector field should look near a vaguely attractive point.

Figures 5.5 and 5.6 allow one to see immediately that in the vicinity of a vaguely attractive fixed point, there should usually appear two fixed points, Fig. 5.5, or a periodic orbit, Fig. 5.6, depending on whether the stability loss takes place, as α passes through α_c, in one real direction or in two complex-conjugate directions.

In fact, this is the essential content of the Hadamard–Perron theorem and of the Hopf theorem which we will discuss in the upcoming sections.

We conclude this section by returning to the problem that we have been using to motivate the analysis of this section: the stability of the stationary solution $\hat{\omega}$ of Eq. (5.1.19).

10 Proposition. *Consider Eq.* (5.1.19). *The stationary solution $\hat{\omega}$ is vaguely attractive near $\alpha_c = \lambda_1 + \lambda_2'''\hat{\omega}_3^2$.*

Observation. The following proof shows that one should not blindly begin to compute mechanically the vague-attractivity constant $\bar{\gamma}_{\alpha_c}$. The reader will note the use of several "tricks" which are not worth being organized in a sequence of propositions refining the criterion of Proposition 8, but which, nevertheless, make the computation possible.

The reader should use these tricks in the exercises at the end of this section.

PROOF. We apply Proposition 8, p. 395. Changing variables to bring the fixed point to the origin, i.e., $(\omega_1, \omega_2, \omega_3) \leftrightarrow (\omega_1, \omega_2, r)$, $r = \omega_3 - \hat{\omega}_3$, Eq. (5.119) becomes

$$\dot{\omega}_1 = (\alpha - \alpha_c)\omega_1 - \hat{\omega}_3\omega_2 - 2\lambda_2'''\hat{\omega}_3 r\omega_1 - \omega_2 r - (\lambda_2'\omega_1^2 + \lambda_2''\omega_2^2 + \lambda_2'''r^2)\omega_1,$$

$$\dot{\omega}_2 = \hat{\omega}_3\omega_1 + (\alpha - \alpha_c)\omega_2 - 2\lambda_2'''\hat{\omega}_3 r\omega_2 + \omega_1 r - (\lambda_2'\omega_1^2 + \lambda_2''\omega_2^2 + \lambda_2'''r^2)\omega_2,$$

$$\dot{r} = -(\lambda_1 + 3\lambda_2'''\hat{\omega}_3^2)r - \hat{\omega}_3(\lambda_2'\omega_1^2 + \lambda_2''\omega_2^2 + 3\lambda_2'''r^2)$$
$$- (\lambda_2'\omega_1^2 + \lambda_2''\omega_2^2 + \lambda_2'''r^2)r. \tag{5.5.48}$$

It is convenient to condense Eq. (5.5.48) by introducing

$$z = \omega_1 + i\omega_2; \qquad \lambda = (\alpha - \alpha_c) + i\hat{\omega}_3; \qquad \tilde{\lambda} = -(\lambda_1 + 3\lambda_2'''\hat{\omega}_3^2),$$
$$E = 2\lambda_2'''\hat{\omega}_3 - i, \tag{5.5.49}$$
$$Q(r, z, \bar{z}) = \lambda_2'(\operatorname{Re} z)^2 + \lambda_2''(\operatorname{Im} z)^2 + \lambda_2'''r^2,$$
$$P(r, z, \bar{z}) = (\lambda_2'(\operatorname{Re} z)^2 + \lambda_2''(\operatorname{Im} z)^2 + 3\lambda_2'''r^2)\hat{\omega}_3.$$

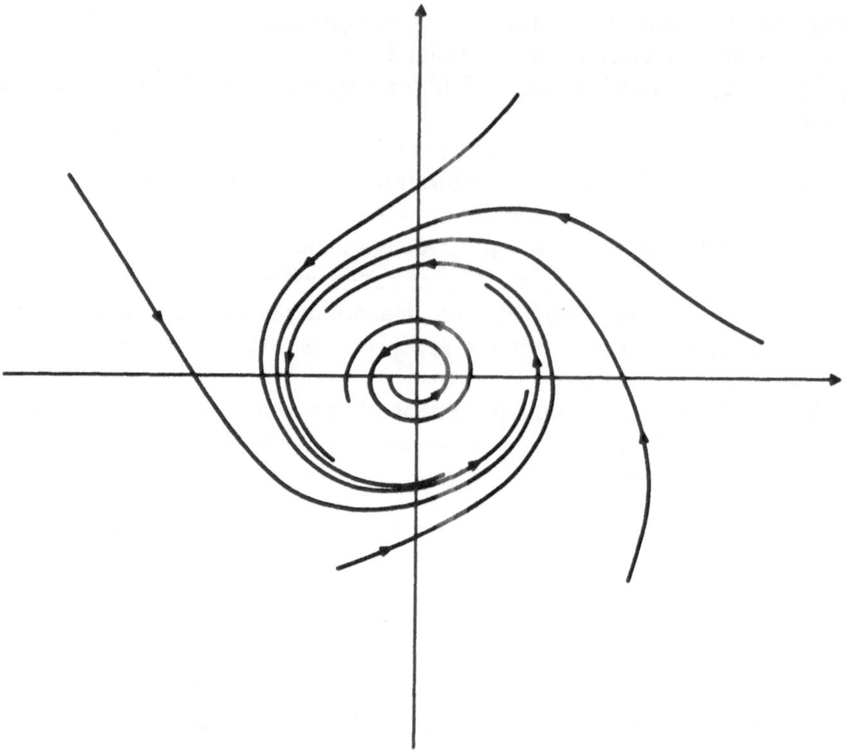

Figure 5.6. In α_c the origin loses stability in two complex directions. The drawing represents only a few curves tangent to the vector field ("trajectories of the field").

Then, multiplying the second of Eqs. (5.5.48) by i and adding it to the first, we find that Eq. (5.5.48) becomes

$$\dot{z} = (\lambda - Er - Q)z, \qquad \dot{r} = \tilde{\lambda}r - P - rQ. \tag{5.5.50}$$

To put the above equation in normal form with respect to λ, $\bar{\lambda}$, we change variables (see Propostion 7):

$$\zeta = z - Azr, \qquad z = \frac{\zeta}{1 - Ar}. \tag{5.5.51}$$

Then the first of Eqs. (5.5.49) becomes

$$\begin{aligned}\dot{\zeta} &= z(\lambda - Er - Q) - Arz(\lambda - Er - Q) - Az(\tilde{\lambda} - P - rQ)\\ &= \zeta\frac{(\lambda - Er - Q)(1 - Ar) - A(\tilde{\lambda}r - P - rQ)}{(1 - Ar)} \equiv \zeta F(\zeta),\end{aligned} \tag{5.5.52}$$

whose linear and quadratic terms are

$$\lambda\zeta + \zeta(-\lambda Ar - Er - \bar{\lambda}Ar + \lambda Ar) \equiv \lambda\zeta + \zeta(-Er - \bar{\lambda}Ar).$$

So we choose $A = -E/\bar{\lambda}$ and Eq. (5.5.48) acquires normal form in the (ζ, r) variables with respect to the eigenvalues $\lambda, \bar{\lambda}$.

From Eq. (5.5.52), it is then easy to compute $\gamma_\alpha(\theta)$: if $\zeta = \rho e^{i\theta}$,

$$\begin{aligned}\gamma_\alpha(\theta) &= \lim_{\rho\to 0}\left[\operatorname{Re}\frac{|\zeta|^2(F(\zeta)-\lambda)}{\rho^4}\right]_{r=0} \\ &= \lim_{\rho\to 0}\operatorname{Re}(\lambda - Q_0 + AP_0 - \lambda)\rho^{-4}\end{aligned} \tag{5.5.53}$$

if Q_0, P_0 are Q, P with $r = 0$. Therefore,

$$\begin{aligned}\gamma_\alpha(\theta) &= -\left(\lambda_2'(\cos\theta)^2 + \lambda_2''(\sin\theta)^2\right) + \frac{2\lambda_2'''\hat{\omega}_3^2\left(\lambda_2'(\cos\theta)^2 + \lambda_2''(\sin\theta)^2\right)}{\lambda_1 + 3\lambda_2'''\hat{\omega}_3^2} \\ &= -\left(\lambda_2'(\cos\theta)^2 + \lambda_2''(\sin\theta)^2\right)\frac{\lambda_1 + \lambda_2'''\hat{\omega}_3^2}{\lambda_1 + 3\lambda_2'''\hat{\omega}_3^2}.\end{aligned} \tag{5.5.54}$$

Therefore,

$$\bar{\gamma}_\alpha = -\frac{\lambda_2' + \lambda_2''}{2}\,\frac{\lambda_1 + \lambda_2'''\hat{\omega}_3^2}{\lambda_1 + 3\lambda_2'''\hat{\omega}_3^2} < 0 \tag{5.5.55}$$

and Proposition 10 follows from Proposition 8. mbe

Exercises and Problems for §5.5

1. Study the vague attractivity of the fixed point $x = 0$ of $\dot{x} = \alpha x + f(x)$, where $f \in C^\infty(\mathbf{R})$, $f(0) = 0$, $f'(0) = 0$. Show that the origin cannot be vaguely attractive unless $f''(0) = 0$.

2. Show, by producing some examples, that if the number $\bar{\gamma}_{\alpha_c}$ of Proposition 8 vanishes, then the fixed point may or may not be vaguely attractive. (Hint: Find examples other than $\dot{z} = (\alpha + i\mu)z \pm z^3\bar{z}^2$, $\alpha \in \mathbf{R}$, $\mu \in \mathbf{R}$, $\alpha_c = 0$, $z \in \mathbf{C}$.)

3. Suppose that the origin is vaguely attractive for $\dot{x} = xf(x^2, \alpha)$ near α_c. Show that the equation

$$\dot{x} = -\mu y + xf(x^2 + y^2, \alpha), \qquad \dot{y} = \mu x + yf(x^2 + y^2, \alpha),$$

also has the origin as a vague attractor near α_c, $\forall \mu \in \mathbf{R}$.

4. Let $a_1 \in C^\infty(\mathbf{R})$. Show that the origin is a vague attractor near α_c for $\dot{x} = -x(x^2 - a_1(\alpha))$ if $a_1(\alpha_c) = 0$.

5. Given $a_1, a_2 \in C^\infty(\mathbf{R})$, show that the origin is vaguely attractive for $\dot{x} = -x(x^2 - a_1(\alpha))(x^2 - a_2(\alpha))$ near α_c if $a_1(\alpha_c) = a_2(\alpha_c) = 0$. (Hint: use Observation (2) to Definition 5, p. 389.)

6. Compute the vague-attractivity indicator $\bar{\gamma} = \bar{\gamma}_{\alpha_c}$ for the origin, see Proposition 8, in the equation

$$\dot{x} = -\mu y - x(x^2 + y^2 - a_1(\alpha)),$$

$$y = +\mu x - y(x^2 + y^2 - a_1(\alpha)),$$

assuming $\mu \in \mathbf{R}$, $a_1(\alpha_c) = 0$, $a_1 \in C^\infty(\mathbf{R})$.

7. Same as Problem 6 for

$$\dot{x} = -\mu y - x(x^2 + y^2 - a_1(\alpha))(x^2 + y^2 - a_2(\alpha)),$$

$$\dot{y} = \mu x - y(x^2 + y^2 - a_2(\alpha))(x^2 + y^2 - a_2(\alpha)),$$

assuming $\mu \in \mathbf{R}$, $a_1, a_2 \in C^\infty(\mathbf{R})$, $a_1(\alpha_c) = a_2(\alpha_c) = 0$, study the vague attractivity. (From Negrini and Salvadori, see references.)

8. Let $z = x_1 + ix_2$, $\lambda = \alpha + i\mu$, α, $\mu \in \mathbf{R}$, and consider the differential equation

$$\dot{z} = \lambda z - \alpha a z\bar{z} - z^2\bar{z}.$$

Apply Proposition 8 to find the vague-attractivity indicator for the origin near $\alpha_c = 0$. (Answer: $\bar{\gamma} = -1$.) What can be said about the vague attractivity of the origin when $\mu = 0$? (Warning: Note that the equation does not have normal form.)

9. Under the assumptions of the first sentence of Proposition 8, only suppose that the equation $\dot{\mathbf{x}} = \mathbf{f}(\mathbf{x}, \alpha)$ has the form

$$\dot{x}_1 = \sigma(\alpha)x_1 - \mu(\alpha)x_2 + S_1(x_1, x_2, \mathbf{y}, \alpha) + N_1(x_1, x_2, \mathbf{y}, \alpha),$$

$$\dot{x}_2 = \mu(\alpha)x_1 + \sigma(\alpha)x_2 + S_2(x_1, x_2, \mathbf{y}, \alpha) + N_2(x_1, x_2, \mathbf{y}, \alpha),$$

$$\mathbf{y} = \tilde{L}(\alpha)\mathbf{y} + \mathbf{F}(x_1, x_2, \mathbf{y}, \alpha),$$

where $\lambda_1(\alpha) = \lambda_2(\alpha) = \sigma(\alpha) + i\mu(\alpha)$, $\mathbf{F}$ has second-order zero at the origin $x_1 = x_2 = 0$, $\mathbf{y} = \mathbf{0}$, for all $\alpha \in I$, N_1 and N_2 have a third order zero at the origin $x_1 = x_2 = 0$, $\mathbf{y} = \mathbf{0}$, $\forall \alpha \in I$, and S_1, S_2 are homogeneous second-order polynomials in $x_1, x_2, \mathbf{y}$. All the functions are supposed to be of class C^∞ in their arguments $(x_1, x_2, \mathbf{y}, \alpha) \in \mathbf{R} \times I$.

Suppose, furthermore, $\mathbf{F}(x_1, x_2, \mathbf{0}, \alpha) \equiv \mathbf{0}$ and also $S_1(x_1, x_2, \mathbf{0}, \alpha_c) \equiv S_2(x_1, x_2, \mathbf{0}, \alpha_c) \equiv 0$ and define $\gamma_\alpha(\theta)$ by Eq. (5.5.24) with the present meaning of the symbols. Show that the origin is vaguely attractive near a point $\alpha_c \in I$, where $\sigma(\alpha_c) = 0$, $\mu(\alpha_c) \neq 0$, if $\bar{\gamma} < 0$, in spite of the presence of the terms S_1, S_2. (Hint: Show that the above equation can be put into normal form with respect to $\lambda_1, \bar{\lambda}_1$ with a change of variables like

$$\xi_j = x_j + \sum_{h=1}^{2} \sum_{k=1}^{d-2} A_{jkh} y_k x_h + \sum_{h,k=1}^{d-2} B_{jhk} y_k y_h, \qquad k, j = 1, 2,$$

and this change of variables does not affect the value of $\gamma_\alpha(\theta)$ because it changes the third-order terms by a quantity vanishing as $\mathbf{y} \to \mathbf{0}$.) This extends Problem 8.

10. Prove that the same conclusions of Proposition 8 hold, replacing the assumption that N_1 and N_2 are of third order with the assumption that $x_1 N_1 + x_2 N_2$ has a fourth-order zero at $x_1 = x_2 = 0$, $\mathbf{y} = \mathbf{0}$, for all $\alpha \in I$. (Hint: Simply go through the proof of Proposition 8.)

11. Same as Problem 9, replacing the assumption that N_1, N_2 are of third order with the assumption that $x_1 N_1 + x_2 N_2$ is of fourth order.

12. Show that the origin is not a vaguely attractive point near $\alpha_c = 0$ for all the values of $E \in \mathbf{C}$ in the equation in $\mathbf{R}^3$:

$$\dot{z} = \lambda z + Ezr - z\bar{z}^2, \qquad r = -r + z\bar{z},$$

where $z = x + iy$, $\lambda = \alpha + i\mu$, $\mu \neq 0$, $x, y, r \in \mathbf{R}$.

13. Put into normal form the equation

$$\dot{\omega}_1 = -\omega_1 - \omega_2 + \omega_1\omega_2, \qquad \dot{\omega}_2 = \omega_1 - \omega_2 + \omega_1^2.$$

(Hint: Introduce $z = \omega_1 + i\omega_2$ and change variables as $\zeta = z + A_2 z^2 + A_1 z\bar{z} + A_0\bar{z}^2$, etc.)

14. Analyze the vague attractivity of the stationary solution $\omega_1 = \omega_2 = 0$, $\omega_3 = 5\alpha$ of the equation

$$\dot{\omega}_1 = -\omega_1 - \omega_2\omega_3, \qquad \dot{\omega}_2 = -\tfrac{1}{3}\omega_2 + \omega_1\omega_3, \qquad \dot{\omega}_3 = -\tfrac{1}{5}\omega_3 + \alpha.$$

15. Consider the equation ("Lorenz equation")

$$\dot{x} = \sigma(-x + y), \qquad \dot{y} = -\sigma x - y - xz, \qquad \dot{z} = -bz + xy - \alpha$$

and study the vague attractivity of its fixed points for $b = \sigma = 1$. (Hint: For the analysis of the fixed point $z = -\alpha/b$ at $\alpha_c = b(1 + \sigma)$, try to use Eqs. (5.5.18) and (5.5.11), with $\alpha = \alpha_c$, $j = 1$, to put the first equation (in the appropriate variables) into the normal form of Eq. (5.5.26); the result will be $\bar{\gamma} = -\sigma/b$. Warning: The analysis of the other fixed points is very cumbersome.)

16. Same as Problem 15 for $b = \frac{8}{3}$, $\sigma = 10$.

17. Suppose that the equation $\dot{\mathbf{x}} = \mathbf{f}(\mathbf{x}, \alpha)$ has a stationary solution $\mathbf{x}(\alpha)$ for $\alpha \leqslant \alpha_c$ depending continuously on α. Let $L(\alpha)$, $\lambda_1(\alpha), \ldots, \lambda_n(\alpha)$ be the stability matrix and its eigenvalues. Assume that, for $\alpha < \alpha_c$, it is $\operatorname{Re}\lambda_j(\alpha) < 0$ and, for $\alpha = \alpha_c$, $\operatorname{Re}\lambda_{j_0}(\alpha_c) = 0$ for some j_0.

Show that if no eigenvalue actually vanishes at $\alpha = \alpha_c$ (i.e., $\operatorname{Im}\lambda_j(\alpha_c) \neq 0$, $j = 1, \ldots, n$), then the solution $\mathbf{x}(\alpha)$ can be continuously continued to $\alpha \gtrsim \alpha_c$ (i.e., there is a continuous function $\alpha \to \mathbf{x}(\alpha)$ defined in the vicinity of α_c and $\mathbf{f}(\mathbf{x}(\alpha)) \equiv \mathbf{0}$).

Show also that if there is an eigenvalue vanishing at $\alpha = \alpha_c$ the solution $\mathbf{x}(\alpha)$ will not admit, in general, a continuation for $\alpha > \alpha_c$. (Hint: Just use the implicit functions theorem for the equation $\mathbf{f}(\mathbf{x}, \alpha) = \mathbf{0}$ near $(\mathbf{x}(\alpha_c), \alpha_c)$; then consider the example $f(x) = \alpha x + x^2 + \alpha$, $x(\alpha) = (-\alpha - (-4\alpha + \alpha^2)^{1/2})/2$, $\alpha_c = 0$, $L(\alpha) \equiv \lambda(\alpha) \simeq -\sqrt{-4\alpha}$).

§5.6. Vague-Attractivity Properties. The Attractive Manifold

Every five years or so, if not more often, someone discovers the theorem of Hadamard and Perron, proving it by Hadamard's method or by Perron's. (Anosov)

The solution $\hat{\omega}$ of Eq. (5.1.19), thought of as a family of differential equations parameterized by a parameter α, is, as shown in Proposition 10, p. 403, §5.5, vaguely attractive near $\alpha_c = \lambda_1 + \lambda_2''' \hat{\omega}_3^2$.

Therefore, the motion $t \to S_t^{(\alpha)}(\omega)$, $t \geqslant 0$, with initial datum close to $\hat{\omega}$ continues to remain quite close to $\hat{\omega}$ if α is near α_c, in spite of the instability of $\hat{\omega}$ for $\alpha > \alpha_c$. We shall see that $\hat{\omega}$ is not only unstable, but it also cannot be an attractor. Hence, the motions which develop from a datum in the vicinity of $\hat{\omega}$, although remaining there, cannot generally have an asymptotic behavior, as $t \to +\infty$, simply given by $S_t(\omega) \to \hat{\omega}$.

In the linear approximation, when the right-hand side of Eq. (5.1.19) is replaced by the function $\omega \to L_\alpha(\omega - \hat{\omega})$, where L_α is the stability matrix (5.5.1) of Eq. (5.1.19) at $\hat{\omega}$, the motion is very simple and $\omega_3 \to \hat{\omega}_3$ exponentially fast ($\simeq \exp - \lambda_1 t$), while $\omega_1^2 + \omega_2^2$ grows exponentially (roughly as $\exp(\alpha - \alpha_c)t$).

The linear approximation is certainly incorrect as soon as $\omega_1^2 + \omega_2^2$ becomes large, just because $\hat{\omega}$ is vaguely attractive [see Observation (2) to Definition 5, p. 389]. However, one can hope that even in the essentially nonlinear motion governed by Eq. (5.1.19), some memory remains of the fact that $\hat{\omega}$ "lost stability" only in two directions, i.e., only for what concerns the components of the motion in the plane $\omega_3 = \hat{\omega}_3$ generated by the eigenvectors $\mathbf{v}^{(1)}, \mathbf{v}^{(2)}$ of the stability matrix of $\hat{\omega}$. We can then think that the motion following a given initial datum ω close to $\hat{\omega}$ develops essentially on a two-dimensional surface, i.e., that the third component $\omega_3(t)$ asymptotically tends to become a function $\varphi(\omega_1(t), \omega_2(t))$ of the first two.

More generally, we can imagine to find ourselves in the following situation, to which the upcoming Proposition 11 will refer.

Let $(\mathbf{x}, \alpha) \to \mathbf{f}(\mathbf{x}, \alpha)$ be a $\mathbf{R}^d$-valued function in $C^\infty(\mathbf{R}^d \times \mathbf{R})$ such that the differential equations

$$\dot{\mathbf{x}} = \mathbf{f}(\mathbf{x}, \alpha), \tag{5.6.1}$$

parameterized by α, have uniformly bounded trajectories as α varies in $I = (\alpha_c - a, \alpha_c + a)$, $a \in (0, 1)$, and have the point $\mathbf{x} = \mathbf{x}_0$ as a stationary solution, $\forall \alpha \in I$,

$$\mathbf{f}(\mathbf{x}_0, \alpha) \equiv \mathbf{0}, \qquad \forall \alpha \in I. \tag{5.6.2}$$

Let L_α be the stability matrix in $\mathbf{x}_0$ and suppose that for $\alpha < \alpha_c$, all its eigenvalues $\lambda_1(\alpha), \ldots, \lambda_d(\alpha)$ have a negative real part, while for $\alpha \in I = (\alpha_c - a, \alpha_c + a)$, only $d - r$ eigenvalues have real parts less or equal to $-\nu_0 < 0$, the others having real parts larger or equal to $-\nu_0' > -\nu_0$ and vanishing for $\alpha = \alpha_c$ (i.e., $\operatorname{Re} \lambda_j(\alpha_c) = 0$ for $j = 1, \ldots, r$.

For simplicity, also suppose that the eigenvalues of L_α are pairwise distinct: so we can choose the eigenvalues $\lambda_1(\alpha), \ldots, \lambda_d(\alpha)$ and the corresponding eigenvectors $\mathbf{v}^{(1)}, \ldots, \mathbf{v}^{(d)}$ so that they are $C^\infty(I)$ functions of α and so that every complex eigenvector appears together with a complex-conjugate eigenvector. We suppose that the eigenvectors and eigenvalues have been so chosen and enumerated.

Under the above assumptions, we may assume without further loss of generality that for $\alpha \in I$, the equation takes the form of Eq. (5.5.10) and that the first r equations describe the evolution of the coordinates relative to the real plane generated by the "unstable" directions $\mathbf{v}^{(1)}, \ldots, \mathbf{v}^{(r)}$ corresponding to the eigenvalues with large real part ($\geqslant -\nu_0'$). Were this not true, we could change the coordinates near $\mathbf{x}_0$ (see Proposition 8, p. 395) to make this true.

Finally, we suppose that $\mathbf{x}_0$ is vaguely attractive for Eq. (5.6.1) near α_c and we denote

$$U, \Gamma(\delta) \tag{5.6.3}$$

a system of neighborhoods associated to $\mathbf{x}_0$ for $\alpha \in I$, whose existence is guaranteed by Definition 5, p. 389, of vague attractivity.

From the discussion of the preceding section, it appears that the just-described situation can be realized for the Eq. (5.1.19), see Proposition 10, p. 403, §5.5, thought of as a family of differential equations parameterized by α: therefore this provides a concrete example to which the following theory can be applied.

We now formulate a proposition giving a positive answer to the conjecture hinted at above, that given $\alpha > \alpha_c$, close enough to α_c, any motion of Eq. (5.6.1) starting close enough to $\mathbf{x}_0$ remains close to $\mathbf{x}_0$ (since $\mathbf{x}_0$ is vaguely attractive) and, furthermore, it can be thought of as developing asymptotically, for $t \to +\infty$, on an invariant surface σ_α.

Such a surface will have dimension r and it will be tangent to the "instability's hyperplane," $x_{r+1} = \cdots = x_d = 0$; furthermore, it will be an attractor for the motions starting in U and its attraction velocity will be exponential and roughly measured, as in the linear case, by the parameter

$$-\nu = \max_{i=r+1,\ldots,d} \operatorname{Re} \lambda_i$$

The surface σ_α will generally be nonunique since, as we shall see, it may contain other smaller attractors. If Λ_α is a minimal attractor in σ_α which has U as its attraction basin, then, clearly, every invariant hypersurface $\sigma' \subset U$ containing Λ_α is an attractor for U; see the exercises at the end of this section.

Finally, the surface σ_α will be described inside the neighborhood $\Gamma(\mathbf{x}_0, \delta) = \{\text{cube centered in } \mathbf{x}_0 \text{ and side } 2\delta\}$ by $d-r$ functions on $\mathbf{R}^r \times \mathbf{R}$ of *preassigned* regularity $C^{(k)}$, $k = 0, 1, \ldots,$ via the equations

$$x_{r+1} = \varphi^{(r+1)}(x_1, \ldots, x_r, \alpha), \ldots, x_d = \varphi^{(d)}(x_1, \ldots, x_r, \alpha) \tag{5.6.4}$$

(where we suppose $\mathbf{x}_0 = \mathbf{0}$) provided δ is small and α is close to α_c. For α close to α_c, this surface will be almost flat: if $k \geqslant 2$, this means that the first derivatives of the functions in Eq. (5.6.4) vanish for $(x_1, \ldots, x_r) = (x_{01}, \ldots, x_{0r})$.

The interest in the above considerations is clear: it will be possible to analyze the asymptotic behavior of some properties of the motions originating near a vaguely attractive point $\mathbf{x}_0$, as $t \to +\infty$, reducing the d equations of Eq. (5.6.1) to the r equations, labeled by $j = 1, 2, \ldots, r$,

$$\dot{x}_j = f^{(j)}\big(x_1, \ldots, x_r, \varphi^{(r+1)}(x_1, \ldots, x_r, \alpha), \ldots, \varphi^{(d)}(x_1, \ldots, x_r, \alpha), \alpha\big). \tag{5.6.5}$$

When d is large and r is small, this may be a very important simplification.

When $r = 1$ or $r = 2$, this may say that the motion near the vaguely attractive point is a "one-dimensional" or "two-dimensional" problem. Figures 5.5–5.6 already suggest that in such cases it will be possible to obtain deeper insights into the theory of the asymptotic behavior of the solutions of the equations starting with initial data close to $\mathbf{x}_0$. They even suggest the results of such a theory (see Figs. 5.5 and 5.6 and §5.7).

In the case of Eq. (5.1.19), we see that $r = 2$ and, therefore, the three equations (5.1.19) can be reduced, for $\alpha - \alpha_c$ small and for the purposes of the analysis of some asymptotic properties, to the first two equations with ω_3 replaced by

$$\omega_3 = \varphi(\omega_1, \omega_2, \alpha) + \hat{\omega}_3, \tag{5.6.6}$$

where φ is a suitable $C^{(k)}$ function (with a preassigned k) having a second-order zero in $\omega_1 = \omega_2 = 0$, i.e., such that there exist $\psi_1, \psi_2, \psi_3 \in C^{(k-2)}$ and

$$\varphi(\omega_1, \omega_2, \alpha) = \omega_1^2\psi_1(\omega_1, \omega_2, \alpha) + \omega_2^2\psi_2(\omega_1, \omega_2, \alpha) + \omega_1\omega_2\psi_3(\omega_1, \omega_2, \alpha), \tag{5.6.7}$$

expressing the tangency of the surfaces σ_α to the instability plane in $\hat{\omega}$ ($\omega_3 = \hat{\omega}_3$, in this case) provided $k \geqslant 2$. For $k < 2$ the near flatness can be expressed by Eq.(5.6.11) (implying Eq.(5.6.7) for $k \geqslant 2$).

A simple consequence of this, as we shall see in §5.7 (see footnote 15 on p. 432), will be that for α close to α_c, $\alpha > \alpha_c$, there is a periodic orbit which is a normal and minimal attractor lying on σ_α with attraction basin $U/\mathcal{C}(\hat{\omega})$, where $\mathcal{C}(\hat{\omega})$ is a one-dimensional curve of points ω through $\hat{\omega}$, whose asymptotic behavior is $S_t^{(\alpha)}(\omega) \to_{t \to +\infty} \hat{\omega}$.

Hence, we shall be able to get a rather complete picture of the motion near $\hat{\omega}$.

We now formulate a precise statement on the above matters as follows.

11 Proposition. *Suppose the validity of the assumptions described in the above text between Eqs.* (5.6.1) *and* (5.6.3). *Consider the symbols introduced there and let, for notational simplicity,* $\mathbf{x}_0 = \mathbf{0}$.

Given $k \geqslant 0$, $C > 0$, *there exist positive constants* $a_+, \delta, \delta_0, \nu \in (0, 1)$ *with* $\delta_0 < \delta$, $a_+ < a$ *and* $d - r$ *functions of class* $C^{(k)}$, *denoted* $\varphi^{(r+1)}, \ldots, \varphi^{(d)}$, *of the* $r + 1$ *variables* $x_1, \ldots, x_r, \alpha$ *defined for*

$$|x_i| < \frac{\delta}{2}, \qquad i = 1, \ldots, r, \qquad \alpha \in I_+ \equiv (\alpha_c - a_+, \alpha_c + a_+) \tag{5.6.8}$$

such that the surfaces $\sigma_\alpha \subset \mathbf{R}^d$ *described by Eqs.* (5.6.4) *have for all* $\alpha \in I_+$ *the properties*:

(i) "*local invariance*":

$$S_t^{(\alpha)}(\sigma_\alpha \cap \Gamma(\delta_0)) \subset \sigma_\alpha, \qquad \forall t \geqslant 0; \tag{5.6.9}$$

(ii) "*local attractivity*": *there exist* $C' > 0$ *such that for all* $\mathbf{w} \in U$ *it is*

$$d\left(S_t^{(\alpha)}(\mathbf{w}), \sigma_\alpha\right) \leqslant C' e^{-\nu t}, \qquad \forall t \geqslant 0 \tag{5.6.10}$$

(iii) "*tangency*" *and* "*flatness*": $\forall j = 1, \ldots, r$,

$$|\varphi^{(j)}(x_1, \ldots, x_r, \alpha)| \leqslant C\left(x_1^2 + \cdots + x_r^2\right)^{3/4}. \tag{5.6.11}$$

Observations.

(1) The reader may be surprised by the fact that, for the first time in this book, an important property is appearing and being considered in class $C^{(k)}$ rather than in class C^{∞}: the reason is due to the fact that in Proposition 11 one cannot choose $k = +\infty$. In fact using the methods of Problems 3 and 4, p. 428, the reader will check that in the equation $\dot{x} = \alpha x - x^3$, $\dot{z} = -z + x^2$ the surface σ_α cannot be of class $C^{(k)}$ for $k \geqslant 1/2\alpha$.

(2) The above proposition is an important part of the "Hadamard and Perron theorem". It is sometimes called the "invariant manifold" or the "attractive manifold" theorem and it had fundamental importance in the development of the qualitative theory of differential equations. It has been intensely studied, undergoing many extensions and generalizations, often trivial but sometimes significant.

(3) The family of surfaces σ_α is generally far from being uniquely determined by Eq. (5.6.1) (see the exercises for §5.6).

(4) The length of the proof and its formulae look quite discouraging. Actually the proof that follows is quite diluted and detailed (to conform to the spirit of this book). The subsections 5.6.A–5.6.D below have only a notational and definitorial character. The first technical step is in subsection 5.6.E with an application of the implicit function theorem with the purpose of stressing some properties of the surfaces $\sigma(\pi_t)$ approximating, as $t \to +\infty$, the surfaces that we are looking for. Subsection 5.6.F collects all the preceding inequalities to obtain further properties of the approximating surfaces $\sigma(\pi_t)$ for "very small" t. Furthermore, it contains the two basic ideas of the proof: (i) the estimates for very short times are possible because the quantity $\nu_0 = \min_{i=r+1,\ldots,d} - \operatorname{Re}\lambda_i > 0$, measuring the attractivity of the stable directions, is much larger than all the other relevant quantities (i.e., for short times, the "strong attractivity of the stable directions prevails over the weak repulsivity of the unstable ones"); and (ii) the long-time estimates, as $t \to +\infty$, can be obtained from the ones for short times because of the equation's autonomy.

These two themes occur again in a more or less repetitive way in subsections 5.6.G–5.6.N, all very similar to each other and which have been included here only for completeness.

The formulae are quite long and they could certainly be simplified and wirtten more compactly. However, they are obtained by applying the procedures suggested in the text and they are left in the form in which they are constructed: in this way, the reader may easily recognize their various parts and their origin and this, perhaps, makes the proof more clear.

The vague attractivity assumption is used at the beginning of the proof to reduce it to an equivalent problem.

The proof says much more than what is stated in Proposition 11 and some of its corollaries are described in the problems at the end of the section.

The proof is adapted from that of Lanford.[12]

PROOF. We shall discuss the proof of this proposition in the apparently particular case when $d = 2$, $r = 1$, and the equation is

$$\dot{x} = \alpha x + P(x,z), \qquad \dot{z} = -\nu_0 z + Q(x,z), \tag{5.6.12}$$

where $\nu_0 > 0$ and P, Q are two $C^\infty(\mathbf{R}^2)$ functions with a second-order zero at the origin:

$$\begin{aligned} P(x,z) &= x^2\overline{P}_1(x,z) + z^2\overline{P}_2(x,z) + xz\overline{P}_3(x,z), \\ Q(x,z) &= x^2\overline{Q}_1(x,z) + z^2\overline{Q}_2(x,z) + xz\overline{Q}_3(x,z) \end{aligned} \tag{5.6.13}$$

and $\overline{P}_i$, $\overline{Q}_i$, $i = 1,2,3$, are in $C^\infty(\mathbf{R}^2)$, see Appendix B.

The stability matrix of $\mathbf{0} \in \mathbf{R}^2$ is

$$L_\alpha = \begin{pmatrix} \alpha & 0 \\ 0 & -\nu_0 \end{pmatrix}, \tag{5.6.14}$$

and we suppose that $\mathbf{x}_0 = \mathbf{0}$ is vaguely attractive near $\alpha_c = 0$.

This case looks quite special; however, its theory forces us to deal with all the difficulties of the general problem whose analysis is a repetition of that relative to Eq. (5.6.12). In the following formulae, it will essentially suffice to think that x and z are vectors with r and $d - r$ components and that α, ν_0 are matrices $r \times r$ or $(d-r) \times (d-r)$, respectively, possibly functions of the parameter α. The first will be a matrix with eigenvalues all having real part not less than $-\nu_0' > -\nu_0$, $\forall \alpha \in I = (\alpha_c - a, \alpha_c + a)$ and vanishing for $\alpha = \alpha_c$, and the second with all the eigenvalues with real part not exceeding $-\nu_0 < 0$, $\forall \alpha \in I$. Furthermore, P, Q also will have to be thought of as depending (smoothly) on α.

Hence, consideration of Eq. (5.6.12) does not diminish the real difficulties of the problem and we prefer to treat it to avoid puzzling the reader with fictitious (mainly notational) difficulties in his first reading of a proof which is complex, although quite natural in its development.

The interested reader will not have difficulties, on a second reading, in interpreting (*mutatis mutandis*) the proofs as relative to the general case (see exercises and problems at the end of this section for some hints and suggestions).

To make the analysis of the proof easier, we have divided it in various basic steps distinguished by alphabetic characters.

[12] See References.

5.6.A. Preliminary Considerations and an Equivalent Problem. Consider Eq. (5.6.12) and let U be the neighborhood introduced in Eq. (5.5.7), whose existence is guaranteed by the vague-attractivity assumption.

Let $\Gamma(\rho) = \{\text{square in } \mathbf{R}^2 \text{ with side size } 2\rho, \text{ centered at the origin}\} = \{\mathbf{w} \mid \mathbf{w} \in \mathbf{R}^2, |w_i| < \rho, i = 1,2\}$.

Choose $k = 0$, first. The case $k > 0$ will be discussed later. Fix $C \in (0,1)$.

Let $\delta, a_+, t_0 \in (0,1)$ be small enough so that $a_+ < a/2$ and the inequalities in Eqs. (5.6.41), (5.6.42), (5.6.43), footnote 13, (5.6.53), (5.6.61), (5.6.76), (5.6.83), and (5.6.84) that we shall meet in the following discussion are satisfied. It is not worth listing them explicitly a priori here. The only fact that we shall really need is that they can all be simultaneously satisfied by choosing δ, a_+, t_0 small enough, once $C < 1$ is given.

Without loss of generality, we also suppose (see Definition 5, p. 389, §5.5) that for all $\alpha \in I = (\alpha_c - a, \alpha_c + a)$,

$$\begin{aligned} &\Gamma(\delta) \subset U, \qquad S_t^{(\alpha)} U \subset \Gamma\left(\frac{\delta}{2}\right), && \forall t \geqslant t_\delta, \\ &S_t^{(\alpha)} \Gamma(\delta_0) \subset \Gamma\left(\frac{\delta}{2}\right), && \forall t \geqslant 0 \end{aligned} \tag{5.6.15}$$

for a suitable choice of $t_\delta > 0$ and of $\delta_0 < \delta$.

Let χ_δ be a $C^\infty(\mathbf{R}^2)$ function which takes values between 0 and 1 and having value 1 on $\Gamma(\delta/2)$ and a value 0 outside $\Gamma(2\delta/3)$. We shall suppose that χ_δ has the form

$$\chi_\delta(x,z) = \chi\left(\frac{x}{\delta}, \frac{z}{\delta}\right), \tag{5.6.16}$$

where $\chi \in C^\infty(\mathbf{R}^2)$ is 1 on $\Gamma(1/2)$ and 0 outside $\Gamma(2/3)$.

So every motion beginning in U enters $\Gamma(\delta/2)$, for good, in a finite time t_δ, and every motion beginning in $\Gamma(\delta_0)$ never leaves $\Gamma(\delta/2)$. This is a consequence of the vague-attractivity assumption.

It is then clear that it will suffice to prove Proposition 11 for the equations

$$\dot{x} = \chi_\delta(x,z)(\alpha x + P(x,z)) \equiv X_\delta(x,z,\alpha), \tag{5.6.17}$$

$$\dot{z} = \nu_0 z + \chi_\delta(x,z) Q(x,z) = -\nu_0 z + Z_\delta(x,z,\alpha). \tag{5.6.18}$$

It is useful to remark, for later use, that for the given values of $a_+, \delta, \delta_0, C, k$, since χ_δ vanishes outside $\Gamma(2\delta/3)$, the solutions $t \to S_t^{(\alpha,\delta)}(\mathbf{w})$ of Eqs. (5.6.17) and (5.6.18) with initial datum $\mathbf{w} \in \Gamma(\delta)$ remain in $\Gamma(\delta)$: just note that

$$S_t^{(\alpha,\delta)}(x,z) = (x, z e^{-\nu_0 t}) \tag{5.6.19}$$

as long as $\chi_\delta(x, z e^{-\nu_0 t}) = 0$.

5.6.B. Some Useful Estimates of Derivatives. We shall need some properties of the solutions of Eqs. (5.6.17) and (5.6.18) thought of as a differential equation, depending on the parameters α and δ and with datum $\mathbf{w} \in \Gamma(\delta)$.

The properties are summarized as follows. There exists a constant $M > 1$ and $t_0 \in (0,1)$ such that, $\forall t \in [-t_0, t_0]$, $\forall \alpha \in (-a, a)$, $\forall \delta \in (0,1)$, $\forall i, j = 1,2$,

$$\left| \frac{\partial \big(S_t^{(\alpha,\delta)}(w_1, w_2) \big)_i}{\partial w_j} - e^{-\mu_i t} \delta_{ij} \right| \leqslant M|t|(|\alpha| + \delta), \tag{5.6.20}$$

$$\left| \frac{\partial \big(S_t^{(\alpha,\delta)}(w_1, w_2) \big)_i}{\partial \alpha} \right| \leqslant M(\delta|t|\delta_{i1} + \delta^2|t|^2\delta_{i2}), \tag{5.6.21}$$

$$\left| \frac{\partial \big(S_t^{(\alpha,\delta)}(w_1, w_2) \big)_i}{\partial t} + \mu_i w_i \right| \leqslant M\delta((|\alpha| + \delta)\delta_{i1} + \delta\delta_{i2}), \tag{5.6.22}$$

where $\mu_1 = 0$, $\mu_2 = \nu_0$, and where we have set $(x, z) = \mathbf{w}$, $\mathbf{w} = (w_1, w_2)$ and have denoted the components of $S_t^{(\alpha,\delta)}(\mathbf{w})$ as $(S_t^{(\alpha,\delta)}(\mathbf{w}))_i$, $i = 1,2$. We shall often use such notations in the following.

The above inequalities follow easily from an analysis of the regularity theorem for differential equations, §2.4, and they will be left to the reader, except Eq. (5.6.20) which is proved, as an example, in Appendix L.

We shall also need the following obvious estimates, immediate consequences of the defintions in Eqs. (5.6.17) and (5.6.18). Let $\mathbf{w} \in \Gamma(\delta)$, $|\alpha| \leqslant a_+$, $\delta < 1$, $i = 1,2$; then

$$\left| \frac{\partial X_\delta(\mathbf{w}, \alpha)}{\partial w_i} \right| \leqslant M(|\alpha| + \delta); \qquad \left| \frac{\partial X_\delta(\mathbf{w}, \alpha)}{\partial \alpha} \right| \leqslant M\delta, \tag{5.6.23}$$

$$\left| \frac{\partial Z_\delta(\mathbf{w}, \alpha)}{\partial w_i} \right| \leqslant M\delta; \qquad \left| \frac{\partial Z_\delta(\mathbf{w}, \alpha)}{\partial \alpha} \right| = 0, \tag{5.6.24}$$

and

$$|Z_\delta(\mathbf{w}, \alpha)| \leqslant M|\mathbf{w}|^2; \qquad |X_\delta(\mathbf{w}, \alpha)| \leqslant M(|\alpha|\,|w_1| + |\mathbf{w}|^2), \tag{5.6.25}$$

where M can be chosen to be the same as before, possibly increasing the latter, which we shall do.

5.6.C. Definition of the Approximate Surfaces. Let π be a $C^\infty(\mathbf{R})$ function such that, $\forall x \in [-\delta, \delta]$, $|\pi(x)| \leqslant \delta$. Interpret this function as defining a surface (a curve in this case, actually) $\sigma(\pi) \subset \Gamma(\delta)$ of parametric equations

$$z = \pi(x), \qquad x \in [-\delta, \delta]. \tag{5.6.26}$$

Also suppose that

$$\left| \frac{\partial \pi}{\partial x} \right| \leqslant C\sqrt{\delta}, \qquad x \in [-\delta, \delta] \tag{5.6.27}$$

(this choice of a bound on $\partial\pi/\partial x$ is quite arbitrary: $C\sqrt{\delta}$ could equally well be replaced by $C\delta^\beta$, $0 < \beta < 1$).

Then by the invariance of $\Gamma(\delta)$, the set $S_t^{(\alpha,\delta)}(\sigma(\pi))$, $t \geqslant 0$, is contained in $\Gamma(\delta)$ and, as we shall see shortly, it is a surface of the form $\sigma(\pi_t)$, where π_t is a new function verifying Eq. (5.6.27) and $|\pi_t| < \delta$.

It is then natural to try to define the surface that we are looking for as the surface $\sigma(\pi_\infty)$, where

$$\pi_\infty = \lim_{t\to+\infty} \pi_t \tag{5.6.28}$$

if this limit exists. In this case, it will in fact be formally true that $S_t^{(\alpha,\delta)}(\sigma(\pi_\infty)) = \sigma(\pi_\infty)$.

5.6.D. Proof that the Approximate Surfaces are Well Defined. First we look for an expression for π_t. This function should be defined by

$$(x, \pi_t(x)) = S_t^{(\alpha,\delta)}(x_0, \pi(x_0)), \tag{5.6.29}$$

where x_0 is a suitable point in $[-\delta, \delta]$ defined, naturally, by Eq. (5.6.29) which should be thought of as an equation defining π_t and x_0 in terms of x and π. Such an equation certainly has a solution since

$$S_t^{(\alpha,\delta)}(\pm\delta, \pi(\pm\delta)) = (\pm\delta, \pi(\pm\delta)e^{-\nu_0 t}) \tag{5.6.30}$$

and, therefore, by continuity, there exists a "function" $A(x,t,\alpha,\pi)$ such that the abscissa of $S_t^{(\alpha,\delta)}(A(x,t,\alpha,\pi), \pi(A(x,t,\alpha,\pi)))$ is just x, i.e.,

$$x_0 = A(x,t,\alpha,\pi) \tag{5.6.31}$$

is the solution of the first of the two equations (5.6.29) and, finally, $\pi_t(x)$ can be defined as the second coordinate of $S^{(\alpha,\delta)}(x_0, \pi(x_0))$ with x_0 given by Eq. (5.6.31).

By Eq. (5.6.19), one naturally sets $A(x,t,\alpha,\pi) \equiv x$ for $|x| \geqslant \delta$.

It is not immediately clear from the above argument that the functions A and π_t are uniquely defined. To this problem we devote the next step.

5.6.E. Alternative Proof of the Existence of π_t: Its Uniqueness for t Small and Estimates of Its Derivatives for t Small. As already noted, the argument in subsection 5.6.D does not prove the uniqueness of π_t, nor does it allow one to estimate its x derivative when one tries to check if it still verifies an inequality of the type of Eq. (5.6.27). In fact, it is a superfluous argument introduced just to help the reader to visualize what is done below.

It is possible to prove constructively the existence and uniqueness of the function A and, at the same time, to obtain an estimate of the derivatives of A with respect to x, t, α by using the implicit function theorem.

To study the function A in this way, we shall write Eq. (5.6.29) as

$$x = x_0 + \int_0^t X_\delta\big(S_\tau^{(\alpha,\delta)}(x_0, \pi(x_0), \alpha)\big)\,d\tau, \tag{5.6.32}$$

$$\pi_t(x) = \pi(x_0)e^{-\nu_0 t} + \int_0^t e^{-\nu_0(t-\tau)} Z_\delta\big(S_\tau^{(\alpha,\delta)}(x_0, \pi(x_0)), \alpha\big)\,d\tau, \tag{5.6.33}$$

obtained from Eqs. (5.6.17) and (5.6.18), pretending that X_δ and Z_δ are "known functions" of t and thinking of them as linear equations.

We write Eq. (5.6.32) in the form $G_\pi(x, x_0, \alpha, t) = 0$, where

$$G_\pi(x, x_0, \alpha, t) = x_0 - x + \int_0^t X_\delta\big(S_\tau^{(\alpha,\delta)}(x_0, \pi(x_0)), \alpha\big)\,d\tau \tag{5.6.34}$$

is a function in $C^\infty(\mathbf{R}^4)$ which we shall mainly consider for $|x| \leqslant \delta$, $|\alpha| \leqslant 2a_+$, $|x_0| \leqslant \delta$, $|t| \leqslant t_0$.

We regard $G_\pi(x, x_0, \alpha, t) = 0$ as an equation for x_0 parameterized by x, α, t at fixed π.

Since the point $(\bar{x}, \bar{x}, \bar{\alpha}, 0)$ is a solution point of our equation, $\forall \bar{x} \in [-\delta, \delta]$, $\forall \bar{\alpha}$, $|\bar{\alpha}| < 2a_+$, we apply the implicit function theorem, see Appendix G, Eq. (G10), to find a square neighborhood with side $2\rho(\bar{x}, \bar{\alpha})$ of $(\bar{x}, \bar{\alpha}, 0)$ in $\mathbf{R}^3$ such that if

$$|x - \bar{x}|, |\alpha - \bar{\alpha}|, |t| < \rho(\bar{x}, \bar{\alpha}), \tag{5.6.35}$$

then $G_\pi(x, x_0, \alpha, t) = 0$ has a solution $x_0 \in [-\delta, \delta]$.

To prove the existence of $\rho(\bar{x}, \bar{\alpha})$, we must study the derivative

$$\frac{\partial G_\pi}{\partial x_0}(x, x_0, \alpha, t). \tag{5.6.36}$$

From Eq. (5.6.34), using Eqs. (5.6.20), (5.6.23), (5.6.26), (5.6.27), and also recalling that $C < 1$, $\delta < 1$ (so that $C\sqrt{\delta} < 1$), one finds

$$\begin{aligned} \left|\frac{\partial G_\pi}{\partial x_0}(x, x_0, \alpha, t) - 1\right| &= \left|\int_0^t \frac{\partial X_\delta\big(S_\tau^{(\alpha,\delta)}(x_0, \pi(x_0)), \alpha\big)}{\partial x_0}\, d\tau\right| \\ &\leqslant 8M(|\alpha| + \delta)(1 + M|t|(|\alpha| + \delta))|t|. \end{aligned} \tag{5.6.37}$$

Furthermore,

$$\left|\frac{\partial G_\pi}{\partial x}(x, x_0, \alpha, t)\right| \equiv 1, \tag{5.6.38}$$

$$\begin{aligned} &\left|\frac{\partial G_\pi}{\partial \alpha}(x, x_0, \alpha, t)\right| \\ &= \left|\int_0^t d\tau \left\{ \frac{\partial X_\delta}{\partial w_1}\big(S_\tau^{(\alpha,\delta)}(x_0, \pi(x_0)), \alpha\big) \frac{\partial\big(S_\tau^{(\alpha,\delta)}(x_0, \pi(x_0))\big)_1}{\partial \alpha} \right.\right. \\ &\qquad + \frac{\partial X_\delta}{\partial w_2}\big(S_\tau^{(\alpha,\delta)}(x_0, \pi(x_0)), \alpha\big) \frac{\partial\big(S_\tau^{(\alpha,\delta)}(x_0, \pi(x_0))\big)_2}{\partial \alpha} \\ &\qquad \left.\left. + \frac{\partial X_\delta}{\partial \alpha}\big(S_\tau^{(\alpha,\delta)}(x_0, \pi(x_0)), \alpha\big) \right\}\right| \\ &\leqslant |t|\big((|\alpha| + \delta)\delta|t|M^2 + \delta^2|t|^2M^2(|\alpha| + \delta) + M\delta\big), \end{aligned} \tag{5.6.39}$$

and, finally,

$$\left|\frac{\partial G_\pi}{\partial t}(x, x_0, \alpha, t)\right| = \left|X_\delta\big(S_t^{(\alpha,\delta)}(x_0, \pi(x_0))\big)\right| \leqslant 2M\delta(\delta + |\alpha|). \tag{5.6.40}$$

The above inequalities for the derivatives are valid for all $|t| \leqslant 1$, $|x| \leqslant \delta$, $|x_0| \leqslant \delta$, $\delta \leqslant 1$.

We now suppose that a_+, δ, t_0 are so small that $\forall |\alpha| < 2a_+$, $|t| \leqslant t_0$:

$$\left|\frac{\partial G_\pi}{\partial x_0}(x, x_0, \alpha, t) - 1\right| < 10M(2a_+ + \delta)|t| < \frac{1}{2}, \tag{5.6.41}$$

$$\left|\frac{\partial G_\pi}{\partial \alpha}(x, x_0, \alpha, t)\right| \leqslant 2M\delta|t| < \frac{1}{10}, \tag{5.6.42}$$

$$\left|\frac{\partial G_\pi}{\partial t}(x, x_0, \alpha, t)\right| \leqslant M\delta(\delta + 2a_+) < \frac{1}{10}. \tag{5.6.43}$$

Here $\frac{1}{2}$ and $\frac{1}{10}$ are arbitrary small numbers, convenient for the upcoming estimates.

Then if $|\bar{\alpha}| < a_+$, $|\bar{x}| < \frac{3}{4}\delta$, and if $\zeta = \min(a_+, t_0, \delta/4)$ the $\rho(\bar{x}, \bar{\alpha})$ just considered can be taken (see Appendix G, Proposition 1) as

$$\rho(\bar{x}, \bar{\alpha}) = \frac{\zeta}{2}\,\frac{\min(|\partial G_\pi/\partial x_0|)}{\max(|\partial G_\pi/\partial x_0| + |\partial G_\pi/\partial x| + |\partial G_\pi/\partial \alpha| + |\partial G_\pi/\partial t|)} \geqslant \frac{\zeta}{10}, \tag{5.6.44}$$

having used Eqs. (5.6.41)–(5.6.43) to get the right-hand side inequality and having considered the maxima and the minima with respect to the parameters t, α, x, x_0 as they vary in $[-t_0, t_0]$, $[-2a_+, 2a_+]$, $[-\delta, \delta]$, $[-\delta, \delta]$.

This shows the existence of A as a function of x, α, t as they vary in[13]

$$|x| \leqslant \delta, \qquad |\alpha| \leqslant a_+, \qquad |t| \leqslant t_+ = \frac{\zeta}{11} \tag{5.6.45}$$

[13]Note that if $G_\pi(x, x_0, \alpha, t) = 0$ has a solution it must be, by Eq. (5.6.25): $|x_0 - x| < 2tM(|\alpha| + \delta)\delta$. We shall suppose the validity of the inequality

$$2t_0 M(a_+ + \delta)\delta < \min(a_+, t_0, \delta/4) = \zeta$$

so that every solution x_0 must verify $|x - x_0| < \zeta$. Then the detailed argument leading to Eqs. (5.6.44) and (5.6.45) is the following.

Note that for $|x| \geqslant \frac{2}{3}\delta$, the determination of A is trivial and $A(x, t, \alpha, \pi) \equiv x$. Then let $\bar{x} \in [-\frac{3}{4}\delta, \frac{3}{4}\delta]$, $\bar{\alpha} \in (-a_+, a_+)$ and note that by Eqs. (5.6.35) and (5.6.44), it is possible to solve uniquely the equation for $A \in [-\zeta, \zeta]$ in the region $|x - \bar{x}| < \zeta/10$, $|\alpha - \bar{\alpha}| < \zeta/10$, $|t| < \zeta/10$. As $\bar{x}, \bar{\alpha}$ vary in $[-3\delta/4, 3\delta/4] \times [-a_+, a_+]$, this parallelepipedal region covers at least a neighborhood V of $[-2\delta/3, 2\delta/3] \times [-a_+, a_+] \times [-\zeta/11, \zeta/11]$.

By the uniqueness of A in each parallelepiped, the functions A thus defined coincide at the points which are common to several parallelepipeds. Furthermore, the functions A have a value equal to x for $|x| > \frac{2}{3}\delta$.

Hence, we have built a continuous piecewise-differentiable solution A of $G_\pi(x, A, \alpha, t) = 0$ in the region of Eq. (5.6.45); and by construction, A is the unique solution, with the property $|A - x| < \zeta$, in this region.

Actually, A must be C^∞ in the region of Eq. (5.6.45), since in each of the parallelepipeds where A has been constructed, A has this property and we have uniqueness.

Finally A is the only solution with $|A| < \delta$ because, as noted above, any such solution must verify $|A - x| < \zeta < \delta/4$ and for $|x| > 2\delta/3$ it is $A \equiv x$.

and it shows, as well, the possibility of estimating the derivatives of A as follows (see Eqs. (5.6.41)–(5.6.43) right-hand sides and Appendix G, Proposition 1):

$$\left|\frac{\partial A}{\partial x}(x,t,\alpha,\pi)-1\right| \equiv \left|-\frac{(\partial G_\pi/\partial x)(x,x_0,\alpha,t)}{(\partial G_\pi/\partial x_0)(x,x_0,\alpha,t)}-1\right| \leqslant 20M|t|(2a_+ +\delta), \tag{5.6.46}$$

$$\left|\frac{\partial A}{\partial \alpha}(x,t,\alpha,\pi)\right| \equiv \left|-\frac{(\partial G_\pi/\partial \alpha)(x,x_0,\alpha,t)}{(\partial G_\pi/\partial x_0)(x,x_0,\alpha,t)}\right| \leqslant 4M|t|\delta, \tag{5.6.47}$$

$$\left|\frac{\partial A}{\partial t}(x,t,\alpha,\pi)\right| \equiv \left|-\frac{(\partial G_\pi/\partial t)(x,x_0,\alpha,t)}{(\partial G_\pi/\partial x_0)(x,x_0,\alpha,t)}\right| \leqslant 4M\delta(2a_+ +\delta) \tag{5.6.48}$$

valid for x,α,t in the region of Eq. (5.6.45). It is important to stress that Eqs. (5.6.46)–(5.6.48) have been obtained independently of the choice of π provided

$$|\pi(x)| \leqslant \delta \quad \text{and} \quad \left|\frac{\partial \pi}{\partial x}(x)\right| \leqslant C\sqrt{\delta}\,, \qquad \forall x \in [-\delta,\delta]. \tag{5.6.49}$$

The above considerations show that the function π_t is well defined at least for $|\alpha| < a_+$, $|t| < t_+$, via Eqs. (5.6.29) and (5.6.31).

The uniqueness of the A function, coming from its construction (see footnote 13, p. 417) allows us to conclude that $\sigma(\pi_t)$ is the $S_t^{(\alpha,\delta)}$ image of $\sigma(\pi)$: $S_t^{(\alpha,\delta)}\sigma(\pi) = \sigma(\pi_t)$. Also note that, by the invariance of $\Gamma(\delta)$, one has $S_t^{(\alpha,\delta)}\sigma(\pi) \subset \Gamma(\delta)$.

The invariance of $\Gamma(\delta)$ for the motions generated by Eqs. (5.6.17) and (5.6.18) also implies that π_t verifies the first of Eqs. (5.6.49) (a property already encountered during the construction of A).

5.6.F. Check of the Validity of Eq. (5.6.49) for π_t, $0 \leqslant t \leqslant t_+$. This check is of fundamental importance since it will allow us to define π_t for all $t \geqslant 0$.

The relation $S_t^{(\alpha,\delta)}\sigma(\pi) = \sigma(\pi_t)$, $t \in [0,t_+]$, will guarantee, taking also into account the group property $S_t^{(\alpha,\delta)}S_{t'}^{(\alpha,\delta)} = S_{t+t'}^{(\alpha,\delta)}$, that if $t \in [0,t_+]$, $t' \in [0,t_+]$, $t+t' \in [0,t_+]$ and if $\pi_t, \pi_{t'}, \pi_{t+t'}$ verify Eq. (5.6.49), then

$$(\pi_t)_{t'} = \pi_{t+t'}\,. \tag{5.6.50}$$

This relation will allow us to uniquely define π_t, $\forall t \geqslant 0$, by dividing the interval $[0,t]$ into intervals with amplitude $\tau < t_+$ and, then, recursively setting

$$\pi_t = (\pi_{t-\tau})_\tau = ((\pi_{t-2\tau})_\tau)_\tau\,. \tag{5.6.51}$$

This definition will necessarily coincide with the one that could be given by setting $S_t^{(\alpha,\delta)}\sigma(\pi) = \sigma(\pi_t)$, $t \geqslant 0$.

Therefore let us verify that, if $0 \leqslant t \leqslant t_+$, π_t fulfills the second of Eqs. (5.6.49) (as noted above, the first has already been checked).

For this purpose, we use Eq. (5.6.33), where instead of x_0, one should imagine $A(x,t,\alpha,\pi)$. Differentiating both sides, one finds

$$\begin{aligned}\frac{\partial \pi_t}{\partial x} &= \frac{\partial \pi}{\partial x_0}(x_0)\,\frac{\partial A}{\partial x}\,e^{-\nu_0 t}\\ &\quad+\sum_{i=1}^{2}\int_0^t\Bigg[e^{-\nu_0(t-\tau)}\frac{\partial Z_\delta}{\partial w_i}\left(S_\tau^{(\alpha,\delta)}(x_0,\pi(x_0)),\alpha\right)\\ &\qquad\times\left\{\frac{\partial\left(S^{(\alpha,\delta)}(x_0,\pi(x_0))\right)_i}{\partial x_0}\right.\\ &\qquad\left.+\frac{\partial\left(S_\tau^{(\alpha,\delta)}(x_0,\pi(x_0))\right)_i}{\partial \pi(x_0)}\,\frac{\partial \pi}{\partial x_0}(x_0)\right\}\frac{\partial A}{\partial x}(x,\tau,\alpha,\pi)\Bigg]d\tau\end{aligned} \tag{5.6.52}$$

with slightly symbolic differentiation notations (hopefully self-explanatory).

By Eqs. (5.6.41), (5.6.46), (5.6.49), and (5.6.20), (5.6.24), Eq. (5.6.52) implies, with some labor, that $\forall t\in[0,t_+]$, $\forall\alpha\in[-a_+,a_+]$, $\forall x\in[-\delta,\delta]$,

$$\begin{aligned}\left|\frac{\partial \pi_t}{\partial x}(x)\right| &\leqslant C\sqrt{\delta}\,e^{-\nu_0 t}(1+20M(2a_+ +\delta)t)\\ &\quad+tM\delta\left\{(1+Mt(a_+ +\delta))+Mt(a_+ +\delta)C\sqrt{\delta}+Mt(a_+ +\delta)\right.\\ &\qquad\left.+C\sqrt{\delta}\,(1+Mt(a_+ +\delta))\right\}\cdot(1+20M(2a_+ +\delta)t)\\ &= C\sqrt{\delta}\Big[e^{-\nu_0 t}(1+20Mt(2a_+ +\delta))+tMC^{-1}\sqrt{\delta}\\ &\qquad\times\left\{(1+Mt(a_+ +\delta))+Mt(a_+ +\delta)C\sqrt{\delta}+Mt(a_+ +\delta)\right.\\ &\qquad\left.+C\sqrt{\delta}\,(1+Mt(a_+ +\delta))\right\}(1+20M(2a_+ +\delta))\Big]\\ &\leqslant C\sqrt{\delta}\left(1-\frac{\nu_0}{2}t\right)\end{aligned} \tag{5.6.53}$$

if δ, a_+, t_0, (recall also that $t_+ \leqslant t_0$), are supposed to have been so chosen that the last inequality in Eq. (5.6.53) holds, $\forall t\in[0,t_+]$.[14]

The above arguments prove that π_t can be defined by $S_t^{(\alpha,\delta)}\sigma(\pi)=\sigma(\pi_t)$, or, equivalently, by Eq. (5.6.51), for $t\geqslant 0$ and show that π_t verifies Eq. (5.6.49) for all $t\geqslant 0$.

[14]One sees that $C\sqrt{\delta}$ could be replaced by $C\delta^\gamma$, $\gamma<1$. The choice $\gamma=1$ could only be made if ν_0 is large enough (or if we decided to allow $C>1$ and C to be large enough).

5.6.G. Proof of the Existence of the Limit as $t \to +\infty$ of π_{nt} for $t \in [0, t_+]$. We shall proceed by recursively evaluating

$$\|\pi_{nt} - \pi_{(n-1)t}\| = \max_{|x|<\delta} |\pi_{nt}(x) - \pi_{(n-1)t}(x)| \tag{5.6.54}$$

and show that the series

$$\sum_{n=0}^{\infty} \|\pi_{nt} - \pi_{(n-1)t}\| < +\infty \tag{5.6.55}$$

converges. This implies that π_{nt} converges uniformly as $n \to +\infty$ to a limit.

To study the series of Eq. (5.6.55), consider two functions π, π' verifying Eq. (5.6.49) and, through them, construct the functions $A(x,t,\alpha,\pi)$ and $A(x,t,\alpha,\pi')$ defined on the set given by Eq. (5.6.45), solving the equations for x_0: $G_\pi(x, x_0, t, \alpha) = 0$ and $G_{\pi'}(x, x_0, t, \alpha) = 0$ as indicated in subsection 5.6.E.

Shortening $A(x,t,\alpha,\pi)$ and $A(x,t,\alpha,\pi')$ in x_0, x_0', respectively, and using Eq. (5.6.33), one then has $\forall t \in [0, t_+]$,

$$\begin{aligned} |\pi_t(x) - \pi_t'(x)| &\leq e^{-\nu_0 t}|\pi(x_0) - \pi'(x_0')| \\ &\quad + \int_0^t e^{-\nu_0(t-\tau)}\Big|Z_\delta\big(S_\tau^{(\alpha,\delta)}(x_0, \pi(x_0)), \alpha\big) \\ &\qquad - Z_\delta\big(S_\tau^{(\alpha,\delta)}(x_0', \pi'(x_0')), \alpha\big)\Big|\, d\tau \end{aligned} \tag{5.6.56}$$

which, by Eqs. (5.6.24), (5.6.49), and (5.6.20), implies

$$\begin{aligned} &|\pi_t(x) - \pi_t'(x)| \\ &\quad \leq e^{-\nu_0 t}\big(|\pi(x_0) - \pi'(x_0)| + |\pi'(x_0) - \pi'(x_0')|\big) \\ &\qquad + \int_0^t M\delta \sum_{i=1}^{2} \Big|\big(S_\tau^{(\alpha,\delta)}(x_0, \pi(x_0))\big)_i - \big(S_\tau^{(\alpha,\delta)}(x_0', \pi'(x_0'))\big)_i\Big|\, d\tau \\ &\quad \leq e^{-\nu_0 t}\big(\|\pi - \pi'\| + C\sqrt{\delta}\,|x_0 - x_0'|\big) \\ &\qquad + \int_0^t 2M\delta(1 + M\delta(a_+ + \delta)\tau)\big(|x_0 - x_0'| + |\pi(x_0) - \pi(x_0')|\big)\, d\tau \\ &\quad \leq \|\pi - \pi'\|\big(e^{-\nu_0 t} + 2M\delta t(1 + M\delta(a_+ + \delta)t)\big) \\ &\qquad + |x_0 - x_0'|\big(C\sqrt{\delta}\, e^{-\nu_0 t} + 2M\delta t(1 + M\delta(a_+ + \delta)t)\big). \end{aligned} \tag{5.6.57}$$

We must therefore estimate $|x_0 - x_0'|$.

Observe that $(x_0, \pi(x_0))$ and $(x_0', \pi'(x_0'))$ are the values of $S_{-t}^{(\alpha,\delta)}(x, \pi_t(x))$ and $S_{-t}^{(\alpha,\delta)}(x, \pi_t'(x))$ and, hence, as in Eq. (5.6.32),

$$\begin{aligned} x_0 &= x - \int_0^t d\tau\, X_\delta\big(S_{-\tau}^{(\alpha,\delta)}(x, \pi_t(x)), \alpha\big), \\ x_0' &= x - \int_0^t d\tau\, X_\delta\big(S_{-\tau}^{(\alpha,\delta)}(x, \pi_t'(x)), \alpha\big) \end{aligned} \tag{5.6.58}$$

and, then, by Eqs. (5.6.23) and (5.6.20),

$$
\begin{aligned}
|x_0 - x_0'| &\leqslant \int_0^t d\tau \left|X_\delta\left(S_{-\tau}^{(\alpha,\delta)}(x,\pi_t(x)),\alpha\right) - X_\delta\left(S_{-\tau}^{(\alpha,\delta)}(x,\pi_t'(x)),\alpha\right)\right| \\
&\leqslant \int_0^t M(a_+ + \delta)2(1 + M\tau(a_+ + \delta))|\pi_t(x) - \pi_t'(x)|\,d\tau \qquad (5.6.59) \\
&\leqslant 2M(a_+ + \delta)t(1 + Mt(a_+ + \delta))|\pi_t(x) - \pi_t(x')|.
\end{aligned}
$$

Hence, Eqs. (5.6.57) and (5.6.59) imply

$$
\begin{aligned}
|\pi_t(x) - \pi_t'(x)| \leqslant \|\pi - \pi'\|&\left(e^{-\nu_0 t} + 2M\delta t(1 + M\delta(a_+ + \delta)t)\right) \\
&+ \left(C\sqrt{\delta}\, e^{-\nu_0 t} + 2M\delta t(1 + M\delta(a_+ + \delta)t)\right) \qquad (5.6.60) \\
&\times (2M(a_+ + \delta)t(1 + Mt(a_+ + \delta)))|\pi_t(x) - \pi_t'(x)|,
\end{aligned}
$$

and from this formula we deduce a bound on $|\pi_t(x) - \pi_t'(x)|$ if a_+, δ, t_0 are so small that the inequality

$$
\begin{aligned}
&\frac{e^{-\nu_0 t} + 2M\delta t(1 + M\delta(a_+ + \delta)t)}{\left\{1 - (2M(a_+ + \delta)t(1 + Mt(a_+ + \delta))) \times \left(e^{-\nu_0 t}C\sqrt{\delta} + 2M\delta t(1 + M\delta(a_+ + \delta)t)\right)\right\}} \qquad (5.6.61) \\
&\leqslant \left(1 - \frac{\nu_0}{2}t\right), \qquad \forall t \in [0, t_0],
\end{aligned}
$$

holds for all t, $0 \leqslant t \leqslant t_0$.

Equations (5.6.61) and (5.6.60) imply $|\pi_t(x) - \pi_t'(x)| \leqslant (1 - (\nu_0/2)t)\|\pi - \pi'\|$; hence, $\forall \alpha \in [-a_+, a_+]$, $\forall t \in [0, t_+]$,

$$
\|\pi_t - \pi_t'\| \leqslant \left(1 - \frac{\nu_0}{2}t\right)\|\pi - \pi'\|. \qquad (5.6.62)
$$

A similar calculation would allow us to show that if π verifies Eq. (5.6.49) and a_+, δ, t_0 are sufficiently small,

$$
\|\pi_t - \pi_{t'}\| \leqslant \gamma|t' - t| \qquad (5.6.63)
$$

for all $\alpha \in [-a_+, a_+]$, $\forall t, t' \in R_+$, $|t - t'| < t_+$ provided γ is suitably chosen.

We shall use this inequality without proof here (see Appendix M where a proof is discussed and an explicit expression for γ is exhibited).

Equation (5.6.62) allows us to recursively estimate Eq. (5.6.54) since it holds under the sole assumption that π and π' verify Eq. (5.6.49) and $t \in [0, t_+]$, $\alpha \in [-a_+, a_+]$. By subsection 5.6.F, one finds

$$
\|\pi_{nt} - \pi_{(n-1)t}\| \leqslant \left(1 - \frac{\nu_0}{2}t\right)^{n-1}\|\pi_t - \pi\| \leqslant 2\delta\left(1 - \frac{\nu_0}{2}t\right)^{n-1} \qquad (5.6.64)
$$

valid for all π verifying Eq. (5.6.49), $\forall n$ integer and $\geqslant 1$.

Hence, the series of Eq. (5.6.55) is uniformly convergent as π varies in the class of the functions verifying Eq. (5.6.49), $\forall t \in [0, t_+]$, $\forall \alpha \in [-a_+, a_+]$.

5.6.H. Independence of the Limit as $n \to +\infty$ of π_{nt} from π and t for $t \in [0, t_+]$. Denote $\pi_{\infty,t,\pi}$ the continuous function defined on $[-\delta, \delta]$, $\forall t \in (0, t_+]$, in terms of a π verifying Eq. (5.6.49), by

$$\lim_{n\to+\infty} \pi_{nt} = \pi_{\infty,t,\pi}, \tag{5.6.65}$$

the continuity being insured by the uniformity of the limit of Eq. (5.6.65), see Eq. (5.6.55).

It is clear that $\pi_{\infty,t,\pi}$ is π independent. In fact, Eq. (5.6.62) recursively implies

$$\|\pi_{nt} - \pi'_{nt}\| \leqslant \left(1 - \frac{\nu_0}{2} t\right)^n \|\pi - \pi'\| \xrightarrow[n\to+\infty]{} 0 \tag{5.6.66}$$

if π, π' verify Eq. (5.6.49). Hence, we shall simply denote $\pi_{\infty,t,\pi}$ as $\pi_{\infty,t}$.

Now let $t', t \in [0, t_+]$ and $t'/t = p/q =$ rational number, p, q integers. Clearly,

$$\pi_{ntp} = \pi_{nt'q}; \tag{5.6.67}$$

hence, in the limit $n \to +\infty$, we see that Eq. (5.6.67) implies

$$\pi_{\infty,t} = \pi_{\infty,t'} \tag{5.6.68}$$

if $t/t' = \{\text{rational number}\}$. Then Eq. (5.6.63) clearly implies that Eq. (5.6.68) holds for all $t, t' \in (0, t_+]$ and $\pi_{\infty,t}$ is t independent.

Denoting π_∞ the function in Eq. (5.6.68), it is obviously true that $(\pi_\infty)_t \equiv \pi_\infty$, $\forall t \geqslant 0$ and this proves the invariance of $\sigma(\pi_\infty)$; hence, Eq. (5.6.9).

5.6.I. Attractivity of $\sigma(\pi_\infty)$. Given $(\bar{x}, \bar{z}) \in \Gamma(\delta)$, let $\bar{\pi}$ be a function verifying Eq. (5.6.49) and $\bar{\pi}(\bar{x}) = \bar{z}$, e.g., $\bar{\pi}(x) \equiv \bar{z}$, $x \in [-\delta, \delta]$. Given $t > 2t_+$, let $\bar{t} \in (0, t_+)$ such that $\bar{t} > t_+/2$ and, furthermore, $t/\bar{t} = N =$ integer.

Then by Eq. (5.6.66) or Eq. (5.6.62),

$$\begin{aligned}
\|\bar{\pi}_t - \pi_\infty\| &\equiv \|\bar{\pi}_t - (\pi_\infty)_t\| = \|(\bar{\pi})_{N\bar{t}} - (\pi_\infty)_{N\bar{t}}\| \\
&= \lim_{n\to\infty} \|(\bar{\pi})_{N\bar{t}} - (\pi_n)_{N\bar{t}}\| \leqslant \lim_{n\to\infty} \left(1 - \frac{\nu_0}{2}\bar{t}\right)^N \|\bar{\pi} - \pi_n\| \qquad (5.6.69) \\
&\leqslant \left(1 - \frac{\nu_0}{2}\bar{t}\right)^N \|\bar{\pi} - \pi_\infty\| \leqslant 2\delta\left(1 - \frac{\nu_0}{4} t_+\right)^{t/t_+}
\end{aligned}$$

which proves that $\sigma(\bar{\pi}_t)$, hence, $S_t^{(\alpha,\delta)}(\bar{x}, \bar{z})$, approaches $\sigma(\pi_\infty)$ with exponential speed so that the attractivity of $\sigma(\pi_\infty)$ is proved in the case of the Eqs. (5.6.17) and (5.6.18) and this immediately leads to Eq. (5.6.10).

5.6.L. Order of Tangency. Let us show that if π is chosen so that it verifies Eq. (5.6.49) as well as

$$|\pi(x)| \leqslant C|x|^{3/2}, \qquad \forall x \in [-\delta, \delta], \tag{5.6.70}$$

then it is also true that

$$|\pi_t(x)| \leqslant C|x|^{3/2}, \qquad \forall x \in [-\delta, \delta], \qquad \forall t \in \mathbf{R}_+ . \tag{5.6.71}$$

Hence, for $x \in [-\delta, \delta]$, we shall have $|\pi_\infty(x)| \leqslant C|x|^{3/2}$, implying Eq. (5.6.11) for $k = 0$.

Suppose that π verifies Eqs. (5.6.49) and (5.6.70), e.g., $\pi(x) \equiv \frac{2}{3} C|x|^{3/2}$.

From Eqs. (5.6.33), (5.6.25), and (5.6.20) and from $S_t^{(\alpha,\delta)}(0,0) \equiv (0,0)$, it follows, with the usual notations, that

$$\begin{aligned}
|\pi_t(x)| &\leqslant e^{-\nu_0 t} C|x_0|^{3/2} + \int_0^t M\left|S_\tau^{(\alpha,\delta)}(x_0, \pi(x_0))\right|^2 d\tau \\
&\leqslant e^{-\nu_0 t} C|x_0|^{3/2} + Mt(1 + 2Mt(a_+ + \delta))^2 \left(x_0^2 + \pi(x_0)^2\right) \\
&\leqslant C|x_0|^{3/2} \Big(e^{-\nu_0 t} + Mt(1 + 2Mt(a_+ + \delta))^2 \\
&\qquad \times \left(C^{-1}\sqrt{|x_0|} + C\sqrt{|x_0|}^{\,3} \right) \Big) \\
&\leqslant C|x_0|^{3/2} \left(e^{-\nu_0 t} + Mt(1 + 2Mt(a_+ + \delta))^2 \left(C^{-1}\sqrt{\delta} + C\sqrt{\delta}^{\,3} \right) \right)
\end{aligned} \tag{5.6.72}$$

for $t \in [0, t_+]$, $\alpha \in [-a_+, a_+]$. From Eqs. (5.6.58), (5.6.25), and (5.6.20) and $S_t^{(\alpha,\delta)}(0,0) \equiv (0,0)$, one finds

$$\begin{aligned}
|x_0| &\leqslant |x| + M \int_0^t \left(a_+ \left| \left(S_{-\tau}^{(\alpha,\delta)}(x, \pi_t(x)) \right)_1 \right| + \left| S_{-\tau}^{(\alpha,\delta)}(x, \pi_t(x)) \right|^2 \right) d\tau \\
&\leqslant |x| + Mt\Big\{ a_+ \left[(1 + Mt(a_+ + \delta))(|x| + |\pi_t(x)|) \right] \\
&\qquad + \left[2(1 + 2Mt(a_+ + \delta))^2 \left(|x|^2 + |\pi_t(x)|^2 \right) \right] \Big\} \\
&\leqslant |x| \Big(1 + Mt\left\{ a_+ [1 + Mt(a_+ + \delta)] + \left[2(1 + 2Mt(a_+ + \delta))^2 \delta \right] \right\} \Big) \\
&\quad + |\pi_t(x)| Mt \Big\{ a_+ [1 + Mt(a_+ + \delta)] \\
&\qquad + \left[2(1 + 2Mt(a_+ + \delta))^2 \delta \right] \Big\}.
\end{aligned} \tag{5.6.73}$$

To simplify the notations, rewrite Eqs. (5.6.72) and (5.6.73) by observing that if $a_+, \delta, t_+ < 1$ (as we have supposing since the beginning of the analysis), there exists $M' > 0$ such that

$$|\pi_t(x)| \leqslant C|x_0|^{3/2}\left(1 - \frac{\nu_0}{2}t + M'\sqrt{\delta}\,t\right), \tag{5.6.72'}$$

$$|x_0| \leqslant |x|(1 + M'(a_+ + \delta)t) + |\pi_t(x)|M'(a_+ + \delta)t. \tag{5.6.73'}$$

Then, taking the $\frac{2}{3}$ power of Eq. (5.6.72') and using Eqs. (5.6.73') and (5.6.72'), one deduces

$$\begin{aligned} |\pi_t(x)|^{2/3} \leqslant C^{2/3}&\left(1 - \frac{\nu_0}{2}t + M'\sqrt{\delta}\,t\right)^{2/3} \\ &\times\left[(1 + M'(a_+ + \delta)t)|x| + M'(a_+ + \delta)t|\pi_t(x)|\right], \end{aligned} \tag{5.6.74}$$

and since $\delta < 1$, $|\pi_t(x)| \leqslant \delta$, one can use $|\pi_t(x)| \leqslant |\pi_t(x)|^{2/3}$ to deduce from Eq. (5.6.74) that

$$|\pi_t(x)|^{2/3} \leqslant C^{2/3}|x|\,\frac{\left(1 - (\nu_0/2)t + M'\sqrt{\delta}\,t\right)^{2/2}(1 + M'(a_+ + \delta)t)}{1 - M'(a_+ + \delta)t\left(1 - (\nu_0/2)t + M'\sqrt{\delta}\,t\right)^{2/3}}. \tag{5.6.75}$$

Hence, let us choose δ, a_+, t_0 to be so small that, $\forall t \in [0, t_0]$, the ratio in Eq. (5.6.75) is bounded by

$$\{\text{ratio in Eq. (5.6.75)}\} \leqslant 1 - \frac{\nu_0 t}{4}, \tag{5.6.76}$$

we see that Eqs. (5.6.75) and (5.6.76) imply, $\forall x \in [-\delta, \delta]$, $\forall t \in [0, t_+]$, $\forall \alpha \in [-a_+ a_+]$:

$$|\pi_t(x)| \leqslant C|x|^{3/2}\left(1 - \frac{\nu_0}{4}t\right)^{3/2} \leqslant C|x|^{3/2}; \tag{5.6.77}$$

hence, the inequality between the left-hand and the right-hand sides holds, $\forall t \geqslant 0$, and this implies Eq. (5.6.11) for $k = 0$.

5.6.M. Regularity in α. This is the last property to check. One proceeds almost exactly in the same way as above. We illustrate the details because in some sense there is here a technical idea new with respect to the ones already met.

Actually we shall prove that π is a Lipshitzian function of α and x for $\alpha \in [-a_+, a_+]$, $x \in [-\delta, \delta]$, i.e., a somewhat stronger result.

Consider a function $(x, \alpha) \to \pi(x, \alpha)$ defined for $x \in [-\delta, \delta]$, $\alpha \in [-a_+, a_+]$, of class $C^{(1)}$ and verifying Eq. (5.6.49) for each α. Define $\pi_t(x, \alpha)$ by thinking of π as a function of x for each α and proceeding as in subsection 5.6.E.

From Eqs. (5.6.33), (5.6.24), (5.6.21), (5.6.20), and (5.6.49) and employing the usual notations, one finds that

$$\pi_t(x,\alpha) = e^{-\nu_0 t}\pi(x_0,\alpha) + \int_0^t e^{-\nu_0(t-\tau)} Z'_\delta\big(S_\tau^{(\alpha,\delta)}(x_0,\pi(x_0,\alpha)),\alpha\big)\,d\tau; \tag{5.6.78}$$

hence, recalling that x_0 is also α dependent and denoting ∂_α the derivative $\partial/\partial\alpha$:

$$\begin{aligned}
|\partial_\alpha \pi_t(x,\alpha)| &\leqslant e^{-\nu_0 t}\left|\partial_\alpha \pi(x_0,\alpha) + \frac{\partial \pi}{\partial x}(x_0,\alpha)\frac{\partial x_0}{\partial \alpha}\right| \\
&\quad + \int_0^t \sum_{i=1}^{2} \left| \frac{\partial Z_\delta}{\partial w_i}\big(S_\tau^{(\alpha,\delta)}(x_0,\pi(x_0,\alpha))\big)\frac{d}{d\alpha}\right. \\
&\qquad \left. \times \Big[\big(S_\tau^{(\alpha,\delta)}(x_0,\pi(x_0,\alpha))\big)_i\Big]\right| \times e^{-\nu_0(t-\tau)}\,d\tau \\
&\leqslant e^{-\nu_0 t}|\partial_\alpha \pi(x_0,\alpha)| + e^{-\nu_0 t} C\sqrt{\delta}\left|\frac{\partial x_0}{\partial\alpha}\right| \\
&\quad + 2M\delta \int_0^t \Big\{ M\tau(\delta+\delta^2 t) + (1+M\tau(a_+ +\delta)) \\
&\qquad \times\left(\left|\frac{\partial x_0}{\partial \alpha}\right| + \left|\frac{\partial \pi}{\partial x_0}(x_0,\alpha)\right|\left|\frac{\partial x_0}{\partial\alpha}\right|\right. \\
&\qquad \left. + \left|\frac{\partial\pi}{\partial\alpha}(x_0,\alpha)\right|\right)\Big\}\,d\tau \\
&\leqslant e^{-\nu_0 t}|\partial_\alpha \pi(x_0,\alpha)| + e^{-\nu_0 t} C\sqrt{\delta}\left|\frac{\partial x_0}{\partial\alpha}\right| + M^2\delta^2(1+\delta t)t^2 \\
&\quad + 2M\delta(1+Mt(a_+ +\delta))t\left((1+C\sqrt{\delta})\left|\frac{\partial x_0}{\partial\alpha}\right| + \left|\frac{\partial\pi}{\partial\alpha}(x_0,\alpha)\right|\right) \\
&\leqslant \big\{e^{-\nu_0 t} + 2M\delta(1+Mt(a_+ +\delta))t\big\}\left|\frac{\partial\pi}{\partial\alpha}(x_0,\alpha)\right| \\
&\quad + M^2\delta^2(1+\delta t)t^2 \\
&\quad + \big\{e^{-\nu_0 t}C\sqrt{\delta} + 2M\delta(1+Mt(a_+ +\delta))t(1+C\sqrt{\delta})\big\}\left|\frac{\partial x_0}{\partial\alpha}\right|.
\end{aligned} \tag{5.6.79}$$

The $\partial x_0/\partial\alpha$ is estimated as in subsection 5.6.G, using Eq. (5.6.58) rewritten as

$$x_0 = x - \int_0^t X_\delta\big(S_{-\tau}^{(\alpha,\delta)}(x,\pi_t(x,\alpha)),\alpha\big)\,d\tau; \tag{5.6.80}$$

hence, proceeding as in the derivation of Eq. (5.6.79) and using Eq. (5.6.23):

$$\left|\frac{\partial x_0}{\partial \alpha}\right| \leqslant \int_0^t d\tau \left\{ M\delta + 2M(a_+ + \delta) \times \left[M\delta\tau(1+\delta\tau) + (1 + M\tau(a_+ + \delta)) \cdot \left|\frac{\partial \pi_t}{\partial \alpha}(x,\alpha)\right| \right]\right\}$$
$$\leqslant M\delta t + 2M^2\delta \frac{t^2}{2}(1+\delta t)(a_+ + \delta) + t(1 + Mt(a_+ + \delta))2M(a_+ + \delta)\left|\frac{\partial \pi_t}{\partial \alpha}(x,\alpha)\right|. \tag{5.6.81}$$

Then Eqs. (5.6.79) and (5.6.81) imply

$$|\partial_\alpha \pi_t(x,\alpha)| \leqslant \frac{A}{1-GS} + \frac{B}{1-GS}|\partial_\alpha \pi(x_0,\alpha)| \tag{5.6.82}$$

with

$$G \equiv \left[e^{-\nu_0 t}C\sqrt{\delta} + 2M\delta(1 + Mt(a_+ + \delta))t(1 + C\sqrt{\delta})\right],$$
$$S \equiv t\left[1 + Mt(a_+ + \delta)2M(a_+ + \delta)\right],$$
$$A \equiv M^2\delta^2(1+\delta t)t^2 + G\left[M\delta t + M^2\delta t^2(1+\delta t)(a_+ + \delta)\right],$$
$$B \equiv e^{-\nu_0 t} + 2M\delta(1 + Mt(a_+ + \delta))t$$

and to understand the essential features of Eq. (5.6.82), we note that if δ, a_+, t_0 are chosen so small that there is $\overline{M}$ such that the first term in Eq. (5.6.82) can be bounded by

$$\left\{\frac{A}{1-GS}\right\} \leqslant \overline{M}\delta\sqrt{\delta}\, t \tag{5.6.83}$$

for all $t \in [0, t_0]$, $\forall \alpha \in [-a_+, a_+]$, and the coefficient of $|\partial_\alpha \pi(x_0,\alpha)|$ in Eq. (5.6.82) can be bounded as

$$\left\{\frac{B}{1-GS}\right\} \leqslant 1 - \frac{\nu_0 t}{2}, \tag{5.6.84}$$

then Eq. (5.6.82) can be simply rewritten, $\forall t \in [0, t_+]$, $\forall \alpha \in [-a_+, a_+]$,

$$\|\partial_\alpha \pi_t\| \equiv \max_{\substack{|x| \leqslant \delta \\ |\alpha| \leqslant a_+}} |\partial_\alpha \pi_t(x,\alpha)| \leqslant \overline{M}\delta\sqrt{\delta}\, t + \left(1 - \frac{\nu_0}{2}t\right)\|\partial_\alpha \pi\| \tag{5.6.85}$$

as is easily seen.

Now we fix π to be a function of the variable x only and verifying Eq. (5.6.49). We can apply Eq. (5.6.85) to the functions $\pi_{nt}, \pi_{(n-1)t}, \ldots$ thought of as functions of x and α. If $t \in [0, t_+]$, $n = 0, 1, 2, \ldots,$

$$\|\partial_\alpha \pi_{nt}\| \leqslant \overline{M}\delta\sqrt{\delta}\, t + \left(1 - \frac{\nu_0}{2}t\right)\|\partial_\alpha \pi_{(n-1)t}\|. \tag{5.6.86}$$

Then, Eq. (5.6.86) implies, recursively,

$$\|\partial_\alpha \pi_{nt}\| \leqslant \bar{M}\delta\sqrt{\delta}\, t\left(1 + \left(1 - \frac{\nu_0}{2}t\right) + \left(1 - \frac{\nu_0}{2}t\right)^2 + \cdots\right) \tag{5.6.87}$$

because $\pi_0 = \pi$ is by hypothesis α independent, so that $\partial_\alpha \pi_0 \equiv 0$, i.e.,

$$\|\partial_\alpha \pi_{nt}\| \leqslant \frac{\bar{M}\delta\sqrt{\delta}\, t}{(\nu_0/2)t} \equiv \frac{2\bar{M}\delta\sqrt{\delta}}{\nu_0}. \tag{5.6.88}$$

The regularity of π_∞ is now easy to prove:

$$\begin{aligned}|\pi_\infty(x,\alpha) - \pi_\infty(x',\alpha')| &= \lim_{n\to\infty} |\pi_{nt}(x,\alpha) - \pi_{nt}(x',\alpha')|\\ &\leqslant \lim_{n\to\infty} (|x - x'| + |\alpha - \alpha'|)\max\left(\left|\frac{\partial \pi_{nt}}{\partial x}\right| + \left|\frac{\partial \pi_{nt}}{\partial \alpha}\right|\right),\end{aligned} \tag{5.6.89}$$

where the maximum is taken on the set $[-\delta,\delta]\times[-a_+,a_+]$ and, by Eq. (5.6.49) (considered for π_{nt}) and Eq. (5.6.88), it can be estimated by $D = \sqrt{\delta}\,(1 + 2M\nu_0^{-1})$.

Hence,

$$|\pi_\infty(x,\alpha) - \pi_\infty(x',\alpha')| \leqslant D(|x - x'| + |\alpha - \alpha'|), \tag{5.6.90}$$

showing that π_∞ is continuous in x and α (i.e., it is in class $C^{(0)}$) and, actually, that it is a Lipshitz function in x and α (with a Lipshitz constant D which can be taken as small as desired by taking δ small).

5.6.N. General Case. To show that π_∞ is k-time differentiable with respect to x if a_+,δ are chosen sufficiently small, one proceeds to estimate $\partial^2\pi_t/\partial x^2$ and, successively, $\partial^3\pi_t/\partial x^2, \ldots, \partial^{k+1}\pi_t/\partial x^{k+1}$ in the same way as in the $k = 0$ case we studied π_t and $\partial\pi_t/\partial x$ to show that π_∞ was $C^{(0)}$, assuming now that π is in $C^{(k+1)}([-\delta,\delta])$ to start with.

Proceeding with the same technique as in subsections 5.6.F, 5.6.G, and 5.6.L, $(1 + k)\delta, (1 + k)a_+, t_0$ are chosen sufficiently small so that inequalities similar to Eqs. (5.6.41), (5.6.42), (5.6.43), and (5.6.53), etc. hold. One finds

$$\left\|\frac{\partial^h \pi_t}{\partial x^h}\right\| \equiv \max_{|x|\leqslant\delta}\left|\frac{\partial^h \pi_t}{\partial x^h}(x)\right| \leqslant \left(1 - \frac{\nu_0}{2}t\right)\left\|\frac{\partial^h \pi}{\partial x^h}\right\| + tR_{k,\delta}\left(\sum_{j=0}^{h-1}\left\|\frac{\partial^j \pi}{\partial x^j}\right\|\right) \tag{5.6.91}$$

for $h = 0, \ldots, k+1$ and $t \in [0, \tilde{t}_+]$ with $\tilde{t}_+$ suitably small provided $\delta, y \to R_{k,\delta}(y)$ is a suitable continous function in the variables δ, y and monotonically increasing in y.

Equation (5.6.91) has the same nature as Eq. (5.6.86), and in the same manner it allows us to show inductively that the Eq. (5.6.49) as well as

$\|\partial^j\pi/\partial x^j\| \leqslant \bar{C}, j=0,\ldots,k+1$, imply

$$\sum_{j=0}^{k+1}\left\|\frac{\partial^j\pi_{tn}}{\partial x^j}\right\| \leqslant (k+1)\frac{R_{k,\delta}\big((k+1)\bar{C}\big)}{\nu_0/2} + (k+1)\bar{C}. \qquad (5.6.92)$$

Equation (5.6.92) means that π_∞ is k-times differentiable.

Along similar lines, it is possible to prove the $C^{(k)}$ regularity in the variable α and, jointly, in α and x for $|\alpha|, |x|$ small. By way of estimates of the $(k+1)$th derivative of π_t with respect to α, of the kth derivative of $\partial\pi_t/\partial x$ with respect to $\alpha, \ldots,$ and of the first derivative with respect to α of $\partial^k\pi_t/\partial x^k$, this regularity property is proved following the ideas and the techniques of subsections 5.6.M and 5.6.N.

The reader who has been determined enough to reach this point shall not have problems in transforming the above hints into a proof. We only stress that from what has been said above, it appears that in order to obtain $C^{(k)}$ regularity, one must impose restrictions on δ, a_+, and t_0 which are k dependent. This means that the above proof cannot be used to prove that the attractive manifold depends in a C^∞ way on x and α: actually, it is an open problem to find whether such a smoothness property can be enjoyed by the attractive manifolds under simple extra assumptions (whose necessity is made clear by the example in Observation 1, p. 411.) mbe

Problems and Complements for §5.6

1. Show that $\mathbf{0} = (0,0)$ is vaguely attractive near $\alpha_c = 0$ for $\dot{x} = \alpha x - x^3$, $\dot{z} = -z$, $(x,z) \in \mathbf{R}^2$.

2. In the context of Problem 1, show that the plane $z = 0$ is an attractive manifold in the sense of Proposition 11.

3. Consider the equation in Problem 1 and the surface σ_α built with three pieces with respective parametric equations

$$\begin{cases} z(\gamma) = \bar{z}e^{-\gamma}; \\ x(\gamma) = \bar{x}(\gamma) = \sqrt{\alpha}\left(1 + \dfrac{\alpha - \bar{x}^2}{\bar{x}^2}e^{-2\alpha\gamma}\right)^{-1/2}, \end{cases} \qquad \gamma \in [0, +\infty)$$

$$\begin{cases} z(\gamma) = \bar{z}'e^{-\gamma}, \\ x(\gamma) = \bar{x}'(\gamma) = -\sqrt{\alpha}\left(1 + \dfrac{\alpha - \bar{x}'^2}{\bar{x}'^2}e^{-2\alpha\gamma}\right)^{-1/2}, \end{cases} \qquad \gamma \in [0, +\infty),$$

$$\begin{cases} z(\gamma) = 0, \\ x(\gamma) = y, \end{cases} \qquad \gamma \in [-\sqrt{\alpha}, \sqrt{\alpha}).$$

Show that σ_α is an attractive manifold $\forall \bar{x}, \bar{x}', \bar{z}, \bar{z}'$ such that $\sqrt{\alpha} \leqslant \bar{x}, -\bar{x}', \alpha > 0$, in the sense of Proposition 11. (Hint: Note that $t \to \bar{x}(t)$ is a solution of $\dot{x} = \alpha x - x^3$ with initial datum $\bar{x}$.)

4. Show that the attractive manifolds in Problem 3 are in class $C^{(k)}$ at fixed α if α is small enough ($2\alpha k < 1$). Show that the equation in Problem 1 admits infinitely many attractive manifolds with are not C^∞ in x. Meditate on how general this nonuniqueness mechanism is.

5. Consider the equation $\dot{x} = \alpha x$, $\dot{z} = -\nu z + x^2$ and determine all the attractive manifolds of the origin for $0 < -\alpha < \nu$. Show that for each $\alpha < 0$ there are infinitely many such manifolds but only one, at most, can be of class C^∞. Find a value of $\alpha < 0$ for which no attractive manifold is of class C^1. (Hint: Note that an attractive manifold must be a union of trajectories of solutions of the differential equation; see also Problem 1. The critical value of α is $\alpha = -\nu/2$.)

6. Using the example of Problem 5 show that the assumption $\operatorname{Re}\lambda_j(\alpha_c) = 0$, $j = 1, \ldots, r$, is essential in Proposition 11. If this assumption is not verified argue that a proposition like Proposition 11 could still hold if the order k of smoothness is restricted as $k < \nu_0/\nu_0'$, at least. See also Problem 7.

7. Prove Proposition 11 for Eq. (5.6.12) when α is near some α_c, $-\nu_0 < \alpha_c < 0$ and $k = 0$. (Hint: Write Eq. (5.6.17) as $\dot{x} = \alpha_c x + \chi_\delta(x,z)((\alpha - \alpha_c)x + P(x,z))$ and proceed as in the proof in §5.6 with the obvious substitution of Eq. (5.6.32), and of the other equations similar to it, with

$$x = e^{\alpha_c t}x_0 + \int_0^t e^{\alpha_c(t-\tau)} X_\delta\big(S_\tau^{(\alpha,\delta)}(x_0, \pi(x_0)), \alpha\big)\, d\tau,$$

etc.)

8. Show the validity of Proposition 11 in the case in which Eq. (5.6.12) is replaced by the equation ($\mu > 0$)

$$\begin{aligned} \dot{x}_1 &= \alpha x_1 - \mu x_2 + P_1(x_1, x_2, z), \\ \dot{x}_2 &= \mu x_1 + \alpha x_2 + P_2(x_1, x_2, z), \\ \dot{z} &= -\nu_0 z + Q(x_1, x_2, z). \end{aligned}$$

(Hint: Write the equation analogous to Eq. (5.6.17) as

$$\begin{aligned} \dot{x}_1 &= -\mu x_2 + \chi_\delta(x_1, x_2, z)(\alpha x_1 + P_1(x_1, x_2, z)), \\ \dot{x}_2 &= \mu x_1 + \chi_\delta(x_1, x_2, z)(\alpha x_2 + P_2(x_1, x_2, z)), \\ \dot{z} &= -\nu_0 z + \chi_\delta(x_1, x_2, z) Q(x_1, x_2, z) \end{aligned}$$

with analogous notations. Then proceed exactly as in the proof in §5.6, substituting Eq. (5.6.32), and the other equations similar to it, with

$$\mathbf{x} = W(t)\mathbf{x}_0 + \int_0^t W(t-\tau)\mathbf{X}_\delta\big(S_\tau^{(\alpha,\delta)}(\mathbf{x}_0, \pi(\mathbf{x}_0)), \alpha\big)\, d\tau,$$

where $W(t) = \begin{pmatrix} \cos t & -\sin t \\ \sin t & \cos t \end{pmatrix}$ is the Wronskian matrix; see, also, problems for §2.5, etc.)

9. Using the same ideas as in Problems 7 and 8, study Proposition 11 in the general case, i.e., for an equation of the form of Eq. (5.5.10).

10. If $\mathbf{x}_0$ is not supposed to be vaguely attractive, recognize that the proof of Proposition 11 can be interpreted as showing the existence of a surface σ_α defined as in Eq. (5.6.4), verifying Eq. (5.6.11) and

(ii′) If $\mathbf{w} \in \Gamma(\delta_0) \cap \sigma_\alpha$ and $S_\tau^{(\alpha)}\mathbf{w} \in \Gamma(\delta/2)$, $\forall \tau \in [0,t]$, then $S_t^{(\alpha)}\mathbf{w} \in \sigma_\alpha$ ("local invariance").

(iii′) If $S_t \mathbf{w} \in \Gamma(\delta/2)$, $\forall t \geqslant 0$, then $d(S_t^{(\alpha)}\mathbf{w}, \sigma_\alpha) \xrightarrow[t \to +\infty]{} 0$ exponentially fast. (Hint: The vague attractivity is used only to reduce the proof to the theory of (5.6.17) and (5.6.18). So just start from them.)

11. Consider the equation $\dot{\mathbf{x}} = \mathbf{f}(\mathbf{x})$ and suppose that $\mathbf{x}_0 = \mathbf{0}$ is a stationary solution for it. Let L be the stability matrix of $\mathbf{x}_0$ and suppose that L has $(d-r)$ eigenvalues with negative real parts and r with zero real parts. Without imposing the vague attractivity of $\mathbf{0}$, interpret the proof of Proposition 11 with $\alpha = 0$ as showing that given $k \geqslant 0$, $C > 0$, there exists δ and δ_0, $\delta > \delta_0 > 0$, and a surface σ of dimension r and described by $(d-r)$ functions $\varphi^{(r+1)}, \dots, \varphi^{(d)}$ of r variables $x_1, \dots, x_r$ with $|x_i| < \delta/2$ and verifying Eq. (5.6.11) as well as the local invariance and attractivity properties of the preceding problem ("theorem of the central manifold").

12. Consider the equation

$$\dot{x} = \lambda x + P(x,z),$$
$$\dot{z} = -\nu z + Q(x,z)$$

with $\lambda, \nu > 0$, and $P, Q \in C^\infty(\mathbf{R}^2)$ with a second-order zero at the origin.

Show that:

$$S_t(x,z) = \big(xe^{\lambda t} + tD(x,z,t), ze^{\nu t} + tE(x,z,t)\big)$$

with D and E of class C^∞ and having a zero of second order in the variables (x,z) at $(0,0)$.

13. Use Problem 12 to show that, in the same context and for all small δ, if π is a $C^{(1)}$ function on $[-\delta,\delta]$ such that

$$|\pi(x)| \leqslant \delta, \qquad \left|\frac{d\pi}{dx}(x)\right| \leqslant \sqrt{\delta} \tag{$*$}$$

and if $\sigma(\pi)$ denotes the curve $z = \pi(x)$, $x \in [-\delta,\delta]$, then $\sigma(\pi)$ is transformed into $S_t\sigma(\pi)$ such that

$$S_t\sigma(\pi) \cap \Gamma(\delta) = \sigma(\pi_t),$$

and π_t verifies $(*)$. (Hint: Use the ideas of the proof of Proposition 11.)

14. In the context of Problems 12 and 13, show that

$$\|\pi_{(n+1)\bar{t}} - \pi_{n\bar{t}}\| < \xi\|\pi_{n\bar{t}} - \pi_{(n-1)\bar{t}}\|,$$
$$\|\pi_{\bar{t}} - \pi'_{\bar{t}}\| \leqslant \xi\|\pi - \pi'\|$$

with $\xi < 1$ (if $\|\cdot\|$ denotes the maximum of a function) provided δ_0 is small enough.

From this, deduce the existence in $\Gamma(\delta)$ of a surface locally invariant for S_{-t} and tangent to the x axis at the origin and such that $S_{-t}\mathbf{w} \to_{t\to+\infty} \mathbf{0}$ exponentially fast in the sense

$$-\lambda = \lim_{t\to+\infty} t^{-1}\log|S_{-t}\mathbf{w}|$$

for all nonzero $\mathbf{w}$ on the surface. Denote this surface by σ_i: it is called the "unstable manifold" through $\mathbf{0}$.

15. In the context of Problem 12, show the existence in $\Gamma(\delta)$ of a surface σ_s locally invariant for S_t, tangent to the z axis, and such that $\forall \mathbf{w} \neq \mathbf{0}$, $\mathbf{w} \in \sigma_s$:

$$-\nu = \lim_{t\to+\infty} t^{-1}\log|S_t\mathbf{w}|$$

("stable manifold through $\mathbf{0}$").

16. Study the generalization of the result of Problems 10–13 to a general equation in $\mathbf{R}^d$, $\dot{\mathbf{x}} = \mathbf{f}(\mathbf{x})$, with $\mathbf{f}(\mathbf{0}) = \mathbf{0}$ and a stability matrix L whose eigenvalues are pairwise distinct and such that none among them has a zero real part, although some of them have a positive real part and others have a negative real part ("hyperbolic unstable point") ("existence of stable and unstable manifolds at a hyperbolic fixed point").

17. Consider the equation

$$\dot{x} = x + \frac{x^2}{\alpha} + \frac{z^2}{\beta},$$

$$\dot{z} = -z + \frac{x^2}{\gamma} + \frac{z^2}{\delta},$$

$\alpha = \delta = 1$, $\beta = -\gamma = 2$, and compute the second derivative at the origin of the function π_s defining (via $x = \pi_s(z)$) the stable manifold of $\mathbf{0}$. (Hint: Write $x = Az^2 + Bz^3 + \cdots$ and insert this expression in the first equation. One finds $A = \beta^{-1}$.)

18. Find some extensions to Problems 14 and 15 to equations in $\mathbf{R}^d$ and study them.

19. Show that if, in the context of Proposition 11, σ_α is regarded as an attractor for the neighborhood U used in the proof [see Eq. (5.6.15)], and if $\tilde{\sigma}_\alpha = \bigcap_{t>0} S_t^{(\alpha)}\sigma_\alpha$ then the function in the left-hand side of Eq. (5.3.21), p. 378, with $A = \tilde{\sigma}_\alpha$ can be estimated by an exponentially decreasing function of t as $t \to +\infty$, i.e., $\tilde{\sigma}_\alpha$ is a normal attractor for U by Proposition 5, §5.3, p. 378. (Hint: Examine the text of Proposition 11 and the discussion around Eq. (5.6.15).)

§5.7. An Application: Bifurcations of the Vaguely Attractive Stationary Points into Periodic Orbits. The Hopf Theorem

After the considerations of §5.5 Proposition 10, p. 403, the theory of § 5.6 can be immediately applied to Eq. (5.1.19).

We can say that with fixed $k, k \geqslant 2$, there is $B > 0$ and a cubic neighborhood $\Gamma(\delta)$ centered at $\hat{\omega}$, with side 2δ, and a family σ_α of $C^{(k)}$ surfaces in $\Gamma(\delta)$ with equations

$$\omega_3 = \hat{\omega}_3 + \varphi_\alpha(\omega_1, \omega_2), \tag{5.7.1}$$

defined for $|\omega_1| \leqslant \delta/2$, $|\omega_2| < \delta/2$ and for α close to α_c, $\alpha \in (\alpha_c - a_+, \alpha_c + a_+)$, and, furthermore,

$$|\varphi_\alpha(\omega_1, \omega_2)| \leqslant B(\omega_1^2 + \omega_2^2), \tag{5.7.2}$$

$\varphi_\alpha \in C^{(k)}([-\delta/2, \delta/2]^2 \times (\alpha_c - a_+, \alpha_c + a_+))$, and for every α close to α_c, the surface σ_α is invariant in the sense of Eq. (5.6.9) and attractive for all the points of $\Gamma(\delta)$ in the sense of Eq. (5.6.10), with exponential speed.

We shall now show that if $(\alpha - \alpha_c) > 0$ is sufficiently small, there is on σ_α a minimal attractor A_α consisting of a periodic orbit with a period approximately $2\pi/\hat{\omega}_3$ and attracting the points on $\sigma_\alpha/\{\hat{\omega}\}$ with exponential speed.

Essentially, by using Proposition 5, §5.3, it will then be easy to show that in the situation of the preceding sentence, $A_\alpha \cup \{\hat{\omega}\}$ is an attractor for which the basin $\Gamma(\delta)$ is normal and $\forall \omega \in \Gamma(\delta)$, $\exists \pi(\omega) \in A_\alpha \cup \{\hat{\omega}\}$ such that

$$|S_t^{(\alpha)}(\omega) - S_t^{(\alpha)}(\pi(\omega))| \xrightarrow[t \to +\infty]{} 0 \tag{5.7.3}$$

exponentially fast.

This statement "completes" the analysis of the asymptotic behavior of the motions of Eq. (5.1.19) with initial datum ω close enough to $\hat{\omega}$ and with a given α slightly above α_c.[15]

To see which is the real motion of the gyroscope corresponding to this asymptotically periodic motion of its angular velocity, it would still be necessary to integrate the "geometric" differential equations connecting the Euler angles with the angular velocity, see Eqs. (5.2.9)–(5.2.11). We shall not discuss this last point.

It will be clear that the preceding statements follow, as a special case, from the following general "Hopf bifurcation theorem" and from the observations to it.

12 Proposition. *Consider a differential equation $\dot{\mathbf{x}} = \mathbf{f}(\mathbf{x}, \alpha)$ in $\mathbf{R}^2$ parameterized by $\alpha \in (-a, a)$ and having the origin $\mathbf{0}$ as a vaguely attractive stationary solution near $\alpha_c = 0$. Suppose that the stability matrix of the origin, denoted $L(\alpha)$, has eigenvalues $\lambda(\alpha) = \alpha + i\mu(\alpha)$, $\overline{\lambda(\alpha)} = \alpha - i\mu(\alpha)$, $\bar{\mu} \equiv \mu(0) \neq 0$. Also suppose that the equation is already put in normal form with respect to $\lambda, \bar{\lambda}$ (see Definition 6, p. 390, §5.5) (this can always be achieved via a change of coordinates, by Proposition 7, p. 391, §5.5):*

$$\begin{aligned} \dot{x} &= \alpha x - \mu(\alpha) y + P(x, y, \alpha), \\ \dot{y} &= \mu(\alpha) x + \alpha y + Q(x, y, \alpha) \end{aligned} \tag{5.7.4}$$

with $P, Q \in C^{(k)}(\mathbf{R}^2 \times (-a, a))$, k being a large enough integer, and with P, Q having a third-order zero in $x = y = 0$, $\forall \alpha \in (-a, a)$.

Finally suppose that the origin is vaguely attractive because the vague-attractivity indicator $\bar{\gamma}_{\alpha_c}$ is negative. Recall that $\bar{\gamma}_{\alpha_c}$ is defined as the average value over θ of $\gamma_{\alpha_c}(\theta)$ with

$$\gamma_\alpha(\theta) = \lim_{\rho \to 0} \frac{xP(x, y, \alpha) + yQ(x, y, \alpha)}{(x^2 + y^2)^2} \tag{5.7.5}$$

if (ρ, θ) are the polar coordinates of (x, y), see (5.5.25).

[15] An even more complete picture, distinguishing the points attracted by A_α from those attracted by $\hat{\omega}$ can be obtained by using the results of Problems 12–19 of §5.6. The outcome would be the one described just before Proposition 11, p. 410.

Then if $\alpha > 0$ is sufficiently small, there is a periodic solution to Eq. (5.7.4) which is an attractor attracting all the points in a small neighborhood of $\mathbf{0}$, *with the exception of* $\mathbf{0}$ *itself, with exponential speed.*

The period T_α of this motion is such that $\lim_{\alpha\to\alpha_c} T_\alpha = 2\pi/\mu(0)$.

Observations.

(1) The requirement on k to be large enough is imposed to guarantee the possibility of further reducing the complexity of Eq. (5.7.4) by changing coordinates so that the function $\gamma_\alpha(\theta)$ in Eq. (5.7.5) becomes θ independent (i.e. $\gamma_\alpha(\theta) \equiv \bar{\gamma}_\alpha$) in the new polar coordinates and, at the same time, so that in the new coordinates the functions

$$\begin{aligned} r(x,y,\alpha) &= xP(x,y,\alpha) + yQ(x,y,\alpha) - \bar{\gamma}_\alpha(x^2+y^2)^2, \\ s(x,y,\alpha) &= xQ(x,y,\alpha) - yP(x,y,\alpha) \end{aligned} \tag{5.7.6}$$

are infinitesimal of fifth order at $x = y = 0$, uniformly in $\alpha \in (-a, a)$, and also have gradients in x, y which are infinitesimal of the fourth order [a property used below in Eqs. (5.7.29) and (5.7.30)].[16] See Observation (8) for more details.

(2) In the application to Eq. (5.1.19), Eq. (5.7.4) is

$$\begin{aligned} \dot{\omega}_1 &= (\alpha - \alpha_c)\omega_1 - \hat{\omega}_3\omega_2 + P(\omega_1, \omega_2, \alpha), \\ \dot{\omega}_2 &= \hat{\omega}_3\omega_1 + (\alpha - \alpha_c)\omega_2 + Q(\omega_1, \omega_2, \alpha); \end{aligned} \tag{5.7.7}$$

where:

$$\begin{aligned} P(\omega_1, \omega_2, \alpha) &= -\omega_1\big(\lambda_2'\omega_1^2 + \lambda_2''\omega_2^2 + \lambda_2'''2\hat{\omega}_3\varphi_\alpha(\omega_1,\omega_2) + \omega_2'''\varphi_\alpha(\omega_1,\omega_2)^2\big) \\ &\quad - \omega_2\varphi_\alpha(\omega_1,\omega_2), \\ Q(\omega_1, \omega_2, \alpha) &= -\omega_2\big(\lambda_2'\omega_1^2 + \lambda_2''\omega_2^2 + \lambda_2'''2\hat{\omega}_3\varphi_\alpha(\omega_1,\omega_2) + \lambda_2'''\varphi_\alpha(\omega_1,\omega_2)^2\big) \\ &\quad + \omega_1\varphi_\alpha(\omega_1,\omega_2), \end{aligned} \tag{5.7.8}$$

φ_α being the function defining the attractive manifold, and it is of class $C^{(k)}$, k chosen (once and for all) as large as desired. Hence,

$$\gamma_\alpha(\theta) = \lim_{\omega_1,\omega_2\to 0} - \frac{\lambda_2'\omega_1^2 + \lambda_2''\omega_2^2 + \lambda_2'''2\hat{\omega}_3\varphi_\alpha(\omega_1,\omega_2)}{(\omega_1^2+\omega_2^2)} \tag{5.7.9}$$

and to evaluate $\bar{\gamma}_{\alpha_c}$ one does not need to know explicitly φ_α. One can proceed as in the proof of Proposition 10, p. 403, setting $r \equiv \varphi_\alpha$; by the same calculation one finds

$$\bar{\gamma}_{\alpha_c} = -\frac{(\lambda_2' + \lambda_2'')(\lambda_1 + \lambda_2'''\hat{\omega}_3^2)}{2(\lambda_1 + 3\lambda_2'''\hat{\omega}_3^2)} < 0 \tag{5.7.10}$$

[16] $k > 5$ will suffice, see Observation (8).

Hence, Eq. (5.1.19) has a periodic attractive solution for $\alpha > \alpha_c$ and $(\alpha - \alpha_c)$ small.

(3) As already noted, the assumption that the equation $\dot{\mathbf{x}} = \mathbf{f}(\mathbf{x}, \alpha)$ has normal form with respect to $\lambda(\alpha), \bar{\lambda}(\alpha)$ is not really restrictive if $\mu(0) \neq 0$, by Proposition 7, §5.5, p. 393.

The assumption $\operatorname{Re}\lambda(\alpha) \equiv \alpha$, is also not too restrictive: if $d\operatorname{Re}\lambda(\alpha)/d\alpha|_{\alpha=\alpha_c} \neq 0$, we can clearly rename $\pm\operatorname{Re}\lambda(\alpha)$ with the name α and fall within the theorem's assumptions. However, pathologies can appear if $\operatorname{Re}\lambda(\alpha)$ has a vanishing derivative at α_c.

(4) The theorem has been formulated in class $C^{(k)}$ rather than in class C^∞ because it is usually applied in connection with the attractive manifold theorem, Proposition 11, p. 410 [as, for instance, in Observation (2)], in which case one cannot take $k = +\infty$.

(5) It is important to stress the rather general situation that the above theorem can cover, if combined with the attractive manifold theorem of §5.6, and with the normal-form theorem (Proposition 7, p. 393, §5.5) when the loss of stability takes place in two nonreal conjugate directions.

One just has to perform the changes of variables (possible if $d\operatorname{Re}\lambda(\alpha)/d\alpha \neq 0$, $\mu(0) \neq 0$) that put the first two equations, among the d equations of the transformed system, into normal form with respect to the two eigenvalues $\lambda(\alpha), \bar{\lambda}(\alpha)$ "responsible for the loss of stability", as

$$\begin{aligned} \dot{x}_1 &= \alpha x_1 - \mu(\alpha)x_2 + \tilde{P}(x_1, x_2, \mathbf{y}, \alpha), \\ \dot{x}_2 &= \mu(\alpha)x_1 + \alpha x_2 + \tilde{Q}(x_1, x_2, \mathbf{y}, \alpha), \end{aligned} \tag{5.7.11}$$

where $\mathbf{y}$ denotes the remaining $(d-2)$ unknowns of the differential equation.

Then one considers the differential equation in $\mathbf{R}^2$ of the form of Eq. (5.7.4) with $P(x_1, x_2, \alpha) = \tilde{P}(x_1, x_2, \mathbf{0}, \alpha)$, $Q(x_1, x_2, \alpha) = \tilde{Q}(x_, x_2, \mathbf{0}, \alpha)$. If this equation verifies the assumptions of Proposition 12, we can infer that the original equation has an attractive periodic orbit for α slightly above α_c.

The proof of this simple criterion is obtained by the obvious extension to $\mathbf{R}^d$ of the discussion in Observation (2) (write $\mathbf{y} = \varphi_\alpha(x_1, x_2)$ and use the fact that φ_α vanishes to second order.

(6) The above theorem has an obvious analogue in one dimension. Consider the equation in $\mathbf{R}$:

$$\dot{x} = \alpha x + p(x, \alpha), \tag{5.7.12}$$

where $p \in C^{(k)}(\mathbf{R}^2)$, k large enough, and p has a third-order zero in $x = 0$, $\forall \alpha \in (-a, a)$, and

$$c(\alpha) = \lim_{x \to 0} \frac{xp(x, \alpha)}{x^4} < 0 \tag{5.7.13}$$

with $c(\alpha) < 0$ and continuous near $\alpha = 0$.

It is easy to see that if k is large, by the implicit function theorem,

Eq. (5.7.12) has two stationary solutions, for $\alpha > 0$ and small ($\bar{x} \simeq \pm\sqrt{-\alpha/c(\alpha)}$).

At such points, the stability matrix turns out to be $\sim -2\alpha < 0$ and, therefore, the two points are attractors with exponential speed for the points in their vicinity.

This observation is sometimes useful in treating cases analogous to the ones discussed in Observation (5), when the stationary solution loses stability because only one real eigenvalue crosses the imaginary axis, as α grows through a critical value α_c, leaving the stationary solution vaguely attractive.

However, it should be stressed that this is a rather rare possibility since it is generally impossible to put a one-dimensional equation into normal form, see Observation (3), p. 396. The existence of normal form can be expected only in systems with "some symmetry".

One can also note that if Eq. (5.7.12) has the property of Eq. (5.7.13), then a small perturbation of it, like

$$\dot{x} = \alpha x + p(x,\alpha) + \epsilon x^2, \tag{5.7.14}$$

can change the vague-attractivity character of $x = 0$ for α near 0, no matter how small ϵ is (exercise). This phenomenon is not possible in equations in which the loss of stability takes place in two complex nonreal directions (essentially just because of the existence of normal forms).

(7) The mechanism of generation of a periodic orbit out of a fixed point when α grows through α_c, described in Proposition 12, is called a "Hopf bifurcation". The solution $\mathbf{x}_0$ loses stability in two complex directions at $\alpha = \alpha_c$ and, if it stays vaguely attractive in the sense of Eq. (5.7.5), it is surrounded by a periodic attractive motion taking place on a curve whose diameter, as we shall see, grows as $\sqrt{\alpha - \alpha_c}$ for $\alpha - \alpha_c > 0$ and small.

(8) As shown in the proof of Proposition 8, p. 395, it is always possible to change smoothly coordinates so as to put Eq. (5.7.4) into a form such that $\gamma_\alpha(\theta)$ is θ independent: $\gamma_\alpha(\theta) \equiv \bar{\gamma}_\alpha$, $\forall \alpha \in (-a,a)$, i.e.,

$$\begin{aligned} \dot{x} &= \alpha x - \mu(\alpha)y + \bar{\gamma}_\alpha x(x^2+y^2) + \tilde{P}(x,y,\alpha), \\ \dot{y} &= \mu(\alpha)y + \alpha x + \bar{\gamma}_\alpha y(x^2+y^2) + \tilde{Q}(x,y,\alpha) \end{aligned} \tag{5.7.15}$$

with $\bar{\gamma}_0 < 0$ and with $\tilde{P}$, $\tilde{Q}$ infinitesimal of fourth order at $x = y = 0$, uniformly in $\alpha \in (-a,a)$ (possibly reducing the value of a); see the change of variables of Eq. (5.5.39) changing Eq. (5.5.38) (i.e., essentially, Eq. (5.7.4) written in complex form) into Eq. (5.5.43) (ie., (5.5.37)).

By Eqs. (5.5.42) and (5.5.38), it is easy to see that the needed change of coordinates involves the third-order Taylor coefficients of P and Q at $x = y = 0$, $\alpha = \alpha_c$, with respect to the variables x, y and it turns out to be of class C^∞ in the variables x, y near $x = y = 0$ and α small (but, in general, only of class $C^{(k-3)}$ in α).

If $k \geqslant 5$, it is easy to see from this that $\tilde{P}$, $\tilde{Q}$ in Eq. (5.7.15) have fifth-order derivatives with respect to x, y continuous in x, y, α near $(0,0,0)$ and also have a fourth-order zero in x, y at $x = y = 0$, $\forall \alpha \in (-a, a)$, if a is small.

Furthermore, the functions r, s of Eq. (5.7.6) are now equal to

$$\begin{aligned} r(x, y, \alpha) &= x\tilde{P}(x, y, \alpha) + y\tilde{Q}(x, y, \alpha), \\ s(x, y, \alpha) &= x\tilde{Q}(x, y, \alpha) - y\tilde{P}(x, y, \alpha) \end{aligned} \tag{5.7.16}$$

by Eq. (5.7.15), and their derivatives in x, y are continuous in x, y, α near $(0,0,0)$ and have a fourth-order zero at $x = y = 0$, $\forall \alpha \in (-a, a)$.

Hence, to fix the ideas, we shall suppose that "k large enough" means $k \geqslant 5$. However, this is not optimal, and one can improve the value of the degree of regularity in x, y, α necessary for P, Q so that a proposition like Proposition 12 will hold. To obtain fine results, one should distinguish the regularity imposed on the α variable and that on the x, y variables.

PROOF. By observation (8), if $k \geqslant 5$, it suffices to treat Eq. (5.7.15) with $\tilde{P}, \tilde{Q}, \partial r, \partial s$ [see Eqs. (5.7.15) and (5.7.16)] being fourth-order infinitesimals in x, y for $x = y = 0$, uniformly in $\alpha \in (-a, a)$ (here ∂ denotes the gradient with respect to the x, y variables).

Let $\bar{\gamma} \equiv -\bar{\gamma}_0$, $\bar{\mu} \equiv \mu(0)$. By the infinitesimality properties of $\tilde{P}$, $\tilde{Q}$, it is possible to find $\bar{\rho} > 0$, $\bar{a} > 0$, $\bar{a} < a$, such that, for all $(x, y) \in C(\bar{\rho})/\{\mathbf{0}\}$, with $C(\bar{\rho}) = \{x, y \mid (x, y) \in \mathbf{R}^2, \sqrt{x^2 + y^2} \leqslant \bar{\rho}\}$ and for all $\alpha \in (-\bar{a}, \bar{a})$,

$$\begin{gathered} \alpha - \tfrac{1}{8}\bar{\gamma}\bar{\rho}^2 < 0, \qquad \tfrac{2}{3}\bar{\mu} < \mu(\alpha) < \tfrac{3}{2}\bar{\mu}, \\ -2\bar{\gamma} < -\bar{\gamma}_\alpha + \frac{r(x, y, \alpha)}{(x^2 + y^2)^2} < -\frac{\bar{\gamma}}{2}, \qquad \frac{|s(x, y, \alpha)|}{(x^2 + y^2)} < \frac{\bar{\mu}}{2}, \end{gathered} \tag{5.7.17}$$

having supposed, for definiteness, that $\bar{\mu} > 0$.

Call $C(\rho', \rho'') = \{\text{annulus with radii } \rho' < \rho''\} = \{x, y \mid (x, y) \in \mathbf{R}^2, \rho' < \sqrt{x^2 + y^2} < \rho''\}$.

We now check that Eq. (5.7.17) allows us to verify that the disk $C(\bar{\rho})$ is $S_t^{(\alpha)}$ invariant and that there is also an invariant annulus $C(\rho'_\alpha, \rho''_\alpha) \subset C(\bar{\rho})$, with $0 < \rho'_\alpha < \rho''_\alpha < \bar{\rho}$ which is an attractor for the points in $C(\bar{\rho})/\{\mathbf{0}\}$, for all $\alpha \in (-\bar{a}, \bar{a})$.

In fact, multiply the first of Eqs. (5.7.15) by x, the second by y, and add the results:

$$\begin{aligned} \frac{d}{dt}\frac{x^2 + y^2}{2} &= \left(\alpha + \bar{\gamma}_\alpha(x^2 + y^2) + \frac{r(x, y, \alpha)}{x^2 + y^2}\right)(x^2 + y^2) \\ &= \begin{cases} < \left(\alpha - \frac{\bar{\gamma}}{2}(x^2 + y^2)\right)(x^2 + y^2) \\ > (\alpha - 2\bar{\gamma}(x^2 + y^2))(x^2 + y^2) \end{cases} \end{aligned} \tag{5.7.18}$$

which shows [see the first of Eqs. (5.7.17)] that the intermediate term in Eq. (5.7.18) is negative on $\partial C(\bar{\rho})$. This means that $C(\bar{\rho})$ is $S_t^{(\alpha)}$ invariant, $\forall t \geqslant 0$, $\forall \alpha \in (-\bar{a}, \bar{a})$, essentially by the same arguments used in justifying Proposition 9, p. 400, §5.5, or as easy to see directly.

Let

$$\bar{\rho}'_\alpha = \sqrt{\frac{\alpha}{2\bar{\gamma}}}, \qquad \bar{\rho}''_\alpha = \sqrt{\frac{2\alpha}{\bar{\gamma}}} \tag{5.7.19}$$

and note that the inequalities in Eq. (5.7.18) show that the intermediate term in Eq. (5.7.18) is positive on $\partial C(\rho'_\alpha)$ and negative on $\partial C(\rho''_\alpha)$; hence, the annulus $C(\rho'_\alpha, \rho''_\alpha)$ is $S_t^{(\alpha)}$ invariant, if α is small so that $\bar{\rho}''_\alpha < \bar{\rho}$.

Equations (5.7.17) and (5.7.18) also show that if $\rho'_\alpha = \frac{1}{2}\bar{\rho}'_\alpha$, $\rho''_\alpha = 2\bar{\rho}''_\alpha < \bar{\rho}$, the annulus $C(\rho'_\alpha, \rho''_\alpha)$ is also invariant and enjoys the property that any initial datum chosen in $C(\bar{\rho})/\{\mathbf{0}\}$ evolves, entering into $C(\rho'_\alpha, \rho''_\alpha)$ in a finite time (see Fig. 5.7), $\forall \alpha \in (-\bar{a}, \bar{a})$.

In fact, if $\bar{\rho} \geqslant (x^2 + y^2)^{1/2} > \rho''_\alpha$, the first inequality in the right-hand side of Eq. (5.7.18) shows that the intermediate term of Eq. (5.7.18) is $\leqslant -6\alpha^2/\bar{\gamma}^2$, so that the "entrance time" in $C(\rho'_\alpha, \rho''_\alpha)$ is finite and can be estimated by $\tau = (\bar{\rho}^2 - \rho''^2_\alpha)\bar{\gamma}^2/12\alpha^2$.

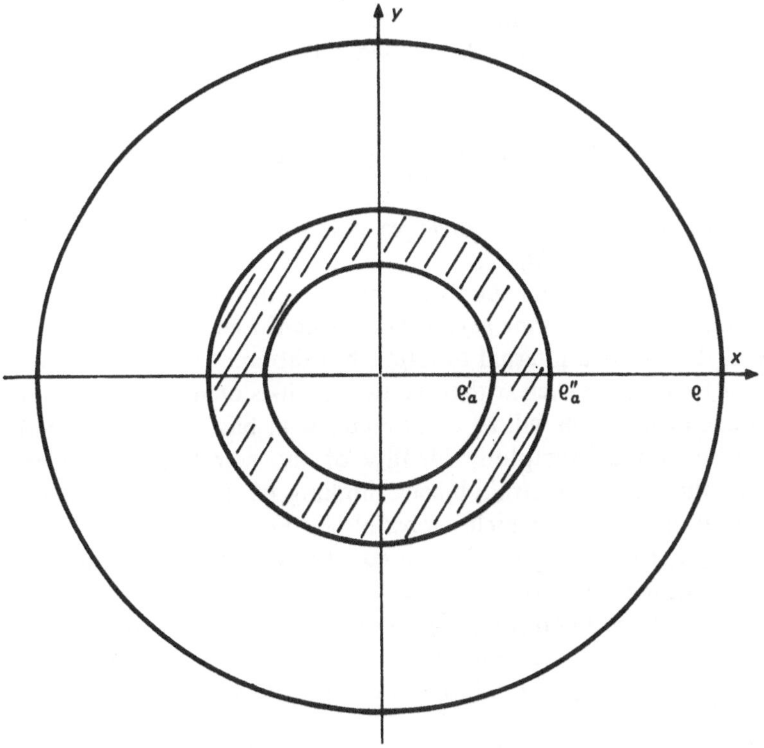

Figure 5.7.

If $0 < \tilde{\rho} = (x^2 + y^2)^{1/2} < \rho'_\alpha$, the intermediate term of Eq. (5.6.18) is not less than $m = \min_{\rho'_\alpha \geqslant \rho \geqslant \tilde{\rho}}(\alpha\rho^2 - 2\bar{\gamma}\rho^4) > 0$ by the second inequality in the right-hand side of Eq. (5.6.18). Hence, the entrance time can now be estimated by $\tau = (\rho'^2_\alpha - \tilde{\rho}^2)/2m$.

This means that every datum close to the origin moves away from the origin until it enters the annulus $C(\rho'_\alpha, \rho''_\alpha)$ in a finite time, while every datum close to $\partial C(\bar{\rho})$ moves towards the origin until it enters the annulus $C(\rho'_\alpha, \rho''_\alpha)$ in a finite time. These motions are spiraling motions, as we now show.

To see that the motions starting in $C(\bar{\rho})/\{\mathbf{0}\}$ are "spiraling motions", it suffices to study them in polar coordinates.

If $S_t^{(\alpha)}(x, y) \equiv (x(t), y(t))$ and if $(\rho(t), \theta(t))$ are the polar coordinates of $(x(t), y(t)) \in C(\bar{\rho})/\{\mathbf{0}\}$,

$$\frac{d\theta}{dt} = \frac{d}{dt}\operatorname{arctg}\frac{y(t)}{x(t)} = \frac{\dot{y}x - y\dot{x}}{x^2 + y^2},$$
$$\frac{d\rho}{dt} = \frac{d}{dt}\sqrt{(x(t))^2 + (y(t))^2} = \frac{x\dot{x} + y\dot{y}}{\sqrt{x^2 + y^2}}. \tag{5.7.20}$$

Note that if $\rho(0) > 0$, $\rho(0) < \bar{\rho}$, then $\rho(t) > 0$ and $\rho(t) < \bar{\rho}$ for all $t \geqslant 0$, because of the above arguments.

Hence, Eq. (5.7.15) and the second and fourth inequalities in (5.7.17) imply

$$\dot{\theta} = \mu(\alpha) + \frac{s(x, y, \alpha)}{x^2 + y^2} \Rightarrow \frac{1}{6}\,\bar{\mu} < \dot{\theta} < 2\bar{\mu}, \tag{5.7.21}$$

i.e., θ is monotonic in t and diverges as $t \to +\infty$. This just means that the motion spirals if $0 < \rho(0) < \bar{\rho}$.

We now show that the spirals associated with the initial data external to $C(\rho''_\alpha)$, but in $C(\bar{\rho})$, become asymptotically confused, as $t \to +\infty$, with those associated with data internal to $C(\rho'_\alpha)$, but different from the origin.

If this happens, it is clear that the two families of spirals are separated by a periodic orbit which will be an attractor with basin containing $C(\bar{\rho})/\{\mathbf{0}\}$.

To discuss the asymptotic identity of the spirals it is convenient to describe them as geometric objects, thinking of them as parameterized in terms of θ instead of t, which is possible by Eq. (5.7.21).

Let $\theta \to \rho_1(\theta)$, $\theta \to \rho_2(\theta)$ be the equations in polar coordinates of two spirals on which two motions of Eq. (5.7.15) run, starting with initial data $\rho_1(0) \geqslant \rho'_\alpha, \theta_1(0) = 0$ and $\rho_2(0) < \rho''_\alpha, \theta_2(0) = 0$ and $\rho_1(0) < \rho_2(0)$.

By the uniqueness theorem for the solutions of the differential equations and by the autonomy of Eq. (5.7.15), it is easy to see that $\rho_2(\theta) - \rho_1(\theta) > 0$, $\forall\theta \geqslant 0$.

We show the existence of $R > 0$, $\epsilon(\alpha) > 0$ such that for α small enough,

$$\rho_2(\theta) - \rho_1(\theta) \leqslant Re^{-\epsilon(\alpha)\theta}. \tag{5.7.22}$$

Then the autonomy of Eq. (5.7.15) and Eqs. (5.7.22) and (5.7.21) plus the attractivity properties of $C(\rho'_\alpha, \rho''_\alpha)$ will easily imply that every datum in $C(\bar{\rho})/\{\mathbf{0}\}$ evolves with exponential speed (with time constant $> \epsilon(\alpha)\bar{\mu}/6$) towards a periodic trajectory of Eq. (5.7.15) which separates geometrically the "outer" spirals (i.e., those originating outside $C(\rho''_\alpha)$) from the "inner" spirals (i.e., those originating inside $C(\rho'_\alpha)$).

To prove Eq. (5.7.22), note that Eqs. (5.7.20), (5.7.19), and (5.7.21) imply

$$\frac{d\rho}{d\theta} = \rho\, \frac{\alpha(x^2+y^2) + \bar{\gamma}_\alpha (x^2+y^2)^2 + r(x,y,\alpha)}{\mu(\alpha)(x^2+y^2) + s(x,y,\alpha)}, \tag{5.7.23}$$

where r, s are infinitesimals of fifth order in x, y at $x = y = 0$, uniformly in $\alpha \in (-\bar{a}, \bar{a})$, while their gradients with respect to x and y have the same property to fourth order.

Equation (5.7.23) will be rewritten as

$$\frac{d}{d\theta}\log\rho = \frac{\alpha + \bar{\gamma}_\alpha \rho^2 + r(x,y,\alpha)/\rho^2}{\mu(\alpha) + s(x,y,\alpha)/\rho^2}. \tag{5.7.24}$$

We now wish to show that the right-hand side of Eq. (5.7.24) is monotonic in ρ for $\rho \in [\rho'_\alpha, \rho''_\alpha]$ at fixed θ and that its ρ derivative stays away from zero.

To estimate the derivative, we just compute it. Basically, the possibility of the bound is due to the fact that to the lowest order in ρ, the right-hand side of Eq. (5.7.24) is $(\alpha + \bar{\gamma}_\alpha \rho^2)/\mu(\alpha)$ whose ρ-derivative is $2\bar{\gamma}_\alpha \rho/\mu(\alpha)$.

So we expect that if $\bar{\rho}$ is small enough [with $\bar{a}$ chosen correspondingly small so that the first of Eqs. (5.7.17) still holds], the ρ derivative of the right-hand side of Eq. (5.7.24) can be estimated, $\forall \rho \in [\rho'_\alpha, \rho''_\alpha]$ (using the orders of infinitesimality of $r, s, \partial r, \partial$ neglect the terms in r, s) to be not larger than:

$$-\frac{\bar{\gamma}}{\bar{\mu}}\sqrt{\frac{1}{2\bar{\gamma}}}\,\sqrt{\alpha} \equiv -\chi\sqrt{\alpha}\,. \tag{5.7.25}$$

A straightforward direct calculation of the ρ derivative of the right-hand side of Eq. (5.7.24) actually proves the above statement by Eq. (5.7.25).

Then recalling that $\rho_2(\theta) > \rho_1(\theta)$, $\forall \theta \geqslant 0$, and writing Eq. (5.7.24) for ρ_2 and ρ_1 and subtracting them, we find, applying the bound on the derivative (5.7.25) (recalling that $\rho'_\alpha \leqslant \rho'_1(\theta)$):

$$\begin{aligned}\frac{d}{d\theta}\log\frac{\rho_2(\theta)}{\rho_1(\theta)} &\leqslant -\chi\sqrt{\alpha}\,(\rho_2(\theta) - \rho_1(\theta)) \\ &= -\chi\sqrt{\alpha}\,\rho_1(\theta)\left(\frac{\rho_2(\theta)}{\rho_1(\theta)} - 1\right) \\ &\leqslant -\chi\sqrt{\alpha}\,\rho'_\alpha\left(\frac{\rho_2(\theta)}{\rho_1(\theta)} - 1\right) = -\frac{\alpha}{2\bar{\mu}}\left(\frac{\rho_2(\theta)}{\rho_1(\theta)} - 1\right)\end{aligned} \tag{5.7.26}$$

which interpreted as a differential inequality for $\rho_2(\theta)/\rho_1(\theta) - 1$, yields

$$\left(1 - \frac{\rho_1(\theta)}{\rho_2(\theta)}\right) \leqslant \left(1 - \frac{\rho_1(0)}{\rho_2(0)}\right) e^{-(\alpha/2\bar{\mu})\theta} \tag{5.7.27}$$

by integration, and this completes the proof. mbe

Exercises and Problems for §5.7

1. The estimate for the coefficient $\epsilon(\alpha)$ in Eq. (5.7.22) is [see Eq. (5.7.27)], $\epsilon(\alpha) = \alpha/2\bar{\mu}$. Is it possible to improve it so that the new estimate $\tilde{\epsilon}(\alpha)$ has the property that $\tilde{\epsilon}(\alpha) \to_{\alpha\to 0} \epsilon > 0$? If not, find a physical interpretation or a motivation of this fact.

2. Consider the differential equation in $\mathbf{R}^2$ written in complex form as

$$\dot{z} = \xi(\alpha)z + P(z, \bar{z}),$$

where $z = x + iy$, $(x, y) \in \mathbf{R}^2$, $\xi(\alpha) = \sigma(\alpha) + i\mu(\alpha)$ and let $\sigma(0) = 0$, $\mu(0) \neq 0$, σ, $\mu \in C^\infty(\mathbf{R})$; suppose P to be a C^∞ function of x, y with a second-order zero at the origin.

In the proof of Proposition 8, p. 395, it was shown [see the change of variables in Eq. (5.5.39)] that in some new coordinates the equation can be given the form $\dot{z} = \xi(\alpha)z + c_2(\alpha)z|z|^2 + O(|z|^4)$, where $O(|z|^4)$ symbolically denotes a function of x, y, α of class C^∞ and with a fourth-order zero at $z = 0$ for all α near zero.

Show that the equation can be given the form:

$$\dot{z} = \xi(\alpha)z + c_2(\alpha)z|z|^2 + O(|z|^5)$$

with the same meaning of the symbols, after a new change of coordinates. (Hint: Again change coordinates as $\zeta = z + \Gamma_4(z, \bar{z})$, where Γ_4 is a homogeneous polynomial in $z, \bar{z}$ of fourth degree, such that the fourth-order terms in the equation cancel, see Eq. (5.5.39)–(5.5.43).)

3. In the context of Problem 2, develop the same ideas to show that, $\forall k \geqslant 0$, the equation can be put, in a suitable coordinate system, in the form

$$\dot{z} = \xi(\alpha)z + c_2(\alpha)z|z|^2 + c_4(\alpha)z|z|^4 + \cdots + c_{2k}z|z|^{2k} + O(|z|^{2k+2}).$$

(Hint: Use induction.)

4. Show that in Problems 2 and 3, the assumption $\sigma(0) = 0$ is not necessary. Actually, if $\sigma(0) \neq 0$, show that, by the same type of arguments, the equation can be given the form

$$\dot{z} = \xi(\alpha)z + O(|z|^k)$$

for all $k \geqslant 0$. (Hint: Note that the reason why one could not eliminate $c_2 z|z|^2$ in Problem 2 was that $\lambda(0) + \overline{\lambda(0)} = 0$).)

5. In Problems 2–4, the parameter α does not play a very essential role. Formulate statements of the same type for α-independent equations. (Hint: Just set $\alpha = 0$ in Problems 2–4 and determine what can be said.)

For information about the problems related to the iterated composition of coordinate transformations transforming the original equations into a fully linear equation $\dot{z} = \xi z$ when $\operatorname{Re}\xi \neq 0$ and $\operatorname{Im}\xi \neq 0$ (by letting $k \to \infty$ in Problem 4), see Moser (References).

6. Discuss the bifurcation pattern, as α grows, for the stationary solutions of the equation

$$\dot{\gamma}_1 = -2\gamma_1 + 4\gamma_2\gamma_3,$$
$$\dot{\gamma}_2 = -9\gamma_2 + 3\gamma_1\gamma_3,$$
$$\dot{\gamma}_3 = -5\gamma_3 - 7\gamma_1\gamma_2 + \alpha.$$

7. Same as Problem 6 for

$$\dot{\gamma}_1 = -2\gamma_1 + 4\gamma_2\gamma_3 + 4\gamma_4\gamma_5,$$
$$\dot{\gamma}_2 = -9\gamma_2 + 3\gamma_1\gamma_3,$$
$$\dot{\gamma}_3 = -5\gamma_3 - 7\gamma_1\gamma_2 + \alpha,$$
$$\dot{\gamma}_4 = -5\gamma_4 - \gamma_1\gamma_5,$$
$$\dot{\gamma}_5 = -\gamma_5 - 3\gamma_1\gamma_4,$$

assuming (without checking it) that when a stationary solution loses stability in one real direction or in two complex ones, it remains vaguely attractive with a negative vague-attractivity indicator [as defined in Eqs. (5.5.24) and (5.5.25)].

8. Find some improvements on the regularity requirements in the variables x, y, and α in Proposition 12, possibly requiring a different order of regularity in x, y, or α.

9. In the context of Proposition 12, suppose that $\bar{\gamma}_{\alpha_c}$, as defined there, is positive. Show that in this case, if $\alpha_c = 0$, there is a repulsive periodic orbit for Eq. (5.7.4) for $\alpha < 0$ small. (Hint: Just change t into $-t$ and apply Proposition 12, noting that the change of t into $-t$ changes the notion of attractivity into that of "repulsivity".)

§5.8. On the Stability Theory for Periodic Orbits and More Complex Attractors (Introduction)

Nondum matura est.

In this section we devote some attention to what happens, as α increases, to the periodic solution of Eq. (5.1.19) whose existence has been established in §5.6 and §5.7.

More generally, one can ask how to establish stability criteria for periodic solutions to differential equations, with uniformly bounded trajectories, of the type:

$$\dot{\mathbf{x}} = \mathbf{f}(\mathbf{x}, \alpha) \tag{5.8.1}$$

with $\mathbf{f} \in C^\infty(\mathbf{R}^d \times \mathbf{R}^d)$ or $C^{(k)}(\mathbf{R}^d \times \mathbf{R})$ with k large enough.

Before examining the evolution of the stability of a periodic orbit of Eq. (5.8.1) with α, it is necessary to investigate the notions of stability of a periodic motion of the equation in $\mathbf{R}^d$:

$$\dot{\mathbf{x}} = \mathbf{f}(\mathbf{x}) \tag{5.8.2}$$

with $\mathbf{f} \in C^\infty(\mathbf{R}^d)$ or $C^{(k)}(\mathbf{R}^d)$ with k large enough and such that Eq. (5.8.2) has bounded trajectories.

Let $t \to \mathbf{x}(t)$, $t \geqslant 0$, be a periodic solution of Eq. (5.8.2) with minimal period $T > 0$.

The stability and the attractivity of this solution is conveniently described in terms of the "Poincaré transformation".

7 Definition. Let $t \to \mathbf{x}(t)$ be a periodic solution of Eq. (5.8.2) with minimal period $T > 0$.

Let ξ_0 be a point on this motion's trajectory, say $\xi_0 = \mathbf{x}(0) \in \mathbf{R}^d$, and let σ be a $(d-1)$-dimensional flat surface element cutting the orbit at the point ξ_0 so that the orbit is not tangent to σ in ξ_0 ("transversal surface element").

It is then possible to define a C^∞ transformation [or a $C^{(k)}$ transformation, if the right-hand side of Eq. (5.8.2) is only of class $C^{(k)}$], on a neighborhood of ξ_0 relative to σ and with values on σ itself, by considering a neighborhood U of ξ_0, on σ so small that the motion, according to Eq. (5.8.2), of the initial datum $\xi \in U$ comes back to intersect σ for the first time after a time $T_\xi \simeq T$ at a point $\Phi_\sigma(\xi) \in \sigma$.

The map of $\sigma \cap U$ into σ associating $\xi \in \sigma \cap U$ the point $\Phi_\sigma(\xi) \in \sigma$ is called the "Poincaré transformation" relative to the given periodic orbit, to the given surface element, and to the given vicinity U.

It is then possible to formulate the following sufficient stability and attractivity criterion (and instability criterion as well) for a periodic orbit. It is the best illustration of the meaning and of the interest of the Poincaré transformations.

13 Proposition. *Let $t \to \mathbf{x}(t)$, $t \geqslant 0$, be a periodic motion for Eq.* (5.8.2) *with minimal period $T > 0$.*

Let σ be a transversal surface element to the trajectory in $\xi_0 = \mathbf{x}(0)$ and introduce on σ Cartesian coordinates $\boldsymbol{\eta} = (\eta_1, \ldots, \eta_{d-1})$ with origin in ξ_0. Denote by $\boldsymbol{\eta}' = \hat{\Phi}_\sigma(\boldsymbol{\eta})$ the Poincaré transformation defined in a suitable neighborhood of ξ_0 on σ. Clearly, $\hat{\Phi}_\sigma(\mathbf{0}) = \mathbf{0}$.

Define the stability matrix of the periodic orbit, relative to σ and to the given system of coordinates on it, as

$$(L_\sigma)_{ij} = \frac{\partial \hat{\Phi}_\sigma^{(i)}}{\partial \eta_j}(\mathbf{0}), \qquad i, j = 1, \ldots, d-1 \tag{5.8.3}$$

Then the periodic orbit is stable and is an attractor, with exponential speed, for the points close enough to it if all the eigenvalues of the matrix L_σ have modulus less than 1.

If at least one among the eigenvalues of L_σ has modulus larger than 1, *the orbit is unstable.*

Observations.

(1) The reader will have no problems in recognizing in this proposition an obvious analogue to Proposition 6, p. 380, §5.4, formulated for maps rather than for differential equations (which can, however, be thought of as "infinitesimal maps").

We shall leave its proof to the reader as an interesting problem [see also Observation (2) below]. To study it, one should first understand the case when $\hat{\Phi}_\sigma$ is a linear map near ξ_0.

Proposition 13 bears the name "stability criterion of Lyapunov" for maps.

(2) Proposition 13 is a special case of a slightly different proposition which could be formulated on the stability of stationary points with respect to the action of repeated applications of a map of $\mathbf{R}^d$ into itself.

The fact that Φ_σ is a Poincaré map plays little role in the proof of Proposition 13. This proof is, in fact, split into two parts: (i) show that the origin is an exponentially attracting (or, alternatively, unstable) point for the iterates of $\hat{\Phi}_\sigma$; (ii) observe that since $\hat{\Phi}_\sigma$ is a Poincaré map relative to a periodic orbit for Eq. (5.8.2), (i) implies that the periodic orbit exponentially attracts the points close enough to it (or is, alternatively, unstable).

Clearly (ii) follows trivially from (i), which could be phrased without reference to the Poincaré map but simply for an arbitrary map of a surface into itself (with a fixed point).

Now consider Eq. (5.8.1) and assume that, $\forall \alpha \in (\alpha', \alpha'') \equiv J$, this equation admits among its solutions a periodic motion $t \to \mathbf{x}_\alpha(t)$, $t \geqslant 0$, with minimal period $T_\alpha > 0$ and such that the function $(\alpha, t) \to \mathbf{x}_\alpha(t)$ is a $C^{(k)}$ function on $J \times [0, +\infty)$, if $C^{(k)}$ is the regularity class in the right-hand side of Eq. (5.8.1).

It will then be possible to consider, $\forall \alpha \in (\alpha', \alpha'')$, the stability matrix $L_\sigma(\alpha)$, see Eq. (5.8.3), relative to a surface element σ which, if $J = (\alpha', \alpha'')$ is a small enough interval, can be supposed to be α independent.

We can choose the Cartesian coordinate system on σ, for each α, with the origin at the point ξ_α at the intersection of σ and the trajectory, and smoothly varying with α so that the Poincaré maps $\hat{\Phi}_{\sigma,\alpha}(\boldsymbol{\eta})$ are defined for $\boldsymbol{\eta} \in U$, where U is a small enough neighborhood of the origin, and $\hat{\Phi}_{\sigma,\alpha}(\boldsymbol{\eta})$ is of class $C^{(k)}$ on $U \times (\alpha', \alpha'')$ in the variables $(\boldsymbol{\eta}, \alpha)$ and

$$\hat{\Phi}_{\sigma,\alpha}(\mathbf{0}) = \mathbf{0}, \qquad \forall \alpha \in J. \tag{5.8.4}$$

We can and shall suppose that $\hat{\Phi}_{\sigma,\alpha}$ is extended arbitrarily to a map of $\mathbf{R}^{d-1}$ into itself, having the same regularity class $C^{(k)}$ (to define this extension it might be first necessary to reduce slightly the size of U).

In analogy with the definitions of stability, attractivity, etc. relative to the solution flows associated with differential equations, we can introduce analogous notions for a single transformation Φ of $\mathbf{R}^d$, or of an open subset of $\mathbf{R}^d$, into itself. What was formerly the family $(S_t)_{t\geq 0}$ of maps associated with the solution of the differential equation now becomes the family $(\Phi^n)_{n\in\mathbf{Z}_+}$ of the iterations of Φ, i.e., one can think of Φ as an "evolution" on $\mathbf{R}^d$ observed at integer times.

We do not repeat the obvious process of setting up the notions of stability, attractivity, vague attractivity, etc. for the iterations of a map Φ, and we just mention that once such definitions are posed in an obvious manner (taking into account the analogous definitions associated with the differential equations), the following proposition on the existence of an attractive manifold and on the Hopf bifurcations holds.

14 Proposition. (i) *Consider Eq.* (5.8.1) *with* $\mathbf{f}\in C^{(k+1)}$, $k\geq 1$, *and suppose that the equation admits a family of periodic orbits verifying the properties illustrated in the above text, following the observations to Proposition* 13.

Suppose that for $\alpha\in J=(\alpha',\alpha'')$, *the stability matrix* $L_\sigma(\alpha)$ *has the eigenvalues* $\lambda_{s+1}(\alpha),\ldots,\lambda_{d-1}(\alpha)$ *with modulus less or equal to* $\nu<1$, *and that for some* $\nu'\in(\nu,1)$, *the other eigenvalues* $\lambda_1(\alpha),\ldots,\lambda_s(\alpha)$ *have modulus larger or equal to* ν'. *Also suppose that the plane generated by the eigenvectors of* $L_\sigma(\alpha)$ *associated with the eigenvalues* $\lambda_1(\alpha),\ldots,\lambda_s(\alpha)$ *coincides with the plane* $\eta_{s+1}=\cdots=\eta_{d-1}=0$.

If the origin is vaguely attractive for the maps $\hat{\Phi}_{\sigma,\alpha}$ *near* $\alpha_c\in J$, *and if* $|\lambda_j(\alpha_c)|=1$, $j=1,\ldots,s$, *there exist* $\epsilon>0$, $\delta>0$, $\delta_0>0$, $\delta_0<\delta$ *and* $d-1-s$ *functions* $\varphi^{(s+1)},\ldots,\varphi^{(d-1)}$ *defined in the neighborhood*[17] $\Gamma_s(\delta/2)\times(\alpha_c-\epsilon,\alpha_c+\epsilon)$ *and there of class* $C^{(k)}$ *such that the equations*

$$\eta_{s+1}=\varphi^{(s+1)}(\eta_1,\ldots,\eta_{d-1},\alpha),\ldots,\eta_{d-1}=\varphi^{(d-1)}(\eta_1,\ldots,\eta_{d-1},\alpha) \tag{5.8.5}$$

define in $\Gamma_{d-1}(\delta/2)$ *a family of surfaces* σ_α *parameterized by* $\alpha\in(\alpha_c-\epsilon,\alpha_c+\epsilon)$ *which are locally invariant, locally attractive, and tangent to the plane* $\eta_{s+1}=\cdots=\eta_{d-1}=0$ *in a sense analogous to Eqs.* (5.6.9)–(5.6.11). *The tangency can be measured as in Eq.* (5.6.11) *in terms of an a priori given constant* $C>0$.

(ii) *Now assume that* $s=2$ *and that* $\lambda_1(\alpha)=\overline{\lambda_2(\alpha)}$ *is the eigenvalue of* L_α *with largest modulus for all* $\alpha\in J$ *and that for* $\alpha=\alpha_c\in J$ *it is* $|\lambda_1(\alpha_c)|=1$, $((d/d\alpha)|\lambda_1(\alpha)|)_{\alpha=\alpha_c}>0$, $\operatorname{Im}\lambda_1(\alpha_c)\neq 0$, $\lambda_1(\alpha_c)^h\neq 1$, $h=1,2,3,4,5$. *Suppose that the vague attractivity of* $\mathbf{0}$ *near* α_c *takes place because a condition analogous to Eq.* (5.5.25), $\bar{\gamma}_{\alpha_c}<0$, *holds. Finally, assume that* k *is large*

[17] As usual, $\Gamma_s(\delta)=\{\mathbf{x}\,|\,\mathbf{x}\in\mathbf{R}^s,|x_i|<\delta,i=1,\ldots,s\}$.

enough and $\alpha - \alpha_c$ *is small enough. Then there is a set on* σ, *which we denote* τ_α, *invariant with respect the to action of* $\hat{\Phi}_{\sigma,\alpha}$ *and homeomorphic to a circle for* $\alpha > \alpha_c$. *Such a set is the intersection between* σ *and a torus which is invariant for the solutions of Eq.* (5.8.1) *and attracts, exponentially fast, all the motions starting close enough to it.*

Observations.

(1) Hence, in a similar way, as the vaguely attractive stationary points may bifurcate, in some circumstances, growing into periodic orbits, the periodic orbits may bifurcate growing into two-dimensional tori.

(2) The proof of the above proposition is parallel to that of Propositions 11, §5.6 and 12, §5.7, and will not be discussed in detail (see problems at the end of this section).

We only mention that the assumptions on the eigenvalues, at $\alpha = \alpha_c$, are needed to be able to put the transformation into a normal form analogous to Eqs. (5.7.4) and (5.7.15), thus allowing us to formulate a vague-attractivity condition like Eq. (5.5.25).

(3) Proposition 14, together with Propositions 7–13 and the problems at the end of the §5.4–§5.8, provide a quite general theory of the stability of the vaguely attractive stationary points and periodic orbits and of their bifurcations, when the regularity class of the differential equation is high enough.

It then becomes natural to ask if it is possible to discuss in a similar fashion the theory of stability and bifurcations (following the loss of stability as a parameter α grows) of attractors or more complex invariant sets.

"Unfortunately", such a question is very difficult, and it seems unsuited to be considered in too general a context. Only within classes of special cases, such a problem can be treated in some detail (e.g., in the case of the theory of the attractors "verifying the axiom A").[18] This is a theme of great interest, which seems to be connected with the theory of many phenomena more general than the ones of a purely mechanical nature, like the theory of turbulence which greatly stimulates research on this subject.

(4) As a comment on the generality of the theory of this and the preceding sections, we must stress that the vague attractivity of a point or of an orbit near a critical value α_c is an interesting hypothesis, mainly for its elegant implications, but is far from being realized always (or even often). It often happens that simple systems of differential equations have stationary points or periodic orbits which are not vaguely attractive near a critical value α_c where they lose stability. In such cases, there is no general theory guiding the theoretical analysis, and various phenomena are possible, like

[18] For a definition, see Smale and, also, Ruelle and Bowen and Ruelle for detailed discussions of some problems (References).

the "sudden" (i.e., for α just above α_c) transition to an asymptotic regime governed by attractors of a nature more complex than a stationary point or a periodic orbit or a two-dimensional torus. Such attractors may be located far from the attractor that lost stability.

In general attractors other than points, periodic orbits or tori run quasi-periodically are called "strange": this qualifies the impossibility of describing these attractors as simple objects, rather than qualifying a well-defined mathematical property.

To illustrate Observations (3), (4) and to get some feeling for how complicated the pattern of the bifurcations may be even for relatively simple differential equations (with quadratic nonlinearities "only"), we give a series of examples.

Some of the results quoted below may be obtained via the theory of the preceding section (like those relative to the stability of the stationary solutions, see §5.4 and §5.5 and the associated problems), possibly using a computer to estimate the eigenvalues of various stability matrices. However, most of the following results can only at present be obtained via the use of numerical experiments (usually absolutely fascinating). They should not be considered as mathematical statements but as empirical observations which may reveal themselves only as first rough approximations to the phenomena that the same nonlinear differential equations may show if studied more carefully.

We leave to the reader, as interesting practical work, the task of checking the following statements analytically (when possible) or numerically (if he has access to a computer).

Example 1: The "Lorenz Model".
Analytically, this is a system of equations that the reader can interpret as equations of motion of a gyroscope subject to suitable forces (following a scheme like the one in §5.1).
The Equations are

$$\dot{x} = -\sigma x + \sigma y, \qquad \dot{y} = -\sigma x - y - xz, \qquad \dot{z} = -bz + yx - \alpha, \quad (5.8.6)$$

$\sigma = 10, b = \frac{8}{3}$. The following items describe the structure of the attractors.
(1) For $0 \leqslant \alpha < \alpha_c^1 = -b(\sigma - 1)$, there is just one stationary point that can be shown to be globally attractive for α small enough. It is locally stable and the eigenvalues of the Lyapunov matrix have a negative real part, $\forall \alpha \in (0, \alpha_c^1)$, and, numerically, it appears to be globally attractive all the way up to α_c^1. The stationary point is stationary for all α, but is unstable for $\alpha > \alpha_c^1$. It is

$$x = y = 0, \qquad z = -\frac{\alpha}{b}. \quad (5.8.7)$$

(2) For $\alpha_c^1 < \alpha < \alpha_c^2 = 2ab(1 + \sigma)/(\sigma - 1 - b) = 83.82$, the preceding point undergoes a bifurcation, losing stability in one real direction but remaining vaguely attractive and it bifurcates in two locally stable stationary solutions. Such solutions exist for all $\alpha > \alpha_c^1$, but lose stability for $\alpha \geqslant \alpha_c^2$.

From a numerical point of view, a randomly chosen initial datum is attracted by one of the above two stationary solutions. The solutions are

$$x = y = \pm\sqrt{\alpha - b(\sigma + 1)}\,, \qquad z = -\sigma - 1. \tag{5.8.8}$$

One should not think, however, that the possible asymptotically different motions consist of the three points of Eqs. (5.8.7) and (5.8.8). For instance, for $\alpha < \alpha_c^2$ and close to it, there are some unstable periodic orbits, as can be rigorously shown.[19] The reason why such asymptotic motions cannot be seen by sampling randomly the initial data space is that they form a set of zero Lebesgue measure.

(3) For $\alpha > \alpha_c^2$, the points of Eq. (5.8.8) lose stability. Such loss of stability takes place in two complex-conjugate directions because two complex-conjugate non-real eigenvalues of the stability matrix cross the imaginary axis from left to right.

However, although the fixed points in Eqs. (5.8.8) still exist, they are *not* vaguely attractive for α near α_c^2. Hence, one cannot apply the Hopf bifurcation theorem to infer the existence of a bifurcation into periodic orbits of each of the points of Eq. (5.8.8). In fact, a strange attractor shows up here, staying alive up to $\alpha \cong 230$, disappearing occasionally only for some small intervals of α when it is replaced by some stable periodic orbits.

(4) For large α, the strange attractor disappears and is replaced by attractors consisting of periodic orbits, as it appears from numerical experiments. The existence of some stable periodic orbits can be proven rigorously for α large (see Robbins in References).

Figures 5.8, 5.9, and 5.10 illustrate the above items.

Example 2: Navier–Stokes equations on a two-dimensional torus with a five-mode truncation.

This is an example in which there are nice Hopf bifurcations. It is, however, more complicated than Example 1. It could also be interpreted mechanically as a system of two coupled rigid bodies with a rather strange looking coupling. However, this mechanical interpretation does not seem to be particularly useful, and we do not discuss it. The physical origin of the model has to be searched for in the theory of fluids. The equations are

$$\begin{aligned}
\dot\gamma_1 &= -2\gamma_1 + 4\gamma_2\gamma_3 + 4\gamma_4\gamma_5,\\
\dot\gamma_2 &= -9\gamma_2 + 3\gamma_1\gamma_3,\\
\dot\gamma_3 &= -5\gamma_3 - 7\gamma_1\gamma_2 + \alpha,\\
\dot\gamma_4 &= -5\gamma_4 - \gamma_1\gamma_5,\\
\dot\gamma_5 &= -\gamma_5 - 3\gamma_1\gamma_4.
\end{aligned} \tag{5.8.9}$$

(1) For α small, the obvious stationary solution, existing $\forall\alpha \geqslant 0$,

$$\gamma_1 = \gamma_2 = \gamma_4 = \gamma_5 = 0, \qquad \gamma_3 = \frac{\alpha}{5} \tag{5.8.10}$$

is stable and globally attractive [this could be proved along the lines of the proof of Eq. (5.2.12) in Proposition 4, §5.2, p. 369].

[19] Applying Problem 16, §5.5, p. 407, to either of the Eqs. (5.8.8) near α_c^2, one computes the vague-attractivity indicator of Proposition 12, Eq. (5.7.5), and shows that it has the "wrong sign", $\bar\gamma > 0$, and then one applies Problem 9, §5.7, p. 441.

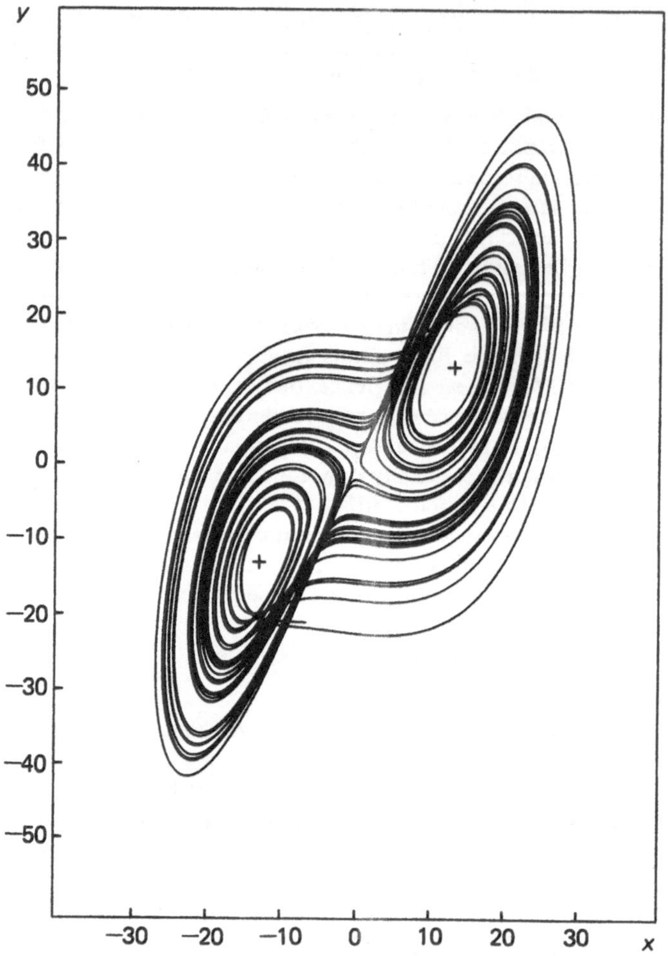

Figure 5.8. Projection on the plane $z = \sigma - 1$ of the fixed points of Eq. (5.8.8) and of a motion corresponding to a given initial datum randomly chosen; $\alpha = 200$. The motion is not periodic.

By the Lyapunov criterion, it remains stable up to $\alpha_c^1 = 5\sqrt{3/2}$. Up to this value, it numerically appears that it is a global attractor.

(2) Near α_c^1, Eq. (5.8.10) is vaguely attractive and loses stability in one real direction, generating two stable attractive solutions

$$\begin{aligned} \gamma_1 &= \epsilon\sqrt{\frac{\sqrt{6}}{7}}\,\sqrt{(\alpha - \alpha_c^1)}, & \gamma_3 &= \sqrt{\frac{3}{2}} \\ \gamma_2 &= \epsilon\sqrt{\frac{1}{7\sqrt{6}}}\,\sqrt{(\alpha - \alpha_c^1)}, & \gamma_4 &= \gamma_5 = 0, \qquad \epsilon = \pm 1. \end{aligned} \tag{5.8.11}$$

Such solutions exist for all $\alpha > \alpha_c^1$ and, numerically, they seem to be globally attractive as long as they are locally stable: this means that randomly chosen initial data are attracted by either of them, see the comment to the point (2) of the Example 1 above.

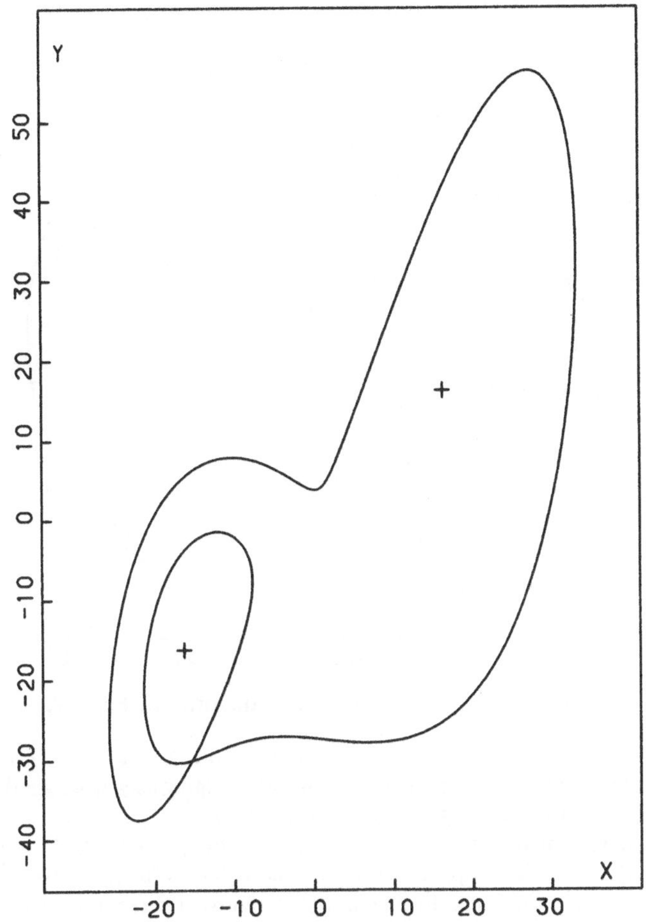

Figure 5.9. $x-y$ projection of a periodic orbit relative to the case $\alpha = 295$. The other periodic orbits that can be experimentally found turn out to be related to the above by the transformation $x \to -x$, $y \to -y$ which is a symmetry of the equation.

They lose stability for $\alpha = \alpha_c^2$:

$$\alpha_c^2 = \frac{80}{9}\sqrt{\frac{3}{2}}\,. \tag{5.8.12}$$

The stability loss takes place in just one real direction again and, again, each of them bifurcates into two new stable solutions which are locally attractive for $\alpha \in (\alpha_c^2, \alpha_c^3)$, but persist for all $\alpha > \alpha_c^2$. If $\epsilon, \sigma = \pm 1$,

$$\begin{gathered}\gamma_1 = \epsilon\sqrt{\frac{5}{3}}\,, \qquad \gamma_2 = \epsilon\,\frac{3}{80}\sqrt{\frac{5}{3}}\,\alpha, \qquad \gamma_3 = \frac{9}{80}\,\alpha, \\ \gamma_4 = \frac{\epsilon\sigma}{3}\sqrt{\left(\frac{9}{80}\alpha\right)^2 - \frac{3}{2}}\,, \qquad \gamma_5 = -\sigma\sqrt{\left(\frac{9}{80}\alpha\right)^2 - \frac{3}{2}}\,\sqrt{\frac{5}{3}}\,,\end{gathered} \tag{5.8.13}$$

and $\alpha_c^3 = 22.8537\ldots$.

At $\alpha = \alpha_c^3$, Eqs. (5.8.13) lose stability in two complex directions and, apparently, they remain vaguely attractive. In fact, one can easily find, numerically,

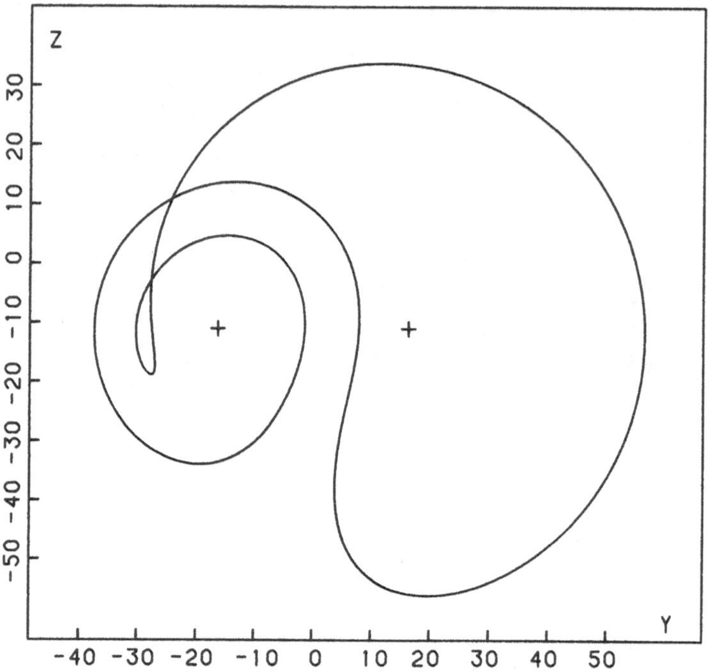

Figure 5.10. $y - z$ projection of the orbit in Fig. 5.9.

that in their vicinity there is a stable periodic orbit, as if a Hopf bifurcation had taken place (in principle, one could even check rigorously whether the vague-attractivity indicator $\bar{\gamma}$ is negative, as it probably is).

The structure of the motions for $\alpha > \alpha_c^3$ is quite fascinating. At various values $\alpha_c^{4,1}, \alpha_c^{4,2}, \alpha_c^{4,3} \ldots$, there appear new periodic orbits bifurcating from the preceding ones because the latter lose stability in one real direction, with the stability matrix of the Poincaré transformation showing the largest eigenvalue crossing the unit circle through -1.

Such cases, although not contemplated in Proposition 14, can nevertheless be theoretically treated under suitable vague-attractivity assumptions, and their theory predicts that the periodic orbit "doubles", doubling also its period,[20] see also Problems 10–13 for §5.8.

The sequence of such bifurcations seems to be infinite and has been observed until the period has reached approximately 2^5 times the initial value. The accumulation point $\lim_{n\to\infty}\alpha_c^{4,n}$, as experimentally measured by a computer, seems to be $\alpha_c^{4,\infty} = 28.6681\ldots$.

For $\alpha = \alpha_c^{5,0} = 28.663\ldots$, there appears a new family of periodic orbits that for $\alpha \in [\alpha_c^{5,0}, \alpha_c^{4,\infty}]$ coexists with the preceding ones, although they are also stable. A randomly chosen initial datum is attracted by one of the stable orbits of the two families.

[20]This can easily be understood intuitively by arguing as in the Observation (6) to Proposition 12, p. 432. Write the Poincaré map as $\hat{\Phi}_{\sigma,\alpha}(x) = (-1 - (\alpha - \alpha_c))x + p(x, \alpha)$, assuming that $xp(x,\alpha)/x^4 \to_{x\to 0} \bar{\gamma} < 0$. One easily finds that there are two points $x_{+,\alpha}, x_{-,\alpha} \simeq \pm\sqrt{-\bar{\gamma}^{-1}(\alpha - \alpha_c)}$ mapped into *each other* by $\hat{\Phi}$. This means that the orbit "doubles".

As α grows beyond $\alpha_c^{5,0}$, these new orbits also undergo the same fate, doubling after losing stability into a double orbit at $\alpha = \alpha_c^{5,1}$ which, in turn, doubles into a double orbit at $\alpha_c^{5,2}$, etc. "indefinitely" with an accumulation point at $\alpha_c^{5,\infty} = \lim_{n\to\infty}\alpha_c^{5,n} = 28.7201\ldots$.

For $\alpha > \alpha_c^{5,\infty}$, it seems that the motion is asymptotically described by a strange attractor up to $\bar{\alpha}_c \cong 34$ with the exception of at least one small interval of values of α, very small, where asymptotic behavior is again ruled by some periodic orbits which, as α grows, lose stability "again through -1" doubling in period infinitely many times.

After $\bar{\alpha}_c$, the motion seems to be governed by periodic and globally attractive orbits whose period and shape vary regularly with α (as before, here global "numerical" attractivity means that if the initial datum is randomly chosen, it converges to one of the above periodic motions).

We stress that the adjective "numerical", referred to some properties of the solutions, means that such properties come out of a computer-assisted study and that they are not mathematically rigorous.

Another exceptionally interesting and marvelous property of the above sequences of bifurcations is that, numerically, the sequences

$$\frac{\alpha_c^{4,n+1} - \alpha_c^{4,n}}{\alpha_c^{4,n} - \alpha_c^{4,n-1}} \quad \text{and} \quad \frac{\alpha_c^{5,n+1} - \alpha_c^{5,n}}{\alpha_c^{5,n} - \alpha_c^{5,n-1}}$$

seem to converge to a limit ρ which is $\simeq 4.67$. This is a numerical value which is conjectured, "Feigenbaum conjecture", to be "universal", i.e., independent of the particular differential equations giving rise to stable periodic orbits which successively grow out of doubling bifurcations when one of them, stable at a given value of α, loses stability as α grows, giving rise to a stable doubled orbit.

However, it is an open problem to formalize in satisfactory generality and to give manageable sufficient conditions for a proof of the validity of this fascinating conjecture which seems to be verified in several cases studied numerically (and different from the above-considered ones). Recently, considerable progress in this direction has been achieved (see Collet, Eckmann, and Campanino, Epstein, Ruelle, and Lanford in References).

The structure of the just discussed bifurcations is illustrated by Figs. 5.11–5.18, representing projections on several planes of trajectories of Eq. (5.8.11).

Example 3: Navier–Stokes equations on a two-dimensional torus with seven modes.

A system exhibiting periodic orbits bifurcating into two-dimensional tori along the scheme suggested by Proposition 14 is the following:

$$\begin{aligned}
\dot{\gamma}_1 &= -2\gamma_1 + 4\sqrt{5}\,\gamma_2\gamma_3 + 4\sqrt{5}\,\gamma_4\gamma_5,\\
\dot{\gamma}_2 &= -9\gamma_2 + 3\sqrt{5}\,\gamma_1\gamma_3,\\
\gamma_3 &= -5\gamma_3 - 7\sqrt{5}\,\gamma_1\gamma_2 + 9\gamma_1\gamma_7 + \alpha,\\
\dot{\gamma}_4 &= -5\gamma_4 - \sqrt{5}\,\gamma_1\gamma_5,\\
\dot{\gamma}_5 &= -\gamma_5 - 3\sqrt{5}\,\gamma_1\gamma_4 + 5\gamma_1\gamma_6,\\
\dot{\gamma}_6 &= -\gamma_6 + 5\gamma_1\gamma_5,\\
\dot{\gamma}_7 &= -5\gamma_7 + 9\gamma_1\gamma_3
\end{aligned} \tag{5.8.14}$$

which can be discussed in a similar way as that of Example 2.

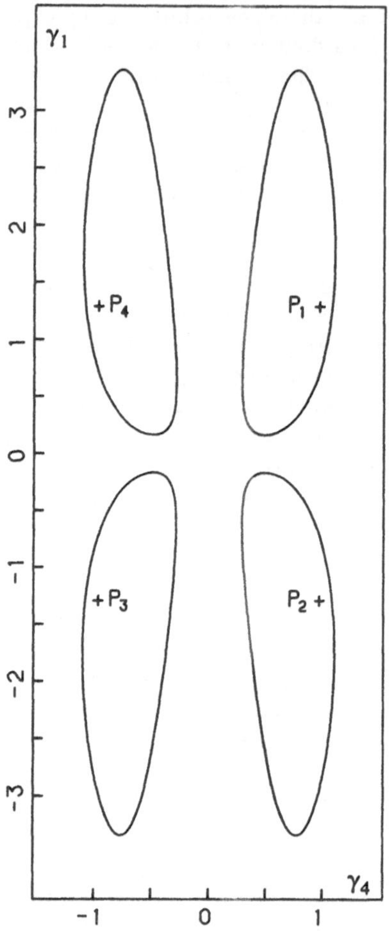

Figure 5.11. $1-4$ projection of the fixed points and periodic orbits after the bifurcation in which the points of Eq. (5.8.13) lose stability; $\alpha = 28$. Equation (5.8.9) has some symmetries which permit us to describe all of the above periodic orbits and fixed points from one of them by applying to it suitable symmetry transformations.

The structure of the bifurcations and attractors is considerably more complicated and interesting. We do not discuss it in detail, feeling that Figs. 5.19–5.25 will, by themselves, excite the reader's curiosity and will stimulate him to read some original papers on the profound theory of Feigenbaum and on Example 3 as well as on Examples 1 and 2 (see Feigenbaum, Franceschini and Tebaldi, Franceschini, Coullet and Tresser in the References).

All the equations of the above examples, as noted in Examples 1 and 2, can be interpreted as equations governing some absurd systems of coupled rigid bodies, but they have been considered in the literature as equations approximating the differential equations describing the motion of simple fluids (like the "Euler" or the "Navier–Stokes" equations or the

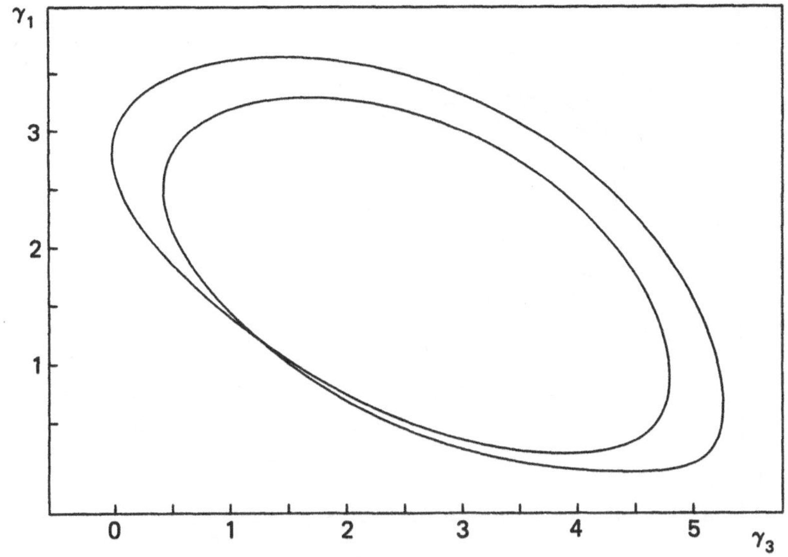

Figure 5.12. 1 − 3 projection of one of the orbits which arise by a doubling bifurcation from one of the orbits of Fig. 5.11 for $\alpha = \alpha_c^{4,1}$; $\alpha = 28{,}60$. The other four doubled periodic orbits are obtained from this symmetry transformation.

"thermofluidodynamics" equations). Their connection with the mechanics of rigid bodies is not surprising, however, if one notes that the above fluid equations can be considered as equations describing infinitely many coupled rigid bodies (with very strange and, perhaps, mechanically unnatural coupling); this remark becomes clearer if one recalls that the equations of

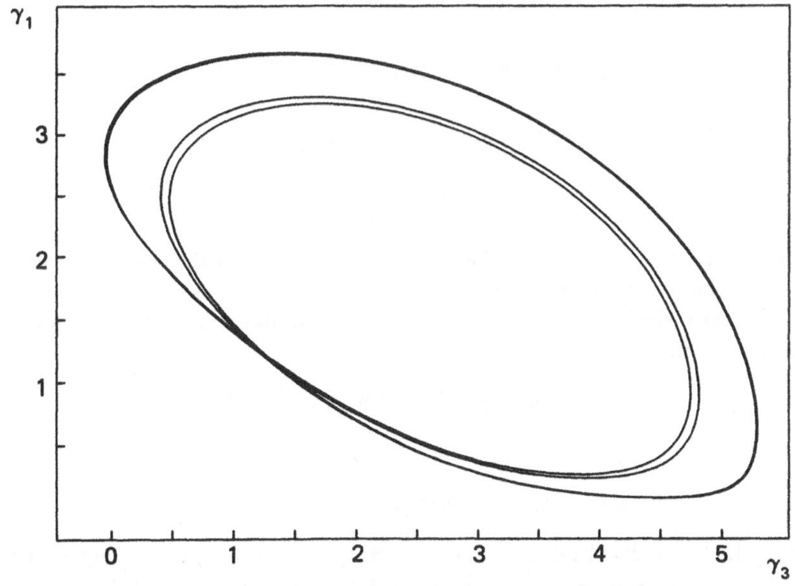

Figure 5.13. Further doubling of the orbit of Fig. 5.12; $\alpha = 28.650\ldots$.

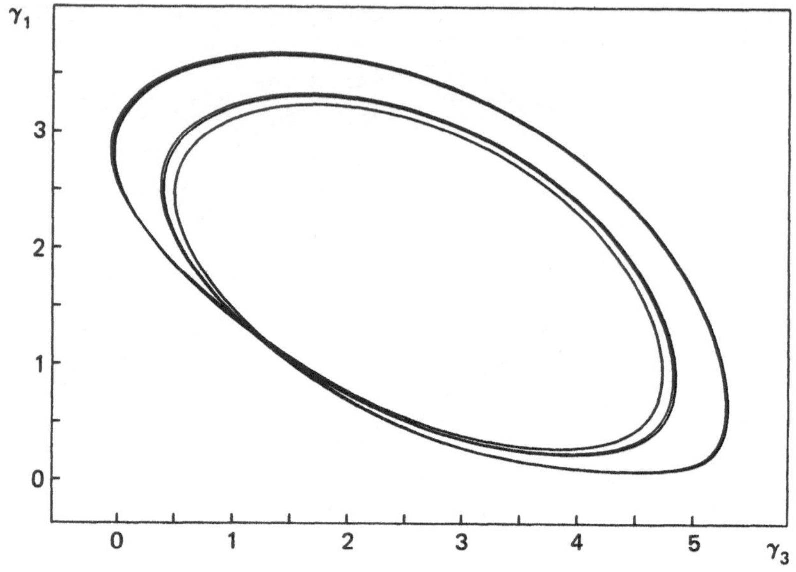

Figure 5.14. Further doubling; $\alpha = 28.666$.

motion of the fluid bodies are usually derived by thinking of them as consisting of many small rigid bodies and applying to each of them the cardinal equations of mechanics.

We shall not further pursue the discussion of the models of dissipative systems and of their stability theory. This is a subject under current intense investigations, and the contents of §5.1–§5.8 provide some introduction to the literature.

Problems and Complements for §5.8

1. Let $\Phi \in C^\infty(\mathbf{R}^d)$ be a map of $\mathbf{R}^d$ into itself with the origin as a fixed point. Write $\mathbf{x}' = \Phi(\mathbf{x})$ as

$$\mathbf{x}' = L\mathbf{x} + \mathbf{F}(\mathbf{x}),$$

where L is a $d \times d$ matrix and $\mathbf{F}$ has a second-order zero at the origin. Suppose that the eigenvalues of L are pairwise distinct.

Show that there is a linear change of coordinates that allows us to put the above map into the form

$$x_1{}^{(j)\prime} = (\operatorname{Re}\lambda_j)x_1{}^{(j)} - (\operatorname{Im}\lambda_j)x_2{}^{(j)} + F_1^{(j)}(\mathbf{x}),$$

$$x_2{}^{(j)\prime} = (\operatorname{Im}\lambda_j)x_1{}^{(j)} + (\operatorname{Re}\lambda_j)x_2{}^{(j)} + F_2^{(j)}(\mathbf{x}),$$

$$x^{(h)\prime} = \lambda_h x^{(h)} + F^{(h)}(\mathbf{x})$$

with $j = 1, \ldots, s$, $h = 2s+1, \ldots, d$, where $\lambda_1, \ldots, \lambda_s$ are the s complex nonreal eigenvalues of L and $\lambda_{2s+1}, \ldots, \lambda_d$ are the $(d-2s)$ real eigenvalues of L; $F_1^{(j)}$,

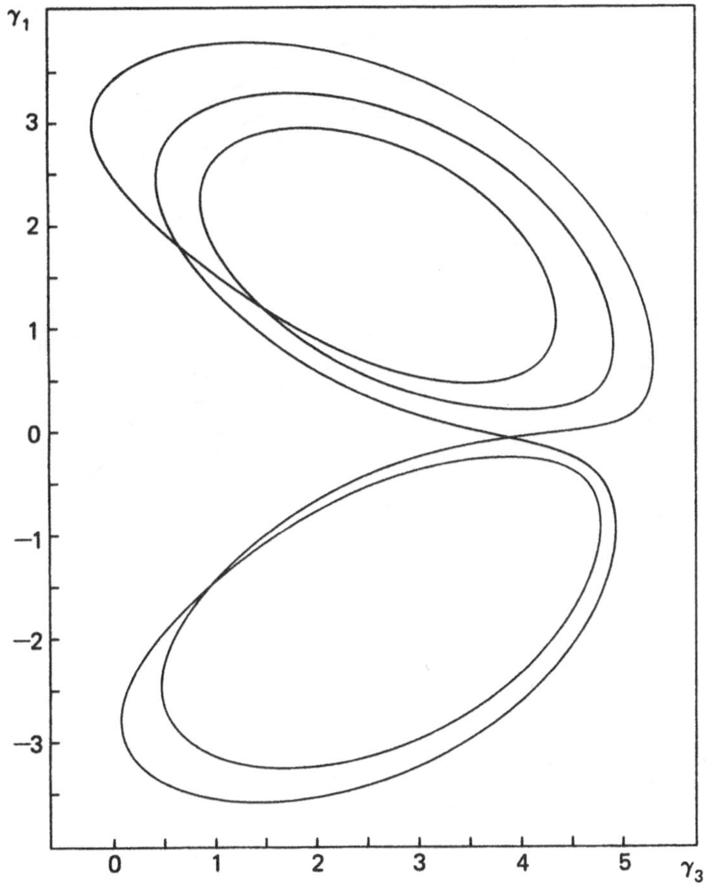

Figure 5.15. One of the four new orbits of the family that is born at $\alpha = \alpha_c^{5,0} =$ 28.663; $\alpha = 28.663$. The other orbits are obtained from this by transforming it with the symmetries of the equation.

$F_2^{(j)}, F^{(h)}$ have a second-order zero at the origin **0**. (Hint: Proceed as in the proof of Proposition 7, p. 391, §5.5.)

2. In the context of Problem 1, suppose that $d = 2$, $\lambda = \lambda_1 = \{\text{complex nonreal}\}$. Let $z = x_1{}^{(1)} + ix_2{}^{(1)}$. Show that the map can be written as a map of **C** into itself:

$$z' = \lambda z + F(z, \bar{z}),$$

where F has a second-order zero at $z = 0$.

3. Show that if $\lambda^3 \neq 1$, $\lambda \neq 0$, the map in Problem 2 can be written in a new coordinate system as

$$\zeta' = \lambda \zeta + N(\zeta, \bar{\zeta}),$$

where N has a third-order zero at the origin $\zeta = 0$. (Hint: Proceed as in the proof of Proposition 8, p. 399, §5.5, i.e., write $F(z, \bar{z}) = a_2 z^2 + a_1 z\bar{z} + a_0 \bar{z}^2 + \tilde{N}(z, \bar{z})$ with $\tilde{N}$ having a third-order zero at $z = 0$. Change variables near $z = 0$ as $\zeta = z + A_2 z^2 +$

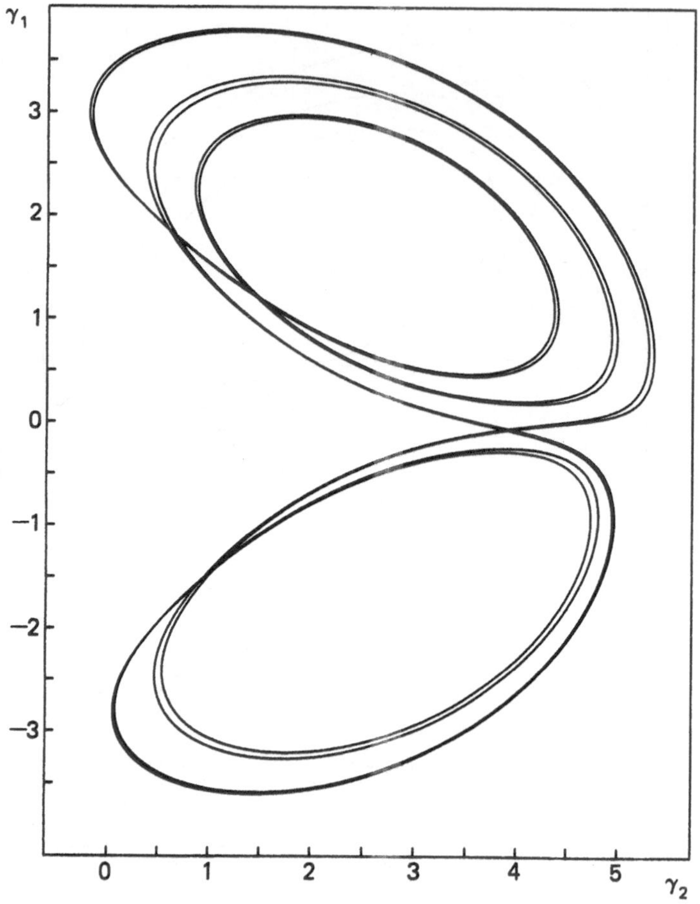

Figure 5.16. Doubling of the orbit in Fig. 5.15 for $\alpha_c^{5,1}$; $\alpha = 28.710$.

$A_1 z\bar{z} + A_0\bar{z}^2$ and choose the A's in order to eliminate the second-order terms from the map in the new coordinates.)

4. Show that if $\lambda^4 \neq 1$, $\lambda \neq 0$, the map in Problem 3, of **C** into itself,

$$\zeta' = \lambda\zeta + N(\zeta, \bar{\zeta})$$

with N having a third-order zero at $\zeta = 0$, can be put into the form

$$z' = \lambda z + bz|z|^2 + Q(z, \bar{z})$$

with Q having a fourth-order zero at $z = 0$, using a change of variables (near the origin) of the form: $z = \zeta + A_3\zeta^3 + A_2\zeta^2\bar{\zeta} + A_1\zeta\bar{\zeta}^2 + A_0\bar{\zeta}^3$.

5. In the context of Problem 4, show that the map can also be written as

$$z' = \lambda z \exp(b|z|^2 + \tilde{Q}(\rho, \theta))$$

near $z = 0$, where $z = \rho e^{i\theta}$ and $\tilde{Q}$ is a C^∞ function of $(\rho, \theta) \in [0, \bar{\rho}) \times T^1$ with a third-order zero at the origin of the ρ variable.

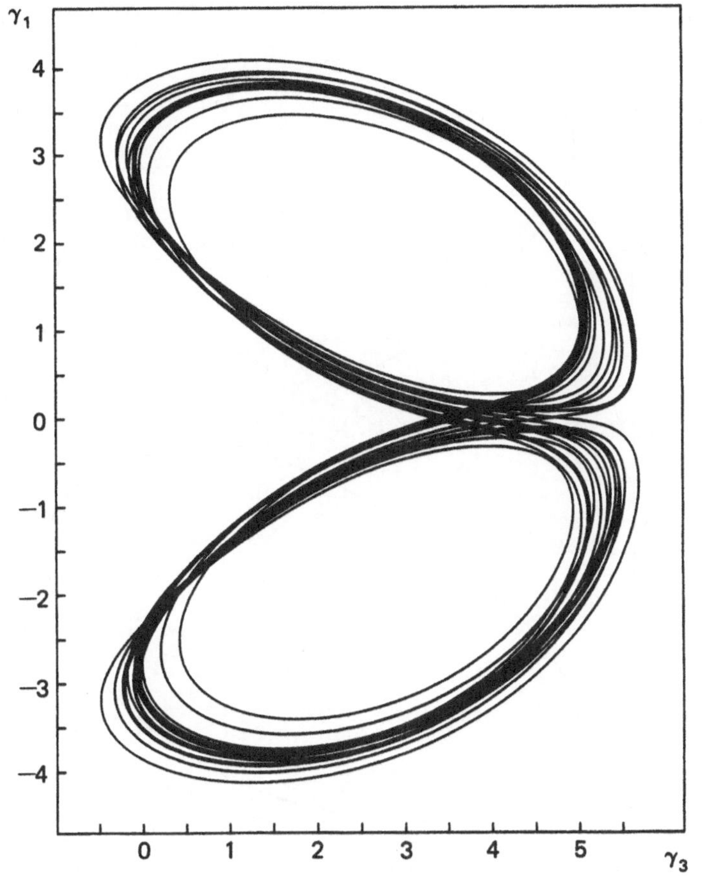

Figure 5.17. Orbit with an asymptotic motion governed, apparently, by a strange attractor; $\alpha = 31$.

6. Consider the map defined as follows: let $z = \rho e^{i\theta}$ and

$$z' = \lambda(\alpha) z \exp\big(b(\alpha)|z|^2 + \tilde{Q}(\rho, \theta, \alpha)\big) \equiv \Phi_\alpha(z),$$

where $\tilde{Q} \in C^\infty([0, \bar{\rho}) \times T^1 \times (-a, a))$, $\lambda, b \in C^\infty((-a, a))$, $|\lambda(0)| = 1$, and $\tilde{Q}$ has a third-order zero at $\rho = 0$, for all $\theta \in \mathbf{T}^1$, for all $\alpha \in (-a, a)$. Show that the origin is vaguely attractive near zero if $\operatorname{Re} b(0) < 0$. (Hint: If $\operatorname{Re} b(0) < 0$, the origin is attractive for $\alpha = 0$; . . . .)

7. Let $\lambda(\alpha) = \exp(\alpha + ib(\alpha))$ and, in the context of Problem 6, let $\operatorname{Re} b(0) < 0$. Show that the maps Φ_α have an attractive invariant set of approximate equation $|z| = \sqrt{\alpha / -\operatorname{Re} b(\alpha)}$ for $\alpha > 0$ small. (Hint: Proceed as in the analysis of the Hopf theorem, performing the analogous steps and estimates.)

Actually (but this is more difficult than the above problem), the invariant set is a curve homeomorphic to a circle. The proof of this could be achieved by writing the equation of the unknown curve as

$$z(\theta) = \sqrt{\frac{-\alpha}{\operatorname{Re} b(\alpha)}}\,(1 + \epsilon(\theta)) e^{i\theta}$$

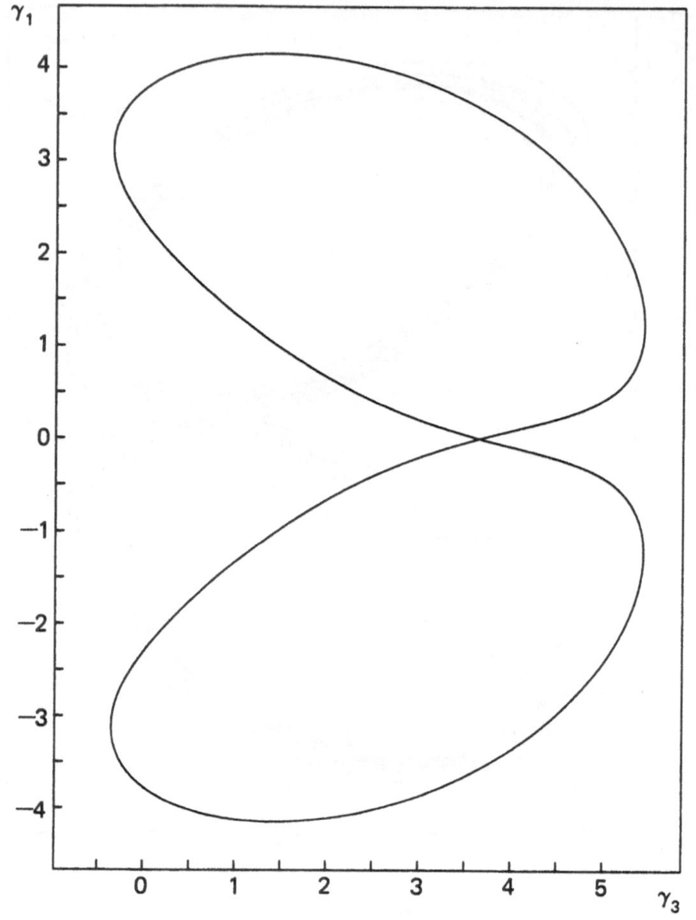

Figure 5.18. $\alpha = 34$; an attractive periodic orbit, $1 - 3$ projection.

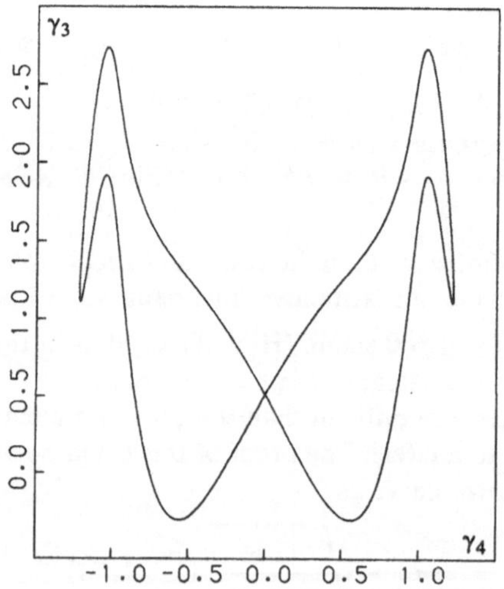

Figure 5.19. A periodic orbit for Eq. (5.8.14) at $\alpha = 71$.

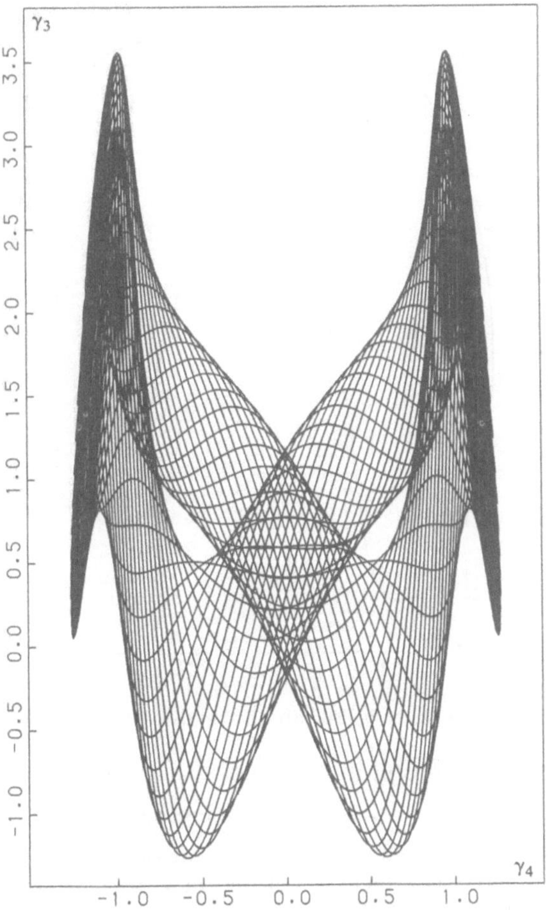

Figure 5.20. $\alpha = 71.60$; the preceding orbit has originated a stable torus (two-dimensional) run quasi-periodically by the motions of Eq. (5.8.14), one of which is shown here.

and trying to determine $\epsilon(\theta)$ by writing the condition that the above curve is Φ_α invariant, i.e.,

$$\theta' = \theta + \beta(\alpha) - \alpha \frac{\operatorname{Im} b(\alpha)}{\operatorname{Re} b(\alpha)} (1 + \epsilon(\theta))^2 + \sqrt{\alpha}^3\, \overline{Q}(1 + \epsilon(\theta), \theta, \alpha),$$

$$1 + \epsilon(\theta') = (1 + \epsilon(\theta)) \exp - \alpha\left[(1 + \epsilon(\theta))^2 - 1\right] + \sqrt{\alpha}^3\, \overline{Q}_1(1 + \epsilon(\theta), \theta, \alpha),$$

where Q, Q_1 are smooth functions of their three arguments. The equation can be solved recursively. The proof, however, is not really straightforward (see Lanford in References).

8. Prove the first part of Proposition 14 for $d = 2$, $s = 1$ (Hint: Proceed as in the proof of Proposition 11, §5.6, p. 410. Here the transformation in Problem 6 plays the role played there by the equation in normal form.)

9. Prove the second part of Proposition 14 for $d = 2$, assuming that the invariant set of Problem 7 is actually homeomorphic to a circle and making use of Problems 2–7 for the reduction to normal form.

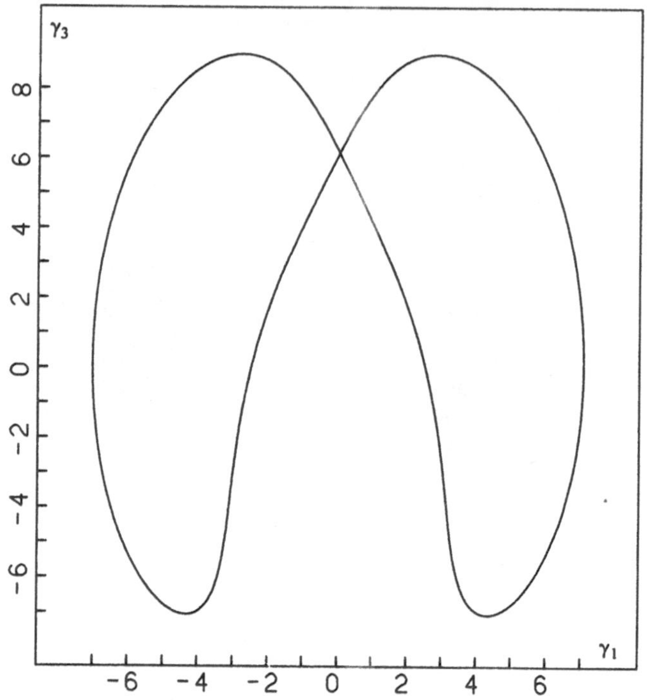

Figure 5.21. $\alpha = 190$; another stable periodic orbit.

10. Consider the C^∞ map Φ of $\mathbf{R}^1$ into itself:

$$x' = \Phi(x) = \lambda x + g(x)$$

with $g \in C^\infty(\mathbf{R})$ having a second-order zero at the origin.

Show that if $\lambda \neq 0$, $\lambda \neq 1$, there is a change of variables transforming the above map into a new one having the form

$$\xi' = \lambda\xi + \xi^3\gamma(\xi)$$

with γ in $C^\infty(\mathbf{R})$, for ξ near 0. (Hint: Let $g(x) = \bar{g}x^2 + \tilde{g}(x)$ with $\tilde{g}$ having a third-order zero at the origin. Set $\xi = x + Gx^2$ and find a suitable G.)

11. In the context of Problem 10, show that if

$$x' = (-1-\alpha)x + x^3\gamma(x,\alpha) \equiv \Phi_\alpha(x)$$

is a family of maps of class C^∞ parameterized by α with $\gamma \in C^\infty(\mathbf{R}^2)$, $\gamma(0,0) > 0$, then there exist two points $x_+(\alpha) \neq x_-(\alpha)$, for $\alpha > 0$ small, such that

$$\Phi_\alpha(x_+(\alpha)) = x_-(\alpha), \qquad \Phi_\alpha(x_-(\alpha)) = x_+(\alpha),$$

i.e., constituting a period 2 orbit ("doubling bifurcation"). Furthermore, show that by the Lyapunov criterion, such an orbit is stable and attractive. (Hint: Use the implicit function theorem to find $x_+(\alpha)$, say, as a root of $\Phi_\alpha^2(x) = x$. Prove the stability by applying the criterion of Lyapunov, Proposition 13, §5.8, p. 442, to $x_+(\alpha)$ and to the map Φ_{α}^2.)

12. Consider a map $\mathbf{x}' = \mathbf{\Phi}(x,\alpha)$ of $\mathbf{R}^d$ into itself, parameterized by $\alpha \in \mathbf{R}$. Let $\mathbf{\Phi} \in C^\infty(\mathbf{R}^d \times \mathbf{R})$, let the origin be a fixed point of the map, for all α near zero, and let $L(\alpha)$ be the origin's stability matrix. Suppose that for $\alpha \in (-a,a)$, all the

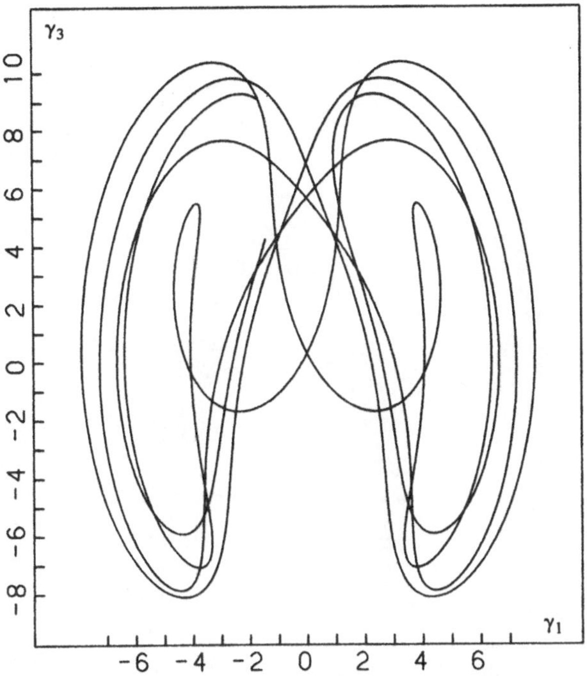

Figure 5.22. $\alpha = 190$; another stable periodic orbit which coexists with that of Fig. 5.21 and with the torus of Fig. 5.23. A randomly chosen initial datum, at this value of α, is attracted either by the periodic motions of Figs. 5.21 and 5.22 [or some of their images by the symmetries of Eq. (5.8.14)] or by the quasi-periodic motion which takes place on the torus of Fig. 5.23.

eigenvalues of $L(\alpha)$ are pairwise distinct and such that $|\lambda_1(0)| = 1 > \nu > |\lambda_2(0)|, \ldots, |\lambda_d(0)|$ with $\nu < 1$.

Using the attractive manifold theorem described in the first part of Proposition 14, p. 444, and Problem 11, show that if $\lambda(\alpha) = -1 - \alpha$, then the origin undergoes a "period doubling bifurcation" as α grows through zero (in the sense of Problem 11). (Hint: Use the attractive manifold theorem to reduce the problem to a one-dimensional problem and then apply Problems 10 and 11.)

13. Prove that Problem 12 implies that if $\boldsymbol{\Phi}(\mathbf{x}, \alpha)$ is (an arbitrary extension of) the Poincaré map for a periodic orbit of a one-parameter family of differential equations in $\mathbf{R}^{d+1}$, then the periodic orbit bifurcates to a stable (exponentially attractive) periodic orbit, as α grows through 0.

14. Study the map $x' = 4\alpha x(1-x), x \in \mathbf{R}$, and show that $[0,1]$ is an invariant set if $\alpha \in [0,1]$. Find the first bifurcation of the fixed points $x = 0$ and $x = x_\alpha > 0$, $x_\alpha = 1 - 1/4\alpha$ (consider the latter only for $\alpha > \frac{1}{4}$). Show that in some sense x_α grows out of a bifurcation of $x = 0$; while when x_α loses stability, it undergoes a doubling bifurcation in the sense of Problem 11.

15. Consider the map Φ in Problem 14 for $\alpha = 1$, restricted to $[0,1]$. Show that the change of variables $y = (2/\pi)\arcsin\sqrt{x}$ transforms this map into the map Ψ:

$$\Psi : y \to \begin{matrix} 2y & \text{if } 0 < y < \frac{1}{2}, \\ 2(1-y) & \text{if } \frac{1}{2} < y < 1. \end{matrix}$$

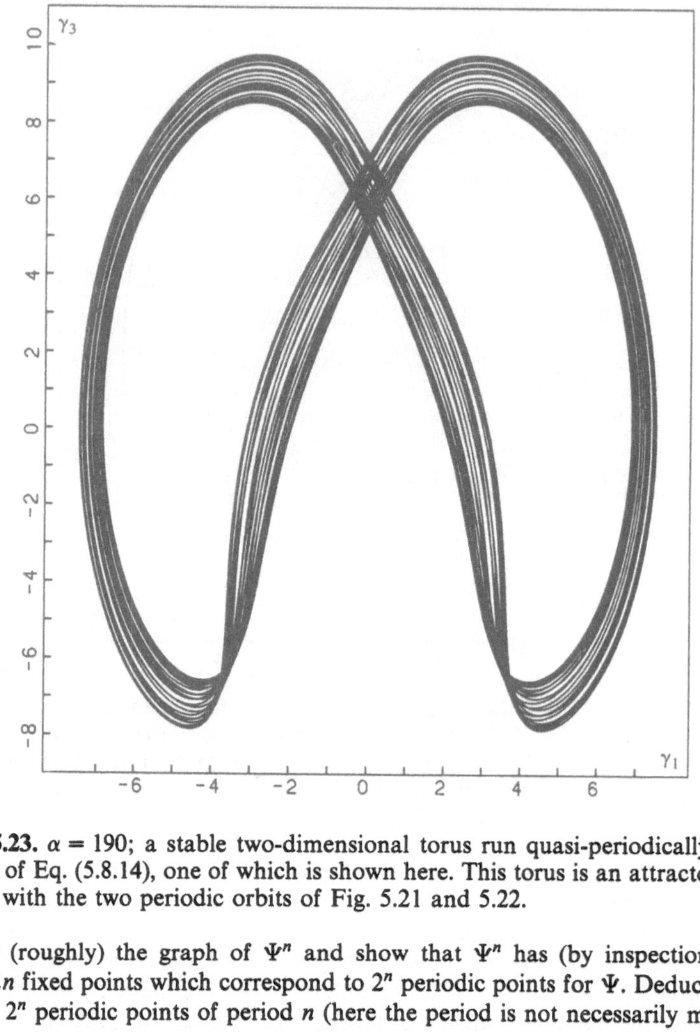

Figure 5.23. $\alpha = 190$; a stable two-dimensional torus run quasi-periodically by the motions of Eq. (5.8.14), one of which is shown here. This torus is an attractor which coexists with the two periodic orbits of Fig. 5.21 and 5.22.

Draw (roughly) the graph of Ψ^n and show that Ψ^n has (by inspection of the graph) $2n$ fixed points which correspond to 2^n periodic points for Ψ. Deduce that Φ also has 2^n periodic points of period n (here the period is not necessarily minimal).

16. Using Problem 15, show that Ψ and Φ have a dense set of periodic points. (Hint: Look at the graph of Ψ^n.)

17. Study the stability of the fixed points of the map of $\mathbf{R}^2 \to \mathbf{R}^2$ parameterized by α, b ("Henon's map"):

$$H(x, y) = (y - \alpha x^2 + 1, bx)$$

with b real and find whether one of its fixed points undergoes, for some fixed value of b, a doubling bifurcation as α grows using Problems 11 and 12.

18. Let $\mathbf{x} \to \mathbf{\Phi}(\mathbf{x})$ be a C^∞ map of the plane into itself which is invertible and area preserving (i.e., area E = area $\mathbf{\Phi}^{-1}(E)$ for all measurable sets). Which relation between the eigenvalues of the Lyapunov stability matrix of a fixed point follows as a consequence of the conservation of the area?

19. Same as Problem 18 for a volume-preserving map of $\mathbf{R}^d$ into itself.

20. In the context of Proposition 13, p. 442, show that the eigenvalues of the stability matrix of a periodic orbit depend neither on the particular system of coordinates introduced on σ nor on the point ξ_0 chosen on the orbit. They are

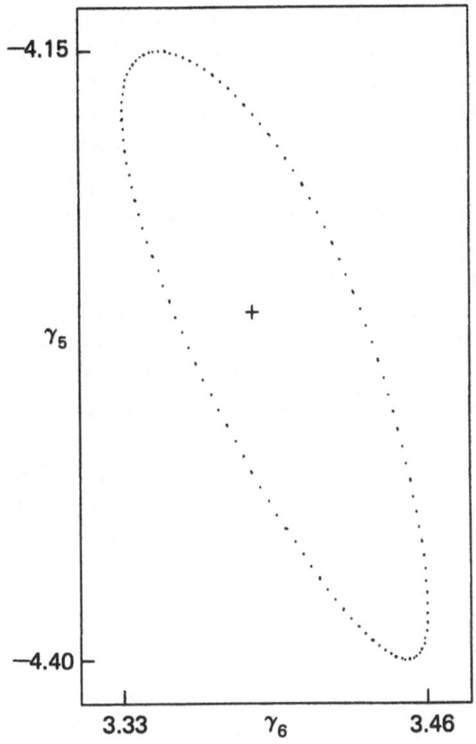

Figure 5.24. A two-dimensional section of the torus of Fig. 5.23.

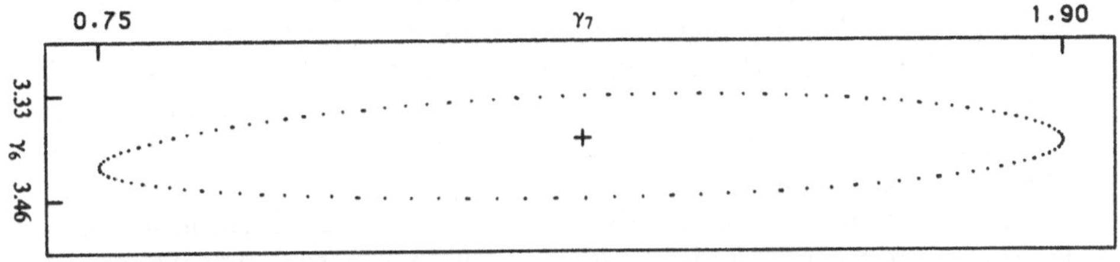

Figure 5.25. A two-dimensional section of the torus of Fig. 5.23.

"characteristic numbers" of the orbit itself. (Hint: This is a problem analogous to Problem 15, p. 387, §5.4. The first statement is proven in exactly the same way. To prove the second, use the equation's trajectories to "transfer" a system of coordinates on σ (through ξ_0) into a system of coordinates on σ' (through ξ'_0), etc.)

§5.9. Stability in Conservative Systems: Introduction

. . . desinas ineptire
et quod perisse vides perditum ducas

This is a very natural problem arising, perhaps for the first time, in the theory of the solar system, where it is still unsolved.

In nature there are many interesting systems which are "quasi-integrable" in the sense that their equations of motion differ, up to "quasi-negligible" terms, from equations of motion of an integrable system.

A nice example is provided by the solar system which we consider via a model in which the solar mass M is $+\infty$, i.e., the Sun is a fixed point mass attracting the planets with a central force with potential energy inversely proportional to the planet's distance and directly proportional to its mass. In the approximation in which the reciprocal attraction among the planets is neglected, it is clear that the solar system is described by as many Hamiltonian integrable systems as the number of planets (i.e., nine), one for each planet. In Chapter 4, §4.9 and §4.10, we saw that such Hamiltonian systems are integrable in the sense of Definitions 10 and 11, §4.8.

It is then attractive to think that the actual motion of the solar system is "close" to this idealized motion followed by nine independent planets.

Keeping for simplicity, the approximation that the Sun is a point mass fixed with respect to the fixed stars, we must compare the solutions of the following two systems of equations: $i = 1, \ldots, 9$,

$$m_i\ddot{\mathbf{x}}^{(i)} = -\frac{Km_i}{|\mathbf{x}^{(i)}|^2}\frac{\mathbf{x}^{(i)}}{|\mathbf{x}^{(i)}|}, \tag{5.9.1}$$

$$m_i\ddot{\mathbf{x}}^{(i)} = -\frac{Km_i}{|\mathbf{x}^{(i)}|^2}\frac{\mathbf{x}^{(i)}}{|\mathbf{x}^{(i)}|} - \epsilon\sum_{j\neq i}^{1,9}\frac{m_i m_j}{(\mathbf{x}^{(i)}-\mathbf{x}^{(j)})^2}\frac{\mathbf{x}^{(i)}-\mathbf{x}^{(j)}}{|\mathbf{x}^{(i)}-\mathbf{x}^{(j)}|}, \tag{5.9.2}$$

at least for initial data which, put into Eq. (5.9.1), give rise to trajectories on which $|\mathbf{x}^{(i)} - \mathbf{x}^{(j)}|$, $i \neq j$, remains so large as to make the second term in the right-hand side of Eq. (5.9.2) small compared to the first. The constant ϵ is the universal gravitation constant, $m_1, \ldots, m_9$ are the masses of the nine main planets, $K = \epsilon M_s$, where M_s is the Sun's real mass; satellites, comets, asteroids, rings, etc. have been disregarded.

Choosing as time origin the flying away instant, the situation in which the solar system is initially found is, as is well known, such that the term in ϵ in Eq. (5.9.2) has a modulus quite a bit smaller than the term representing the Sun's attraction. The first question, preliminary to the comparison between the solutions of Eqs. (5.9.1) and (5.9.2), is whether this situation remains unchanged as time goes by.

This property can be easily verified through the explicit solution of the various Kepler problems in the case of Eq. (5.9.1). Hence, we see that, in fact, this question is intimately related to the comparison between Eqs. (5.9.1) and (5.9.2).

From the general results of the theory of ordinary differential equations, it is evident that "close equations yield close solutions"; however, this closeness is not uniform over time. It does not, indeed, follow from the regularity theorems and the initial data and parameters' dependence that close equations with close initial data produce solutions which stay close forever or solutions whose trajectories, as sets, remain close. The first possibility is almost always false.

One then asks if the corrections to the equations of motion (5.9.1) due to the presence of the term in ϵ in Eq. (5.9.2), though small, may lead to changes in the motions which, in the long run, result in a motion very different from the one foreseen in Eq. (5.9.1).

A priori, one could even consider "unthinkable" or undesirable catastrophic events, like interplanetary collisions or capture of a planet by the burning Sun.

Of course, one wishes to have analytic instruments for the solutions of Eq. (5.9.2) and of comparison with those of Eq. (5.9.1), allowing not only the exclusion of such catastrophes, but even to show that it is true, or essentially true, that the planets' movements are described by Eq. (5.9.1) and that, if needed, one can compute or estimate the deviations between the motions of Eq. (5.9.1) and those of Eq. (5.9.2) with equal initial data at least for long times, i.e., of astronomical magnitude, long compared with the revolution periods of the various planets.

In other words, one wishes to use Eq. (5.9.1) for "rough" astronomical predictions and to have algorithms to compute the corrections at least for times of the order of magnitude of several thousand years.

That this is a delicate problem can be deduced from the fact that rough estimates, too pessimistic, of the errors lead to the conclusion that the reciprocal influence between the planets may become important within a few years.

For instance, the time necessary for a collision between two heavenly bodies of the size of Venus and Earth, assuming that at time zero they are standing still (relative to the fixed stars) at a distance $d(T, V)$, equal to the actual Earth-Venus maximal observed distance, could be estimated by T_{coll} such that (accelerated motion estimate)

$$\frac{\epsilon}{2} \frac{(m_T + m_V)}{d(T, V)^2} T_{\text{coll}}^2 = d(T, V) \Rightarrow T_{\text{coll}} \lesssim 370 \text{ years.} \tag{5.9.3}$$

Hence, we see that even to establish some accurate predictions for times of a few centuries, a remarkable precision is needed, i.e., it is necessary to take into account the fact that the planets' motion at the initial time is very far from a situation bound to a collision and that, obviously, the corrections to the motion described by Eq. (5.9.1), originated by the additional terms in ϵ in Eq. (5.9.2), are not always favorable to collisions (or escapes, etc.). Think of the two-body problem where the systematic attraction results only in providing a curvature to the trajectory. On the average, the effects favorable to catastrophic events may be much smaller or even totally absent with respect to the above pessimistic calculation.

This and similar problems, which may obviously be formulated for systems very different from the solar system (like harmonic oscillators with conservative anharmonic additional perturbing forces or, more generally, for systems "close" to integrable systems), are typical stability problems for conservative systems.

To the above problems, one adds analogous problems of stability of integrable systems perturbed by the addition, among the active forces, of external forces varying with simple time laws ("nonautonomous Hamiltonian systems").

All of the above problems are much more difficult than one might imagine, perhaps naively. Only recently have some techniques apt to provide some answers been developed (and are being developed), although we are still quite far from a "satisfactory" theory even for very small perturbations.

The main result on this theme is the following theorem ("Kolmogorov–Arnold–Moser theorem") which we shall analyze in some particularly interesting cases in the §5.12. The reader who wishes to obtain deeper insights may consult Moser (see References).

15 Proposition. *Consider a mechanical system in* $\mathbf{R}^d$ *with* l *degrees of freedom, subject to conservative forces with potential energy* $\Phi_0 \in C^\infty(\mathbf{R}^d)$ *bounded from below and subject to ideal constraints.*

Suppose that the system is canonically integrable on some open set W *of the phase space (see Definition* 11, *p.* 289, §4.8) *and call* H_0 *its Hamiltonian.*

If $I: W \leftrightarrow V \times \mathbf{T}^l$ *is the integrating transformation and if we set* $(\mathbf{A}, \boldsymbol{\varphi}) = I(\mathbf{p}, \mathbf{q})$ *the system's motion in* $(\mathbf{A}, \boldsymbol{\varphi})$ *coordinates is, by definition,*

$$\hat{S}_t(\mathbf{A}, \boldsymbol{\varphi}) = I\big(S_t(I^{-1}(\mathbf{A}, \boldsymbol{\varphi}))\big) = (\mathbf{A}, \boldsymbol{\varphi} + \boldsymbol{\omega}(\mathbf{A})t), \tag{5.9.4}$$

where $\boldsymbol{\omega}(\mathbf{A}) = (\omega_1(\mathbf{A}), \ldots, \omega_l(\mathbf{A}))$ *are* l *pulsations corresponding to the* l *prime integrals* $\mathbf{A} = (A_1, \ldots, A_l)$, *and* $\boldsymbol{\omega}_0(\mathbf{A}) = \partial h_0(\mathbf{A})$ *if* $h_0(\mathbf{A}) = H_0(I^{-1}(\mathbf{A}, \boldsymbol{\varphi}))$ [$\boldsymbol{\varphi}$ *independent because of the integrating character of* I, *see Observation* (1), *p.* 289].

Assume that the matrix

$$J_{ij} = \frac{\partial \omega_i}{\partial A_j} \tag{5.9.5}$$

has nonvanishing determinant on all of V ("*nonisochrony*" *of the system*) *and suppose* V *is bounded.*

Then, if $\Psi \in C^\infty(\mathbf{R}^d)$ *is a uniformly bounded potential energy, the mechanical system with the same constraints but with an active force with potential energy*

$$\Phi_0 + \epsilon\Psi \tag{5.9.6}$$

has various remarkable properties which will be described calling $S_t^{(\epsilon)}$ *and* $\hat{S}_t^{(\epsilon)}$, $t \in \mathbf{R}$, *the transformations generating the motions, corresponding to Eq.* (5.9.6) *and to the given constraints, in the coordinates* $(\mathbf{p}, \mathbf{q})$ *and* $(\mathbf{A}, \boldsymbol{\varphi})$, *respectively; if* $(\mathbf{A}, \boldsymbol{\varphi}) = I(\mathbf{p}, \mathbf{q})$, *then*

$$\hat{S}_t^{(\epsilon)}(\mathbf{A}, \boldsymbol{\varphi}) = I\big(S_t^{(\epsilon)}(\mathbf{p}, \mathbf{q})\big) \tag{5.9.7}$$

(*and* $\hat{S}_t^{(\epsilon)}(\mathbf{A}, \boldsymbol{\varphi})$ *is only defined for those pairs* $(\mathbf{A}, \boldsymbol{\varphi})$ *for which Eq.* (5.9.7) *makes sense*).

(i) *There is a subset* $W^{(\epsilon)} \subset W$ *invariant for the transformations* $S_t^{(\epsilon)}$ *and a map* $\mathbf{F}^{(\epsilon)} : W^{(\epsilon)} \leftrightarrow V^{(\epsilon)} \times \mathbf{T}^l$, $V^{(\epsilon)} \subset V$, *invertible and continuous, denoted*

$$\mathbf{F}^{(\epsilon)}(\mathbf{A}, \boldsymbol{\varphi}) = (\mathbf{a}(\mathbf{A}, \boldsymbol{\varphi}, \epsilon), \boldsymbol{\Psi}(\mathbf{A}, \boldsymbol{\varphi}, \epsilon)). \tag{5.9.8}$$

Furthermore, there is a continuous function $\boldsymbol{\Omega}^{(\epsilon)} : W^{(\epsilon)} \to \mathbf{R}^l$ *such that*

$$\mathbf{F}^{(\epsilon)}\big(\hat{S}_t^{(\epsilon)}(\mathbf{A}, \boldsymbol{\varphi})\big) = \big(\mathbf{a}(\mathbf{A}, \boldsymbol{\varphi}, \epsilon), \boldsymbol{\Psi}(\mathbf{A}, \boldsymbol{\varphi}, \epsilon) + \boldsymbol{\Omega}^{(\epsilon)}(\mathbf{A}, \boldsymbol{\varphi})t\big). \tag{5.9.9}$$

Therefore, the motions with initial datum in $W^{(\epsilon)}$ *can be thought of as rotations of an l-dimensional torus.*

(ii) *The set* $W^{(\epsilon)} \subset W$ *is generally only measurable in the sense of Lebesgue and not necessarily in the usual sense of Riemann, and its measure is such that*

$$\frac{\text{volume } W^{(\epsilon)}}{\text{volume } W} \underset{\epsilon \to 0}{\to} 1. \tag{5.9.10}$$

(iii) *The functions* $(\epsilon, \mathbf{A}, \boldsymbol{\varphi}) \to \mathbf{F}^{(\epsilon)}(\mathbf{A}, \boldsymbol{\varphi})$ *can be extended to* $C^{(k)}$ *functions with arbitrary preassigned k on* $(-1, 1) \times V \times \mathbf{T}^l$ *and the same can be said of the functions* $(\mathbf{A}, \boldsymbol{\varphi}) \to \boldsymbol{\Omega}^{(\epsilon)}(\mathbf{A}, \boldsymbol{\varphi})$. *Furthermore, such extensions have the property*

$$\mathbf{F}^{(0)}(\mathbf{A}, \boldsymbol{\varphi}) \equiv (\mathbf{A}, \boldsymbol{\varphi}), \qquad \boldsymbol{\Omega}^{(0)}(\mathbf{A}, \boldsymbol{\varphi}) \equiv \frac{\partial h_0}{\partial \mathbf{A}}(\mathbf{A}). \tag{5.9.11}$$

(iv) *If the original system is an analytic analytically integrable system and* $\boldsymbol{\Psi}$ *is also analytic, then one can take* $k = +\infty$ *in* (iii).

Observations.

(1) This theorem tells us that, in some sense, perturbing an integrable system with proper pulsations "really" variable, see Eq. (5.9.5), i.e., "non-isochronous", one obtains a system that can still be thought of as a system moving essentially in the same way as the unperturbed system, see Eq. (5.9.11).

(2) $W^{(\epsilon)}$ can be thought of as foliated into invariant l-dimensional tori with equations

$$(\mathbf{A}, \boldsymbol{\varphi}) = (\mathbf{F}^{(\epsilon)})^{-1}(\mathbf{a}, \boldsymbol{\psi}), \qquad \boldsymbol{\psi} \in \mathbf{T}^l \tag{5.9.12}$$

parameterized by l parameters $\mathbf{a} \in V^{(\epsilon)}$. By Eq. (5.9.11), each of such tori is a slight deformation of the torus described by $\{\mathbf{a}\} \times \mathbf{T}^l$ in the original variables.

(3) One interprets Observation (2) as saying that the foliation of the phase space into invariant tori (characteristic of the integrable systems) is at least in the canonically integrable cases, preserved under small perturbations, provided one disregards a subset of phase space with small measure.

(4) The fact that $V^{(\epsilon)}$ can only be shown to be Lebesgue measurable (and it probably cannot be chosen Riemann measurable) is quite unpleasant because it means that $W^{(\epsilon)}$, although containing many points (for ϵ

small) cannot be approximated by "nice sets" and, therefore, it becomes difficult to decide constructively whether a given point is or is not in $W^{(\epsilon)}$.

However, a little thought shows that (iii) partially solves this problem from a practical point of view.

(5) Note that a l-dimensional torus in a $2l$-dimensional space does not split the space $\mathbf{R}^{2l}$ into "interior" and "exterior" parts, unless $l = 1$.[21] This is perhaps what makes clearer the incompleteness of the result (iii). In fact, a point beginning its motion in $W/W^{(\epsilon)}$, i.e., outside the invariant tori, may "sneak" through the tori of the foliations very far from the vicinity of the unperturbed torus on which it would move it $\epsilon = 0$. This phenomenon, called "Arnold's diffusion", is not at all understood, (see Nekhorossiev, references).

It would be nice to understand criteria sufficient for the existence of a Riemann-measurable set of initial data which does not undergo the Arnold diffusion, i.e., implying that even though only a Lebesgue measurable set of points in phase space moves essentially as if the perturbation were not present (i.e., quasi-periodically, on tori close to the unperturbed ones), there is a Riemann measurable set of points moving (perhaps not quasi-periodically) close to the unperturbed tori.

In fact, this is what the numerical experiments sometimes seem to suggest. It can be rigorously proved for some nonautonomous 1-degree-of-freedom systems (once the above theorem is extended, as can be done, to the nonautonomous system with external periodic forces of Hamiltonian type) or for 2-degrees-of-freedom autonomous systems.

In such cases, however, the entire problem disappears as the motion takes place on a three-dimensional set (because, in the first case, the "phase space" is three dimensional (p, q, t) and in the second, although the phase space is four dimensional, the motion takes place on the three-dimensional surface of constant energy), and in $\mathbf{R}^3$ a two-dimensional torus has an interior and an exterior.

(6) The above theorem cannot be applied to perturbations of harmonic oscillators since the nonisochrony condition of Eq. (5.9.5) is manifestly violated.

Nevertheless, if the l pulsations $\omega_0 = (\omega_1, \ldots, \omega_l)$ of the harmonic oscillator verify a "nonresonance" condition: $\exists C < \infty$, $\alpha < \infty$ and

$$|\omega_0 \cdot \nu|^{-1} \leqslant C|\nu|^{\alpha}, \qquad \forall \nu \neq \mathbf{0}, \tag{5.9.13}$$

where $\nu = (\nu_1, \ldots, \nu_l) \in \mathbf{Z}^l$ is an "integer vector", then Eq. (5.9.5) can be replaced by a condition on Ψ. Namely, if $f(\mathbf{A}, \varphi)$ is the function Ψ in the $(\mathbf{A}, \varphi)$ variables, $f(\mathbf{A}, \varphi) = \Psi(I^{-1}(\mathbf{A}, \varphi))$, $(\mathbf{A}, \varphi) \in V \times \mathbf{T}^l$, and if we define

$$f_0(\mathbf{A}) = \frac{1}{(2\pi)^l} \int_{\mathbf{T}^l} f(\mathbf{A}, \varphi)\, d\varphi, \tag{5.9.14}$$

and if the matrix $(\partial^2 f_0/\partial A_i \partial A_j)(\mathbf{A})$, $i, j = 1, \ldots, l$, has nonvanishing deter-

[21] nor $\mathbf{R}^{2l-1}$, if energy conservation is taken into account, unless $l \leqslant 2$.

minant $\forall \mathbf{A} \in V$, the theorem's results (i), (ii), (iii), and (iv) hold without change.

(7) Even worse is the situation of the solar system, i.e., if one tries to apply the above theorem to Eq. (5.9.2) as a perturbation to Eq. (5.9.1).

The problem lies not so much in the unboundedness of the potentials in the Kepler motions. In fact, in a vicinity W of the Kepler motions of the actual planets, there are no collisions, so the perturbation is bounded there (W has to be thought of as a subset in the nine planets' phase space $\mathbf{R}^{27} \times \mathbf{R}^{27}$).

The difficulty lies in the fact that for the unperturbed system described by Eq. (5.9.1), Kepler's laws hold and say that each planet moves periodically with pulsation ω_i and, therefore, the system moves quasi-periodically with nine independent pulsations instead of the 27 that should be present if the system were really anisochronous and the condition (5.9.5) cannot hold (since two of the three pulsations of each planet i have to be integer multiples of ω_i).

Nevertheless, it is possible to find a version of Proposition 15 covering this problem at least in some nontrivial cases of N gravitating point masses attracted by a fixed center and attracting each other (see also p. 499).

Without quoting the exact results, we mention one of their consequences: there exist quasiperiodic motions of the planets (i.e., solutions of Eq. (5.9.2)] which take place on almost circular, almost closed, and almost coplanar orbits of distinct radii, provided the masses are very small; hence, there are motions of Eq. (5.9.2) quasi-periodic and without collisions or escapes.

The last statement and result solves a problem which for centuries fascinated physicists, mathematicians, and astronomers. Newton's universal gravitation law is not incompatible, by itself, with the stability of the solar system, a fact empirically observed since millennia and hoped for by everybody. Nevertheless, it remains an open question whether or not our own solar system, modeled by Eq. (5.9.2), is actually stable: the initial data on the positions and velocities of the planets and their masses seem too far from values to which the above-mentioned extensions of Proposition 15 can be applied.

(8) The proof of Proposition 15 gives much more information than its text expresses. It might even be possible to extract from it, and from its extensions mentioned in Observation (6), some astronomically interesting results. However, much work has to be done, since the results of Proposition 15 and its extensions are seldom obtained in "optimal" form. Actually, to my knowledge, careful estimates based on the proof of the theorem and taking full advantage of the peculiarities of a given equation of interest are still lacking even in the simplest cases.

(9) The ideas for the proof of the above theorem arise from perturbation theory for classical Hamiltonian systems. To this we devote the next section, and in §5.11 and §5.12 we shall show how the ideas of perturbation theory may be applied to prove Proposition 15 in the simplest case of a canonically analytically integrable system, analytically perturbed.

It is a shame that the old classical perturbation theory, which gave rise to analytical mechanics and to the Hamilton–Jacobi method, is nowadays almost forgotten since many people seem to know or care only for the quantum-mechanical perturbation theory. This fact is largely responsible for the aura of mystery which still seems to surround the above theorem.

§5.10. The Formal Theory of Perturbations by the Hamilton–Jacobi Method

Or ti riman, lettor, sovra 'l tuo banco,
Dietro pensando a ciò che si preliba,
S'esser vuoi lieto assai prima che stanco.
Messo t'ho innanzi: omai per te ti ciba;
Ché a se torce tutta la mia cura
Quella materia ond'io son fatto scriba.[22]

Consider an l-degree-of-freedom system described by a Hamiltonian function H on an open set W in phase space.

Denote $(\mathbf{p}, \mathbf{q})$ the points in W and denote

$$(\mathbf{p}, \mathbf{q}) \to H(\mathbf{p}, \mathbf{q}), \qquad (\mathbf{p}, \mathbf{q}) \in W \tag{5.10.1}$$

the Hamiltonian function H.

We shall suppose that this system is canonically analytically integrable via an analytic canonical transformation I integrating it by transforming W into $V \times \mathbf{T}^l$ with $V \subset \mathbf{R}^l$, open and bounded.[23]

The transformation I transforms the Hamiltonian H into a function of the first l variables of $(\mathbf{A}, \boldsymbol{\varphi}) = I(\mathbf{p}, \mathbf{q}) : h(\mathbf{A}) = H(I^{-1}(\mathbf{A}, \boldsymbol{\varphi}))$, $\forall (\mathbf{A}, \boldsymbol{\varphi}) \in V \times \mathbf{T}^l$.

We recall that the variables $\mathbf{A}$ are called "action variables", while the $\boldsymbol{\varphi}$ variables are called "angle variables".

Let F be an analytic function on W and consider the Hamiltonian system described on W by the Hamiltonian function

$$H(\mathbf{p}, \mathbf{q}) + \epsilon F(\mathbf{p}, \mathbf{q}) \tag{5.10.2}$$

or, in the action-angle variables, $(\mathbf{A}, \boldsymbol{\varphi}) \in \mathbf{T}^l$,

$$h(\mathbf{A}) + \epsilon f(\mathbf{A}, \boldsymbol{\varphi}) \tag{5.10.3}$$

[22] In basic English:

Now stay, o reader, on your bench,
thinking about what is foreshadowed
if you wish to be happy before being tired.
I did initiate you: now proceed by yourself;
as my whole thoughts are absorbed
by the matter about which I am scribe.
(Dante, *Paradiso*, Canto X)

[23] See Definition 11, p. 289, §4.8.

with

$$h(\mathbf{A}) = H\big(I^{-1}(\mathbf{A},\boldsymbol{\varphi})\big), \qquad f(\mathbf{A},\boldsymbol{\varphi}) = F\big(I^{-1}(\mathbf{A},\boldsymbol{\varphi})\big). \qquad (5.10.4)$$

Perturbation theory proposes to compare, for ϵ small, the motions of the system with Hamiltonian h and those of the system with Hamiltonian $h + \epsilon f$, usually with the same initial data.

As stated in §5.9 the comparison methods for solutions of a differential equation depending on a parameter (Lyapunov criterion, attractive manifold theorem, Hopf theorem, etc.) often reveal themselves to be inadequate in the analysis of the problems and difficulties connected with the stability of conservative systems. Such problems appear quite different from those arising in the theory of dissipative systems, at least at the beginning (although the advanced theory ultimately may conceptually coincide).

However, the special form of the Hamiltonian equations permits the use of a simple algorithm, of great interest for applications, for the analysis of the motions of quasi-integrable systems.

The idea is to change variables via a completely canonical transformation $(\mathbf{A},\boldsymbol{\varphi})\to(\mathbf{A}',\boldsymbol{\varphi}')$, arranging things so that the "old" Hamiltonian (5.10.3) takes the form

$$h_\epsilon^{(n)}(\mathbf{A}') + \epsilon^{n+1} f_\epsilon^{(n)}(\mathbf{A}',\boldsymbol{\varphi}') \qquad (5.10.5)$$

in the new variables, where $(\mathbf{A}',\boldsymbol{\varphi}')$ denote the new variables and $h_\epsilon^{(n)}$, $f_\epsilon^{(n)}$ are analytic functions of ϵ near 0, of $\boldsymbol{\varphi}'$ in $\mathbf{T}^l$ and of $\mathbf{A}'$ in a suitable open set.

Hence, for ϵ small, the error that would be made supposing that in the variables $(\mathbf{A}',\boldsymbol{\varphi}')$ the system is integrable and described by the Hamiltonian $h_\epsilon^{(n)}$ is very much smaller than the one that would be made assuming the system as integrable in the original variables $(\mathbf{A},\boldsymbol{\varphi})$ simply setting $\epsilon = 0$ in Eq. (5.10.3) (provided, as we suppose as an extra essential requirement of construction, the canonical transformation itself is not singular at $\epsilon = 0$).

Intuitively, neglecting in Eq. (5.10.5), or, better, in the Hamiltonian equations associated with Eq. (5.10.5), the $\boldsymbol{\varphi}'$-dependent term produces an error of the order $\epsilon^{n+1}T$ in the equations' solutions, if these solutions are observed up to a time T. Hence, given an approximation η, it will be possible to retain it, though neglecting the influence of $f_\epsilon^{(n+1)}$ on the motions of Eq. (5.10.5), for a time of the order $T_{\epsilon,\eta} \propto \eta\epsilon^{-(n+1)}$.

For ϵ small, this may be a substantially better result for $n > 0$ than the one corresponding to the simple, but often too rough, analogous approximation with $n = 0$ (i.e., $\epsilon = 0$ in Eq. (5.10.3)).

The reader will easily realize that the method that will be used for the "reduction to higher order" of the perturbation via a canonical transformation is nothing more than a method for constructing successive approximations to the time independent solutions of the Hamilton–Jacobi equation (3.11.68), p. 225.

There are two remarkable cases which can actually be treated along the above lines, building a completely canonical transformation changing Eq.

(5.10.3) into Eq. (5.10.5) at least for $(\mathbf{A}, \varphi)$ in a neighborhood of the form $S_\rho(\mathbf{A}_0) \times \mathbf{T}^l \subset V \times \mathbf{T}^l$, where $S_\rho(\mathbf{A}_0)$ is a sphere with radius ρ in $\mathbf{R}^l$ around a preassigned point $\mathbf{A}_0$, and for some $n > 0$ and ϵ small.

The first case arises when

$$h(\mathbf{A}) = \omega_0 \cdot (\mathbf{A} - \mathbf{A}_0) + h(\mathbf{A}_0) \tag{5.10.6}$$

with $\omega_0 \in \mathbf{R}^l$ such that there are $C, \alpha > 0$, for which

$$C = \sup_{\nu \in \mathbf{Z}^l, |\nu| \neq 0} \frac{|\omega_0 \cdot \nu|^{-1}}{|\nu|^\alpha} < +\infty. \tag{5.10.7}$$

The second case arises when the Fourier coefficients of the development of f:

$$\begin{aligned} f(\mathbf{A}, \varphi) &= \sum_{\nu \in \mathbf{Z}^l} f_\nu(\mathbf{A}) e^{i\nu \cdot \varphi}, \\ f_\nu(\mathbf{A}) &= \frac{1}{(2\pi)^l} \int_{\mathbf{T}^l} f(\mathbf{A}, \varphi) e^{-i\nu \cdot \varphi} \, d\varphi, \end{aligned} \tag{5.10.8}$$

vanish for $|\nu| > N$ and, setting

$$\omega(\mathbf{A}) = \frac{\partial h}{\partial \mathbf{A}}(\mathbf{A}), \tag{5.10.9}$$

one has

$$|\omega(\mathbf{A}_0) \cdot \nu| > 0, \qquad \forall \nu \in \mathbf{Z}^l, \qquad 0 < |\nu| \leqslant N. \tag{5.10.10}$$

In the first case, it is even possible to put the Hamiltonian into the form of Eq. (5.10.5), $\forall n = 0, 1, \ldots,$ provided ϵ is small enough (depending, however, on the choice of n).

The above statements are illustrated in the following classical propositions.

16 Proposition. *Consider the Hamiltonian* (5.10.3) *on* $V \times \mathbf{T}^l$ *with*

$$f(\mathbf{A}, \varphi) \equiv \sum_{\substack{\nu \in \mathbf{Z}^l \\ |\nu| \leqslant N}} f_\nu(\mathbf{A}) e^{i\nu \cdot \varphi} \tag{5.10.11}$$

analytic on $V \times \mathbf{T}^l$, *with* $N > 0$, *and suppose that*, $\forall \mathbf{A}_0 \in V$, *the function* h *is such that*

$$|\omega(\mathbf{A}_0) \cdot \nu| > 0, \qquad \nu \in \mathbf{Z}^l, \quad 0 < |\nu| \leqslant N. \tag{5.10.12}$$

Then there exist $\rho_1 > 0$, $\epsilon_1 > 0$ *and*, $\forall \epsilon \in (-\epsilon_1, \epsilon_1)$, *a completely canonical transformation* $(\mathbf{A}, \varphi) \leftrightarrow (\mathbf{A}', \varphi')$ *defined for* $(\mathbf{A}, \varphi) \in W_\epsilon$, *with* $V \times \mathbf{T}^l \supset W_\epsilon \supset S_{\rho_1/2}(\mathbf{A}_0) \times \mathbf{T}^l$ *and with values onto* $S_{\rho_1}(\mathbf{A}_0) \times \mathbf{T}^l$ *smoothly depending on* ϵ *and transforming the Hamiltonian* (5.10.3) *into*

$$h_\epsilon^{(1)}(\mathbf{A}') + \epsilon^2 f_\epsilon^{(1)}(\mathbf{A}', \varphi'), \tag{5.10.13}$$

where $h_\epsilon^{(1)}, f_\epsilon^{(1)}$ *are analytic in* $\epsilon, \mathbf{A}', \varphi'$. *Furthermore* $h_\epsilon^{(1)}$ *can be given a simple expression; see Eq.* (5.10.25).

Observation. As mentioned above, the reader should interpret the proof that follows as a "perturbative solution to order ϵ" of the Hamilton–Jacobi equation in the time-independent case, i.e., when H in Eq. (3.11.68), p. 225, does not explicitly depend on t. Actually, the above proposition is the basic example of how the method of Hamilton–Jacobi concretely works. Most applications of the Hamilton–Jacobi's method are based on this proposition.

PROOF. The canonical transformation will be determined by looking for a generating function Φ, see §3.11 and §3.12 from p. 220 on.

Since we expect that such a transformation is close to the identity up to infinitesimals of order ϵ, we shall try to write the unknown generating function as

$$\mathbf{A}' \cdot \boldsymbol{\varphi} + \Phi(\mathbf{A}', \boldsymbol{\varphi}), \tag{5.10.14}$$

where $(\mathbf{A}', \boldsymbol{\varphi}) \to \mathbf{A}' \cdot \boldsymbol{\varphi}$ is the generating function of the identity map and Φ is infinitesimal in ϵ.

The function Φ will be determined by requiring that the Hamiltonian in the new variables $(\mathbf{A}', \boldsymbol{\varphi}')$ defined by the formal map

$$\begin{aligned} \mathbf{A} &= \mathbf{A}' + \frac{\partial \Phi}{\partial \boldsymbol{\varphi}}(\mathbf{A}', \boldsymbol{\varphi}), \\ \boldsymbol{\varphi}' &= \boldsymbol{\varphi} + \frac{\partial \Phi}{\partial \mathbf{A}'}(\mathbf{A}', \boldsymbol{\varphi}), \end{aligned} \tag{5.10.15}$$

i.e., the function

$$h\left(\mathbf{A}' + \frac{\partial \Phi}{\partial \boldsymbol{\varphi}}(\mathbf{A}', \boldsymbol{\varphi})\right) + \epsilon f\left(\mathbf{A}' + \frac{\partial \Phi}{\partial \boldsymbol{\varphi}}(\mathbf{A}', \boldsymbol{\varphi}), \boldsymbol{\varphi}\right), \tag{5.10.16}$$

is $\boldsymbol{\varphi}$ independent up to terms infinitesimal of higher order in ϵ.

Since, as already said, we expect that $\Phi \simeq O(\epsilon)$, we can heuristically find, by developing Eq. (5.10.16) in series with respect to $\partial\Phi/\partial\mathbf{A}'$, that the equation for Φ (the "Hamilton–Jacobi equation to first order in ϵ") is

$$\frac{\partial h}{\partial \mathbf{A}'}(\mathbf{A}') \cdot \frac{\partial \Phi}{\partial \boldsymbol{\varphi}}(\mathbf{A}', \boldsymbol{\varphi}) + \epsilon f(\mathbf{A}', \boldsymbol{\varphi}) = \{\boldsymbol{\varphi}\text{-independent function}\} \tag{5.10.17}$$

which, written in terms of the Fourier components of Φ, means that if $\boldsymbol{\omega}(\mathbf{A}') \equiv (\partial h/\partial \mathbf{A}')(\mathbf{A}')$,

$$i(\boldsymbol{\omega}(\mathbf{A}') \cdot \boldsymbol{\nu})\Phi_{\boldsymbol{\nu}}(\mathbf{A}') + \epsilon f_{\boldsymbol{\nu}}(\mathbf{A}') = 0, \qquad \forall \boldsymbol{\nu} \in \mathbf{Z}^l, \quad |\boldsymbol{\nu}| > 0. \tag{5.10.18}$$

This equation is really a soluble equation if $|\mathbf{A}' - \mathbf{A}_0| \leqslant \bar{\rho}_1$ with $\bar{\rho}_1$ so small that the closure of $S_{\bar{\rho}_1}(\mathbf{A}_0)$ is such that $\overline{S_{\bar{\rho}_1}(\mathbf{A}_0)} \subset V$ and therefore

$$\boldsymbol{\omega}(\mathbf{A}') \cdot \boldsymbol{\nu} \neq 0, \qquad \forall \boldsymbol{\nu} \in \mathbf{Z}^l, \quad 0 < |\boldsymbol{\nu}| \leqslant N, \tag{5.10.19}$$

see Eq. (5.10.12).

Then we can define in $S_{\bar{\rho}_1}(\mathbf{A}_0) \times \mathbf{T}^l$ the analytic function

$$\Phi(\mathbf{A}', \boldsymbol{\varphi}) = \epsilon \sum_{0 < |\boldsymbol{\nu}| \leqslant N} \frac{f_{\boldsymbol{\nu}}(\mathbf{A}') e^{i\boldsymbol{\nu} \cdot \boldsymbol{\varphi}}}{-i\boldsymbol{\omega}(\mathbf{A}') \cdot \boldsymbol{\nu}} \tag{5.10.20}$$

and we observe, as follows from the implicit function theorem, see Appendix G, Corollaries 3 and 4, that the second of Eqs. (5.10.15) can be uniquely inverted with respect to $\boldsymbol{\varphi}$ and the first of Eqs. (5.10.15) can be inverted with respect to $\mathbf{A}'$ in the respective forms

$$\begin{aligned} \boldsymbol{\varphi} &= \boldsymbol{\varphi}' + \boldsymbol{\Delta}(\mathbf{A}',\boldsymbol{\varphi}'), &\quad \boldsymbol{\Delta} &\in C^\infty\left(\overline{S_{\bar\rho_1}(\mathbf{A}_0)} \times \mathbf{T}^l\right), \\ \mathbf{A}' &= \mathbf{A} + \boldsymbol{\Xi}'(\mathbf{A},\boldsymbol{\varphi}), &\quad \boldsymbol{\Xi}' &\in C^\infty\left(\overline{S_{\bar\rho_1/2}(\mathbf{A}_0)} \times \mathbf{T}^l\right) \end{aligned} \tag{5.10.21}$$

if ϵ is small enough,[24] i.e., if $|\epsilon| < \tilde\epsilon_1$ with $\tilde\epsilon_1$ suitably chosen; and also there is $B > 0$ such that

$$\left|\frac{\partial \Phi}{\partial \boldsymbol{\varphi}}(\mathbf{A}',\boldsymbol{\varphi})\right| = |\boldsymbol{\Xi}'(\mathbf{A},\boldsymbol{\varphi})| < B|\epsilon|, \tag{5.10.22}$$

so that if $B|\epsilon| < \bar\rho_1/8$, the maps $(\mathbf{A}',\boldsymbol{\varphi}') \to \mathcal{C}(\mathbf{A}',\boldsymbol{\varphi}') = (\mathbf{A},\boldsymbol{\varphi})$:

$$\begin{aligned} \mathbf{A} &= \mathbf{A}' + \frac{\partial \Phi}{\partial \boldsymbol{\varphi}}(\mathbf{A}',\boldsymbol{\varphi}' + \boldsymbol{\Delta}(\mathbf{A}',\boldsymbol{\varphi}')), \\ \boldsymbol{\varphi} &= \boldsymbol{\varphi}' + \boldsymbol{\Delta}(\mathbf{A}',\boldsymbol{\varphi}'), \end{aligned} \tag{5.10.23}$$

and $(\mathbf{A}',\boldsymbol{\varphi}') \to \mathcal{C}'(\mathbf{A},\boldsymbol{\varphi})$:

$$\begin{aligned} \mathbf{A}' &= \mathbf{A} + \boldsymbol{\Xi}'(\mathbf{A},\boldsymbol{\varphi}), \\ \boldsymbol{\varphi}' &= \boldsymbol{\varphi} + \frac{\partial \Phi}{\partial \mathbf{A}'}(\mathbf{A} + \boldsymbol{\Xi}'(\mathbf{A},\boldsymbol{\varphi}),\boldsymbol{\varphi}) \end{aligned} \tag{5.10.24}$$

are well defined on $S_{\bar\rho_1/2}(\mathbf{A}_0) \times \mathbf{T}^l$ and take values in $S_{\bar\rho_1}(\mathbf{A}_0) \times \mathbf{T}^l$. Furthermore, $\mathcal{C}$ and $\mathcal{C}'$ map $S_{\bar\rho_1/4}(\mathbf{A}_0) \times \mathbf{T}^l$ into $S_{\bar\rho_1/2}(\mathbf{A}_0) \times \mathbf{T}^l$ and $\mathcal{C}\mathcal{C}' \equiv \mathcal{C}'\mathcal{C} \equiv$ {identity map} on $S_{\bar\rho_1/4}(\mathbf{A}_0) \times \mathbf{T}^l$ by construction (and by the uniqueness part of the implicit function theorem).

Therefore, the Jacobian determinants of $\mathcal{C}$ or $\mathcal{C}'$ on $S_{\rho_1/4}(\mathbf{A}_0) \times \mathbf{T}^l$ cannot vanish and, hence, by Proposition 21, §3.11, p. 220, $\mathcal{C}$ is a completely canonical map of $S_{\bar\rho_1/4}(\mathbf{A}_0) \times \mathbf{T}^l$ onto its image W_ϵ in $S_{\bar\rho_1/2}(\mathbf{A}_0) \times \mathbf{T}^l \subset V \times \mathbf{T}^l$ and $W_\epsilon \supset S_{\bar\rho_1/8}(\mathbf{A}_0) \times \mathbf{T}^l$. So we take $\epsilon_1 = \min(\tilde\epsilon_1, \bar\rho_1/8B)$ and $\rho_1 = \bar\rho_1/4$. By the construction of Φ, it is clear [see Eq. (5.10.17)] that the Hamiltonian function in the $(\mathbf{A}',\boldsymbol{\varphi}')$ variables has the form of Eq. (5.10.13). By substituting Eq. (5.10.20) into Eq. (5.10.17), one, in fact, also obtains

$$h_\epsilon^{(1)}(\mathbf{A}') = h(\mathbf{A}') + \epsilon f_0(\mathbf{A}'), \tag{5.10.25}$$

where $f_{\mathbf{0}}$ is the $\mathbf{0}$th Fourier coefficient of f, see Eq. (5.10.8).

The analyticity of the canonical maps $\mathcal{C}$ and $\mathcal{C}'$ will not be discussed here. It follows immediately if Eqs. (5.10.21) are obtained via the application of analytic implicit function theorems that will be discussed in the next section; see Propositions 18–20. mbe

The above discussion is the basis for the most common algorithms in the calculations of the perturbed Hamiltonian motions; it leads to the natural

[24] $\boldsymbol{\Xi}'$ and $\boldsymbol{\Delta}$ are C^∞ also in ϵ, jointly with $(\mathbf{A},\boldsymbol{\varphi})$ or $(\mathbf{A}',\boldsymbol{\varphi}')$, by the implicit functions' theorem.

idea of iterating the procedure by reducing the perturbation from $O(\epsilon^2)$ to $O(\epsilon^4)$, etc.

The difficulty lies in the fact that, in general, the new Hamiltonian (5.10.16) which, to first order in ϵ reduces to Eq. (5.10.25), no longer has the form necessary for applicability of Proposition 16. In fact, the perturbation of order ϵ^2 will be a function of $(\mathbf{A}', \boldsymbol{\varphi}')$ which has *all* the harmonic components in $\boldsymbol{\varphi}'$ nonvanishing, disregarding exceptional cases. One can easily convince oneself of this with some thought, noting that $(\partial\Phi/\partial\boldsymbol{\varphi})(\mathbf{A}', \boldsymbol{\varphi}' + \boldsymbol{\Delta}(\mathbf{A}', \boldsymbol{\varphi}'))$ contains terms like $\exp i\boldsymbol{\Delta}(\mathbf{A}', \boldsymbol{\varphi}') \cdot \boldsymbol{\nu}$ which, unless some miraculous cancellations take place, will no longer be trigonometric polynomials in $\boldsymbol{\varphi}'$.

The following proposition, valid in the other case considered in the introduction to Proposition 16, is quite interesting because it shows that with a slight modification of the method of the above proof but under different assumptions, one can "remove" the perturbation to an arbitrary order in ϵ.

17 Proposition. *Consider the Hamiltonian function given by Eq.* (5.10.3) *on* $V \times \mathbf{T}^l$ *with* h *verifying Eqs.* (5.10.6) *and* (5.10.7) *and* f *analytic. There is* $\rho > 0$ *such that*:

(1) *For each* $n = 0, 1, \ldots$ *there exists* $\epsilon_n > 0$ *and,* $\forall |\epsilon| < \epsilon_n$, *functions* $\Phi_{\epsilon,n}$ *defined on* $S_\rho(\mathbf{A}_0) \times \mathbf{T}^l$ *and analytic in* ϵ *and in the other arguments* $(\mathbf{A}, \boldsymbol{\varphi})$, *generating completely canonical transformation* $(\mathbf{A}, \boldsymbol{\varphi}) \leftrightarrow (\mathbf{A}', \boldsymbol{\varphi}')$ *such that*

$$\mathbf{A} = \mathbf{A}' + \frac{\partial \Phi_{\epsilon,n}}{\partial \boldsymbol{\varphi}}(\mathbf{A}', \boldsymbol{\varphi}),$$
$$\boldsymbol{\varphi}' = \boldsymbol{\varphi} + \frac{\partial \Phi_{\epsilon,n}}{\partial \mathbf{A}'}(\mathbf{A}', \boldsymbol{\varphi}) \tag{5.10.26}$$

which map a subset $W_{\epsilon,n} \supset S_{\rho/2}(\mathbf{A}_0) \times \mathbf{T}^l$, *such that* $W_{\epsilon,n} \subset V \times \mathbf{T}^l$, *onto* $S_\rho(\mathbf{A}_0) \times \mathbf{T}^l$.

(2) *The map of Eq* (5.10.26) *transforms the Hamiltonian into the form*

$$h_{\epsilon,n}(\mathbf{A}') + \epsilon^{n+1} f_\epsilon^{(n)}(\mathbf{A}', \boldsymbol{\varphi}'), \tag{5.10.27}$$

where $h_{\epsilon,n}(\mathbf{A}')$ *is analytic in* $\epsilon, \mathbf{A}'$ *and* $f_\epsilon^{(n)}$ *is also analytic in* $\epsilon, \mathbf{A}', \boldsymbol{\varphi}'$. *An explicit expression for* $h_{\epsilon,n}(\mathbf{A}')$ *is Eq.* (5.10.41).

Observation. The construction described in the proof of this proposition is often referred to as the "Birkhoff transformation".

PROOF. Define heuristically:

$$\Phi_{\epsilon,n}(\mathbf{A}', \boldsymbol{\varphi}) = \sum_{k=1}^{n} \epsilon^k \Phi^{(k)}(\mathbf{A}', \boldsymbol{\varphi}) \tag{5.10.28}$$

and consider the Hamiltonian in the new variables $(\mathbf{A}', \boldsymbol{\varphi})$, Eq. (5.10.26):

$$h\left(\mathbf{A}' + \frac{\partial \Phi_{\epsilon,n}}{\partial \boldsymbol{\varphi}}(\mathbf{A}', \boldsymbol{\varphi})\right) + \epsilon f\left(\mathbf{A}' + \frac{\partial \Phi_{\epsilon,n}}{\partial \boldsymbol{\varphi}}(\mathbf{A}', \boldsymbol{\varphi}), \boldsymbol{\varphi}\right). \tag{5.10.29}$$

Developing this expression in powers of ϵ using the analyticity of f and h (the latter is actually linear) in $\mathbf{A}$, impose that the resulting series in ϵ,

$$\sum_{k=0}^{\infty} \psi^{(k)}(\mathbf{A}', \varphi)\epsilon^k, \tag{5.10.30}$$

has all the coefficients $\psi^{(k)}$, $k = 0, 1, \ldots, n$, φ independent.

One easily finds that this condition allows one to determine recursively $\Phi^{(1)}, \ldots, \Phi^{(n)}$ [and, for instance, it appears that $\epsilon\Phi^{(1)}$ is given by Eq. (5.10.20), of course].

Then, once the expressions for $\Phi^{(1)}, \ldots, \Phi^{(n)}$ are found, one shall write Eq. (5.10.26), and by taking ϵ small, proceeding exactly as in the proof of Proposition 16, the implicit function theorem will be used to guarantee that Eq. (5.10.26) actually defines a canonical transformation between $S_\rho(\mathbf{A}_0) \times \mathbf{T}^l$ and some $W_{\epsilon,n} \subset V \times \mathbf{T}^l$ and $W_{\epsilon,n} \supset S_{\rho/2}(\mathbf{A}_0) \times \mathbf{T}^l$. The invertibility conditions will depend on n. By construction, Eq. (5.10.27) will then follow, with $h_{\epsilon,n}, f_\epsilon^{(n)}$ of class C^∞ in $\epsilon, \mathbf{A}', \varphi'$. They are actually analytic and this point can be commented as at the end of the proof of Proposition 16, see p. 474.

Hence, the whole problem is to show that one can find $\Phi^{(1)}, \ldots, \Phi^{(n)}$ so that the formal series of Eq. (5.10.30) has the first $(n+1)$ coefficients vanishing. This is a purely algebraic problem.

As we shall see in the following section where the question will be more systematically treated, the analyticity assumption on f implies that it can be developed in the Taylor series about $\mathbf{A}_0$ and in the Fourier series in φ in the form

$$f(\mathbf{A}, \varphi) = \sum_{\mathbf{a}\in\mathbf{Z}^l_+} \sum_{\boldsymbol{\nu}\in\mathbf{Z}^l} f_{\boldsymbol{\nu}}^{(\mathbf{a})}(\mathbf{A}-\mathbf{A}_0)^{\mathbf{a}} e^{i\boldsymbol{\nu}\cdot\varphi} \equiv \sum_{\mathbf{a}\in\mathbf{Z}^l_+} f^{(\mathbf{a})}(\mathbf{A}_0, \varphi)(\mathbf{A}-\mathbf{A}_0)^{\mathbf{a}}, \tag{5.10.31}$$

where, see Definition 13, p. 333, $\mathbf{a} = (a_1, \ldots, a_l) \in \mathbf{Z}^l_+$ and $\boldsymbol{\nu} = (\nu_1, \ldots, \nu_l) \in \mathbf{Z}^l$ and

$$(\mathbf{A}-\mathbf{A}_0)^{\mathbf{a}} = \prod_{i=1}^{l} (A_i - A_{0i})^{a_i}, \qquad \boldsymbol{\nu}\cdot\varphi = \sum_{i=1}^{l} \nu_i \varphi_i \tag{5.10.32}$$

and, furthermore, there is $R > 0$, $\rho_0 > 0$, $\xi_0 > 0$, such that

$$|f_{\boldsymbol{\nu}}^{(\mathbf{a})}| \leqslant R\rho_0^{-|\mathbf{a}|} e^{-\xi_0|\boldsymbol{\nu}|}, \qquad \forall \mathbf{a} \in \mathbf{Z}^l_+, \quad \forall \boldsymbol{\nu} \in \mathbf{Z}^l, \tag{5.10.33}$$

if

$$|\mathbf{a}| = \sum_{i=1}^{l} a_i, \qquad |\boldsymbol{\nu}| = \sum_{i=1}^{l} |\nu_i|.$$

This inequality is not immediately obvious and it will be discussed in §5.11; for the time being, we suppose and use Eq. (5.10.33) without discussion.

Developing Eq. (5.10.29) in powers of ϵ and collecting the terms of equal order in ϵ and setting

$$f^{(\mathbf{a})}(\mathbf{A}',\boldsymbol{\varphi}) \equiv \frac{1}{a_1!\ldots a_l!}\,\frac{\partial^{|\mathbf{a}|}f}{\partial(\mathbf{A}')^{\mathbf{a}}}(\mathbf{A}',\boldsymbol{\varphi})$$

(it is the $\mathbf{a}$th coefficient of the Taylor expansion of f around $\mathbf{A}'$ at fixed $\boldsymbol{\varphi}$), one finds [using Eq. (5.10.31)]

$$\psi^{(k)}(\mathbf{A}',\boldsymbol{\varphi}) = \Bigg\{\sum_{\mathbf{a}\in\mathbf{Z}^l_+} f^{(\mathbf{a})}(\mathbf{A}',\boldsymbol{\varphi}) \sum_{n^1_1,\ldots,n^1_{a_1}} \cdots \sum^{*}_{n^l_1,\ldots,n^l_{a_l}} \prod_{j=1}^{l}\left(\prod_{s=1}^{a_j}\frac{\partial\Phi^{(n^j_s)}}{\partial\varphi_j}(\mathbf{A}',\boldsymbol{\varphi})\right)\Bigg\}$$
$$+\,\boldsymbol{\omega}_0\cdot\frac{\partial\Phi^{(k)}}{\partial\boldsymbol{\varphi}}(\mathbf{A}',\boldsymbol{\varphi}) \equiv \left\{N^{(k)}(\mathbf{A}',\boldsymbol{\varphi})\right\} + \boldsymbol{\omega}_0\cdot\frac{\partial\Phi^{(k)}}{\partial\boldsymbol{\varphi}}(\mathbf{A}',\boldsymbol{\varphi}) \tag{5.10.34}$$

for $k = 1,2,\ldots,$ and the $*$ means that the sum is performed subject to the constraint $\sum_{j=1}^{l}\sum_{s=1}^{a_j} n^j_s = k-1$. Furthermore, we set

$$\psi^{(0)}(\mathbf{A}',\boldsymbol{\varphi}) \equiv h(\mathbf{A}'). \tag{5.10.35}$$

The condition that $\psi^{(1)}$ is $\boldsymbol{\varphi}$ independent (hence, $\boldsymbol{\varphi}'$ independent) becomes, by Eq. (5.10.34),

$$f(\mathbf{A}',\boldsymbol{\varphi}) + \boldsymbol{\omega}_0\cdot\frac{\partial\Phi^{(1)}}{\partial\boldsymbol{\varphi}}(\boldsymbol{\varphi},\mathbf{A}') = \{\boldsymbol{\varphi}\text{-independent function}\} \tag{5.10.36}$$

and it determines $\Phi^{(1)}$, up to a function of $\mathbf{A}'$ alone, as (say):

$$\Phi^{(1)}(\mathbf{A}',\boldsymbol{\varphi}) = \sum_{0<|\boldsymbol{\nu}|,\,\boldsymbol{\nu}\in\mathbf{Z}^l} \frac{f_{\boldsymbol{\nu}}(\mathbf{A}')e^{i\boldsymbol{\nu}\cdot\boldsymbol{\varphi}}}{-i\boldsymbol{\omega}_0\cdot\boldsymbol{\nu}}, \tag{5.10.37}$$

where $f_{\boldsymbol{\nu}}(\mathbf{A}')$ is the $\boldsymbol{\nu}$th Fourier coefficient of $f(\mathbf{A}',\boldsymbol{\varphi})$ at $\mathbf{A}'$ fixed:

$$f_{\boldsymbol{\nu}}(\mathbf{A}') = \sum_{\mathbf{a}\in\mathbf{Z}^l_+} f^{(\mathbf{a})}_{\boldsymbol{\nu}}(\mathbf{A}_0)(\mathbf{A}'-\mathbf{A}_0)^{\mathbf{a}}. \tag{5.10.38}$$

Replacing $f_{\boldsymbol{\nu}}(\mathbf{A}')$ in Eq. (5.10.37) by Eq. (5.10.38) and using Eqs. (5.10.33) and (5.10.7), one easily sees that the series of Eqs. (5.10.37) and converges and defines a C^∞ function of $(\mathbf{A}',\boldsymbol{\varphi}) \in S_{\rho_0}(\mathbf{A}_0)\times\mathbf{T}^l$ (actually such a function is analytic, as could be shown).

Then, from Eq. (5.10.34), one sees that

$$N^{(2)}(\mathbf{A}',\boldsymbol{\varphi}) = \sum_{j=1}^{l} f^{(\mathbf{e}_j)}(\mathbf{A}',\boldsymbol{\varphi})\,\frac{\partial\Phi^{(1)}}{\partial\varphi_j}(\mathbf{A}',\boldsymbol{\varphi}) \tag{5.10.39}$$

with $\mathbf{e}_1 = (1,0,\ldots,0)$, $\mathbf{e}_2 = (0,1,\ldots,0)\ldots.$

From what has been said above, it follows that $N^{(2)}$ is a $C^\infty(S_{\rho_0}(\mathbf{A}_0)\times\mathbf{T}^l)$ function (actually analytic), and if $N^{(2)}_{\boldsymbol{\nu}}(\mathbf{A}')$ denotes its $\boldsymbol{\nu}$th Fourier coefficient, the condition that $\psi^{(2)}$ given by Eq. (5.10.34) is $\boldsymbol{\varphi}$ independent yields

$$\Phi^{(2)}(\mathbf{A},\varphi') = \sum_{\nu\neq 0} \frac{N_\nu^{(2)}(\mathbf{A}')e^{i\nu\cdot\varphi}}{-i\omega_0\cdot\nu} \tag{5.10.40}$$

which, again from Eq. (5.10.33) and from Eqs. (5.10.31), (5.10.37), and (5.10.38), turns out to be a C^∞ function on $S_{\rho_0}(\mathbf{A}_0)\times\mathbf{T}^l$ (actually analytic), etc., inductively.

It is clear that

$$h_{\epsilon,n}(\mathbf{A}') = h(\mathbf{A}') + \sum_{k=1}^{n} \epsilon^k N_0^{(k)}(\mathbf{A}'). \tag{5.10.41}$$

mbe

Observations.

(1) Equations (5.10.37) and (5.10.40) and their generalizations to higher k show that $N^{(k)}(\mathbf{A}',\varphi)$ can be chosen to be n independent. It becomes natural to try to consider the limit as $n\to\infty$. In this limit, the perturbation would disappear and the system would be transformed into a system with Hamiltonian

$$h_\epsilon(\mathbf{A}') = h_0(\mathbf{A}') + \sum_{k=1}^{\infty} N_0^{(k)}(\mathbf{A}')\epsilon^k,$$

and it would therefore be integrable. However, the estimates on ϵ_n that can be derived by applying the scheme suggested in the above proof appear to be such that $\epsilon_n\to 0$ as $n\to\infty$, save some exceptional cases. Therefore, nothing can be concluded about the limit $n\to\infty$.

It is known that it cannot happen, in general, that the series ("Birkhoff's formal series").

$$\sum_{k=1}^{\infty} \epsilon^k N_0^{(k)}(\mathbf{A}'), \qquad \sum_{k=1}^{\infty} \epsilon^k \Phi^{(k)}(\mathbf{A}',\varphi) \tag{5.10.42}$$

converge, defining analytic functions of $(\mathbf{A}',\varphi,\epsilon)$ in $(\mathbf{A}',\varphi)\in S_\rho(\mathbf{A}_0)\times\mathbf{T}^l$ and in ϵ near zero and, at the same time, $\epsilon^{n+1}\partial f_\epsilon^{(n)} \to_{n\to\infty} 0$ uniformly in the same region of $(\mathbf{A}',\varphi,\epsilon)$.

This would, in fact, imply the existence of l prime integrals analytic in $\epsilon,\mathbf{A},\varphi$ for ϵ close to 0, $\mathbf{A}$ close to $\mathbf{A}_0$ and $\varphi\in\mathbf{T}^l$: namely, $(A_1',\ldots,A_l')$, and via such integrals ("uniform integrals"), the system would be analytically integrable with a canonical transformation leading one to take such integrals as the new action variables. This property has been shown to be impossible in a number of interesting cases.

A simple example in which the series (5.10.42) can be explicitly computed is in Problem 16 at the end of this section: in the example the second of (5.10.42) does not converge. However if $N_k(\mathbf{A})$ depend on $\mathbf{A}$ via $\omega\cdot\mathbf{A}$, only, the series converge: this is a nice criterion of Rüssmann (see References).

(2) Various algorithms used in practice to study perturbations of integrable motions are based on the two propositions illustrated above. The simplest is the following.

First, develop f in a Fourier series. This usually causes great problems. In fact, it is often possible to compute only a few Fourier coefficients for f. However, on the other hand, such coefficients often decrease, as $|\boldsymbol{\nu}| \to \infty$, very quickly. Then, if f is written as

$$f = f^{[\leqslant N]} + f^{[>N]}, \tag{5.10.43}$$

where, for f given by Eq. (5.10.31), we set [see Eq. (5.10.38)]

$$f^{[\leqslant N]}(\mathbf{A}, \boldsymbol{\varphi}) = \sum_{\mathbf{a} \in \mathbf{Z}_+^l} \sum_{\substack{\boldsymbol{\nu} \in \mathbf{Z}^l \\ |\boldsymbol{\nu}| \leqslant N}} f_{\boldsymbol{\nu}}^{(\mathbf{a})}(\mathbf{A} - \mathbf{A}_0)^{\mathbf{a}} e^{i\boldsymbol{\nu} \cdot \boldsymbol{\varphi}} \equiv \sum_{\substack{\boldsymbol{\nu} \in \mathbf{Z}^l \\ |\boldsymbol{\nu}| \leqslant N}} f_{\boldsymbol{\nu}}(\mathbf{A}) e^{i\boldsymbol{\nu} \cdot \boldsymbol{\varphi}}, \tag{5.10.44}$$

one has that $\epsilon f^{[>N]}$ is very small even for N not too large and its contribution to the Hamiltonian equation produces an error, in a fixed given time, much smaller than $O(\epsilon)$, say $O(\epsilon\eta)$ with $\eta \ll 1$.

It is then possible to apply Proposition 16 to the system with Hamiltonian $h + \epsilon f^{[\leqslant N]}$ and remove the perturbation to $O(\epsilon^2)$. In the new variables, neglecting the perturbation of $O(\epsilon^2)$ will cause an error, over a fixed time, of order $O(\epsilon^2 + \epsilon\eta)$ on the original equations' solutions. This is often a very good approximation if $\boldsymbol{\omega}(\mathbf{A}) \cdot \boldsymbol{\nu} \neq 0$, $\forall 0 < |\boldsymbol{\nu}| < N$, $\forall \mathbf{A} \in \{$set of interesting initial actions$\}$.

(3) A special case of great importance to which, however, the above algorithm cannot be applied directly is that of the perturbations of the motion of the Kepler system when, in defining the unperturbed system, one neglects the reciprocal attractions between the planets [i.e., one takes Eq. (5.9.2) as a perturbation of Eq. (5.9.1)].

As we saw, the Kepler motions are rigorously periodic, and to every planet a single pulsation is associated rather than three: the other two vanish (or are integer multiples of the first, depending on which variables are chosen to integrate the motion) as a consequence of the conservation of angular momentum and of the wonderful nature of the Newtonian force which singles it out among the central forces as the most impressive, see §4.9 and §4.10.

It is therefore certainly impossible to satisfy Eq. (5.10.10) with reasonable N. Hence, the above approximation scheme cannot be applied.

Nevertheless, a very similar scheme can be applied.

Consider the planets' motions in action-angle coordinates $(\mathbf{A}, \boldsymbol{\varphi})$, where $\mathbf{A} = (\mathbf{A}^{(1)}, \ldots, \mathbf{A}^{(n)})$, $\boldsymbol{\varphi} = (\boldsymbol{\varphi}^{(1)}, \ldots, \boldsymbol{\varphi}^{(n)})$ and $(\mathbf{A}^{(i)}, \boldsymbol{\varphi}^{(i)})$ are the natural variables for the systems Sun–ith planet, in terms of which the Hamiltonian takes the form (if M_s = Sun's mass), m_i = ith planet's mass, $\epsilon_{ij} = \sqrt{m_i m_j}\,/M_s$, see p. 464:

$$h_0(\mathbf{A}) - \sum_{i<j}^{1,n} \frac{\epsilon m_i m_j}{|x_i - x_j|} \equiv h_0(\mathbf{A}) - \sum_{i<j} \epsilon_{ij} \frac{K\sqrt{m_i m_j}}{|x_i - x_j|}, \tag{5.10.45}$$

$$h_0(\mathbf{A}) = \sum_{i=1}^{n} \bar{h}_0(A_1^{(i)}), \tag{5.10.46}$$

having denoted $A_1^{(j)}$ the first component of $\mathbf{A}^{(j)} = (A_1^{(j)}, A_2^{(j)}, A_3^{(j)})$, and we recall that $\mathbf{A}^{(j)}$ can be chosen as follows (see problems for §4.10):

$$
\begin{aligned}
A_1^{(j)} &= m_j \omega_j a_j^2 = \frac{(\epsilon M_s)^{3/2} m_j}{(-2E_j)^{1/2}} \equiv L_j, \\
A_2^{(j)} &= m_j A(j) \equiv G_j, \qquad\qquad (5.10.47) \\
A_3^{(j)} &= m_j A(j) \cos i(j) \equiv \Theta_j,
\end{aligned}
$$

where $A(j)/2$ is the areal velocity of the jth planet, E_j its energy, a_j is the major semiaxis of its orbit, and $i(j)$ is the inclination of the jth orbit on the ecliptic plane (the ecliptic plane is traditionally the plane of the Earth's orbit or more precisely a reference plane fixed with the stars).

The angle variables associated with such action variables are $\varphi_1^{(j)} \equiv l^{(j)}$, $\varphi_2^{(j)} \equiv g^{(j)}$, $\varphi_3^{(j)} \equiv h^{(j)}$ known in astronomy as the "average anomaly", the "major semiaxis longitude" and the "node-line longitude" with respect to the fixed axes established on the ecliptic plane (i.e., on the xy plane of the chosen inertial frame); see Problem 16 to §4.10, p. 305, for a discussion of these variables.

Equation (5.10.46) shows that in the unperturbed motions, $g^{(j)}, h^{(j)}$ are constants (i.e., $\omega_2^{(j)} = \omega_3^{(j)} = 0$) since h_0 only depends on the variables L_j, $j = 1, \ldots, n$.

One can then proceed to write the perturbation in Eq. (5.10.45) in terms of the $(\mathbf{A}, \varphi)$ coordinates (a nontrivial task, in practice; see Problem 15, p. 305, §4.10 for the similar question in the case of the planar problem), and afterwards one can try to apply the scheme seen in the proof of Proposition 16 to build a canonical map $(\mathbf{A}, \varphi) \to (\mathbf{A}', \varphi')$ transforming Eq. (5.10.45) into a function independent on the $\varphi_1^{(j)}$ variables, $j = 1, \ldots, n$, to first order in $\bar{\epsilon} = \max_{ij} \epsilon_{ij}$. One shall proceed as prescribed in the proof of Proposition 16, considering $\varphi_2^{(j)}, \varphi_3^{(j)}$ as parameters.

If we call $f(\mathbf{A}, \varphi)$ the perturbation term of Eq. (5.10.45), when expressed in the action-angle variables apt to describe the unperturbed system, we introduce the new canonical variables $(\mathbf{A}', \varphi')$ via the generating function $\mathbf{A}' \cdot \varphi + \Phi(\mathbf{A}', \varphi)$ with

$$
\Phi(\mathbf{A}', \varphi) = \sum_{\substack{\nu \in \mathbf{Z}^n \\ 0 < |\nu| < N}} \frac{f_\nu(\mathbf{A}') e^{i\nu \cdot \varphi}}{-i \sum_{j=1}^n \nu_1^j \omega_j (A_1'^{(1)})}, \tag{5.10.48}
$$

where N is a "large number" which we imagine here to have chosen such that for some a priori given purposes, neglecting $f^{[>N]}$ in Eq. (5.10.45), produces a negligible error.

In this way we obtain a Hamiltonian having the form

$$
\begin{aligned}
&h(A_1'^{(1)}, \ldots, \varphi_1'^{(1)}, \ldots) \\
&\quad = h_0(\mathbf{A}') + \epsilon h_1(A_1'^{(1)}, \ldots, A_3'^{(n)}; \varphi_2'^{(2)}, \ldots, \varphi_n'^{(3)}) + O(\epsilon^2), \qquad (5.10.49)
\end{aligned}
$$

and the equations of motion will become

$$\dot{A}_1^{(j)} = 0, \qquad j = 1, \ldots, n,$$

$$\begin{cases} \dot{\varphi}_\sigma^{(j)} = \epsilon \dfrac{\partial h_1}{\partial A_\sigma^{(j)}} (A_1'^{(1)}, \ldots, A_3'^{(n)}; \varphi_2'^{(2)}, \ldots, \varphi_n'^{(3)}), \\ \qquad\qquad\qquad\qquad\qquad j = 1, \ldots, n; \quad \sigma = 2, 3, \\ \dot{A}_\sigma^{(j)} = -\epsilon \dfrac{\partial h_1}{\partial |_\sigma'^{(j)}} (A_1'^{(1)}, \ldots, A_3'^{(n)}; \varphi_2'^{(2)}, \ldots, \varphi_n'^{(3)}), \end{cases}$$

$$j = 1, \ldots, n; \quad \sigma = 2, 3, \tag{5.10.50}$$

$$\dot{\varphi}_1^{(j)} = \frac{\partial h_0}{\partial A_1'^{(j)}} (A_1'^{(1)}, \ldots, A_1'^{(n)}) + \epsilon \frac{\partial h_1}{\partial A_1'^{(j)}} (A_1'^{(1)}, \ldots, A_3'^{(n)}; \varphi_2'^{(2)}, \ldots, \varphi_n'^{(3)}), \qquad j = 1, \ldots, n,$$

up to $O(\epsilon^2)$.

Since h_1 is $\varphi_1'^{(j)}$ independent, the equations in curly brackets form a system of Hamiltonian equations parameterized by the initial data of $A_1'^{(j)}$ and with $2n$ degrees of freedom. Once they are "solved", the last of Eqs. (5.10.50) is an ordinary differential equation expressing $\dot{\varphi}_1'^{(j)}$ in terms of a known function of t and, therefore, it is "trivial".

In celestial mechanics, it sometimes happens that inside the neighborhood $W = V \times \mathbf{T}^l$, of interesting initial data, h_1 itself can be written as $\bar{h}_1 + \mu \tilde{h}_1$, where $\bar{h}_1$ is an integrable Hamiltonian and μ is a "small" parameter.

It will then be possible to apply again perturbation theory, Proposition 16, to study the motion of the Hamiltonian system in Eqs. (5.10.50), described by the equations in curly brackets, as a perturbation of a simple motion, (see problem 2 at the end of this section where a similar but simpler situation occurs).

A very interesting case when this happens is the case when the unperturbed motion of the planets that one considers is a motion in which the planets wander around orbits with small eccentricity and small inclination. The resulting parameter μ is of an order of magnitude related to the maximum eccentricity and to the maximum inclination.

(4) The representation of the planet's motion thus obtained is very suggestive: the planet keeps moving with roughly the same revolution period on the same elliptic orbit [in Eqs. (5.10.50), the first and the last equations say that the average anomalies rotate with about the same unperturbed pulsations up to $O(\epsilon)$]; but the node lines and the major semiaxis longitude have a movement developing on a very slow time scale of $O(\epsilon^{-1})$ [because of the factor ϵ in the curly bracket equations in Eqs. (5.10.50)] called "precession" which is quasi-periodic with the periods

characteristic of the Hamiltonian $\bar{h}_1$.[25] In the same quasi-periodic way vary the inclinations of the orbits and the areal velocities. The main motion obtained by neglecting $O(\epsilon)$ in Eqs. (5.10.50) should be called a "deferent motion", while the $O(\epsilon)$ corrections expressed by the curly-bracket terms in Eqs. (5.10.50) should be called the "epicyclical" motions, to do some justice to the Greek astronomers and to Ptolemy, in particular.

The above "Ptolemaic" description is accurate only to $O(\epsilon^2 + \epsilon\mu)T$ if T is the time for which one wishes to make astronomical predictions.

The reader should consult books on celestial mechanics to see concrete applications of the procedures and approximation schemes to some astronomical problems (among which the simplest is the theoretical calculation of Mercury's perihelion precession).

Exercises and Problems for §5.10

1. Apply the idea of the proof of Proposition 17 to study the Hamiltonian system

$$\omega_0 \cdot \mathbf{A} + \epsilon g(\boldsymbol{\varphi}), \qquad (\mathbf{A}, \boldsymbol{\varphi}) \in \mathbf{R}^l \times \mathbf{T}^l$$

with ω_0 verifying Eq. (5.10.7). Deduce that the system is integrable for small ϵ (for an alternative solution to this problem, see Problem 1, p. 290, §4.8).

2. Apply the scheme suggested in Observation (3), p. 479, to discuss to higher order the motion associated with the system in $\mathbf{R}^{l+1} \times \mathbf{T}^{l+1}$:

$$A + \epsilon(\mathbf{B} \cdot \omega_0 + \mu g(A, \mathbf{B}, \varphi, \boldsymbol{\psi}))$$

if $(A, \mathbf{B}; \varphi, \boldsymbol{\psi})$ are canonical action-angle variables $A \in \mathbf{R}$, $\mathbf{B} \in \mathbf{R}^l$, $\varphi \in \mathbf{T}^1$, $\boldsymbol{\psi} \in \mathbf{T}^l$ (note that for $\epsilon = 0$, this system has only 1 frequency rather than $l + 1$). Explicitly calculate the "daily" and "secular" components of the motions to $O(\epsilon^2 + \epsilon\mu)$ after finding the secular Hamiltonian $\bar{h}_1$, see Observation (3), p. 479, and assuming a "nonresonance" condition on ω_0 like Eq. (5.10.7).

3. Same as Problem 2 for the system in $\mathbf{R}^2$:

$$\tfrac{1}{2}(p_1^2 + q_1^2) + \tfrac{1}{2}(p_2^2 + q_2^2) + \epsilon(2q_1 + q_2)^4.$$

(Hint: Find the action-angle variables $A_1, A_2, \varphi_1, \varphi_2$ when $\epsilon = 0$ (just polar coordinates) for the two oscillators; then completely canonically change variables $A = (A_1 + A_2)/2$, $B = (A_1 - A_2)/2$, $\varphi = \varphi_1 - \varphi_2$, $\psi = \varphi_1 + \varphi_2$, and then apply the method of Observation (3), p. 479.)

4. Same as Problem 2 for the system in $\mathbf{R}^2$:

$$\frac{p_1^2}{2} + \frac{p_2^2}{2} - \frac{1}{\sqrt{q_1^2 + q_2^2}} + \epsilon(q_1 - q_2).$$

5. Consider the "restricted three-body problem" in $\mathbf{R}^2$:

$$H(\mathbf{p}_1, \mathbf{p}_2, \mathbf{q}_1, \mathbf{q}_2) = \frac{\mathbf{p}_1^2}{2m_1} + \frac{\mathbf{p}_2^2}{2m_2} - \frac{km_1}{|\mathbf{q}_1|} - \frac{km_2}{|\mathbf{q}_2|} - \epsilon\frac{m_1 m_2}{|\mathbf{q}_1 - \mathbf{q}_2|},$$

[25] It is called a "secular motion" since in some simple cases this time scale is of the order of centuries.

$\mathbf{p}_i, \mathbf{q}_i \in \mathbf{R}^2$, $i = 1, 2$. Using the results of Problem 15, p. 305, §4.10, write (with patience) up to second order in the eccentricities of the two bodies the Hamiltonian in the Kepler coordinates (i.e., in the action-angle variables corresponding to $\epsilon = 0$); see Problem 11, p. 303, §4.10.

Show that if the eccentricities are neglected, together with quantities of order $O(\epsilon^2)$, the secular motion [in the language of Observation (3), p. 479] is described by the Hamiltonian

$$h_0 + \epsilon \bar{h}_1 = -\frac{m_1^3 k^2}{2L_1'^2} - \frac{m_2^3 k^2}{2L_2'^2} - \epsilon m_1 m_2 \int_0^{2\pi} \frac{d\alpha}{2\pi} \left(a_1'^2 + a_2'^2 - 2a_1' a_2' \cos\alpha\right)^{-1/2}$$

(where $L = m\sqrt{k}\, a$, $a =$ major semiaxis; see Problem 11, p. 303, §4.10) ("0th order in the eccentricity").

6. Show that in the context of Problem 5, the secular Hamiltonian $\bar{h}_1$ of the Hamiltonian in Problem 5 is eccentricity independent even to first order in the eccentricity.

Does this mean that, to first order in the eccentricities, the Kepler ellipses remain fixed in space? (Answer: no.) Show that they move quasi-periodically "without full precession" (i.e., g_1', g_2' vary continuously with a small amplitude of oscillation, i.e., $< 2\pi$) to first order in the eccentricities.

7. Show that to second order in the eccentricities, the secular Hamiltonian of Problems 5 and 6 depends both on the L's and on the e's (i.e., on the G's) and has the form (without explicitly computing f_{ij}) $h_1 = \bar{h}_1^{(0)}(L_1', L_2') + e_1'^2 f_{11}(L_1', L_2', g_2' - g_1') + 2e_1' e_2' f_{12}(L_1', L_2', g_2' - g_1') + e_2'^2 f_{22}(L_1', L_2', g_2' - g_1')$. Show that the above secular Hamiltonian is integrable and that it says that, if f_{ij} are nontrivial, the relative position of the perihelions precesses to $O(e^2)$. (Hint: Use Problem 15, p. 305, §4.10. Canonically change variables as

$$\gamma = g_1 + g_2, \qquad G = \frac{G_1 + G_2}{2},$$

$$\tilde{\gamma} = g_2 - g_1, \qquad \tilde{G} = \frac{G_2 - G_1}{2}$$

and note that the Hamiltonian "effectively" takes the form of a Hamiltonian for a one-dimensional system (integrable by quadratures or by the Hamilton-Jacobi method).)

8. Attempt a concrete computation of f_{ij} and of the angular velocity of the precession, assuming that the unperturbed motions take place on ellipses of small eccentricity and with semiaxes a_1, a_2 such that $a_2 - a_1$ is "of the order" of a_1 and a_2 (i.e., with quite different semiaxes).

9. (i) Let $\Gamma(L) \subset \mathbf{R}^l$ be a cube centered at the origin and with side $2L$. Let $\boldsymbol{\nu} \in \mathbf{Z}^l$, $|\boldsymbol{\nu}| > 0$ and let $\Gamma_\epsilon(L)$ be the set of the points $\boldsymbol{\omega} \in \Gamma(L)$ such that $|\boldsymbol{\omega} \cdot \boldsymbol{\nu}/|\boldsymbol{\nu}|| < \epsilon$. Show that the measure of $\Gamma_\epsilon(L)$ does not exceed $2\epsilon\sqrt{l}\,(2L\sqrt{l}\,)^{l-1}$. (Hint: Just look at the geometrical meaning of the inequality $|\boldsymbol{\omega} \cdot \boldsymbol{\nu}/|\boldsymbol{\nu}|| < \epsilon$, the $\sqrt{l}$ arises from $|\boldsymbol{\nu}| = \sum_{i=1}^l |\nu_i| \leqslant \sqrt{l}\,(\sum_{i=1}^l |\nu_i|^2)^{1/2}$.)

(ii) Deduce that the measure of the set Γ_C of the points $\boldsymbol{\omega} \in \Gamma(L)$ such that $|\boldsymbol{\omega} \cdot \boldsymbol{\nu}|^{-1} \leqslant C|\boldsymbol{\nu}|^l$, $\forall |\boldsymbol{\nu}| > 0$, has a complement with Lebesgue measure not exceeding

$$2C^{-1}\left(2L\sqrt{l}\,\right)^{l-1}\sqrt{l} \sum_{|\boldsymbol{\nu}|>0} |\boldsymbol{\nu}|^{-l-1}$$

(see, also, Problem 11).

10. Using Problem 9 show that $\bigcup_C \Gamma_C = \tilde{\Gamma} \subset \Gamma(L)$ has the same Lebesgue measure of $\Gamma(L)$, i.e., $(2L)^l$, although its complement is dense.

11. Without using the Lebesgue-measure theory, infer from the inequalities of Problem 9 above that $\tilde{\Gamma}$, in Problem 10, is a dense set in $\Gamma(L)$.

12. Consider a time-dependent Hamiltonian with one degree of freedom: $h_0(A) + \epsilon f_0(A, \varphi, t)$, where $(A, \varphi) \in \mathbf{R}^1 \times \mathbf{T}^1$ and $t \in \mathbf{T}^1$ is interpreted as the time appearing in a 2π-periodic time-dependent perturbation to the system with Hamiltonian h_0.

Develop a formal perturbation theory for the above system proving propositions analogous to Propositions 16 and 17 of this section. (Hint: Use a time-dependent canonical transformation with generating function $A'\varphi + \Phi(A', \varphi, t)$ and proceed, as in this section, using the Hamilton–Jacobi method.)

13. Consider the time-dependent system on $\mathbf{R}^1 \times \mathbf{T}^1$,

$$\frac{A^2}{2} + \epsilon(\cos\varphi + \cos(\varphi - t)),$$

and applying the results of Problem 12, remove the perturbation to $O(\epsilon^2)$ near the points with $A = \omega_0 = (1 + \sqrt{5})/2$ (see exercises and problems to §2.20 for the theory of the number ω_0).

14. Same as in Problem 13, but to $O(\epsilon^4)$. (Warning: The calculations are quite long.)

15. Let $h(A)$ be a C^∞ function defined on a sphere $S_\rho(\mathbf{A}_0) \subset \mathbf{R}^l$ with gradient $\omega(\mathbf{A}) = (\partial h/\partial \mathbf{A})(\mathbf{A})$ bounded by $|\omega(\mathbf{A})| < E$ and such that the matrix $M_{ij} = \partial^2 h/\partial A_i \partial A_j$ is invertible for all $\mathbf{A} \in S_\rho(\mathbf{A}_0)$ and

$$\sum_{i,j=1}^{l} |(M^{-1})_{ij}| \leqslant \eta < +\infty.$$

Suppose that the correspondence $\mathbf{A} \to \omega(\mathbf{A})$ is one to one between $S_\rho(A_0)$ and $\omega(S_\rho(\mathbf{A}_0))$. Denote, for $C > 0$:

$$S_\rho(\mathbf{A}_0, C) = \{\mathbf{A} \mid \mathbf{A} \in S_\rho(\mathbf{A}_0),\ |\omega(\mathbf{A}) \cdot \boldsymbol{\nu}|^{-1} < C|\boldsymbol{\nu}|^l,\ \forall \boldsymbol{\nu} \neq 0\}.$$

Show that there is $B > 0$, depending only on l, such that

$$1 \geqslant \frac{\operatorname{vol} S_\rho(\mathbf{A}_0, C)}{\operatorname{vol} S_\rho(\mathbf{A}_0)} \geqslant \left(1 - B\frac{(E\eta\rho^{-1})^l}{EC}\right).$$

(Hint: Use the change of variable formula:

$$\int_{S_\rho(\mathbf{A}_0, C)} d\mathbf{A} \equiv \int_{S_\rho(\mathbf{A}_0)} d\mathbf{A} - \int_{S_\rho(\mathbf{A}_0)/S_\rho(\mathbf{A}_0, C)} d\mathbf{A}$$

$$= \operatorname{vol} S_\rho(\mathbf{A}_0) - \int_{\omega(S_\rho(\mathbf{A}_0)/S_\rho(\mathbf{A}_0, C))} \left|\det \frac{\partial \mathbf{A}}{\partial \omega}\right| d\omega$$

$$\geqslant \operatorname{vol} S_\rho(\mathbf{A}_0) - \eta^l \int_{\omega(S_\rho(\mathbf{A}_0)/S_\rho(\mathbf{A}_0, C))} d\omega$$

$$\geqslant \operatorname{vol} S_\rho(\mathbf{A}_0) - \eta^l \sum_{\boldsymbol{\nu} \neq \mathbf{0}} \int_{\substack{|\omega| < E \\ |\omega \cdot \boldsymbol{\nu}/|\boldsymbol{\nu}|| < C^{-1}|\boldsymbol{\nu}|^{-l-1}}} d\omega \geqslant \qquad \text{(by Problem 9)}$$

$$\geqslant \operatorname{vol} S_\rho(\mathbf{A}_0) - \eta^l C^{-1} (2E\sqrt{l})^{l-1} \sqrt{l} \sum_{|\boldsymbol{\nu}| > 0} |\boldsymbol{\nu}|^{-l-1}$$

and then recall that $\operatorname{vol} S_\rho(\mathbf{A}_0) = \operatorname{const} \rho^l$.)

16. Let $\omega_0 = (\omega, 1) \in \mathbf{R}^2$ be such that $|\omega\nu_1 + \nu_2|^{-1} < C(|\nu_1| + |\nu_2|)^\alpha$ for some α, $C > 0$. Let f be a function on $\mathbf{T}^1$ with Fourier coefficients $f_\nu \neq 0$, $\forall \nu \neq 0$, e.g.,

$$f(\varphi) = 2 \sum_{n=1}^{\infty} e^{-\xi n} \cos n\varphi, \qquad \xi > 0.$$

Consider the Hamiltonian system on $\mathbf{R}^2 \times \mathbf{T}^2$

$$H_\epsilon = (\omega A_1 + A_2) + \epsilon(A_2 + f(\varphi_1) f(\varphi_2)).$$

Show that the Birkhoff formal series (5.10.42) are

$$h_\epsilon(\mathbf{A}') = (\omega A_1 + A_2) + \epsilon A_2,$$

$$\Phi_\epsilon(\mathbf{A}', \varphi) = \sum_{k=1}^{\infty} \epsilon^k \left(\sum_{\nu \in \mathbf{Z}^2} \frac{e^{-\xi|\nu|} e^{i\nu\cdot\varphi}}{-i(\omega\nu_1 + \nu_2)} \left(\frac{-\nu_2}{\omega\nu_1 + \nu_2} \right)^{k-1} \right),$$

and prove that the series for Φ_ϵ does not converge. (Hint: Using the explicit solubility of the equations for H_ϵ, see Problem 1, §4.8, p. 290, one sees that the passage to action-angle variables for H_ϵ must be singular for a dense set of values of ϵ: the singularities arise in correspondence of the values of ϵ for which the formal sum of the Φ_ϵ-series makes no sense (to sum formally the series permute them).)

17. In the context of Problem 16, show that the function Φ_ϵ, obtained by permuting $\sum_k$ and $\sum_\nu$ and summing the geometric series, makes sense and is analytic in $\mathbf{A}, \varphi$ for many values of ϵ and, whenever this happens, H_ϵ is indeed integrable by the canonical map generated by Φ_ϵ. [Hint: Use Problem 9 above to identify the values of ϵ which allow bounds of the type $|\omega\nu_1 + (1+\epsilon)\nu_2|^{-1} < C|\nu|^\alpha$, $C, \alpha > 0$.]

§5.11. Some Simple Properties of the Holomorphic Functions. Analytic Theorems for the Implicit Functions

In §5.10, we mentioned, without discussion, some properties of the analytic functions. Such properties can be derived in the more general context of the theory of holomorphic functions.

Such functions are basically defined as analytic functions of complex variables, i.e., a $\mathbf{C}^p$-valued function f defined on an open subset $W \subset \mathbf{C}^l$ is holomorphic if it can be developed in an absolutely convergent power series around each point of W.

For a more detailed discussion of some perturbation theory problems, it is convenient to state the following definition which is general enough for our purposes. It is a definition that is provided more with the aim of fixing some notations rather than with the objective of developing the part of the holomorphic functions theory that we need. In this and in the following sections, we suppose that the reader is familiar with the basic properties of holomorphic functions, i.e., the Cauchy integral formula, the theory of the Taylor–Laurent expansions in power series, and the identity principle. Such properties will be repeatedly used in §5.11 and §5.12.

Let l, p, q be positive integers.

8 Definition. (i) We introduce the following notations: $\forall \mathbf{a} = (a_1, \ldots, a_l) \in \mathbf{Z}_+^l$, $\forall \boldsymbol{\nu} = (\nu_1, \ldots, \nu_l) \in \mathbf{Z}^l$,

$$|\mathbf{a}| = \sum_{i=1}^{l} a_i, \qquad |\boldsymbol{\nu}| = \sum_{i=1}^{p} |\nu_i|, \tag{5.11.1}$$

while if $\mathbf{w} \in \mathbf{C}^q$, $\mathbf{w} = (w_1, \ldots, w_q)$,

$$|\mathbf{w}| = \max_{1 \leqslant i \leqslant q} |w_i|, \qquad \|\mathbf{w}\| = \sum_{i=1}^{q} |w_i|. \tag{5.11.2}$$

(ii) For $\mathbf{A}_0 \in \mathbf{C}^l$, $\rho > 0$, $\xi > 0$, we set

$$\begin{aligned} \hat{\mathbb{S}}_\rho(\mathbf{A}_0) &= \{\mathbf{A} \mid \mathbf{A} \in \mathbf{C}^l, |\mathbf{A} - \mathbf{A}_0| < \rho\} \\ C(\xi) &= \{\mathbf{z} \mid \mathbf{z} \in \mathbf{C}^p, e^{-\xi} < |z_j| < e^{\xi}, \forall j = 1, \ldots, p\}, \\ C(\rho, \xi; \mathbf{A}_0) &= \hat{\mathbb{S}}_\rho(\mathbf{A}_0) \times C(\xi). \end{aligned} \tag{5.11.3}$$

The first two such sets will be called, respectively, the "complex multisphere", with center $\mathbf{A}_0$ and radius ρ, and the "complex multiannulus", with inner radius $e^{-\xi}$ and outer radius e^{ξ}. If $\mathbf{A}_0$ is real, we define

$$\mathbb{S}_\rho(\mathbf{A}_0) = \{\mathbf{A} \mid \mathbf{A} \in \mathbf{R}^l, |A_i - A_{0i}| < \rho, i = 1, \ldots, l\}, \tag{5.11.4}$$

calling it the "real multisphere" with center $\mathbf{A}_0$ and radius ρ.

The set $\mathbb{S}_\rho(\mathbf{A}_0) \times \mathbf{T}^p$ will be identified to a subset of $C(\rho, \xi; \mathbf{A}_0)$ via the map

$$(\mathbf{A}, \boldsymbol{\varphi}) \to (\mathbf{A}, \mathbf{z}), \qquad z_j = e^{i\varphi_j}, \qquad j = 1, \ldots, l. \tag{5.11.5}$$

(iii) If $W \subset \mathbf{C}^q$ is open and if F is a $\mathbf{C}^p$-valued function, we say that F has a convergent power series expansion around $\mathbf{w}_0 \in W$ if there is a family of $\mathbf{C}^p$-vectors $\{F^{(\mathbf{a})}(\mathbf{w}_0)\}_{\mathbf{a} \in \mathbf{Z}_+^q}$ such that for some $\tilde{\rho} > 0$:

$$F(\mathbf{w}) = \sum_{\mathbf{a} \in \mathbf{Z}_+^q} F^{(\mathbf{a})}(\mathbf{w}_0)(\mathbf{w} - \mathbf{w}_0)^{\mathbf{a}}, \qquad \forall |\mathbf{w} - \mathbf{w}_0| < \tilde{\rho}, \tag{5.11.6}$$

having set

$$(\mathbf{w} - \mathbf{w}_0)^{\mathbf{a}} = \prod_{j=1}^{q} (w_j - w_{0j})^{a_j} \tag{5.11.7}$$

for $\mathbf{a} = (a_1, \ldots, a_q)$, and

$$\sum_{\mathbf{a} \in \mathbf{Z}_+^q} |F^{(\mathbf{a})}(\mathbf{w}_0)| \rho^{|\mathbf{a}|} < +\infty, \qquad \forall \rho < \tilde{\rho}. \tag{5.11.8}$$

(iv) A function F is holomorphic in the open subset $W \subset \mathbf{C}^q$ if it has a convergent power series around every point $\mathbf{w} \in W$. In this case, one defines the derivatives of F as

$$\frac{\partial^{|\mathbf{a}|} F}{\partial \mathbf{w}^{\mathbf{a}}}(\mathbf{w}_0) = \mathbf{a}!\, F^{(\mathbf{a})}(\mathbf{w}_0), \qquad \forall \mathbf{w}_0 \in W \tag{5.11.9}$$

if $\mathbf{a}! = a_1! \ldots a_q!$.

Observation. One calls Eq. (5.11.6) the Taylor series of F at $\mathbf{w}_0$, because of Eq. (5.11.9).

9 Definition. Let l, p, q be positive integers and $\mathbf{A}_0 \in \mathbf{R}^l$ and use the notations of Definition 8 above.

(i) Let f, g, h be three functions defined, respectively, on $\mathbb{S}_\rho(\mathbf{A}_0)$, $\mathbb{S}_\rho(\mathbf{A}_0) \times \mathbf{T}^p$, $\mathbf{T}^p$ with values in $\mathbf{R}^q$. We shall say that they are holomorphic in $\hat{\mathbb{S}}_\rho(\mathbf{A}_0)$, $C(\rho, \xi; \mathbf{A}_0)$, $C(\xi)$ respectively if, identifying $\mathbb{S}_\rho(\mathbf{A}_0)$, $\mathbb{S}_\rho(\mathbf{A}_0) \times \mathbf{T}^p$, $\mathbf{T}^p$ as subsets of $\hat{\mathbb{S}}_\rho(\mathbf{A}_0)$, $C(\rho, \xi; \mathbf{A}_0)$, $C(\xi)$, respectively, as explained in Definition 8 (ii) above, they can be extended to holomorphic functions $\bar{f}, \bar{g}, \bar{h}$ on the larger sets $\hat{\mathbb{S}}_\rho(\mathbf{A}_0)$, $C(\rho, \xi; \mathbf{A}_0)$, $C(\xi)$.

The functions of the type $\bar{f}, \bar{g}, \bar{h}$ will be called "holomorphic in $\hat{\mathbb{S}}_\rho(\mathbf{A}_0)$, $C(\rho, \xi; \mathbf{A}_0)$, $C(\xi)$", respectively, and "real on $\mathbb{S}_\rho(\mathbf{A}_0)$, $\mathbb{S}_\rho(\mathbf{A}_0) \times \mathbf{T}^p$, $\mathbf{T}^p$", respectively.

Sometimes the extensions $\bar{f}, \bar{g}, \bar{h}$ will still be called f, g, h, dropping the bar.

(ii) If F is holomorphic on $C(\rho, \xi; \mathbf{A}_0)$ or on $C(\xi)$, we define its "$\boldsymbol{\varphi}$ derivatives" by setting $\partial/\partial\varphi_k = iz_k \partial/\partial z_k$, $k = 1, \ldots, p$.

Observations.

(1) It is easy to deduce from the definition of an analytic function on $V \times \mathbf{T}^l$ (see Definitions 13 and 14, p. 333 and p. 334, §4.13) that if f, g, and h are analytic on V or on $V \times \mathbf{T}^p$, $\mathbf{T}^p$, respectively, then given $\mathbf{A}_0 \in V$, there exist $\rho > 0$, $\xi > 0$ such that f is holomorphic in $\hat{\mathbb{S}}_\rho(\mathbf{A}_0)$, g in $C(\rho, \xi; \mathbf{A}_0)$, and h in $C(\xi)$. In general, however, ρ and ξ may be very small even if V is large.

(2) This definition is particularly useful because it provides a simple description of an important class of functions on $\mathbf{T}^p$ or on $\mathbb{S}_\rho(\mathbf{A}_0) \times \mathbf{T}^p$, thinking of $\mathbf{T}^p$ as a subset of $C(\xi)$ via the natural correspondence

$$\boldsymbol{\varphi} = (\varphi_1, \ldots, \varphi_p) \in \mathbf{T}^p \leftrightarrow \mathbf{z} = (e^{i\varphi_1}, \ldots, e^{i\varphi_p}) \tag{5.11.10}$$

already pointed out several times.

The classical theorems on the theory of the holomorphic functions (Taylor and Laurent expansions, Cauchy's formula, identity principle, etc.) imply the following proposition which we do not prove since it can be found, with other symbols, in any elementary textbook on holomorphic functions.

18 Proposition. *Let f, g, h be holomorphic functions on $\hat{\mathbb{S}}_\rho(\mathbf{A}_0)$, $C(\rho, \xi; \mathbf{A}_0)$, $C(\xi)$, respectively, see Eq.* (5.11.4), *with values in $\mathbf{C}^q$.*

Using the notation of Definitions 8 *and* 9 *and setting, for*

$$\boldsymbol{\nu} = (\nu_1, \ldots, \nu_p) \in \mathbf{Z}^p, \qquad \mathbf{z} \in C(\xi)$$
$$\mathbf{z}^{\boldsymbol{\nu}} = \prod_{j=1}^{p} z_j^{\nu_j}: \tag{5.11.11}$$

(i) *There exist sequences of vectors in* $\mathbf{C}^q$

$$\{f^{(\mathbf{a})}\}_{\mathbf{a}\in\mathbf{Z}^l_+}, \qquad \{g_\nu^{(\mathbf{a})}\}_{\mathbf{a}\in\mathbf{Z}^l_+,\nu\in\mathbf{Z}^p}, \qquad \{h_\nu\}_{\nu\in\mathbf{Z}^p}$$

such that

$$\begin{aligned}\bar{f}(\mathbf{A}) &= \sum_{\mathbf{a}\in\mathbf{Z}^l_+} f^{(\mathbf{a})}(\mathbf{A}-\mathbf{A}_0)^{\mathbf{a}},\\ \bar{g}(\mathbf{A},z) &= \sum_{\mathbf{a}\in\mathbf{Z}^l_+}\sum_{\nu\in\mathbf{Z}^p} g_\nu^{(\mathbf{a})}(\mathbf{A}-\mathbf{A}_0)^{\mathbf{a}}z^{\nu},\\ h(\mathbf{z}) &= \sum_{\nu\in\mathbf{Z}^p} h_\nu \mathbf{z}^{\nu}.\end{aligned} \tag{5.11.12}$$

(ii) *Denoting* $g(\mathbf{A},\boldsymbol{\varphi})\equiv g(\mathbf{A},\mathbf{z})$, $h(\boldsymbol{\varphi})\equiv h(\mathbf{z})$ *if* $\mathbf{z}=(e^{i\varphi_1},\dots,e^{i\varphi_p})$, $\boldsymbol{\varphi}\in\mathbf{T}^p$, then

$$\begin{aligned}f^{(\mathbf{a})} &= \frac{1}{\mathbf{a}!}\frac{\partial^{|\mathbf{a}|}f}{\partial\mathbf{A}^{\mathbf{a}}}(\mathbf{A}_0),\\ g_\nu^{(\mathbf{a})} &= \frac{1}{\mathbf{a}!}\int_{\mathbf{T}^p}\frac{\partial^{|\mathbf{a}|}g}{\partial\mathbf{A}^{\mathbf{a}}}(\mathbf{A}_0,\boldsymbol{\varphi})e^{-i\nu\cdot\boldsymbol{\varphi}}\frac{d\boldsymbol{\varphi}}{(2\pi)^p},\\ h_\nu &= \int_{\mathbf{T}^p} h(\boldsymbol{\varphi})e^{-i\nu\cdot\boldsymbol{\varphi}}\frac{d\boldsymbol{\varphi}}{(2\pi)^p}.\end{aligned} \tag{5.11.13}$$

(iii) *Setting*

$$\begin{aligned}|f|_\rho &= \sup|f(\mathbf{A})|,\\ |g|_{\rho,\xi} &= \sup|g(\mathbf{A},\mathbf{z})|,\\ |h|_\xi &= \sup|h(\mathbf{z})|,\end{aligned} \tag{5.11.14}$$

where the suprema are taken over the functions' respective domains of definition [*and Eq.* (5.11.2) *is used for* $|\cdot|$], *it is*

$$\begin{aligned}|f^{(\mathbf{a})}| &\le |f|_\rho\rho^{-|\mathbf{a}|},\\ |g_\nu^{(\mathbf{a})}| &\le |g|_{\rho,\xi}\rho^{-|\mathbf{a}|}e^{-\xi|\nu|},\\ |h_\nu| &\le |h|_\xi e^{-\xi|\nu|}.\end{aligned} \tag{5.11.15}$$

(iv) *If the coefficients of the series in Eq.* (5.11.12) *can be bounded by a constant times, respectively,* $\rho^{-|\mathbf{a}|}$, *or* $\rho^{-|\mathbf{a}|}e^{-\xi|\nu|}$ *or* $e^{-\xi|\nu|}$, *then the sums of the series of Eq.* (5.11.12) *define holomorphic functions on* $\hat{S}_\rho(\mathbf{A}_0)$, $C(\rho,\xi;\mathbf{A}_0)$, $C(\xi)$, *respectively.*

(v) *The second of Eqs.* (5.11.12) *can also be written*

$$g(\mathbf{A},\mathbf{z}) = \sum_{\nu\in\mathbf{Z}^p} g_\nu(\mathbf{A})\mathbf{z}^{\nu} \tag{5.11.16}$$

with $g_\nu(A)$ *holomorphic in* $\hat{S}_\rho(\mathbf{A}_0)$ *and such that its Taylor series around* $\mathbf{A}_0$ *is obtained by inspecting the second of Eqs.* (5.11.12) *and considering the sum over* $\mathbf{a}$ *only.*

(vi) *If f, g, h are real on $\mathbb{S}_\rho(\mathbf{A}_0)$, $\mathbb{S}_\rho(\mathbf{A}_0)\times\mathbf{T}^p$, $\mathbf{T}^p$ for $\mathbf{A}_0\in\mathbf{R}^l$, the $f^{(\mathbf{a})}$ coefficients are also real, while $g_\nu^{(\mathbf{a})}$ and h_ν are complex conjugates to $g_{-\nu}^{(\mathbf{a})}$, $h_{-\nu}$, respectively, and vice versa.*

Observations.

(1) Note that the convergence of the series of Eq. (5.11.12) stated in (i) follows from Eq. (5.11.15) only if $|f|_\rho$, $|g|_{\rho,\xi}$, and $|h|_\xi$ are finite. This is, however, not necessarily true in general so that (iii) and (iv) are not reciprocal statements.

(2) If F is holomorphic in a region $W\subset C^q$ and $\tilde{\mathbf{w}}\in W$, $\hat{\mathbb{S}}_\rho(\tilde{\mathbf{w}})\subset W$, and if we wish to estimate the derivatives $(\partial F/\partial w_k)(\tilde{\mathbf{w}})$, we can use Eqs. (5.11.15) and (5.11.13) as follows.

Here and below, we regard a matrix-valued function with values on the matrices $l\times q$ as a $\mathbf{C}^{lq}$-valued function,[26] and we consider F as a holomorphic function on $\hat{\mathbb{S}}_\rho(\tilde{\mathbf{w}})$. To bound the $l\times q$ matrix $\partial F/\partial\mathbf{w}$ (assuming that F is $\mathbf{C}^l$ valued), consider the first of Eqs. (5.11.13) and (5.11.15) written for $|\mathbf{a}|=1$. It gives

$$\left|\frac{\partial F}{\partial\mathbf{w}}(\tilde{\mathbf{w}})\right|\leqslant\Big(\sup_{\mathbf{w}\in\hat{\mathbb{S}}_\rho(\tilde{\mathbf{w}})}|F(\mathbf{w})|\Big)\rho^{-1}\leqslant\frac{\sup|F(\mathbf{w})|}{\rho},\tag{5.11.17}$$

where the second supremum is over W.

From this remark, it immediately follows that

$$\begin{aligned}&\left|\frac{\partial f}{\partial\mathbf{A}}\right|_{\rho'}\leqslant\frac{|f|_\rho}{\rho-\rho'},\qquad\left|\frac{\partial g}{\partial\mathbf{A}}\right|_{\rho',\xi}\leqslant\frac{|g|_{\rho,\xi}}{\rho-\rho'},\\&\left|\frac{\partial g}{\partial\mathbf{z}}\right|_{\rho,\xi'}\leqslant\frac{|g|_{\rho,\xi}}{e^{-\xi'}-e^{-\xi}}\leqslant|g|_{\rho,\xi}\frac{e^\xi}{\delta},\\&\left|\frac{\partial g}{\partial\varphi_k}\right|_{\rho,\xi'}\equiv\left|iz_k\frac{\partial g}{\partial z_k}\right|_{\rho,\xi'}\leqslant|g|_{\rho,\xi}\frac{e^{2\xi}}{\delta}\end{aligned}\tag{5.11.18}$$

for $\rho'<\rho$, $\xi'<\xi$, if $\delta=\xi-\xi'$. Analogous inequalities hold for the higher-order derivatives, e.g., $|\partial^2 f/\partial\mathbf{A}\partial\mathbf{A}|_{\rho'}\leqslant 2|f|_\rho/(\rho-\rho')^2$ [see Eq. (5.11.9)].

These simple estimates will be called "dimensional estimates". In physics, one says that a "dimensional estimate" is any estimate of the derivative of a function F at a given point in terms of the function's maximum in a region divided by the distance of the point to the region's boundary ("characteristic magnitude of F" divided by a "characteristic length"). Recall that physicists rightly believe that all functions (with, possibly, some exceptions) are analytic.

We now possess the terminology necessary to formulate the analytic implicit function theorem. This theorem is a particularly simple and strong version of the ordinary implicit function theorem valid when the defining

[26] so that $|M(\mathbf{w})|=\sup_{i,j}|M_{ij}(\mathbf{w})|$, $\|M(\mathbf{w})\|=\sum_{i,j}|M_{ij}|$

function is analytic. This theorem will play a key role in the proof of Proposition 22 which, in turn, is the heart of the proof of Proposition 15 in the analytic case.

The proof of the propositions that follow uses very elementary aspects of the theory of holomorphic functions and it will be discussed in Appendix N. Propositions 19–21 are "analytic implicit function theorems".

19 Proposition. *Let $l > 0$ be an integer, $\mathbf{A}_0 \in \mathbf{R}^l$, and $\mathbf{f}$ be a $\mathbf{C}^l$-valued function holomorphic in the complex multisphere $\hat{\mathbb{S}}_\rho(\mathbf{A}_0) \subset \mathbf{C}^l$ and real on $\mathbb{S}_\rho(\mathbf{A}_0)$.*

Consider the equation for $\mathbf{A}$:

$$\mathbf{A} - \mathbf{A}_0 + \mathbf{f}(\mathbf{A}) = \mathbf{0}. \tag{5.11.19}$$

There exists a constant γ (e.g., $\gamma = 2^8$) *such that if*

$$\gamma|\mathbf{f}|_\rho < \rho, \tag{5.11.20}$$

Eq. (5.11.19) *admits a unique solution* $\mathbf{A}_1 \in \mathbb{S}_\rho(\mathbf{A}_0)$, i.e., *such that* $\mathbf{A}_1 \in \mathbf{R}^l$ *and*

$$|\mathbf{A}_1 - \mathbf{A}_0| < \rho. \tag{5.11.21}$$

20 Proposition. *Let $p, l > 0$ be integers, let $\mathbf{A}_0 \in \mathbf{R}^l$, and $\rho, \xi, \delta > 0$, $\delta < \xi$, $\delta < 1$.*

Let $\mathbf{g}$ be an $\mathbf{R}^p$-valued analytic function on $\mathbb{S}_\rho(\mathbf{A}_0) \times \mathbf{T}^p$ holomorphic in $C(\rho, \xi; \mathbf{A}_0)$.

Consider the equation

$$\boldsymbol{\varphi}' = \boldsymbol{\varphi} + \mathbf{g}(\mathbf{A}, \boldsymbol{\varphi}) \tag{5.11.22}$$

thought of as an equation on $\mathbf{T}^p$ parametrized by $\boldsymbol{\varphi}' \in \mathbf{T}^p$ and $\mathbf{A} \in \mathbb{S}_\rho(\mathbf{A}_0)$. Then there exists a constant γ (e.g., $\gamma = 2^8$) *such that*:

(i) *Equation* (5.11.22) *is soluble if*

$$\gamma|\mathbf{g}|_{\rho,\xi} e^{2\xi} \delta^{-1} < 1 \tag{5.11.23}$$

and admits a solution of the form

$$\boldsymbol{\varphi} = \boldsymbol{\varphi}' + \boldsymbol{\Delta}(\mathbf{A}, \boldsymbol{\varphi}') \tag{5.11.24}$$

with $\boldsymbol{\Delta}$ being an $\mathbf{R}^p$-valued analytic function on $\mathbb{S}_\rho(\mathbf{A}_0) \times \mathbf{T}^p$ holomorphic in $C(\rho, \xi - \delta, \mathbf{A}_0)$.

(ii) *The function $\boldsymbol{\Delta}$ can be bounded as*

$$|\boldsymbol{\Delta}|_{\rho,\xi-\delta} \leqslant |\mathbf{g}|_{\rho,\xi}. \tag{5.11.25}$$

(iii) *The only function inverting Eq.* (5.11.22) *and enjoying the properties* (i) *and* (ii) *above is* $\boldsymbol{\Delta}$.

Observations.

(1) The reader should note that the above two implicit function theorems have "dimensional nature", i.e., they just say what can be naively guessed.

In fact, in order to invert an implicit equation "close to the identity" like Eq. (5.11.22), one expects to have to impose that the derivatives of g are small compared to the derivatives of the identity map (i.e., small compared to 1). This is precisely the meaning of Eq. (5.11.23): if we wish to invert inside the annulus with external radius $e^{\xi-\delta}$ and internal radius $e^{-(\xi-\delta)}$, we estimate the gradient of $\mathbf{g}$ in the region by $|\mathbf{g}|_{\rho,\xi}\delta^{-1}e^{\xi}$, see Eq. (5.11.18). For $\xi \gg 1$, this is still not the same as Eq. (5.11.23) (while it is such for $\xi \leqslant 1$). However, if ξ is large, we are asking for the inversion of Eq. (5.11.16) in a very large region and some extra conditions stem out of the requirement of global invertibility[27] (see the proof).

(2) Proposition 19 is an infinite-dimensional version of the implicit function theorem, since one can consider all the Taylor coefficients of f at $\mathbf{A}_0$ as parameters in Eq. (5.11.19). Also, Eq. (5.11.22) is susceptible to such an interpretation.

(3) Note that the constant γ in Propositions 19 and 20 is l- and p-independent. It is also the same in Propositions 19 and 20: but this has been arranged so as to avoid introducing too many constants. The numerical value of γ is not optimal.

(4) Proposition 20 is remarkable because it is a "global inversion" theorem. The equation is posed on all of $\mathbf{T}^p$ and not just locally.

A proposition analogous to Proposition 20 holds for the equation

$$\mathbf{w}' = \mathbf{w} + \mathbf{G}(\mathbf{w}, \mathbf{z}), \tag{5.11.26}$$

where $\mathbf{G}$ is a $\mathbf{C}^l$-valued function holomorphic on $C(\rho, \xi; \mathbf{A}_0)$ and real on $\mathbb{S}_\rho(\mathbf{A}_0) \times \mathbf{T}^p \subset C(\rho, \xi; \mathbf{A}_0)$:

21 Proposition. *Let $l, p > 0$ be integers, $\mathbf{A}_0 \in \mathbf{R}^l$, and $\rho, \xi, \tau > 0$, $\tau < 1$.*

Let $\mathbf{G}$ be an $\mathbf{R}^l$-valued analytic function on $\mathbb{S}_\rho(\mathbf{A}_0) \times \mathbf{T}^p$ holomorphic in $C(\rho, \xi; \mathbf{A}_0)$.

Consider Eq. (5.11.26) *as an equation for* $\mathbf{w}$ *parameterized by* $\mathbf{w}', \mathbf{z}$.

(i) *There is a constant* γ (e.g., $\gamma = 2^8$) *such that if*

$$\gamma |\mathbf{G}|_{\rho,\xi}\rho^{-1}\tau^{-1} < 1, \tag{5.11.27}$$

Eq. (5.11.26) *is soluble,* $\forall \mathbf{w}' \in \hat{\mathbb{S}}_{\rho e^{-\tau}}(\mathbf{A}_0)$, *and admits a solution of the form*

$$\mathbf{w} = \mathbf{w}' + \mathbf{D}(\mathbf{w}', \mathbf{z}) \tag{5.11.28}$$

with $\mathbf{D}$ *holomorphic on* $C(\rho e^{-\tau}, \xi; \mathbf{A}_0)$ *real on* $\mathbb{S}_{\rho e^{-\tau}}(\mathbf{A}_0) \times \mathbf{T}^p$.

(ii) *The following bound can be put on* $\mathbf{D}$:

$$|\mathbf{D}|_{\rho e^{-\tau},\xi} \leqslant |\mathbf{G}|_{\rho,\xi}. \tag{5.11.29}$$

(iii) $\mathbf{D}$ *is the only function inverting Eq.* (5.11.26) *and enjoying the properties* (i) *and* (ii) *above.*

[27] Also, it makes a difference to bound $\partial/\partial\boldsymbol{\varphi}$ rather than $\partial/\partial\mathbf{z}$ for ξ large.

(iv) *Fixing* $\mathbf{w} \in C(\rho e^{-\tau}, \xi; \mathbf{A}_0)$, *Eq.* (5.11.28) *yields the only* $\mathbf{w} \in C(\rho, \xi; \mathbf{A}_0)$ *verifying Eq.* (5.11.26).

Observations.

(1) The above proposition makes sense, and is true, in a natural way if $p = 0$ (just drop everywhere $\mathbf{z}$ and the index ξ). Likewise, Proposition 20 makes sense in a natural way if $l = 0$ (just drop $\mathbf{A}$ and the index ρ everywhere).

(2) Setting $\mathbf{w}' = 0$ in Eq. (5.11.26) as well as $p = 0$, $\mathbf{A}_0 = \mathbf{0}$ and applying Proposition 21, one deduces Proposition 19 with $\mathbf{A}_0 = \mathbf{0}$. Since $\mathbf{A}_0 = \mathbf{0}$ is clearly not restrictive, we see that Proposition 19 is a corollary of Proposition 21.

(3) This proposition is clearly analogous to Proposition 20 and has, also, a "dimensional nature", see Observation (1), p. 490; more generally, the comments made on Proposition 20 can be repeated with obvious modifications for Proposition 21. An analogue of item (iv) in Proposition 21 could also be formulated for Proposition 20, but we do not need it.

Problems and Exercises for §5.11

1. After studying the proof in Appendix N of Proposition 20, find the best value for the constant γ or, at least, a better value for it.

2. Same as Problem 1 for Proposition 21 and for its corollary, Proposition 19.

3. Apply Proposition 20 to invert the equation

$$l = \xi - e \sin \xi,$$

$\xi \in \mathbf{T}^1$, $l \in \mathbf{T}^1$, appearing in the theory of the two-body problem, see Problem 13, p. 304, §4.10. e is a parameter, $0 < e < 1$. Find for which values of e the above equation can be globally inverted.

4. Same as Problem 3 with the new γ computed in Problem 1.

5. Show that the equation in Problem 3 can be inverted for all $e \in [0, 1)$ in the sense that there is a function g analytic on $\mathbf{T}^1$ such that $\xi = l - g(l)$, for each given $e \in [0, 1)$. (Hint: Do not use Proposition 20 directly.)

§5.12. Perturbations of Trajectories. The Small Denominators Theorem

Another perturbative problem that could be studied is the following. Let $(\mathbf{A}, \boldsymbol{\varphi}) \to h_0(\mathbf{A})$ be an analytic Hamiltonian on $V \times \mathbf{T}^l$ which we suppose such that the matrix

$$M_0(\mathbf{A})_{ij} = \frac{\partial^2 h_0}{\partial A_i \, \partial A_j}(\mathbf{A}) \tag{5.12.1}$$

has nonvanishing determinant on $V \times \mathbf{T}^l$ ("nonisochronous integrable system").

It is clear that given $\mathbf{A}_0 \in V$, the torus $\{\mathbf{A}_0\} \times \mathbf{T}^l$ is an l-dimensional torus invariant for the motion associated with the Hamiltonian h_0. The Hamiltonian flow on the phase space $V \times \mathbf{T}^l$ induces on the torus a quasi-periodic flow $\boldsymbol{\varphi} \to \boldsymbol{\varphi} + \boldsymbol{\omega}_0 t$, $t \geqslant 0$, with pulsations

$$\boldsymbol{\omega}_0 = \frac{\partial h_0}{\partial \mathbf{A}}(\mathbf{A}_0) \equiv \boldsymbol{\omega}(\mathbf{A}_0). \tag{5.12.2}$$

If f_0 is an analytic function on $V \times \mathbf{T}^l$, it is natural to ask whether the motions on $V \times \mathbf{T}^l$ associated with the perturbed Hamiltonian,

$$H_0(\mathbf{A}, \boldsymbol{\varphi}) = h_0(\mathbf{A}) + f_0(\mathbf{A}, \boldsymbol{\varphi}), \tag{5.12.3}$$

leave a torus invariant, inducing on it a quasi-periodic flow with pulsations $\boldsymbol{\omega}(\mathbf{A}_0)$, i.e., "with the same spectrum" as before. One could call this problem "the spectrum-conservation problem".

Intuitively, one could expect that a torus on which a quasi-periodic motion with pulsations $\boldsymbol{\omega}(\mathbf{A}_0)$ takes place will continue to exist but it will be "deformed" inside $V \times \mathbf{T}^l$ if compared to the one relative to the $f_0 = 0$ case, at least if $M_0(\mathbf{A}_0)$ is invertible[28] and f_0 is small.

One can note that this perturbation problem differs from the one of the preceding sections; the latter was in fact concerned with the study of the perturbations of motions with given initial datum.

Now we are considering a whole family of motions enjoying a certain common property, namely, quasi-periodicity with pulsations $\boldsymbol{\omega}(\mathbf{A}_0)$, and we ask whether a family of motions with the same property still exists after perturbation.

Proposition 15 of §5.9 provides an anwer, in some sense affirmative, to the above question.

We now formulate and prove a proposition that, as it appears from the observations that follow it, also proves important parts of Proposition 15 and essentially gives all the ingredients necessary for a full proof of Proposition 15 in the analytic case.

The proof of Proposition 22 that follows is taken from Arnold and is specifically fit for the analytic case under examination. The proof of the analogous proposition in the $C^{(k)}$-differentiable case (with k large enough) is due to Moser and is based on a method which is technically different from the one that we shall discuss.

Before stating Proposition 22, which we shall call the "small-denominators theorem" for reasons which are obvious from its proof (or "Arnold's theorem", see References), we set up some notational conventions, see also Eqs. (5.11.1)–(5.11.3), (5.11.14).

[28] In the case $\boldsymbol{\omega}(\mathbf{A}) \equiv \boldsymbol{\omega}_0$, $\forall \mathbf{A} \in V$ and, hence, $M_0(\mathbf{A}) \equiv 0$, it is easy to give a counterexample. Let $l = 1$, $h_0(A) = A$ so that $\omega(A) \equiv 1$ and $M_0 \equiv 0$. Let $f(A, \varphi) \equiv \epsilon A$. Then the unperturbed motions have pulsation 1, while the perturbed ones have pulsation $(1 + \epsilon) \neq 1$, if $\epsilon \neq 0$.

10 Definition. (i) If $\mathbf{a} \in \mathbf{Z}_+^l$, $\boldsymbol{\nu} \in \mathbf{Z}^l$, we set

$$|\mathbf{a}| = \sum_{i=1}^{l} a_i, \qquad |\boldsymbol{\nu}| = \sum_{i=1}^{l} |\nu_i|.$$

(ii) If $\mathbf{w} \in \mathbf{C}^q$, we set

$$|\mathbf{w}| = \max_{1 < i < q} |w_i|, \qquad \|\mathbf{w}\| = \sum_{i=1}^{q} |w_i|.$$

A matrix M, $l \times l$, will be regarded as an element of $\mathbf{C}^q$ with $q = l^2$ so that it will make sense to write $|M|, \|M\|$.

(iii) If f is holomorphic in $C(\rho, \xi; \mathbf{A}_0)$[29] and takes its values in $\mathbf{C}^q$ and if h is holomorphic in $\hat{\mathbb{S}}_\rho(\mathbf{A}_0)$[29] and takes its values in $\mathbf{C}^q$, we set [see Eqs. (5.11.11) and (5.11.3)]

$$|f|_{\rho,\xi} = \sup|f(\mathbf{A}, \mathbf{z})|, \qquad \|f\|_{\rho,\xi} = \sup\|f(\mathbf{A}, \boldsymbol{\xi})\|,$$
$$|h|_\rho = \sup|h(\mathbf{A})|, \qquad \|h\|_\rho = \sup\|h(\mathbf{A})\|,$$

where the suprema are taken over the domains of definition of the various functions.

The small-denominators theorem can then be formulated as follows.

22 Proposition. *Let h_0, f_0 be two real analytic functions on $\mathbb{S}_{\rho_0}(\mathbf{A}_0) \times \mathbf{T}^l$, $\rho_0 > 0$, holomorphic in $C(\rho_0, \xi_0; \mathbf{A}_0)$, $\xi_0 < 1$. Assume that h_0 depends only on the action variables $\mathbf{A}$ in $(\mathbf{A}, \boldsymbol{\varphi}) \in \mathbb{S}_{\rho_0}(\mathbf{A}_0) \times \mathbf{T}^l$ and that the matrix M_0 of Eq. (5.12.1) is nonsingular. Suppose that $\omega_0 = (\partial h_0/\partial \mathbf{A})(\mathbf{A}_0)$ has the "nonresonance" property:*

$$|\omega_0 \cdot \boldsymbol{\nu}|^{-1} \leqslant C|\boldsymbol{\nu}|^l, \qquad \forall \boldsymbol{\nu} \in \mathbf{Z}^l, \ |\boldsymbol{\nu}| > 0 \tag{5.12.4}$$

for some $C > 0$ ("resonance parameter"). Let E_0, η_0, ϵ_0 be such that

$$E_0 > \left|\frac{\partial h_0}{\partial \mathbf{A}}\right|_{\rho_0,\xi_0}, \qquad \eta_0 > \|M_0^{-1}\|_{\rho_0,\xi_0},$$
$$\epsilon_0 > \left|\frac{\partial f_0}{\partial \mathbf{A}}\right|_{\rho_0,\xi_0} + \rho_0^{-1}\left|\frac{\partial f_0}{\partial \boldsymbol{\varphi}}\right|_{\rho_0,\xi_0}. \tag{5.12.5}$$

Then there exist constants $B, a, b, c > 0$, only depending upon the number l of degrees of freedom,[30] *such that if*

$$q \equiv BC\epsilon_0(CE_0)^a\left(\eta_0 E_0 \rho_0^{-1}\right)^b \xi_0^{-c} < 1, \tag{5.12.6}$$

one can find in $\mathbb{S}_{\rho_0}(\mathbf{A}_0) \times \mathbf{T}^l$ a torus $\mathcal{T}(\omega_0)$ with parametric equations

$$\begin{aligned} \mathbf{A} &= \mathbf{A}_0 + \boldsymbol{\alpha}(\boldsymbol{\varphi}'), \\ \boldsymbol{\varphi} &= \boldsymbol{\varphi}' + \boldsymbol{\beta}(\boldsymbol{\varphi}'), \end{aligned} \qquad \boldsymbol{\varphi}' \in \mathbf{T}^l \tag{5.12.7}$$

[29] See (5.11.3) for this symbol's meaning.

[30] e.g., a rather rough, though not "totally absurd", estimate says that one can take $a = b = 14$, $c = 2(10l + 6)$, $B = (12l)!\, 10^{40l}$ (very far from optimal).

and such that:

(i) $\mathcal{T}(\omega_0)$ *is invariant for the evolution in* $\mathcal{S}_{\rho_0}(\mathbf{A}_0) \times \mathbf{T}^l$ *associated with the Hamiltonian* (5.12.3). *On* $\mathcal{T}(\omega_0)$ *the evolution is described by the map*

$$\varphi' \to \varphi' + \omega_0 t, \qquad t \in \mathbf{R}_+ \tag{5.12.8}$$

and is therefore quasi-periodic with pulsations ω_0.

(ii) *The functions* $\boldsymbol{\alpha}$, $\boldsymbol{\beta}$ *are analytic on* $\mathbf{T}^l$ *and*

$$\rho_0^{-1}|\boldsymbol{\alpha}(\varphi')| + |\boldsymbol{\beta}(\varphi')| \leqslant q. \tag{5.12.9}$$

Observations.

(1) Using the notations of Proposition 18, p. 487, §5.11, we see by Proposition 18, Eqs. (5.11.16) and (5.11.12), that f_0 can be written as

$$\begin{aligned} f_0(\mathbf{A}, \mathbf{z}) &= \sum_{\boldsymbol{\nu} \in \mathbf{Z}^l} f_{0\boldsymbol{\nu}}(\mathbf{A}) \mathbf{z}^{\boldsymbol{\nu}} \\ &\equiv \sum_{\mathbf{a} \in \mathbf{Z}^l_+} \sum_{\boldsymbol{\nu} \in \mathbf{Z}^l} f_{0\boldsymbol{\nu}}^{(\mathbf{a})} (\mathbf{A} - \mathbf{A}_0)^{\mathbf{a}} \mathbf{z}^{\boldsymbol{\nu}}, \end{aligned} \tag{5.12.10}$$

where $f_{0\boldsymbol{\nu}}(\mathbf{A})$ is the sum of the series in $\mathbf{a}$ in the right-hand side of Eq. (5.12.10) and is holomorphic in $\hat{\mathcal{S}}_\rho(\mathbf{A}_0)$.

Then the derivatives $\partial/\partial\varphi$ appearing in Eq. (5.12.5) can be simply defined, when $z_j \neq e^{i\varphi_j}$, for some $j = 1, \ldots, l$, as $\partial/\partial\varphi_k \equiv iz_k \partial/\partial z_k$ (see Definition 9 (ii), p. 487).

(2) It follows from the theory of the Taylor–Laurent expansions for holomorphic functions that if $g_1, \ldots, g_r$ are r real analytic functions on $V \times \mathbf{T}^l$, $V \subset \mathbf{R}^l$ open, it is possible to find two functions $\mathbf{A} \to \rho(\mathbf{A})$, $\mathbf{A} \to \xi(\mathbf{A})$ positive and continuous on V such that $\mathcal{S}_{\rho(\mathbf{A})}(\mathbf{A}) \subset V$, $\forall \mathbf{A} \in V$, and, furthermore, such that $g_1, \ldots, g_r$ are holomorphic in $C(\rho(\mathbf{A}), \xi(\mathbf{A}); \mathbf{A})$, see Definition 8, p. 486, and $|g_j|_{\rho(A), \xi(\mathbf{A})}$ are continuous functions of $\mathbf{A}$ in V, $j = 1, \ldots, r$.

It is therefore clear that Proposition 22 could be formulated in an apparently more general form by only requiring the analyticity of h_0, f_0, M_0^{-1} in $\mathcal{S}_\rho(\mathbf{A}_0) \times \mathbf{T}^l$ rather than their holomorphy in $C(\rho, \xi; \mathbf{A}_0)$.

(3) An elementary result of measure theory would allow us to deduce that, given $C > 0$ and supposing V bounded, the set $V(C)$ of the points $\mathbf{A} \in V$ such that

$$|\boldsymbol{\nu} \cdot \omega(\mathbf{A})|^{-1} \equiv \left| \boldsymbol{\nu} \cdot \frac{\partial h_0}{\partial \mathbf{A}}(\mathbf{A}) \right|^{-1} < C|\boldsymbol{\nu}|^l, \qquad \forall \boldsymbol{\nu} \in \mathbf{Z}^l, \ |\boldsymbol{\nu}| > 0$$

has a Lebesgue measure $\mu(V(C)) \to \mu(V)$ as $C \to +\infty$ if $M_0(\mathbf{A})^{-1}$ exists, $\forall \mathbf{A} \in V$; see Problems 9 and 10 to §5.10, p. 483.

In particular, this implies that for all $\mathbf{A}$'s outside a set of zero Lebesgue measure, there is a number C, depending on $\mathbf{A}$, such that the above inequality holds.

The question of determining or estimating a number C such that $C \geqslant \sup_{\boldsymbol{\nu} \in \mathbf{Z}^l, |\boldsymbol{\nu}| \neq 0} |\boldsymbol{\nu} \cdot \omega|^{-1} |\boldsymbol{\nu}|^{-l}$ is, for a given ω, an interesting and difficult number-theoretic problem. Some of its aspects are discussed in detail in the problems of §2.20.

(4) Suppose $f_0 = \lambda \tilde{f}_0$, with $\tilde{f}_0$ λ-independent, and fix a set $\tilde{V} \subset V$, bounded and closed. Using the notations of the Observation (2) let

$$\rho_0 = \min_{\mathbf{A} \in \tilde{V}} \rho(\mathbf{A}), \qquad \xi_0 = \min_{\mathbf{A} \in \tilde{V}} \xi(\mathbf{A}),$$

$$E_0 = \max_{\mathbf{A} \in \tilde{V}} \left| \frac{\partial h_0}{\partial \mathbf{A}} \right|_{\rho(\mathbf{A})}, \qquad \eta_0 = \max_{\mathbf{A} \in \tilde{V}} \| M_0^{-1} \|_{\rho(\mathbf{A})}.$$

Then apply Proposition 22 to the Hamiltonian system described by Eq. (5.12.3) in $\mathbb{S}_{\rho_0}(\mathbf{A}_0) \times \mathbf{T}^l$ with $\mathbf{A}_0 \in \tilde{V}(C)$, see Observation (3) above. By Eq. (5.12.6), one immediately deduces that for λ small, the perturbed Hamiltonian system admits simultaneously coexisting invariant tori $\mathfrak{T}(\omega(\mathbf{A}_0))$, $\forall \mathbf{A}_0 \in \tilde{V}(C)$. Such tori will be located geometrically close to the unperturbed tori, by Eq. (5.12.9).

This means that the "less resonant" the pulsations ω of the unperturbed quasi-periodic motions,[31] the larger the perturbations' intensity λ has to become before it can possibly suceed in destroying these motions and the invariant tori on which they take place.

(5) Observations (1)–(4) above show that the statement (i) of Proposition 15, p. 466, §5.9, and the statement that $W^{(\epsilon)} \neq \emptyset$ follow from Proposition 22. From the proof of Proposition 22, however, all of Proposition 15 follows with some effort, in the analytic case. We shall not discuss this problem, (see for instance Pöschel, Chierchia–Gallavotti, Gallavotti in the reference list).

(6) The condition (5.12.6) involves only the derivatives of h_0 and f_0: this is natural since only such functions appear in the Hamiltonian equations of motion.

Also, it should be noted that the nature of the condition (5.12.6) is quite simple: given $h_0, f_0, \mathbf{A}_0$, one can form the quantities $E_0, \epsilon_0, C, \rho_0, \xi_0, \eta_0$ and, with them, the "dimensionless quantities" $C\epsilon_0, CE_0, \eta_0\rho_0^{-1}E_0, \xi_0$ in terms of which all the other dimensionless quantities can be formed. It can be seen that

$$CE_0 \geqslant 1, \qquad \eta_0\rho_0^{-1}E_0 \geqslant 1, \tag{5.12.11}$$

see Problem 1 at the end of the section.

Then Eq. (5.12.6) just says that the perturbation strength $C\epsilon_0$ has to be small compared to the other "small" dimensionless quantities $(CE_0)^{-1}$, $(\eta_0\rho_0^{-1}E_0)^{-1}$, and ξ_0 which are relevant to the problem.

Note that in the above argument, the parameters "relevant to the problem" are just $E_0, \epsilon_0, C, \rho_0, \xi_0, \eta_0$: this is, in fact, not obvious and, a priori, one might expect that other quantities may be relevant, like $F_0 = |\partial^2 h_0/\partial \mathbf{A}\, \partial \mathbf{A}|_{\rho_0}$ or $\tilde{F}_0 = |\partial^3 h_0/\partial \mathbf{A}\, \partial \mathbf{A}\, \partial \mathbf{A}|_{\rho_0}$, etc.

All that the above argument says is that if the results of Proposition 22 hold under conditions that just involve $E_0, \epsilon_0, C, \rho_0, \xi_0, \eta_0$, then it is not

[31] i.e., the smaller C is.

surprising that such conditions can take the form of Eq. (5.12.6), i.e., the simplest imaginable form.

(7) The condition $\eta_0 < +\infty$ or something like it must be necessary: in fact, for isochronous systems the above theorem cannot hold. Just consider $l = 1$, $h_0(A) = A$, $f_0(A, \varphi) = \epsilon A$; in this case all the perturbed motions have pulsations $\omega = 1 + \epsilon$ and none $\omega = \omega_0 = 1$.

We call η_0 the "anisochrony parameter" and a system for which $\eta_0 < +\infty$ is said to be "anisochronous" near $\mathbf{A}_0$.

The systems of harmonic oscillators are strictly isochronous, $\eta_0 = +\infty$, and the theorem does not directly apply to them.

However, if $h_0(\mathbf{A}) = \boldsymbol{\omega}_0 \cdot \mathbf{A}$ and if $\boldsymbol{\omega}_0$ verifies Eq. (5.12.4), then the theorem can still be indirectly applied under some additional assumptions.

In fact, let $f_0 = \lambda \tilde{f}_0$ with $\tilde{f}_0$ λ-independent. Assume that

$$\tilde{\eta}_0 = \left\| \left(\frac{\partial^2 \tilde{f}_{00}}{\partial \mathbf{A}\, \partial \mathbf{A}} \right)^{-1} \right\|_{\rho_0} < +\infty, \tag{5.12.12}$$

where $\tilde{f}_{00}$ denotes the average of $\tilde{f}_0$ over $\mathbf{T}^l$, i.e., its Fourier coefficient with $\boldsymbol{\nu} = \mathbf{0}$ [see also Eq. (5.11.16)].

Now apply Proposition 17, p. 475, to change variables completely canonically and to transform the problem into that of the analysis of the Hamiltonian systems with Hamiltonian

$$h_0'(\mathbf{A}) + \lambda^{n+1} f_0'(\mathbf{a}, \boldsymbol{\varphi}), \tag{5.12.13}$$

where h_0', f_0' are holomorphic in $C(\rho_0/2, \xi_0/2; \mathbf{A}_0)$ and in the variable λ for λ close to zero, see Eq. (5.10.27); choose n as $n = al + b + l$, a and b being the constant in Eq. (5.12.6).

Also, from Eq. (5.10.41), we see that

$$h_0'(\mathbf{A}) = h_0(\mathbf{A}) + \lambda \tilde{f}_{00}(\mathbf{A}) + \lambda^2 \tilde{h}(\mathbf{A}), \tag{5.12.14}$$

where $\tilde{h}$ is analytic in λ (actually, it is a polynomial) and in $\mathbf{A}$, near $\mathbf{A}_0$.

Therefore, if λ is small enough, the quantities $E_0', \eta_0', \epsilon_0'$ such that

$$E_0' \geq \left| \frac{\partial h_0'}{\partial \mathbf{A}} \right|_{\rho_0/2}, \qquad \eta_0' \geq \left\| \left(\frac{\partial^2 h_0'}{\partial \mathbf{A}\, \partial \mathbf{A}} \right)^{-1} \right\|_{\rho_0/2}, \tag{5.12.15}$$

$$\epsilon_0' \geq \lambda^{bl+1} \left(\left| \frac{\partial f_0'}{\partial \mathbf{A}} \right|_{\rho_0/2, \xi_0/2} + \frac{2}{\rho_0} \left| \frac{\partial f_0'}{\partial \boldsymbol{\varphi}} \right|_{\rho_0/2, \xi_0/2} \right)$$

can be chosen so that for a suitable $K > 0$, depending on $E_0, \epsilon_0, \rho_0, \xi_0$, but not on λ, and $\forall \lambda$ small:

$$E_0' \leq 2E_0, \qquad \eta_0' \leq 2\tilde{\eta}_0 \lambda^{-1}, \qquad \epsilon_0' \leq K \lambda^{bl+1}. \tag{5.12.16}$$

We now consider the points $\mathbf{A}_0' \in S_{\rho_0/4}(\mathbf{A}_0)$ for which $\boldsymbol{\omega}'(\mathbf{A}_0') = (\partial h_0'/\partial \mathbf{A})(\mathbf{A}'_0)$ is such that

$$|\boldsymbol{\omega}'(\mathbf{A}_0') \cdot \boldsymbol{\nu}|^{-1} \leq C \lambda^{-l} |\boldsymbol{\nu}|^l, \qquad \forall \boldsymbol{\nu} \in \mathbf{Z}^l, \ |\boldsymbol{\nu}| > 0. \tag{5.12.17}$$

Using the results of the Problems 9 and 15, §5.10, p. 483 and 484, and the estimate on η_0', it is possible to see that such points actually exist and fill a considerable part of $\mathbb{S}_{\rho_0/4}(\mathbf{A}_0')$ (in fact, their ensemble forms a set whose measure approaches that of $\mathbb{S}_{\rho_0/4}(\mathbf{A}_0')$ itself as $C \to \infty$ uniformly in λ).

Then we apply Proposition 20 to the Hamiltonian $h_0' + \lambda^{bl+l+1} f_0'$ regarded as holomorphic on $C(\rho_0/4, \xi_0/4; \mathbf{A}_0')$ with $\mathbf{A}_0'$ verifying Eq. (5.12.17).

Equation (5.12.6) now becomes

$$BC\lambda^{-l}K\lambda^{al+b+l+1}\left(2\lambda^{-l}E_0\right)^a\left(2\tilde{\eta}_0\lambda^{-1}4\rho_0^{-1}2E_0\right)^b\left(4\xi_0^{-1}\right)^c < 1$$

which can be fulfilled for λ small.

This could be interpreted as saying that the quasi-periodic motions with $\mathbf{A}_0'$ such that Eq. (5.12.17) holds are not destroyed by the perturbation, but survive with a slightly modified pulsation (since $\boldsymbol{\omega}'(\mathbf{A}_0') = \boldsymbol{\omega}_0 + O(\lambda)$), running on slightly deformed tori.

(8) So Observation (7) shows that the nonisochrony condition, $\eta_0 < +\infty$, can be essentially weakened. One can ask whether this is the case for the "nonresonance" condition $C < +\infty$ as well. The whole discussion of perturbation theory, §5.10, suggests that this is not the case.

In fact, by considering some extreme cases, it is easy to show that one cannot go too far toward weakening the conditions.

Consider an harmonic isochronous resonating oscillator in $\mathbf{R}^3$:

$$H(\mathbf{p},\mathbf{q}) = \frac{\mathbf{p}^2}{2} + \frac{\mathbf{q}^2}{2}\,. \tag{5.12.18}$$

We can pass to action-angle coordinates $(\mathbf{A},\boldsymbol{\varphi}) \in (\mathbf{R}_+/\{0\})^3 \times \mathbf{T}^3$ to describe most of the oscillators' motions via the Hamiltonian[32]

$$h_0(\mathbf{A}) = A_1 + A_2 + A_3$$

on $(0,+\infty)^3 \times \mathbf{T}^3$. A further completely canonical change of coordinates, $\mathbf{A} \to \tilde{\mathbf{A}}$, $\boldsymbol{\varphi} \to \tilde{\boldsymbol{\varphi}}$:

$$\begin{aligned} \tilde{A}_1 &= A_1 + A_2 + A_3, & \tilde{\varphi}_1 &= \varphi_1, \\ \tilde{A}_2 &= A_2, & \tilde{\varphi}_2 &= \varphi_2 - \varphi_1, \\ \tilde{A}_3 &= A_3, & \tilde{\varphi}_3 &= \varphi_3 - \varphi_1 \end{aligned} \tag{5.12.19}$$

(see Problem 33, §3.11, p. 232) transforms the Hamiltonian (5.12.18) into the form

$$\tilde{h}_0(\tilde{\mathbf{A}}) = \tilde{A}_1, \qquad (\tilde{\mathbf{A}},\tilde{\boldsymbol{\varphi}}) \in (0,+\infty)^3 \times \mathbf{T}^3. \tag{5.12.20}$$

Let $f_0(\tilde{A}_2, \tilde{A}_3, \tilde{\varphi}_2, \tilde{\varphi}_3)$ be an analytic nonintegrable Hamiltonian on $V \times \mathbf{T}^2 \subset (0,+\infty)^2 \times \mathbf{T}^2$: its existence is not obvious, but we state without proof that it exists and that it can be chosen so that it produces nonquasiperiodic motions.[33]

[32] See exercises for §4.1.

[33] An example could be constructed on the basis of Observation (3), p. 332, but the discussion is quite long.

Clearly, the system

$$\tilde{h}_0(\tilde{\mathbf{A}}) + \epsilon f_0(\tilde{A}_2, \tilde{A}_3, \tilde{\varphi}_2, \tilde{\varphi}_3) \tag{5.12.21}$$

cannot be integrable as, manifestly, the coordinates corresponding to the degrees of freedom with indices 2 and 3 verify the equations with Hamiltonian ϵf_0 which gives rise, for $\epsilon \neq 0$, to motions coinciding with those of f_0 up to a change of scale in time and which are not quasiperiodic, i.e., not integrable by criterion (i), p. 350.

This example clearly shows why resonances can be important. In a resonance situation, it happens that some degrees of freedom of the system "do not move at all" as can be seen by suitable changes of coordinates. Hence, upon perturbation, their motion will be entirely governed by the perturbation and it will therefore become important whether or not the perturbation by itself is integrable.

If the perturbation by itself describes an integrable system in the phase-space region around a resonant torus of the unperturbed system, the above argument suggests that something could, nevertheless, be done. This is in fact the situation found in celestial mechanics in the vicinity of the unperturbed tori corresponding to orbits of small eccentricity and small inclination. As shown in §5.10, Observation (3), p. 479, in this situation one can set up some perturbation scheme to compute the secular perturbations. The scheme can lead to a rigorous proof of the tori's conservation (under suitable assumptions on the phase-space region which is considered). This proof is in a celebrated paper by Arnold, see References.

Of course, in the above discussion, one could have directly started from Eqs. (5.12.20) and (5.12.21), but we thought that starting from a physical system would be easier for the reader. On the other hand, the choice of $\mathbf{R}^3$ is essential to the argument: if we had chosen $\mathbf{R}^2$, the argument could have failed since only $\tilde{A}_2, \tilde{\varphi}_2$ would have been present, i.e., f_0 would have described a one-degree-of-freedom system (which is "necessarily"[34] integrable).

(9) The above observation shows that the nonresonance condition is essential in a case in which the resonance is very manifest, i.e., the unperturbed system is isochronous and resonating.

However, one could think that the nonintegrability phenomenon might be only related to isochronous resonances: if a system is anisochronous one might argue that the perturbation will cause the motion to wander around in phase space, keeping it away from the resonances most of the time. The fallaciousness of this way of reasoning is made clear by an example that goes back to Poincaré.

Consider the system on $\mathbf{R}^2 \times \mathbf{T}^2$:

$$H(A_1, A_2, \varphi_1, \varphi_2) = \frac{A_1^2 + A_2^2}{2} + \epsilon f(\varphi_1, \varphi_2) \tag{5.12.22}$$

[34] See the statement (19), p. 362, and §2.7 for general conditions of integrability.

with

$$g(\varphi_2-\varphi_1)=\int_0^{2\pi} f(\varphi_1+\psi,\varphi_2+\psi)\frac{d\psi}{2\pi}=\text{not constant.}\qquad(5.12.23)$$

To fix the ideas we shall take $f(\varphi_1,\varphi_2)=(1-\cos(\varphi_2-\varphi_1))$: in this case (5.12.22) has a simple physical meaning, as it describes two points ideally bound to a unit circle attracting each other via a harmonic force. The reader should, as an exercise, understand the physical meaning of the argument below and why it can be immediately extended to the general case if (5.12.23) holds. For $\epsilon=0$ all the motions on the torus $\{\mathbf{A}_0\}\times\mathbf{T}^2$, $\mathbf{A}_0=(1,1)$, are periodic with pulsations $\boldsymbol{\omega}_0=\mathbf{A}_0=(1,1)$, so the torus is resonant.

Suppose, per absurdum, that the torus is not destroyed for small ϵ, in the sense that there exists an invariant torus (i.e., invariant with respect to the perturbed motion) with parametric equations:

$$\begin{aligned}\mathbf{A}&=\mathbf{A}_0+\boldsymbol{\alpha}(\boldsymbol{\varphi}'),\\ \boldsymbol{\varphi}&=\boldsymbol{\varphi}'+\boldsymbol{\beta}(\boldsymbol{\varphi}'),\end{aligned}\qquad(5.12.24)$$

where $\boldsymbol{\alpha}$, $\boldsymbol{\beta}$ are $\mathbf{R}^2$-valued functions in $C^\infty(\mathbf{T}^2)$, and that the torus given by Eq. (5.12.24) is close to the unperturbed torus for ϵ small:

$$\gamma(\epsilon)=\max|\boldsymbol{\alpha}(\boldsymbol{\varphi}')|+\max|\boldsymbol{\beta}(\boldsymbol{\varphi}')|\underset{\epsilon\to0}{\to}0\qquad(5.12.25)$$

and that the motion on the torus in Eq. (5.12.24) is described by $\boldsymbol{\varphi}'\to\boldsymbol{\varphi}'+\boldsymbol{\omega}_0t$, $\boldsymbol{\omega}_0=(1,1)=\mathbf{A}_0$, i.e., we suppose that the perturbed torus is run periodically with the same spectrum as that corresponding to the unperturbed torus $\{\mathbf{A}_0\}\times\mathbf{T}^2$.

We write the Hamiltonian equations for $h_0+\epsilon f$ and subtract the two equations for A_1 and A_2:

$$\dot{A}_2-\dot{A}_1=-2\epsilon\sin(\varphi_2-\varphi_1).\qquad(5.12.26)$$

Then we integrate both sides between $t=0$ and $t=2\pi$, assuming that we have computed them on a motion developing on the torus of Eq. (5.12.24) with initial datum corresponding to $\boldsymbol{\varphi}'\in\mathbf{T}^2$.

Since the motion is periodic, by assumption, with period 2π ($\omega_0=(1,1)$), we find

$$\begin{aligned}0&=-2\epsilon\int_0^{2\pi}\sin(\varphi_2-\varphi_1)\,dt\\ &=-2\epsilon\int_0^{2\pi}\sin\big[\varphi_2'-\varphi_1'+\beta_2(\varphi_1'+t,\varphi_2'+t)-\beta_1(\varphi_1'+t,\varphi_2'+t)\big]\,dt\\ &=-4\pi\epsilon\big(\big[\sin(\varphi_2'-\varphi_1')\big]+2\tilde{\gamma}(\epsilon)\big)\underset{\epsilon\to0}{\rightrightarrows}-4\pi\epsilon\sin(\varphi_2'-\varphi_1'),\end{aligned}\qquad(5.12.27)$$

where $\tilde{\gamma}(\epsilon)\in[-\gamma(\epsilon),\gamma(\epsilon)]$ is suitably chosen. This is absurd if $\varphi_2'-\varphi_1'\neq0$, π and shows that the torus of Eq. (5.12.24) cannot exist as an invariant torus run periodically with pulsation $\omega_0=(1,1)$. The resonating torus corresponding to $\mathbf{A}_0 = (1, 1)$ is "destroyed" upon perturbation, no matter how small.

The above argument shows that the torus is "destroyed", but does not show that all the periodic motions with period 2π are destroyed, e.g., if $\varphi_1 = \varphi_2$ or $\varphi_1 = \varphi_2 + \pi$, we form, together with $A_1 = A_2 = 1$, two sets of initial data, evolving periodically with period 2π and, topologically, such sets are two circles (i.e., like $\mathbf{T}^1$ instead of $\mathbf{T}^2$).

This example is interesting because it considers a case in which all the assumptions of Proposition 22 hold except the nonresonance condition (5.12.4), thereby showing its necessity. However, it does not provide an example as "shocking" as the one of Observation (8) above, since the perturbed system still exhibits only quasi-periodic motions or motions with rather trivial asymptotic behavior.[35] Much more interesting in this respect would be the case when f in Eq. (5.12.23) is replaced by a function really depending on both φ_1 and φ_2, not only on $\varphi_2 - \varphi_1$. In such a case, one expects to find some motions with very complex asymptotic behavior near a resonating unperturbed torus.

(10) Observations (7)–(9) above clarify the necessity of the assumptions in Proposition 22. They can be summarized as follows: nonresonating quasi-periodic motions on l-dimensional tori are preserved, in anisochronous systems, in the presence of small perturbations; they are also preserved in isochronous nonresonating systems for all the nonisochronous small perturbations. Resonating motions on l-dimensional tori are generally destroyed by small perturbations in both the isochronous and the nonisochronous cases.

(11) It is important that the reader who is about to read the following proof realizes that all the very numerous inequalities that he will meet can easily be guessed on "dimensional grounds", i.e., using what we called in §5.11, Observation (2), p. 489, "dimensional estimates". In this way, one can easily check the calculations (which we give in great detail only for completeness since this book is supposed to be elementary).

The possibility of simple dimensional estimates is what makes the proof in the analytic case easy to visualize.

In the upcoming proof no attention is paid to optimal estimates, nor to the evaluation of the various constants. However, in principle, the proof below does not contain any crude approximation, and if the constants are evaluated with care it should give results which are optimal in the given generality of the assumptions.

This, of course, does not mean that in particular cases the estimates could not be greatly improved.

Finally let us point out to the reader familiar with present trends in statistical mechanics and field theory that the proof below yields a nice example of a vast class of theorems which can be proved by what has become known in physics as the "renormalization group method".

[35] as can be seen by the completely canonical change of variables $A = (A_1 + A_2)/2$, $B = (A_1 - A_2)/2$, $\varphi = \varphi_1 + \varphi_2$, $\psi = \varphi_1 - \varphi_2$ (exercise).

PROOF. We think of the unperturbed Hamiltonian h_0 and the perturbation f_0 as a pair of holomorphic functions on $C(\rho_0, \xi_0; \mathbf{A}_0) \subset \mathbf{C}^{2l}$, real on $\mathbb{S}_{\rho_0}(\mathbf{A}_0) \times \mathbf{T}^l$, see Definition 9, p. 487, §5.11. To this pair we associate the "characteristic numbers" $E_0, \eta_0, \xi_0, \rho_0, \epsilon_0$ verifying Eq. (5.12.5).

We have already noted that [see Eq. (5.12.11)]

$$\eta_0 E_0 \rho_0^{-1} \geqslant 1, \qquad CE_0 \geqslant 1. \tag{5.12.28}$$

In the course of the proof, we shall have to "give up" some analyticity in the $\mathbf{A}$ and $\boldsymbol{\varphi}$ variables in order to make dimensional estimates. The amount of analyticity that is given up is, to a great extent, arbitrary: we introduce some "analyticity loss" parameters $\delta_0 > \delta_1 > \cdots$ which we shall use to describe precisely the analyticity loss.

To be definite, we fix

$$\delta_k = 2^{-4}(1+k)^{-2}\xi_0 \tag{5.12.29}$$

so that $5\sum_{k=0}^{\infty}\delta_k < \xi_0 < 1$. For simplicity, we shall assume that $C\epsilon_0 < 1$.

We shall identify $\boldsymbol{\varphi} = (\varphi_1, \ldots, \varphi_l) \in \mathbf{T}^l$ with $\mathbf{z} = (e^{i\varphi_1}, \ldots, e^{i\varphi_l}) \in \mathbf{C}^l$ and freely use an "angular notation" for $\mathbf{z}$ even if $\mathbf{z}$ is not on the product of the l unit circles. In this case, $\partial/\partial\varphi_k$ means $iz_k\partial/\partial z_k$, see Definition 9 (ii), p. 487. Also, it will be convenient to write $e^{i\boldsymbol{\Delta}} \equiv (e^{i\Delta_1}, \ldots, e^{i\Delta_l})$ and $\mathbf{z}e^{i\boldsymbol{\Delta}} = (z_1 e^{i\Delta_1}, \ldots, z_l e^{i\Delta_l})$ for $\boldsymbol{\Delta} \in \mathbf{C}^l$. Such conventions greatly simplify the notations.

The proof proceeds by applying perturbation theory along the lines of §5.10. Since the first problem is that f_0 does not fulfill the assumptions of Proposition 17, we shall divide f_0 into two parts: one very small $\sim O(\epsilon_0^2)$ and the other fulfilling the assumptions of Proposition 17, i.e., with only finitely many Fourier components ("Arnold's regularization").

Then we shall apply Proposition 17 to find a canonical transformation changing the Hamiltonian into a "renormalized" one with an integrable part $h_1(\mathbf{A})$ plus a perturbation $f_1(\mathbf{A}, \boldsymbol{\varphi})$ with f_1 of $O(\epsilon_0^2)$. Afterwards, we proceed to find a point $\mathbf{A}_1$ such that $(\partial h_1/\partial \mathbf{A})(\mathbf{A}_1) = \omega_0$, and we shall again be in a position to begin the procedure all over again, provided we control the new characteristic parameters $E_1, \epsilon_1, \rho_1, \xi_1, \eta_1$. Basically, the whole argument is reduced to the research of an expression of $E_1, \epsilon_1, \ldots$ in terms of $E_0, \epsilon_0, \ldots$ ("Kolmogorov's iteration").

To reduce f_0 to a trigonometric polynomial plus a small remainder, we introduce the "ultraviolet cut off":

$$N_0 = 2\delta_0^{-1}\log\left(C\epsilon_0\delta_0^l\right)^{-1} > 1 \tag{5.12.30}$$

and define the "regularized perturbation"

$$f_0^{[\leqslant N_0]}(\mathbf{A}, \boldsymbol{\varphi}) = \sum_{\substack{\boldsymbol{\nu}\in\mathbf{Z}^l \\ |\boldsymbol{\nu}| \leqslant N_0}} f_{0\boldsymbol{\nu}}(\mathbf{A}) e^{i\boldsymbol{\nu}\cdot\boldsymbol{\varphi}}, \tag{5.12.31}$$

using for f_0 the notation of Eqs. (5.11.16), p. 488. We set $f_0^{[>N_0]} = f_0 - f_0^{[\leqslant N_0]}$.

The choice of N_0 has been made so that $f^{[>N_0]}$ is indeed of $O(\epsilon_0^2)$. This can be seen immediately by applying the estimates of Eqs. (5.11.15) to the functions $\partial f_0/\partial \mathbf{A}$ and $\rho_0^{-1}\partial f_0/\partial \boldsymbol{\varphi}$, holomorphic in $C(\rho_0, \xi_0; \mathbf{A}_0)$, regarded as function on $C(\xi)$ parameterized by $\mathbf{A} \in \hat{\mathbb{S}}_{\rho_0}(\mathbf{A}_0)$:

$$\left|\frac{\partial f_{0\boldsymbol{\nu}}}{\partial \mathbf{A}}\right| \leqslant \epsilon_0 e^{-\xi_0|\boldsymbol{\nu}|},$$
$$|\nu_i f_{0\boldsymbol{\nu}}| \leqslant \epsilon_0 \rho_0 e^{-\xi_0|\boldsymbol{\nu}|}, \qquad \forall \boldsymbol{\nu} \in \mathbf{Z}^l, \quad i = 1, \ldots, l \tag{5.12.32}$$

by the third of Eqs. (5.11.15),

Also note that by item (v) Proposition 18, p. 487, the functions $\partial f_{0\boldsymbol{\nu}}/\partial \mathbf{A}$ and $\nu_i f_{0\boldsymbol{\nu}}$ are, $\forall \boldsymbol{\nu} \in \mathbf{Z}^l$, $\forall i = 1, \ldots, l$, holomorphic on $\hat{\mathbb{S}}_{\rho_0}(\mathbf{A}_0)$. Just apply (v) to the functions $g = \partial f_0/\partial \mathbf{A}$ and $g = \partial f_0/\partial \boldsymbol{\varphi}$.

Equation (5.12.32) allow us to bound $f_0^{[>N_0]}$ and $f_0^{[\leqslant N_0]}$ as follows. There exist B_1, B_2, $1 \leqslant B_1 \leqslant B_2$ such that

$$\left|\frac{\partial f_0^{[\leqslant N_0]}}{\partial \mathbf{A}}\right|_{\rho_0, \xi_0 - \delta_0} + \rho_0^{-1}\left|\frac{\partial f_0^{[\leqslant N_0]}}{\partial \boldsymbol{\varphi}}\right|_{\rho_0, \xi_0 - \delta_0} \leqslant B_1 \epsilon_0 \delta_0^{-l},$$
$$\left|\frac{\partial f_0^{[>N_0]}}{\partial \mathbf{A}}\right|_{\rho_0, \xi_0 - \delta_0} + \rho_0^{-1}\left|\frac{\partial f_0^{[>N_0]}}{\partial \boldsymbol{\varphi}}\right|_{\rho_0, \xi_0 - \delta_0} \leqslant B_2 \epsilon_0^2 C. \tag{5.12.33}$$

These estimates follow by substituting the bounds given by Eq. (5.12.32) into Eq. (5.12.31) or into the analogous expression for $f^{[>N_0]}$, after the appropriate differentiations. For instance, consider the second of Eqs. (5.12.33). One has

$$\left|\frac{\partial f_0^{[>N_0]}}{\partial \mathbf{A}}(\mathbf{A}, \boldsymbol{\varphi})\right| = \left|\sum_{\substack{\boldsymbol{\nu}\in\mathbf{Z}^l\\ |\boldsymbol{\nu}|>N_0}} \frac{\partial f_{0\boldsymbol{\nu}}}{\partial \mathbf{A}}(\mathbf{A}) e^{i\boldsymbol{\nu}\cdot\boldsymbol{\varphi}}\right| \equiv \left|\sum_{\substack{\boldsymbol{\nu}\in\mathbf{Z}^l\\ |\boldsymbol{\nu}|>N_0}} \frac{\partial f_{0\boldsymbol{\nu}}}{\partial \mathbf{A}}(\mathbf{A}) \mathbf{z}^{\boldsymbol{\nu}}\right|$$
$$\leqslant \sum_{\substack{\boldsymbol{\nu}\in\mathbf{Z}^l\\ |\boldsymbol{\nu}|>N_0}} \epsilon_0 e^{-\delta_0|\boldsymbol{\nu}|} \leqslant \epsilon_0 e^{-(\delta_0/2)N_0} \sum_{\boldsymbol{\nu}\in\mathbf{Z}^l} e^{-(\delta_0/2)|\boldsymbol{\nu}|} \tag{5.12.34}$$
$$= C\epsilon_0^2 \delta_0^l \left(\frac{1 + e^{-\delta_0/2}}{1 - e^{-\delta_0/2}}\right)^l \leqslant B' C\epsilon_0^2,$$
$$\forall (\mathbf{A}, \mathbf{z}) \in C(\rho_0, \xi_0 - \delta_0; \mathbf{A}_0),$$

where in the first equality, we use the symbolic but suggestive "angular notation" for $\mathbf{z}$, and $B' > 0$ is a suitable constant.

Similarly,

$$\left|\frac{\partial f_0^{[>N_0]}}{\partial \boldsymbol{\varphi}}(\mathbf{A}, \mathbf{z})\right| = \left|\sum_{\substack{\boldsymbol{\nu}\in\mathbf{Z}^l\\ |\boldsymbol{\nu}|>N_0}} \boldsymbol{\nu} f_{0\boldsymbol{\nu}}(\mathbf{A}) \mathbf{z}^{\boldsymbol{\nu}}\right| \leqslant \epsilon_0 \rho_0 \sum_{|\boldsymbol{\nu}|>N_0} e^{-\delta_0|\boldsymbol{\nu}|} \leqslant B' C\epsilon_0^2 \rho_0. \tag{5.12.35}$$

Hence, the second of Eqs. (5.12.33) follows from Eqs. (5.12.34) and (5.12.35). The first of Eqs. (5.12.33) follows from the same type of arguments.[36]

Following the ideas of perturbation theory, we shall now construct a canonical change of variables using, as in the proof to Proposition 16, §5.10, p. 472, a generating function of the form $(\mathbf{A}',\boldsymbol{\varphi})\to(\mathbf{A}'\cdot\boldsymbol{\varphi})+\Phi_0(\mathbf{A}',\boldsymbol{\varphi})$, where Φ_0 is defined on a suitable set $\mathbb{S}_{\tilde\rho_0}(\mathbf{A}_0)\times\mathbf{T}^l$ as

$$\Phi_0(\mathbf{A}',\boldsymbol{\varphi})=\sum_{\substack{\boldsymbol{\nu}\in\mathbf{Z}^l\\ 0<|\boldsymbol{\nu}|<N_0}}\frac{f_{0\boldsymbol{\nu}}(\mathbf{A})\mathbf{z}^{\boldsymbol{\nu}}}{-i\omega(\mathbf{A})\cdot\boldsymbol{\nu}} \tag{5.12.36}$$

which defines a holomorphic function of $(\mathbf{A}',\mathbf{z})\in C(\tilde\rho_0,\xi_0-\delta_0;\mathbf{A}_0)$ if $\tilde\rho_0$ is chosen so small that, by consequence of Eq. (5.12.4), $|\omega(\mathbf{A})\cdot\boldsymbol{\nu}|>0$, $\forall\boldsymbol{\nu}\in\mathbf{Z}^l$, $0<|\boldsymbol{\nu}|\leqslant N_0$, $\forall\mathbf{A}\in\hat{\mathbb{S}}_{\tilde\rho_0}(\mathbf{A}_0)$.

Actually, we can easily find a simple choice for $\tilde\rho_0$, good enough for our purposes. In fact, $\forall\mathbf{A}\in\hat{\mathbb{S}}_{\tilde\rho_0}(\mathbf{A}_0)$ and if $\tilde\rho_0<\rho_0/2$, it is

$$\begin{aligned}|\omega(\mathbf{A}')\cdot\boldsymbol{\nu}|^{-1}&\equiv|(\omega_0+(\omega(\mathbf{A}')-\omega(\mathbf{A}_0)))\cdot\boldsymbol{\nu}|^{-1}\\ &\leqslant|\omega_0\cdot\boldsymbol{\nu}|^{-1}\left|1-\frac{|(\omega(\mathbf{A}')-\omega(\mathbf{A}_0))\cdot\boldsymbol{\nu}|}{|\omega_0\cdot\boldsymbol{\nu}|}\right|^{-1}\\ &\leqslant C|\boldsymbol{\nu}|^l\left|1-2C|\boldsymbol{\nu}|^{l+1}l\frac{E_0}{\rho_0}\tilde\rho_0\right|^{-1}\end{aligned} \tag{5.12.37}$$

because we can bound $|\omega(\mathbf{A}')-\omega(\mathbf{A}_0)|$ as

$$\begin{aligned}|\omega(\mathbf{A}')-\omega(\mathbf{A}_0)|&\equiv\left|\int_0^1 dt\,\frac{d}{dt}\,\omega(\mathbf{A}_0+t(\mathbf{A}'-\mathbf{A}_0))\right|\\ &=\left|\int_0^1\left(\sum_{j=1}^{l}\frac{\partial\omega}{\partial A_j}(\mathbf{A}_0+t(\mathbf{A}'-\mathbf{A}_0))(A'_j-A_{0j})\right)dt\right|\\ &\leqslant l\frac{E_0}{\rho_0-\tilde\rho_0}\tilde\rho_0\leqslant 2lE_0\frac{\tilde\rho_0}{\rho_0}\end{aligned} \tag{5.12.38}$$

by a dimensional estimate like the first of Eqs. (5.11.18); hence, if we define

$$\tilde\rho_0=\rho_0\left(2lCE_0N_0^{l+1}\right)^{-1}, \tag{5.12.39}$$

we see, since $CE_0\geqslant 1$, $N_0\geqslant 1$, $\tilde\rho_0<\rho_0/2$, that Eq. (5.12.37) implies

$$|\omega(\mathbf{A}')\cdot\boldsymbol{\nu}|^{-1}<2C|\boldsymbol{\nu}|^l,\qquad\forall\boldsymbol{\nu},\quad 0<|\boldsymbol{\nu}|\leqslant N_0 \tag{5.12.40}$$

for $\mathbf{A}'\in\hat{\mathbb{S}}_{\tilde\rho_0}(\mathbf{A}_0)$.

[36]One could take, say, $B_1=B_2=(4\sqrt{e})^l 2$, because $(1+e^{-\delta_0/2})/(1-e^{-\delta_0/2})\leqslant 4\sqrt{e}\,\delta_0^{-1}$, if $\delta_0<1$.

Hence, Eq. (5.12.36) implies that Φ_0 is holomorphic in $C(\tilde{\rho}_0, \xi_0 - \delta_0; \mathbf{A}_0)$ and that, using the second of Eqs. (5.12.32),[37]

$$\begin{aligned}|\Phi_0(\mathbf{A}',\mathbf{z})| &\leqslant \left| \sum_{0<|\boldsymbol{\nu}|\leqslant N_0} \frac{|f_{0\boldsymbol{\nu}}(\mathbf{A}')|}{|\omega(\mathbf{A}')\cdot\boldsymbol{\nu}|} e^{(\xi_0-\delta_0)|\boldsymbol{\nu}|} \right| \\ &\leqslant \sum_{0<|\boldsymbol{\nu}|\leqslant N_0} 2C|\boldsymbol{\nu}|^{l} |f_{0\boldsymbol{\nu}}(\mathbf{A}')| e^{(\xi_0-\delta_0)|\boldsymbol{\nu}|} \\ &\leqslant \sum_{|\boldsymbol{\nu}|\neq 0} 2C|\boldsymbol{\nu}|^{l-1} l \epsilon_0 \rho_0 e^{-\delta_0|\boldsymbol{\nu}|} \leqslant B_3 \epsilon_0 C \rho_0 \delta_0^{-2l+1}\end{aligned} \tag{5.12.41}$$

for all $(\mathbf{A}',\mathbf{z}) \in C(\tilde{\rho}_0, \xi_0 - \delta_0; \mathbf{A}_0)$, with $B_3 > 2$.[38]

Hence, by the dimensional estimates of Eq. (5.11.18),

$$\begin{aligned}\left|\frac{\partial\Phi_0}{\partial\mathbf{A}'}\right|_{\tilde{\rho}_0/2,\xi_0-\delta_0} &\leqslant 2B_3\epsilon_0 C\rho_0\delta_0^{-2l+1}\tilde{\rho}_0^{-1}, \\ \left|\frac{\partial\Phi_0}{\partial\boldsymbol{\varphi}}\right|_{\tilde{\rho}_0,\xi_0-2\delta_0} &\leqslant B_3\epsilon_0 C\rho_0\delta_0^{-2l}e^{2\xi_0}\delta_0^{-1}.\end{aligned} \tag{5.12.42}$$

Therefore, it makes sense to consider the map

$$\begin{aligned}\mathbf{A} &= \mathbf{A}' + \frac{\partial\Phi_0}{\partial\boldsymbol{\varphi}}(\mathbf{A}',\mathbf{z}), \\ z_j' &= z_j \exp i \frac{\partial\Phi_0}{\partial A_j'}(\mathbf{A}',\mathbf{z}),\end{aligned} \qquad j = 1, \dots, l \tag{5.12.43}$$

defined for $(\mathbf{A}',\mathbf{z}) \in C(\tilde{\rho}_0, \xi_0 - \delta_0; \mathbf{A}_0)$ with values in $\mathbf{C}^{2l}$. Here we regard the second of Eqs. (5.12.43) as the complex version of

$$\boldsymbol{\varphi}' = \boldsymbol{\varphi} + \frac{\partial\Phi_0}{\partial\mathbf{A}'}(\mathbf{A}',\mathbf{z}). \tag{5.12.43'}$$

Now arises the problem of inverting the first of Eqs. (5.12.43) or the second of Eqs. (5.12.43) in the respective forms

$$\begin{aligned}\mathbf{A}' &= \mathbf{A} + \boldsymbol{\Xi}'(\mathbf{A},\mathbf{z}), \\ z_j &= z_j' \exp i\Delta_j(\mathbf{A}',\mathbf{z}'),\end{aligned} \qquad j = 1, \dots, l, \tag{5.12.44}$$

where the second should be regarded as the complex extension of

$$\boldsymbol{\varphi} = \boldsymbol{\varphi}' + \boldsymbol{\Delta}(\mathbf{A}',\boldsymbol{\varphi}'). \tag{5.12.45}$$

[37] Recall that $|\boldsymbol{\nu}|$, $\boldsymbol{\nu} \in \mathbf{Z}^l$, and $|\mathbf{w}|$, $\mathbf{w} \in \mathbf{C}^l$, have a different meaning by our conventions, Eqs. (5.11.1) and (5.11.2). This explains the factor l.

[38] Using

$$\sum_{\boldsymbol{\nu}} |\boldsymbol{\nu}|^a e^{-\delta|\boldsymbol{\nu}|} \leqslant \max_{0\leqslant y\leqslant\infty}(y^a e^{-\delta y/2})\left(\sum_{\boldsymbol{\nu}} e^{-\delta|\boldsymbol{\nu}|/2}\right) \leqslant (\max_{y\geqslant 0} y^a e^{-y})(2/\delta)^a(4\sqrt{e})^l \delta^{-l},$$

one can take, say, $B_3 = l!\,2^l(4\sqrt{e})^l > 2$.

For this purpose, we use, respectively, Proposition 21, p. 491, and Proposition 20, p. 490, §5.11 (choosing, say, $\tau = \log 2$). They guarantee that the above inversions can indeed be made in the desired form, via Eqs. (5.12.39) and (5.12.42), if

$$\frac{1}{2l} B_4 \epsilon_0 C \frac{\rho_0}{\tilde{\rho}_0} \delta_0^{-2l} \equiv B_4 \epsilon_0 C E_0 C N_0^{l+1} \delta_0^{-2l} < 1, \tag{5.12.46}$$

where B_4 is a suitable constant determined by imposing Eqs. (5.11.23) and (5.11.27).[39] In this case, $\boldsymbol{\Xi}'$ is holomorphic on $C(\tilde{\rho}_0/2, \xi_0 - 2\delta_0; \mathbf{A}_0)$ as well as $\boldsymbol{\Delta}$ and they verify the bounds

$$\begin{aligned} |\boldsymbol{\Xi}'|_{\tilde{\rho}_0/2,\, \xi_0 - 2\delta_0} &< B_3 \epsilon_0 C \rho_0 \delta_0^{-2l} e^{2\xi_0} < \frac{\tilde{\rho}_0}{8}, \\ |\boldsymbol{\Delta}|_{\tilde{\rho}_0/2,\, \xi_0 - 2\delta_0} &< 2 B_3 \epsilon_0 C \rho_0 \delta_0^{-2l+1} \tilde{\rho}_0^{-1} < \delta_0, \end{aligned} \tag{5.12.47}$$

where the first right-hand-side inequalities follow from Eq. (5.11.29) or Eq. (5.11.25), while the second right-hand-side inequalities follow from Eq. (5.12.46) if B_4 is chosen as in the footnote.[39]

Equations (5.12.47) permit us to define on $C(\tilde{\rho}_0/2, \xi_0 - 2\delta_0; \mathbf{A}_0)$, say, the functions

$$\begin{aligned} \boldsymbol{\Xi}(\mathbf{A}', \mathbf{z}') &= \frac{\partial \Phi_0}{\partial \boldsymbol{\varphi}} (\mathbf{A}', \mathbf{z}' e^{i\boldsymbol{\Delta}(\mathbf{A}', \mathbf{z}')}), \\ \boldsymbol{\Delta}'(\mathbf{A}, \mathbf{z}) &= \frac{\partial \Phi_0}{\partial \mathbf{A}'} (\mathbf{A} + \boldsymbol{\Xi}'(\mathbf{A}, \mathbf{z}), \mathbf{z}) \end{aligned} \tag{5.12.48}$$

and, by Eqs. (5.12.42) and (5.12.39) they verify

$$\begin{aligned} |\boldsymbol{\Xi}|_{\tilde{\rho}_0/2, \xi - 3\delta_0} &\leqslant B_3 \epsilon_0 C \rho_0 \delta_0^{-2l} e^{2\xi_0} < \frac{\tilde{\rho}_0}{8}, \\ |\boldsymbol{\Delta}'|_{\tilde{\rho}_0/4, \xi_0 - 2\delta_0} &\leqslant 4 l B_3 \epsilon_0 C E_0 C N_0^{l+1} \delta_0^{-2l+1} < \delta_0. \end{aligned} \tag{5.12.49}$$

Therefore, we define the maps $\mathscr{C}_0$:

$$\begin{aligned} \mathbf{A} &= \mathbf{A}' + \boldsymbol{\Xi}(\mathbf{A}', \mathbf{z}'), \\ \mathbf{z} &= \mathbf{z}' \exp i\boldsymbol{\Delta}(\mathbf{A}', \mathbf{z}'), \end{aligned} \qquad (\mathbf{A}', \mathbf{z}') \in C\left(\frac{\tilde{\rho}_0}{2}, \xi_0 - 2\delta_0; \mathbf{A}_0\right), \tag{5.12.50}$$

[39] One can take $B_4 = 4\gamma e^2 B_3 l$. Note that, not surprisingly, both inversions require the *same* condition up to a constant factor, adjusted in Eq. (5.12.46) to be the same. This is basically so because the implicit function theorems impose conditions on $\partial^2\Phi_0/\partial\mathbf{A}'\,\partial\boldsymbol{\varphi}$ for the first inversion or on $\partial^2\Phi_0/\partial\boldsymbol{\varphi}\,\partial\mathbf{A}'$ for the second.

Actually, it would be easy to check that Eqs. (5.12.42) and (5.12.46) automatically imply that the matrix $J_{ij} = \delta_{ij} + \partial^2\Phi_0/\partial A_j'\,\partial\varphi_i$ is invertible in $C(\tilde{\rho}_0/8, \xi_0 - 4\delta_0; \mathbf{A}_0)$ if B_3, B_4 are chosen as in footnote 37 to p. 505 and as above. Hence, the general theory of the canonical transformations, §3.11, Problems 9–11, p. 228, shows that under the condition of Eq. (5.12.46), the map of Eq. (5.12.43) locally generates a completely canonical transformation defined on $S_{\tilde{\rho}_0/8}(\mathbf{A}_0) \times \mathbf{T}^l$, changing $(\mathbf{A}', \boldsymbol{\varphi}')$ into $(\mathbf{A}, \boldsymbol{\varphi})$. This map is actually a globally canonical map, as we shall see.

and $\tilde{\mathcal{C}}_0$:

$$\begin{aligned}\mathbf{A}' &= \mathbf{A} + \boldsymbol{\Xi}'(\mathbf{A},\mathbf{z}),\\ \mathbf{z}' &= \mathbf{z}\exp i\boldsymbol{\Delta}'(\mathbf{A},\mathbf{z}),\end{aligned} \qquad (\mathbf{A},\mathbf{z}) \in C\left(\frac{\tilde{\rho}_0}{2}, \xi_0 - 2\delta_0; \mathbf{A}_0\right) \tag{5.12.51}$$

which have the properties [by Eqs. (5.12.47) and (5.12.49)]

$$\begin{aligned}\mathcal{C}_0 &: C\left(\frac{\tilde{\rho}_0}{4}, \xi_0 - 3\delta_0; \mathbf{A}_0\right) \to C\left(\frac{\tilde{\rho}_0}{2}, \xi_0 - 2\delta_0; \mathbf{A}_0\right),\\ \tilde{\mathcal{C}}_0 &: C\left(\frac{\tilde{\rho}_0}{4}, \xi_0 - 3\delta_0; \mathbf{A}_0\right) \to C\left(\frac{\tilde{\rho}_0}{2}, \xi_0 - 2\delta_0; \mathbf{A}_0\right).\end{aligned} \tag{5.12.52}$$

Hence, it makes sense to consider $\mathcal{C}_0\tilde{\mathcal{C}}_0$ and $\tilde{\mathcal{C}}_0\mathcal{C}_0$ on $C(\tilde{\rho}_0/4, \xi_0 - 3\delta_0; \mathbf{A}_0)$. By construction, $\mathcal{C}_0$ and $\tilde{\mathcal{C}}_0$ are inverses of each other:

$$\mathcal{C}_0\tilde{\mathcal{C}}_0 \equiv \tilde{\mathcal{C}}_0\mathcal{C}_0 \equiv \{\text{identity map}\} \tag{5.12.53}$$

on $C(\tilde{\rho}_0/4, \xi_0 - 3\delta_0; \mathbf{A}_0)$.

It follows from the general theory of canonical transformations that $\mathcal{C}_0$ and $\tilde{\mathcal{C}}_0$ are completely canonical maps, inverse to each other, of $\mathbb{S}_{\tilde{\rho}_0/4} \times \mathbf{T}^l$ onto its image.

If a motion takes place in $\mathcal{C}_0(\mathbb{S}_{\tilde{\rho}_0/4} \times \mathbf{T}^l)$, it can be described in the $(\mathbf{A}', \boldsymbol{\varphi}')$ variables as a motion developing under the equations associated with Hamiltonian:

$$H_1(\mathbf{A}', \boldsymbol{\varphi}') = h_0(\mathbf{A}' + \boldsymbol{\Xi}(\mathbf{A}', \boldsymbol{\varphi}')) + f_0(\mathbf{A}' + \boldsymbol{\Xi}(\mathbf{A}', \boldsymbol{\varphi}'), \boldsymbol{\varphi}' + \boldsymbol{\Delta}(\mathbf{A}', \boldsymbol{\varphi}')) \tag{5.12.54}$$

which, following the perturbation theory, we write as

$$\begin{aligned}H_1(\mathbf{A}', \boldsymbol{\varphi}') &\equiv \{h_0(\mathbf{A}') + f_{00}(\mathbf{A}')\}\\ &\quad + \{h_0(\mathbf{A}' + \boldsymbol{\Xi}(\mathbf{A}', \boldsymbol{\varphi}')) - h_0(\mathbf{A}')\\ &\qquad + f_0(\mathbf{A}' + \boldsymbol{\Xi}(\mathbf{A}', \boldsymbol{\varphi}'), \boldsymbol{\varphi}' + \boldsymbol{\Delta}(\mathbf{A}', \boldsymbol{\varphi}')) - f_{00}(\mathbf{A}')\}\\ &\equiv \{h_1(\mathbf{A}')\} + \{f_1(\mathbf{A}', \boldsymbol{\varphi}')\},\end{aligned} \tag{5.12.55}$$

where h_1 and f_1 are implicitly defined, respectively, as the first and second curly-bracket terms in the intermediate equality in Eq. (5.12.55), $\forall(\mathbf{A}', \mathbf{z}') \in C(\tilde{\rho}_0/4, \xi_0 - 3\delta_0; \mathbf{A}_0)$.

We shall henceforth regard $C(\tilde{\rho}_0/4, \xi_0 - 3\delta_0; \mathbf{A}_0)$ as the domain of definition and holomorphy of h_1 and f_1: however we shall further reduce it, later, for the purpose of using dimensional estimates or for other needs. This basic choice of domain is convenient since we control well $\mathcal{C}_0$ on this set, see Eq. (5.12.52).

Our next task, according to the program of the proof, is to find a point $\mathbf{A}_1 \in \mathbb{S}_{\tilde{\rho}_0/4}(\mathbf{A}_0)$ such that

$$\frac{\partial h_1}{\partial \mathbf{A}'}(\mathbf{A}_1) = \boldsymbol{\omega}_0. \tag{5.12.56}$$

Recalling that $\omega_0 = (\partial h_0/\partial \mathbf{A})(\mathbf{A}_0)$, this equation can be elaborated as

$$\begin{aligned}
&\omega(\mathbf{A}') - \omega(\mathbf{A}_0) + \frac{\partial f_{00}}{\partial \mathbf{A}'}(\mathbf{A}') = \mathbf{0} \\
&\quad \Rightarrow M(\mathbf{A}_0)(\mathbf{A}' - \mathbf{A}_0) \\
&\qquad + \left[\omega(\mathbf{A}') - \omega(\mathbf{A}_0) - M(\mathbf{A}_0)(\mathbf{A}' - \mathbf{A}_0) + \frac{\partial f_{00}}{\partial \mathbf{A}'}(\mathbf{A}')\right] = \mathbf{0} \qquad (5.12.57) \\
&\quad \Rightarrow (\mathbf{A}' - \mathbf{A}_0) + M(\mathbf{A}_0)^{-1}\Big[\omega(\mathbf{A}') - \omega(\mathbf{A}_0) \\
&\qquad\qquad - M(\mathbf{A}_0)(\mathbf{A}' - \mathbf{A}_0) + \frac{\partial f_{00}}{\partial \mathbf{A}'}(\mathbf{A}')\Big] \\
&\quad \equiv (\mathbf{A}' - \mathbf{A}_0) + \mathbf{n}(\mathbf{A}') = \mathbf{0},
\end{aligned}$$

where $\mathbf{n}$ is defined on $\hat{\mathbb{S}}_{\rho_0}(\mathbf{A}_0)$ by the term within square brackets in the third relation.

We can apply Proposition 19, p. 490, to the last equation to deduce that if $\gamma|n|_\rho < \rho$ for some $\rho < \tilde{\rho}_0/4$, then the equation admits a unique solution $\mathbf{A}_1 \in \mathbb{S}_\rho(\mathbf{A}_0)$. Hence, we must estimate $|\mathbf{n}|_\rho$ for $\rho < \tilde{\rho}_0/4 < \rho_0/2$.

For this purpose we note that

$$\begin{aligned}
&|\omega(\mathbf{A}') - \omega(\mathbf{A}_0) - M(\mathbf{A}_0)(\mathbf{A}' - \mathbf{A}_0)| \\
&\quad \equiv \left|\int_0^1 d\tau \int_0^\tau d\theta \frac{d^2}{d\theta^2}\omega(\mathbf{A}_0 + \theta(\mathbf{A} - \mathbf{A}_0))\right| \\
&\quad \equiv \left|\sum_{i,j=1}^{l} \int_0^1 d\tau \int_0^\tau d\theta \frac{\partial^2 \omega}{\partial A_i \partial A_j}(\mathbf{A}_0 + \theta(\mathbf{A}' - \mathbf{A}_0))(A_i' - A_{0i})(A_j' - A_{0j})\right| \\
&\quad \leqslant \rho^2 l^2 2 \frac{E_0}{(\rho_0 - \rho)^2} \leqslant 8 l^2 E_0 \left(\frac{\rho}{\rho_0}\right)^2,
\end{aligned}$$

having estimated the second derivative of $\boldsymbol{\omega}$ by a dimensional estimate; see Eqs. (5.11.9) and (5.11.18). Hence, if $\rho < \tilde{\rho}_0/4$ and if the first of Eqs. (5.12.32) is used with $\boldsymbol{\nu} = \mathbf{0}$:

$$|\mathbf{n}|_\rho \leqslant \eta_0\left(8 l^2 E_0\left(\frac{\rho}{\rho_0}\right)^2 + \epsilon_0\right) \qquad (5.12.58)$$

so that if we choose (recalling that $CE_0 > 1$, $C\epsilon_0 < 1$)

$$\rho = \frac{\tilde{\rho}_0}{8l}\sqrt{\frac{\epsilon_0}{E_0}} < \frac{\tilde{\rho}_0}{8},$$

we see that $|\mathbf{n}|_\rho < 2\epsilon_0\eta_0$. Applying Proposition 19, p. 490, we see that if $2\gamma\eta_0\epsilon_0 < \tilde{\rho}_0/8$, then Eq. (5.12.56) admits a solution $\mathbf{A}_1 \in \mathbb{S}_{\tilde{\rho}_0/8}(\mathbf{A}_0)$. The condition $2\gamma\eta_0\epsilon_0 < \tilde{\rho}_0/8$ can be rewritten, recalling the expression (5.12.39) for $\tilde{\rho}_0$, $16\gamma\epsilon_0 C(\eta_0\rho_0^{-1}E_0)N_0^{l+1} < 1$, and it can be implied together with Eq. (5.12.46) by requiring

$$B_5\epsilon_0 C E_0 C\big(\eta_0\rho_0^{-1}E_0\big)N_0^{l+1}\delta_0^{-2l} < 1, \tag{5.12.59}$$

having used Eq. (5.12.28) and having chosen B_5 suitably,[40] $B_5 > 4$.

Hence, if Eq. (5.12.59) holds, we can consider the Hamiltonian h_1 and its perturbation f_1 in Eq. (5.12.55) as functions defined and holomorphic in $C(\tilde{\rho}_0/8, \xi_0 - 3\delta_0; \mathbf{A}_1)$ with $\mathbf{A}_1$ so chosen that Eq. (5.12.56) holds.

We are now in a position where we can try to iterate the argument. We can in fact associate with the Hamiltonians h_1, f_1 in $C(\tilde{\rho}_0/16, \xi_0 - 4\delta_0; \mathbf{A}_1)$[41] the characteristic parameters $\rho_1 = \tilde{\rho}_0/16$, $\xi_1 = \xi_0 - 4\delta_0$ and E_1, η_1, ϵ_1, where E_1, h_1, ϵ_1 are estimates of

$$\left|\frac{\partial h_1}{\partial \mathbf{A}'}\right|_{\rho_1}, \qquad \left\|\left(\frac{\partial^2 h_1}{\partial \mathbf{A}'\partial \mathbf{A}'}\right)^{-1}\right\|_{\rho_1}, \qquad \left|\frac{\partial f}{\partial \mathbf{A}'}\right|_{\rho_1,\xi_1} + \rho_1^{-1}\left|\frac{\partial f_1}{\partial \boldsymbol{\varphi}'}\right|_{\rho_1,\xi_1}.$$

To find E_1, η_1, ϵ_1, we apply, as usual, some dimensional estimates.

The E_1 estimate is based on the first of Eqs. (5.12.32):

$$\left|\frac{\partial h_1}{\partial \mathbf{A}'}(\mathbf{A}')\right| = \left|\frac{\partial h_0}{\partial \mathbf{A}'}(\mathbf{A}') + \frac{\partial f_{00}}{\partial \mathbf{A}'}(\mathbf{A}')\right| \leqslant E_0 + \epsilon_0. \tag{5.12.60}$$

The η_1 estimate is based on the dimensional estimate, Eq. (5.11.18), for $\sigma_{ij}(\mathbf{A}') = (\partial^2 f_{00}/\partial \mathbf{A}'_i \partial A'_j)(\mathbf{A}')$ as $|\sigma_{ij}(\mathbf{A}')| \leqslant \epsilon_0/(\rho_0 - \tilde{\rho}_0/4) \leqslant 2\epsilon_0\rho_0^{-1}$, $\forall \mathbf{A}' \in \mathcal{S}_{\rho_1}(\mathbf{A}_1)$; in fact,

$$\begin{aligned}
M_1(\mathbf{A}')^{-1} &= \big(M_0(\mathbf{A}') + \sigma(\mathbf{A}')\big)^{-1}\\
&= \Big(M_0(\mathbf{A}')\big(1 + M_0(\mathbf{A}')^{-1}\sigma(\mathbf{A}')\big)\Big)^{-1}\\
&= \big(1 + M_0(\mathbf{A}')^{-1}\sigma(\mathbf{A}')\big)^{-1} M_0(\mathbf{A}')^{-1}\\
&\equiv M_0(\mathbf{A}')^{-1} + \Big[\big(1 + M_0(\mathbf{A}')^{-1}\sigma(\mathbf{A}')\big)^{-1} - 1\Big]M_0(\mathbf{A}')^{-1},
\end{aligned} \tag{5.12.61}$$

and (since given two $l \times l$ matrices R and S it is $\|RS\| \leqslant \|R\|\,\|S\|$ and $\|(1+R)^{-1} - 1\| \leqslant 2\|R\|$, if $\|R\| < \frac{1}{2}$)[42] we see that $\|M_0(\mathbf{A}')^{-1}\sigma(\mathbf{A}')\| \leqslant 2\epsilon_0\eta_0\rho_0^{-1} < \frac{1}{2}$ [by Eq. (5.12.59)] and

$$\|M_1(\mathbf{A}')\| \leqslant \eta_0 + 4\epsilon_0\eta_0^2\rho_0^{-1}. \tag{5.12.62}$$

The estimate of ϵ_1 is slightly more complicated because it involves the derivatives of $\boldsymbol{\Xi}$ and $\boldsymbol{\Delta}$ which, however, can be easily estimated dimensionally.

[40] e.g., one could take $B_5 = 2B_4 16\gamma = 2^{32} l!\,(8\sqrt{e}\,)^l \leqslant 2^{32} l!\, 2^{4l}$, if $\gamma = 2^8$.

[41] We further restrict the domain in which we consider h_1, f_1 to be able to perform dimensional estimates later.

[42] See Appendix E, Eqs. (E.2) and (E.10), p. 526.

We first elaborate the formal expression of f_1 by adding and subtracting suitable terms:

$$\begin{aligned}
f_1(\mathbf{A}',\mathbf{z}') &= h_0(\mathbf{A}' + \Xi(\mathbf{A}',\mathbf{z}')) - h_0(\mathbf{A}') + f_0^{[\leqslant N_0]}\big(\mathbf{A}' + \Xi(\mathbf{A}',\mathbf{z}'),\mathbf{z}'e^{i\Delta(\mathbf{A}',\mathbf{z}')}\big)\\
&\quad - f_{00}(\mathbf{A}') + f^{[>N_0]}\big(\mathbf{A}' + \Xi(\mathbf{A}',\mathbf{z}'),\mathbf{z}'e^{i\Delta(\mathbf{A}',\mathbf{z}')}\big)\\
&= \{h_0(\mathbf{A}' + \Xi(\mathbf{A}',\mathbf{z}')) - h_0(\mathbf{A}') - \omega(\mathbf{A}')\cdot\Xi(\mathbf{A}',\mathbf{z}')\} \qquad (5.12.63)\\
&\quad + \big\{f_0^{[\leqslant N_0]}\big(\mathbf{A}' + \Xi(\mathbf{A}',\mathbf{z}'),\mathbf{z}'e^{i\Delta(\mathbf{A}',\mathbf{z}')}\big) - f_{00}(\mathbf{A}') + \omega(\mathbf{A}')\cdot\Xi(\mathbf{A}',\mathbf{z}')\big\}\\
&\quad + \big\{f_0^{[>N_0]}\big(\mathbf{A}' + \Xi(\mathbf{A}',\mathbf{z}'),\mathbf{z}'e^{i\Delta(h\mathbf{A}',\mathbf{z}')}\big)\big\},
\end{aligned}$$

where the addition and subtraction of $\omega\cdot\Xi$ is suggested by the formal perturbation theory and by the fact that, if $(\mathbf{A},\mathbf{z}) = \mathscr{C}_0(\mathbf{A}',\mathbf{z}')$, it is $\Xi(\mathbf{A}',\mathbf{z}') = (\partial\Phi_0/\partial\varphi)(\mathbf{A}',\mathbf{z})$ so that the various terms in curly brackets formally have size $O(\epsilon_0^2)$. In fact, the first term is manifestly of $O(|\Xi|^2)$ and Ξ is formally of $O(\epsilon_0)$; the third term is by construction of formal order $O(\epsilon_0^2)$, while the second term can be rewritten as

$$f_0^{[\leqslant N_0]}\big(\mathbf{A}' + \Xi(\mathbf{A}',\mathbf{z}'),\mathbf{z}'e^{i\Delta(\mathbf{A}',\mathbf{z}')}\big) - f_0^{[\leqslant N_0]}\big(\mathbf{A}',\mathbf{z}'e^{i\Delta(\mathbf{A}',\mathbf{z}')}\big) \qquad (5.12.64)$$

because by the definition of Φ_0,

$$\omega(\mathbf{A}')\cdot\Xi(\mathbf{A}',\mathbf{z}') \equiv \omega(\mathbf{A}')\cdot\frac{\partial\Phi_0}{\partial\varphi}(\mathbf{A}',\mathbf{z}) = f_0^{[\leqslant N_0]}(\mathbf{A}',\mathbf{z}) - f_{00}(\mathbf{A}')$$

and, therefore, Eq. (5.12.64) is formally of $O(\epsilon_0)O(\Xi)$, i.e., $O(\epsilon_0^2)$ (here we use that $\mathbf{z}'e^{i\Delta(\mathbf{A}',\mathbf{z}')} \equiv \mathbf{z}$, also).

Writing the three terms in curly brackets in Eq. (5.12.63) as $f_0^{\mathrm{I}}, f_0^{\mathrm{II}}, f_0^{\mathrm{III}}$, respectively, we now show rigorously that they have the right order of magnitude.

Dropping the $(\mathbf{A}',\mathbf{z}')$ in the arguments of Ξ,Δ, for simplicity, and using Eq. (5.12.64) and the Taylor–Lagrange formulae, we find

$$\begin{aligned}
f_1^{\mathrm{I}} &= \int_0^1 dt_1 \int_0^{t_1} dt_2 \frac{d^2}{dt_2^2} h_0(\mathbf{A}' + t\Xi)\\
&\equiv \int_0^1 (1-t)\left(\sum_{j,k=1}^{l} \frac{\partial^2 h_0}{\partial A_j \partial A_k}(\mathbf{A}' + t\Xi)\Xi_j\Xi_k\right) dt,\\
f_1^{\mathrm{II}} &= \int_0^1 dt \frac{d}{dt} f_0^{[\leqslant N_0]}(\mathbf{A}' + t\Xi,\mathbf{z}'e^{i\Delta}) \qquad (5.12.65)\\
&\equiv \int_0^1 dt \left(\sum_{j=1}^{l} \frac{\partial f^{[\leqslant N_0]}}{\partial A_j}(\mathbf{A}' + t\Xi,\mathbf{z}'e^{i\Delta})\Xi_j\right),\\
f_1^{\mathrm{III}} &= f_0^{[>N_0]}(\mathbf{A}' + \Xi,\mathbf{z}'e^{i\Delta}).
\end{aligned}$$

Easy bounds can now be found for $f_1^{\mathrm{I}}, f_1^{\mathrm{II}}, f_1^{\mathrm{III}}$ by dimensional estimates. In fact, combining Eq. (5.11.18) with the Eq. (5.12.33), the second of Eqs. (5.12.47) and the first of Eqs. (5.12.49), we deduce the following inequalities, except the fourth:

$$\begin{aligned}
&\left|\frac{\partial^2 h_0}{\partial \mathbf{A}\,\partial \mathbf{A}}\right|_{\tilde{\rho}_0} \leqslant \frac{E_0}{\rho_0-\tilde{\rho}_0} \leqslant 2E_0\rho_0^{-1},\\
&\left|\frac{\partial^3 h_0}{\partial \mathbf{A}\,\partial \mathbf{A}\,\partial \mathbf{A}}\right|_{\tilde{\rho}_0} \leqslant \frac{2!\,E_0}{(\rho_0-\tilde{\rho}_0)^2} \leqslant 8E_0\rho_0^{-2},\\
&\left|\frac{\partial f_0^{[\leqslant N_0]}}{\partial \mathbf{A}}\right|_{\tilde{\rho}_0,\xi_0-\delta_0} \leqslant B_1\epsilon_0\delta_0^{-l},\\
&|f_0^{[>N_0]}|_{\tilde{\rho}_0,\xi_0-2\delta_0} \leqslant lB_2\epsilon_0^2C\rho_0,\\
&|\boldsymbol{\Xi}|_{\tilde{\rho}_0/2,\xi_0-3\delta_0} \leqslant B_3\epsilon_0C\rho_0\delta_0^{-2l}e^{-2\xi_0} < \frac{\tilde{\rho}_0}{8},\\
&|\boldsymbol{\Delta}|_{\tilde{\rho}_0/2,\xi_0-3\delta_0} \leqslant 4lB_3\epsilon_0CE_0CN_0^{l+1}\delta_0^{-2l+1} \leqslant \delta_0.
\end{aligned}\tag{5.12.66}$$

To prove the fourth inequality, we use the second of Eqs. (4.12.33):

$$\begin{aligned}
|f_0^{[>N_0]}(\mathbf{A},\mathbf{z})| &= \left|\sum_{|\boldsymbol{\nu}|>N_0} f_{0\boldsymbol{\nu}}(\mathbf{A})\mathbf{z}^{\boldsymbol{\nu}}\right|\\
&\leqslant \sum_{|\boldsymbol{\nu}|>N_0} |f_{0\boldsymbol{\nu}}(\mathbf{A})|e^{(\xi_0-\delta)|\boldsymbol{\nu}|}\\
&\leqslant \sum_{|\boldsymbol{\nu}|>N_0} |\boldsymbol{\nu}||f_{0\boldsymbol{\nu}}(\mathbf{A})|e^{(\xi_0-\delta_0)|\boldsymbol{\nu}|}\\
&\leqslant \epsilon_0\rho_0 l\sum_{|\boldsymbol{\nu}|>N_0} e^{-\delta_0|\boldsymbol{\nu}|} \leqslant \epsilon_0\rho_0 l e^{-N_0\delta_0/2}\sum_{|\boldsymbol{\nu}|>N_0} e^{-(\delta_0/2)|\boldsymbol{\nu}|}\\
&\leqslant B_2 l\epsilon_0^2C\rho_0
\end{aligned}\tag{5.12.67}$$

[see also Eq. (5.12.35)].

For $(\mathbf{A}',\mathbf{z}') \in C(\tilde{\rho}_0/4, \xi_0-3\delta_0; \mathbf{A}_0)$, we see that $(\mathbf{A}'+t\boldsymbol{\Xi}, \mathbf{z}'e^{i\boldsymbol{\Delta}}) \in C(\tilde{\rho}_0/2, \xi_0-2\delta_0; \mathbf{A}_0)$; so we can insert the bounds of Eq. (5.12.66) into Eq. (5.12.65) using $e^{\xi} < e < 4$ for simplicity, to obtain

$$\begin{aligned}
|f_1^{\mathrm{I}}|_{\tilde{\rho}_0/4,\xi_0-3\delta_0} &\leqslant 2E_0\rho_0^{-1}l^2\left(B_3\epsilon_0C\rho_0\delta_0^{-2l}e^2\right)^2\\
&\leqslant 2^9B_3^2l^2\epsilon_0^2CE_0C\rho_0\delta_0^{-4l},\\
|f_1^{\mathrm{II}}|_{\tilde{\rho}_0/4,\xi_0-3\delta_0} &\leqslant 2^4B_1B_3l\epsilon_0^2C\delta_0^{-3l}\rho_0,\\
|f_1^{\mathrm{III}}|_{\tilde{\rho}_0/4,\xi_0-3\delta_0} &\leqslant B_2l\epsilon_0^2C\rho_0
\end{aligned}\tag{5.12.68}$$

so that

$$|f_1|_{\tilde\rho_0/4,\xi_0-3\delta_0} \leq B_6\epsilon_0^2 CE_0C\delta_0^{-4l}\rho_0, \tag{5.12.69}$$

where B_6 is suitably chosen.[43]

Now we note that $\mathbf{A}_1 \in \mathbb{S}_{\tilde\rho_0/8}(\mathbf{A}_0)$ so that $C(\rho_1,\xi_1;\mathbf{A}_1) \subset C(\frac{3}{16}\tilde\rho_0,\xi_0-4\delta_0;\mathbf{A}_0)$, and we see that the boundary of $C(\rho_1,\xi_1;\mathbf{A}_1)$ is quite far from that of $C(\tilde\rho_0/4,\xi_0-3\delta_0;\mathbf{A}_0)$ and we can use Eq. (5.12.69) to estimate dimensionally

$$\sup\left|\frac{\partial f_1}{\partial \mathbf{A}'}(\mathbf{A}',\mathbf{z}')\right| + \rho_1^{-1}\sup\left|\frac{\partial f_1}{\partial \boldsymbol{\varphi}'}(\mathbf{A}',\mathbf{z}')\right|, \tag{5.12.70}$$

where the supremum is taken over $C(\rho_1,\xi_1;\mathbf{A}_1)$. Calling $\bar\epsilon_1$ Eq. (5.12.70), we immediately find

$$\begin{aligned}\bar\epsilon_1 &\leq |f_1|_{\tilde\rho_0/4,\xi_0-3\delta_0}\left(\frac{1}{\tilde\rho_0/4-\frac{3}{16}\tilde\rho_0}+\frac{1}{\rho_1}\frac{e^{2\delta_0}}{\delta_0}\right)\\ &\leq \frac{2^5e^2}{\delta_0\tilde\rho_0}|f_1|_{\tilde\rho_0/4,\xi_0-3\delta_0} \leq B_7\epsilon_0^2C(E_0C)^2N_0^{l+1}\delta_0^{-4l-1}\end{aligned} \tag{5.12.71}$$

recalling that $\rho_1 = \tilde\rho_0/16$ and suitably choosing B_7.[44]

So, collecting all the above inequalities (5.12.71), (5.12.62), and (5.12.60) and the definitions of ρ_1,ξ_1, we see that we can take the following quantities as characteristic parameters for the Hamiltonians h_1, f_1 in $C(\rho_1,\xi_1;\mathbf{A}_1)$:

$$\begin{aligned}\rho_1 &= \frac{\rho_0}{32lE_0CN_0^{l+1}}, \qquad N_0 = 2\delta_0^{-1}\log(C\epsilon_0\delta_0^l)^{-1},\\ \xi_1 &= \xi_0-4\delta_0,\\ E_1 &= E_0+\epsilon_0,\\ \eta_1 &= \eta_0+4\epsilon_0\eta_0^2\rho_0^{-1},\\ \epsilon_1 &= B_8C\epsilon_0^2(CE_0)^2\left(\log(C\epsilon_0\delta_0^l)^{-1}\right)^{l+1}\delta_0^{-5l-2},\end{aligned} \tag{5.12.72}$$

having replaced N_0 in Eq. (5.12.71) with its expression, and $B_8 = 2^{l+1}B_7$ *provided* the condition of Eq. (5.12.59) holds.

We now study the mappings K_n which map $(\rho_n,\xi_n,E_n,\eta_n,\epsilon_n)$ into $(\rho_{n+1},\xi_{n+1},E_{n+1},\eta_{n+1},\epsilon_{n+1})$ defined by Eq. (5.12.72) in which δ_0 is replaced by δ_n and $0\to n$, $1\to n+1$, forgetting the condition of Eq. (5.12.59).

We note that

$$(\rho_n,\xi_n,E_n,\eta_n,\epsilon_n) = K_{n-1}\cdots K_0(\rho_0,\xi_0,E_0,\eta_0,\epsilon_0) \tag{5.12.73}$$

and we wish to show that if $C\epsilon_0$ is small enough (so that the inequality in

[43] e.g., $B_6 = 2^9l^2B_3^2 < 2^{4l}l^2(l!)^2 2^9$.

[44] e.g., $B_7 = 2^9B_6 = 2^{18+4l}l^2(l!)^2$.

Eq. (5.12.85) below holds), then:

$$\begin{aligned}
&\xi_n > \xi_\infty = \xi_0 - 4\sum_{j=0}^{\infty}\delta_j,\\
&E_n \leqslant 2E_0,\\
&\eta_n \leqslant 2\eta_0,\\
&(\epsilon_0 C)^{(2+1/2)^n} \leqslant \epsilon_n C \leqslant (\epsilon_0 C)^{(2-1/2)^n},\\
&\rho_n \geqslant \rho_0\Big\{(E_0C)^n\Big[\xi_0^{-n}(n!)^2 2^{n^2/2}\big(\log(C\epsilon_0)^{-1}\big)^{2n}\Big]^{l+1}\Big\}^{-1}.
\end{aligned} \tag{5.12.74}$$

An inductive proof of the validity of Eq. (5.12.74) under a condition of the form of Eq. (5.12.85) is described below, between Eqs. (5.12.75) and (5.12.86), for completeness. The reader should, however, first realize that Eqs. (5.12.74) and (5.12.86) are quite obviously valid under a condition of the type of Eq. (5.12.85) below.

The first equation is obvious by our choice of δ_j. The second and third follow from the last two if, say,

$$\begin{aligned}
&\epsilon_0 < \frac{E_0}{2},\\
&\sum_{n=0}^{\infty}(C\epsilon_0)^{(3/2)^n-1} < 2,\\
&\epsilon_0\eta_0\rho_0^{-1} < \tfrac{1}{8},\\
&\sum_{n=0}^{\infty}(C\epsilon_0)^{(3/2)^n-1}\Big[(E_0C)^n\xi_0^{-n}(n!)^2 2^{(n/2)^2}\big(\log(C\epsilon_0)^{-1}\big)^{2n}\Big]^{l+1} < 2\log 2.
\end{aligned} \tag{5.12.75}$$

The fourth inequality in (5.12.74) is proved by noting that for $x<1$, one has $\sup_{0\leqslant x\leqslant 1}x^a(\log x^{-1})^{l+1} \leqslant a^{-l-1}(l+1)!$, $\forall a>0$; hence, from Eq. (5.12.72),

$$\begin{aligned}
&\epsilon_n C\delta_n^l \geqslant B_8(CE_0)^2(\epsilon_{n-1}C\delta_{n-1})^2\delta_{n-1}^{-6l-2}2^{-l},\\
&\epsilon_n C\delta_n^l \leqslant B_8(2CE_0)^2(\epsilon_{n-1}C\delta_{n-1})^{5/3}\delta_{n-1}^{-6l-2}3^{l+1}(l+1)!,
\end{aligned} \tag{5.12.76}$$

where we have bounded the ratio $(\delta_n/\delta_{n-1})^l$ by 2^{-l} or 1 (below or above) and we have applied the above elementary inequality with $a=\frac{1}{3}$.

Since B_8 is very large, e.g., $B_8 2^{-l}>1$, and $CE_0>1$, $\delta_{n-1}^{-(6l+2)}>1$, we conclude from the first of Eqs. (5.12.76) that

$$\epsilon_n C\delta_n^l \geqslant (\epsilon_{n-1}C\delta_{n-1}^l)^2 \geqslant (\epsilon_0 C\delta_0^l)^{2^n} \tag{5.12.77}$$

which immediately implies the lower bound in Eq. (5.12.74) for $C\epsilon_n$ if $C\epsilon_0$ is small enough. Taking into account the explicit expression (5.12.29) for δ_n, the condition turns out to be

$$(C\epsilon_0)^{-1/4}\delta_0^l > 1. \tag{5.12.78}$$

Recalling the form of Eq. (5.12.29), the second inequality in Eqs. (5.12.76) gives

$$C\epsilon_n\delta_n^l \leqslant \left(C\epsilon_0\delta_0^l\right)^{(5/3)^n}$$

$$\times \prod_{k=1}^{n}\left[B_8(2E_0C)^2 e^{l+1}(l+1)!\left(\frac{\xi_0}{16}\right)^{-6l-2}(1+(n-k))^{12l+4}\right]^{(5/3)^{k-1}}$$

$$\equiv \left(C\epsilon_0\delta_0^l\right)^{(5/3)^n}\left[B_8(2E_0C)^2 3^{l+1}(l+1)!\left(\frac{\xi_0}{16}\right)^{-6l-2}\right]^{((5/3)^n-1)(3/2)}$$

$$\times \exp(12l+4)\left(\frac{5}{3}\right)^{n-1}\sum_{k=0}^{n-1}\left(\frac{3}{5}\right)^k\log(1+k)$$

$$\leqslant \left(C\epsilon_0\delta_0^l(E_0C)^3\xi_0^{-9l-3}B_9\right)^{(5/3)^n} \tag{5.12.79}$$

if B_9 is suitably chosen.[45]

Since for all $n \geqslant 0$, $\delta_0^{(5/3)^n} \leqslant \delta_n$, we deduce from Eq. (5.12.79):

$$\begin{aligned}(C\epsilon_n) &\leqslant (C\epsilon_0)^{(3/2)^n}\left[(C\epsilon_0)^{1-(9/10)^n}(E_0C)^3\xi_0^{-9l-3}B_9\right]^{(5/3)^n}\\ &\leqslant (C\epsilon_0)^{(3/2)^n},\end{aligned} \tag{5.12.80}$$

provided (the worst case being $n = 1$)

$$(C\epsilon_0)^{1/10}(E_0C)^3\xi_0^{-9l-3}B_9 < 1. \tag{5.12.81}$$

Finally we consider the last of Eqs. (5.12.74). Using the recursive definition of ρ_n,

$$\rho_n = \rho_0\frac{(\delta_{n-1}\cdots\delta_0)^{l+1}}{(2^5E_0Cl)^n 2^{n(l+1)}\left(\prod_{k=1}^n\log\left(C\epsilon_{n-k}\delta_{n-k}^l\right)^{-1}\right)^{l+1}}, \tag{5.12.82}$$

and using Eq. (5.12.77) and the explicit form of δ_n, this becomes

$$\rho_n \geqslant \rho_0\frac{1}{\left(l2^{5l+10}E_0C\right)^n}\left[\frac{(n!)^{-2}\xi_0^n}{2^{n(n-1)/2}\left(\log\left(C\epsilon_0\delta_0^l\right)^{-1}\right)^n}\right]^{l+1}. \tag{5.12.83}$$

So if $C\epsilon_0$ is small enough, the last inequality in Eqs. (5.12.74) holds. More precisely, it holds if

$$\begin{gathered}C\epsilon_0 < \delta_0^l,\\ \frac{\left(\log(C\epsilon_0)^{-1}\right)^{l+1}}{2^{6l+11}l} > 1.\end{gathered} \tag{5.12.84}$$

Note that the conditions (5.12.84), (5.12.81), (5.12.78), and (5.12.75) can all

[45] e.g., $B_9 = 8B_8 3^{(3/2)(l+1)}(l+1)!^{3/2}2^6\exp\frac{3}{5}(12l+4)\sum_{h=0}^{\infty}(\frac{3}{5})^h\log(1+h)$.

be satisfied by imposing a single condition which also permits us to satisfy Eq. (5.12.59):

$$B_{10}\epsilon_0 C(E_0 C)^{\bar{a}}(E_0\eta_0\rho_0^{-1})^{\bar{b}}\xi_0^{-\bar{c}} < 1, \tag{5.12.85}$$

where B_{10} is a suitable constant and so are $\bar{a},\bar{b},\bar{c}$.

It is now easy to see that if Eq. (5.12.85) holds, then, $\forall n \geqslant 0$, the analogue of Eq. (5.12.59)

$$B_5\epsilon_n C E_n C(E_n\eta_n\rho_n^{-1})N_n^{l+1}\delta_n^{-2l} < 1 \tag{5.12.86}$$

holds if $C\epsilon_0$ is small enough. In fact, Eq. (5.12.85) implies Eq. (5.12.74), as just shown, and Eq. (5.12.74) inserted into Eq. (5.12.86) just gives a condition like Eq. (5.12.85) with possibly new values for the constants $B_{10},\bar{a},\bar{b},\bar{c}$.

So under a condition of the form of Eq. (5.12.6), we can guarantee that the sequence of numbers $(\rho_n,\xi_n,E_n,\eta_n,\epsilon_n)$ recursively defined in Eq. (5.12.73) verifies Eqs. (5.12.74) and (5.12.86) as well.

Clearly, this means that under the condition (5.12.6), with B,a,b,c suitably chosen (and l dependent), we can define a sequence of completely canonical transformations, $\mathcal{C}_0,\mathcal{C}_1,\ldots,$ having the form

$$\begin{aligned} \mathbf{A} &= \mathbf{A}' + \boldsymbol{\Xi}^{(n)}(\mathbf{A}',\mathbf{z}'),\\ \mathbf{z} &= \mathbf{z}'\exp i\boldsymbol{\Delta}^{(n)}(\mathbf{A}',\mathbf{z}') \end{aligned} \tag{5.12.87}$$

and such that, $\forall j = 0,1,\ldots,$

$$\mathcal{C}_j : C(\rho_{j+1},\xi_{j+1};\mathbf{A}_{j+1}) \to C(\rho_j,\xi_j;\mathbf{A}_j)$$

and [see Eq. (5.12.52) and the discussion following it and Eqs. (5.12.49) and (5.12.47)] $\boldsymbol{\Xi}^{(j)},\boldsymbol{\Delta}^{(j)}$ can be bounded in $C(\tilde{\rho}_j/4,\xi_j-3\delta_j;\mathbf{A}_j)\supset C(\rho_{j+1},\xi_{j+1};\mathbf{A}_{j+1})$ by

$$\begin{aligned} |\mathbf{A}_{j+1}-\mathbf{A}_j| &\leqslant \tilde{\rho}_j,\\ |\boldsymbol{\Xi}^{(j)}(\mathbf{A}',\mathbf{z}')| &\leqslant B_3\epsilon_j C\delta_j^{-2l}\rho_j,\\ |\boldsymbol{\Delta}^{(j)}(\mathbf{A}',\mathbf{z}')| &\leqslant 2B_3\epsilon_j C\delta_j^{-2l}\frac{\rho_j}{\tilde{\rho}_j}, \end{aligned} \tag{5.12.88}$$

where $\tilde{\rho}_j$ is defined as $\tilde{\rho}_0$ with the index j replacing 0 everywhere.

The maps $\mathcal{C}_j$ are very close to the identity map on the very small set on which they are defined.

In fact, setting $|(\mathbf{A},\mathbf{z})-(\mathbf{A}',\mathbf{z}')| \equiv |\mathbf{A}-\mathbf{A}'| + \rho_0|\mathbf{z}-\mathbf{z}'|$, one finds for every pair $(\mathbf{A}',\mathbf{z}'),(\mathbf{A}'',\mathbf{z}'')\in C(\rho_{j+1},\xi_{j+1};\mathbf{A}_{j+1})$:

$$|\mathcal{C}_j(\mathbf{A}',\mathbf{z}')-\mathcal{C}_j(\mathbf{A}'',\mathbf{z}'')| \leqslant (1+\theta_j)|(\mathbf{A}',\mathbf{z}')-(\mathbf{A}'',\mathbf{z}'')|, \tag{5.12.89}$$

where θ_j is a small number that can be taken to be

$$\theta_j = B_{11}\epsilon_j C\delta_j^{-2l-1}\frac{\rho_j}{\tilde{\rho}_j}\frac{\rho_0}{\tilde{\rho}_j} \tag{5.12.90}$$

which is implied by a simple calculation based on the dimensional estimates of the derivatives of $\boldsymbol{\Xi}, \boldsymbol{\Delta}$ on the set $C(\rho_{j+1}, \xi_{j+1}; \mathbf{A}_j)$ (possible since $\boldsymbol{\Xi}, \boldsymbol{\Delta}$ are holomorphic on a much larger set, i.e., $C(\tilde{\rho}_j/4, \xi_j - 3\delta_j; \mathbf{A}_j)$).

Clearly, Eqs. (5.12.74) imply that $\theta_j \to 0$ very fast, in particular, $\sum_{j=0}^{\infty}\theta_j < \infty$. They also imply that $\sum_{j=0}^{\infty}\theta_j \to 0$ as $\epsilon_0 \to 0$.

Then we can define parametrically a torus

$$(\mathbf{A}, \mathbf{z}) = \mathcal{C}_0 \cdots \mathcal{C}_{n-1}\mathcal{C}_n(\mathbf{A}_{n+1}, \mathbf{z}'), \qquad \mathbf{z}' \in \mathbf{T}^l \tag{5.12.91}$$

which can be written more explicitly as

$$\begin{aligned} \mathbf{A} &= \mathbf{A}_0 + \boldsymbol{\alpha}^{(n)}(\boldsymbol{\varphi}') \\ \boldsymbol{\varphi} &= \boldsymbol{\varphi}' + \boldsymbol{\beta}^{(n)}(\boldsymbol{\varphi}'), \end{aligned} \qquad \boldsymbol{\varphi}' \in \mathbf{T}^l, \tag{5.12.92}$$

where $\boldsymbol{\alpha}^{(n)}$, $\boldsymbol{\beta}^{(n)}$ are defined by comparison between the right-hand sides of Eqs. (5.12.91) and (5.12.92). By construction, $\boldsymbol{\alpha}^{(n)}$ and $\boldsymbol{\beta}^{(n)}$ are holomorphic on the multiannulus $C(\xi_n) \supset C(\xi_\infty)$ and also

$$\begin{aligned} &|\boldsymbol{\alpha}^{(n)}(\mathbf{z}') - \boldsymbol{\alpha}^{(n-1)}(\mathbf{z}')| + \rho_0|\boldsymbol{\beta}^{(n)}(\mathbf{z}') - \boldsymbol{\beta}^{(n-1)}(\mathbf{z}')| \\ &\quad = |\mathcal{C}_0 \cdots \mathcal{C}_n(\mathbf{A}_{n+1}, \mathbf{z}') - \mathcal{C}_0 \cdots \mathcal{C}_{n-1}(\mathbf{A}_n, \mathbf{z}')| \\ &\quad \leqslant \left(\prod_{j=1}^{\infty}(1+\theta_j)\right)|\mathcal{C}_n(\mathbf{A}_{n+1}, \mathbf{z}') - (\mathbf{A}_n, \mathbf{z}')| \\ &\quad \leqslant \left(\prod_{j=1}^{\infty}(1+\theta_j)\right)(|\mathbf{A}_{n+1} - \mathbf{A}_n| + |\boldsymbol{\Xi}^{(n)}| + \rho_0 e^{\xi_0}|\boldsymbol{\Delta}^{(n)}|) \equiv \sigma_n, \end{aligned} \tag{5.12.93}$$

where $|\boldsymbol{\Xi}^{(n)}|, |\boldsymbol{\Delta}^{(n)}|$ denote the right-hand sides of the second and third of Eqs. (5.12.88). Clearly, $\sigma_n \to_{n\to\infty} 0$ by Eqs. (5.12.88) and (5.12.74), and the right-hand side of Eq. (5.12.93) is summable over n.

Hence, the limits

$$\boldsymbol{\alpha}_\infty(\boldsymbol{\varphi}') = \lim_{n\to\infty}\boldsymbol{\alpha}^{(n)}(\boldsymbol{\varphi}'), \qquad \boldsymbol{\beta}_\infty(\boldsymbol{\varphi}') = \lim_{n\to\infty}\boldsymbol{\beta}^{(n)}(\boldsymbol{\varphi}') \tag{5.12.94}$$

exist and define (by the convergence theorem of Vitali on the sequences of holomorphic functions) two holomorphic functions of $\boldsymbol{\varphi}'$ in $C(\xi_\infty)$.

Via the parametric equations:

$$\begin{aligned} \mathbf{A} &= \mathbf{A}_0 + \boldsymbol{\alpha}_\infty(\boldsymbol{\varphi}'), \\ \boldsymbol{\varphi} &= \boldsymbol{\varphi}' + \boldsymbol{\beta}_\infty(\boldsymbol{\varphi}'), \end{aligned} \qquad \boldsymbol{\varphi}' \in \mathbf{T}^l, \tag{5.12.95}$$

one defines a torus $\mathcal{T}(\boldsymbol{\omega}_0) \subset \mathcal{S}_{\rho_0}(\mathbf{A}_0) \times \mathbf{T}^l$.

It is easy to see from Eqs. (5.12.88) and (5.12.74) that $\boldsymbol{\alpha}_\infty$ and $\boldsymbol{\beta}_\infty$ are small if ϵ_0 is small, i.e., a property like Eq. (5.12.9) holds (possibly redefining B, a, b, c).

So it remains to prove that $\mathcal{T}(\boldsymbol{\omega}_0)$ is an invariant torus run "quasi-periodically" with spectrum $\boldsymbol{\omega}_0$.

The Hamiltonian flow $S_t^{(n)}$, which describes in the coordinates defined by the canonical transformation $\mathcal{C}_0 \cdots \mathcal{C}_{n-1}$ the perturbed Hamiltonian flow S_t associated with Eq. (5.12.1), is such that the coordinates of $S_t^{(n)}(\mathbf{A}_n,$

φ') are

$$\begin{aligned} &\mathbf{A}_n + \rho_n O(\epsilon_n t), \\ &\varphi' + \omega_0 t + O((1+E_n t)\epsilon_n t) \end{aligned} \tag{5.12.96}$$

because the Hamiltonian f_n contributes terms of $O(\epsilon_n)$ to the equations of motion. Of course, Eq. (5.12.96) holds only as long as the point in Eq. (5.12.96) is inside $C(\rho_n, \epsilon_n; \mathbf{A}_n)$.[46]

If $t > 0$ is fixed, it is clear that $\rho_n O(\epsilon_n t) \ll \rho_n$ for n large, by Eq. (5.12.74), and, therefore, by Eqs. (5.12.89) and (5.12.96), we get

$$\begin{aligned} &|\mathcal{C}_0 \cdots \mathcal{C}_{n-1}(S_t^{(n)}(\mathbf{A}_n, \varphi')) - \mathcal{C}_0 \cdots \mathcal{C}_{n-1}(\mathbf{A}_n, \varphi' + \omega_0 t)| \\ &\quad \leqslant \left(\prod_{j=1}^{\infty} (1+\theta_j) \right) (\rho_n O(\epsilon_n t) + \rho_0 O(\epsilon_n t(1+E_n t))) \underset{n\to\infty}{\to} 0. \end{aligned} \tag{5.12.97}$$

Hence,

$$\begin{aligned} \lim_{n\to\infty} S_t \mathcal{C}_0 \cdots \mathcal{C}_{n-1}(\mathbf{A}_n, \varphi') &\equiv \lim_{n\to\infty} \mathcal{C}_0 \cdots \mathcal{C}_{n-1}(S_t^{(n)}(\mathbf{A}_n, \varphi')) \\ &= \lim_{n\to\infty} \mathcal{C}_0 \cdots \mathcal{C}_{n-1}(\mathbf{A}_n, \varphi' + \omega_0 t); \end{aligned} \tag{5.12.98}$$

but the first and third limit exist by Eqs. (5.12.91) and (5.12.94) and their equality means

$$\begin{aligned} &S_t(\mathbf{A}_0 + \boldsymbol{\alpha}_\infty(\varphi'), \varphi' + \boldsymbol{\beta}_\infty(\varphi')) \\ &\quad = (\mathbf{A}_0 + \boldsymbol{\alpha}_\infty(\varphi' + \omega_0 t), \varphi' + \omega_0 t + \boldsymbol{\beta}_\infty(\varphi' + \omega_0 t)) \end{aligned} \tag{5.12.99}$$

which just says that $\mathcal{T}(\omega_0)$ is an invariant torus for the perturbed motion on which quasi-periodic motions with spectrum ω_0 take place ($t \geqslant 0$ being arbitrary). mbe

Problems for §5.12

1. Let $\mathbf{A} \in \mathcal{S}_\rho(\mathbf{A}_0)$ and write Eq. (5.12.4) for $\boldsymbol{\nu} = \mathbf{e}^{(1)} = (1, 0, 0, \ldots, 0)$: $|\omega_1(\mathbf{A})|^{-1} < C$. Deduce that this implies $E_0 C > 1$, with the assumptions and notations of, say, Proposition 22.

In the above context show that

$$|M_{ij}(\mathbf{A}_0)| \leqslant \frac{E_0}{\rho_0},$$

[46] One finds this as follows: let $S_t^{(n)}(\mathbf{A}_n, \varphi') \equiv (\mathbf{A}(t), \varphi'(t))$ so that

$$\dot{\mathbf{A}} = -\frac{\partial f_n}{\partial \varphi}, \qquad \dot{\varphi}' = \omega_n(\mathbf{A}) + \frac{\partial f_n}{\partial \mathbf{A}} \equiv \omega_0 + (\omega_n(\mathbf{A}) - \omega_n(\mathbf{A}_n)) + \frac{\partial f_n}{\partial \mathbf{A}}$$

Therefore, by Taylor's theorem and by a dimensional estimate, one finds after integration on t:

$$|\mathbf{A}(t) - \mathbf{A}_n| < \epsilon_n \rho_n t, \qquad |\varphi'(t) - \varphi' - \omega_0 t| \leqslant \left(\frac{E_n}{\rho_n} \rho_n \epsilon_n t \right) t + \epsilon_n t$$

which implies (5.12.96).

by a dimensional estimate [see Eq. (5.11.18)], and deduce from this that $l|M^{-1}|_{\rho_0}E_0 > \rho_0$, $\|M^{-1}\|_{\rho_0}E_0 > \rho_0$. (Hint: $1 \equiv |(M(\mathbf{A})^{-1}M(\mathbf{A}))_{11}| = |\sum_{k=1}^{l} M(\mathbf{A})^{-1}_{1k} M_{k1}(\mathbf{A})| \leqslant (E_0/\rho_0) l |M(\mathbf{A})^{-1}|$ or $\leqslant (E_0/\rho_0)\|M(\mathbf{A})^{-1}\| \ldots$.)

2. Consider the Hamiltonian system on $\mathbf{R}^{d+1} \times \mathbf{R}^{d+1}$:

$$\frac{A^2}{2} + \frac{\mathbf{B}^2}{2} + \epsilon f(\varphi, \psi), \qquad (\varphi, \psi) \in \mathbf{T}^1 \times \mathbf{T}^d; \quad (A, \mathbf{B}) \in \mathbf{R}^1 \times \mathbf{R}^d.$$

Consider the motions near the resonating torus $A = 1$, $\mathbf{B} = \mathbf{0}$ and write

$$A = 1 + \sqrt{\epsilon}\, a_\epsilon(t\sqrt{\epsilon}\,), \qquad \varphi = \delta_\epsilon(t\sqrt{\epsilon}\,),$$

$$\mathbf{B} = \sqrt{\epsilon}\, \mathbf{b}_\epsilon(t\sqrt{\epsilon}\,), \qquad \psi = \gamma_\epsilon(t\sqrt{\epsilon}\,)$$

for the solution to the Hamiltonian equations with initial datum

$$a_\epsilon(0) = a_0, \qquad \mathbf{b}_\epsilon(0) = \mathbf{b}_0, \qquad \gamma_\epsilon(0) = \gamma_0, \qquad \delta_\epsilon(0) = \delta_0.$$

Show that the solutions to the Hamiltonian equations are such that $a_\epsilon, \mathbf{b}_\epsilon, \gamma_\epsilon$ (but not δ_ϵ) have a limit as $\epsilon \to 0$ and this limit verifies the equations

$$\dot{a} = 0,$$
$$\dot{\gamma} = \mathbf{b},$$
$$\dot{\mathbf{b}} = -\frac{\partial \bar{f}}{\partial \gamma}(\gamma),$$

where $\bar{f}(\gamma) = \int_0^{2\pi} f(\theta, \gamma)\, d\theta / 2\pi$. Show that the limit is approached with a speed $O(\epsilon t)$ at fixed t. (Hint: Write the Hamiltonian equations and note that, after dividing them by $\sqrt{\epsilon}$, they converge formally to the above equations for $a, \mathbf{b}, \gamma$. Then apply the ideas of the proof of Proposition 13, p. 187, §3.8, and of §3.7 and §3.8.)

3. In the context of Problem 2, take $d = 1$. Show that "up to a time $O(1/\epsilon)$", the motion is quasi-periodic with pulsations $\omega_1 = 1$, $\omega_2 = (2\pi / T_{b_0, \gamma_0})\sqrt{\epsilon}$, where

$$T_{b_0, \gamma_0} = 2 \int_{\gamma_-}^{\gamma_+} \big(2(E_0 - \bar{f}(\gamma))\big)^{-1/2} d\gamma,$$

where $E_0 = b_0^2/2 + \bar{f}(\gamma_0)$ and γ_-, γ_+ are 0 and 2π if the equation $E_0 = \bar{f}(\gamma)$ has no roots; otherwise, they are two suitably chosen roots of this equation. Consider only the case $T_{b_0, \gamma_0} < +\infty$ (however, the data for which $T_{b_0, \gamma_0} = +\infty$ are exceptional).

4. Find a result analogous to the one of Problem 2 near a general torus with rational pulsations for the solution flow of the equations associated with the Hamiltonian

$$\frac{\mathbf{A}^2}{2} + \epsilon f(\varphi).$$

(Hint: First extend Problem 2 to the case when f depends on $A, \mathbf{B}$ also; then canonically change variables so that the torus under analysis appears to be run with pulsations $\omega = (\omega_0, 0, 0, \ldots, 0)$.)

5. Consider a time-dependent Hamiltonian with one degree of freedom: $h_0(A) + f_0(A, \varphi, t)$; see Problems 12–14, p. 484, §5.10.

Suppose that h_0 is holomorphic in $\hat{S}_{\rho_0}(\mathbf{A}_0)$ and that f_0 is holomorphic in $C(\rho_0, \xi_0, \mathbf{A}_0) = \hat{S}_{\rho_0}(\mathbf{A}_0) \times C(\xi_0) = \{A, \varphi, t \,|\, (A, z, \zeta) \in \mathbf{C}^3,\ |A - A_0| < \rho,\ e^{-\xi_0} < |z| < e^{\xi_0},\ e^{-\xi_0} < |\zeta| < e^{\xi_0}\}$, where $z = e^{i\varphi}$, $\zeta = e^{it}$.

Using the formal perturbation theory of Problem 12, §5.10, p. 484, prove that if $dh_0/dA \neq 0$ and f_0 is "small" and

$$|\omega(A_0)\nu_1 + \nu_2|^{-1} \leqslant C(|\nu_1| + |\nu_2|)^{\alpha}, \qquad \forall \nu \in \mathbf{Z}^2;$$

for some $C, \alpha > 0$, then the perturbed motion, regarded as taking place on the space of the variables (A, φ, t), leaves invariant a torus on which a quasi-periodic motion with pulsations $(\omega(A_0), 1)$ takes place in the following sense. There exist two holomorphic functions α_∞, β_∞ on $\mathbf{T}^2$ such that setting

$$A = A_0 + \alpha_\infty(\varphi' t'),$$
$$\varphi = \varphi' + \beta_\infty(\varphi', t'),$$
$$t = t',$$

the solution of the equations of motion with datum assigned at time t' and given by $A = A_0 + \alpha_\infty(\varphi', t')$, $\varphi = \varphi' + \beta_\infty(\varphi', t')$ for some $\varphi' \in \mathbf{T}^1$ evolves at time $t' + \tau$ into

$$A(\tau) = A_0 + \alpha_\infty(\varphi' + \omega\tau, t' + \tau),$$
$$\varphi(\tau) = \varphi' + \omega\tau + \beta_\infty(\varphi' + \omega\tau, t' + \tau),$$

i.e., regarding this problem's phase space as $\mathbf{R} \times \mathbf{T}^2$, the above motions can be regarded as taking place on a two-dimensional torus in $\mathbf{R} \times \mathbf{T}^2$ and having pulsations $(\omega, 1)$. (Hint: Just repeat the proof of Proposition 22. No real simplification arises in this apparently simpler case.)

6. Consider the system ("Duffing oscillator")

$$\frac{p^2}{2} + \frac{q^4}{4} + \epsilon q \sin t.$$

Fix an initial datum p_0, q_0. Show that if ϵ is small enough, the trajectory with datum (p_0, q_0) at any initial time t_0 is uniformly bounded in time. (Hint: Show that p_0, q_0 is between two unperturbed tori in the phase space $\mathbf{R}^1 \times \mathbf{T}^2$, of the system with $\epsilon = 0$, having pulsations $(\omega_1, 1), (\omega_2, 1)$ (see preceding problem) nonresonant and with finite resonance parameter C. Use Problem 5 to show that for ϵ small, such tori are slightly deformed but remain invariant. Then use the fact that a two-dimensional torus in a three-dimensional space has an "interior" and an "exterior".)

7. In the context of Problem 5, define $\boldsymbol{\varphi} = (\varphi, t)$ and

$$E_0 \geqslant \left|\frac{dh}{dA}\right|_{\rho_0}, \qquad \eta_0 \geqslant \left|\left(\frac{d^2h}{dA^2}\right)^{-1}\right|_{\rho_0},$$

$$\epsilon_0 \geqslant \left|\frac{\partial f}{\partial A}\right|_{\rho_0, \xi_0} + \frac{1}{\rho_0}\left|\frac{\partial f}{\partial \boldsymbol{\varphi}}\right|_{\rho_0, \xi_0}.$$

Then the condition of smallness of ϵ_0 for the property envisioned there is implied by the following condition, as can be proven:[47]

$$10^{10}\left(\eta_0 E_0 \rho_0^{-1}\right)^4 (CE_0)^4 C\epsilon_0 < 1.$$

Derive a similar formula (i.e., prove the statement in Problem 5, explicitly computing the constants) and try to improve it.

[47] G. Inglese, see references.

8. Consider the system on $\mathbf{R} \times \mathbf{T}^2$ ("Escande–Doveil pendulum")

$$\frac{A^2}{2} + \epsilon(\cos\varphi + \cos(\varphi - t)),$$

where t is the time; see Problems 12–14, §5.10, p. 484. Apply the result of Problem 5 with the estimate in Problem 7 to place a bound on how large ϵ must be in order that one cannot guarantee the "stability of the quasi-periodic motions with pulsations $(\omega_0, 1)$" with $\omega_0 = (\sqrt{5} - 1)/2 = \{\text{golden section}\}$.

9. Same as Problem 8, but applying the results of Problems 5 and 7 to the system obtained from that in Problem 8 by first removing the perturbation to $O(\epsilon^2)$ by ordinary perturbation theory; see Problems 12–14, §5.10, p. 484.

10. Same as Problem 9, but first removing the perturbation to $O(\epsilon^4)$. Show that in this way one obtains much better results.

11. Suppose that in observing the motions of the system in Problem 8, one is able to see them with an absolute precision η of four digits (in decimal basis) and for an observation time T about equal to 50 periods of the forcing term, $T = 50 \cdot 2\pi$.

Note that (see (5.12.86)) to achieve a given accuracy for a given time, one only needs to "remove the perturbation" to an order n such that $O(\epsilon_n T(1 + E_n T)) < \eta$. Using this remark, estimate a threshold for the "survival" of motions which look quasi-periodic, within the error η up to time T, with pulsations $(\omega_0, 1)$, $\omega_0 = (\sqrt{5} - 1)/2$, and compare the result with the experimental value of the "threshold of disappearance" of the quasi-periodic motion in question: $\epsilon \simeq 0.75$.

12. Try to compute the constants B, a, b, c in Eq. (5.12.6), explicitly improving the values of the constants B_1–B_9 suggested in the proof of Proposition 22. An example of a rigorous result is[48]

$$l^{12l}10^{40l}\left(\eta_0 E_0 \rho_0^{-1} C E_0\right)^{14} \xi_0^{-2(10l+6)} < 1.$$

[48] L. Chierchia, see references.

APPENDIX A

The Cauchy-Schwartz Inequality

1 Proposition. *Let Ω be a closed bounded Riemann-measurable (or Lebesgue-measurable) set in $\mathbf{R}^d$. Let $f, g \in C^{(0)}(\Omega)$ be two $\mathbf{R}$-valued functions. Then*

$$\left|\int_\Omega f(\xi)g(\xi)\,d\xi\right| \leqslant \sqrt{\int_\Omega f(\xi)^2\,d\xi}\;\sqrt{\int_\Omega g(\xi)^2\,d\xi}\,. \tag{A1}$$

PROOF. In fact, $\forall \lambda \in \mathbf{R}$:

$$0 \leqslant \int_\Omega (f(\xi)+\lambda g(\xi))^2\,d\xi = \int_\Omega f(\xi)^2\,d\xi + 2\lambda\int_\Omega f(\xi)g(\xi)\,d\xi + \lambda^2\int_\Omega g(\xi)^2\,d\xi\,.$$

Hence, this polynomial of second degree in λ must have a non-negative discriminant. Its discriminant is simply the difference between the square of the right-hand side of (A1) and the square of the left-hand side. mbe

Exercise.

Prove (A1) by observing that

$$\int_\Omega f(\xi)g(\xi)\,d\xi = \lim_{\delta\to 0}\sum_i f(\xi_i)g(\xi_i)\operatorname{vol}\Delta_i\,, \qquad \int_\Omega f(\xi)^2\,d\xi = \lim_{\delta\to 0}\sum_i f(\xi_i)^2\operatorname{vol}\Delta_i\,,$$

$$\int_\Omega g(\xi)^2\,d\xi = \lim_{\delta\to 0}\sum\nolimits_i g(\xi_i)^2\operatorname{vol}\Delta_i\,,$$

where $\Delta_1, \Delta_2, \ldots, \Delta_n$ are a pavement of Ω with parallel cubes with side δ and $\xi_i \in \Delta_i \cap \Omega$. Then apply the "ordinary Cauchy inequality" $|\Sigma_i a_i b_i| \leqslant (\Sigma_i a_i^2)^{1/2} \times (\Sigma_i b_i^2)^{1/2}$ to the sequences

$$a_i = f(\xi_i)\sqrt{\operatorname{vol}\Delta_i}\,, \qquad b_i = g(\xi_i)\sqrt{\operatorname{vol}\Delta_i}$$

By $\operatorname{vol}\Delta_i$, we mean the volume of Δ_i.

APPENDIX B

The Lagrange-Taylor Expansion

1 Proposition. *Let $f \in C^{(k)}(\mathbf{R}^d)$ and suppose that f has a zero of order $(m+1)$, $m < k$, in $\mathbf{x}_0$. Then*

$$f(\mathbf{x}) = \sum_{\substack{\alpha_1, \ldots, \alpha_d \\ \alpha_i \geq 0, \sum_i \alpha_i = m+1}} \bar{f}_{x_0, \alpha_1, \ldots, \alpha_d}(\mathbf{x}) \prod_{i=1}^{d} \frac{(x_i - (\mathbf{x}_0)_i)^{\alpha_i}}{\alpha_i!} \tag{B1}$$

and the functions $\bar{f}_{\mathbf{x}_0, \alpha_1, \ldots, \alpha_d} \in C^{(k-(m+1))}(\mathbf{R}^d)$. If they are regarded as functions of $(\mathbf{x}_0, \mathbf{x}) \in \mathbf{R}^{2d}$, then they are in $C^{(k-(m+1))}(\mathbf{R}^{2d})$.

PROOF. Consider the function $\lambda \to f(\mathbf{x}_0 + \lambda(\mathbf{x} - \mathbf{x}_0))$ which has in $\lambda = 0$ a zero of order $m+1$, i.e., it has the first m derivatives vanishing. Then

$$\begin{aligned} f(\mathbf{x}) &= \int_0^1 d\lambda_1 \frac{d}{d\lambda_1} f(\lambda_1(\mathbf{x} - \mathbf{x}_0) + \mathbf{x}_0) \\ &= \int_0^1 d\lambda_1 \int_0^{\lambda_1} d\lambda_2 \frac{d^2}{d\lambda_2^2} f(\lambda_2(\mathbf{x} - \mathbf{x}_0) + \mathbf{x}_0) \\ &= \int_0^1 d\lambda_1 \int_0^{\lambda_1} d\lambda_2 \cdots \int_0^{\lambda_m} d\lambda_{m+1} \frac{d^{m+1}}{d\lambda^{m+1}} f(\lambda_{m+1}(\mathbf{x} - \mathbf{x}_0) + \mathbf{x}_0) \\ &= \int_0^1 \frac{\lambda^m}{m!} \frac{d^{m+1}}{d\lambda^{m+1}} f(\lambda(\mathbf{x} - \mathbf{x}_0) + \mathbf{x}_0)\, d\lambda. \end{aligned} \tag{B2}$$

Expressing the derivative with respect to λ in terms of the derivatives with respect to the x coordinates, one inductively finds

$$\frac{d^{m+1}}{d\lambda^{m+1}} f(\lambda(\mathbf{x}-\mathbf{x}_0)+\mathbf{x}_0)$$

$$= (m+1)! \sum_{\substack{\alpha_1,\ldots,\alpha_d \\ \alpha_i \geq 0, \sum_i \alpha_i = m+1}} \frac{\partial^{m+1} f}{\partial x_1^{\alpha_1} \ldots \partial x_d^{\alpha_d}} (\lambda(\mathbf{x}-\mathbf{x}_0)+\mathbf{x}_0) \quad \text{(B3)}$$

$$\times \prod_{i=1}^{d} \frac{(x_i - (\mathbf{x}_0)_i)^{\alpha_i}}{\alpha_i!}$$

and this proves the proposition, showing that one can take

$$\bar{f}_{\mathbf{x}_0,\alpha_1 \ldots \alpha_d}(\mathbf{x}) = \int_0^1 \frac{(m+1)!}{m!} \lambda^m \frac{\partial^{m+1} f}{\partial x_1^{\alpha_1} \ldots \partial x_d^{\alpha_d}} (\lambda(\mathbf{x}-\mathbf{x}_0)+\mathbf{x}_0)\, d\lambda. \quad \text{(B4)}$$

mbe

Observation. The same proof holds if $f \in C^{(k)}(\Omega)$ and Ω is a convex open set.

Corollary. *If $f \in C^{(k)}(\mathbf{R}^d \times \mathbf{R}^n)$ has a zero of order $m+1$, $m < k$, in $\mathbf{x}_0$ for each $\mathbf{y} \in \mathbf{R}^n$:*

$$f(\mathbf{x},\mathbf{y}) = \sum_{\substack{\alpha_1,\ldots,\alpha_d \\ 0 \leq \alpha_i, \sum_i \alpha_i = m+1}} \bar{f}_{\mathbf{x}_0,\alpha_1,\ldots,\alpha_d}(\mathbf{x},\mathbf{y}) \prod_{i=1}^{d} \frac{(x_i - (\mathbf{x}_0)_i)^{\alpha_i}}{\alpha_i!} \quad \text{(B5)}$$

and the functions $\bar{f}$, thought of as functions of $(\mathbf{x}_0,\mathbf{x},\mathbf{y}) \in \mathbf{R}^d \times \mathbf{R}^d \times \mathbf{R}^n$, are in $C^{(k-(m+1))}(\mathbf{R}^d \times \mathbf{R}^d \times \mathbf{R}^n)$.

Proof. It is a repetition of the above proof.

3 Proposition. *If $f \in C^{(k)}(\mathbf{R}^d \times \mathbf{R}^n)$, the function*

$$f(\mathbf{x},\mathbf{y}) - \sum_{\substack{\alpha_1,\ldots,\alpha_d \\ 0 \leq \alpha_i, \sum_i \alpha_i \leq m}} \frac{\partial^{\alpha_1+\cdots+\alpha_d} f}{\partial x_1^{\alpha_1} \ldots \partial x_d^{\alpha_d}} (\mathbf{x}_0,\mathbf{y}) \prod_{i=1}^{d} \frac{(x_i - (\mathbf{x}_0)_i)^{\alpha_i}}{\alpha_i!} \quad \text{(B6)}$$

has, for each $m < k$, a zero in $\mathbf{x}$ of order m at $\mathbf{x}_0$ for all $\mathbf{y} \in \mathbf{R}^n$. Furthermore, the function in Eq. (B6) *has a representation like the right-hand side of Eq.* (B5) *with functions $\bar{f}$ having the same properties as those of Corollary 2 above.*

Proof. One first checks that Eq. (B6) has all the x derivatives vanishing in $(\mathbf{x}_0,\mathbf{y})$ up to order m. Then one applies Corollary 2 or repeats the proof of Proposition 1. This time,

$$\bar{f}_{\mathbf{x}_0,\alpha_1,\ldots,\alpha_d}(\mathbf{x},\mathbf{y}) = \int_0^1 \frac{(m+1)!}{m!} \lambda^m \frac{\partial^{m+1} f}{\partial x_1^{\alpha_1} \cdots \partial x_d^{\alpha_d}} (\lambda(\mathbf{x}-\mathbf{x}_0)+\mathbf{x}_0,\mathbf{y})\, d\lambda. \quad \text{(B7)}$$

mbe

APPENDIX C

C^∞-Functions with Bounded Support and Related Functions

1. There is a nonzero function $\psi_a \in C^\infty(\mathbf{R})$, $\psi_a \geqslant 0$ with support in $[0, a]$, $a > 0$, and one can take

$$\psi_a(t) = 0 \qquad \text{if} \quad t \notin (0, a),$$

$$\psi_a(t) = \exp - \frac{1}{t^2(a-t)^2} \qquad \text{if} \quad t \in (0, a).$$

2. There exists a nondecreasing function $g \in C^\infty(\mathbf{R})$ vanishing for $t \leqslant 0$ and equal to 1 for $t \geqslant a > 0$. For instance,

$$\chi(t) = C \int_{-\infty}^{t} \psi_a(\tau)\, d\tau,$$

where

$$C^{-1} = \int_{-\infty}^{+\infty} \psi_a(\tau)\, d\tau.$$

3. The function

$$g_{\alpha,\beta}(t) = \chi(t - \alpha + a)\chi(\beta - t + a)$$

has value 1 if $t \in [\alpha, \beta]$, 0 if $t \leqslant \alpha - a$ or $t \geqslant \beta + a$ and is non-negative.

4. The function in $C^\infty(\mathbf{R}^d)$,

$$g(\xi_1, \ldots, \xi_d) = \prod_{i=1}^{d} g_{\alpha_i, \beta_i}(\xi_i),$$

has value 1 on the parallelipiped $[\alpha_1, \beta_1] \times \cdots \times [\alpha_d, \beta_d]$; it is non-negative and has bounded support.

APPENDIX D

Principle of the Vanishing Integrals

1 Proposition. *Let $f \in C^\infty([\alpha, \beta])$ and suppose*

$$\int_\alpha^\beta f(t)z(t)\,dt = 0$$

for all $z \in C_0^\infty([\alpha, \beta])$. Then $f \equiv 0$.

PROOF. If $f \not\equiv 0$, there is $t_0 \in (\alpha, \beta)$, where $f(t_0) \neq 0$. Let $[\bar{\alpha}, \bar{\beta}] \subset (\alpha, \beta)$ be an interval around t_0 such that $|f(t)| \geqslant |f(t_0)|/2$, $\forall t \in [\bar{\alpha}, \bar{\beta}]$. Let $t \to \chi(t)$, $t \in [\alpha, \beta]$ be a C^∞ function positive in t_0 and vanishing outside $[\bar{\alpha}, \bar{\beta}]$. Then $t \to f(t_0)\chi(t)$ is in $C_0^\infty([\alpha, \beta])$ and

$$0 = \int_\alpha^\beta f(t)\chi(t)f(t_0)\,dt \geqslant \frac{1}{2}\int_{\bar{\alpha}}^{\bar{\beta}} |f(t_0)|^2 \chi(t)\,dt > 0.$$

mbe

APPENDIX E

Matrix Notations. Eigenvalues and Eigenvectors. A List of Some Basic Results in Algebra

The reader who wishes more details (or proofs) on the subjects discussed below may consult Fano, Chaps. 1 and 2; see references.

1. Given a matrix J, $l \times m$ and a matrix L, $m \times p$, JL denotes the $l \times p$ matrix obtained by multiplying "rows by columns" the matrices J and L.

2. If J is an $l \times m$ matrix and $\mathbf{x} \in \mathbf{C}^m$, we denote $\mathbf{y} = L\mathbf{x}$ the vector of $\mathbf{C}^l$ with components

$$y_i = \sum_{k=1}^{m} J_{ik} x_k, \qquad i = 1, \ldots, l. \tag{E1}$$

3. The determinant, $\det J$, of matrix J is defined for all the square matrices. If J and L are two $d \times d$ square matrices, $\det JL = \det J \det L$. The determinant is a linear combination of products of matrix elements.

4. The sum of two $l \times m$ matrices is an $l \times m$ matrix with matrix elements given by the sums of the homonymous matrix elements of the two matrices. The matrix λJ, $\lambda \in \mathbf{C}$ is the matrix whose elements are those of J multiplied by λ.

The modulus of an $l \times m$ matrix J is

$$|J| = \sum_{i=1}^{l} \sum_{j=1}^{m} |J_{ij}|. \tag{E2}$$

In §5.12 (only) we use the symbol $\|J\|$ for the right-hand side of (E2) and $|J|$ for $\max|J_{ij}|$. If J is an $l \times m$ matrix and L is a $m \times p$ matrix,

$$|JL| \leqslant |J|\,|L|. \tag{E3}$$

5. The identity matrix, $d \times d$, will usually be simply denoted by 1 and similarly, the product of $\lambda \in \mathbf{C}$ with the identity matrix will be denoted λ.

6. The eigenvalues of a square matrix are the solutions of the algebraic equation in λ ("secular or characteristic equation"):

$$\det(J - \lambda) = 0. \tag{E4}$$

7. The inverse matrix to a square matrix J exists if and only if $\det J \neq 0$ and it will be denoted J^{-1}: it is characterized by the property $JJ^{-1} = J^{-1}J = 1$. Its matrix elements are expressible as ratios of determinants of submatrices of J by the determinant of J.

8. If $f(z) = \sum_{n=0}^{\infty} c_n z^n$ is a power series with radius of convergence $\rho > 0$ and if J is a square matrix such that $|J| < \rho$ and if $J^0 \equiv 1$, the series

$$f_{ij} = \sum_{n=0}^{\infty} c_n (J^n)_{ij} \tag{E5}$$

are absolutely convergent since

$$|f_{ij}| \leqslant \sum_{n=0}^{\infty} c_n |J^n| \leqslant \sum_{n=0}^{\infty} c_n |J|^n < \sum_{n=0}^{\infty} c_n \rho^n < \infty. \tag{E6}$$

They define a matrix that will be denoted $f(J)$.

If $(P(z), Q(z)$ are two polynomials and $PQ(z)$ is their product polynomial, it is clear that

$$P(J)\,Q(J) = PQ(J) \tag{E7}$$

(if one thinks of the definition of the product of polynomials and of the fact that the product of matrices is distributive).

Similarly, if $f(z)$, $g(z)$ are two powers series with radius of convergence ρ, their product power series $fg(z)$ has the same radius of convergence and the above relation is generalized by

$$f(J)\,g(J) = fg(J). \tag{E8}$$

In particular, if $|J| < 1$, $f(z) = 1 - z$, $g(z) = (1 - z)^{-1} = \sum_{n=0}^{\infty} z^n$, $fg(z) \equiv 1$, so that $g(J)$ is the inverse to $(1 - J)$; i.e.,

$$(1 - J)^{-1} = \sum_{n=0}^{\infty} J^n \tag{E9}$$

and

$$|(1 - J)^{-1} - 1| = \left|\sum_{n=1}^{\infty} J^n\right| \leqslant \frac{|J|}{1 - |J|}\,. \tag{E10}$$

9. A real square matrix J is said to be "orthogonal" if $J^{-1} = J^T$, where $(J^T)_{ij} = J_{ji}, \forall i, j$. The orthogonal $d \times d$ matrices can also be thought of as "rotations of $\mathbf{R}^d$". The rotation of $\mathbf{R}^d$ corresponding to the orthogonal matrix J will be the map of $\mathbf{R}^d$ into itself:

$$\mathbf{x} \to J\mathbf{x} \tag{E11}$$

10. If J is a $d \times d$ matrix and if $\mathbf{y}, \mathbf{x} \in \mathbf{C}^d$,

$$\mathbf{x} \cdot J\mathbf{y} = J^T\mathbf{x} \cdot \mathbf{y} \tag{E12}$$

11. The eigenvalues of a matrix enjoy remarkable properties.

1 Proposition. *If J is a $d \times d$ matrix with pairwise-distinct eigenvalues $\lambda_1, \ldots, \lambda_d$, there are d vectors $\mathbf{v}^{(1)}, \ldots, \mathbf{v}^{(d)} \in \mathbf{C}^d$, generally complex even if J is a real matrix, such that*

$$J\mathbf{v}^{(i)} = \lambda_i \mathbf{v}^{(i)}, \qquad i = 1, \ldots, d$$

and they are linearly independent.

If J is a real matrix, the eigenvalues and the eigenvectors can be arranged so that they appear in complex-conjugate pairs.

If the matrix J varies in the neighborhood (in the sense that $|J - J_0|$ is small) of a matrix J_0 with pairwise-distinct eigenvalues, then the eigenvalues and the corresponding eigenvectors can be chosen and labeled so that they vary smoothly with J, i.e., so that the eigenvalue λ_j of J and the corresponding eigenvector's components $(\mathbf{v}^{(j)})_k, j, k = 1, \ldots, d$, are C^∞ functions of the matrix elements of J.

APPENDIX F

Positive-Definite Matrices. Eigenvalues and Eigenvectors. A List of the Basic Properties

The reader who wishes more details (or proofs) on the subjects discussed below may consult Fano, Chaps. 1 and 2, in references.

Definition. A real matrix $V = (V_{ij})_{i,j=1,\ldots,d}$ is "positive definite" if

(i) $V_{ij} = V_{ji}$, $i, j = 1, \ldots, d$; (F1)

(ii) For all $\boldsymbol{\alpha} = (\alpha_1, \ldots, \alpha_d) \in \mathbf{R}^d$,

$$(V\boldsymbol{\alpha} \cdot \boldsymbol{\alpha}) = \sum_{i,j=1}^{d} V_{ij}\alpha_i\alpha_j > 0. \tag{F2}$$

We now collect the main properties of the positive-definite matrices in two propositions.

First, note that $\det V \neq 0$; otherwise, there would be $\boldsymbol{\alpha}_0 \neq \mathbf{0}$ such that $V\boldsymbol{\alpha}_0 = \mathbf{0}$, contradicting (ii) above.

The following proposition states the "existence of an orthonormal basis on which V is diagonal".

1 Proposition. *If V is a $d \times d$ positive-definite matrix, there exist d positive numbers $\lambda_1, \ldots, \lambda_d$ and an orthonormal basis $\mathbf{v}^{(1)}, \ldots, \mathbf{v}^{(d)}$ in $\mathbf{R}^d$ such that*

$$V\mathbf{v}^{(j)} = \lambda_j \mathbf{v}^{(j)}, \qquad j = 1, \ldots, d \tag{F3}$$

and the orthogonal matrix

$$J_{ij} = (\mathbf{v}^{(i)})_j, \qquad i, j = 1, \ldots, d \tag{F4}$$

is such that

$$JVJ^T = \Lambda, \qquad V = J^T\Lambda J, \tag{F5}$$

where Λ *is the diagonal* $d \times d$ *matrix with diagonal elements given by* $\lambda_1, \ldots, \lambda_d$.

Observation. Eq. (F5) implies that $\lambda_1, \ldots, \lambda_d$ are the eigenvalues of V counted according to multiplicity. In fact,

$$\det(V - \lambda) = \det(J^T\Lambda J - \lambda) \equiv \det J^T(\Lambda - \lambda)J$$
$$= \det(\Lambda - \lambda) = \prod_{i=1}^{d} (\lambda_i - \lambda)$$

(since $J^TJ \equiv 1$, $\det J \det J^T = \det JJ^T = 1$).

2 Corollary. *If* V *is a positive-definite* $d \times d$ *matrix, there is a positive-definite matrix* $\sqrt{V}$ *such that* $(\sqrt{V})^2 = V$.

More generally, if $a \in \mathbf{R}$, *there is a positive-definite matrix* V^a *such that,* $\forall a, b \in \mathbf{R}$, $V^aV^b = V^{a+b}$ *and* $V^1 = V$, $V^0 = 1$.

PROOF. If Λ is a diagonal $d \times d$ matrix such that Eq. (F5) holds, we set $\Lambda^a = \{$diagonal matrix and diagonal elements $\lambda_1^a, \ldots, \lambda_d^a\}$. Then $\Lambda^a\Lambda^b = \Lambda^{a+b}$, $\forall a, b \in \mathbf{R}$, $\Lambda^0 = 1$, $\Lambda^1 = \Lambda$; so we set

$$V^a = J^T\Lambda^a J \tag{F7}$$

and V^a verifies the desired properties. $V^{1/2} = \sqrt{V}$ by definition. mbe

3 Corollary. *If* V *is a positive-definite* $d \times d$ *matrix, there exists a continuous function* $\mu(V) > 0$ *depending on the matrix elements of* V *such that*

$$V\boldsymbol{\alpha} \cdot \boldsymbol{\alpha} \geqq \mu(V)|\boldsymbol{\alpha}|^2. \tag{F8}$$

In fact, $\mu(V) = \min_i \lambda_i$, *if* $\lambda_1, \ldots, \lambda_d$ *are the eigenvalues of* V.

PROOF.

$$V\boldsymbol{\alpha} \cdot \boldsymbol{\alpha} = J^T\Lambda J\boldsymbol{\alpha} \cdot \boldsymbol{\alpha} = \Lambda J\boldsymbol{\alpha} \cdot J\boldsymbol{\alpha} = \sum_{i=1}^{d} \lambda_i (J\boldsymbol{\alpha})_i^2$$
$$\geqslant \mu(V) \sum_{i=1}^{d} (J\boldsymbol{\alpha})_i^2 = \mu(V)J\boldsymbol{\alpha} \cdot J\boldsymbol{\alpha} \tag{F9}$$
$$= \mu(V)J^TJ\boldsymbol{\alpha} \cdot \boldsymbol{\alpha} = \mu(V)\boldsymbol{\alpha} \cdot \boldsymbol{\alpha}.$$

mbe

We conclude with a generalization of the above results.

4 Proposition. *Let* G, V *be two positive-definite* $d \times d$ *matrices. There are* d *independent vectors* $\mathbf{v}^{(1)}, \ldots, \mathbf{v}^{(d)} \in \mathbf{R}^d$ *and* d *positive numbers*

$\lambda_1, \ldots, \lambda_d$ *such that*

$$V\mathbf{v}^{(j)} = \lambda_j G\mathbf{v}^{(j)}, \qquad j = 1, \ldots, d, \tag{F10}$$

$$G\mathbf{v}^{(i)} \cdot \mathbf{v}^{(j)} = \delta_{ij}, \qquad i, j = 1, \ldots, d. \tag{F11}$$

The numbers $\lambda_1, \ldots, \lambda_d$ *are the solutions repeated with multiplicity of*

$$\det(V - \lambda G) = 0. \tag{F12}$$

There is a function $\mu(V, G) > 0$ *continuously dependent on the matrix elements of* V, G *such that*

$$(V\boldsymbol{\alpha} \cdot \boldsymbol{\alpha}) \geqslant \mu(V, G)(G\boldsymbol{\alpha} \cdot \boldsymbol{\alpha}). \tag{F13}$$

Observation. This proposition is easily reduced to the preceding ones. If $\mathbf{w}^{(1)}, \ldots, \mathbf{w}^{(d)}$ are the eigenvectors of the positive-definite matrix $W = G^{-1/2} V G^{-1/2}$, the $\mathbf{v}^{(j)}$ are

$$\mathbf{v}^{(j)} = G^{-1/2}\mathbf{w}^{(j)}, \qquad j = 1, \ldots, d. \tag{F14}$$

APPENDIX G

Implicit Function Theorems

Let $\mathbf{f} \in C^{\infty}(\mathbf{R}^m \times \mathbf{R}^d)$ be a function with values in $\mathbf{R}^d$ associating to $(\mathbf{x}, \mathbf{y}) \in \mathbf{R}^m \times \mathbf{R}^d$ the value $\mathbf{f}(\mathbf{x}, \mathbf{y})$.

Consider the equation for $\mathbf{y} \in \mathbf{R}^d$ parameterized by $\mathbf{x}$:

$$\mathbf{f}(\mathbf{x}, \mathbf{y}) = \mathbf{0} \tag{G1}$$

which is a system with d equations in d unknowns $y_1, \ldots, y_d$.

Suppose that $(\mathbf{x}_0, \mathbf{y}_0) \in \mathbf{R}^m \times \mathbf{R}^d$ verifies Eq. (G1). By the Taylor theorem, see Appendix B,

$$\mathbf{f}(\mathbf{x}, \mathbf{y}) = J(\mathbf{y} - \mathbf{y}_0) + L(\mathbf{x} - \mathbf{x}_0) + \mathbf{N}(\mathbf{x}, \mathbf{y}), \tag{G2}$$

where J, L are $d \times d$ and $d \times m$ matrices built with the derivatives of f:

$$J_{ij} = \frac{\partial f^{(i)}}{\partial y_j}(\mathbf{x}_0, \mathbf{y}_0), \qquad i, j = 1, \ldots, d, \tag{G3}$$

$$L_{ij} = \frac{\partial f^{(i)}}{\partial x_j}(\mathbf{x}_0, \mathbf{y}_0) \qquad i = 1, \ldots, d; j = 1, \ldots, m \tag{G4}$$

and $\mathbf{N}$ is an $\mathbf{R}^d$-valued C^{∞} function with a second-order zero in $(\mathbf{x}_0, \mathbf{y}_0)$, see Appendix B, Proposition 3 with $m = 1$, $k = +\infty$.

The implicit function theorem compares the solution of Eq. (G1), written as

$$J(\mathbf{y} - \mathbf{y}_0) + L(\mathbf{x} - \mathbf{x}_0) + \mathbf{N}(\mathbf{x}, \mathbf{y}) = \mathbf{0}, \tag{G5}$$

with that of the linear equations ($d \times d$ linear system)

$$J(\mathbf{y} - \mathbf{y}_0) + L(\mathbf{x} - \mathbf{x}_0) = \mathbf{0}. \tag{G6}$$

If $\det J \neq 0$, the matrix J^{-1} exists and Eq. (G6) has the unique solution

$$\mathbf{y}-\mathbf{y}_0 = -J^{-1}L(\mathbf{x}-\mathbf{x}_0). \tag{G7}$$

Therefore, it becomes natural to think that Eq. (G5) admits a solution differing from Eq. (G7) "by higher-order infinitesimals in $\mathbf{x}-\mathbf{x}_0$", since such is the difference between Eq. (G5) and Eq. (G6). More precisely, one can hope that there exists in a vicinity U of $\mathbf{x}_0$ a function $\boldsymbol{\varphi}(\mathbf{x})$ such that

$$\mathbf{f}(\mathbf{x},\boldsymbol{\varphi}(\mathbf{x})) \equiv \mathbf{0}, \qquad \mathbf{x}\in U, \tag{G8}$$

$$\boldsymbol{\varphi}(\mathbf{x}) = -J^{-1}L(\mathbf{x}-\mathbf{x}_0)+\boldsymbol{\Phi}(\mathbf{x}), \qquad \mathbf{x}\in U, \tag{G9}$$

where $\boldsymbol{\Phi}\in C^\infty(U)$ and has a second-order zero at $\mathbf{x}_0$.

This is, in fact, the content of the implicit function theorems. Since we shall also need explicit estimates of the size of the set U, on which $\boldsymbol{\Phi}$ can be defined, and on the size of $\boldsymbol{\Phi}(U)$, its $\boldsymbol{\Phi}$ image, it is more appropriate to describe the proof in notations which are convenient for us rather than to refer to a standard book.

We first treat the $d=1$ case, denoting $\Gamma_n(\mathbf{x},\rho)\subset\mathbf{R}^n$ the closed cube with center $\mathbf{x}\in\mathbf{R}^n$ and side 2ρ.

1 Proposition. *Given $\delta>0$, $\alpha>0$, $\alpha\geqslant\delta$, define*

$$\rho_{\delta,\alpha} = \frac{\delta}{2}\,\frac{\min|\partial f/\partial y|}{\max\left(\sum_{j=1}^m|\partial f/\partial x_j|+|\partial f/\partial y|\right)}, \tag{G10}$$

where the minimum and the maximum are considered as $\mathbf{x}$ varies in $\Gamma_m(\mathbf{x}_0,\alpha)$ and as $y-y_0$ varies in $[-\delta,\delta]$ and we suppose that $\mathbf{f}(\mathbf{x}_0,\mathbf{y}_0)=\mathbf{0}$.

If $\rho_{\delta,\alpha}>0$, it is possible to define a function φ in $C^\infty(\Gamma_m(\mathbf{x}_0,\rho_{\delta,\alpha}))$ verifying Eq. (G8) *for every $\mathbf{x}\in\Gamma_m(\mathbf{x}_0,\rho_{\delta,\alpha})$.*

Furthermore, all the solutions of Eq. (G1) *in $\Gamma_m(\mathbf{x}_0,\rho_{\delta,\alpha})\times[y_0-\delta,y_0+\delta]$ have the form $(\mathbf{x},\varphi(\mathbf{x}))$ and*

$$\frac{\partial\varphi}{\partial x_i}(\mathbf{x}) = -\frac{(\partial f/\partial x_i)(\mathbf{x},\varphi(\mathbf{x}))}{(\partial f/\partial y)(\mathbf{x},\varphi(\mathbf{x}))}, \qquad \mathbf{x}\in\Gamma_m(\mathbf{x}_0,\rho_{\delta,\alpha}). \tag{G11}$$

PROOF. Let $(\mathbf{x},y_0+\delta)$ be a point on the upper face of the parallelepiped $\Gamma_m(\mathbf{x}_0,\rho_{\delta,\alpha})\times[y_0-\delta,y_0+\delta]$.

We show that on this face f has a well-defined sign, opposite to the one it has on the lower face.

Since $\partial f/\partial y$ cannot vanish in the parallelepiped, by the choice of $\rho_{\delta,\alpha}$ and because $\rho_{\delta,\alpha}>0$, this will imply that for each $\mathbf{x}\in\Gamma_m(\mathbf{x}_0,\rho_{\delta,\alpha})$ there is one and only one point $\varphi(\mathbf{x})\in[y_0-\delta,y_0+\delta]$ such that $f(\mathbf{x},\varphi(\mathbf{x}))=0$ (note that $\partial f/\partial y\neq0\Rightarrow$ strict monotonicity).

To show that f takes opposite signs on the opposite faces suppose, to be definite, $\partial f/\partial y>0$ in $\Gamma_m(\mathbf{x}_0,\alpha)\times[y_0-\delta,y_0+\delta]$. Then

$$\begin{aligned} f(\mathbf{x},\mathbf{y}_0+\delta) &\equiv f(\mathbf{x},\mathbf{y}_0+\delta)-f(\mathbf{x}_0,\mathbf{y}_0)\\ &= f(\mathbf{x},y_0+\delta)-f(\mathbf{x},y_0)+f(\mathbf{x},y_0)-f(\mathbf{x}_0,y_0) \end{aligned} \tag{G12}$$

and we apply the Lagrange theorem to find $\tilde{\mathbf{x}}$ and $\tilde{y}$, intermediate between $\mathbf{x}$ and $\mathbf{x}_0$ and between y_0 and $y_0+\delta$, such that the right-hand side of Eq. (G12) can be written

$$\begin{aligned} f(\mathbf{x}, y_0+\delta) &= \frac{\partial f}{\partial y}(\mathbf{x}, \tilde{y})\delta + \sum_{j=1}^{m} \frac{\partial f}{\partial x_j}(\tilde{\mathbf{x}}, y_0)(\mathbf{x}-\mathbf{x}_0)_j \\ &\geqslant \left(\min\left|\frac{\partial f}{\partial y}\right|\right)\delta - \left(\max \sum_{j=1}^{m}\left|\frac{\partial f}{\partial x_j}\right|\right)\rho_{\delta,\alpha} \qquad \text{(G13)} \\ &\geqslant \left(\min\left|\frac{\partial f}{\partial y}\right|\right)\frac{\delta}{2} > 0. \end{aligned}$$

Similarly, one proves that $f(\mathbf{x}, y_0-\delta) < 0$, $\forall \mathbf{x} \in \Gamma_m(\mathbf{x}_0, \rho_{\delta,\alpha})$. This proves the existence of $\varphi(\mathbf{x})$ and its uniqueness.

To show the differentiability in the direction of the axis $\mathbf{e} = (e_1, \ldots, e_d)$ observe that, given $\mathbf{x} \in \Gamma_m(\mathbf{x}_0, \rho_{\delta,\alpha})$ and given ϵ such that $\mathbf{x}_0 + \epsilon\mathbf{e} \in \Gamma_m(\mathbf{x}_0, \rho_{\delta,\alpha})$ and if $\tilde{\mathbf{x}}, \tilde{y}$ are suitable intermediate points between $\mathbf{x}$ and $\mathbf{x}+\epsilon\mathbf{e}$ or $\varphi(\mathbf{x})$ and $\varphi(\mathbf{x}+\epsilon\mathbf{e})$, one finds

$$0 \equiv f(\mathbf{x}+\epsilon\mathbf{e}, \varphi(\mathbf{x}+\epsilon\mathbf{e})) - f(\mathbf{x}, \varphi(\mathbf{x})), \qquad \text{(G14)}$$

$$0 = \sum_{i=1}^{m} e_i \frac{\partial f}{\partial x_i}(\tilde{\mathbf{x}}, \tilde{y})\epsilon + \frac{\partial f}{\partial y}(\tilde{\mathbf{x}}, \tilde{y})(\varphi(\mathbf{x}+\epsilon\mathbf{e}) - \varphi(\mathbf{x})) \qquad \text{(G15)}$$

by the Lagrange theorem, and this shows that

$$|\varphi(\mathbf{x}+\epsilon\mathbf{e}) - \varphi(\mathbf{x})| \leqslant \frac{\max\sum_{i=1}^{m}|\partial f/\partial x_i|}{\min|\partial f/\partial y|}\epsilon \leqslant \frac{\delta\epsilon}{2\rho_{\delta,\alpha}}, \qquad \text{(G16)}$$

i.e., φ is continuous.

Equation (G15) also shows, dividing it by ϵ and letting $\epsilon \to 0$, that

$$\lim_{\epsilon\to 0} \frac{\varphi(\mathbf{x}+\epsilon\mathbf{e}) - \varphi(\mathbf{x})}{\epsilon} = \frac{-\sum_{i=1}^{m} e_i(\partial f/\partial x_i)(\mathbf{x}, \varphi(\mathbf{x}))}{(\partial f/\partial y)(\mathbf{x}, \varphi(\mathbf{x}))}, \qquad \text{(G17)}$$

proving the differentiability of φ and Eq. (G11).

By the composed-function chain-differentiation theorem, it is clear that eq. (G11) implies that $\partial\varphi/\partial x_j$ are differentiable in $\mathbf{x}$ and their derivatives can be expressed in terms of φ, of its first derivative and of f and its first two partial derivatives. Therefore, $\partial^2\varphi/\partial x_j \partial x_k$ are differentiable, etc., i.e., $\varphi \in C^{\infty}(\Gamma_m(\mathbf{x}_0, \rho_{\delta,\alpha}))$. mbe

Observation. It appears from the proof that the same results hold if no relation is assumed a priori between α and δ, provided $\rho_{\delta,\alpha}$ is replaced by

$$\bar{\rho}_{\delta,\alpha} = \left\{ \text{minimum between } \alpha \text{ and } \frac{\delta}{2}\,\frac{\min|\partial f/\partial y|}{\max\sum_{j=1}^{m}|\partial f/\partial x_j|} \right\} \qquad \text{(G18)}$$

which is a better result; see Eq. (G13).

To deal with the general case, introduce, given a matrix M,

$$|M| = \sum_{i,j} |M_{ij}| \qquad \text{(G19)}$$

and note that $|M \cdot N| \leqslant |M||N|$ if $M \cdot N$ makes sense, i.e., if the number of columns of M equals that of the rows of N.

Also define the matrices

$$J(\mathbf{x},\mathbf{y}) = \frac{\partial \mathbf{f}}{\partial \mathbf{y}}(\mathbf{x},\mathbf{y}), \qquad L(\mathbf{x},\mathbf{y}) = \frac{\partial \mathbf{f}}{\partial \mathbf{x}}(\mathbf{x},\mathbf{y}) \tag{G20}$$

2 Proposition. *Given $\delta, \alpha > 0$, define*

$$\rho_{\delta,\alpha} = \frac{1}{2}\,\frac{\delta - 2(\max|J^{-1}|)(\alpha \max|\partial \mathbf{N}/\partial \mathbf{x}| + \delta \max|\partial \mathbf{N}/\partial \mathbf{y}|)}{\max|J^{-1}L|} \tag{G21}$$

with the maxima taken on $\Gamma_m(\mathbf{x}_0,\alpha) \times \Gamma_d(\mathbf{y}_0,\delta)$ and set $\rho_{\delta,\alpha} = 0$ if J^{-1} does not exist at some point of this set. Suppose $\mathbf{f}(\mathbf{x}_0,\mathbf{y}_0) = \mathbf{0}$, and $\alpha > \rho_{\delta,\alpha} > 0$.

It is then possible to find $\boldsymbol{\varphi} \in C^\infty(\Gamma_m(\mathbf{x}_0,\rho_{\delta,\alpha}))$ with values in $\Gamma_d(\mathbf{y}_0,\delta)$ verifying Eq. (G8).

Furthermore, all the solutions of Eq. (G1) *in $\Gamma_m(\mathbf{x}_0,\rho_{\delta,\alpha}) \times \Gamma_d(\mathbf{y}_0,\delta)$ have the form $(\mathbf{x},\boldsymbol{\varphi}(\mathbf{x}))$ and*

$$\frac{\partial \varphi_i}{\partial x_j}(\mathbf{x}) = -\sum_{k=1}^{d}\left(J(\mathbf{x},\boldsymbol{\varphi}(\mathbf{x}))^{-1}\right)_{ik}\frac{\partial f_k}{\partial x_j}(\mathbf{x},\boldsymbol{\varphi}(\mathbf{x})). \tag{G22}$$

Observations.

(1) Note that the above proposition is nonempty. Using the fact $\mathbf{N}$ has a second-order zero at $(\mathbf{x}_0,\mathbf{y}_0)$, given $B > (|J(\mathbf{x}_0,\mathbf{y}_0)^{-1}L(\mathbf{x}_0,\mathbf{y}_0)|)^{-1}$, we see that for δ small enough (depending on B) and $\alpha = B\delta$ it is:

$$0 < \frac{1}{2}\,\frac{1}{\max|J^{-1}L|}\,\delta < \rho_{\delta,B\delta} < \frac{1}{\max|J^{-1}L|}\,\delta < B\delta \tag{G23}$$

(2) There are two methods to prove a theorem like the above. The most natural would be to deduce it as a corollary of Proposition 1. One would just proceed by substitution as in the solution of the linear systems.

The assumption $\det J \neq 0$ implies that there is at least one derivative $(\partial f^{(i_1)}/\partial y_1)(\mathbf{x}_0,\mathbf{y}_0) \neq 0$. Then we apply Proposition 1 to the function $f = f^{(i_1)}$ with $y = y_1$ and $\mathbf{x}$ replaced by $(\mathbf{x}, y_2, \ldots, y_d)$ and call $\varphi^1(\mathbf{x}, y_2, \ldots, y_d)$ its solution defined close enough to $\mathbf{x}_0, (\mathbf{y}_0)_2, (\mathbf{y}_0)_3, \ldots, (\mathbf{y}_0)_d$. Then, supposing $i_1 = 1$, consider

$$\begin{aligned} &f^{(2)}(\mathbf{x}, \varphi_1(\mathbf{x}, y_2, \ldots, y_d), y_2, \ldots, y_d) = 0 \\ &\qquad\vdots \\ &f^{(d)}(\mathbf{x}, \varphi_1(\mathbf{x}, y_2, \ldots, y_d), y_2, \ldots, y_d) = 0. \end{aligned} \tag{G24}$$

The determinant of the Jacobian matrix J_1 of the left-hand side of Eq. (G.24) with respect to $y_2, \ldots, y_d$ cannot vanish in $\mathbf{x}_0, (\mathbf{y}_0)_2, \ldots, (\mathbf{y}_0)_d$ because it can be shown to coincide with the determinant of the linear system of equations obtained from the system $J(\mathbf{x}_0,\mathbf{y}_0)\boldsymbol{\xi} = \boldsymbol{\eta}$ by solving its first equation with respect to ξ_1 and substituting into the others.

Therefore, we can again apply Proposition 1, expressing, say, y_2 as a function of $\mathbf{x}, y_3, \ldots, y_d$ close enough to $\mathbf{x}_0, (\mathbf{y}_0)_3, \ldots, (\mathbf{y}_0)_d$, etc. The only difficulty is that the left-hand side of Eq. (G24) is only defined, and C^∞, in a small vicinity of $\mathbf{x}_0, (\mathbf{y}_0)_2, \ldots, (\mathbf{y}_0)_d$ and not on all of $\mathbf{R}^m \times \mathbf{R}^{d-1}$, as would be required by Proposition 1. This is, however, an obviously trivial difficulty.

What is more difficult in this method is to keep track of the size of the neighborhoods involved, in order to obtain an explicit formula like Eq. (G21). Therefore, here we shall adopt another classical method of proof. The triumph of the naive substitution method will appear in Appendix N where, however, additional assumptions on f are made.

PROOF. Write Eq. (G1) as Eq. (G5) and let

$$\mathbf{y}' - \mathbf{y}_0 = -J^{-1}L(\mathbf{x} - \mathbf{x}_0) - J^{-1}\mathbf{N}(\mathbf{x}, \mathbf{y}) \tag{G25}$$

for $(\mathbf{x}, \mathbf{y}) \in \Gamma_m(\mathbf{x}_0, \alpha) \times \Gamma_d(\mathbf{y}_0, \delta)$.

Note that $|\mathbf{y}' - \mathbf{y}_0| < \delta$ if $(\mathbf{x}, \mathbf{y}) \in \Gamma_m(\mathbf{x}_0, \rho_{\delta,\alpha}) \times \Gamma_d(\mathbf{y}_0, \delta)$ and if (as supposed) $\alpha > \rho_{\delta,\alpha} > 0$. In fact, by the Lagrange theorem and $\mathbf{N}(\mathbf{x}_0, \mathbf{y}_0) = \mathbf{0}$, it follows that

$$\begin{aligned} |\mathbf{y}' - \mathbf{y}_0| &\leqslant |J^{-1}L|\rho_{\delta,\alpha} + |J^{-1}(\mathbf{N}(\mathbf{x}, \mathbf{y}) - \mathbf{N}(\mathbf{x}_0, \mathbf{y}_0))| \\ &\leqslant |J^{-1}L|\rho_{\delta,\alpha} + \max|J^{-1}|\left(\left|\frac{\partial \mathbf{N}}{\partial \mathbf{x}}\right|\alpha + \left|\frac{\partial \mathbf{N}}{\partial \mathbf{y}}\right|\delta\right) < \frac{\delta}{2} \end{aligned} \tag{G26}$$

Therefore, at $\mathbf{x}$ fixed in $\Gamma_m(\mathbf{x}_0, \rho_{\delta,\alpha})$, Eq. (G25) yields a map of $\Gamma_d(\mathbf{y}_0, \delta)$ into itself.

We can, therefore, recursively define, for each fixed $\mathbf{x} \in \Gamma_m(\mathbf{x}_0, \rho_{\delta,\alpha})$,

$$\mathbf{y}_n - \mathbf{y}_0 = -J^{-1}L(\mathbf{x} - \mathbf{x}_0) - J^{-1}\mathbf{N}(\mathbf{x}, \mathbf{y}_{n-1}), \tag{G27}$$

$n = 1, 2, \ldots$. Then,

$$\begin{aligned} |\mathbf{y}_n - \mathbf{y}_{n-1}| &= |J^{-1}(\mathbf{N}(\mathbf{x}, \mathbf{y}_{n-1}) - \mathbf{N}(\mathbf{x}, \mathbf{y}_{n-2}))| \\ &\leqslant (\max|J^{-1}|)\left(\max\left|\frac{\partial \mathbf{N}}{\partial \mathbf{y}}\right|\right)|\mathbf{y}_{n-1} - \mathbf{y}_{n-2}| \\ &\leqslant \frac{1}{2}|\mathbf{y}_{n-1} - \mathbf{y}_{n-2}|, \end{aligned} \tag{G28}$$

having used in the last step the hypothesis $\rho_{\delta,\alpha} > 0$ which implies that $(\max|J^{-1}|)(\max|\partial\mathbf{N}/\partial\mathbf{y}|) < \frac{1}{2}$.

Therefore, $|\mathbf{y}_n - \mathbf{y}_{n-1}| \leqslant 2^{-(n-1)}|\mathbf{y}_1 - \mathbf{y}_0|$ and there exists the limit

$$\boldsymbol{\varphi}(\mathbf{x}) = \lim_{n\to\infty} \mathbf{y}_n = \mathbf{y}_0 + \sum_{k=1}^{\infty} (\mathbf{y}_k - \mathbf{y}_{k-1}). \tag{G29}$$

If $(\mathbf{x}, \tilde{\mathbf{y}})$ is another solution to Eq. (G1) in $\Gamma_m(\mathbf{x}_0, \rho_{\delta,\alpha}) \times \Gamma_d(\mathbf{y}_0, \delta)$, we can write Eq. (G1) in the form of Eq. (G5) for $\tilde{\mathbf{y}}$ and $\boldsymbol{\varphi}(\mathbf{x})$ and subtract

$$|\tilde{\mathbf{y}} - \boldsymbol{\varphi}(\mathbf{x})| = |J^{-1}(\mathbf{N}(\mathbf{x}, \tilde{\mathbf{y}}) - \mathbf{N}(\mathbf{x}, \boldsymbol{\varphi}(\mathbf{x}))| \leqslant \tfrac{1}{2}|\tilde{\mathbf{y}} - \boldsymbol{\varphi}(\mathbf{x})|, \tag{G30}$$

i.e., $\tilde{\mathbf{y}} = \boldsymbol{\varphi}(\mathbf{x})$, proving uniqueness.

The differentiability statement is proved as in Proposition 1. mbe

3 Corollary. *Under the assumptions of Proposition* 2, *let* $m = d$ *and, see Eq.* (G23), *give* $B, C > 0$ *such that* $B, C > 1$ *and*

$$B > (\min|J^{-1}L|)^{-1},$$
$$C > (\min|L^{-1}J|)^{-1}, \tag{G31}$$

where the minima are taken over $\Gamma_d(\mathbf{x}_0, \bar{\alpha}) \times \Gamma_d(\mathbf{y}_0, \bar{\delta})$ *with given* $\bar{\alpha}, \bar{\delta} > 0$. *Suppose that* $\delta > 0$ *is so small that* $\delta, B\delta < \bar{\alpha}, \bar{\delta}$.

Define $\rho_{\delta,\alpha}$ *as in Eq.* (G21) *and* $\tilde{\rho}_{\alpha,\delta}$ *as*

$$\tilde{\rho}_{\alpha,\delta} = \frac{1}{2}\,\frac{\alpha - 2(\max|L^{-1}|)(\alpha \max|\partial\mathbf{N}/\partial\mathbf{x}| + \delta \max|\partial\mathbf{N}/\partial\mathbf{y}|)}{\max|L^{-1}J|}, \tag{G32}$$

where the maxima are now considered on $\Gamma_d(\mathbf{x}_0, \bar{\alpha}) \times \Gamma_d(\mathbf{y}_0, \bar{\delta})$ *both for* $\rho_{\delta,\alpha}$ *and* $\tilde{\rho}_{\alpha,\delta}$.

Then if δ *is so small that*

$$0 < \rho \equiv \rho_{\delta, B\delta} < B\delta \quad \text{and} \quad 0 < \tilde{\rho} \equiv \tilde{\rho}_{\rho/BC, \rho/B} < \delta \tag{G33}$$

(*which is possible by Observation* (1), *p.* 535), *the* φ *image of* $\Gamma_d(\mathbf{x}_0, \rho)$ *covers* $\Gamma_d(\mathbf{y}_0, \tilde{\rho})$.

Observations.

(1) This means that if the Jacobians of $\mathbf{f}$ with respect to $\mathbf{x}$ and with respect to $\mathbf{y}$ have nonvanishing determinant at $(\mathbf{x}_0, \mathbf{y}_0)$, the $\mathbf{f}$ sets up a correspondence between $\mathbf{x}, \mathbf{y}$ near $\mathbf{x}_0, \mathbf{y}_0$ of class C^∞ with the inverse of class C^∞ and sending open sets onto open sets (it is a local "C^∞ diffeomorphism").

(2) Since Corollary 3 is quantitative, it says much more: it gives, in fact, estimates of the size of the regions where $\mathbf{f}$ can be inverted.

PROOF. Just apply Proposition 2 twice, to express y in terms of x and vice versa (make a two-dimensional drawing to better understand the situation). mbe

Another important application of Proposition 2 is the following corollary used in §5.10.

4 Corollary. *Let* $\mathbf{f} \in C^\infty(\mathbf{T}^l)$ *with values in* $\mathbf{R}^l$. *Consider the equation for* $\varphi \in \mathbf{T}^l$:

$$\varphi' = \varphi + \epsilon\mathbf{f}(\varphi) \tag{G34}$$

with $\epsilon \in \mathbf{R}_+$ *and suppose* $\max|\mathbf{f}(\varphi)| \leqslant 1$, $\max|(\partial\mathbf{f}/\partial\varphi)(\varphi)| \leqslant 1$.

There is $\epsilon_l > 0$, *depending only on* l *and not on* $\mathbf{f}$, *such that,* $\forall \epsilon < \epsilon_l$, *the above equation can be solved uniquely in the form*

$$\varphi = \varphi' + \epsilon\mathbf{g}(\varphi', \epsilon) \tag{G35}$$

with $\mathbf{g} \in C^\infty(\mathbf{T}^l)$ *at fixed* ϵ *and* $\max_\varphi|\mathbf{g}(\varphi, \epsilon)| \leqslant 1$.

Furthermore, if φ *verifies Eq.* (G34), *then it is given by Eq.* (G35) *up to* $2\pi\boldsymbol{\nu}$, $\boldsymbol{\nu} \in \mathbf{Z}^l$.

Observation. This is a "global theorem" involving an inversion on a large set, namely, $\mathbf{T}^l$.

It can be improved to cover the case when $\mathbf{f}$ depends parametrically on some $\mathbf{A} \in \mathbf{R}^p$ so that $(\mathbf{A}, \varphi) \to \mathbf{f}(\mathbf{A}, \varphi)$ is a C^∞ function on $\mathbf{R}^p \times \mathbf{T}^l$. Then if $\mathbf{f}$ verifies the assumptions of the corollary for each $\mathbf{A} \in V \subset \mathbf{R}^p$, one easily sees that $\mathbf{g} \in C^\infty(V \times \mathbf{T}^l)$, $\forall \epsilon < \epsilon_l$.

PROOF. Let $0 \leqslant \epsilon < \frac{1}{4}$. The Jacobian matrices L, J of Eq. (G34) regarded as an implicit equation $\mathbf{F}(\varphi, \varphi') = \mathbf{0}$ in $\mathbf{R}^l \times \mathbf{R}^l$ near the solution $(\varphi_0, \varphi_0 + \epsilon\mathbf{f}(\varphi_0))$, with φ_0 given in $\mathbf{T}^l$, are

$$L_{ij} = \delta_{ij}, \qquad J_{ij} = \delta_{ij} + \epsilon \frac{\partial f_i}{\partial \varphi_j}, \tag{G36}$$

and by assumption [see Eqs. (E2), (E3), and (E10)] and since $\epsilon < \frac{1}{4}$:

$$(l - \tfrac{1}{4}) < |J| < (l + \tfrac{1}{4}), \qquad (l - \tfrac{1}{2}) < |J^{-1}| < (l + \tfrac{1}{2}) \tag{G37}$$

so that the constants B, C in Eq. (G31) can be chosen $B, C \geqslant (l - \frac{1}{2})^{-1}$. We now apply Corollary 3 to our equation near $(\varphi_0, \varphi_0 + \epsilon\mathbf{f}(\varphi_0))$ by choosing $\delta = \sqrt{\epsilon}$, say, and noting that from Eqs. (G21) and (G32), it easily follows that for ϵ small enough,

$$\frac{\delta}{4(l - \frac{1}{2})^2} \leqslant \rho, B\tilde{\rho} \leqslant B\delta. \tag{G38}$$

Noting that $\delta \gg \epsilon$, we see that Corollary 3 implies that as φ_0 varies on $\mathbf{T}^l$, the point $\varphi_0 + \epsilon\mathbf{f}(\varphi_0)$ also varies covering $\mathbf{T}^l$ if ϵ is small enough.

Furthermore, the map of Eq. (G34) is one to one, for ϵ very small, as a map of $\mathbf{T}^l$ onto itself. In fact, if $\varphi_1, \varphi_2 \in \mathbf{T}^l$ and if the segment σ given by $t \to \varphi_1 t + \varphi_2(1 - t)$, $t \in [0, 1]$, is the shortest segment on $\mathbf{T}^l$ connecting φ_1 and φ_2, we see that the points $\varphi_1' = \varphi_1 + \epsilon\mathbf{f}(\varphi_1)$ and $\varphi_2' = \varphi_2 + \epsilon\mathbf{f}(\varphi_2)$ can coincide mod 2π only if $\varphi_1' = \varphi_2'$, if ϵ is small.[1]
But

$$\begin{aligned} |\varphi_1' - \varphi_2'| &= |\varphi_1 - \varphi_2 + \epsilon\mathbf{f}(\varphi_1) - \epsilon\mathbf{f}(\varphi_2)| \\ &\geqslant |\varphi_1 - \varphi_2| - \epsilon\left(\max_{\varphi} \sum_{i,j} \left|\frac{\partial f_i}{\partial \varphi_j}(\varphi)\right|\right)|\varphi_1 - \varphi_2| \\ &\geqslant (1 - \epsilon)|\varphi_1 - \varphi_2| > 0 \end{aligned}$$

if $\varphi_1 \neq \varphi_2$.

Since $\mathbf{f}$ is periodic, the assumption that σ is the shortest path on $\mathbf{T}^l$ leading from φ_1 to φ_2 cannot be restrictive and, therefore, the map $\varphi \to \varphi + \epsilon\mathbf{f}(\varphi)$ is one to one for $\epsilon < 1$.

[1] In fact, $(\varphi_1 - \varphi_2)_i$ cannot be too large ($\leqslant \pi$) if σ is the shortest segment joining φ_1 and φ_2 on $\mathbf{T}^l$.

So the map of Eq. (G34) can be inverted on $\mathbf{T}^{\prime}$ and its inverse map $\boldsymbol{\varphi}' \to \mathbf{F}_\epsilon(\boldsymbol{\varphi}')$ is C^∞ near every point if ϵ is small enough. Clearly, Eq. (G35) holds with $\mathbf{g}(\boldsymbol{\varphi}', \epsilon) = -\mathbf{f}(\mathbf{F}_\epsilon(\boldsymbol{\varphi}'))$ which also proves $|\mathbf{g}| < 1$. mbe

Concluding Remark

The above proofs do not really make use of the fact that $\mathbf{f}$ is of class C^∞. If $\mathbf{f}$ is only supposed to be of class $C^{(k)}$, $k \geqslant 1$, the ideas of the proofs still work, and the only difference will be that the inverse function φ will not turn out to be of class C^∞, of course, but only of class $C^{(k)}$. We use the above "$C^{(k)}$-version" of the implicit function theorems only in §5.7.

Exercise

In the context of Proposition 1, compute the second derivative of $\varphi(\mathbf{x})$ in terms of φ and of its first derivative $\partial\varphi/\partial\mathbf{x}$ and in terms of f and of its first two derivatives (Answer:

$$\frac{\partial^2\varphi}{\partial x_j \partial x_i} = -\frac{(\partial^2 f/\partial x_i \partial x_j)(\mathbf{x},\varphi) + (\partial^2 f/\partial x_i \partial y)(\mathbf{x},\varphi)\cdot \partial\varphi/\partial x_j}{(\partial f/\partial y)(\mathbf{x},\varphi)}$$

$$+ \frac{(\partial f/\partial x_i)(\mathbf{x},\varphi)\big((\partial^2 f/\partial x_i \partial y)(\mathbf{x},\varphi) + (\partial^2 f/\partial y^2)(\mathbf{x},\varphi)\partial\varphi/\partial x_i\big)}{((\partial f/\partial y)(\mathbf{x},\varphi))^2}.$$

APPENDIX H

The Ascoli-Arzelà Convergence Criterion

The following elegant proposition is famous.

1 Proposition. *Let Ω be a closed bounded set in $\mathbf{R}^d$. Let $(f_n)_{n=0}^{\infty}$ be a sequence of continuous functions defined on Ω such that:*

(i) *The sequence $(f_n)_{n=0}^{\infty}$ is "equibounded"*, i.e., *there exists M such that*

$$\|f_n\| = \max_{\xi\in\Omega} |f_n(\xi)| \leqslant M; \tag{H1}$$

(ii) *the sequence $(f_n)_{n=0}^{\infty}$ is "equicontinuous"*, i.e., *given $\epsilon > 0$ there exists $\delta_\epsilon > 0$ such that*

$$\sup_{n,|\xi-\xi'|<\delta_\epsilon} |f_n(\xi) - f_n(\xi')| < \epsilon. \tag{H2}$$

Then there is a subsequence $(f_{n_i})_{i=0}^{\infty}$ such that the limit

$$f(\xi) = \lim_{i\to\infty} f_{n_i}(\xi) \tag{H3}$$

exists, uniformly, $\forall\xi \in \Omega$.

Observations.

(1) Hence, f is continuous on Ω.

(2) The most interesting aspect of this theorem is, perhaps, the uniformity of the convergence.

PROOF. Let $\Omega_0 \subset \Omega$ be a denumerable dense subset of Ω (to be concrete, think of the case when Ω is a square and Ω_0 is the set of its points with rational coordinates). We shall write $\Omega_0 = \{\xi_0, \xi_1, \ldots\}$.

By the equiboundedness condition, it will be possible to find a subsequence $(f_{n_i})_{i=0}^{\infty}$ of $(f_n)_{n=0}^{\infty}$ such that the limits

$$\lim_{i\to\infty} f_{n_i}(\xi_j) \equiv f(\xi_j), \qquad j = 1, 2, \ldots \tag{H4}$$

exist. For instance, one can use the Cantor diagonal method; f is defined by the right-hand side of Eq. (H4).

Without loss of generality, we may and shall assume that the subsequence $(n_i)_{i=0}^{+\infty}$ coincides with $(0, 1, 2, \ldots)$, i.e., that the limits $\lim_{n\to\infty} f_n(\xi_j) \equiv f(\xi_j)$ exist without passing to a subsequence. This will now be used to show that the function f defined on Ω_0 can be extended to Ω by showing that the limit $\lim_{n\to\infty} f_n(\xi)$ exists.

In fact, we show that $(f_n(\xi))_{n=0}^{\infty}$ is a Cauchy sequence for all $\xi \in \Omega$.

So let $\xi \in \Omega$. Given $\epsilon > 0$ let $\tilde{\xi} \in \Omega_0$ be such that $|\xi - \tilde{\xi}| < \delta_\epsilon$; then, by Eq. (H2):

$$\begin{aligned} |f_n(\xi) - f_m(\xi)| &\leqslant |f_n(\xi) - f_n(\tilde{\xi})| + |f_n(\tilde{\xi}) - f_m(\tilde{\xi})| + |f_m(\tilde{\xi}) - f_m(\xi)| \\ &\leqslant 2\epsilon + |f_n(\tilde{\xi}) - f_m(\tilde{\xi})| \underset{n,m\to\infty}{\longrightarrow} 2\epsilon \end{aligned} \tag{H5}$$

because $f_n(\tilde{\xi}))_{n=0}^{\infty}$ is a Cauchy sequence. Hence, by the arbitrariness of ϵ, we see that $(f_n(\xi))_{n=0}^{\infty}$ is also a Cauchy sequence and we can define, $\forall \xi \in \Omega$, $f(\xi) = \lim_{n\to\infty} f_n(\xi)$.

If $\xi, \eta \in \Omega$, $|\xi - \eta| < \delta_\epsilon$, it then follows from Eq. (H2) that

$$|f(\xi) - f(\eta)| = \lim_{n\to\infty} |f_n(\xi) - f_n(\eta)| \leqslant \epsilon. \tag{H6}$$

It remains to show that the limit given by Eq. (H3) is uniform on Ω.

Otherwise, we could find $\epsilon > 0$, a sequence $n_i \to_{i\to\infty} \infty$ and points $\mathbf{x}_i \in \Omega$ such that

$$|f_{n_i}(\mathbf{x}_i) - f(\mathbf{x}_i)| > \epsilon, \qquad i = 1, 2, \ldots. \tag{H7}$$

Of course, we may and shall assume without loss of generality that $n_i = i$, i.e.,

$$|f_n(\mathbf{x}_n) - f(\mathbf{x}_n)| > \epsilon, \qquad n = 1, 2, \ldots. \tag{H8}$$

This is impossible because there would be an accumulation point $\bar{\mathbf{x}} \in \Omega$ for the sequence $\mathbf{x}_n$, $n = 1, \ldots$ and again we may assume, without loss of generality, that $\lim \mathbf{x}_n = \bar{\mathbf{x}}$. Then if $|\bar{\mathbf{x}} - \mathbf{x}_n| < \delta_{\epsilon/4}$, using Eqs. (H6) and (H2),

$$\begin{aligned} \epsilon < |f_n(\mathbf{x}_n) - f(\mathbf{x}_n)| &\leqslant |f_n(\mathbf{x}_n) - f_n(\bar{\mathbf{x}})| + |f_n(\bar{\mathbf{x}}) - f(\bar{\mathbf{x}})| + |f(\bar{\mathbf{x}}) - f(\mathbf{x}_n)| \\ &\leqslant \frac{2\epsilon}{4} + |f_n(\bar{\mathbf{x}}) - f_m(\bar{\mathbf{x}})| \underset{n\to\infty}{\longrightarrow} \frac{\epsilon}{2} \end{aligned} \tag{H9}$$

which is absurd. mbe

2 Corollary. *Under the assumptions of Proposition* 1, *aside from that of boundedness (or of closure or both) for* Ω, *the same conclusions hold with the exception of the uniformity of the convergence of* $f_{n_i}(\xi)$ *to* $f(\xi)$. *Nevertheless,* f *is uniformly continuous on* Ω.

PROOF. By insepection of the proof of Proposition 1.

Exercises

1. Let $(f_n)_{n=1}^{\infty}$ be a sequence of $C^1(\Omega)$ functions on a convex set which is the closure of its interior. If there is M such that

$$\sup_n \max_{\xi \in \Omega} \left| \frac{\partial f_n}{\partial \xi}(\xi) \right| \leqslant M,$$

then $(f_n)_{n=1}$ is an equicontinuous family on Ω. (Hint: Express the variation of f_n as the integral of its derivative along a segment joining two points.)

2. Define $C^{(\epsilon)}(\Omega)$, $\epsilon \in (0,1)$, to be the set of the functions such that

$$|f|_\epsilon = \sup_{\mathbf{x} \neq \mathbf{y}} \frac{|f(\mathbf{x}) - f(\mathbf{y})|}{|\mathbf{x} - \mathbf{y}|^\epsilon} + \sup_{\mathbf{x}} |f(\mathbf{x})| < +\infty.$$

Then any sequence $(f_n)_{n=0}^{\infty}$, $f_n \in C^{(\epsilon)}(\Omega)$, such that $|f_n|_\epsilon \leqslant M < \infty$, $\forall n$, is an equicontinuous equibounded sequence.

APPENDIX I

Fourier Series for Functions in $\bar{C}^{\infty}([0, L])$

Here we prove Lemma 11, §4.5, p. 266.

If $u \in \bar{C}^{\infty}([0, L])$, we set

$$\begin{aligned} u^*(x) &= u(x), && x \in [0, L], \\ u^*(L + x) &= -u(L - x), && x \in [0, L] \end{aligned} \tag{I1}$$

and by the assumption on the even derivatives of u in 0 and in L, i.e., that they should vanish, it is clear that the function thus defined on $[0, 2L]$ is in $C^{\infty}([0, 2L])$ and is periodic, together with all its derivatives, with period $2L$.

By the Fourier theorem, we set

$$\hat{u}_h^* = \frac{1}{2L} \int_0^{2L} u^*(x) e^{-i(2\pi h/2L)x}\, dx \tag{I2}$$

for $h \in \mathbf{Z}$ and observe that, $\forall h \geqslant 0$,

$$\begin{aligned} \hat{u}_h^* &= \frac{1}{2L} \int_0^L \left(u(x) e^{-i(\pi h/L)x} - u(L - x) e^{-i(\pi h/L)(L+x)} \right) dx \\ &= \frac{1}{2L} \int_0^L u(x) (e^{-i(\pi h/L)x} - e^{i(\pi h/L)x})\, dx \\ &= \frac{-i}{L} \int_0^L u(x) \sin \frac{\pi h x}{L}\, dx \\ &\equiv -\frac{i}{2} \bar{u}(h) = -\hat{u}_{-h}^*, \end{aligned} \tag{I3}$$

having used the change of variables $x \to L - x$.

Therefore, for $x \in [0, 2L]$,

$$u^*(x) = \sum_{h=-\infty}^{+\infty} \hat{u}_h^* e^{i\pi h x/L} = \sum_{h=1}^{\infty} \bar{u}_h \sin \frac{\pi h}{L} x. \tag{I4}$$

Hence, for $x \in [0, L]$,

$$u(x) = \sum_{h=1}^{\infty} \bar{u}_h \sin \frac{\pi h}{L} x, \tag{I5}$$

where $\bar{u}_h$ defined in Eq. (I3) coincides with Eq. (4.5.20). Equation (4.5.21) follows from Eq. (I3) and from the decay properties as $h \to \infty$ of the Fourier coefficients for C^{∞}-periodic functions. Equation (I5) gives Eq. (4.5.22).

mbe

APPENDIX L

Proof of Eq. (5.6.20)

Let $(S_t^{(\alpha,\delta)}(w_1, w_2))_i = \sigma_i(t, w)$, $i = 1, 2$; $t \in [0, 1]$. From Eqs. (5.6.17) and (5.6.18), one sees that

$$\sigma_1(t, \mathbf{w}) = w_1 + \int_0^t \chi_\delta(\boldsymbol{\sigma}(\tau, \mathbf{w}))(\alpha\sigma_1(\tau, \mathbf{w}) + P(\boldsymbol{\sigma}(\tau, \mathbf{w})))\, d\tau,$$

$$\sigma_2(t, \mathbf{w}) = e^{-\nu_0 t} w_2 + \int_0^t e^{-\nu_0(t-\tau)} \chi_\delta(\boldsymbol{\sigma}(\tau, \mathbf{w})) Q(\boldsymbol{\sigma}(\tau, \mathbf{w}))\, d\tau.$$

Consider, for instance, $\partial\boldsymbol{\sigma}/\partial w_1$ and, dropping the $\mathbf{w}$ variables in the arguments of $\boldsymbol{\sigma}$ for simplicity, note that

$$\frac{\partial \sigma_1}{\partial w_1}(t) = 1 + \int_0^t \Bigg\{ \partial\chi_\delta(\boldsymbol{\sigma}(\tau)) \cdot \frac{\partial \boldsymbol{\sigma}}{\partial w_1}(\tau)\,(\alpha\sigma_1(\tau) + P(\boldsymbol{\sigma}(\tau))) + \chi_\delta(\boldsymbol{\sigma}(\tau))\left(\alpha \frac{\partial \sigma_1}{\partial w_1}(\tau) + \partial P(\boldsymbol{\sigma}(\tau)) \cdot \frac{\partial \boldsymbol{\sigma}}{\partial w_1}\right)\Bigg\}\, d\tau$$

$$\frac{\partial \sigma_2}{\partial w_1}(t) = \int_0^t e^{-\nu_0(t-\tau)} \Bigg\{ \partial\chi_\delta(\boldsymbol{\sigma}(\tau)) \cdot \frac{\partial \boldsymbol{\sigma}}{\partial w_1}(\tau) Q(\boldsymbol{\sigma}(\tau)) + \chi_\delta(\boldsymbol{\sigma}(\tau)) \partial Q(\boldsymbol{\sigma}(\tau)) \cdot \frac{\partial \boldsymbol{\sigma}}{\partial w_1}(\tau)\Bigg\}\, d\tau,$$

where ∂g denotes $(\partial g/\partial w_1, \partial g/\partial w_2)$ if g is a function of w_1, w_2 and other variables.

Hence, using Eq. (5.6.15) and the fact that P and Q have a second-order zero at the origin, we see that there are two constants p, q such that

$$\left|\frac{\partial\sigma_1}{\partial w_1}(t) - 1\right| \leqslant p\int_0^t \left\{ \frac{1}{\delta}\left|\frac{\partial\boldsymbol{\sigma}}{\partial w_1}(\tau)\right|(|\alpha|\delta + \delta^2) + |\alpha|\left|\frac{\partial\boldsymbol{\sigma}}{\partial w_1}(\tau)\right| + \delta\left|\frac{\partial\boldsymbol{\sigma}}{\partial w_1}(\tau)\right|\right\} d\tau$$

(since $|\boldsymbol{\sigma}(\tau)| \leqslant \delta\sqrt{2}$) and

$$\left|\frac{\partial\sigma_2}{\partial w_1}(t)\right| \leqslant q\int_0^t \left\{ \frac{1}{\delta}\left|\frac{\partial\boldsymbol{\sigma}}{\partial w_1}(\tau)\right|\delta^2 + \delta\left|\frac{\partial\boldsymbol{\sigma}}{\partial w_1}(\tau)\right|\right\} d\tau.$$

Therefore, adding and subtracting 1 appropriately:

$$\left|\frac{\partial\sigma_1}{\partial w_1}(t) - 1\right| \leqslant 2p(|\alpha| + \delta)t + 2p(|\alpha| + \delta) \times \int_0^t \left\{\left|\frac{\partial\sigma_1}{\partial w_1}(\tau) - 1\right| + \left|\frac{\partial\sigma_2}{\partial w_1}(\tau)\right|\right\} d\tau,$$

$$\left|\frac{\partial\sigma_2}{\partial w_2}(t)\right| \leqslant 2q\delta t + 2q\delta\int_0^t \left\{\left|\frac{\partial\sigma_1}{\partial w_1}(\tau) - 1\right| + \left|\frac{\partial\sigma_2}{\partial w_1}(\tau)\right|\right\} d\tau.$$

Setting $y(t) = |(\partial\sigma_1/\partial w_1)(t) - 1| + |(\partial\sigma_2/\partial w_1)(t)|$, the preceding inequalities, added up, imply

$$y(t) \leqslant 2(p + q)(|\alpha| + \delta)t + 2(p + q)(|\alpha| + \delta)\int_0^t y(\tau)\, d\tau$$

and $y(0) = 0$. The above integral inequality implies

$$y(t) \leqq \bar{y}(t), \qquad \forall t \geqslant 0,$$

where

$$\bar{y}(t) = 2(p + q)(|\alpha| + \delta)t + 2(p + q)(|\alpha| + \delta)\int_0^t \bar{y}(\tau)\, d\tau$$

and $\bar{y}(0) = 0$ (see Problems 8 and 9, §2.5). Hence,

$$\bar{y}(t) = (e^{2(p+q)(|\alpha|+\delta)t} - 1) \leqslant Mt(|\alpha| + \delta)$$

for $0 \leqslant t \leqslant 1$, $|\alpha| \leqslant 1$, $\delta \leqslant 1$ (and M could be $2(p + q)e^{4(p+q)}$).

An identical argument could be given for $t \in (-1, 0)$.

APPENDIX M

Proof of Eq. (5.6.63)

Let

$$(x, \pi_t(x)) = S_t^{(\alpha,\delta)}(x_0, \pi(x_0)), \qquad (x, \pi_{t'}(x)) = S_{t'}^{(\alpha,\delta)}(x'_0, \pi(x'_0)). \quad \text{(M1)}$$

Then from Eq. (5.6.33), it follows that

$$|\pi_t(x) - \pi_{t'}(x)| \leqslant |e^{-\nu_0 t}\pi(x_0) - e^{-\nu_0 t'}\pi(x'_0)| + \left| \int_0^t e^{-\nu_0(t-\tau)} Z_\tau\left(S_\tau^{(\alpha,\delta)}(x_0, \pi(x_0), \alpha)\right) d\tau - \int_0^{t'} e^{-\nu_0(t'-\tau)} Z_\delta\left(S_\tau^{(\alpha,\delta)}(x'_0, \pi(x'_0)), \alpha\right) d\tau \right|. \quad \text{(M2)}$$

Using Eqs. (5.6.25), (5.6.24), and (5.6.49) and supposing $0 \leqslant t' < t \leqslant t_+$, the right-hand side of Eq. (M2) is

$$\begin{aligned}
&\leqslant |e^{-\nu_0 t} - e^{-\nu_0 t'}|\,|\pi(x_0)| + e^{-\nu_0 t'}|\pi(x_0) - \pi(x'_0)| \\
&\quad + M\delta^2|t - t'| + \int_0^{t'} |e^{-\nu_0(t-\tau)} - e^{-\nu_0(t'-\tau)}| M\delta^2\, d\tau \\
&\quad + \int_0^{t'} e^{-\nu_0(t'-\tau)} |Z_\delta\left(S_\tau^{(\alpha,\delta)}(x_0, \pi(x_0), \alpha)\right) - Z_\delta\left(S_\tau^{(\alpha,\delta)}(x'_0, \pi(x'_0), \alpha)\right)| \\
&\leqslant \delta\nu_0|t - t'| + |\pi(x_0) - \pi(x'_0)| + (1 + \nu_0 t)M\delta^2|t - t'| \\
&\quad + 2Mt\delta(1 + M(a_+ + \delta)t)(|x_0 - x'_0| + |\pi(x_0) - \pi(x'_0)|) \\
&\leqslant \left(\delta\nu_0 + (1 + \nu_0 t)M\delta^2\right)|t - t'| + \Big\{\left(C\sqrt{\delta} + 2Mt\delta(1 + M(a_+ + \delta)t)\right. \\
&\qquad \left. + 2Mt\delta(1 + M(a_+ + \delta)t)C\sqrt{\delta}\right)\Big\}|x_0 - x'_0|.
\end{aligned} \quad \text{(M3)}$$

To estimate $|x_0 - x_0'|$, one proceeds as in subsection 5.6.G, p. 420, using the expressions analogous to Eq. (5.6.58):

$$\begin{aligned} x_0 &= x - \int_0^t d\tau\, X_\delta\big(S_{-\tau}^{(\alpha,\delta)}(x,\pi_t(x)),\alpha\big),\\ x_0' &= x - \int_0^{t'} d\tau\, X_\delta\big(S_{-\tau}^{(\alpha,\delta)}(x,\pi_t(x)),\alpha\big), \end{aligned} \tag{M4}$$

By Eqs. (5.6.25), (5.6.23), and (5.6.20),

$$\begin{aligned} |x_0 - x_0'| &\leqslant M|t-t'|(a_+\delta + \delta^2)\\ &\quad + \int_0^{t'} d\tau \big|X_\delta\big(S_{-\tau}^{(\alpha,\delta)}(x,\pi_t(x)),\alpha\big) - X_\delta\big(S_{-\tau}^{(\alpha,\delta)}(x,\pi_{t'}(x)),\alpha\big)\big|\\ &\leqslant M(a_+\delta+\delta^2)|t-t'|\\ &\quad + 2M(a_+ + \delta)t(1+M(\alpha_+ + \delta)t)|\pi_t(x) - \pi_{t'}(x)|. \end{aligned} \tag{M5}$$

The restrictions imposed on a, t_0, by the second of Eqs. (5.6.41) imply (recall that $C, \delta < 1$)

$$\begin{aligned} \theta &= \big(C\sqrt{\delta} + 2tM\delta(1+M(a_+ + \delta)t)(1+C\sqrt{\delta})\big)\\ &\quad \times 2M(a_+ + \delta)t(1+M(a_+ + \delta)t)\\ &\leqslant (1+\tfrac{1}{10}(1+\tfrac{1}{20}))(1+1)\tfrac{1}{10}(1+\tfrac{1}{10}) < \tfrac{1}{2}. \end{aligned} \tag{M6}$$

By combining the last of Eqs. (M5) with the last of Eqs. (M3), it follows that

$$(1-\theta)|\pi_t(x) - \pi_{t'}(x)| \leqslant \big(\delta\nu_0 + (1+\nu_0 \mathbf{t})M\delta^2\big)|t-t'|, \tag{M7}$$

so that, since $\theta < \frac{1}{2}$:

$$|\pi_t(x) - \pi_{t'}(x)| \leqslant 2\big(\delta\nu_0 + (1+\nu_0 t_+)M\delta^2\big)|t-t'|, \tag{M8}$$

$\forall t, t' \in [0, t_+]$, and, therefore, by Eq. (5.6.51), for all $t, t' \in \mathbf{R}_+$, $t' \leqslant t$, $|t-t'| \leqslant t_+$.

APPENDIX N

The Analytic Implicit Function Theorem

The proofs of Propositions 20 and 21, §5.11, are based on the following idea.

Let F be a holomorphic function of a single complex variable $z \in \Omega \subset \mathbf{C}$. Assume that its complex derivative, denoted by a prime in this section, $F'(z)$, does not vanish in Ω.

It is an elementary consequence of the theory of power series that as z' varies in a small vicinity of $z'_0 = F(z_0)$, and z varies close to z_0 the equation $z' = F(z)$ can be uniquely solved for z and its solution is a function I defined in a neighborhood U of z'_0 and holomorphic in U;

$$F(I(z')) \equiv z' \tag{N1}$$

for all z' in U, and

$$I(F(z)) \equiv z \tag{N2}$$

for all z in a suitable neighborhood of z_0. The function I has Taylor coefficients in z'_0 which can be computed via an easy algorithm from those of F in z_0.

The function F will be invertible on the whole $F(\Omega)$ if and only if $F(z) \neq F(z')$ whenever $z \neq z'$. In this case the inverse function I will be holomorphic on $F(\Omega)$ and it will be the unique inverse of F defined on $F(\Omega)$.

A simple criterion implying that $F(z) \neq F(z')$ for $z \neq z'$ is the following. Suppose that for every pair $z, z' \in \Omega$ there is a smooth curve $\Lambda(z, z') \subset \Omega$ with length $|\Lambda(z, z')|$ bounded by

$$\Lambda(z, z') < \beta(\Omega)|z - z'|, \tag{N3}$$

where $\beta(\Omega)$ is a suitable constant. Then F will be a one to one map between Ω and $F(\Omega)$ if

$$\sigma \equiv \beta(\Omega) \sup_{z\in\Omega} |F'(z) - 1| < 1. \tag{N4}$$

In fact (N4) implies

$$\begin{aligned}|F(z) - F(z')| &\equiv \left|\int_{\Lambda(z,z')} F'(\zeta)\, d\zeta\right| \equiv \left|\int_{\Lambda(z,z')} d\zeta + \int_{\Lambda(z,z')} (F'(\zeta) - 1)\, d\zeta\right| \\ &\geqslant |z - z'| - \sigma|z - z'| = (1-\sigma)|z - z'|.\end{aligned} \tag{N5}$$

Let us prove Proposition 20 using the above remarks.

First consider the inversion problem for the equation

$$\varphi' = \varphi + g(\varphi) \qquad \bmod 2\pi \tag{N6}$$

with $\varphi \in \mathbf{T}^1$ and g holomorphic on $C(\xi)$. Let $\bar{g}$ be the holomorphic extension of g to $C(\xi)$.

Equation (N6) can be written

$$z' = ze^{i\bar{g}(z)} \equiv F(z), \qquad z \in \mathbf{T}^1. \tag{N7}$$

Let $\delta \in (0,1)$, $\delta < \xi/2$ (say); we regard (N7) as an equation for $z \in C(\xi - \delta)$, i.e., $\Omega = D(\xi - \delta)$ in the language of the above discussion.

Between any two points $z, z' \in C(\xi)$ we can draw a line $\Lambda(z,z')$ contained in $C(\xi)$ with length bounded by $2\pi|z - z'|$: therefore we see that we can take $\beta(C(\xi - \delta)) = 2\pi$.

Hence Eq. (N7) can be inverted in $C(\xi - \delta)$ under the condition

$$2\pi \sup_{z\in C(\xi-\delta)} |e^{i\bar{g}(z)} - 1 + iz\bar{g}'(z)e^{i\bar{g}(z)}| < 1 \tag{N8}$$

which also ensures that $F'(z) \neq 0$ because

$$F'(z) \equiv 1 + (e^{i\bar{g}(z)} - 1) + iz\bar{g}'(z)e^{i\bar{g}(z)}. \tag{N9}$$

The supremum in inequality (N8) can be estimated by using a dimensional estimate like (5.11.18): one finds

$$2\pi\left((e^{|g|_\xi} - 1) + e^{\xi}|g|_\xi e^{|g|_\xi} e^{\xi}\delta^{-1}\right) < 2\pi e^{2\xi} e^{|g|_\xi} |g|_\xi \delta^{-1} \tag{N10}$$

and the above discussion implies that we can define a function $I(z')$ on $F(C(\xi - \delta))$ such that

$$F(I(z')) = z' \qquad \text{for all} \quad z' \in F(C(\xi - \delta)) \tag{N11}$$

provided

$$4\pi e^{2\xi} e^{|g|_\xi} |g|_\xi \delta^{-1} < 1. \tag{N12}$$

The form of F,

$$F(z) = ze^{i\bar{g}(z)}, \tag{N13}$$

implies that

$$F(C(\xi - \delta)) \subset C(\xi - \delta - |g|_\xi) \tag{N14}$$

because the F-image of $\partial C(\xi - \delta)$ consists of two lines outside $C(\xi - \delta - |g|_\xi)$ and the boundary of $F(C(\xi - \delta))$ is $F(\partial C(\xi - \delta))$. (The latter property follows from general properties of holomorphic functions but it can also be seen directly in our case as follows. If $z_0' \in F(C(\xi - \delta))$, there is a sequence

$z_n \in C(\xi - \delta)$ such that

$$z_0' = \lim_{n\to\infty} F(z_n), \tag{N15}$$

and, without loss of generality, we may suppose that the sequence z_n converges to a limit z_0. If $z_0 \in \partial D(\xi - \delta)$, then $z_0' \in F(\partial C(\xi - \delta))$; if $z_0 \in C(\xi - \delta)$, then the local invertibility of F implies that z_0' is interior to $F(C(\xi - \delta))$ which is impossible.)

Therefore if Eq. (N12) holds the function I inverse to F is holomorphic at least in $C(\xi - 2\delta)$, because Eq. (N12) implies $|g|_\xi < \delta$.

Assuming the validity of the inequality in Eq. (N12), set

$$\bar{\Delta}(z') = -\bar{g}(I(z')), \qquad z' \in C(\xi - 2\delta). \tag{N16}$$

This defines a holomorphic function on $C(\xi - 2\delta)$ such that

$$|\bar{\Delta}|_{\xi-2\delta} < |g|_\xi \tag{N17}$$

and

$$I(z') = z' e^{i\bar{\Delta}(z')}. \tag{N18}$$

As z varies on the unit circle, the point $z' = F(z)$ also varies on the unit circle so that $\bar{\Delta}$ is real on the F-image of the unit circle: since $|g|_\xi < \delta$ and F is given by Eq. (N13) it is clear (by a continuity argument) that as z varies on the unit circle z' varies covering the entire unit circle. This means that $\bar{\Delta}$ is real on $\mathbf{T}^1$ and we can define

$$\Delta(\varphi) = \bar{\Delta}(e^{i\varphi}) \tag{N19}$$

and Δ is analytic and real on $\mathbf{T}^1$.

Since δ is arbitrary in $(0, \xi/2)$ we replace 2δ by δ and find that the theorem is proved, in the case considered, under the condition

$$8\pi e^{2\xi} e^{|g|_\xi} |g|_\xi \delta^{-1} < 1. \tag{N20}$$

Next we study the inversion problem for the equation

$$\varphi' = \varphi + g(\mathbf{A}, \varphi), \tag{N21}$$

where g is holomorphic on $C(\rho, \xi; \mathbf{A}_0)$.

We write Eq. (N21) as

$$z' = z \exp i\bar{g}(\mathbf{A}, z). \tag{N22}$$

Clearly, $\forall \mathbf{A} \in \hat{\mathbb{S}}_\rho(\mathbf{A}_0)$, we can repeat the above argument, once more keeping in all the formulae an explicit $\mathbf{A}$ dependence which will, however, play no role whatsoever. So Eq. (N22) will be invertible in the form

$$z = z' \exp i\bar{\Delta}(\mathbf{A}, z') \tag{N23}$$

with $\bar{\Delta}$ holomorphic on $C(\rho, \xi; \mathbf{A}_0)$ if Eq. (N20) holds with $|g|_\xi$ replaced by $|g|_{\rho,\xi}$. The function $\bar{\Delta}$ will also turn out to be real for $\mathbf{A} \in \mathbb{S}_\rho(\mathbf{A}_0)$, i.e., for $\mathbf{A}$ real and for $|z'| = 1$.

The same conclusions clearly hold if g is defined and holomorphic on a more general set of the form $W \times \mathbf{T}^1$ with $W \subset \mathbf{C}^l$ open. Equation (N22) is inverted by Eq. (N23) if Eq. (N20) holds with $|g|_\xi$ replaced by the supremum of g in $W \times C(\xi)$.

With these remarks in mind, we can quickly conclude the proof of Proposition 20.

Consider the case contemplated in Proposition 20:

$$\boldsymbol{\varphi}' = \boldsymbol{\varphi} + \mathbf{g}(\mathbf{A}, \boldsymbol{\varphi}) \tag{N24}$$

with $\mathbf{g}$ extending to a holomorphic function on $C(\rho, \xi; \mathbf{A}_0)$. Write the system of Eq. (N24) as

$$z_k' = z_k \exp i g_k(\mathbf{A}, \mathbf{z}), \qquad k = 1, \ldots, p \tag{N25}$$

and consider the first equation for z_1:

$$z_1' = z_1 \exp i g_1(\mathbf{A}, z_1, z_2, \ldots, z_p). \tag{N26}$$

If Eq. (N20) holds with $|g|_\xi$ replaced by $|\mathbf{g}|_{\rho,\xi}$, we can invert Eq. (N26) as

$$z_1 = z_1' \exp i\tilde{\Delta}_1(\mathbf{A}, z_1', z_2, \ldots, z_p) \tag{N27}$$

with $\tilde{\Delta}_1$ holomorphic for $\mathbf{A} \in \hat{\mathbb{S}}_\rho(\mathbf{A}_0)$, $z_1' \in C(\xi - \delta)$, and $z_k \in C(\xi)$ for all $k = 2, \ldots, p$. Also, $|\mathbf{g}|_{\rho,\xi} < \delta$.

Furthermore, Eq. (N27) inverts Eq. (N26) on the same set $\hat{\mathbb{S}}_\rho(\mathbf{A}_0) \times C(\xi - \delta) \times C(\xi)^{l-1}$, and

$$|\tilde{\Delta}_1| \leqslant |g_1|_{\rho,\xi} \leqslant |\mathbf{g}|_{\rho,\xi} < \delta, \tag{N28}$$

where $|\tilde{\Delta}_1|$ denotes the supremum of $\tilde{\Delta}_1$ on its domain of definition.

Finally, $\tilde{\Delta}_1$ is real if $\mathbf{A} \in \mathbb{S}_\rho(\mathbf{A}_0), |z_1'| = |z_2| = \cdots = |z_k| = 1$.

Now substitute Eq. (N27) into the Eq. (N25) for $k = 2, \ldots, p$ and set

$$g_k^{(1)}(\mathbf{A}, z_1', z_2, \ldots, z_p) = g_k\left(\mathbf{A}, z_1' e^{i\tilde{\Delta}_1(\mathbf{A}, z_1', z_2, \ldots, z_p)}, z_2, \ldots, z_p\right) \tag{N29}$$

which are defined and holomorphic for $\mathbf{A} \in \hat{\mathbb{S}}_\rho(\mathbf{A}_0), z_1' \in C(\xi - \delta), z_2, \ldots, z_p \in C(\xi)$ and, of course, the supremum of $g_k^{(1)}$ on its domain of definition can be estimated as

$$\sup |g_k^{(1)}| \leqslant |g_k|_{\rho,\xi} \leqslant |\mathbf{g}|_{\rho,\xi}. \tag{N30}$$

Hence, we can take as parameters $\mathbf{A}, z_1', z_3 \ldots z_p$ and solve the equation

$$z_2' = z_2 \exp i g_2^{(1)}(\mathbf{A}, z_1', z_2, z_3, \ldots, z_p) \tag{N31}$$

for z_2 as before, etc.

After p steps, we will have inverted the full system, in the desired form, on the set $C(\rho, \xi - \delta; \mathbf{A}_0)$ under the sole condition

$$8\pi e^{2\xi} e^{|\mathbf{g}|_{\rho,\xi}} |\mathbf{g}|_{\rho,\xi} \delta^{-1} < 1 \tag{N32}$$

which, if $\delta \leqslant 1$, and, hence, $|\mathbf{g}|_{\rho,\xi} < \delta$, can be put into the form of Eq. (5.11.19) with $\gamma < 2^8$.

With some care, one could find smaller values for γ. mbe

In the same way, one can prove the implicit function theorem mentioned in Proposition 21. Since this is a "local theorem", the proof is actually slightly easier than the above.

APPENDIX O

A Simple Algorithm for the Solutions of the Differential Equations. The Finite-Difference Method

Consider $f \in C^{\infty}(\mathbf{R}^d)$ and the equation

$$\dot{\mathbf{x}} = \mathbf{f}(\mathbf{x}), \qquad \mathbf{x}(0) = \mathbf{x}_0. \tag{O1}$$

We wish to estimate $\mathbf{x}(\tau)$ for a given $\tau > 0$.

Let $\eta = \tau/N$, $N \in \mathbf{Z}_+$, and inductively define

$$\begin{aligned} \mathbf{x}_0 &= \mathbf{x}(0), \\ \mathbf{x}_n &= \mathbf{x}_{n-1} + \eta f(\mathbf{x}_{n-1}), \qquad n = 1, 2, \ldots, N. \end{aligned} \tag{O2}$$

Let

$$C = \sup_{x \in \Omega} \sum_{i=1}^{d} |f^{(i)}(\mathbf{x})|, \qquad L = \sup_{x \in \Omega} \sum_{i,j=1}^{d} \left| \frac{\partial f^{(i)}}{\partial x_j}(\mathbf{x}) \right|, \tag{O3}$$

where $\Omega \subset \mathbf{R}^d$ is some convex region where one can a priori guarantee that $\mathbf{x}(t)$, $\forall t \in [0, \tau]$, and $\mathbf{x}_n$, $\forall n = 0, 1, \ldots, N$, will fall ($\Omega$ has to be found in each case: out of desperation, one could always take $\Omega = \mathbf{R}^d$).

Then

$$|\mathbf{x}_N - \mathbf{x}(\tau)| \leqslant \frac{C\tau}{2N}(e^{L\tau} - 1). \tag{O4}$$

This formula gives an a priori estimate of the error that would be committed if one iteratively solved Eq. (O1) with the method of Eq. (O2) ("finite-difference algorithm"). It can be used in many of the exercises proposed in this book, where the use of a computer is suggested.

The proof of Eq. (O4) is a simple consequence of the considerations and proofs given in §2.2–§2.4.

PROOF. Let $\mathbf{d}_k = \mathbf{x}_k - \mathbf{x}(k\eta)$, $k = 0, 1, \ldots, N$.
One finds

$$\begin{aligned}\mathbf{d}_k &= \mathbf{x}_k - \mathbf{x}(k\eta)\\ &= \mathbf{x}_{k-1} + \eta f(\mathbf{x}_{k-1}) - \mathbf{x}((k-1)\eta) - \int_0^\eta f(\mathbf{x}((k-1)\eta + \theta))\, d\theta \qquad (\mathrm{O}5)\\ &= \mathbf{d}_{k-1} + \int_0^\eta (f(\mathbf{x}((k-1)\eta + \theta)) - f(\mathbf{x}_{k-1}))\, d\theta.\end{aligned}$$

Hence, applying Taylor's formula and adding and subtracting suitable terms:

$$\begin{aligned}|\mathbf{d}_k| &\leqslant |\mathbf{d}_{k-1}| + L\int_0^\eta |\mathbf{x}((k-1)\eta + \theta) - \mathbf{x}_{k-1}|\, d\theta\\ &\leqslant |\mathbf{d}_{k-1}| + L\int_0^\eta (|\mathbf{x}((k-1)\eta + \theta) - \mathbf{x}((k-1)\eta)| + |\mathbf{d}_{k-1}|)\, d\theta \qquad (\mathrm{O}6)\\ &\leqslant |\mathbf{d}_{k-1}| + L\eta|\mathbf{d}_{k-1}| + LC\int_0^\eta \theta\, d\theta,\end{aligned}$$

where in the last inequality, we have estimated the derivative of x by recalling that $\dot{\mathbf{x}} = \mathbf{f}(\mathbf{x})$ and $|\mathbf{f}(\mathbf{x})| \leqslant C$. If $\Omega \neq \mathbf{R}^d$ the Taylor's formula can still be applied by the convexity assumption on Ω: this easily follows from the proofs of appendix A.

Then

$$|\mathbf{d}_k| \leqslant (1 + L\eta)|\mathbf{d}_{k-1}| + \frac{LC}{2}\eta^2 \qquad (\mathrm{O}7)$$

which, by iteration, yields (recall $d_0 = 0$)

$$\begin{aligned}|\mathbf{d}_k| &\leqslant \frac{LC}{2}\eta^2\left[1 + (1 + L\eta) + (1 + L\eta)^2 + \cdots + (1 + L\eta)^{k-1}\right]\\ &= \frac{LC}{2}\eta^2 \frac{(1 + L\eta)^k - 1}{L\eta} = \frac{C\eta}{2}\left[(1 + L\eta)^k - 1\right]\end{aligned} \qquad (\mathrm{O}8)$$

which for $k = N$, recalling that $\eta = \tau/N$, becomes

$$|\mathbf{x}_N - \mathbf{x}(\tau)| \leqslant \frac{C\tau}{2N}\left[\left(1 + \frac{L\tau}{N}\right)^N - 1\right] \leqslant \frac{C\tau}{2N}(e^{L\tau} - 1). \qquad (\mathrm{O}9)$$

APPENDIX P

Some Astronomical Data

(1) Gravitational constant $k = 6.67 \times 10^{-8}\ \mathrm{cm}^3/\mathrm{g}(\mathrm{sec})^2$.
(2) Radius of the Sun: $R_s = 6.96 \times 10^5$ Km.
Mass of the Sun: $M_s = 1.99 \times 10^{33}$ g.
Density of the Sun: $\rho_s = 1.41\ \mathrm{g}/\mathrm{cm}^3$.

(3) Elements of the Planets' Orbits.

Planet	Major semiaxis of the orbit u.a.	Major semiaxis of the orbit 10^6km	Period of sideral revolution Days	Eccentricity of orbit	Ecliptic inclination	Longitude Ascendent node	Longitude Perigee
Mercury	0.387099	57.91	87.969	0.206625	7° 0′13″.8	47°44′66″	76°40′32″
Venus	0.723332	108.21	224.700	0.006793	3 23 39.3	76 14 11	130 51 20
Earth	1.000000	149.60	365.257	0.016729			102 04 41
Mars	1.52369	227.94	686.980	0.093357	1 51 0.0	49 10 25	335 58 19
Jupiter	5.2028	778.34	4332.587	0.048417	1 18 21.2	99 56 55	13 31 33
Saturn	9.540	1427.2	10759.21	0.055720	2 29 26.1	113 13 37	92 04 39
Uranus	19.18	2869.3	30685.	0.0471	0 46 22.0	73 43 36	169 51
Neptune	30.07	4498.5	60188.	0.0087	1 46 28.1	131 13 51	44 10
Pluto	39.44	5900.	90700.	0.247	17 08 24	109 38 02	223 30

[1]For the year 1950.

(4) Physical Characteristics of the Planets.

Planet	Equatorial radius km	Equatorial radius In earth Radii	Mass In earth Mass	Mass 10^{27}g	Density g/cm^3	Equatorial gravitational acceleration	Escape velocity Km/s	Period of sideral rotation	Equator's inclination on the orbit's plane
Mercury	2437	0.382	0.055	0.330	5.5	372	4.3	$58^d.65$	7°
Venus	6050	0.950	0.816	4.87	5.2	887	10.4	$243^d, 2$**	3°24′
Earth	6378	1.000	1.000	5.98	5.5	981	11.2	$23^h56^m4^s.1$	23°27′
Mars	3394	0.531	0.107	0.64	3.9	376	5.0	$24^h37^m22^s.6$	24°56′
Jupiter	71400	11.2	318.	1900.	1.3	2500	61.	$9^h50^m.5$	3°07′
Saturn	60400	9.5	95.1	568.	0.7	1100	36.	10^h14^m	26°45′
Uranus	24800	3.9	14.6	87.	1.6	950	22.	10^h49^m**	82°
Neptune	25050	3.9	17.2	103.	1.7	1150	24.	$15^h.8 \pm 1$	29°
Pluto	< 2900	0.45	0.9*	5.5*	–	–	–	$6^d.4$	–

* Approximate date
** Retrograde

(5) Satellites of the Planets.

Planet	Satellite	Average distance from the planet's center 10^3 km	Period of the sideral revolution in days	Period of synodic revolution
Earth	Moon	384.4	27.321661	$29^d12^h44^m02^s.8$
Mars	1. Phobos	9.4	0.318910	07 39 26 .65
	2. Deimos	43.5	1.262441	1 06 21 15 .68
Jupiter	1. Io	421.8	1.769138	1 18 28 35 .95
	2. Europa	671.4	3.551181	3 13 17 53 .74
	3. Ganymede	1071.	7.154553	7 03 59 35 .86
	4. Callisto	1884.	16.689018	16 18 05 06 .92
	5. Amalthea	181.	0.498179	11 57 27 .6
	6.	11500.	250.62	260 .0
	7.	11750.	259.8	276 .10
	8.	23500.	738.9	631 .05
	9.	23700.	755.	626
	10.	11750.	260.	276
	11.	22500.	696.	599
	12.	21000.	625.	546
Saturn	1. Mimas	185.7	0.942422	22 37 12 .4
	2. Encelado	238.2	1.370218	1 08 53 21 .9
	3. Tethys	294.8	1.887802	1 21 18 54 .8
	4. Dione	377.7	2.736916	2 17 42 09 .7
	5. Rhea	527.5	4.517503	4 12 27 56 .2
	6. Titan	1223.	15.945452	15 23 15 25
	7. Hyperion	1484.	21.276665	21 07 39 06
	8. Iapetus	3563.	79.33082	79 22 04 56
	9. Phoebe	12950.	550.45	536 16
	10. Themis	157.5	0.749	
Uranus	1. Ariel	191.8	2.52038	2 12 29 40
	2. Umbriel	267.3	4.14418	4 03 28 25
	3. Titania	438.7	8.70588	8 17 00
	4. Oberon	586.6	13.46326	13 11 15 36
	5. Miranda	130.1	1.414	
Neptune	1. Triton	353.6	5.87683	5 21 03 27
	2. Nereid	6000 ?.	500.	

E. on the ecliptic plane
P. on the plane of the planet's equator
B. on the plane of the planet's orbit
R. retrograde rotation

Inclination of the orbit	Orbit eccen-tricity	Radius of the Satellite km	Mass (Planet's mass): (Satellite's mass)	10^{24}g
5°1E	0.0549	1738	81.3	73.4
1.8P	0.019	14		
1.4P	0.003	8		
0P	Small	1660	24 000	79
0P	and	1440	39 800	47.8
0P	variable	2470	12 400	153
0P		2340	21 000	90
0P	0.003	80		
28.5B	0.155	60		
28.0B	0.207	20		
R33B	0.38	20		
R24B	0.25	11		
28.3B	0.140	10		
R16.6B	0.207	12		
R	0.13	10		
1.5P	0.0196	260	15 000 000	0.038
0.0P	0.0045	300	8 000 000	0.07
1.1P	0.0000	600	870 000	0.65
0.0P	0.0021	650	555 000	1.03
0.3P	0.0009	900	250 000	2.3
0.3P	0.0289	2500	4 150	137
0.6P	0.110	200	5 000 000	0.11
14.7P	0.029	600	100 000	5
R30P	0.166	150		
		300		
0P	0.007	300		
0P	0.008	200		
0P	0.023	500		
0P	0.010	400		
R20P	0.000	2000	700	150
	0.7	150	3 000 000	0.05

From F. Bakouline, E. Kononovitch, V. Moroz, *Astronomie Générale*, MIR, Moscow, 1975.

Definitions and Symbols

Si T est un ensemble, et A une partie de T, on notera φ_A la fonction charactéristique de A, si cela n'entraîne pas de confusion.

(Bourbaki, ch. IX)

$C^\infty(A)$: if $A \subset \mathbf{R}^d$ is an open set: the set of the functions on A continuous, together with their partial derivatives of all orders; shortened often as C^∞ when A is understood.

$C_0^\infty(A)$: if $A \subset \mathbf{R}^d$ is an open set: subset of $C^\infty(A)$ consisting of the functions vanishing outside a closed bounded set contained in A.

$C^{(k)}(A)$: if $A \subset \mathbf{R}^d$ is an open set: it is the set of the functions on A with partial derivatives of order $\leqslant k$ continuous on A, k being a non-negative integer.

$C^\infty(Q)$: with $Q \subset \mathbf{R}^d$ arbitrary set with dense interior Q_0: set of the functions in $C^\infty(Q_0)$ which can be extended to continuous functions on Q, together with all their derivatives.

$C_0^\infty(Q)$: with $Q \subset \mathbf{R}^d$ arbitrary set with dense interior Q_0: set of the functions in $C^\infty(Q_0)$ vanishing outside some closed bounded set contained in Q_0.

$C^{(k)}(Q)$: with $Q \subset \mathbf{R}^d$ arbitrary set with dense interior: defined as $C^\infty(Q)$, considering only the first k derivatives.

$C^\infty(\mathbf{T}^d)$: functions of class C^∞ on the d-dimensional torus $\mathbf{T}^d$ (see Definition 12, p. 100, and Definition 13, p. 101, §2.21).

$\overline{C}^{\infty}([0, L])$: functions in $C^{\infty}([0, L])$ vanishing in 0 and L together with all the even-order derivatives.

$\mathbf{C}^d$: complex d-dimensional space and (or) complex d-dimensional vector space.

$\mathbf{R}^d$: real d-dimensional space and (or) real d-dimensional vector space.

$\mathbf{T}^d$: d-dimensional torus with side 2π (see p. 101).

$\mathbf{R}, \mathbf{R}^1$: real line.

$\mathbf{C}, \mathbf{C}^1$: complex plane.

$\mathbf{R}_+$: interval $[0, +\infty)$.

S_t: solution flow for an autonomous differential equation.

$\mathbf{Z}^d$: lattice of the d-tuples of integers.

$\mathbf{Z}, \mathbf{Z}^1$: integer numbers.

$\mathbf{Z}_+$: non-negative integers.

$\boldsymbol{\xi}, \boldsymbol{\eta}, \ldots$: points or vectors in $\mathbf{R}^d$, $\mathbf{C}^d$.

$\boldsymbol{\varphi}, \boldsymbol{\psi}, \ldots$: points in $\mathbf{T}^d$.

$(X^{(\alpha)})_{\alpha \in J}$: family of objects $X^{(\alpha)}$ parameterized by an index varying in an index set.

t: real parameter with the interpretation of time.

$\dot{}$: t-derivative.

$\ddot{}$: second t-derivative.

$O(\xi)$: quantity of the order of magnitude of ξ: it means that there is $C > 0$, $\xi_c > 0$ such that $|O(\xi)| \leq C|\xi|$ if $|\xi| < \xi_c$. Used only when ξ is an "infinitesimal" variable.

$o(\xi)$: quantity infinitesimal of higher order compared to ξ: it means $\lim_{\xi \to 0} |\xi|^{-1} |o(\xi)| = 0$.

mbe: end-of-proof symbol.

$\mathbf{x} \cdot \mathbf{y}$: scalar product of vectors in $\mathbf{R}^d$.

$\mathbf{x} \wedge \mathbf{y}$: vector product of two vectors in $\mathbf{R}^3$.

$P - Q$: vector whose components in a given frame of reference are the differences of the homonymous coordinates of P and Q in the same frame of reference.

$\equiv$: identity or, often, implicit definition.

$\wedge$: vector product symbol.

Re, Im: real or imaginary part of a complex number.

$/, \backslash$: symbols for the set theoretic difference.

∂: partial derivative or boundary of a set

∂: gradient

References

A. Suggested Books

Arnold, V.: *Equations Differentielles Ordinaires*. Mir, Moscow, 1976.

Arnold, V.: *Chapitres supplementaires de la théorie des equations differentielles ordinaires*. Mir, Moscow, 1980.

Arnold, V.: *Methodes mathematiques de la mécanique classique*. Mir, Moscow, 1978.

Boyer, C.: *History of Mathematics*. Wiley, New York, 1968.

Dreyer, J.: *A History of Astronomy from Thales to Kepler*. Dover, New York, 1953.

Galilei, G.: *Dialogues Concerning Two New Sciences*. Dover, New York, 1982.

Landau L., Lifschitz, E.: *Mécanique*. Mir, Moscow, 1966.

Landau, L., Lifschitz, E.: *Mécanique des fluides*. Mir, Moscow, 1966.

Levi-Civita, T., Amaldi, V.: *Lezioni di Meccanica Razionale*. Zanichelli, Bologna, 1949.

Mach, E.: *The Science of Mechanics*. Open Court, La Salle-London, 1942.

Newton, I.: *The Mathematical Principles of Natural Philosophy*. Transl. by A. Motte, ed. F. Cajori, Univ. of California Press, Berkeley, 1930.

Sommerfeld, A.: *Atomic Structure and Spectral Lines*. London, 1934.

Truesdell, C.: *Essays in the History of Mechanics*. Springer-Verlag, New York, 1974.

B. Works Developing Themes Introduced in this Book

Arnold, V.: Small denominators and problems of stability of motion in classical and celestial mechanics. *Russ. Math. Surveys* **18**(6) 85, (1963).

Arnold, V., Avez, A.: *Ergodic Problems in Classical Mechanics*. Benjamin, New York, 1963.

Gallavotti, G.: *Aspetti della teoria ergodica, qualitativa e statistica del moto*, Quaderni dell'. Unione Matematica Italiana, vol. 21. Pitagora Editrice, Bologna, 1981.

Gnedenko, B.: *The Theory of Probability*. Mir, Moscow, 1973.

Huang, K.: *Statistical Mechanics*. Wiley, New York, 1963.

Khintchin, A.: *Mathematical Foundations of Statistical Mechanics*. Dover, New York, 1957.

Khintchin, A.: *Mathematical Foundations of Information Theory*. Dover, New York, 1963.

Kornfeld, I., Sinai, J. Fomin, S.: *Ergodicescaia Teoria*. Nauka, Moscow, 1980.

Lagrange, L.: *Mécanique analytique*. Paris, 1788.

Lanford, O.: *Entropy and Equilibrium States in Classical Statistical Mechanics*, Lecture Notes in Physics 20. ed. A. Lenard, Springer-Verlag, Berlin, 1972.

Marsden, J., McCracken, M.: *The Hopf Bifurcation and Its Applications*. Springer-Verlag, Berlin, 1976.

Miranda, C.: *Partial Differential Equations of Elliptic Type*. Springer-Verlag, Berlin, 1970.

Moser, J.: *Lectures on Hamiltonian Systems*. Mem. Am. Math. Soc., vol. 81, 1968.

Moser, J.: *Stable and random motions in dynamical systems*. Annals of mathematical studies, Princeton Univ. Press, Princeton, 1973.

Poincaré, H.: *Les méthodes nouvelles de la mécanique celeste*, vols. 1 and 2. Gauthier-Villars, Paris, 1897.

Reed, M.: *Abstract nonlinear wave equations*, in Lecture Notes in Mathematics 507. Springer-Verlag, Berlin, 1975.

Ruelle, D.: *Statistical Mechanics. Rigorous Results*. Benjamin, New York, 1969.

C. Quoted References.

The numbers in square brackets at the end of each reference refer to the page of this book where the quotation is made. When the reference concerns works already listed in §A, B above it will be identified by the first word in its title.

Arnold, V.: *Méthodes* . . . [211, 362].

Arnold, V.: A proof of a theorem of A. N. Kolmogorov on the invariance of quasiperiodic motions under small perturbations of the Hamiltonian.: *Russian Math. Surveys* **18**(5), 9 (1963). [493]

Arnold, V.: Small denominators . . . [499]

Arnold, V., Avez, A.: *Ergodic* . . . [353, 357].

Bakouline, P., Kononovitch, E., Monoz, V.: *Astronomie generale*. Mir, Moscow, 1975 [555, 559].

Berry, M.: Regular and irregular motions, in *AIP Conference proceedings, Vol. 46*, ed. S. Jorna. American Institute of Physics, New York, 1978 [302].

Bowen, R., Ruelle, D.: The ergodic theory of axiom *A* flows. *Inventiones Mat.* **29**, 181 (1975) [445].

Campanino, M., Epstein, H., Ruelle, D.: On Feigenbaum's functional equation $g \circ g(\lambda x) + \lambda g(x) = 0$. *Topology* **21**, 125, 1982 [451].

Collet, P., Eckmann, J.P.: *Iterated Maps of the Interval as Dynamical Systems*, Birkhauser, Boston, 1980 [451].

Courant, R., Hilbert, D.: *Methods of Mathematical Physics, Vol. II*. Interscience, New York, 1962 [329].

Chierchia, L.: Thesis, Universitá di Roma, 1981 [520].

Chierchia, L., Gallavotti, G.: Smooth prime integrals for quasi-integrable Hamiltonian systems. Nuovo Cimento **67B**, 277 (1982) [496].

De Prit, A.: Free rotations of a rigid body studied in phase space. *Am. J. Phys.* **35**, 424 (1967) [318, 319].

Fano, G.: *Mathematical Methods of Quantum Mechanics*. McGraw–Hill, New York, 1971, [526, 529].

Feigenbaum, M.: Quantitative universality for a class of nonlinear transformations. *J. Stat. Phys.* **19**, 25 (1978) [452].

Finzi, B., Udeschini, P.: *Esercizi di Mecconica razionale*. Tamburini, Milano, 1974 [244].

Franceschini, V.: A Feigenbaum sequence of bifurcations in the Lorenz model. *J. Stat. Phys.* **22**, 397 (1980) [452].

Franceschini, V., Tebaldi, C.: Sequences of infinite bifurcations and turbulence in a five mode truncation of the Navier-Stokes Equations. *J. Stat. Phys.* **21**, 707 (1979) [452].

Franceschini, V., Tebaldi, C.: A seven mode truncation of the plane incompressible Navier-Stokes equations. *J. Stat. Phys.* **25**, 397 (1981) [452].

Galilei, G.: "Il Saggiatore," in *Opere*, Edizione Nazionale, Firenze, 1896, vol. VI [2].

Gallavotti, G.: Perturbation theory for classical Hamiltonian systems, in *Progress in Physics*, ed. J. Frölich. Birkhauser, Boston, 1982. [496].

Hadamard, H.: Sur l'iteration et les solutions asymptotiques des équations differentielles, Bulletin Societé Mathématique de France, **29**, 224, 1901. [411]

Inglese, G.: Thesis, Universitá di Roma, 1981 [519].

Khintchin, A.: *Mathematical . . . Information . . .* [357].

Khintchin, A.: *Continued Fractions*. Noordhoff, Gröningen, 1963 [97].

Kobussen, J.: Some comments on the lagrangian formalism for systems with general velocity dependent forces. *Acta Physica Austrioca* **51**, 293 (1979) [139, 140].

Landau, L., Lifschitz, E.: *Mécanique . . .* [48, 231, 311, 327].

Lanford, O.: *Bifurcations of periodic solutions into invariant tori. The work of Ruelle and Takens*, in Lecture Notes in Mathematics 322. Springer-Verlag, Berlin, 1973 [399, 412, 459].

Lanford, O.: A computer assisted proof of Feigenbaum's conjecture, preprint, Berkeley, 1981 [451].

Mach, E.: *The Science . . .* [9, 211, 243].

Metcherskij, L.: *Récueil de problèmes de mecanique rationelle*. Mir, Moscow, 1973 [244].

Moser, J.: *Stable* . . . [441, 466].

Moser, J.: Lectures . . . [466, 493].

Negrini, P., Salvadori, L.: Attractivity and Hopf bifurcations, *Nonlinear Analysis*, **3**, 87, 1978, [406].

Nekhorossiev, V.: An exponential estimate of the time of stability of nearly-integrable Hamiltonian systems, *Russian Mathematical Surveys*, **32**(6), 1, 1972, [468].

Newton, I.: *The Mathematical* . . . [9, 11, 299].

Poincaré, H.: *Les méthodes* . . . [17, 330].

Pöschel, J.: Integrability of Hamiltonian systems on Cantor sets, *Comm. Pure Appl. Math.* **35**, 653 (1982). ETH-Zürich, 1981 [496].

Robbins, K.: Periodic solutions and bifurcation structure at high *R* in the Lorenz model. *SIAM J. Appl. Math.* **36**, (1979) [447].

Ruelle, D.: A measure associated with the axiom **A** attractors. *Am. J. Math.* **98**, 619 (1976) [445].

Rüssmann, H.: Über das Verhalten analytischer hamiltonscher Differentialglechungen in der Mähe einer Gleichgewichtslösung, *Math. Annalen* **154**, 285 (1964) [362].

Rüssmann, H.: Über die Normalform analytischer Hamiltonscher Differentialgleichungen in der Nähe einer gleichgewichtslösung, *Math. Annalen* **169**, 55 (1967), [478].

Smale, S.: Differentiable Dynamical Systems, *Bulletin Am. Math. Soc.*, **73**, 747, 1967 [445].

Sommerfeld, A.: *Atomic* . . . [330].

Tresser, C., Coullet, P.: Itérations d'endomorphismes et groupe de renormalization, *C. R. Acad. Sci.*, Paris, **287A**, 577 (1978) [452].

Truesdell, C.: "*Essays* . . . ," [9].

Wittaker, E.: *A treatise on the analytical dynamics of the rigid body*. Cambridge University Press, 1937 [328].

Index